T0303863

Autonomic
Pharmacology

Autonomic Pharmacology

KENNETH J. BROADLEY

Welsh School of Pharmacy,
University of Wales, Cardiff, UK

CRC Press
Taylor & Francis Group
Boca Raton London New York

CRC Press is an imprint of the
Taylor & Francis Group, an **informa** business

UK   Taylor & Francis Ltd, 1 Gunpowder Square, London EC4A 3DE
USA   Taylor & Francis Inc., 1900 Frost Road, Suite 101, Bristol, PA 19007

British Library Cataloguing in Publication Data

A catalogue record for this book is available from the British Library
ISBN 0-7484-0556-9 (formerly 013 052390 9)

Library of Congress Cataloguing Publication data are available

The publisher assumes no responsibility for any injury or damage to persons or property as a matter of product liability, negligence or otherwise, or from any use or operation of any methods, products or dosage regimes contained in this book. Independent verification of diagnosis and drug dosages should be obtained.

Cover design by Jim Wilkie

Typeset in Times 10/12pt by MCS Ltd, Salisbury, Wiltshire
Printed in Great Britain by T.J. Press (Padstow) Ltd

Contents

Note: see the front of each individual chapter for a detailed contents listing.

Preface

The inspiration to write this book came from a desire to update the book of the same name written over 15 years ago by my friend and former colleague Dr Michael Day. In the intervening years autonomic pharmacology has made considerable advances and a new book on the subject was felt to be overdue. Whereas Mike Day's book was intended to serve as a student textbook, nowadays most major student textbooks of pharmacology would cover the area of autonomic pharmacology quite adequately. This book was therefore written to serve as a definitive reference text on the subject with a comprehensive overview of the current state of knowledge.

My intention was not to attempt to deal with every aspect of the area of autonomic pharmacology with every minor action of the drugs considered nor the more obscure ideas on mechanism of drug action. I have tried to present the consensus view on drugs acting on the autonomic nervous system and the receptor classification. Sometimes this may be my personal interpretation of the data.

In writing this book I have made several assumptions. First, there may be readers with little pharmacology background who may want to find out about the actions and uses of drugs acting on the autonomic nervous system. As a result there may be sections of the book where I may appear to have been overenthusiastic in defining the terms used or the background. I hope that the experienced pharmacologist among the books readers will overlook this indulgence. The second assumption is that few people will read this book from cover-to-cover, and I have therefore tried to make each chapter relatively self-contained with appropriate reference to relevant chapters for more information on related areas. This has necessitated some repetition in definitions between chapters.

In spite of carrying out research now for over 20 years in the general area of autonomic pharmacology, writing this book has shown me how some areas have moved very rapidly during that time whereas others have remained quite static. It has also become clear to me that autonomic pharmacology cannot be considered in isolation. The subject has taken me into a wide range of areas of science – medicinal chemistry, immunology, endocrine biology, haematology, neuropharmacology, pharmaceutics, pharmacognosy, and so on. The emphasis, however, is on pharmacology – the science of the *action* of drugs on the body. I believe that the study of a drug's action in the intact organism or isolated tissues is the cornerstone of drug evaluation and essential for a better understanding of the mechanism of action. I have therefore tried to illustrate the action of the drugs acting on the autonomic nervous system and stressed the need for adequate experimental design. This is not a manual of practical pharmacology but I hope it gives some useful guidelines. I frequently persuade my graduate students of the value of showing a typical trace, the 'thesis trace'. This was my opportunity to put theory into practice and to obtain a few 'textbook traces' of my own, which are liberally scattered throughout this book.

Finally, I must express my gratitude to several people who have helped during the task of writing this book, which turned out to be more demanding and time-consuming than I had imagined. Several of my colleagues at the University of Wales in Cardiff have willingly helped in various ways. Professor Paul Nicholls, Dr John Smith and Dr Bob Sewell have provided helpful discussion on matters that I could not cope with. I especially want to thank Dr Terry Wall of the Bute Library for so much assistance with literature searches and tracking down obscure chemical structures. I am also grateful to Brenda Williams for producing the figures and tables of chemical structures. I also thank Professor David Brown of University College London, Professor Steve Nahorski of the University of Leicester and Dr Carlo Maggi of Menarini Pharmaceuticals, Florence, for kindly reading and commenting on three chapters of the book.

My final and most sincere thanks go to my family. First, to my sons, Simon and Duncan, and their wives, Charlotte and Sheila, with whom I have not spent as much time as I should have liked over the past year or so and to Duncan for help with drawing the diagrams. My special appreciation and love goes to Pat, my wife, who has patiently tolerated many hours of being a 'textbook widow' while the writing has been in progress. Without her understanding and immense help in the typing of the entire manuscript, this task would not have been possible.

Ken Broadley
Cardiff, UK

1

Anatomy and Physiology of the Autonomic Nervous System

1.1 Introduction

The autonomic nervous system controls the internal involuntary functions of the body. These are the functions concerned with the maintenance of a constant internal environment, that is, homeostasis. The major activities and systems under the control of the autonomic nervous system include digestion, the cardiovascular system (such as the blood pressure), blood chemistry, breathing and body temperature. Control of these functions is below the level of consciousness, hence the term involuntary nervous system. The major cellular structures that are innervated by the autonomic nervous system are smooth muscle, cardiac muscle, glandular tissue and adipocytes (fat cells). Thus, the digestion of food occurs by the secretion of digestive enzymes from the intestinal glands of the intestinal mucosa and rhythmic contraction and relaxation of circular and longitudinal smooth muscle arranged in the intestinal wall propels the chyme along and churns it up to aid digestion. We are unaware of this process and have little conscious control over it. The autonomic nervous system modulates this digestive activity by either speeding or attenuating.

The autonomic nervous system has been subdivided into the sympathetic and parasympathetic systems based upon their physiological functions and their anatomy. The word 'sympathetic' is derived from the term *nervi sympathici majores* used by Winslow (1732) to describe the nerves which he thought carried out 'sympathies' and co-ordinate various visceral functions. The sympathetic nervous system is concerned with adaption to stressful situations, which Cannon (1929) described as preparing the body for 'flight and fight'. In contrast, the parasympathetic division's role is to conserve and restore energy. Thus, in general they have opposing actions upon tissues, the heart being a good example. The rate and force of cardiac contractions are elevated by an increase in sympathetic nerve discharge in response to a stressful stimulus. In contrast, raised parasympathetic nerve activity to the heart causes slowing. Digestion is a function not essential for survival of immediate danger and is therefore retarded by sympathetic nervous activity, whereas parasympathetic activity stimulates digestion during rest and recovery from stress.

The role of the autonomic nervous systems in controlling smooth muscle activity differs from that of the somatic nervous system which is responsible for the contraction of skeletal (striated, voluntary) muscle. The contraction of mammalian skeletal muscle depends almost entirely upon the arrival of a nerve impulse along the somatic nerve and release of the neurotransmitter, acetylcholine (Ach). If the somatic nerve is destroyed, then the muscle becomes paralysed and eventually atrophies. In contrast, smooth muscle continues to function in the absence of an autonomic innervation but without its modulating influence.

A feature common to both the autonomic and somatic nervous systems is that both operate as a reflex arc. Some authorities regard the autonomic nervous system solely as an efferent system carrying motor impulses to the smooth muscle. However, afferent fibres dispatch sensory information regarding the body's internal environment via autonomic or somatic nerves to the central nervous system (CNS) where it is integrated before efferent impulses are sent to the effector organ. In this book, the autonomic nervous system is regarded both as an afferent and efferent system, since it is important to understand the autonomic reflexes when considering the pharmacology of drugs acting on the autonomic nervous system and its effectors. Furthermore, the sensory nerves are increasingly being recognized as sites of drug action in their own right.

Unlike the efferent somatic nerves which leave the spinal cord and pass directly to the skeletal muscle, autonomic efferents synapse once outside the spinal cord or brain stem before arriving at the innervated smooth muscle. These synapses occur with groups of cell bodies which are the autonomic ganglia.

1.2   Smooth Muscle

Smooth muscle fibres are arranged in sheets in the walls of the hollow organs such as the intestine and bladder. The individual fibres may be arranged in parallel, such as in the longitudinal muscle layer of the intestine which surrounds an inner circular muscle layer. In other organs, such as the uterus and bladder, the fibres form a network in which they run randomly in different directions. Smooth muscle has been classified by Bozler (1948) into two types: single-unit and multi-unit muscle (Figure 1.1).

1.2.1   *Single-unit Smooth Muscle*

The individual muscle cells are arranged so that adjacent fibres are in contact with each other through specialized gap junctions, which allows the spread of a wave of contraction throughout the muscle. The point of contact probably provides a low resistance pathway for ephaptic transmission, whereby local circuit currents are set up between adjacent cells. Chemical or mechanical transmission between smooth muscle cells is unlikely to occur.

The muscles develop tone or display rhythmic contractile activity. This tone or myogenic activity is not dependent upon the presence of the autonomic nerves but the nerve impulse to the smooth muscle modifies this activity either by increasing or decreasing the magnitude of contractions and their frequency. Not all fibres of

3

(a)　　　　　　　Single–unit type　　　　　　　　　　Multi–unit type

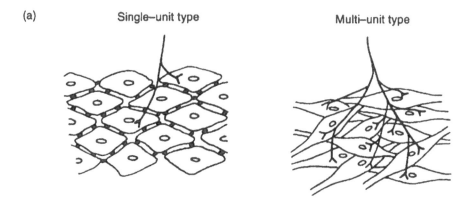

(b)　Skeletal muscle

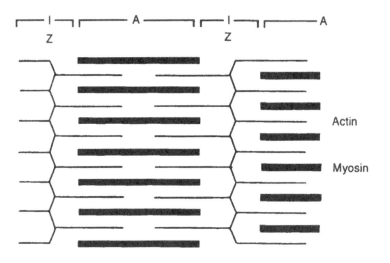

(c)　Smooth muscle cell

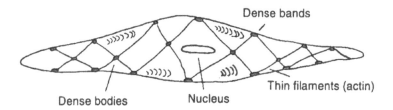

Figure 1.1　(a) Arrangement and innervation patterns of smooth muscle cells of the single-unit and multi-unit types. (b) Contractile filaments arrangement in skeletal muscle showing the A and I bands and the Z line forming the point of attachment of actin myofilaments. (c) A smooth muscle cell showing contractile myofilaments and their points of attachment to the cell membrane at dense bodies.

single-unit smooth muscle may receive an autonomic innervation, but its influence will spread readily to adjacent fibres. Single-unit smooth muscle does not respond to single electrical stimuli of the innervating nerves, although a small depolarization of each cell membrane occurs. With trains of stimuli, depolarizations summate to finally induce an action potential and contraction of the muscle.

The membranes of smooth muscle cells are polarized, the inside being negative with respect to the outside. The resting membrane potential is ~-60 mV. This is due to the combined effects of the sodium pump which expels Na^+ from the cell, the relatively greater permeability of the cell membrane to K^+ which reaches a higher concentration on the inside of the cell, and to the presence of large organic anions on the inside of the cell. Reduction of the resting membrane potential to a critical level generates a propagated action potential whereby the membrane is depolarized and contraction ensues.

The spontaneous rhythmic activity of smooth muscle is generated by pacemaker potentials or slow depolarizations of the cell membrane. These have been classified by Kuriyama (1970) into three types.

1 *Type I.* A slow potential of short duration acting as a prepotential for an individual spontaneous spike. These may be due to oscillations in internal Ca^{2+} concentration, $[Ca]_i$ (Tomita & Watanabe 1973).

2 *Type II.* An excitatory potential generated by nervous activity which may occur at a local level. For example, the intramural plexus of the intestine contains a local nerve loop triggered by a rise in intraluminal pressure (see later). This is therefore neurogenic in origin and responsible for peristalsis.

3 *Type III.* Slow depolarization leading to a train of spike discharges and known as slow waves. Each depolarization and train of superimposed spikes is accompanied by a contraction of the muscle. The frequency ranges from 1 to 18 per minute in different muscles. Their generation and propagation are not prevented by tetrodotoxin (TT_X), the inhibitor of Na^+ channels in excitable cells, with specificity for nervous tissue. Slow waves are not therefore neurogenic but myogenic. The frequency of myogenic activity is increased by stretching. Slow waves and spontaneous tone are often abolished by indomethacin, an inhibitor of cyclo-oxygenase, and are therefore attributed to endogenous arachidonic acid products (eg guinea-pig tracheal and intestinal smooth muscle) (Tomita 1989).

Special pacemaker cells may lie between longitudinal and circular muscle layers of the gastrointestinal tract. They may affect both layers equally, or one layer more than the other depending on species and location. Slow waves in the region of highest frequency will drive cells in neighbouring areas. In some tissues the pacemaker is in a fixed position, for example, in the ureter it is at the end nearest the kidney. In other tissues it may vary as different regions have the highest frequency. Smooth muscle slow waves consist of an initial component which is probably the driving force originating in specialist pacemaker cells, followed by a secondary component with superimposed spike potentials. These latter two aspects are probably due to Ca^{2+} influx, since they are abolished by the calcium channel antagonists, verapamil and nifedipine, along with the rhythmic contraction (Figure 1.2) (Golenhofen 1981). The mechanism of generation of these slow waves may involve oscillations in the electrogenic Na^+–Ca^{2+} exchange mechanism, since they are blocked by ouabain, an

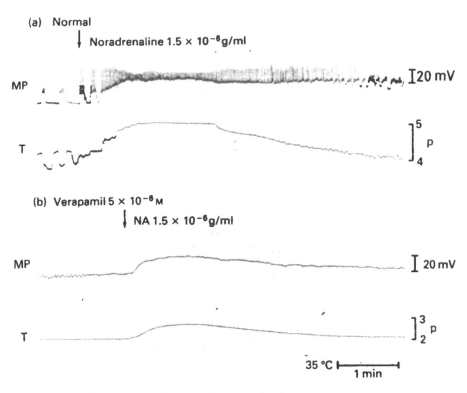

Figure 1.2 Membrane potential (MP) and tension development (T) of guinea-pig portal vein smooth muscle (a) under normal conditions and (b) in the presence of the Ca^{2+} channel antagonist, verapamil. Noradrenaline (NA) was added at the arrow, inducing an increase in the firing rate of spike potentials, sustained rather than slow wave depolarizations and a sustained contraction in place of phasic contractions. Verapamil blocks the slow waves. Reproduced with permission from Golenhofen *et al.* (1973).

inhibitor of Na^+, K^+-ATPase which is responsible for the Na^+ pump (Tomita 1981).

The degree of spontaneous myogenic activity varies from muscle to muscle; even different blood vessels display varying amounts of myogenic activity. For example, the guinea-pig and rat portal veins show spontaneous activity whereas rabbit aorta and ear artery do not.

Addition of contractile agents, such as Ach or noradrenaline, induces an increase in firing rate of the spike potentials, a sustained continuous depolarization and a sustained tonic contraction of the muscle (Figure 1.2) whether it displays myogenic activity or not. These contractions to noradrenaline are often seen to be biphasic (eg rabbit ear artery). An initial rapid rise in tension is followed by a more slowly developing contraction. These are also referred to as *phasic* and *tonic* contractions, respectively. The former, however, is apparently less sensitive to Ca^{2+} channel antagonism and thus contrasts with the phasic rhythmic slow waves. The contractions are often resistant to removal of extracellular Ca^{2+}. Contractions induced by these agents therefore have only a minor role for Ca^{2+} influx and may occur with minimal depolarization (from -60 to -55 mV). They have therefore been attributed to pharmacomechanical coupling which can be partly explained by release of

6

intracellular Ca^{2+} following activation of an extracellular receptor by the contractile agent (Tomita 1989).

Depolarization of smooth muscle cells may also be brought about by stretch, by raising the extracellular K^+, $[K^+]_e$, or by passing a depolarizing current via an electrode. Stretch produces a depolarization and a quick stretch may induce a contraction in some tissues. Raising extracellular K^+ induces a positive charge to the inside of the cell membrane causing it to become depolarized. This ion is particularly effective because the resting membrane potential is largely due to the concentration gradient of K^+ across the cell membrane. The resulting depolarization is more complete than with contractile drugs and the contraction, being sensitive to Ca^{2+} channel blockade, is therefore due primarily to the influx of Ca^{2+} through voltage-dependent calcium channels (L channels) (Karaki 1987, Tomita 1989). Addition of Ca^{2+} to the extracellular space will increase the action potential amplitude and enhance contractility, but the effect is dampened by an opposing membrane stabilizing action. Barium, however, can replace the Ca^{2+} and on addition to isolated smooth muscle tissues induces a powerful contraction. Conversely, manganese blocks the influx of Ca^{2+}. Application of an electrical stimulus to smooth muscle causes depolarization, the membrane adjacent to the cathode being depolarized; the membrane at the anode is hyperpolarized. To stimulate smooth muscle to contract by direct application of an electrical impulse, pulse widths of >10 msec are required. However, to stimulate the autonomic nerve components in isolated smooth muscle preparations, lower pulse widths of 0.1–2 msec will suffice.

1.2.2 Multi-unit Smooth Muscle

This type of smooth muscle occurs less commonly and is characterized by a lack of connections between individual fibres. They are separated by gaps of ≥1 μm. Each fibre may thus receive an autonomic innervation (Figure 1.1). This arrangement occurs in muscles that undergo relatively rapid contractions such as the ciliary muscles of the eye, that adjust lense curvature, and in the vas deferens, which contracts during ejaculation. The piloerector muscles of the skin and circular muscle bands around certain blood vessels are also of the multi-unit type. Such muscles appear to constrict only in response to the arrival of autonomic nerve impulses. Multi-unit smooth muscle responds to a single stimulus of the nerve and trains of pulses yield bigger responses. It is usual for both multi-unit and single-unit types to occur in most smooth muscle tissues.

The smooth muscle cells (Figure 1.1) are usually described as elongated and tapered at each end, although variations in this general shape are more usual. The size is in the range 2–3 μm diameter and 15–20 μm long in blood vessels, 5–6 μm diameter and 30–40 μm long in the intestine, and up to 0.5 mm long in the uterus. The smooth muscle fibres contain a single centrally located nucleus and, unlike skeletal muscle, have no transverse striations. In skeletal muscle the contractile filaments, myosin and actin, are arranged longitudinally in alternating layers as thick and thin myofilaments, respectively. Their alternating parallel arrangement produces the characteristic dense transverse bands (A bands) where they are both present, and the thin bands (I bands) where only actin occurs (Figure 1.1). In contrast, the actin and myosin of smooth muscle fibres are very thin and arranged more randomly so that no striations are seen. The contractile filaments are seen as fine longitudinal striations. Dense bodies can be observed in the sarcoplasm and these are probably connected to the dense bands attached to the cell membrane. These probably form the point of attachment of the thin filaments. A cytoskeleton of intermediate filaments probably exists, serving as a means of force transduction between the dense bodies and its membrane attachment (Gabella 1981).

Smooth muscle contraction is thought to occur in a similar fashion to that in skeletal muscle, although at a slower rate, probably because of a reduced availability of ATP as an energy source. According to the sliding-filament theory, actin and myosin filaments form cross-bridges as a result of which the two myofilaments slide past each other. Unlike skeletal muscle, however, there appears to be no troponin, the calcium-binding protein on the actin myofilament that regulates actin–myosin cross-bridging. Calmodulin is the calcium-binding protein of smooth muscle. Smooth muscle cells have an abundant sarcoplasmic reticulum (SR). The rough or granular endoplasmic reticulum (ER) is the site of synthesis of new membranes, filaments and glycogen. The smooth ER (SR) is probably a site of storage of Ca^{2+} and release. Although control of contractile mechanisms is dependent upon intracellular Ca^{2+}, it appears that only a minor contribution comes from this intracellular storage site. The transverse or T tubules, which form a continuation of the sarcolemma and pass into skeletal muscle fibres to connect with the SR, are absent in smooth muscle. The roles of myosin light-chain kinase, calmodulin and Ca^{2+} in the contractile responses of smooth muscle are described in Chapter 13.

Smooth muscle cells receive a blood supply usually via capillaries. In blood vessel walls, the tunica media smooth muscle is not vascularized in small animal species, but the blood vessels are confined to the advential layer where they are known as *vasa vasorum*. In man, the vasa vasorum does penetrate to the medial layer but the inner media is probably supplied by diffusion through the intima.

The autonomic nervous system also controls the activities of cardiac muscle and glandular tissue. Cardiac muscle has similar features to skeletal muscle in that it is striated, the calcium-binding protein is troponin, and each cell has a centrally located nucleus. However, the muscle fibres consist of a network of branching cells which interconnect to enable the spread of contraction throughout the myocardium. The cells are joined end to end by means of intercalated discs which form a strong bond between adjacent cells and facilitate the transmission of the electrical impulse quickly from cell to cell. The spontaneous activity of the heart is maintained by the rhythmic electrical discharge of cells in the sinuatrial node which acts as the pacemaker. The pacemaker cells are densely innervated by the autonomic nerves which control their rate of firing. However, as with all cells under autonomic control, they continue to function when the autonomic influence has been removed by denervation or by drugs.

1.3 Structure of Autonomic Nerves

Like all nervous tissue, autonomic nerves (neurones) consist of a cell body (perikaryon) from which originate two types of process – axons and dendrites. Efferent autonomic nerves are typically multipolar having many short dendrites each forming an extension of the cell body. These conduct incoming nerve impulses towards the cell body. There is usually a single axon extending from the cell body which carries the nerve impulse to another nerve cell or to the effector cell. Axons of the autonomic nervous system may or may not be myelenated. The lipid sheath of myelenated nerves consists of rolled up layers of Schwann cells (neurolemmocytes) which protect and insulate the axon. The myelin sheath is interrupted at regular intervals by the nodes of Ranvier, where voltage-gated Na$^+$ channels are highly concentrated. Nerve transmission occurs by local circuit currents whereby there is forward flow in the cytoplasm to the next node where it leaves and returns to the previous node via the extracellular fluid. The nerve action potential is transmitted virtually instantaneously between nodes because it is via electrolyte. The myelin therefore accelerates the conduction velocity and reduces the energy requirement of the sodium pump which is confined to the nodes of Ranvier.

Unmyelenated neurones are surrounded by neurolemmocytes which do not form myelin. These nerves are protected and nourished by the tissues that surround them. Their rate of conduction is relatively slow (0.7–2.3 m/sec) compared with myelenated fibres (3–120 m/sec). Nerve fibres are classified as A, B and C fibres according to their diameter, conduction velocity and myelenation. A and B fibres are myelenated, B fibres being smaller and conducting more slowly, while C fibres are unmyelenated. Preganglionic autonomic nerve fibres are of the B type with a diameter of ~3 μm, while postganglionic fibres are C fibres with a diameter of 1 μm.

The cell bodies of peripheral autonomic nerves located outside the CNS are grouped together in ganglia. The cell bodies are encapsulated in satellite or glial cells which are equivalent to the Schwann cells. The outer surface of the cell body does not therefore come into direct contact with the surrounding connective tissue or extracellular space. Under electron microscopy, the cell body can be seen to contain the nucleus and several organelles essential for metabolism, growth, repair and synthesis of neurotransmitters. These organelles include the Nissl bodies (chromatophilic substance) which consists of the rough ER. This is a network of tubules that branch thoughout the cytoplasm to circulate materials and store enzymes. Being rough ER, attached to the outer surface are ribosomes which are the sites of protein synthesis. Also present are mitochondria, neurotubules (microtubules), neurofilaments and the Golgi apparatus. The latter is connected to the Nissl bodies and is the site to which proteins are transferred, sorted and packaged in vesicles. They are transported to the axon terminal along the neurotubules together with vesicles containing neurotransmitter precursors. The neurofilaments in the cell body extend into the axon and provide a semi-rigid framework.

Synaptic transmission between pre- and postganglionic nerves is axodendritic, that is, from the preganglionic neurone branches to the postganglionic cell body dendrites. The neurotransmitter is Ach which on release from the presynaptic axon terminal interacts with its receptor to cause depolarization or an excitatory postsynaptic potential (EPSP). This may be insufficient to trigger an action potential in the axon of the postsynaptic neurone. However, summation of the EPSPs by either repeated firing or by actions at multiple sites (eg different dendrites) will lead to generation of a propagated action potential at the axon hillock. The axon hillock at the junction between cell body and axon prior to myelenation has the lowest excitability threshold and is the site of action potential generation. Some neurotransmitter–receptor interactions result in hyperpolarization or an inhibitory postsynaptic potential (IPSP). This stabilizes the postsynpatic membrane making the neurone less likely to trigger an action potential. Examples are noradrenaline or dopamine stimulating α-adrenoceptors on ganglionic cell bodies.

As a general rule, the predominant neurotransmitters at the axon terminal of postganglionic neurones are Ach for parasympathetic nerves and noradrenaline for sympathetic nerves.

1.4 Structure of the Autonomic Nervous System

The description of the autonomic nervous system is here divided into two components – the efferent pathways leaving the spinal cord or brain stem and the afferent pathways transmitting sensory data to the CNS.

1.5 The Efferent Autonomic Pathways

The efferent pathways originate in the brain stem (midbrain and medulla) and spinal cord. The cell bodies of those neurones emerging from the brain stem are aggregated in groups called nuclei. These are located at the levels of the respective cranial nerve roots through which the bundles of axons emerge. The cell bodies of neurones leaving the spinal cord are located in the lateral portion of the intermediate zone of the grey matter throughout the thoracic and the first two or three lumbar segments and in the second, third and fourth sacral segments (Figure 1.3). Whether sympathetic or parasympathetic, all of the neurones then synapse with a second neurone before reaching the effector organ. This synapse occurs at the autonomic ganglia which contain the cell bodies of the postganglionic nerves. The preganglionic neurones of the sympathetic and parasympathetic efferent pathways emerge from separate locations (Netter 1974, Gabella 1976).

1.5.1 *Sympathetic pathways*

The sympathetic system constitutes the thoracolumbar outflow from the spinal cord. The preganglionic fibres pass from both sides of the spinal cord across the anterior roots and leave the vertebral column with the spinal nerve at each segment from the first thoracic to second or third lumbar segment. They then synapse in ganglia of two types – paravertebral ganglia or peripheral (collateral, prevertebral) ganglia.

Paravertebral (vertebral) ganglia form a chain running down either side of the vertebral column. Each ganglion is associated with a spinal cord segment but they are connected with adjacent segments to form the sympathetic chain. The preganglionic fibres leave the spinal nerve and reach the vertebral ganglia associated with the corresponding segment via short myelinated nerve trunks called *white rami communicantes* (Figure 1.4). The preganglionic fibres may synapse in the vertebral ganglion. Alternatively, it may branch and run up or down the sympathetic chain to synapse in a more distant vertebral ganglion.

The preganglionic axon provides many synapses with the ganglion cell body. The fibres run parallel with the dendrites to form axodendritic synapses. The preganglionic nerve endings contain round agranular synaptic vesicles which contain the transmitter, Ach. The presence of catecholamine in the sympathetic ganglion cell body is indicated by formaldehyde-induced fluorescence. This represents neurotransmitter which has been synthesized ready for transfer to the nerve terminal. However, there are also some non-fluorescent cell bodies which display a strong reaction for

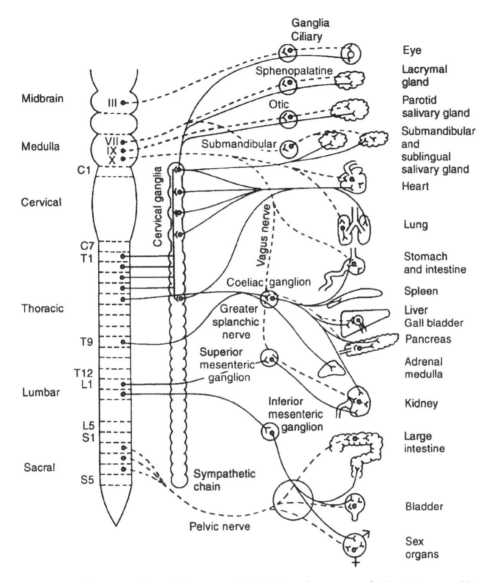

Figure 1.3 Efferent outflows of the sympathetic (—) and parasympathetic (---) nerves. Note that some efferent pathways leaving the spinal cord, together with their paths through the sympathetic chain, have been omitted for simplicity.

choline-acetylase, the enzyme responsible for synthesis of Ach. These cell bodies are probably of the minor group of sympathetic neurones which have Ach as their transmitter and which provide vasodilator fibres to the limbs and sweat-promoting (sudomotor) fibres to the skin.

Also observed in ganglia are chromaffin cells which have a high-intensity staining for catecholamines including dopamine and noradrenaline and have a granular appearance in the electron microscope. These extramedullary chromaffin cells differ from those in the medulla in not containing adrenaline. This is because of the absence of the phenylethanolamine-N-methyltransferase which is induced by the high levels of corticosteroids found in the adrenal gland (see Chapter 11). Some of these are described as small intensely fluorescent (SIF) cells. They have short processes and thus differ from the neurone cell bodies of the ganglion. They are sheathed by satellite cells, have a central nucleus and the cytoplasm has abundant ribosomes and mitochondria. Nerve endings originating from the spinal cord synapsing with chromaffin cells of the rat superior cervical ganglion have been identified. The precise role of these chromaffin cells is uncertain. They may release catecholamines which exert a modulatory role upon ganglionic transmission by inhibition of Ach release or they may form interneurones between sympathetic ganglionic cell bodies. The released catecholamine may induce IPSPs on the dendrites of the postganglionic nerve and thus inhibition of transmission.

The postganglionic axons may leave the vertebral ganglion as a postganglionic nerve trunk and pass directly to an effector such as the heart or lungs. However, others may return to the spinal nerve via *grey rami communicantes* and are then distributed to smooth muscle of blood vessel walls, sweat glands and erector pili muscles of the skin which make the hair stand on end (Figure 1.4). Grey rami are usually unmyelenated.

Peripheral ganglia lie more distally from the spinal cord, within the body cavities. The preganglionic sympathetic fibres emerging from the ventral root of the spinal cord (first thoracic to second or third lumbar) cross the white rami, do not synapse in

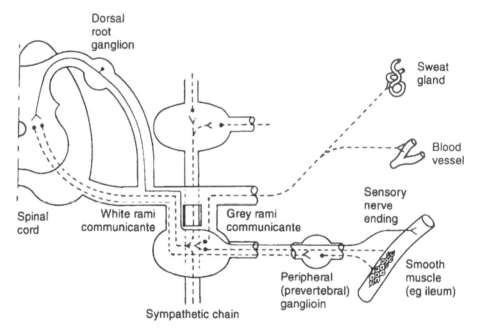

Figure 1.4 Sympathetic nerve pathways through one spinal segment and its associated vertebral ganglion in the sympathetic chain. Afferent sensory pathway illustrated by solid line (—). Alternative efferent pathways (---): (a) dividing in the vertebral ganglion and passing up or down the chain to adjacent segments; (b) synapsing in the vertebral ganglion of the sympathetic chain and then passing through a peripheral ganglion to the smooth muscle; (c) synapsing in the vertebral ganglion and returning via the grey rami communicante to the spinal nerve to innervate blood vessels and sweat glands; or (d) synapsing more distally in a peripheral ganglion (eg coeliac ganglion) after passing through the sympathetic chain.

12

the vertebral ganglion but pass straight through and synapse more distally in the peripheral ganglia (Figure 1.4). These are often not bilaterally symmetrical and often form a plexus such as the abdominal plexus which contains the coeliac, superior mesenteric and inferior mesenteric ganglia. The coeliac and superior mesenteric ganglia constitute a crescent-shaped structure on both sides of the coeliac artery known as the solar plexus. This structure gives rise to several nerve trunks running in all directions including the gastric, hepatic and splanchnic nerves which supply the stomach, liver and small intestine (Figure 1.3). These nerves are usually unmyelenated, but there are thinly myelenated fibres, for example, running from the superior cervical ganglion to the eye. This is the uppermost ganglion of the body, receiving preganglionic fibres from the first to fourth thoracic segments, and forms an extension of the sympathetic chain.

There are many fewer preganglionic fibres than ganglion neurones (1:63 in man) which indicates *divergence* of each preganglionic fibre onto several ganglionic cells. However, most ganglionic neurones also receive synapses from many preganglionic fibres (*convergence*).

1.5.2 *The adrenal medullae*

The adrenal medullae may be regarded as specialized sympathetic ganglia. The chromaffin cells receive a sympathetic preganglionic fibre from thoracic segments 5 to 9, which form the greater splanchnic nerve supplying the coeliac ganglion. However, the nerves supplying the adrenal medullae do not synapse in the coeliac ganglion but pass through and as they enter the medullae lose their myelin sheath. The nerve terminals form close junctions with the chromaffin cells. These are cholinergic nerve endings which release Ach onto micotinic receptors. Denervation of these nerves does not result in a loss of the catecholamine synthesis and storage capacity of the chromaffin cells. The chromaffin cells are of two types, noradrenaline-containing, and adrenaline-containing which exist in a ratio of 1:7. The latter arise because the enzyme that converts noradrenaine to adrenaline (phenethanolamine-N-methyltransferase, PNMT) is induced by corticosteroids. These cells receive the adrenocortical hormones in the blood from the adrenal cortex. The release of catecholamines is by exocytosis and is Ca^{2+}-dependent. Stress and exercise increase the output of adrenaline into the circulation by at least 10-fold to reinforce the action of sympathetic nerve discharge. Chromaffin cells have also been shown to contain peptides which are mainly the opioid peptides, the enkephalins. Also found are neuropeptides including neuropeptide Y and vasoactive intestinal peptide (VIP). These may also be located in the neurone terminals, however it has been suggested that they are co-released with catecholamines during stress. The enkephalins may also be released into the circulation to play an important role in adaptation to the environment in response to stress (Carmichael 1986).

1.5.3 *Parasympathetic pathways*

There are two regions from which parasympathetic efferents arise; these are the cranial outflow and the sacral outflow, which are therefore anatomically above and below the sympathetic outflows, respectively. The preganglionic fibres of the cranial outflow are present in the IIIrd cranial nerve (oculomotor nerve) arising from the midbrain, and the VIIth (facial nerve), IXth (glossopharyngeal nerve) and Xth (vagus nerve) cranial nerves arising from the medulla oblongata. These nerves innervate the eye, salivary glands, lacrymal glands, the heart, lungs and upper gastrointestinal tract as shown in Table 1.1. The preganglionic fibres of the sacral outflow leave the spinal cord via the ventral roots of sacral segments 2,3 and 4.

Table 1.1 Innervation of organs by the sympathetic and parasympathetic divisions of the autonomic nervous system and the physiological response to nerve activity

Organ	Tissue	Sympathetic nerves	Receptor[a]	Parasympathetic nerves
Eye	Iris			
	Dilator pupillae (radial muscle)	Contraction, dilated pupil (mydriases)	α_1	NI
	Constrictor pupillae (circular muscle)	NI		Contraction, constricted pupil (miosis)
	Ciliary body	Relaxation for far vision	β_2	Contracts for near vision
	Eyelids smooth muscle	Eyelids raised	α	NI
	Conjunctival blood vessels	Vasoconstriction	α_1	
	Lacrymal glands			
	Vasculature	Vasoconstriction	α_1	NI
	Secretory cells	NI	—	Secretion
	Nictitating membrane	Contraction	α_1	NI
Salivary glands	Blood vessels and mucous cells	Vasoconstriction, thick mucinous secretion	α_1	Vasodilatation
	Acinar cells	Secretion of amylase	β_1	Profuse watery secretion
Gastro-intestinal tract	Circular/longitudinal muscle tone and motility	Relax, reduced motility, delayed passage	α_1/β_1	Contraction, increased motility and passage
	Sphincters	Contract	α_1	Relax
	Digestive enzymes	Inhibition	α_2	Increased secretion
Gall bladder and duct	Smooth muscle	Relaxation	β_2	Contraction
Pancreas	Acini	Decreased enzyme secretion	α	Increased secretion
	β-Cells of islet	Insulin secretion increased	β_2	Release of insulin
		Insulin secretion decreased	α_2	
Spleen	Capsule, arteries	Contracts to expell blood	α_1	NI
Kidneys	Renal arterioles	Vasoconstriction, reduced flow	α_1	NI
	Juxtaglomerular apparatus	Renin release	β_1	NI
Bladder	Detrusor muscle	Relaxation? Neck contracts	—	Contracts
	Trigone and internal sphincter	Contraction	α_1	Relaxes (micturition)
Ureter	Motility and tone	Increased	α_1	NI?

14

Table 1.1 (*Continued*)

Organ	Tissue	Sympathetic nerves	Receptor[a]	Parasympathetic nerves
Heart	SA node, AV node and bundle	Increased rate (positive chronotropy) and conduction	β_1	Slowed (negative chronotrophy)
	Atria	Increased contractility (positive inotropy)	β_1	Reduced contractility (negative inotropy)
	Ventricular myocardium	Increased contractility	β_1	NI
Blood vessels	Head, thorax, mucous membranes, skin, viscera	Vasoconstriction	α_1	Localized vasodilatation mostly NI
	Face	Vasodilatation (blush)		—
	Skeletal muscle	Vasodilatation predominates[b]	M	NI
	Coronary	Vasoconstriction, reduced flow[b]	α_1	Vasodilatation, increased flow
Blood pressure	Cardiac output × peripheral resistance	Increases		Falls
Skin	Blood vessels	Vasoconstriction (pallor)	α_1	NI
	Eccrine sweat gland	Dilute secretion	M	NI
	Apocrine sweat gland	NI		NI
	Pilomotor muscles	Contraction (piloerection)	α_1	NI
				NI
Lungs	Smooth muscle	Relax, bronchodilatation	β_2	Contract, bronchoconstriction
	Mucous glands	—		Increased secretion
Genital tract: male	Penile blood vessels	NI	—	Vasodilatation (erection)
	Vasa deferentia, seminal vesicles, prostate	Contraction (ejaculation)	α_1	NI
Genital tract: female	Uterus	Contraction, but depends on hormonal influences and species	α_1	NI?
Metabolic	White adipose tissue	NI		NI
	Brown adipose tissue	Lipolysis and thermogenesis	β_3	NI
	Liver	Glycogenolysis, rise in blood sugar	β_2	Minor, ↓ glucose output
	Thyroid gland	Thyroid secretion	β	
		Vasoconstriction	α	Vasodilatation

Notes: [a] Receptor types involved in responses mediated via nerve stimulation. α, β_1, β_2 and β_3 are adrenoceptors, M are muscarinic receptors. Other receptors may be present but are not involved in the response to nerve activity.
[b] Metabolic vasodilatation usually has overriding control together with β_2-mediated effects of circulating adrenaline.
NI, Not innervated; SA, sinuatrial; AV, atrioventricular.

These combine to form the pelvic nerve which supplies the bladder, lower abdominal viscera and external genitalia.

The preganglionic parasympathetic fibres synapse in ganglia, which unlike sympathetic ganglia lie very close to or within the organ that they innervate. The postganglionic parasympathetic fibres are therefore short and non-myelenated. Often the cell bodies of the postganglionic neurones are distributed throughout the tissue so that no discrete ganglia occur but instead they form a network of interconnecting fibres known as a plexus. For example, the cardiac ganglia of the vagus nerve form a plexus around the great vessels or in the atrial wall which give rise to the postganglionic fibres to the heart. Also running through this plexus are sympathetic postganglionic fibres and sensory afferents. The vagus nerve also contains somatic motor fibres which innervate the striated muscle of the pharynx and larynx, which controls swallowing. The ciliary ganglion lies behind the eyeball, receiving preganglionic fibres from the oculomotor nerve, and innervates the ciliary muscle and smooth muscles of the sphincter pupillae (circular muscle) of the iris.

1.5.4 The Enteric Nervous System

The vagal efferents to the gastrointestinal tract are unique in that they synapse with ganglia located within the walls of the intestine. They are thus exposed directly to the mechanical activity of the gut. These ganglion cells form two major plexuses: the myenteric (Auerbach's) plexus situated between the inner circular and outer longitudinal muscle layers and the submucosal (Meisner's) plexus. These plexuses are interconnected and stretch uninterrupted the length of the gastrointestinal tract. Together with sympathetic postganglionic neurones which synapse with them, they form what is known as the enteric nervous system. This is defined as the intrinsic innervation of the gastrointestinal tract. Although originally classified by Langley (1921) as a third division of the autonomic nervous system, it is in fact comprised of components of the sympathetic and parasympathetic divisions (Gershon 1981). However, in the gut the autonomic innervation appears to be different from that of other organs in the degree of independence from the CNS that it displays. Peristaltic activity of the gut is a local nervous reflex involving the enteric nervous system which can occur *in vitro* without any central connections.

The enteric nervous system consists of a diverse group of neurone types characterized by their electrophysiological properties and neurotransmitter content. Based upon their presence and pharmacological activities, substances with a putative transmitter role in the peristaltic reflex include Ach, noradrenaline, dopamine, 5-hydroxytryptamine (5-HT), adenosine triphosphate (ATP, purinergic neurones), enkephalin and peptides such as vasoactive intestinal peptide (VIP), somatostatin, cholecystokinin, substance P (a tachykinin) and bombesin.

The role of these substances in non-adrenergic non-cholinergic pathways will be discussed in Chapter 12. Of these, Ach has been shown unequivocally to be released from the myenteric plexus (Paton & Zar 1968). Longitudinal muscle, from which the myenteric plexus had been removed, failed to release Ach. These cholinergic neurones are involved in the control of intestinal motility, the short non-myelenated postganglionic neurones innervating both longitudinal and circular muscle. The peristaltic reflex is initiated by distension of the intestinal lumen, or by application of pressure or of 5-HT to the mucosa. This can be demonstrated in an isolated segment of guinea-pig intestine such as the Trendelenburg preparation (1917). In this experiment the pressure is raised within a sealed segment of intestine to induce waves of phasic pressure changes.

The parasympathetic ganglia of the enteric plexuses are vital to the peristaltic reflex since it can be inhibited by ganglion blockade with hexamethonium and partially by antagonism of Ach at the postganglionic nerve endings with atropine. The precise location of the sensory nerve endings is unclear; they may lie in the neuronal processes in the myenteric plexus or processes reaching into the circular muscle or submucosa. Stimulation of these local afferent neurones activates three distinct neural pathways. First, an ascending excitatory pathway is activated to contract circular muscle behind the bolus. These are short processes and probably utilize Ach as the transmitter. At the same time, a descending inhibitory pathway running along the myenteric plexus is activated, which relaxes circular muscle in advance of the bolus. This probably has two components. One is activated after a short latency (1 sec) and is associated with fast EPSPs. The transmitter is non-cholinergic and non-adrenergic and may involve ATP or VIP. Another component of this descending inhibition is thought to involve a short interneurone and inhibition of the cholinergic excitation. The final phase of the peristaltic reflex is a descending excitation. This is a longer latency (2–11 sec) excitation pathway and may receive sensory inputs from the submucous plexus. This is probably responsible for the propulsive wave of peristalsis involving longitudinal muscle (Figure 1.5).

The sympathetic nervous system impinges upon the enteric nervous system through postganglionic neurones which synapse with the intramural parasympathetic ganglia. Thus, sympathetic nerve activity induces intestinal relaxation predominantly by inhibition of the parasympathetic excitatory activity, and to a lesser extent by direct inhibition of intestinal smooth muscle. The site of action is predominantly a presynaptic inhibition of Ach release onto the parasympathetic ganglia. The

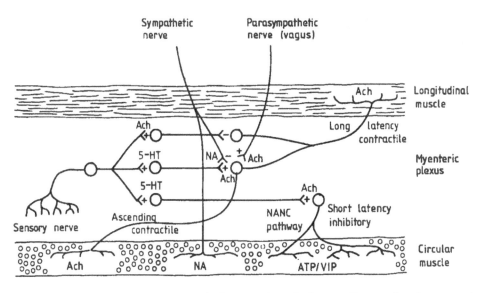

Figure 1.5 Enteric nervous system of the gastrointestinal tract. Proposed pathways and neurotransmitters are shown, Ach (Ach), noradrenaline (NA), 5-hydroxytryptamine (5-HT), adenosine triphosphate (ATP) and vasoactive intestinal peptide (VIP). Positive and inhibitory transmission indicated by + and − signs, respectively.

sympathetic innervation of the intestine is regarded as extrinsic in nature since the cell bodies lie outside the intestinal wall. The nerve endings and neurotransmitters (noradrenaline and dopamine) disappear after extrinsic denervation. Some intrinsic nerves, however, show evidence of containing 5-HT which may serve a neurotransmitter role. Initiation of the peristaltic reflex by mucosal stimulation has been shown to release 5-HT from the myenteric plexus. The presence of 5-HT has been demonstrated in association with neurones, although the major source is the 5-HT storage cells (enterochromaffin cells) of the mucosa. 5-HT has also been shown to have complex pharmacological activity on the intestine. It has excitatory activity on enteric ganglia causing release of Ach and contraction of smooth muscle but it also activates the non-adrenergic non-cholinergic inhibitory pathways. The induction of peristalsis by application of 5-HT is unlikely to be due to stimulation of the distension-sensitive receptors (Gershon 1981, North 1986).

1.5.5 Summary of Efferent Autonomic Pathways (Figure 1.6)

1 The sympathetic nerves leave the spinal cord via the thoracolumbar segments, with ganglia lying in the sympathetic chain or more peripherally in the body cavities.

2 The parasympathetic nerves emerge from the brain stem or the sacral segments of the spinal cord. The ganglia lie close to or within the organ of innervation, with very short postganglionic fibres.

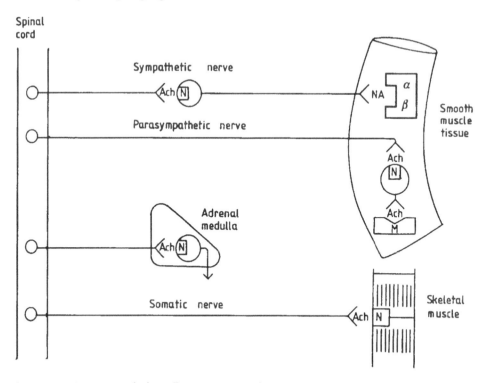

Figure 1.6 Summary of the efferent nerve pathways showing the location of synapses, transmitters (Ach, acetylcholine; NA, noradrenaline) and receptors (M, muscarinic; N, nicotinic; α- and β-adrenoceptors).

3 The neurotransmitter at both sympathetic and parasympathetic ganglia is Ach, where it interacts with *nicotinic* receptors.

4 The neurotransmitter at postganglionic parasympathetic nerve endings in contact with the effector is Ach, where it acts on *muscarinic* receptors.

5 The neurotransmitter at the postganglionic sympathetic nerve endings is usually noradrenaline, where it acts on α- or β-adrenoceptors. Some sympathetic postganglionic nerves release Ach in sweat glands and the blood vessels supplying skeletal muscle.

1.6 Physiological Responses of the Innervated Organs to Autonomic Nerve Stimulation

Generally, the organs are innervated by both divisions of the autonomic nervous systems, which have opposing effects. However, because the parasympathetic nerves are less widely distributed and limited to the thoracic, abdominal and pelvic viscera, some organs receive only a sympathetic innervation. The physiological effects of sympathetic nerve activity to an organ can be anticipated by considering its role in preparing the body for flight or fight. So, for example, the rate and force of cardiac contractions increase to improve the blood flow to skeletal muscle where vascular smooth muscle is dilated. Sympathetic activity is generally more widespread which is reflected in the greater branching of sympathetic fibres in the sympathetic chain. The origins of sympathetic fibres supplying a particular organ are therefore often difficult to identify since there is considerable overlap between adjacent ganglia. A general scheme is presented in Figure 1.3.

1.6.1 *The Eye*

The sympathetic fibres originate from the upper three or four thoracic segments of the spinal cord. They pass through the sympathetic chain and ascend to the superior cervical ganglion where they synapse with postganglionic fibres running to various tissues innervated in the head and neck. The sympathetic nerves innervate the dilator pupillae muscle (outer radial muscle) of the iris, the fibres spreading over the whole iris. They also innervate the blood vessels of the conjunctiva and retina and the smooth muscles of the eyelids (the superior and inferior tarsal muscles).

Sympathetic stimulation causes dilatation of the pupil (mydriasis) to permit more light into the eye, blanching of the conjunctiva due to vasoconstriction and raising of the upper eyelid to produce a staring gaze. In certain species, such as the cat, the nictitating membrane, or third eyelid, is well developed and receives a sympathetic innervation. Stimulation causes contraction of the smooth muscle which retracts the cartilagenous membrane into the corner of the eye.

The parasympathetic nerves pass in the IIIrd cranial nerve via the ciliary ganglion. These innervate the constrictor pupillae muscles (inner circular muscle) of the iris and the smooth muscle of the ciliary body. The ciliary body forms a ring of smooth muscle around the lens, which is held in place by suspensory ligaments attached to the ciliary body. Stimulation of parasympathetic nerves contracts the ciliary body moving it inwards towards the lens. The suspensory ligaments become less taut and the lens assumes a greater curvature for focusing on near objects (Figure 1.7). This

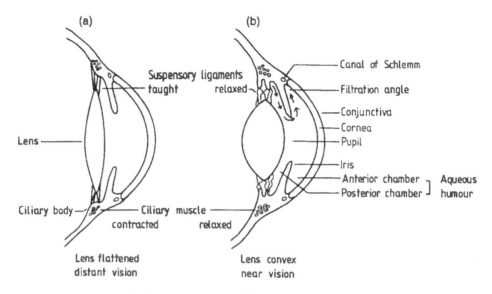

Figure 1.7 Diagram of the anterior part of the eye. (a) Ciliary muscle contracted under parasympathetic innervation to tighten suspensory ligaments and pull the lens into a flattened shape for distant vision. (b) Ciliary muscle relaxed, allowing suspensory ligaments to relax. The lens becomes convex for near vision. Arrows indicate direction of flow of aqueous humour from site of formation in ciliary body to its drainage via the filtration angle into the canal of Schlemm.

process is termed *accommodation* and is accompanied by parasympathetic-mediated constriction of the pupil, which increases the resolving power of the lens. For distant vision, the ciliary body relaxes, the suspensory ligaments become taut and flatten the lens.

At the junction between the cornea and the base of the iris and ciliary body is the canal of Schlemm. This is the conventional route of drainage of aqueous humour from the posterior and anterior chambers of the eye back into the venous and lymphatic system; it reaches the canal of Schlemm via a trabecular network. The aqueous humour is formed by capillaries of the ciliary body and provides nourishment to the capillary-free lens and cornea. When the iris and ciliary body constrict under parasympathetic stimulation, they move away from the canal of Schlemm and widen the filtration angle and facilitate drainage (Figure 1.7) (Kaufman *et al.* 1984).

1.6.2 *Salivary Glands*

The secretion of saliva is solely under the control of the autonomic nervous system with no hormonal influence. The postganglionic sympathetic fibres to the salivary glands originate in the superior cervical ganglion. These end mainly in blood vessels of the glands but also in close contact with the secretory acinar cells. Species differences occur in the degree of sympathetic innervation; the submandibular glands of the guinea-pig, rat and mouse receive none, other than to the blood vessels. Acinar cells of the rat parotid gland, however, are in intimate contact with sympathetic nerve endings. The effect of sympathetic nerve stimulation upon salivation is often described as producing a thick, mucous-rich and viscous saliva. This is probably mainly a secondary consequence of the reduced blood flow through the glands due to vasoconstriction. As a result, the mouth feels dry during stressful situations. The sympathetic nerves or circulating adrenaline exert their effects on electrolyte and water secretion by activating α-adrenoceptors. In contrast, secretion of enzymes including amylase and peroxidase by exocytosis from acinar cells is stimulated by sympathetic activation through β-adrenoceptors (Danielsson *et al.* 1982).

The parasympathetic innervation arises from the medulla, the VIIth cranial nerve synapsing in the submandibular ganglia before reaching the *submandibular* and *sublingual* salivary glands. The IXth cranial nerve synapses in the otic ganglion before innervating the largest of the salivary glands, the *parotid* gland. The nerve endings penetrate the basal membrane and come into direct contact with the acinar cells. Parasympathetic stimulation produces a profuse watery saliva rich in enzymes to aid digestion. Although Ach is the primary parasympathetic neurotransmitter, the neuropeptide VIP is also found in parasympathetic neurones supplying the salivary glands, where it is involved in the vasodilator response (see Chapter 12).

1.6.3 *Lacrymal Glands*

The lacrymal or tear glands are located under each upper lateral eyelid. The tears pass through lacrymal ducts into the conjunctival sac under the upper eyelid. They flow across the eye and drain into two small lacrymal ducts on either side of the inner canthus. These remove excess tears into a lacrymal sac which is a dilated portion of the nasolacrymal duct running into the nasal cavity. Hence excessive production of lacrymal secretion due to irritation or allergy (eg hay fever) often results in the need to blow the nose. When the production of lacrymal secretion exceeds the drainage capacity, tears become visible.

The lacrymal glands receive sympathetic fibres via the superior cervical ganglia which mainly innervate the vasculature. Stimulation produces vasoconstriction. The parasympathetic innervation is via the VIIth cranial nerve which synapses in the sphenopalatine ganglion with postganglionic fibres which terminate with the secretory cells. Stimulation produces profound lacrymal secretion.

1.6.4 *Thyroid Gland*

The sympathetic innervation of the thyroid gland is via the middle and superior cervical ganglia. The parasympathetic preganglionic fibres arise from the dorsal nucleus of the vagus nerve. Both sympathetic and parasympathetic fibres innervate primarily the vasculature of the gland, the sympathetic nerves causing vasoconstriction and the parasympathetic causing vasodilatation. Stimulation of the sympathetic fibres also causes an increase in secretion of thyroid hormone (Melander *et al.* 1974). In addition, β-adrenoceptor stimulation increases the synthesis of thyroid hormone (Melander *et al.* 1973). As will be seen later (Chapter 4), there is a close interrelationship between the thyroid gland and sympathetic activity, since raised thyroid hormone levels have been found to increase the number of β-adrenoceptor binding sites in organs such as the heart (Williams *et al.* 1977).

1.6.5 *Heart*

Sympathetic nerves to the heart leave the spinal cord at the upper five or six thoracic segments. Two ganglionic synapses are then involved. First, the preganglionic fibres synapse within the upper five or six ganglia of the sympathetic chain and second, fibres pass through these vertebral ganglia and ascend to the three cervical ganglia (intermediate, middle and superior cervical). (NB. The first thoracic is sometimes fused with an inferior cervical sympathetic ganglion to form the stellate ganglion.) The postganglionic fibres arising from this wide area run together and merge at the cardiac plexus along with the parasympathetic fibres. The sympathetic nerves terminate predominantly in the sinuatrial node pacemaker tissue, the atrioventricular node and the conducting tissue. The distribution of sympathetic fibres to the ventricular and atrial myocardium is less conspicuous. Stimulation of the sympathetic nerves increases pacemaker activity resulting in speeding of the heart (tachycardia, positive chronotropy). The force of contraction also increases (positive inotropy) resulting in a raised cardiac output.

The parasympathetic supplies to the heart are the vagus nerves arising from the medulla. These innervate the sinuatrial node (right vagus), atrioventricular node (left vagus) and atrioventricular bundle (Bundle of His) with only minor distribution of parasympathetic fibres to ventricular muscle. Vagal nerve activity causes slowing of heart rate (bradycardia, negative chronotropy) and slowing of conduction through the Bundle of His (negative dromotropy). The stroke volume is in fact increased but cardiac output falls because of the fall in heart rate (Higgins *et al.* 1973).

1.6.6 *Blood Vessels*

It is the smooth muscle of predominantly arterioles that receive the autonomic innervation, since these vessels control the peripheral resistance and hence blood pressure. The innervation is usually confined to the outer adventitial layer so that only the outer smooth muscle cells are directly innervated. Arteries are innervated but capillaries are not, and veins are only sparsely supplied. Most of the vascular sympathetic fibres originate from the ganglia of the sympathetic chain, the postganglionic fibres returning to the spinal nerve via the grey rami communicantes before passing to regions adjacent to the spinal segment (Figure 1.4). Head and sacral regions are supplied via the cervical ganglia and lower sacral ganglia, respectively, while visceral regions also receive preganglionic fibres in the splanchnic nerve which synapse in peripheral ganglia of the coeliac and mesenteric plexuses.

The sympathetic innervation to the blood vessels to the head, thorax, skin and visceral regions produce vasoconstriction. The skin becomes pale and heat loss is prevented, hence the term 'cold feet' to describe the reaction to fear. In some regions, such as the skin of the face, sympathetic stimulation can also cause vasodilatation, as observed in blushing. The blood vessels of mucous membranes, for example in the nasopharyngeal passages, constrict to reduce blood flow and allow mucous secretion to dry up. The visceral regions require less blood flow since digestion is slowed in stressful situations. Since this vascular bed contributes a large proportion to the total peripheral resistance, the vasoconstriction here, combined with the raised cardiac output, results in the increase in blood pressure that occurs in stress. The blood vessels of skeletal muscle, in contrast, dilate in response to sympathetic stimulation, increasing blood flow for increased muscle activity. The postganglionic fibres, however, differ from the other sympathetic nerves in that they release Ach as the neurotransmitter and are therefore sympathetic cholinergic nerves. Clearly the sympathetic nervous system redistributes blood from the skin and viscera to skeletal muscle with a net increase in blood pressure.

Parasympathetic nerves innervate only few blood vessels. In the upper part of the body, parasympathetic fibres in the cranial outflows from the medulla produce vasodilatation on stimulation. There is no parasympathetic nerve supply to the vessels of the skin or visceral region.

1.6.6.1   *Coronary circulation*

The circulation of the heart represents a special case, since it is greatly affected not only by the direct autonomic innervation but also indirectly by the autonomic effects upon the mass of myocardium that it supplies. Thus, it is difficult to separate these indirect and direct effects when autonomic nerves to the heart are stimulated. The indirect actions arise from the physical compression of the coronary vessels by the surrounding myocardium so that sympathetic-mediated positive inotropy and chronotropy results in reduced coronary flow. The myocardial activity also affects coronary flow through the local autoregulatory mechanisms. Increases in myocardial activity cause local hypoxia, hypercapnia (raised CO_2 levels) and a concomitant acid pH, all of which are associated with coronary vasodilatation. The local regulation of coronary flow by changes in O_2 level is thought to be through the release of adenosine (the adenosine hypothesis), first proposed by Berne (1980). There is strong evidence in favour of this hypothesis but it still remains to be irrefutably accepted. The precise link between oxygen levels, CO_2, adenosine and other possible mediators such as K^+ ions in the control of coronary flow have been extensively studied, however detailed consideration of this topic is beyond the scope of this book (Feigl 1983).

Both large and small coronary vessels do receive an innervation by the sympathetic and parasympathetic nerves independent of that to the myocardium. The parasympathetic innervation is via the vagus, the nerve endings of which terminate in the adventitia–media border. The ganglia probably lie at the base of the heart. The postganglionic sympathetic fibres arise from the stellate ganglion and have varicose swellings with dense granules innervating the adventitia–media border with a junction of $0.1–0.5$ μm.

The responses to sympathetic and parasympathetic nerve stimulation are complicated by indirect myocardial effects. However, if the heart is arrested by electrically-induced fibrillation or K^+-induced depolarization, parasympathetic stimulation has been shown to cause coronary vasodilatation. Stimulation of the stellate sympathetic ganglion to beating hearts causes an increase in coronary flow indicative of vasodilatation. However, the majority of this effect is due to the metabolic autoregulatory consequences of the concomitant tachycardia. In arrested hearts, the predominant effect of sympathetic nerve stimulation is vasoconstriction via α-adrenoceptors. Vasodilator β-adrenoceptors are present but they do not appear to be activated by nerve stimulation (Feigl 1983).

1.6.7 *Skin*

The innervation and responses of the cutaneous blood vessels have been described above. However, it is in the autonomic innervation of the sweat (sudoriferous) glands that most confusion occurs. Sweat glands of humans are of two types: *eccrine* and *apocrine* glands. The eccrine glands are widely distributed in the skin of the body and secrete a dilute sweat in response to sympathetic nerve activity and a rise in body temperature. Eccrine glands that respond to psychological stress are most numerous on the palms of the hands and soles of the feet. They are innervated by sympathetic fibres, the postganglionic neurones originating in the adjacent ganglia of the sympathetic chain (Figure 1.4). In the head region, the postganglionic fibres leave the superior cervical ganglion to innervate sweat glands, cutaneous blood vessels and erector pili muscles. The sympathetic postganglionic nerves to sweat glands utilize Ach and not noradrenaline as the neurotransmitter. It is this that causes confusion in certain textbooks since sweating is variously described as a cholinergic, sympathetic or even parasympathetic response. However, it is clearly a *sympathetic* response.

The apocrine glands, in contrast, are larger and distributed mainly around the external genitalia and the axillae (armpits). The apocrine sweat glands become active in adolescence and appear to be involved in sexual function through the release of pheromones in the secretions. These airborne products may stimulate the olfactory senses resulting in release of gonadotrophic hormones, LH (luteinizing hormone) and FSH (follicle-stimulating hormone), from the anterior pituitary gland. These control sexual activity through their release of sex hormones (testosterone and oestradiol) from the gonads. Body odour also arises primarily from the bacterial decomposition of apocrine gland secretions. These glands are not innervated by autonomic nerves but may respond via α-adrenoceptors to circulating catecholamines (adrenaline) released from the adrenal medullae in response to stress.

Two types of sweating are recognized. *Thermoregulatory* sweating occurs over the entire body in response to changes in environmental temperature but does not involve the apocrine glands. *Emotional* sweating is confined to the palms, soles and axillae. In hot surroundings, mental stress may induce sweating in the rest of the body. Thermoregulatory sweating is initiated by an increase in body temperature detected either by peripheral heat-sensitive receptors or by the hypothalamus. Another sympathetic response of the skin is piloerection – the hair stands on end. This is a thermoregulatory response to cold since the air trapped close to the skin in the hair of certain animals serves as an insulating layer. In certain animals this also serves a defense role to attack, since it creates the impression of a larger size, for example, the hair on a cat's back stands up. In humans, where body hair is minimal, this response to cold or emotion is seen as 'goose-flesh'. Again, there is no parasympathetic supply to piloerector muscles.

The relationship between sweating and cutaneous blood flow in response to stress or temperature changes is complex. In stress situations or exposure to extreme cold, the sympathoadrenal system is activated to cause cutaneous vasoconstriction and piloerection to minimize heat loss. Shivering also occurs. Simultaneously, adrenaline release causes vasodilatation of skeletal muscle blood vessels to redistribute blood to the interior of the body. It also induces lipolysis (see later) and glycogenolysis for extra thermogenesis. Exposure to heat causes firstly cutaneous vasodilatation and this is followed by sweating.

The anterior region of the hypothalamus appears to be the predominant centre of the brain for thermoregulation. Thermosensitive neurones of the anterior region respond to a rise in blood temperature causing vasodilatation and sweating. The anterior region is often regarded as a parasympathetic centre of the hypothalamus since its stimulation electrically is associated with a wide range of parasympathetic responses. However, one of these responses is sweating, a sympathetic response. The relationship between cutaneous blood flow and sweating and their control by the hypothalamus is therefore complex and poorly understood. While sweating is the sympathetic response to heat and stress, it is clearly not accompanied by a sympathetic-mediated vasoconstriction to aid heat loss, but by vasodilatation. Part of the cutaneous vasodilatation may be due to the release of bradykinin from sweat glands.

1.6.8 *Lungs*

Sympathetic preganglionic fibres originate from the upper four or five thoracic segments of the spinal cord and synapse in the corresponding thoracic ganglia of the sympathetic chain before reaching the lungs. Some synapse in the cervical and stellate ganglia. The postganglionic fibres innervate the trachea, bronchi and extend as far as the bronchioles. Sympathetic innervation appears to become less dense on descending to the central airways in humans and guinea-pigs (Richardson 1979). Sympathetic stimulation causes relaxation of the circular smooth muscle of the respiratory tract leading to bronchodilatation. An excitatory component (α-adrenoceptor-mediated) exists in the trachea of many species including man, but not the guinea-pig (Gabella 1987). The parasympathetic innervation of the airways is via the vagus nerve which synapses in the plexus around the hilum. The postganglionic nerves innervate the bronchial musculature and mucous glands, stimulation causing bronchoconstriction and increased secretion, respectively. A non-adrenergic non-cholinergic excitatory response has also been reported (see Chapter 12) (Gabella 1987).

1.6.9 *Gastrointestinal Tract*

Sympathetic preganglionic fibres innervating the oesophagus arise from the fourth to sixth thoracic segments, while those innervating the stomach, small intestine and ascending and transverse colon arise from the fifth or sixth to the eleventh thoracic segments. The nerves from the upper segments synapse in the adjacent vertebral ganglia of the sympathetic chain, whereas lower down they pass through the vertebral ganglia emerging as the splanchnic nerves to synapse in the coeliac ganglion or mesenteric ganglia. Postganglionic fibres pass together with blood vessels of the mesentery to the smooth muscle of the small intestine. The descending colon, sigmoid colon and rectum receive sympathetic innervation via synapses in the inferior mesenteric, hypogastric and pelvic plexuses. Sympathetic stimulation inhibits tone and motility of the gastrointestinal tract. The tone in the circular muscle of the sphincters, such as the cardiac and pyloric sphincters at the entrance and exit, respectively, of the stomach is, however, increased resulting in delayed emptying (enterogastric reflex). It is questionable whether sympathetic innervation decreases gastric secretion but it probably inhibits secretion of other intestinal secretions. Certainly, noradrenaline exerts antisecretory effects in the rat jejunum (Williams *et al.* 1990).

The parasympathetic innervation of the gastrointestinal tract from oesophagus to transverse colon is via the vagus nerve. It passes through the coeliac and mesenteric plexuses to synapse with intrinsic neurones of the myenteric (Auerbach's) plexus (Figure 1.5). The descending colon and rectum receive preganglionic parasympathetic fibres from the second to fourth sacral segments. These run in the pelvic nerves to again synapse in the myenteric plexus. Stimulation of these extrinsic parasympathetic nerves increases motility of the gut and the secretion of gastric and other digestive enzymes, and of gastric acid from the parietal (oxyntic) cells of the stomach. The tone of the sphincters is reduced. Vagal nerve stimulation also causes the release of gastrin from the enteroendocrine cells of the pyloric region of the stomach. This involves the intermediate release of various peptides which are found in the endocrine cells and enteric neurones of the gastrointestinal tract. These include somatostatin, cholecystokinin (CCK-8) and gastrin-releasing peptide (GRP). The relationships between these and gastric acid secretion are described more thoroughly in Chapters 8 and 12 (see Figure 8.3) (Dockray 1992). Gastrin is carried by the blood to the parietal cells which are stimulated to secrete gastric acid. Thus digestion and passage of contents is accelerated. The peristaltic reflex, however, involves a local reflex of the intrinsic enteric nervous system which has already been described. The extrinsic autonomic nerves appear to exert a modulating effect upon this local reflex.

1.6.10 Liver

The liver receives predominantly sympathetic fibres which arise from thoracic segments 5 to 9. These pass via the splanchnic nerve to synapse in the coeliac ganglion. Sympathetic nerve stimulation promotes glycogenolysis leading to raised blood sugar levels. A minor parasympathetic innervation via the vagus and hepatic parasympathetic nerve, on stimulation, decreases hepatic glucose output and increases glycogen deposition. Blockade by drugs or section of these parasympathetic nerves reduces sensitivity to the hypoglycaemic effect of insulin (insulin resistance), suggesting a role of the parasympathetic in the control of insulin sensitivity (Lautt 1980).

1.6.11 Pancreas

This organ is innervated by both the sympathetic (via the coeliac ganglion) and parasympathetic (vagus) divisions. Sympathetic stimulation causes vasoconstriction and a dual effect upon insulin release from the β cells of the pancreatic islets. Activation of sympathetic nerves causes inhibition of glucose-stimulated insulin secretion via α-adrenoceptors, whereas β-adrenoceptors stimulate insulin release. Sympathetic stimulation also increases the glucagon output from the α cells of the islets. Vagal stimulation releases insulin. Stimulation of the pancreatic nerve after vagal blockade with atropine, however, causes inhibition of insulin release. Thus both sympathetic and parasympathetic divisions play an important role in the control of insulin release (Porte & Robertson 1973).

Parasympathetic stimulation to the pancreas also promotes the exocrine secretion of pancreatic digestive juices into the duodenum.

1.6.12   Gall Bladder and Bile Duct

The sympathetic innervation of the gall bladder and bile duct is via the splanchnic nerves and coeliac ganglion, stimulation inhibiting the musculature. Parasympathetic innervation via the vagus increases tone and motility of the gall bladder and relaxes the sphincter.

1.6.13   Spleen

The capsule and trabeculae of the spleen contain smooth muscle fibres which are under sympathetic innervation only, via the coeliac ganglion. No fibres appear in the white (lymphoid tissue) and red pulp (rich in stored erythrocytes), but both arteries and veins within the spleen are richly innervated. Sympathetic stimulation causes contraction of the splenic capsule and release of erythrocytes and lymphocytes into the bloodstream. Thus, in stress, bursts of physical exercise, or haemorrhage, the spleen releases substantial amounts of blood into the circulation.

1.6.14   Kidneys

The kidneys are under sympathetic innervation, the preganglionic fibres emerging from the fourth thoracic segment to the second lumbar. The preganglionic fibres synapse in the coeliac ganglion, the superior and inferior mesenteric ganglia and the renal plexus. The postganglionic fibres supply primarily the vasculature, stimulation causing vasoconstriction, reduced renal blood flow and thus reduced urine output. The role of dopaminergic sympathetic nerves is described in Chapter 12. A further effect of sympathetic nerve stimulation is to enhance the release of renin from the juxtaglomerular apparatus. This initiates the renin–angiotensin–aldosterone cascade of renal blood flow and urine volume regulation. The raised aldosterone levels promote water reabsorption in the kidney tubules and therefore also reduce urine production. This reflex occurs in response to stressful stimuli and to a loss of blood or fall in blood pressure.

1.6.15   Urinary Bladder

The bladder has three principal components: the body, which consists mainly of detrusor smooth muscle; the trigone, which is a triangular sheet of smooth muscle through which the ureters and urethra pass; and the neck. At the base of the bladder the smooth muscle fibres pass around the upper urethra for several centimetres, forming an *internal sphincter*. The parasympathetic innervation of the bladder is from the sacral outflow via the pelvic nerves. Stimulation causes contraction of the detrusor muscle and relaxation of the internal sphincter with consequent micturition. At the same time, the external sphincter, which surrounds the urethra and consists of skeletal muscle continuous with the perineal muscle, is under conscious control by the pudendal nerve and is relaxed.

The sympathetic innervation of the bladder arises from the lower thoracic and upper two or three lumbar segments and synapses in the inferior mesenteric, hypogastric and visical ganglia. Activation of the sympathetic nervous system naturally retards micturition, a reflex likely to be inconvenient during flight and fight situations. The precise innervation is not clear but appears to be mainly at the base of the bladder, stimulation closing the neck by pulling down the base.

1.6.16   Genital Tract

The sympathetic innervation of the genital tract arises from the lower thoracic and upper lumbar segments, most of the fibres synapsing in the inferior mesenteric, hypogastric and preaortic plexuses. In the male, the postganglionic fibres innervate the vasa deferentia, seminal vesicles, blood vessels and prostate gland. Stimulation causes vasoconstriction and contraction of the smooth muscle of the prostate, vas deferens and seminal vesicles, resulting in ejaculation of semen. Somatic nerves also play an important role in the conscious control of ejaculation through contraction of the ischio- and bulbo-cavernous muscles. These cause rhythmic compression of the urethra and are also responsible for expelling the last drops of urine during micturition. The sympathetic nerves also increase the tone in the internal sphincter of the urinary bladder and thereby prevent semen from entering the bladder.

In the female, the sympathetic nerves innervate the uterus and fallopian tubes which contract on stimulation. Mucous-secreting glands that open into the vestibule of the vagina are also stimulated to secrete. The density of the sympathetic innervation of the uterus is species-dependent and in the human is richer in the cervix than the corpus or fundus. The innervation also appears to decline during pregnancy. The ovaries also receive a sympathetic innervation along the blood vessels.

The parasympathetic innervation arises from the sacral outflow and passes via the pelvic nerves to the diffusely arranged pelvic plexuses. In the male, postganglionic fibres supply the blood vessels of the corpus cavernosum of the penis causing arteriolar vasodilatation. The sinusoids of the corpus become engorged with blood causing penile erection. Distension of the sinusoids compresses the veins which additionally prevents drainage of blood. In the female, a similar response of the clitoris occurs. Thus, the parasympathetic innervation causes erection while the sympathetic is responsible for ejaculation. The uterus receives a parasympathetic innervation but its function is uncertain.

1.6.17   Adipocytes

Adipose or fat tissue is of two types. Brown adipose tissue (BAT) has primarily a thermoregulatory function and is sparsely located in adult humans to an interscapular fat pad. It receives a rich sympathetic innervation with thin, naked axons found closely attached to the fat cells. Stimulation induces lipolysis, the breakdown of triglyceride to fatty acids, and thermogenesis, measured *in vivo* as an increase in metabolic rate or oxygen consumption. White adipose tissue (WAT), making up $\geq 10\%$ of body mass, is widely distributed but receives only a sparse sympathetic innervation. It is not involved in thermogenesis. The adipocytes are not regarded as being innervated but lipolysis and mobilization of plasma free fatty acids (FFA) can be stimulated by circulating adrenaline or noradrenaline released by distant sympathetic nerve endings. The major effect of sympathetic innervation to WAT is on its blood vessels. Stimulation causes vasoconstriction and reduced blood flow which is followed by a reactive hyperaemia or increase in flow. The reduced blood flow paradoxically inhibits the release of FFA from WAT, probably because of reduced availability of albumen to which FFA binds (Fredholm 1985).

1.7 Afferent Sensory Pathways

Sensory information concerning the internal environment of the body is transmitted to the spinal cord or brain via afferent autonomic nerves. In the spinal cord or brain this information is integrated and appropriate motor impulses are passed down the efferent pathways already considered, to make the necessary adjustments to maintain constancy (homeostasis). In this way, involuntary functions such as blood pressure, heart rate, blood pO_2 and pCO_2, body temperature and digestion are controlled. Traditionally, we think of the senses of taste, smell, sight, hearing and touch; these are the *somatic* senses. However, the autonomic sensory pathways detect the internal environment and are termed the *visceral* senses. The afferent impulses of visceral origin reach CNS through primary sensory neurones that do not differ significantly from those involved in somatic senses.

The sense organs or receptors are the peripheral ends of the dendrites of afferent neurones. These are unipolar neurones in which the single process of the cell body divides. One arm is the dendrite, which has a branched terminal and extends to the periphery carrying nerve impulses towards the cell body; the other branch forms the axon which carries impulses away from the cell body to the brain or spinal cord. The sensory nerve endings convert environmental stimuli as various forms of energy (eg pressure, heat, chemical) into nerve action potentials.

There is clear histochemical evidence that the afferent neurones pass along the same nerve tracts as do the efferent nerves. The dendrites cross peripheral sympathetic ganglia and parasympathetic plexuses uninterrupted. The cell bodies of afferent neurones in the region of the sympathetic thoracolumbar system are generally regarded to lie in the dorsal root ganglion of T1 to L2 or L3 spinal nerves. The dendrites of these cells pass peripherally to enter the sympathetic trunk via white rami communicante. They then run along the cardiac, pulmonary and splanchnic nerves to the organs of the thorax and abdomen (Figure 1.4).

The axons of these cell bodies enter the spinal cord where they synapse with neurones of the intermediolateral cell columns of the grey matter. An interneurone in the spinal cord completes the reflex arc with the pre- and postganglionic efferent neurones. Certain intestinal reflexes may be abolished by section of the dorsal roots. However, there are central connections which reach the hypothalamus via the reticular formation of the brain stem. Afferent pathways are also associated with the parasympathetic division, the cranial section of which includes the vagus and glossopharyngeal afferent nerves. The cell bodies of the afferent vagal nerves lie in the nodose ganglion and lying above it and smaller, the jugular ganglion. The afferents from the distal colon, rectum and bladder run in the second to fourth sacral parasympathetic nerves, with cell bodies in the dorsal root ganglion.

The homeostatic reflexes and the centres for their control have been identified from the loss of control after section of spinal cord or brain level. It is convenient to consider these areas from the simplest spinal level in an ascending order.

1.7.1 Spinal Reflexes

The simplest form of reflex occurs at the spinal level, without the need for central connections. These reflexes include those for emptying the bladder and rectum. Filling of these organs causes stimulation of stretch receptors in the walls, resulting in an increase in traffic of afferent impulses to the sacral spinal cord. On reaching a specific intensity, there is reflex contraction of the detrusor muscle of the bladder and increased peristalsis of the descending colon. The rectum and sigmoid colon contract vigorously, the internal anal sphincter (smooth muscle) relaxes and defaecation occurs. Normally, ascending pathways to the medulla transmit this information on fullness of the rectum and a voluntary decision can be made to inhibit defaecation. This is achieved by contraction of the external anal sphincter (skeletal muscle) and relaxation of the rectum and sigmoid colon which reduces the pressure and thus stimulation of stretch receptors. This may continue until the rectum is distended to the point where defaecation is unavoidable. In very young children this voluntary control of defaecation does not occur because the necessary nerve pathways have yet to reach full maturity. Similarly, in adults whose spinal cord is severed at or above the sacral level (paraplegics), voluntary control is lost and the reflex is purely spinal. The micturition reflex of the bladder operates in a similar fashion.

The sex organs of both male and female can respond by a spinal reflex. Stimulation of the vagina of a bitch in which the spinal cord is transected results in movement of the tail into a copulatory position. However, it is clear that the whole copulatory act involves a complex interaction of physical and emotional sensory inputs and an integral role is played by higher centres.

1.7.2 Medullary Reflexes

The medulla oblongata lies in the brain stem and is concerned with autonomic reflexes of the cardiovascular, respiratory and gastrointestinal systems.

1.7.2.1 Cardiovascular reflexes

The heart and arterioles are controlled by sensory signals detecting arterial blood pressure and blood chemistry. Arterial blood pressure is detected by the terminal dendrites of afferent nerves located in the aortic arch and the carotid sinus, which is at the bifurcation of the common carotid artery. These serve as baroreceptors and are stimulated by pressures >60 mmHg. The aortic arch is supplied by the vagus nerve whose cell bodies lie within the nodose ganglion, while the carotid sinus is supplied by the glossopharyngeal nerve. The central axons of the vagus and glossopharyngeal nerves terminate in the nucleus of the solitary tract (tractus solitarius) in the medulla. From here fibres pass to the cardioinhibitory and vasomotor centres of the medullary reticular formation. These centres are in turn connected to the dorsal motor nucleus of the efferent vagal nerves and to the sympathetic efferents via the intermediolateral cell column of the spinal cord. A rapid increase in arterial blood pressure results in increased afferent traffic to the vagal cardioinhibitory centre, which sends out impulses to the heart resulting in vagal slowing. At the same time, the sympathetic vasomotor centre is inhibited, resulting in reduction of sympathetic vasoconstrictor tone. Conversely, falls in arterial blood pressure are compensated for by escape from this sympathetic inhibition, with a consequent rise in cardiac output and vasoconstriction. Sympathetic stimulation also passes to the spleen and kidneys to increase blood volume. The spleen capsule contracts to release more blood into the circulation and renin is released from the kidney to activate the renin–angiotensin–aldosterone system to promote water retention.

It is important to remember these cardiovascular reflexes when considering the effects of drugs which interfere with the transmission along autonomic pathways. For example, a side-effect of several antihypertensive drugs is *orthostatic hypotension*. When changing from a supine to erect position, the blood tends to pool under gravity to the lower part of the body resulting in a temporary fall in pressure in the carotid sinus. The reduction in baroreceptor stimulation and consequent escape from sympathetic inhibition results in an increased sympathetic vasomotor outflow to restore blood pressure. This occurs rapidly with only transient dizziness occurring with abrupt change of body position. However, if the sympathetic pathways are blocked by drugs, the reflex is blunted, resulting in a more prolonged period of dizziness or even syncope (fainting) due to the reduced venous return and resultant compromised cardiac output to the cerebral circulation. This is orthostatic hypotension. A useful measure of the cardiovascular reflex and its inhibition by drugs in experimental animals is the carotid occlusion response. The two common carotid arteries are clamped, resulting in a fall in pressure in the carotid sinus and a compensatory rise in peripheral blood pressure. A similar test in human subjects is Valsalva's manoeuvre. In this test the subject is asked to make a forced expiration against a closed glottis. This raises intrathoracic pressure which restricts venous return to the heart and consequently cardiac output. The fall in blood pressure is detected by baroreceptors, resulting in compensatory vasoconstriction. This persists after release of the pressure and is seen as an overshoot rise in blood pressure (Faulkner & Sharpey-Schafer 1959).

Baroreceptors of vagal afferents are also found in the right atria and are stimulated during atrial filling. This results in reflex sympathetic discharge to the heart, increasing its rate and lowering venous pressure (Bainbridge reflex). Additionally, there is a reflex increase in urine flow (Hainsworth 1991).

Also affecting blood pressure is the pO_2, pCO_2 and pH of the blood. These are detected by chemoreceptors of the carotid bodies which are located at the bifurcation of the common carotid arteries. They are innervated by the efferent sympathetic fibres which control the blood flow through them, but more important are the sensory nerve endings which are extremely sensitive to a reduction in pO_2, a rise in pCO_2 or a fall in pH. These result in increased nerve traffic to the medulla and increased activity in the efferent vagal nerves and a decrease in sympathetic activity. Thus heart rate slows and blood pressure falls.

Various neurotransmitters have been identified in the carotid body chemoreceptors, but the precise mechanism whereby these may stimulate the sensory receptors is uncertain. Veratrum alkaloids (veratridine) and nicotine have been shown to induce reflex vagally-mediated bradycardia and hypotension through stimulation of an afferent vagal pathway, which is known as the von Bezold–Jarische reflex. The chemoreceptors for this reflex are located in the coronary circulation. They are also stimulated by bradykinin and prostaglandin and are probably activated by these substances when released during myocardial ischaemia or infarction (Hainsworth 1991). 5-Hydroxytryptamine also induces this reflex and a possible role for 5-HT both at the sensory nerve endings and the central level of cardiovascular control has been suggested (Pires & Ramage 1990, Hainsworth 1991).

1.7.2.2   Respiratory reflexes

The medullary respiratory centre consists of the inspiratory centre located medially in the reticular formation and the expiratory centre located more laterally. A higher pneumotaxic centre in the reticular formation of the pons regulates the rhythmic nature of inspiration and expiration by turning off inspiration. The most important regulator of the respiratory centres is discharge from chemoreceptors located in the medulla, a rise in pCO_2 causing an increase in the rate of ventilation. A rise in H^+ concentration also stimulates ventilation, as does a fall in pO_2 detected by carotid body chemoreceptors. The activity of the respiratory centre is also modified by afferent vagal impulses from stretch receptors in the bronchial tree, especially the smaller branches. When the lungs expand during inspiration the nerve traffic to the inspiratory centre increases until inspiration is brought to a halt. This is known as the Hering–Breuer reflex. In anaesthetized animals, section of the vagal nerves results in prolonged inspiration.

Many other reflexes are associated with breathing and the respiratory centre. For example, the *cough reflex* is initiated by stimulation of vagal tracheal mechanoreceptors or chemoreceptors deeper in the airways. A forced expiration ensues against a closed glottis which on opening results in expulsion of the offending material. A *sneeze* is a similar reflex involving stimulation of sensory nerve endings in the nasal passages, with the forced expiration directed through the nose with the mouth closed. The act of swallowing also inhibits respiration.

1.7.2.3   Swallowing and vomiting reflexes

When food is voluntarily pushed by the tongue into the pharynx, the glossopharyngeal nerve endings are stimulated and the involuntary swallowing reflex is initiated by the passage of afferent nerve impulses to the medulla. The swallowing centre of the medulla coordinates a series of precisely timed events via the cranial nerves which propel the food bolus to the oesophagus, where vagally-modulated peristaltic activity takes over.

The vomiting reflex is initiated by irritation of sensory vagal nerve endings in the stomach or duodenal wall by noxious materials in the chyme. These impulses pass to the medullary vomiting centre. Circulating chemicals including drugs such as the cardiac glycosides, nicotine, opiates and chemotherapeutic agents also cause nausea and vomiting by stimulation of the chemosensitive trigger zone (CTZ) in the area postrema in the floor of the fourth ventricle. The blood–brain (BB) barrier in the area postrema is poorly developed and the CTZ is readily reached by these circulating substances. The vomiting centre receives input from the CTZ and from the vestibular apparatus of the inner ear, resulting in motion sickness. The ensuing vomiting reflex is mediated via the vagus nerve, phrenic nerve to the diaphragm and spinal nerves to the abdominal muscles. The upper region of the stomach relaxes while the pylorus contracts and the co-ordinated contractions of the skeletal muscle of the diaphragm and abdominal wall onto the stomach forces the contents out. Nausea and vomiting are associated with other autonomic responses including palor, sweating, salivation, irregular heart rate and hypotension.

1.7.3 Midbrain

The midbrain or mesencephalon is a portion of the brain stem lying above the pons and below the thalamus. The superior colliculus receives and integrates incoming impulses from the eye for which branches of the optic nerve (cranial nerve II) serve as the afferent conductor. Emerging from the midbrain are the two oculomotor nerves (cranial nerve III) which innervate the iris and ciliary body. Thus, the amount of light entering the eye through the pupil, focusing and involuntary eye movements are controlled.

1.7.4 Hypothalamus

The specific role of the hypothalamus in the regulation of body temperature through the autonomic nervous system has already been considered (p. 24). Additionally, it controls the osmotic pressure of the blood which is often regarded as homeostatic control of the internal environment in the context of autonomic control. A 'thirst' and 'thirst satiety' centre regulate the intake of water through drinking and its output via the kidneys and sweat glands. Nuclei within these centres at the supraoptic nuclei act as osmoreceptors monitoring the osmotic concentration of the blood. Unmyelenated neurones from these nuclei pass to the posterior pituitary where they release antidiuretic hormone (ADH, vasopressin) that has formed in the cell body. An increase in osmotic pressure due to water loss, emotional or physical stress and drugs such as nicotine increases the secretion of vasopressin. This passes to the kidneys where the permeability of the kidney tubules is increased and more water is reabsorbed, urine formation is inhibited and blood volume is raised. Simultaneously, vasopressin constricts peripheral arterioles to raise blood pressure.

The hypothalamus clearly has a central role in autonomic control and the body's response to stress. Stimulation of the anterior and medial regions generally gives rise to parasympathetic responses, whereas stimulation of sites in the posterior and lateral hypothalamus is associated with a wide range of sympathetic responses. In addition to the activation of the sympathetic nervous system and release of adrenaline from the adrenal medullae during stress, the hypothalamus is responsible for promoting secretions of the anterior and posterior pituitary gland. The anterior pituitary (adenohypophysis) secretes adrenocorticotrophic hormone (ACTH) in response to ACTH-releasing factor released by the hypothalamus. ACTH is carried in the circulation to the adrenal cortex to release steroids which assist the body's response to stress by mobilizing blood sugar and promoting tissue repair. The hypothalamus receives impulses from higher centres such as the cerebral cortex, limbic lobe and hippocampus to enable expression of emotions such as anxiety, fear, rage or pleasure through the autonomic nervous system.

1.7.5 **Thalamus**

This area of the brain is located above the hypothalamus and below the cerebrum and forms the lateral walls of the third ventricle. Amongst its several activities is the receipt of sensory pain impulses. The sensory neurones for pain of visceral structures are associated primarily with the sympathetic afferent pathways, in particular the splanchnic nerves. The cell bodies lie in the dorsal root ganglia of the thoracic and upper two or three lumbar segments of the spinal cord. Vagotomy has no effect on pain induced by intestinal distension. Pain impulses from the pelvic viscera (bladder) are transmitted along the sacral parasympathetic afferents via the dorsal root ganglion.

The pain nerve endings or nociceptors are naked endings of dendrites having no myelin or neurolemmocytes. They are similar to those that detect cutaneous pain forming the ends of somatic nerves. In fact, pain from the upper organs of the body is transmitted via cranial nerves whose cell bodies lie in the brain stem. The localization of these receptors and the pain threshold is variable. The intestine, for example, can be cut or burned without causing pain but if inflammed, the threshold is raised and pain is experienced. The type of stimulus is also important since distension of the intestine by obstruction or gas can cause severe pain (intestinal colic). These nociceptors display the phenomenon of adaptation which is relatively slow.

In the spinal cord the pathways are common to those for somatic pain. The afferent axons of the sensory fibres synapse with ascending pathways in the lateral spinothalamic tract of the spinal cord. The impulses reach the lateral nucleus of the thalamus through the spinal medial and trigeminal lemnisci relays of the reticular formation. From here, further neurones radiate to the postcentral gyrus of the cerebral cortex. This radiation is such that the areas of the body are 'projected' in order, with the feet at the top, then the head and abdominal viscera at the bottom. There is no discrimination between somatic and visceral pain at this stage. Indeed, pain arising from visceral structures is often perceived in cutaneous areas not always immediately above the organ. This known as *referred pain*. The cutaneous zone coincides with the segment of the spinal cord through which both the somatic and autonomic nerves enter. For example, the pain of angina originates from stimulation of cardiac sensory nerve endings of the sympathetic afferents to thoracic segments one to four. These coincide with the entry of primary sensory neurones from the left side so that the pain is referred to the chest and inner left arm. Similarly, gall bladder distention is felt under the shoulder blades. Thus the site of superficial pain sensation is extremely useful for identifying the origin of visceral pain.

In common with somatic pain, visceral pain causes reflex contraction of adjacent skeletal muscle seen as spasm of the abdominal wall. Also, neural connections pass to the medullary vomiting, respiratory and cardiovascular centres. Gastrointestinal pain is therefore accompanied by nausea, vomiting, sweating and a fall in blood pressure.

The efferent pathways of the sympathetic nervous system may also have a role in chronic pain induced by injury and inflammation. Pain after nerve injury has been shown to be relieved by sympathectomy and by α-adrenoceptor blockade (phentolamine and the α_2-selective antagonist, yohimbine – see Chapter 5). The damaged afferent nerves are sensitive to mechanical, thermal and chemical stimuli, but unlike intact nociceptive afferents are also sensitive to noradrenaline. This is attributed to the possible release of noradrenaline from the efferent neurones which sensitizes nearby sensory nerve endings to the painful stimuli. The precise mechanism remains to be elucidated, but the process may have particular relevance to inflammatory conditions (Koltzenburg & McMahon 1991).

1.8 Efferent Functions of Afferent Neurones

Sensory afferent nerves have been known for a long time to release neurotransmitters from their peripheral nerve endings. As long ago as 1901, Bayliss showed that electrical stimulation of the dorsal roots induced vasodilatation and that this was unaffected by section between the dorsal root and the spinal cord or by removal of the efferent sympathetic ganglia. It was, however, eliminated by removal of the dorsal root ganglion and a subsequent degeneration of the afferent nerves. Thus, the vasodilator response must have been due to impulses travelling in reverse or antidromic direction along the afferent nerve. Local reflexes involving antidromic nerve impulses are known to occur without central connections to the spinal cord. The 'triple response' following scratching the skin or local release of histamine by a nettle sting or insect bite is an example of such an axon reflex. The sensory nerve endings are stimulated by histamine and nerve impulses pass up the dendrites to nearby branches down which they travel in an antidromic direction. The neurotransmitter(s) released at these sensory branch endings induces a variety of responses including vasodilatation, which gives rise to the flare component of the triple response. The mediators also cause plasma protein extravasation due to increased venular permeability, degranulation of mast cells and changes in inflammatory cells (lymphocytes, granulocytes and macrophages). This response to local tissue injury or exposure to irritants is known as *neurogenic inflammation.* Stimulants of these sensory nerve endings include capsaicin, the pungent ingredient of capsicum peppers, 5-HT, bradykinin (note that bradykinin was a vasodilator mediator associated with sweating) and histamine (Maggi 1991a). The mediators released from the sensory nerve endings that induce the tissue response are the tachykinins (eg substance P and neurokinin A) and calcitonin gene-related peptide (CGRP). The pharmacology of these peptides and of ATP, which may also be released, will be discussed in Chapter 12.

1.9 Relationships Between Higher Centres and Autonomic and Somatic Nervous Systems

The somatic nervous system is involved in voluntary control of skeletal muscle whereas the autonomic nervous system is said to control internal functions below the level of consciousness. The latter is therefore under control of the hypothalamus and medullary centres. However, it is clear that higher centres of the brain such as the cortex can have a substantial influence on autonomic function. This is illustrated by the response to anticipation of a stressful event such as an interview, public speaking or performance. The palms of the hands sweat, the pulse rate accelerates and there is a desire to empty the bladder. It is also possible to exert some conscious control over internal functions such as blood pressure and heart rate, not through somatic nerve activity, but by mental effort. Yoga and Zen masters have long demonstrated their ability to regulate involuntary body functions, and more recently similar approaches have been applied through the technique of biofeedback. This involves an acquired skill to control a recorded body function such as heart rate.

Certain functions involve both autonomic and somatic nervous systems operating in harmony. For example, control of body temperature involves purely involuntary control of cutaneous blood flow through autonomic regulation of vascular smooth muscle, together with shivering which is skeletal muscle contraction through somatic nerve activity. Both responses are controlled by the posterior hypothalamus. Defaecation and ejaculation are also examples of basic autonomic functions that may be controlled by conscious effort. The basic defaecation reflex is autonomic, involving peristaltic contractions of the smooth muscle of the rectum and sigmoid colon and relaxation of the internal anal sphincter, but only in young children or adults whose spinal cord is severed above the sacral region. Above the age of ~2 years, the reflex may be controlled voluntarily by somatic nerve (pudental nerve, second, third and fourth sacral) impulses from the brain to the external anal sphincter (skeletal muscle). This nerve also innervates the bulbo-cavernosus skeletal muscle, voluntary control of which aids propulsion of semen in the urethra during ejaculation.

Although the sympathetic and parasympathetic divisions usually operate antagonistically, there are occasions where co-ordinated activity of both divisions is required. For example, the copulatory activity of the male involves involuntary responses of both the parasympathetic (erection) and sympathetic (ejaculation), together with the somatic nerve activity and a substantial emotional involvement from higher centres.

1.10   Considerations in the Classification of Autonomic Nerve Pathways

The autonomic nervous system has been divided into the sympathetic and parasympathetic divisions, but what is the basis of this classification? Three different criteria can be used.

1   *Functional basis*: The American physiologist, Cannon (1929), defined the sympathetic nervous system as being concerned with stress situations and the parasympathetic for restoration and conservation of energy.

2   *Neurotransmitters*: In general, the sympathetic nerves utilize noradrenaline and are termed *noradrenergic*, while the parasympathetic nerves release Ach onto the effector and are therefore termed *cholinergic*.

3   *Location of outflow from the spinal cord and brain stem*: The cranial and sacral preganglionic neurones and the ganglionic neurones with which they synapse constitutes the parasympathetic division. The thoracolumbar outflow from the spinal cord is the sympathetic division.

The *functional classification* is satisfactory but does depend upon the ability to induce a physiological response by stimulating the autonomic nerves. Langley (1892) determined the course of autonomic nerves by selectively stimulating single thoracic roots in anaesthetized cats and measuring the responses of the pupil and salivary glands. Functional classification has helped pharmacologists to develop drugs that mimic the physiological response, but as more knowledge is acquired, it is becoming clear that stimulation of autonomic nerves may induce more than one functional response of an organ because of the presence of multiple receptors to the neurotransmitter and indeed multiple neurotransmitters. The functional effect observed is often the resultant of two or more actions. For example, although sympathetic stimulation to the lung causes bronchodilatation via β-adrenoceptors, an underlying bronchoconstriction via α-adrenoceptors also occurs, which is revealed by blockade of the β-adrenoceptors.

The physiological roles of autonomic nerves have also been examined by their selective destruction. Partial loss of both parasympathetic and sympathetic innervation is not fatal in most species but homeostatic control is greatly impaired, so that the animal is less able to tolerate severe changes in environmental conditions such as raised or lowered ambient temperature. When the sympathetic chains were removed from cats, the animals survived and reproduced normally, providing that they were not unduly stressed (Cannon 1929). It is of interest that reproductive function of males failed after abdominal sympathectomy because of ejaculation failure. The survival from gross sympathectomy does not imply that the sympathetic nervous system is unimportant but that compensatory mechanisms must occur to adjust for the lack of sympathetic control. For example, the adrenal medullary function persists and up-regulation of the effector sensitivity to the reduced transmitter level soon occurs (see Chapter 14). The missing functions from a sympathectomized organ correspond relatively closely to the physiological role of the sympathetic nerve, although factors such as multiple innervation and compensatory effects due to denervation may have an influence. However, the response of selected organs to sympathectomy can indicate whether sympathetic nerves exert a tonic influence and, if so, the nature of that response. For example, the nictitating membrane relaxes after sympathectomy.

Sympathectomy has been produced by surgically severing the nerve, in which case the precise location can be identified in terms of the organs innervated. More generalized sympathectomy can be achieved by immunosympathectomy or by treatment with the neurotoxin, 6-hydroxydopamine, or by guanethidine (see Chapter 6). Immunosympathectomy is produced by injection of an antiserum to nerve growth factor (NGF) into neonate animals. NGF in neonates increases sympathetic neurone numbers and the antiserum causes permanent destruction of $\leqslant 90\%$ of sympathetic ganglion cells (Levi-Montalcini 1972). In the adult, the effects are less severe and partially reversible. NGF antiserum appears to have a selective effect, since nerves associated with the genital organs are relatively resistant, and in brown adipose tissue the fibres supplying blood vessels are destroyed whereas those to the adipocytes are unaffected (Derry *et al.* 1969). The effects on sympathetic nerves are more readily achieved than on parasympathetic nerves because the sympathetic innervation is not fully developed at birth and is thus susceptible to these interventions during early life.

Classification by neurotransmitter into noradrenergic and cholinergic nerves causes the most confusion. The nature of the neurotransmitter in autonomic nerves has been identified principally by histochemical techniques. For example, catecholamine-containing neurones are identified by formaldehyde-induced fluorescence, a technique that can distinguish between noradrenaline and other biogenic amines. Cholinergic fibres are identified histologically by their choline acetyltransferase or acetylcholinesterase reactivity (Chapter 7). The nature of transmitter vesicles in neurones can be examined under electron microscopy. To a lesser extent, the release of transmitter into perfusates after nerve stimulation has been determined, but the precise source is not always easy to identify. Using such techniques, the nerves have been defined as cholinergic or noradrenergic. In most cases cholinergic and parasympathetic nerves are synonomous, as are noradrenergic and sympathetic nerves. However, the sympathetic nerves supplying sweat glands and blood vessels of skeletal muscle are cholinergic.

The courses taken by autonomic neurones has been traced by combining these histochemical techniques with denervation studies. *Denervation* is section of a postganglionic sympathetic neurone which leads to degeneration of the distal part of the axon. Noradrenaline levels fall and its storage vesicles disappear. The axon breaks down and is taken up by Schwann cells. This does not occur immediately, otherwise experiments on isolated tissues, such as Finkleman's (1932) preparation of rabbit isolated jejunum or Paton's (1957) preparation of transmurally stimulated guinea-pig ileum, would not yield a response when the autonomic nerve endings are stimulated. However, ~2 days after severing the postganglionic cervical sympathetic fibres of the cat, stimulation of the distal stump elicits no response of the nictitating membrane. Regeneration of the axon then commences, providing that the cell body is intact. The axon stump sends out new roots supported by increased activity of the cell body seen as breakdown of chromatophilic substance (chromatolysis) which provides new protein for cell growth. The terminal sprout grows along the tract formed by the basal lamina of Schwann cells which arrange themselves into a continuous cord of cells extending to the nerve terminal. *Sympathectomy*, by contrast, involves destruction of the ganglion cell body, the neurone from which cannot regenerate. *Decentralization* involves section of the preganglionic nerve. This again results in loss of the preganglionic axon and choline acetyltransferase. However, the postganglionic cell body is virtually unaffected. The neurone retains its usual features and noradrenaline levels and electrical properties are unchanged (see Figure 14.4).

These techniques have been used to determine the course of nerves and whether they are pre- or postganglionic or have afferent or efferent functions. For example, intracranial vagotomy leads to degeneration of many vagal axons distal to the interruption, indicating their efferent role. There is also chromatolysis of cell bodies in the nodose ganglion lying below the section, which shows that the nerves also carry afferent fibres. These methods have therefore shown the presence of both afferent and efferent fibres and of both cholinergic and noradrenergic fibres in common nerve tracts. These observations make the classification of autonomic nerve pathways based on anatomical features such as transmitter content open to misinterpretation.

Classification according to the site of outflow is based upon the location of the cell bodies of the preganglionic nerves and is therefore purely anatomical. In terms of the sympathetic and parasympathetic divisions of the autonomic nervous system, this is the only sound method of classification since it does not suffer the pitfalls of the other two classifications, that is, anomalies in the transmitters or multiple physiological effects. This method is, however, of little help to the pharmacologist. A sound knowledge of the physiological responses of the body's organs to sympathetic and parasympathetic nerve activity is an essential prerequisite if the pharmacology of drugs that interact with the autonomic nervous system is to be understood. However, the presence or absence of an innervation by either division of the autonomic nervous system, and therefore of a physiological response, does not preclude a pharmacologic action of drugs that mimic sympathetic or parasympathetic activity. As will be seen in the following chapters, the neurotransmitters for the parasympathetic (Ach) and sympathetic (noradrenaline) divisions exert their activity by binding to specific receptors. These receptors may exist in tissues, although there is no appropriate innervation. Thus, the apocrine sweat glands are not innervated but increase their activity in response to circulating adrenaline. Similarly, blood vessels such as the rat aorta receive no sympathetic innervation but constrict in response to noradrenaline (Patil *et al.* 1972). The nictitating membrane of the cat receives no parasympathetic supply but contracts in response to Ach. Blood vessels devoid of a parasympathetic innervation exhibit relaxation responses due to release of endothelium-derived relaxant factors (EDRF) such as nitric oxide (NO). Other tissues that are clearly non-innervated but are responsive to transmitters include the formed elements of the blood. Adrenoceptors have been identified on lymphocytes and platelets, the latter being involved in promoting or inhibiting aggregation as a prelude to coagulation of blood. The pharmacologist therefore additionally requires a knowledge of the receptors present in each organ or tissue and their responses to stimulation. One overriding response of each organ is usually produced by stimulation of the innervating sympathetic or parasympathetic nerves. The receptor type mediating that response to sympathetic nerve stimulation is shown in Table 1.1.

Sympathetic Neurotransmission

2.1 Introduction

The responses of the sympathetic nervous system are now universally accepted to be mediated by the release of noradrenaline from the sympathetic nerve terminals. The concept of neurohumoral transmission between a nerve and either another nerve or the effector smooth muscle cell was first introduced by Elliott in 1905. The similarity in pharmacological activity of adrenaline extracted from the adrenal gland and of sympathetic nerve stimulation had already been noted (Langley 1901). Elliott showed that this extract had little activity on smooth muscle that received no sympathetic innervation but produced effects on sympathetically innervated tissues even after postganglionic section and degeneration of the sympathetic nerve. Thus, adrenaline was proposed to be the sympathetic neurotransmitter which Langley (1906) suggested was released from sympathetic nerves onto areas which he called 'receptive substances'. Thus grew the concept of a drug receptor.

The first demonstration of the release of a substance from sympathetic neurones was made by Loewi in experiments started in 1921. His pioneering work involved stimulation of the vago-sympathetic nerve of a frog isolated perfused heart, from which the perfusate passed to a recipient frog heart. On stimulation of the nerve trunk, the donor heart slowed and this was followed by slowing of the recipient. This was the first indication that nerve stimulation caused release of a chemical transmitter which Loewi referred to as *Vagusstoff* (vagus-substance), and was subsequently identified as acetylcholine (Ach) (Loewi & Navratil 1926). In later experiments Loewi used frogs at a different time of year (summer) and showed that the sympathetic nerve predominated when the vago-sympathetic nerve was stimulated. The donor heart increased in rate and force of contraction and this effect was also transferred to the recipient heart. This cardiac stimulant substance was termed *Acceleranstoff* and later suggested to be adrenaline. At the same time Cannon & Uridil (1921) reported that stimulation of the sympathetic hepatic nerve to the liver of anaesthetized cats caused the release of an adrenaline-like substance into the circulation. This substance increased blood pressure and heart rate and was originally called 'sympathin'.

From its pharmacological properties, sympathin closely resembled adrenaline but there were certain differences. Cannon & Rosenbleuth (1933) showed that hepatic nerve stimulation caused a rise in blood pressure which persisted after administration of ergotoxin. In contrast, the adrenaline-induced pressor response was converted to a fall in blood pressure after ergotoxin. Hepatic nerve stimulation also caused a contraction of the nictitating membrane, but no effect on the non-pregnant uterus, which was relaxed by adrenaline. To explain these differences, they postulated that a single transmitter, thought to be adrenaline, was released by sympathetic nerves and combined with another substance at or near the effector, to produce either an excitatory or inhibitory transmitter. These were termed sympathin E, which was responsible for the excitatory effects (pressor), and sympathin I, which exerted the inhibitory effects of adrenaline. A criticism of this hypothesis is that the sympathin released by stimulation of tissues showing a excitatory response such as the cardioaccelerator nerves to the heart (sympathin E) produced only a weak relaxation of the non-pregnant uterus. If the theory had been correct, it should have caused an excitatory response. Clearly, the conversion of adrenaline to sympathin would have to occur at the nerve terminal since injected adrenaline would also owe its pharmacological activity to its conversion to sympathin. As we now know, the results of Cannon & Rosenblueth can be explained by the fact that sympathetic nerves release noradrenaline. The differences observed between stimulation of hepatic, splanchnic (with adrenals tied off) and cardioaccelerator nerves, however, appear to this writer to remain unexplained, since all should release a common transmitter, namely noradrenaline. Why did the non-pregnant uterus relax in response to adrenaline and splanchnic nerve stimulation but not after hepatic nerve stimulation? The possibility of adrenaline release from the adrenal medullae by splanchnic nerve stimulation should have been excluded by their ligation.

Bacq (1975) persisted with the earlier proposals that noradrenaline was the sympathetic neurotransmitter and conclusive evidence to support this hypothesis was finally provided by von Euler (1946). He demonstrated that highly purified extracts of sympathetically innervated tissues contained a substance that mimicked the effects of sympathetic nerve stimulation which resembled noradrenaline both chemically and pharmacologically. It is now accepted that noradrenaline is the major neurotransmitter released from the postganglionic sympathetic nerve endings of mammals. The earlier proposal from the work of Loewi that adrenaline was the sympathetic neurotransmitter of the frog has in fact been proved to be correct (Stene-Larsen & Helle 1978).

2.2 Criteria for Neurohumoral Transmission

In spite of the growing evidence to support the concept of neurohumoral transmission, there remained a few sceptics who were unwilling to accept the idea. The alternative was a direct electrical connection between nerve and effector. A major objection was that the time of transmission was considered to be too rapid for the release of a chemical substance and its passage across a synapse to the effector. However, more recent electrophysiological recordings in pre- and postjunctional cells have demonstrated time lags of between 1 and 3 msec (Brock & Cunnane 1988) or between 6 and 40 msec (Holman 1970) for the spontaneous junction potentials in a variety of noradrenergically innervated smooth muscle. Another problem was the discrepancy between amounts of transmitter recovered during nerve stimulation and the amount required exogenously to produce a similar response. Clearly, this can be accounted for by the uptake and metabolism of endogenous transmitter at the synapse before it can leave in the effluent from the tissue. The most convincing evidence, however, against electrogenic transmission is that retrograde transmission across synapses does not occur. That is, stimulation of the postjunctional cell does not cause depolarization of the prejunctional cell; neurotransmitters are only released from the prejunctional neurone.

Several criteria have been laid down for identification of a putative neurotransmitter. These have been proposed over the years by several authorities (Eccles 1964, Triggle & Triggle 1976) for the classical neurotransmitters. It might be thought that there is no place these days for such criteria with the major transmitters of the autonomic nervous system now recognized. However, there may still be transmitters, co-transmitters or mediators of physiological responses awaiting identification or verification of their transmitter role in various organs. For example, an inhibitory response of the gastrointestinal tract in response to nerve stimulation persists when the sympathetic and parasympathetic effects are blocked. The inhibitory neurotransmitter of this response has yet to be definitely identified, although ATP (Burnstock

1972) or a peptide (Daniel 1985), such as vasoactive intestinal peptide (VIP) or neurotensin, may have claims to neurotransmitter roles (Lefebvre 1986) (see Chapter 12). The following criteria are therefore still relevant.

1 Administration of the putative transmitter must produce an identical response to that caused by stimulation of the appropriate nerve.

2 The proposed transmitter or its precursor and the appropriate synthetic enzyme(s) must be present in the nerve.

3 The transmitter must be demonstrated in the effluent from the innervated tissue during nerve stimulation and absent (or at a greatly reduced level) when not stimulating.

4 Mechanisms for the termination of the transmitter action (metabolism and/or uptake) should be present.

5 The responses to the putative transmitter and to nerve stimulation should be modified in a similar manner by drugs.

This latter criterion in fact fails for several drugs with regard to sympathetic nerve stimulation and noradrenaline. For example, the adrenergic neurone blockers such as guanethidine block the effect of nerve stimulation but potentiate the actions of nopradrenaline (see Chapter 5).

2.3 Sympathetic Neurotransmission

As the postganglionic sympathetic neurone enters the terminal organ it branches extensively and takes on a beaded appearance (Figure 2.1). These thickenings are the varicosities (1 μm diameter and 1–3 μm long). They are brightly fluorescent, globular structures joined by narrow lengths of fibre with a low level of fluorescence. Although not the end of the nerve fibre, these form the synapse with the smooth muscle or glandular cells. There may be several thousand varicosities for each sympathetic neurone. Varicosities increase the number of regions per nerve fibre where transmitter release can occur. Thus sympathetic neurones differ from somatic nerves, in which each branch ends in a discrete motor end plate. The varicose fibres run parallel with the smooth muscle cells, each axon making several contacts with the same cell via a varicosity and with several different cells.

2.3.1 *Synaptic Cleft*

The junction between the varicosity and effector cell is the synaptic cleft which varies in distance in different tissues. In the vas deferens, a single axon supplies each muscle fibre (multi-unit) and is deeply embedded in the muscle cell. The Schwann cell covering of the axon becomes incomplete in the varicose region, the naked varicosities making close neuroeffector contact with the smooth muscle cells. The synaptic cleft is only 10–20 nm, known as close junctions. Some muscle cells receive no direct innervation. In contrast, the synaptic cleft of blood vessel smooth muscle is larger at 200 nm or $\geq$4 μm (Bevan & Su 1973). Terminal sympathetic axons are confined to the adventitio-medial junction and ramify over the smooth muscle surface, rarely penetrating more than two to three cells into the smooth muscle. The released noradrenaline therefore has to penetrate into the muscle layers

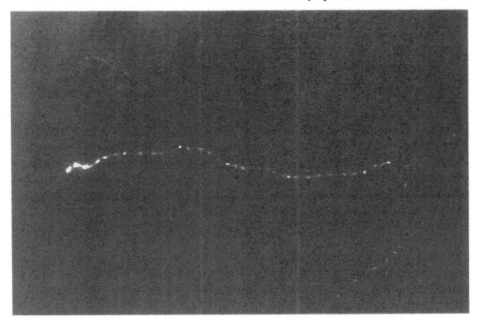

Figure 2.1 The terminal portion of a noradrenergic axon showing the characteristic beaded appearance of the varicosities. These are revealed by formaldehyde-induced fluorescence of the catecholamines in a whole-mount preparation of the mesentery of a guinea-pig. Magnification ×225. Reproduced with permission from Gabella (1992); the provision of an original photomicrograph by Professor Gabella is gratefully acknowledged.

slowly. Excitation of the inner layers of smooth muscle probably therefore occurs by propagation of the impulse from cell to cell rather than by diffusion of noradrenaline. More recent studies suggest that varicosities do form closer contact (<100 nm) in certain arterioles (Luff & McLachlan 1989). The synaptic cleft distance is affected by the contractile activity of innervated tissues.

The synaptic cleft distance has important implications for the concentration of noradrenaline at the effector cell surface and the actions of endogenously and exogenously added noradrenaline. The narrow cleft and dense innervation of the vas deferens favours retention of released noradrenaline within the confines of the cleft, with little diffusion to extracellular spaces. There is rapid removal by reuptake into the neurone. With larger nerve–muscle junctions, effective neurotransmitter concentrations at the receptor will be lower due to dissipation. The concentration of noradrenaline in the synapses of various blood vessels has been estimated by comparing the reduction of the response to nerve stimulation and to exogenous noradrenaline by the α-adrenoceptor antagonist, phenoxybenzamine (Bevan & Su 1973). These figures showed a higher concentration of intrasynaptic noradrenaline as the cleft distance became smaller. The relationship between synaptic cleft distance and neuronal uptake is illustrated by the effects of the uptake inhibitor, cocaine (see later). Inhibition of neuronal uptake potentiates the effects of exogenously added noradrenaline because it diverts a higher concentration into the vicinity of the receptor. The potentiating effect of cocaine is greater with smaller neuromuscular intervals. Data of this type can be used as evidence for whether or not adrenergic receptors are closely associated with the nerve terminals. That is, whether they are truly innervated. For example, the production of supersensitivity by cocaine in the rat portal vein indicates that the functionally important α-adrenoceptors are located close to the nerve terminals (Ljung *et al.* 1973). An analysis of β-adrenoceptors has also shown that inhibition of uptake potentiates responses mediated via β_1-adrenoceptors in the heart but not β_2-adrenoceptors, for example, in the airways. This suggests wide synaptic clefts or lack of innervation of the β_2-adrenoceptors (Hawthorn & Broadley 1982).

The width of the synaptic cleft might be expected to influence the latency of transmission of junction potentials, however, there does not appear to be a relationship. The vas deferens, arteries and nictitating membrane all appear to have latencies between 10 and 200 msec (Triggle & Triggle 1976).

2.3.2   Noradrenaline Storage Vesicles

The highly fluorescent varicosities of sympathetic neurones can be seen to contain aggregations of vesicles under the electron microscope. These are characterized by having an electron-dense core. The varicosities contain only storage vesicles and mitochondria, whereas the intervaricose portions of the axon contain the subcellular structures seen higher up the axon, including the microtubules and endoplasmic reticulum. Vesicles also appear less densely distributed in the axon, cell body and dendrites. These probably represent storage vesicles in transit from their site of origin in the cell body.

The vesicles may be divided into two principle types according to size – small vesicles containing electron-dense granules with a diameter of 30–60 nm and larger dense-core vesicles of 85 nm diameter. The ratio of large to small vesicles depends upon the tissue and the species. The percentage of large vesicles varies from only 2–4 % in the rat iris and vas deferens to 33–40 % in ox spleen and human blood vessels (Thureson-Klein 1982). The noradrenaline storage vesicles have been separated out from tissue homogenates by differential and density-gradient centrifugation. The sizes of vesicles correlate well with the small and large dense-core vesicles seen under the electron microscope.

Storage granules or vesicles are transported along the axon from the cell body to the nerve terminal varicosities. This was first demonstrated by Dahlström (1965) by ligating the noradrenergic neurone which led to accumulation of fluorescent noradrenaline-containing granules above the restriction. The transport is mainly in the form of large dense-core vesicles since it is these that predominate above the restriction. The small dense-core granules that accumulate are probably formed from the large vesicles (Fillenz 1990). The vesicles are transported at a rate of 5–10 mm/hr via the neurotubules. The role of the neurotubules is indicated by the fact that vesicle transport is inhibited by colchicine and vinblastine which destroy neurotubules.

The *large dense-cored vesicles* are probably formed in the Golgi apparatus of the cell body; they are found in the cell body and this is the major vesicle type in preterminal axons. As the large dense-core vesicles are transported along the axon, noradrenaline synthesis occurs and the level progressively increases towards the axon terminal (Klein 1982). The *small dense-cored vesicles* are suggested to be formed by two processes, both from the large vesicles after release of their noradrenaline by exocytosis, and by budding from the ER (Triggle & Triggle 1976, Fillenz 1990). Which of these processes is more important is a matter of debate (Fillenz 1992).

The fact that there are differences in the composition of the membranes of large and small vesicles has been used as an argument against the small vesicles representing recycled large vesicular membranes (De Camilli & Jahn 1990). If this were indeed the case then some small vesicles would travel down the nerve independently of large dense-core vesicles. Both large and small vesicles contain noradrenaline as the granular electron-dense core material. The large vesicles appear uniform in density and core size whereas the small vesicles appear to show considerable variations.

There are probably two pools of noradrenaline within each vesicle: a small, readily released, *mobile* pool and a larger *reserve* pool. The readily released pool represents only a few percent of releasable noradrenaline. It contains newly synthesized noradrenaline and is replaced from the stable pool (Figure 2.2). Radiolabelled noradrenaline can be taken up by the small vesicles, suggesting that they are not completely filled by endogenously generated noradrenaline. A third pool of noradrenaline occurs in the cytoplasm. This pool is normally relatively small due to uptake into the vesicles and enzymatic breakdown via monoamine oxidase (MAO). This *labile* pool is not released by nerve action potentials but is the store released by the indirectly acting sympathomimetic amines (see later and Chapter 4).

In addition to the noradrenaline, the core of both small and large vesicles contain adenosine triphosphate (ATP), which has been identified histochemically by use of the ATP-specific uranaffin reaction (Richards & Da Prada 1977). It has been suggested that noradrenaline is bound in the vesicle core as a noradrenaline : ATP complex in a ratio of ~4 : 1. ATP, however, is released in addition to noradrenaline from the nerve terminal following a nerve action potential. There is now strong

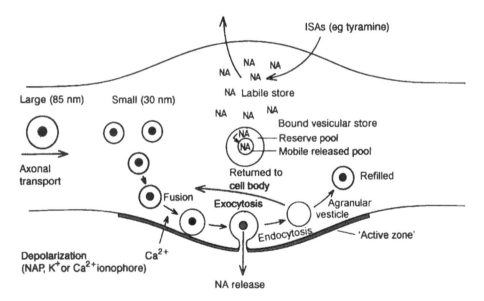

Figure 2.2 Diagram of a terminal varicosity of a noradrenergic neurone. The storage pools of noradrenaline (NA) in the cytosol (labile store) and in large and small vesicles (reserve and mobile pools) are shown. The release of noradrenaline by exocytosis is shown. This process involves Ca^{2+} influx following depolarization by the nerve action potential (NAP), by increased extracellular K^+ levels or by Ca^{2+} ionophore. Indirectly acting sympathomimetic amines (ISAs) release noradrenaline from the labile store.

evidence that ATP serves as a co-transmitter in the autonomic nerves to several organs (see Chapter 12). The cores of the large vesicles also contain soluble proteins, peptides and lipids. The major soluble protein is dopamine-β-hydroxylase (DBH), the enzyme responsible for the stage of noradrenaline synthesis from dopamine to noradrenaline. The other proteins present are the chromogranins/ secretogranins. These may serve as regulatory proteins involved in the packaging of the vesicle. They are also found in peptidergic nerves and may serve as precursors of the peptide transmitters (Huttner *et al.* 1988). Large vesicles also contain opioid peptides, somatostatin, substance P and neuropeptide Y (Fried *et al.* 1985, Fillenz 1992).

The membranes of the vesicles appear to consist mainly of lipid. Also present in the membrane is DBH which is therefore found in both the core (20%) and membrane of the vesicle and constitutes the particulate fraction. It is also probably free in the cytoplasm, the soluble component. However, reserpine, which blocks uptake of noradrenaline and dopamine into the vesicles, prevents conversion of dopamine to noradrenaline indicating that the vesicle is the major source of DBH. Its presence in the releasable core is illustrated by the fact that DBH is released together with noradrenaline on stimulation of the nerve. Indeed, in the adrenal medullary chromaffin cells, which are often used as a model for transmitter release, the release of soluble proteins, including DBH, and catecholamine was in the same proportion as found in the chromaffin granules. DBH is transported along the axon with the large dense-core vesicles. Since small vesicles contain DBH, it is clear that these vesicles are capable of synthesizing noradrenaline. DBH immunoreactivity has also been located to the ER, indicating that this is a probable origin of the small vesicles (Fillenz 1992).

2.3.3 Exocytosis

The release of noradrenaline from the sympathetic nerve varicosities by a nerve action potential is now widely believed to occur by a process of exocytosis. The small vesicles are concentrated at the axon endings at 'active zones' of the synaptic junction. The arrival of the nerve action potential at the nerve terminal leads to an influx of Ca^{2+} into the cytoplasm through specific channels. This occurs through open voltage-dependent channels.

The Ca^{2+} channels of excitable cell membranes have been subdivided into four subtypes according to their electrophysiological properties, location and the specificity of antagonists (Table 2.1). These are the L-, N-, T- and P-type Ca^{2+} channels (Tsien *et al.* 1991). The L-type Ca^{2+} channels are the best characterized and are defined as high-voltage-activated channels sensitive to the 1,4-dihy-dropyridine Ca^{2+} channel blockers, such as nifedipine. This distinguishes them from the T-, N- and P-type channels which are generally dihydropyridine-insensitive. L-Type channels are the major pathways for voltage-gated influx of Ca^{2+} into cardiac and smooth muscle cells and in some endocrine cells. They are also blocked by other calcium channel antagonists such as verapamil and diltiazem. T-Type Ca^{2+} channels are also known as low-voltage-activated Ca^{2+} channels because they are opened with small depolarization from negative membrane potentials. They are 'fast' channels because they exhibit rapid inactivation. T-Type channels are less sensitive to Cd^{2+}, but more sensitive to Ni^{2+} and amiloride than are L-type channels. They are found in

Table 2.1 Ca^{2+} channel types and their locations and blockers

Type	Blockers	Activation	Location/function
L	Dihydropyridines (DHP, eg nifedipine), verapamil, diltiazem	High voltage activated, long-lasting current, slow inactivation	Cardiac and smooth muscle contractions, endocrine cell secretion
T	DHP-insensitive Ni^{2+}, amiloride, flunarizine	Low voltage activated by small depolarizations, transient current	Absent in sympathetic neurones, support cardiac pacemaker function
P	DHP-insensitive funnel web spider toxin (FT_X)	Moderately high voltage (<-50 mV) non-inactivating	CNS neurones, adrenal chromaffin cells (?)
N	DHP-insensitive ω-conotoxin (also blocks neuronal L-channel)	High voltage activated, moderate rate of inactivation	Neurones only, activation of Ca^{2+}-dependent transmitter release

a wide range of cell types but are absent from adrenal chromaffin cells and sympathetic neurones, their main function being to support pacemaker activity.

P-Type Ca^{2+} channels are a novel class of high-voltage-activated Ca^{2+} channel. They are prominent in cerebellar Purkinje cells and are activated over potential ranges less negative than -50 mV. P-Type channels are insensitive to dihydropyridines and ω-conotoxin but are potentially blocked by FT_X, a toxin purified from the funnel web spider, *Agelenopsis aperta*.

The N-type Ca^{2+} channels are high-voltage-activated but are resistant to dihydropyridines, and exhibit smaller single channel conductance than do the L-type channels. They have been demonstrated exclusively in neuronal tissue. Patch-clamp studies have demonstrated that ω-conotoxin GVIA, a 27 amino acid peptide from the marine snail, *Conus geographus*, inhibits the L- and N-type calcium channels in dorsal root ganglia, sensory and sympathetic neurones but not in smooth or cardiac muscle cells (McCleskey *et al.* 1987). The responses of several tissues, such as the vas deferens (Brock *et al.* 1989), mesenteric arteries (Pruneau & Angus 1990) and atria (De Luca *et al.* 1990), to sympathetic nerve stimulation have been shown to be blocked by ω-conotoxin, without block of the smooth muscle response to exogenously added noradrenaline (Figure 2.3). The dihydropyridine L-type channel blockers inhibit vascular smooth muscle contraction (Godfraind *et al.* 1986), but require considerably higher concentrations to inhibit noradrenaline release from sympathetic nerves (Göthert *et al.* 1979). This, and the fact that ω-conotoxin does not inhibit the smooth muscle response, which is dependent upon Ca^{2+} influx through L-type channels, suggests that the release of noradrenaline is induced by the influx of Ca^{2+} through N-type channels. Blockade of the propagation of the action potential by ω-conotoxin was not involved in its inhibitory action. This contrasts with the action of tetrodotoxin (TT_X) which also abolishes the response to sympathetic nerve stimulation but by selectively blocking membrane Na^+ channels of nerves rather than smooth or cardiac muscle (Figure 2.3). The high voltage Ca^{2+}

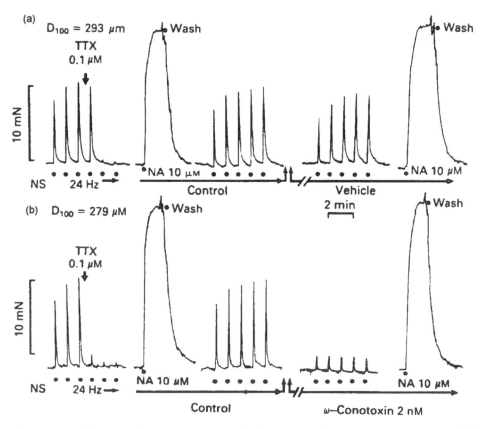

Figure 2.3 Inhibition of the contractions of the rat small mesenteric arteries to field stimulation of sympathetic nerves (●) but not of exogenously added noradrenaline (NA) by the N-type Ca^{2+} channel blocker, ω-conotoxin. Vehicle (upper panel) has no effect. Also shown is the inhibitory effect of tetrodoxin (TTX) upon responses to field stimulation, proving that it is mediated via stimulation of sympathetic nerves. Reproduced with permission from Pruneau & Angus (1990).

channels of adrenal chromaffin cells are insensitive to both dihydropyridines and ω-conotoxin and are therefore neither L-, N- or T-channels. It is not certain yet whether the release of adrenaline from chromaffin cells is therefore due to influx of Ca^{2+} through the P-type channel and susceptible to blockade by the funnel web spider toxin, FT_X.

The N-type channels of sympathetic neurones are probably located close to the active zone of the synaptic membrane where the small vesicles cluster. Large dense-core vesicles are more diffusely distributed in the nerve terminal. In the vas deferens they represent ~3% of the total vesicular population (Fillenz 1990). Their exocytosis appears to take place away from the synaptic junction and may explain occasional large spontaneous excitatory junction potentials (SEJPs). However, more recent studies suggest that EJPs of certain smooth muscle such as the rodent vas deferens are not due to noradrenaline release but to the ATP which is co-released (Astrand *et al.* 1988, Sneddon & Machaly 1992). The frequency of stimulation of nerves affects the ratio of released peptide and monoamine, with high frequency favouring the release of peptides. It is inferred that exocytosis from large vesicles therefore requires higher frequency stimulation. L-Type Ca^{2+} channels may be involved in release from large vesicles, and higher Ca^{2+} levels may be required for their exocytosis and consequently higher stimulation frequencies (De Camilli & Jahn 1990, Trimble *et al.* 1991).

Following the voltage-gated Ca^{2+} influx, the local cytosolic calcium concentration ($[Ca^{2+}]_i$) may rise to several hundred micromolar. As a result, the vesicles become fused with the nerve terminal plasma membrane. The interior of the vesicle opens to the outside, allowing the soluble contents to escape into the synaptic cleft. This is exocytosis (Figure 2.2). It is regarded as quantal in that the entire contents of a number of vesicles are released. It is dependent on the extracellular Ca^{2+} levels. The lipid components of the vesicle membrane and proteins of the cytoplasm are not released. However, vesicular DBH and chromogranin A are released. In addition to raising $[Ca^{2+}]_i$ by depolarization from an action potential, it may also be raised by depolarizing extracellular levels of K^+ and by application of the Ca^{2+} ionophore, A23187, the latter method bypassing the voltage-dependent Ca^{2+} channels. Both of these methods have been shown to induce release of radiolabelled noradrenaline from synaptosomes. Synaptosomes are resealed nerve terminals and varicosities formed after homogenization of tissues, usually from the brain. Differential centrifugation of such homogenates separates out at progressively higher speeds, unbroken cells and nuclei, a crude mitochondrial fraction also containing synaptosomes, and finally the microsomal pellet containing noradrenaline storage vesicles. Potassium-induced depolarization of nerve endings reduces the noradrenaline level and the number of small dense-core vesicles in parallel, indicating a common process of exocytosis (Pollard *et al.* 1982).

The vesicles may be bound to the cytoskeleton of the neurone and only released for exocytosis by the activation of a vesicle protein. Synapsin I and II have been identified. These can be phosphorylated by cAMP-dependent protein kinase and Ca^{2+}-calmodulin-dependent protein kinase (type I). This process may release vesicles from the actin-based cytoskeleton of the neurone and increase the pool available for exocytosis (Augustine *et al.* 1987, De Camille & Jahn 1990). However, activation of these protein kinases (protein kinase C and cAMP-dependent) may be involved in modulation of transmitter release through prejunctional α_2-adrenoceptors rather than being directly responsible for release (see Chapter 6). The precise mechanism for the triggering of exocytosis by the rise in cytosolic Ca^{2+} levels in the vicinity of the synaptic vesicle remains obscure. It is thought that a protein located on the cytoplasmic side of the vesicle membrane or in the cytoplasm itself serves as a Ca^{2+} binding trigger molecule. Several candidates have been identified including calmodulin which is a 148 amino acid protein with four Ca^{2+} binding domains. Binding to at least three of these induces a conformational change in its structure. Its addition to isolated synaptic vesicles causes the release of transmitter. Trifluoroperazine, a calmodulin antagonist, blocks the release of transmitter from synaptosomes (Schwietzer & Kelly 1982). Trifluoroperazine, however, has other actions such as blockade of Ca^{2+} channels and activation of voltage-gated Na^+ channels (Clapham & Neher 1984) and results obtained from its use must be treated with caution. The long half-life of Ca^{2+}-calmodulin (10 ms) makes it unlikely that it is the sole regulator of exocytosis in the rapid response of neurotransmission. It may operate by activating Ca^{2+}-calmodulin-dependent protein kinases which may modulate transmitter release (Augustine *et al.* 1987). Further elucidation of the role of calmodulin in transmitter release must, however, await the development of more selective antagonists of calmodulin.

Synaptophysin may be a further vesicle membrane-spanning, Ca^{2+} binding protein, with the Ca^{2+} binding domain on the cytoplasmic side. It could serve as a Ca^{2+}-regulated pore or junction between the vesicle and plasma membrane. Thus, the initial step for exocytosis may be the formation of a pore through which the vesicle contents pass across the neuronal membrane (Trimble *et al.* 1991). Another group of cytoplasmic Ca^{2+} binding proteins are the *annexins*. These can promote the fusion of chromaffin granules of the adrenal medulla in the presence of low Ca^{2+} levels and fatty acids. One of these, *calpactin*, is a strong candidate for linking the increase in $[Ca^{2+}]_i$ to exocytosis by promoting the fusion of the vesicle with the inner surface of the plasma membrane (Trimble *et al.* 1991). Finally, it is suggested that specific GTP-binding proteins are also involved in the fusion of the vesicle with the plasma membrane (De Camilli & Jahn 1990).

Following exocytosis and release of neurotransmitter, the vesicle membrane becomes incorporated into the neuronal membrane. However, it must be retrieved from the plasma membrane quite rapidly otherwise there would be considerable expansion of the membrane and there is no evidence that this occurs during repeated nerve activity. There is also a need for rapid reuse of the vesicles since their loss during nerve activity would otherwise require provision of new vesicles by transport from the cell body. Their reformation is termed endocytosis. A decrease in the number of vesicles may occur if after exocytosis they are unable to refil with transmitter and therefore cannot be used. Release of noradrenaline from the vas deferens by electrical stimulation leads to decreased numbers of small dense-core vesicles and an increase in the number of small agranular vesicles (Pollard *et al.* 1982). These probably represent unfilled vesicles. Electrical stimulation of sympathetic nerves to guinea-pig heart causes depletion of noradrenaline. After 10 min of stimulation, storage capacity is unchanged; while after 70 min of stimulation, storage capacity is reduced (Wakade & Wakade 1982). The unchanged status at 10 min may be due to reuse of vesicles.

Vesicles appear to have the ability to pass retrogradely from the nerve terminal to the cell body, probably when not refilled. The retrograde movement of the membrane-bound DBH has been measured, but its enzymatic activity is considerably reduced. When measured from immunofluorescence, the DBH moving retrogradely and orthogradely was similar. Thus DBH appears to move back to the cell body in an inactive form, probably in empty vesicles (Fillenz 1990).

2.3.4 Effects of Repeated Nerve Stimulation

Repeated stimulation of sympathetic nerves normally does not cause a loss of noradrenaline because the synthesis of noradrenaline and replenishment of vesicular content is capable of compensating for the release. Only after prolonged or severe stimulation, or when synthesis is inhibited, is there a gradual depletion of noradrenaline content. Repeated nerve stimulation has been shown to cause a gradual decline in the output of both noradrenaline and the vesicular enzyme, DBH. When synthesis was inhibited, the rate of decline of both noradrenaline and DBH was monophasic suggesting release from a common pool. When synthesis was not inhibited, noradrenaline showed a biphasic decline which suggests release from two pools, one of which contained newly synthesized noradrenaline (Fillenz 1990). The number of small dense-core vesicles in the vas deferens has been shown to be reduced by nerve stimulation (Cote *et al.* 1970).

2.3.5 *Excitatory and Inhibitory Junction Potentials and Noradrenaline Release*

The release of transmitter from sympathetic nerves in multimolecular packets or quanta was first suggested by the recording of spontaneous excitatory junction potentials (SEJPs) (Burnstock & Holman 1962). Stimulation of excitatory sympathetic nerves induces transient depolarization, termed excitatory junction potentials (EJPs). In inhibitory nerves to intestinal smooth muscle, for example, a single stimulus elicits hyperpolarization or an inhibitory junction potential (IJP) which is mediated by β-adrenoceptor stimulation. Generally, contraction and relaxation of the smooth muscle are associated with depolarization and hyperpolarization, respectively, of the smooth muscle membrane. However, in some muscles contraction can occur without changes in membrane potential. In these muscles Ca^{2+} may be released intracellularly by second messengers or raised by receptor-operated channels by addition of noradrenaline, for example (Van Breemen & Saida 1989). Similarly, relaxation may not necessarily be associated with hyperpolarization, as in tracheal smooth muscle (Cameron *et al.* 1983).

The depolarization (EJPs) elicited by sympathetic nerve stimulation to some tissues (guinea-pig mesenteric veins and arteries, rat and mouse anococcygeus muscle) is blocked by antagonists of noradrenaline (α-adrenoceptor antagonists) but is resistant to α-adrenoceptor blockade in other tissues (rodent vas deferens) (Brock & Cunnane 1992). This suggests that the EJPs recorded in the vas deferens are not due to noradrenaline release. Since noradrenaline and ATP are co-stored in sympathetic nerves and the release of ATP has been demonstrated (Kirkpatrick & Burnstock 1987), it is suggested that these EJPs are in fact due to the ATP. A component of the contraction of these tissues is, however, probably elicited by the release of noradrenaline. In this case the noradrenaline would not produce a change in membrane potential. Contractions mediated through activation of α-adrenoceptors are not always associated with muscle action potentials (Minneman 1988). In the rat anococcygeus muscle, the EJPs induced by sympathetic nerve stimulation *are* blocked by α-adrenoceptor antagonists and therefore due to released noradrenaline, which when added exogenously also induces membrane depolarization.

Calculation of the amount of noradrenaline released from each varicosity per stimulus has been made from the total noradrenaline content of a tissue and the number of varicosities estimated from histochemical analysis. From the estimated vesicular content of a varicosity and the molecular weight of noradrenaline, it has been suggested that only 1–3 % of the vesicular content was released per stimulus. These calculations are obviously prone to many errors and assumptions, but it does appear that the amount of noradrenaline released from each varicosity is less than the content of a single vesicle. This has been attributed to the fact that suprathreshold nerve stimulation frequently fails to induce transmitter release from an individual varicosity. This *intermittent* release of transmitter was demonstrated in the vas deferens with intracellular records of the EJPs (Blakeley & Cunnane 1979). By differentiation of the rising phase of the EJP due to electrical stimulation or of SEJPs, a peak was identified which was termed a 'discrete event'. This was attributed to the release of transmitter from a single varicosity. The height and duration of discrete events from a single smooth muscle cell could be matched up and were considered to reflect the release from a single vesicle. The conclusions were that transmitter is released from a varicosity in quanta, that the release is intermittent, and that release causing SEJPs and EJPs is from identical sites. To determine whether the failure of individual varicosities to release transmitter was due to a failure of the nerve action potential to invade the entire axon terminal, it was necessary to examine transmitter release from individual varicosities. Brock & Cunnane (1987) have applied suction electrodes to the smooth muscle surface, which allowed the electrical activity generated by a small group of varicosities (~10) to be examined. It was concluded that the action potential actively invades the entire nerve and that intermittency of transmitter release is due to a low probability of transmitter release by the invaded varicosity. An increase in the rate of arrival of action potentials at the nerve terminal increases the number of active varicosities. Graded responses to sympathetic nerve activity are therefore achieved by varying the rate of nerve traffic (Brock & Cunnane 1992). Clearly, these electrophysiological observations in the vas deferens do not necessarily provide information on the mechanism of noradrenaline release, since the action potential in this tissue is probably due to the ATP release rather than the noradrenaline. However, the same principles may apply.

2.3.6 The 'Burn & Rand' Hypothesis

In 1959, Burn & Rand proposed that Ach served as an intermediate transmitter in the release of noradrenaline from the sympathetic postganglionic neurone, possibly by raising the neuronal membrane permeability to Ca^{2+} (Burn & Rand 1965). This 'cholinergic link' or 'Burn and Rand' Hypothesis was supported by the observation that Ach and nicotine have effects in certain tissues that are identical to sympathetic stimulation and are prevented when noradrenaline is depleted with reserpine (Burn & Rand 1965). Furthermore, anticholinesterases, which prevent the breakdown of Ach, potentiate the actions of sympathetic nerve stimulation (Malik 1970). When responses to sympathetic nerve stimulation are blocked with the adrenergic neurone blocking agents, guanethidine and bretylium, further stimulation produces a parasympathetic response (Day & Rand 1961). This is attributed to the stimulation of cholinergic fibres in the sympathetic nerves which release Ach onto muscarinic receptors, since the response is blocked by atropine. Finally, drugs such as hemicholinium, which interferes with Ach synthesis, and botulinum toxin, which reduces Ach release from cholinergic nerves, reduce the effects of sympathetic stimulation such as to the nictitating membrane (Ambache 1951). Histochemical evidence is supportive of the hypothesis since acetylcholinesterase has been found closely associated with sympathetic neurones. Cholinergic-sympathetic fibres supplying sweat glands and certain blood vessels, where Ach is the major transmitter of the sympathetic nerves, have already been described (Chapter 1). It was therefore not surprising that the suggestion should have been made that all sympathetic nerves might release Ach.

In spite of convincing evidence and the persuasive arguments of Burn (1975), support for the Burn & Rand Hypothesis has wained. The pharmacological data could be explained by the stimulation of a mixture of sympathetic and parasympathetic nerves. The cholinergic fibres in many sympathetic nerves could be parasympathetic in origin or even postganglionic cholinergic sympathetic fibres similar to those supplying sweat glands. For the hypothesis to regain any credibility, it is necessary to demonstrate the presence of both Ach and noradrenaline stored within the same sympathetic nerve fibre and this has yet to be done (Day 1979). The sympathetic effects of nicotine are probably due to an indirect action at the noradrenergic nerve terminal causing the release of noradrenaline (Chapter 11). One possible mechanism is the induction of an action potential in the terminal axon since antidromic impulses have been detected in sympathetic nerve trunks (Ferry 1966). However, in the presence of TT_x to block propagation of action potentials along the nerve, higher concentrations of nicotine still cause release of noradrenaline in isolated atria. This indicates that nicotine induces a local depolarization of the axon terminal with consequent influx of Ca^{2+} and release of transmitter (Sarantos-Laska *et al.* 1981).

2.4 Synthesis of Noradrenaline

The synthetic pathways for noradrenaline and adrenaline were first proposed by Blaschko (1939). These were subsequently verified initially with chromaffin cells of the adrenal medulla (Gurin & Delluva 1947) and then for noradrenergic nerves (Goodall & Kirshner 1958). In mammalian tissues, the biosynthesis of cate-cholamines results in the formation of noradrenaline in sympathetic nerves. In chromaffin cells of the adrenal medulla, it continues with the N-methylation of noradrenaline to adrenaline, by the action of phenethanolamine-N-methyltransferase (PNMT) but not in those chromaffin cells producing noradrenaline. Separate adrenaline-containing neurones carrying the PNMT enzyme are also suggested to occur in the brain (Fuller 1982). In non-mammalian peripheral sympathetic nerves, including the frog, the final product of synthesis is also adrenaline, due to the presence of the N-methyltransferase (Stene-Larsen & Helle 1978). Each biosynthetic step occurs in an identifable compartment of the neurone where the appropriate enzyme is located (Figure 2.4).

The dietry amino acid precursor in the blood stream is tyrosine. In addition, dietry phenylalanine may be utilized by conversion to tyrosine outside the neurone, by the action of tyrosine hydroxylase or a closely related enzyme. It is likely that *phenylalanine hydroxylase* is a separate enzyme since its absence occurs in the inborn error of metabolism, phenylketonuria. In this congenital defect, phenylalanine is not converted to tyrosine but diverted to phenylpyruvic acid through transamination. The levels of noradrenaline are reduced but not abolished, suggesting some conversion of dietry tyrosine to dopa by tyrosine hydroxylase. The reduced activity can be attributed to the inhibition of tyrosine hydroxylase by the build-up of phenylalanine. Phenylalanine hydroxylase can be inhibited by fluorophenylalanine and by esculetin (Bowman & Rand 1980) (Figure 2.5).

2.4.1 *Tyrosine Hydroxylase (Tyrosine-3-monooxygenase)*

Tyrosine is transported across the neuronal membrane by active transport. Inside the neurone, tyrosine is converted to dihydroxyphenylalanine (dopa) by tyrosine hydroxylase (EC 1.14.3a) (Udenfriend 1966). This enzyme has a cytoplasmic location, as shown by its distribution in the supernatant after differential centrifuga-tion of tissue homogenates, and is associated with neurotubules, as determined by immunofluorescence. Tyrosine hydroxylase consists of subunits each having a

55

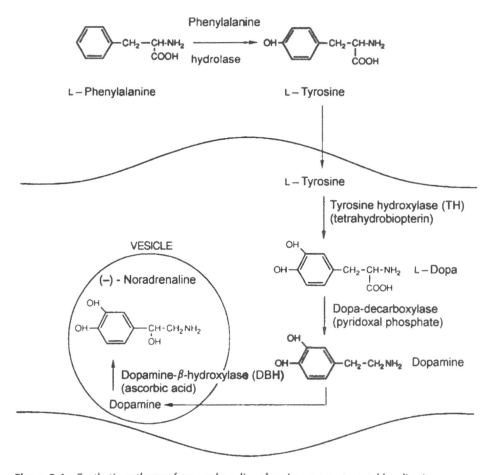

Figure 2.4 Synthetic pathways for noradrenaline showing compartmental localization.

relative molecular mass $(M_r) = 60\ 000$. Each subunit has binding sites for tyrosine, for molecular oxygen and for reduced pteridine co-factor, each of which is a requirement for activation of this enzyme.

Tyrosine hydroxylase is specific for catecholamines and is the rate-limiting step since inhibition of this, but not subsequent stages, results in depletion of noradrenaline. Changes in the activity of this enzyme therefore regulate the synthesis and levels of noradrenaline in the neurone. Thus, the levels remain relatively constant with changes in nerve activity. Tyrosine hydroxylase may be activated, inhibited, or the *amount* of enzyme may be increased (induction) in response to changes in sympathetic nerve activity. An increase in nerve activity is associated with an increase in the rate of noradrenaline synthesis. An increase in nerve traffic in the superior cervical ganglion of the rat has been shown to stimulate tyrosine hydroxylation without the release of noradrenaline. This suggests that synthesis may be regulated independently of the release (Anden & Grabowska-Anden 1985). The amount of enzyme is also increased by raised sympathetic activity such as during stress (Thoenen 1970). Induction of the enzyme in the postganglionic neurone can also be produced by an increased firing rate of the preganglionic nerve, an effect that

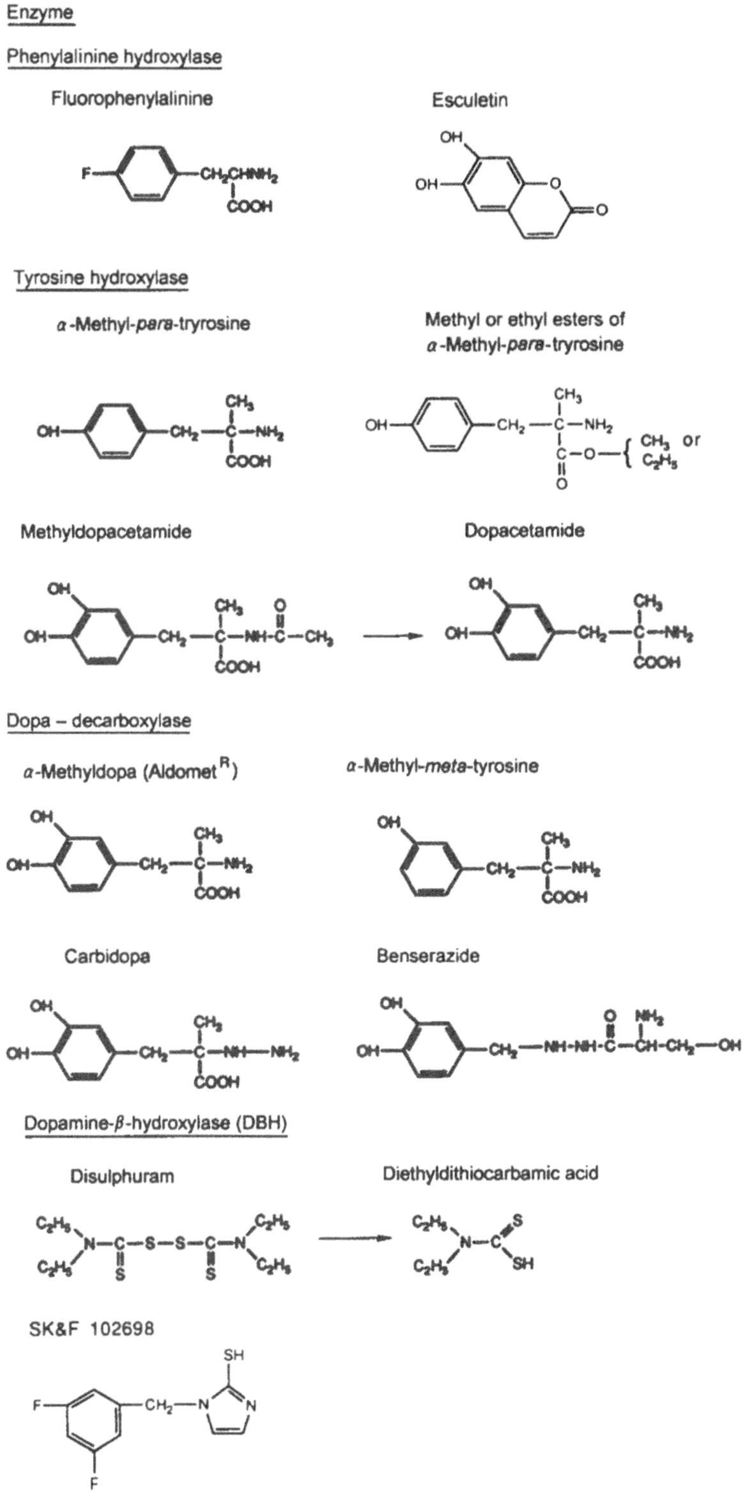

Enzyme

Phenylalinine hydroxylase

Fluorophenylalinine

Esculetin

Tyrosine hydroxylase

a-Methyl-*para*-tryrosine

Methyl or ethyl esters of
a-Methyl-*para*-tryrosine

Methyldopacetamide

Dopacetamide

Dopa – decarboxylase

a-Methyldopa (Aldomet[R])

a-Methyl-*meta*-tyrosine

Carbidopa

Benserazide

Dopamine-β-hydroxylase (DBH)

Disulphuram

Diethyldithiocarbamic acid

SK&F 102698

Figure 2.5 Inhibitors of noradrenaline synthetic enzymes.

can be prevented by blockade of ganglionic transmission with a nicotinic receptor antagonist (Mueller *et al.* 1970).

The stimulation of noradrenaline synthesis by increased nerve activity is thought to be initiated by the influx of Ca^{2+} that causes the release of noradrenaline. As measured by the increased accumulation of noradrenaline into synaptosomes, with its breakdown inhibited by a MAO inhibitor, tyrosine hydroxylase can be activated by raising intracellular Ca^{2+} with the Ca^{2+} ionophor, A23187, and by raising extracellular K^+. The Ca^{2+} ionophor causes an influx of calcium by creating channels that are voltage-independent. Agents that raise intraneuronal cyclic adenosine monophosphate (cAMP) also stimulate synthesis. For example, the stable analogue, 8-bromo-cAMP, activates tyrosine hydroxylase in the absence of extracellular Ca^{2+} and is therefore probably independent of the influx of Ca^{2+}. Presynaptic receptors on the neuronal terminal, in addition to regulating release of transmitter (see Chapter 6), control its synthesis. Adenosine and isoprenaline, through prejunctional A_2 purinoceptors and β-adrenoceptors, respectively, enhance synthesis by elevating cAMP levels through the stimulation of adenylyl cyclase. This effect of adenosine appears to contrast with its inhibition of transmitter release (see Chapter 6) which is mediated via A_1-purinoceptors through *inhibition* of adenylyl cyclase. Prejunctional α_2-adrenoceptors are negatively coupled to adenylyl cyclase and their stimulation by the selective α_2-agonist, clonidine, results in inhibition of Ca^{2+}- and K^+-induced noradrenaline synthesis in brain synaptosomes. This depression of synthesis is blocked by selective α_2-antagonists. The effect does not, however, appear to be related to the changes in cAMP levels since α_2-agonists inhibit synthesis induced by cAMP-independent (A23187 and K^+) and not cAMP-dependent (isoprenaline and adenosine) mechanisms. The α_2-agonists appear to inhibit synthesis at a step beyond Ca^{2+} influx, which contrasts with their action on release of transmitter which is due to modulation of the voltage-dependent Ca^{2+} channel. Whether α_2-adrenoceptors modulate noradrenaline synthesis in intact peripheral nerve endings remains to be established (Fillenz 1990, 1992).

The activation of tyrosine hydroxylase by cAMP-dependent processes is probably through phosphorylation of the enzyme (Fillenz 1992). Phosphorylation can occur by three different classes of protein kinases which may have different effects upon enzyme activity. These classes of protein kinases are: (1) cAMP-dependent protein kinase which increases the affinity for the co-factor pteridine with no change in V_{max}; (2) cAMP-independent protein kinase which increases V_{max} but which has no effect on affinity for the co-factor; and (3) Ca^{2+}-calmodulin-dependent protein kinase. The enzyme appears to exist in two affinity states for the naturally occuring reduced form of the co-factor, tetrahydrobiopterin (BH_4). Phosphorylation converts the low-affinity to the high-affinity form, thus allowing lower concentrations of co-factor to activate it (Fillenz 1990).

A model for the immediate regulation of tyrosine hydroxylase by nerve activity and by activation of β, α_2 and A_2-receptors is shown in Figure 2.6. Other prejunctional receptors have also been identified which appear to modulate the synthesis of noradrenaline, including inhibitory muscarinic receptors for Ach.

Longer-term regulation of the enzyme also occurs. For example, depletion of noradrenaline stores by reserpine is followed by an increase in tyrosine hydroxylase (Mueller *et al.* 1969a). The time course shows a delay of between 10 and 18 hr and a duration of several days. In adrenal medullary cells, the induction of tyrosine hydroxylase by reserpine is also accompanied by an increase in GTP cyclohydrolase,

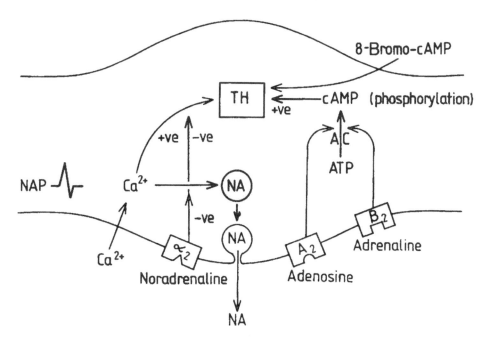

Figure 2.6 Control of noradrenaline (NA) synthesis at the level of tyrosine hydroxylase (TH) by the nerve action potential (NAP)-mediated influx of Ca^{2+} and by prejunctional α_2- and β_2-adrenoceptors and A_2 adenosine receptors. AC represents adenylyl cyclase.

the enzyme that synthesizes the pteridine co-factor. The induction of tyrosine hydroxylase is blocked by actinomycin and cycloheximide (Mueller *et al.* 1969b), evidence that it requires protein synthesis through RNA transcription and translation. The increase in activity is first seen in the cell body but later occurs in the nerve terminal, presumably as a result of transport of the enzyme along the axon. It is therefore a delayed, long-lasting increase in enzyme protein, distinct from the immediate regulation at the nerve terminal described above.

2.4.1.1 *Inhibition of tyrosine hydroxylase*

Importantly, tyrosine hydroxylase is subject to end-product inhibition, that is, it is inhibited by noradrenaline. Another potent inhibitor is α-methyl-*p*-tyrosine, which effectively reduces the vesicular content of noradrenaline since this is the rate-limiting step of synthesis. Noradrenaline levels are effectively reduced by α-methyltyrosine, which is ultimately converted to the 'false transmitter', α-methylnoradrenaline (Spector *et al.* 1965). α-Methyl-*p*-tyrosine reduces synthesis of adrenaline and lowers blood pressure in patients with adrenal medullary tumour – phaeochromocytoma. It is useful in patients prior to surgery and may control symptoms for many months but is ineffective in essential hypertension. It is effectively combined with phenoxybenzamine and β-adrenoceptor blockade (see Chapter 5) (Brogden *et al.* 1981). Other inhibitors of tyrosine hydroxylase include α-methylphenylalanine, the methyl and ethyl esters of α-methyl-*p*-tyrosine and dopacetamide (Figure 2.5). The esters of these compounds act as prodrugs, effective by oral administration and after absorption they are hydrolysed to the active drug.

The tyrosine analogues inhibit by competing for the active centre of the enzyme, while the catechols compete for the pteridine co-factor.

2.4.2 *Dopa-decarboxylase*

The next step in the synthesis of noradrenaline is decarboxylation of dopa to dopamine. This occurs in the cytoplasm of the neurone by the action of dopa-decarboxylase (Figure 2.4). This enzyme is a relatively non-specific aromatic L-amino acid decarboxylase (EC 4.1.12.5) which also removes the carboxyl group from *m*-tyrosine, *p*-tyrosine, phenylalanine, methyldopa, the 5-hydroxytryptamine (serotonin) precursor, 5-hydroxytryptophan, and the histamine precursor, histidine. It has a K_m for conversion of dopa to dopamine of 4×10^{-4} M. In common with tyrosine hydroxylase, it decarboxylates L-amino acids only, thus it is laevodopa that is administered for the treatment of Parkinson's disease to raise central dopamine levels.

Since the three amino acid precursors of noradrenaline (L-phenylalanine, L-tyrosine and L-dopa) are all capable of decarboxylation, they can yield alternative intermediates to the major ones shown in Figure 2.4. For example, low levels of tyramine and octopamine (norsynephrine) occur naturally (Figure 2.7). Octopamine does not appear to serve a transmitter role in vertebrates, however, it fulfils many of the functions of adrenaline and noradrenaline in arthropods. It is a circulating hormone in crustaceans and insects. In the locust, octopaminergic nerves have been described which modulate the contractions of the extensor-tibiae muscle of the hind leg. These responses are mimicked by the action of octopamine on specific octopamine receptors. These receptors are said to be different from adrenoceptors, but from agonist and antagonist potencies they resemble α_2-adrenoceptors (Evans 1981).

Dopa-decarboxylase requires the presence of pyridoxal 5'-phosphate as a co-factor. The conversion of dopa to dopamine is rapid, thus dopa does not accumulate in the sympathetic neurone. It is therefore not a rate-limiting step and inhibition of dopa-decarboxylase has little effect on tissue noradrenaline levels.

Dopa-decarboxylase inhibitors include the substrates α-methyldopa (Aldomet[R]) and α-methyl-*m*-tyrosine (Figure 2.5). These agents are themselves converted to α-methyldopamine and α-methyltyramine, respectively, by dopa-decarboxylase and then to α-methylnoradrenaline and metaraminol (see Figure 6.4 and Table 4.1), respectively. These serve as false transmitters, replacing noradrenaline in the storage vesicles on a mole for mole basis. It is this, rather than enzyme inhibition, that probably explains the minor depletion of noradrenaline.

Other inhibitors are shown in Figure 2.5 and include the clinically useful carbidopa and benserazide. These are peripherally acting since they poorly penetrate the blood–brain barrier. They are combined with L-dopa (Madopar[R] and Sinemet[R], respectively) in the treatment of Parkinson's disease to prevent its peripheral decarboxylation to dopamine thus allowing lower doses of L-dopa to be used.

2.4.3 *Dopamine-β-hydroxylase*

The final step in the synthesis of noradrenaline occurs in the storage vesicle where the enzyme dopamine-β-hydroxylase (DBH) (EC 1.14.17.1) is mainly located. Its

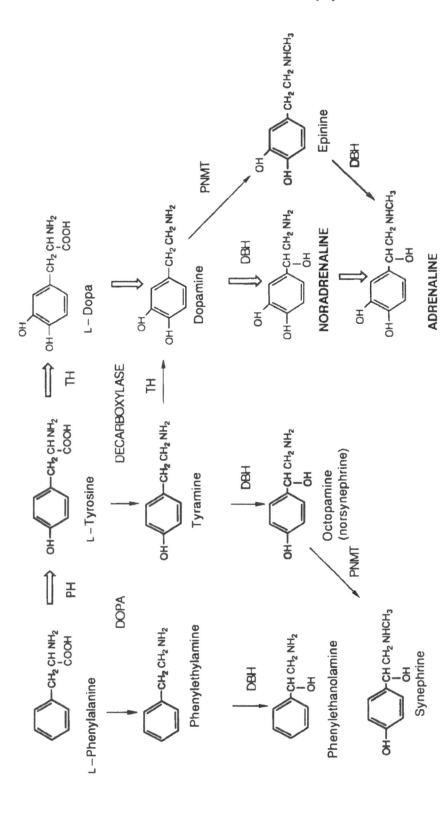

Figure 2.7 Alternative minor pathways of synthesis of catecholamines. Major route is shown by bold arrow. PH, Phenylalanine hydrolase; DBH, dopamine β-hydroxylase; PNMT, phenethanolamine-N-methyltransferase (in chromaffin cells of adrenal medulla).

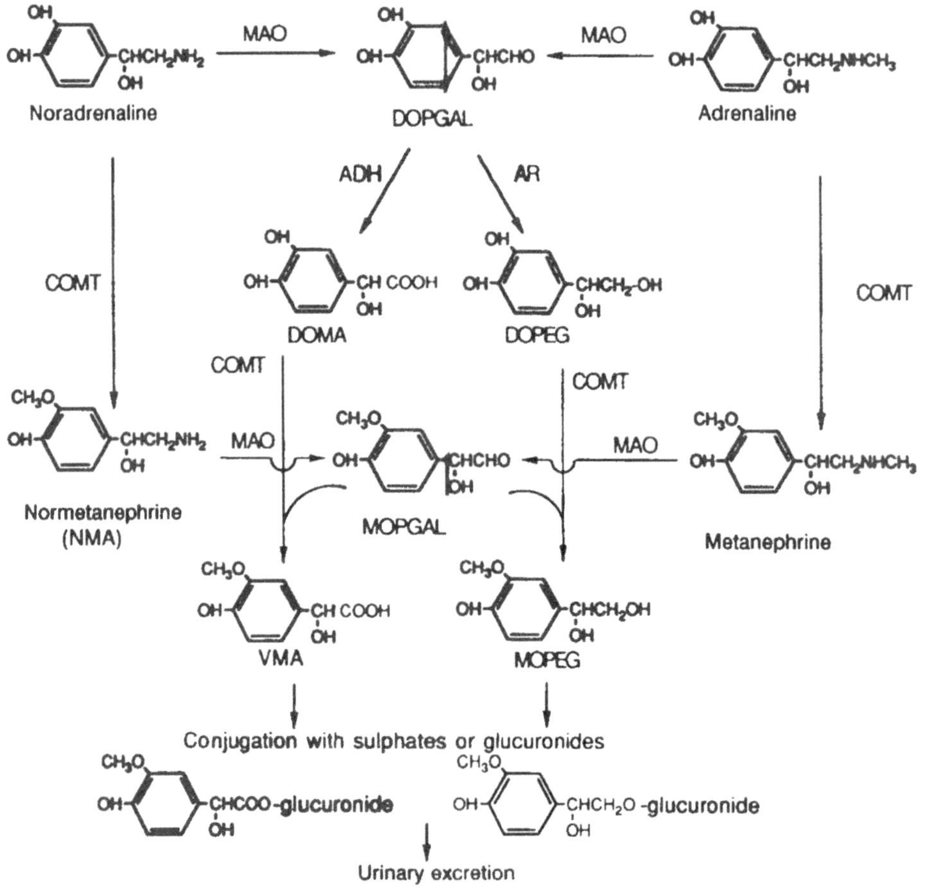

Figure 2.8 Metabolic pathways for noradrenaline and adrenaline. Enzymes: ADH, aldehyde dehydrogenase; AR, aldehyde reductase; COMT, catechol-O-methyl transferase; MAO, monoamine oxidase. Metabolites; DOMA, 3,4-dihydroxymandelic acid; DOPEG, 3,4-dihydroxyphenylglycol; DOPGAL, 3,4-dihydroxyphenylglycolic aldehyde; MOPEG, 3-methoxy-4-hydroxyphenylethyleneglycol; VMA, 3-methoxy-4-hydroxymandelic acid ('vanillylmandelic acid').

distribution in vesicles, transport both down the axon and retrogradely in inactive form, and release during exocytosis have already been described. It is therefore a useful marker of the release from sympathetic neurones, a possible advantage being that, unlike noradrenaline, it is not taken up. However, plasma noradrenaline and DBH levels are not well correlated (Weinshilboum 1979).

DBH is therefore found in the vesicular fraction after differential centrifugation (microsomal fraction) but also in the final supernatant, indicating its release from the endoplasmic reticulum. A soluble form is found in the core of the large vesicles and released by nerve activity. This is the most active form. A membrane-bound DBH is less active and is the precursor of the highly active soluble form. This appears to have a hydrophobic tail which anchors it to the vesicular membrane (Fillenz 1990, 1992). DBH is a copper-containing glycoprotein consisting of four subunits each of $M_r = 75\ 000$. It requires the presence of ascorbic acid as a co-factor. The enzyme is relatively specific for phenylethylamines and converts tyramine to octopamine and phenylethylamine to phenylethanolamine (Figure 2.7). The products have a chiral centre and are always the $(-)$-isomers.

Inhibitors of DBH include sulphydryl compounds and tropolone which complex with the copper residue of the enzyme. An example is disulphuram (Antabuse^R) which is converted to the active metabolite, diethyldithiocarbamic acid (Figure 2.5). Since this is not the rate-limiting step, these inhibitors do not have a marked effect upon the production of noradrenaline or adrenaline and cause only a small decrease in noradrenaline levels, with an accompanying rise in dopamine content. Disulphuram itself is an aldehyde dehydrogenase inhibitor and is used as a disincentive to alcohol consumption by alcoholics. Ethanol is normally converted to acetaldehyde by alcohol dehydrogenase in the liver and then further oxidized by aldehyde dehydrogenase. In the presence of disulphuram, however, the aldehyde accumulates and gives rise to the acetaldehyde syndrome. This is most unpleasant and includes nausea, sweating, vertigo, hypotension (probably because of DBH inhibition and a fall in noradrenaline levels in sympathetic nerves) and weakness. A more recent DBH inhibitor is SK&F 102698 (Figure 2.5) which elevates renal dopamine levels three-fold (Kinoshita *et al.* 1990) (see also Chapter 12, dopaminergic neurones).

2.5 Metabolism and Uptake of Noradrenaline

Following the release of noradrenaline into the synaptic cleft and its generation of a physiological response, it must be rapidly removed from the vicinity of the receptor. This will prevent the repeated stimulation of the receptor sites and permit fine control of the sympathetic innervation of an effector. It will also conserve transmitter. The principal route for removing from the synaptic cleft, noradrenaline released by sympathetic nerve activity, is by reuptake into the nerve terminal. This process is the neuronal uptake or uptake$_1$ and accounts for the removal of ~70% of the noradrenaline from the biophase. A secondary uptake mechanism is into extraneuronal or uptake$_2$ sites. These two uptake compartments contain the two major metabolic enzymes for noradrenaline, monoamine oxidase (MAO) and catechol-O-methyltransferase (COMT). MAO is present mainly in the noradrenergic nerve terminals but also occurs extraneuronally, the highest levels being found in the liver and kidney. Most of the activity of COMT takes place in extraneuronal tissues, although a minor source is the sympathetic nerve ending. The removal of noradrenaline from the neuronal cytoplasm, either following its uptake or spontaneous leakage, is therefore by uptake into the storage vesicles and by the activity of MAO.

Two further enzymes are involved in the metabolism of noradrenaline following the activity of MAO. The aldehyde product of MAO activity (3,4-dihydroxyphenylglycolic aldehyde, DOPGAL) undergoes extraneuronal oxidation by aldehyde dehydrogenase (ADH), when MAO activity has occurred outside neurones such as in the liver or intestine. Inside the neurone, DOPGAL is reduced by aldehyde reductase (AR). The pathways for the metabolism of noradrenaline by these enzymes are shown in Figure 2.8 and a schematic diagram of their locations is shown in Figure 2.9.

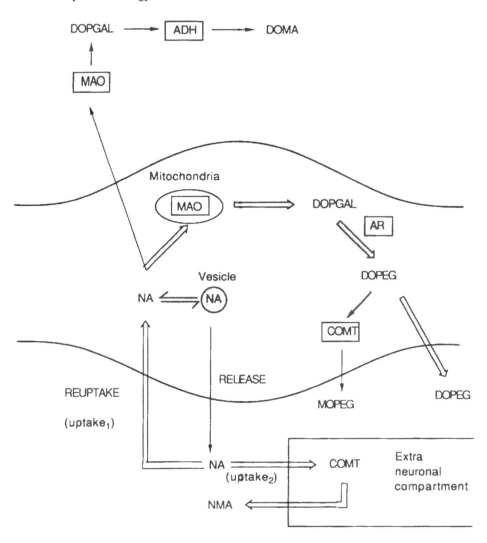

Figure 2.9 Metabolic pathways and compartments for noradrenaline. Enzymes (shown in boxes): ADH, aldehyde dehydrogenase; AR, aldehyde reductase; COMT, catechol-O-methyl transferase; MAO, monoamine oxidase. Metabolites (unboxed): DOMA, 3,4-dihydroxy-mandelic acid; DOPEG, 3,4-dihydroxyphenylglycol; DOPGAL, 3,4-dihydroxyphenylglycolic aldehyde; MOPEG, 3-methoxy-4-hydroxyphenylethyleneglycol; NMA, normetanephrine. Major routes are shown in bold arrows.

2.5.1 Monoamine Oxidase

Monoamine oxidase (monoamine oxygen oxidoreductase, EC 1.4.3.4) is tightly bound to the outer membrane of mitochondria. It deaminates primary and secondary amines that are free in the neuronal cytoplasm but not those bound in storage vesicles. In 1928, Hare first described an oxidase enzyme that deaminated tyramine (*p*-hydroxyphenylethylamine) to *p*-hydroxyphenylacetic acid. This enzyme was subsequently found to be widely distributed and capable of deaminating a number of amines of both dietary and endogenous neurotransmitter origin including noradrenaline, adrenaline, dopamine, octopamine and serotonin (5-hydroxytryptamine, 5-HT). In most sympathetically innervated tissues previously labelled with [^{3}H]noradrenaline, it is the metabolites of noradrenaline that make up the majority of the radiolabelled products of spontaneous outflow. Unchanged noradrenaline represents <10% and the major metabolite is DOPEG, the product of deamination by MAO to DOPGAL immediately followed by reduction by aldehyde reductase (Figure 2.8).

Generally, the levels of the deaminated acid (DOMA) formed by extraneuronal oxidation of noradrenaline are low, except in a few tissues such as the guinea-pig atria where MAO is regarded as extraneuronally located (Langer 1974). DOPEG also represents the major metabolite produced after nerve stimulation, indirectly acting sympathomimetic amines and depolarization by raised extracellular K$^+$. It is formed after reuptake of the released noradrenaline, which is then metabolized intraneuronally via MAO and aldehyde reductase. This is supported by the fact that inhibition of neuronal uptake by cocaine prevents the formation of DOPEG during nerve stimulation. The increases in O-methylated deaminated metabolites, such as MOPEG, that occur during nerve stimulation are also attributed to reuptake of noradrenaline followed by intraneuronal deamination followed in turn by intraneuronal methylation. However, the increased level of normetanephrine during nerve stimulation is due to extraneuronal postjunctional COMT activity (Figure 2.8). This is elevated by cocaine because of greater levels of noradrenaline reaching the extraneuronal sites of O-methylation. Thus, DOPEG levels provide a useful index for increased sympathetic nerve activity and increased noradrenaline release.

These observations have been taken to indicate that the primary purpose of neuronal uptake is to inactivate the neurotransmitter, with conservation of transmitter by restorage in vesicles being a secondary role. Thus, enhanced synthesis of noradrenaline during nerve stimulation via tyrosine hydroxylase may be more important for maintenance of noradrenaline stores, as discussed earlier. An increase in tyrosine hydroxylase activity may be induced by the elevated levels of intraneuronal deaminated metabolites. In fact, the opposite occurs, with DOPEG being equipotent with noradrenaline at *inhibiting* tyrosine hydroxylase (Rubio & Langer 1973). However, one explanation for stimulation of tyrosine hydroxylase has been that the conversion of DOPGAL to DOPEG by aldehyde reductase involves the simultaneous oxidation of NADPH to NADP$^+$ (Figure 2.10). This elevated NADP$^+$ activates the pentose shunt for oxidation of glucose and may be responsible for maintaining high levels of NADPH. The latter may be involved in reduction of the pteridine co-factor of tyrosine hydroxylase (q.v.). The presence of NADPH is necessary for the activity of dihydropteradin reductase which is responsible for maintaining the co-factor in its active reduced state, tetrahydrobiopterin (BH$_4$) (Figure 2.10) (Langer 1974).

2.5.1.1 *MAO-A and MAO-B*

Two isozymes of MAO having different substrate specificities and inhibitor sensitivities have now been identified. Johnston (1968) found that only a portion of any MAO was inhibited by the then newly synthesized inhibitor, clorgyline; this portion was called MAO-A and the clorgyline-resistant fraction was termed MAO-B. At about the same time, the potent inhibitor, selegiline [(−)-deprenyl], was

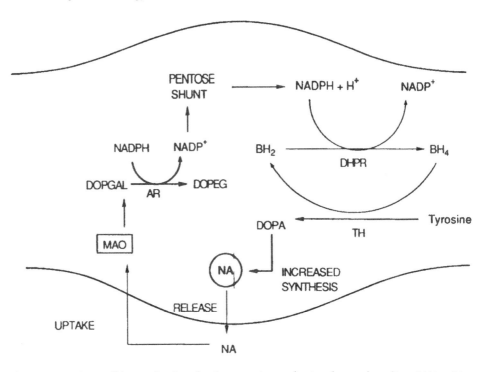

Figure 2.10 A possible mechanism for increase in synthesis of noradrenaline (NA) arising from its release. Reuptake is followed by deamination to DOPGAL. The conversion of DOPGAL to DOPEG produces NADP$^+$ which enters the pentose shunt. This activity may be responsible for maintenance of tetrahydrobiopterin (BH$_4$) in its active reduced state by means of dihydropteridine reductase (DHPR). The BH$_4$ would then activate tyrosine hydroxylase (TH) to increase synthesis of noradrenaline (NA). Abbreviations are otherwise as in Figures 2.8 and 2.9.

synthesized and found to be selective for MAO-B. Since then, the distribution of the two isozymes in different tissues and species has been established and their substrate and inhibitor specificities identified.

The A form predominates in the stomach, intestine and placenta (~70 %) (Tipton 1990). MAO-A is the subtype found in peripheral sympathetic neurones but is also located extraneuronally. The preferred substrates for MAO-A are the polar aromatic amines including noradrenaline, adrenaline, 5-HT and octopamine. 5-HT is a substrate only for the A form. MAO-B is found exclusively in platelets and in serotonergic neurones and predominates in the brain. MAO-B selectively inactivates non-polar aromatic amines such as benzylamine and phenylethylamine. Dopamine is a relatively selective substrate for MAO-B. It is converted to the intermediate aldehyde and then via aldehyde reductase to DOPET or via aldehyde dehydrogenase to DOPAC (Figure 2.11).

The existence of two different enzymes has been proven by the cloning and sequencing of the cDNA coding for human liver MAO-A and MAO-B (Bach *et al.* 1988). The active site of MAO consists of a flavine residue, flavine adenine diphosphate (FAD), which is covalently bound to a sequence of five amino acids (Tyr-Cys-Gly-Gly-Ser) through a thioester linkage to the cysteinyl residue (Figure 2.12). This five amino acid chain forms a portion of a sequence of 20 amino acids that is identical for both isozymes (Figure 2.12).

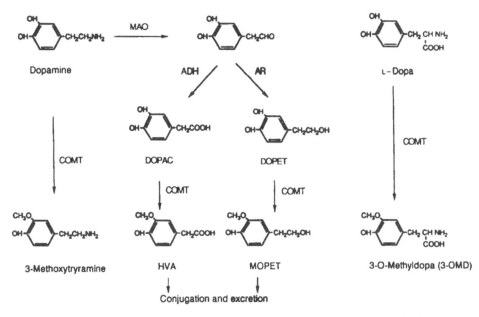

Figure 2.11 Metabolic routes for dopamine and L-dopa. DOPAC, 3,4-Dihydroxypheny-lacetic acid; DOPET, 3,4-dihydroxyphenylethanol; HVA, 3-methoxy-4-hydroxyphenylacetic acid (homovanillic acid); MOPET, 3-methoxy-4-hydroxyphenylethanol. Other abbreviations as in Figure 2.9.

The reaction of MAO with its substrates requires molecular O_2 and is represented by the equation:

$$R\text{-}CH_2NH_2 + O_2 + H_2O \rightarrow R\text{-}CHO + NH_3 + H_2O_2.$$

The amine ($R\text{-}CH_2NH_2$) binds to the enzyme (E), the FAD being reduced. An intermediate complex is formed which is then converted to an imine product (R-CH = NH):

$$R\text{-}CH_2NH_2 + E(FAD) \rightleftharpoons (R\text{-}CH_2NH_2)\text{-}E(FAD) \rightarrow (R\text{-}CH = NH)\text{-}E(FADH_2)$$
$$\rightleftharpoons R\text{-}CH = NH + E(FADH_2).$$

The imine intermediate reacts with water to yield the aldehyde and ammonia: .

$$R\text{-}CH = NH + H_2O \rightarrow R\text{-}CHO + NH_3.$$

The reduced enzyme reacts with molecular oxygen to regenerate free enzyme and release hydrogen peroxide:

$$E(FADH_2) + O_2 \rightleftharpoons O_2\text{-}E(FADH_2) \rightarrow H_2O_2\text{-}E(FAD) \rightleftharpoons H_2O_2 + E(FAD).$$

It should be noted that a methyl group in the 2 position of the side-chain renders the amine immune to MAO.

2.5.1.2 Monoamine oxidase inhibitors

Iproniazid and isoniazid were first introduced for the treatment of tuberculosis, and the variable tuberculostatic action of iproniazid resulted in its rejection in favour of

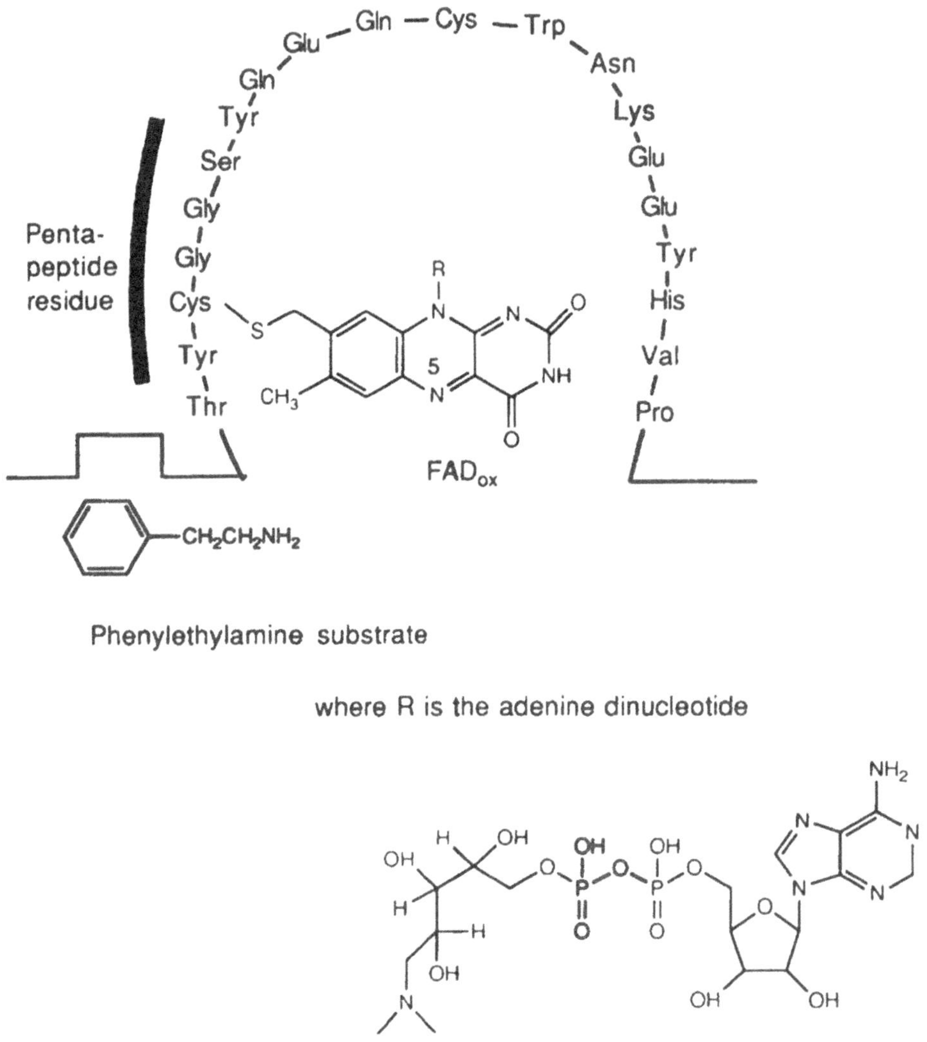

Phenylethylamine substrate

where R is the adenine dinucleotide

Figure 2.12 Amino acid sequence and attachment of FAD to MAO-A and MAO-B.

isoniazid. Although central effects were noted in both drugs, it was iproniazid that produced most euphoria and hyperexcitability. In 1952, Zeller & Barsky found that iproniazid but not isoniazid was a potent inhibitor of MAO in rats. This led to the conclusion that the mental stimulation seen in patients taking this drug was related to inhibition of MAO and the subsequent rise in neurotransmitter amine levels such as noradrenaline and 5-HT in the brain. Subsequently, other MAO inhibitors based on the hydrazine structure of iproniazid were developed as antidepressant drugs, including nialamide, isocarboxazide, phenelzine and pheniprazine (Figure 2.13). To eliminate potential hepatotoxicity of the hydrazines, in 1962 the first non-hydrazine MAO inhibitor, tranylcypromine, was introduced (Cole 1964). This was followed soon afterwards by the propargylamines such as pargyline. Clorgyline and selegiline [(−)-deprenyl] are also members of this group but were the first to show selectivity towards MAO-A and MAO-B, respectively (Jarrott & Vajda 1987).

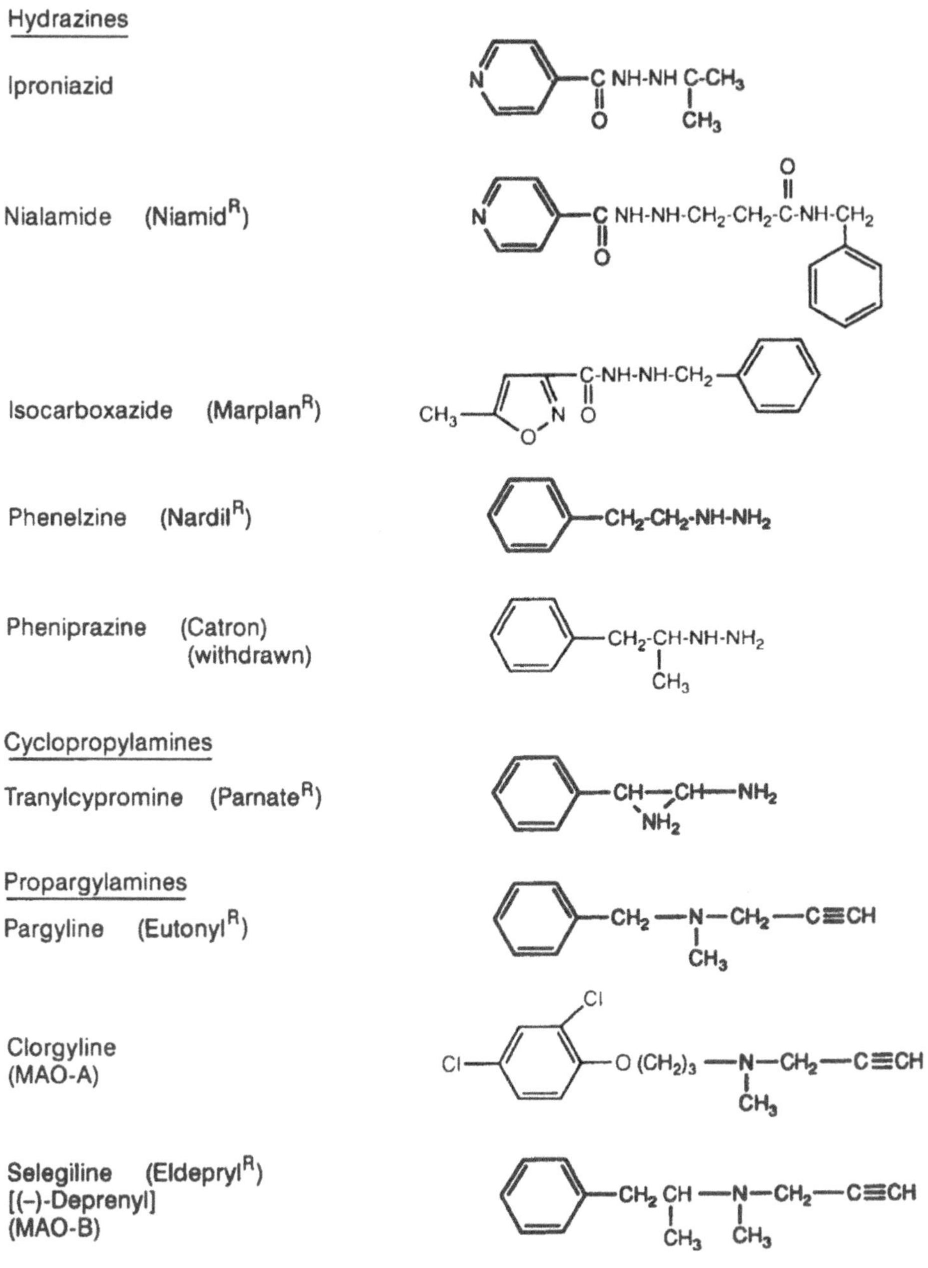

Hydrazines

Iproniazid $-C(=O)NH\text{-}NH\text{-}CH(CH_3)\text{-}CH_3$ (attached to pyridine)

Nialamide   (Niamid[R]) $-C(=O)NH\text{-}NH\text{-}CH_2\text{-}CH_2\text{-}C(=O)\text{-}NH\text{-}CH_2\text{-}$(phenyl) (attached to pyridine)

Isocarboxazide   (Marplan[R]) CH_3-(isoxazole)$-C(=O)\text{-}NH\text{-}NH\text{-}CH_2\text{-}$(phenyl)

Phenelzine   (Nardil[R])   (phenyl)$-CH_2\text{-}CH_2\text{-}NH\text{-}NH_2$

Pheniprazine   (Catron) (withdrawn)   (phenyl)$-CH_2\text{-}CH(CH_3)\text{-}NH\text{-}NH_2$

Cyclopropylamines

Tranylcypromine   (Parnate[R])   (phenyl)$-CH\text{--}CH\text{--}NH_2$ (cyclopropyl, with NH_2)

Propargylamines

Pargyline   (Eutonyl[R])   (phenyl)$-CH_2\text{--}N(CH_3)\text{--}CH_2\text{--}C{\equiv}CH$

Clorgyline (MAO-A)   (2,4-dichlorophenyl)$-O(CH_2)_3\text{--}N(CH_3)\text{--}CH_2\text{--}C{\equiv}CH$

Selegiline   (Eldepryl[R]) [(−)-Deprenyl] (MAO-B)   (phenyl)$-CH_2\text{--}CH(CH_3)\text{--}N(CH_3)\text{--}CH_2\text{--}C{\equiv}CH$

Figure 2.13 Irreversible monoamine oxidase (MAO) inhibitors.

This group of MAO inhibitors are regarded as irreversible antagonists. They are also known as mechanism-based or 'suicide' inhibitors. They bind to the enzyme and are themselves oxidized in the same way as are substrate amines, with the enzyme–FAD complex being converted to the reduced form (E-FADH$_2$). During this reaction, however, they are converted to reactive electrophilic intermediates which then interact irreversibly with the FAD component of the enzyme usually at the N-5 portion (Figure 2.14). This results in irreversible inhibition of the enzyme.

The propargylamine (selegiline) intermediate reacts with the N-5 position. The hydrazine, phenylhydrazine, through the highly reactive diazene intermediate reacts at the 4a position of FAD_{ox}. The cyclopropylamines inhibit the enzyme in different ways depending on their structure. Tranylcypromine is shown in Figure 2.14 reacting through a radical intermediate, not with FAD, but with the FAD binding site, probably at the cysteine residue (Kyburz 1990).

More recently, competitive reversible and short-acting MAO inhibitors have been developed. The effect of these is readily reversed by any procedure that reduces inhibitor concentration. Furthermore, the duration of action will be controlled by its rate of removal by metabolism and elimination because no covalent bond formation is involved. Substrate and inhibitor binding to the enzyme is mutually exclusive and high concentrations of substrate will displace a competitive inhibitor from the enzyme. Examples include brofaromine (CGP 11305A), moclobemide, cimoxatone and toloxatone, which are selective for MAO-A. Ro 19-6327 is a highly potent and reversible selective MAO-B inhibitor (Figure 2.15) (Kyburz 1990).

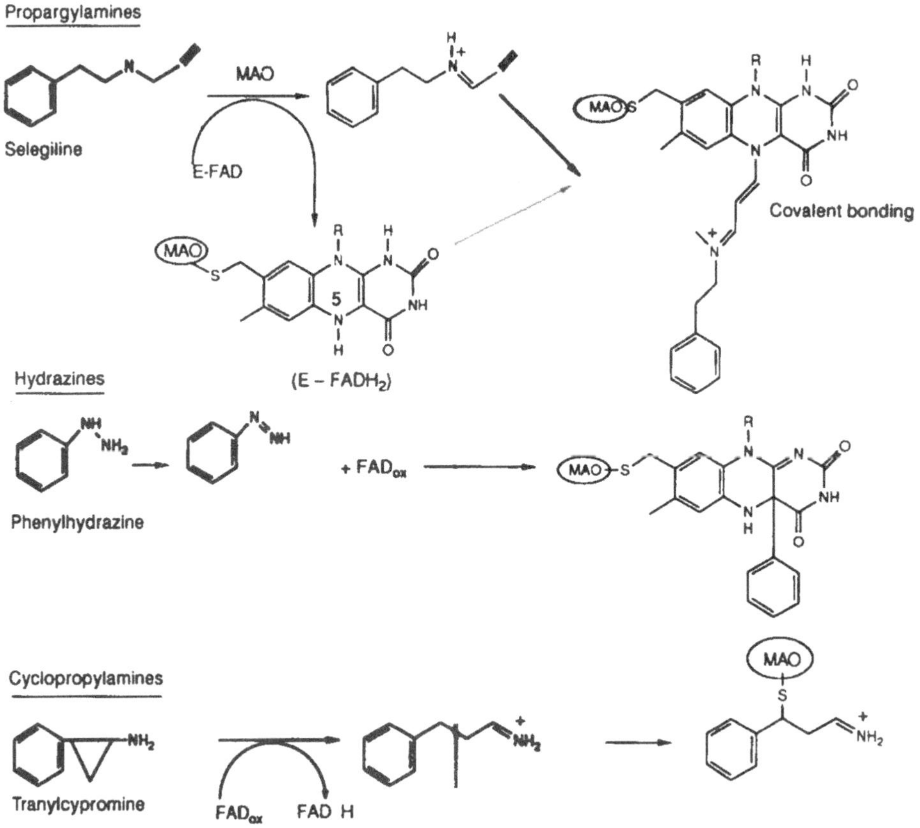

Figure 2.14 Mechanisms of irreversible inhibition of MAO by propargylamines, hydrazines and a cyclopropylamine. R is the adenine dinucleotide residue (see Figure 2.12) and E represents the enzyme, MAO. After Kyburz (1990).

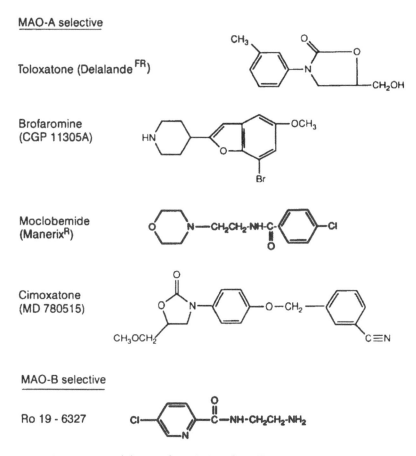

MAO-A selective

Toloxatone (Delalande ^{FR})

Brofaromine
(CGP 11305A)

Moclobemide
(Manerix^R)

Cimoxatone
(MD 780515)

MAO-B selective

Ro 19 - 6327

Figure 2.15 Competitive inhibitors of MAO-A and MAO-B.

2.5.1.3 *Pharmacology and clinical uses*

The initial use of MAO inhibitors was in the treatment of depression. Their effectiveness is attributed primarily to the inhibition of MAO-A in the CNS, which permits elevation of 5-HT levels. This is supported by the finding that the selective MAO-B inhibitor, selegiline, was not an effective antidepressant in patients with endogenous depression. When combined with the 5-HT precursor, 5-hydroxytryptophan, however, selegiline had a good antidepressant action. Differences in the cellular location of MAO-A and MAO-B in the brain have been identified. MAO-B appears to be located in the cell bodies of the raphe nucleus in serotonergic neurones, and in astrocytes. MAO-A has been found in cell bodies of the locus coeruleus and in the terminals of serotonergic nerves. The significance of this different distribution to the antidepressant action of MAO inhibitors remains to be elucidated (Youdim & Finberg 1990).

The initial enthusiasm for the early irreversible and non-selective MAO inhibitors (iproniazid and tranylcypromine) was soon dampened by the occurrence of serious side-effects. The hepatotoxicity could be overcome by avoiding hydrazine

71

MAO inhibitors. However, the occurrence of life-threatening hypertensive crisis associated with the consumption of tyramine-containing foods has been the cause of most concern in the use of MAO inhibitors. This is known as the 'cheese effect' because of the high tyramine content of cheeses, although other foods rich in tyramine include yeast and meat extracts (Bovril, Marmite, Vegemite), red wines (Chianti), broad bean pods and pickled herrings. MAO in the intestine and liver normally deaminates dietary tyramine to *p*-hydroxyphenylacetic acid so that little ingested tyramine reaches peripheral sympathetic neurones. However, during MAO inhibitor therapy, the ingestion of these foods results in high levels of tyramine reaching sympathetic nerve terminals where it can enter and cause release of noradrenaline (see indirectly acting sympathomimetic amines, Chapter 4). The released noradrenaline causes vasoconstriction and a transient hypertension which in certain individuals has been fatal due to intracranial haemorrhage. This is a potential problem with any indirectly acting sympathomimetic amine, in particular those that may be used in over-the-counter (OTC) cough and cold remedies, such as the decongestant phenylpropanolamine (eg Contac 400[R]), amphetamine, ephedrine and fenfluramine (see Tables 4.1 and 4.4.). Thus, MAO inhibitors potentiate the peripheral autonomic activity of indirectly acting sympathomimetic amines (Figure 2.16) (Ryall 1961).

The potentiation is in fact irrespective of whether the amine is a substrate for MAO. Sympathomimetic amines methylated at the α-position in the side-chain are immune to breakdown by MAO, since the basic reaction is not possible. This is unimportant for exogenously administered catecholamines, such as α-methyl-noradrenaline (cobefrine), which are metabolized principally by COMT. However, with non-catechols deamination is the only route of metabolism and if these are α-methylated they are neither O-methylated nor deaminated (amphetamine, ephedrine; Table 4.1) and have a longer plasma half-life. Thus, indirectly acting sympathomimetic amines such as ephedrine and amphetamine, that are not

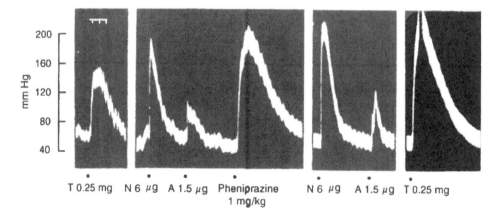

Figure 2.16 Effects of the irreversible monoamine amine oxidase inhibitor, pheniprazine, on the pressor responses to an indirectly acting sympathomimetic amine (tyramine, T) and directly acting amines (noradrenaline, N and adrenaline, A) of the spinal cat. Time scale is in minutes and all drugs are injected intravenously. Note that tyramine is substantially potentiated whereas noradrenaline and adrenaline are only slightly enhanced. Reproduced with permission from Ryall (1961).

substrates for either MAO-A or MAO-B, would not be deaminated in the gut anyway. Similarly β-phenylethylamine would not be deaminated since it is a selective substrate for MAO-B which is virtually absent in the intestine. However, in the sympathetic nerve terminals MAO inhibition increases the cytoplasmic pool of noradrenaline. It is this pool that is released by indirectly acting sympathomimetic amines, which are therefore potentiated (Youdim & Finberg 1991).

The responses to exogenously administered, directly acting sympathomimetic amines are not markedly potentiated by MAO inhibitors because MAO is not important for their inactivation. However, there may be some potentiation because the build-up of cytosolic noradrenaline exerts a neuronal uptake inhibitory effect (Figure 2.16).

Although the MAO inhibitors were widely used as antidepressants between 1957 and 1962, their use has since declined rapidly because of the life-threatening hypertensive crisis and the ever increasing list of dietary restrictions. Other side-effects including anticholinergic actions such as dry mouth, constipation, blurred vision, hesitancy of micturition and dysuria. These, combined with a past history of doubtful efficacy together with the demonstration that tricyclic antidepressants were superior, has limited MAO inhibitors to only 1.5% of all antidepressant prescriptions. It appears that the risks of death from a sudden hypertensive crisis in patients using MAO inhibitors has been exaggerated and the levels of tyramine in foods is probably lower than early reports suggested. Thus, in recent years there has been considerable renewed interest in MAO inhibitors which avoid these problems (Kyburz 1990).

The newer MAO inhibitors moclobemide (Manerix[R]) and toloxatone (Delalande[FR]), which are selective for MAO-A and used as antidepressants do not produce the 'cheese effect'. High levels of dietary tyramine are able to displace these inhibitors from the enzyme, making it available for the degradation of at least some of the tyramine (Tipton 1990).

The second major use of MAO inhibitors is as an adjunct to L-dopa in the treatment of Parkinson's disease. Here it is MAO-B inhibitors such as (−)-selegiline (Eldepryl[R]) that are of value. The rationale for their use is that dopamine formed by the decarboxylation of L-dopa (Figure 2.8) would be deaminated peripherally by MAO before substantial levels could be reached in the brain. As dopamine is metabolized preferentially by MAO-B (Figure 2.11), inhibition of this subtype would allow brain levels of dopamine to build up. MAO-B activity is in fact increased with age and in Parkinson's disease. In the latter case, this contributes to the degradation of dopamine and its depletion from the nigrostriatal brain region (Strolin Benedetti & Dostert 1989). The selective MAO-B inhibitors are devoid of the 'cheese effect' because of the absence of MAO-B in sympathetic neurones, where MAO-A will deaminate any tyramine taken up (Youdim & Finberg 1990).

The effectiveness of selegiline in the management of Parkinson's disease and delaying the onset of disability has also recently been attributed to an elevation of superoxide dismutase (SOD) levels in the striatum (Clow *et al.* 1991). This enzyme protects against the neurotoxicity of reactive oxygen species such as hydroxyl radicals. The oxidation of amine substrates by MAO results in the production of H_2O_2, and in the absence of adequate levels of antioxidants (reduced glutathione and ascorbic acid) leads to the production of hydroxyl radical ($^\cdot OH$). Further support for a neurotoxicity mechanism comes from the observation that the contaminant of designer heroin, N-methyl-4-phenyl-1,2,3,6-tetrahydropyridine (MPTP), induced a

Parkinson's-like syndrome in addicts. MPTP is a substrate for MAO-B which converts it to MPDP plus H_2O_2 and thence to MPP^+. This toxicity is prevented by selegiline (Youdim & Finberg 1991).

Orthostatic hypotension is a side-effect of MAO inhibitors. This fall in blood pressure is attributed to inhibition of dietary tyramine breakdown which is then converted to octopamine by dopamine-β-hydroxylase (Figure 2.7). This acts as a false transmitter producing weaker sympathetic vasoconstriction. The fall in blood pressure has been utilized with pargyline (Eutonyl[R]) which has been used as an antihypertensive agent.

2.5.2 Semi-carbazide-sensitive Amine Oxidase

A separate enzyme responsible for the deamination of histamine was initially termed histaminase but later renamed diamine oxidase (DAO; EC 1.4.3.6) to distinguish it from MAO. This is a member of a group of copper-containing enzymes which are sensitive to inhibition by semi-carbazide and hence termed semi-carbazide-sensitive amine oxidases (SSAOs). The preferred substrate of the SSAOs not deaminating histamine appears to be benzylamine. This is not a naturally occurring amine and the physiological significance remains to be established. However, SSAO enzymes do deaminate putricine, spermidine and spermine. The latter two amines are produced by herbivores from cellulose fermentation in the rumen. SSAOs may in fact serve a general role for inactivating and detoxifying dietary amines. A further distinction from MAO is the location of SSAO in smooth muscle cells of blood vessel walls, in addition to adipose tissue, and its absence from the brain. The close association with the vessel wall makes SSAO ideally suited to scavenging circulating amines. These may be endogenous amines (eg adrenaline), dietary amines or xenobiotics arising from industrial or environmental sources.

The deamination by SSAO occurs by a similar mechanism to that for MAO, with the generation of an aldehyde, H_2O_2 and ammonia. SSAO is not sensitive to acetylenic MAO inhibitors (eg clorgyline). It is inhibited by semi-carbazide and by the compound MDL 72145, but the latter cannot be used when MAO-B is present. The 4-picolylamine derivative of MDL 72145, B24, has recently been found to be a highly selective, reversible inhibitor of SSAO (Callingham *et al.* 1991).

2.5.3 Catechol-O-methyl Transferase (COMT)

COMT (EC 2.1.1.6) is a non-membrane-bound, soluble enzyme found in the cytoplasm of glial cells in the CNS and in many peripheral tissues, the highest level being found in the kidney and liver. It is largely located in the extraneuronal compartment so that denervation of a tissue has little or no effect upon COMT activity (Axelrod 1971). Immunofluorescence indicates its presence in non-neuronal tissue.

This enzyme catalyses the transfer of a methyl group from the methyl donor S-adenosyl-L-methionine (SAM) , to the hydroxyl group of substrates containing catechol groupings. Mg^{2+} serves as an activator. O-Methylation occurs almost exclusively in the *meta*-position (3-position) of catecholamines *in vivo* (Figure 2.8). There may be a small fraction of O-methylated in the *para*-position *in vitro*. There is no O-methylation of amines carrying a single hydroxyl group in the *meta*-position, such as phenylephrine (Table 4.1). There is no stereospecificity since both D and L isomers are O-methylated. In addition to noradrenaline and adrenaline, their catecholamine precursors (dopamine and L-dopa) and the deaminated metabolites of noradrenaline (DOPEG and DOMA) and dopamine (DOPAC) are all substrates (Figures 2.8 and 2.11). The O-methylated products of noradrenaline and adrenaline are normetanephrine and metanephrine, respectively, and dopamine is converted to methoxytyramine. O-Methylation does not appear to be a primary mechanism for inactivation of the neurotransmitter since both normetanephrine and metanephrine retain some pharmacological activity at α-adrenoceptors, although O-methylation does markedly reduce their activity at β-adrenoceptors (Langer 1974). In fact, normetanephrine and metanephrine have weak blocking activity at β-adrenoceptors, a property that is even more pronounced with the methylated product of isoprenaline, O-methylisoprenaline. The metabolites of noradrenaline and adrenaline that are both O-methylated and deaminated are 3-methoxy-4-hydroxyphenylglycol (MOPEG) and 3-methoxy-4-hydroxymandelic acid (VMA). These represent the second most important fraction released by nerve activity or spontaneously. They are the products of intraneuronal deamination following reuptake of the released noradrenaline which is in turn followed by intraneuronal O-methylation (Figure 2.9). These metabolites are virtually inactive. When neuronal uptake is blocked, normetanephrine becomes the principal O-methylated metabolite.

COMT therefore appears to be of major importance for the termination of action of catecholamines in sparsely innervated tissues where MAO levels are low and those released from the adrenal medullae or administered by injection into the blood stream. Generally very few of the drugs based upon noradrenaline (sympathomimetic amines) are catecholamines and they are therefore stable to degradation by COMT. For example, the isomer of isoprenaline, orciprenaline (metaproterenol, Alupent[R]), has hydroxyl groups in the 3- and 5-positions of the benzene ring instead of the 3- and 4-positions and is therefore not inactivated by COMT and has a longer duration of action than isoprenaline (isoproterenol) (Table 4.1).

Measurement of the metabolism of noradrenaline by COMT following its extraneuronal uptake is complicated by the additional presence of a neuronal uptake site and MAO. A useful alternative substrate is isoprenaline which is not taken up neuronally or metabolized by MAO. On perfusion or incubation of tissues with [³H]isoprenaline, [³H]-O-methylisoprenaline ([³H]OMI) accumulates in the tissue and medium (Figure 2.17). This is a saturable process with a $K_m = 2.9 \pm 1.5$ μM in rat perfused hearts. Discrepancies exist between perfused hearts and isolated purified enzymes, the latter displaying a higher K_m value (200–400 μM) for a range of substrates. This difference may be due to separation of the enzyme from a tissue component that may increase the affinity of the enzyme (Trendelenburg 1976). The values obtained with purified COMT vary considerably, possibly because of instability of the enzyme, different levels of Mg^{2+}, and the formation from SAM of S-adenosylhomocysteine, which is a potent inhibitor of COMT. There was also early evidence that COMT may exist in several isoforms which may vary between tissues (Guldberg & Marsden 1975).

There is also a discrepancy between whole perfused organs and isolated incubated tissues since, for example, the O-methylating capacity of the perfused heart was 50-fold that of isolated ventricular strips. This has been attributed to the presence of at least two O-methylating compartments in the perfused heart: one in the vascular

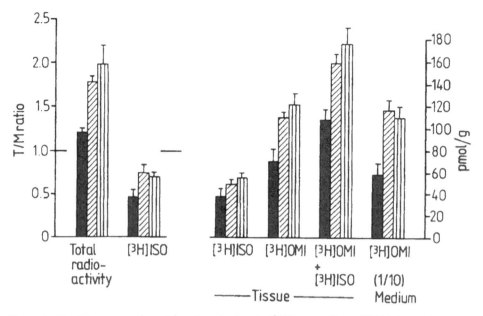

Figure 2.17 Extraneuronal uptake (uptake$_2$) of [^{3}H]isoprenaline ([^{3}H]ISO) and its O-methylation in guinea-pig isolated lung strips. Tissues were incubated with [^{3}H]isoprenaline (0.1 μM) for 40 (■), 60 (■) or 80 min (▥). Tissue levels of total radioactivity and [^{3}H]ISO are expressed as the tissue/medium ratio (T/M ratio). Individual and combined levels of [^{3}H]ISO and its O-methylated metabolite, O-methylisoprenaline, in the tissue or medium are expressed as picomoles per gram ($n = 4$). Reproduced with permission from Broadley & Paton (1990).

smooth muscle which is poorly supplied with substrate in ventricular stips and a second site possibly in the myocardial cells (Trendelenberg 1976).

Since O-methylation by COMT and the extraneuronal uptake site appear to be closely associated, their interrelationship will be discussed further when extraneuronal uptake of catecholamines is considered.

2.5.3.1 COMT inhibitors

One of the first COMT inhibitors to be described was pyrogallol. Many of the inhibitors of COMT are either catechols or, like pyrogallol, have three hydroxyl substituents on the benzene ring (Figure 2.18). Pyrogallol and catechol derivatives serve as competitive substrates for COMT and inhibit the enzyme both *in vitro* and *in vivo*. This author showed the COMT inhibitory properties of catechol, pyrogallol and two acid degradation products of noradrenaline, noradnamine and diadrenaline ether (Abbs *et al.* 1967). Pyrogallol, however, has a high toxicity *in vivo* attributed to inhibition of other enzyme systems and to the formation of methaemoglobin (Guldberg & Marsden 1975) and thus has no clinical use. Subsequently, tropolone and U-0521 (3,4-dihydroxy-2-methylpropiophenone) have been shown to have COMT inhibitory properties. Tropolone appears to inhibit by formation of an enzyme–Mg^{2+}–tropolone complex; it is toxic and like U-0521 is ineffective after oral administration and therefore of no clinical use. U-0521, however, is a widely used tool drug for examining the role of COMT. Newer orally active and potent

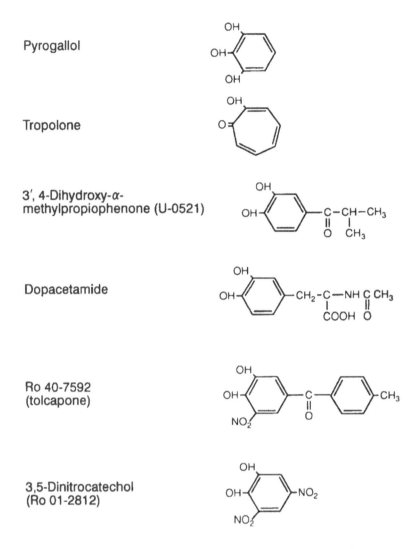

Pyrogallol

Tropolone

3', 4-Dihydroxy-α-
methylpropiophenone (U-0521)

Dopacetamide

Ro 40-7592
(tolcapone)

3,5-Dinitrocatechol
(Ro 01-2812)

Figure 2.18 Structures of some catechol-O-methyl transferase (COMT) inhibitors.

COMT inhibitors with low toxicity are being developed, including Ro 40-7592 (tolcapone) (Zürcher *et al.* 1990).

2.5.3.2 *Pharmacology and clinical uses*

The increase in blood pressure in response to an injection of adrenaline was shown many years ago to be enhanced and prolonged after inhibition of COMT by pyrogallol. U-0521 potentiates the responses of isolated atria to adrenaline and isoprenaline (Figure 2.19). These results indicate that COMT is an important route for their inactivation. Noradrenaline, however, is not potentiated in the atria unless neuronal uptake is inhibited by cocaine. Thus, when the major route of removal of noradrenaline is no longer available, COMT takes on a significant role at low physiological concentrations (Kaumann 1972).

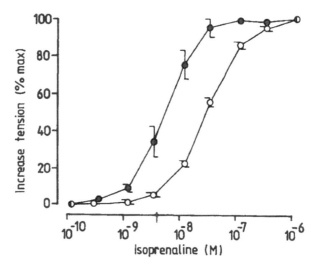

Figure 2.19 Potentiation of the positive inotropic responses of the guinea-pig isolated left atria to isoprenaline by the catechol-O-methyl transferase (COMT) inhibitor, U-0521 (10^{-4} M). The concentration–response curve for isoprenaline in the presence of U-0521 (●) is displaced to the left of that obtained before exposure to U-0521 (O). Mean ($n = 4$) increases in left atrial tension caused by cumulative addition of isoprenaline are plotted as a percentage of the pre-U-0521 maximum. Atria were removed from reserpine pretreated guinea-pigs (0.5 mg/kg ip 24 hr before use) and paced at 2 Hz. Unpublished results.

COMT inhibitors currently have no therapeutic applications of their peripheral action, however, they are being developed as adjuncts to L-dopa in the treatment of Parkinson's disease. The reasoning behind this is as follows. Endogenous and administered L-dopa are normally converted to dopamine by dopa-decarboxylase (Figure 2.4). When L-dopa is combined with a carboxylase inhibitor such as benserazide (in Madopar[R]) or carbidopa (in Sinemet[R]), the peripheral levels of dopa increase. Dopa is then substantially O-methylated to 3-O-methyldopa (3-OMD) (Figure 2.11). This is inactive and has a very long half-life (~15 hr). It accumulates in the plasma and is then transported into the brain by the same carrier that transports dopa. The advantages of combining with MAO inhibitors have already been described, however the further addition of an orally active COMT inhibitor would also permit a reduction of the dose and frequency of administration of dopa. The formation of 3-OMD in the periphery would be reduced, resulting in improved bioavailability of dopa for the brain. An example of this type of inhibitor is tolcapone (Figure 2.18).

2.5.4   *Neuronal Uptake of Noradrenaline (Uptake₁)*

The major route of removal of noradrenaline from the synapse, and thus its inactivation by removal from the vicinity of the receptor site, is by its transport back into the sympathetic neurone. Evidence that noradrenaline was taken up into sympathetic nerves was provided by the demonstration that after infusion of [³H]noradrenaline, stimulation of sympathetic nerves to the spleen resulted in the appearance of [³H]noradrenaline in the effluent (Hertting & Axelrod 1961).

Furthermore, examination of tissues by electron microscopy combined with autoradiography, following such infusions, suggested that the noradrenaline was taken up and incorporated into intraneuronal storage vesicles. Similarly, after depletion of noradrenaline with reserpine and inhibition of MAO with nialamide, fluorescence histochemistry revealed that administration of noradrenaline caused the reappearance of fluorescent axons and nerve terminals. Without MAO inhibition, the fluorescence was weaker because the incorporation of noradrenaline into the storage vesicles is prevented by reserpine and it is rapidly deaminated by MAO. This reappearance of noradrenaline fluorescence was prevented by surgical sympathectomy prior to the noradrenaline administration (Hamberger *et al.* 1964). Thus, there are two mechanisms for concentrating noradrenaline intraneuronally – uptake across the neuronal membrane and subsequent uptake into the storage vesicles. The administration of [^{3}H]noradrenaline leads to its accumulation in the microsomal particulate fraction of cardiac tissue homogenates. Autoradiography after incubation of tissues with radiolabelled noradrenaline has shown that the vesicles closest to the surface of the varicosity are preferentially loaded. Furthermore, it is the varicosities nearest to the surface of the tissue that are preferentially labelled (Trendelenburg 1991).

The characteristics of the noradrenaline uptake process (uptake$_1$) were examined in the perfused heart (Iversen 1967). It has a powerful concentrating effect, with the concentration in the heart perfused with low concentrations of noradrenaline being 30 times that in the perfusion solution. The accumulation of noradrenaline by perfused tissues is a net accumulation rather than exchange for endogenous amine. The uptake of noradrenaline or adrenaline is a saturable process that for concentrations up to 1 μg/ml obeys Michaelis–Menten kinetics (Figure 2.20). The affinity of noradrenaline for the uptake site is high, ranging from 0.4 to 1.6 $\times$ 10^{-6} M in various tissues (Iversen 1971). The process depends upon the presence of an intact

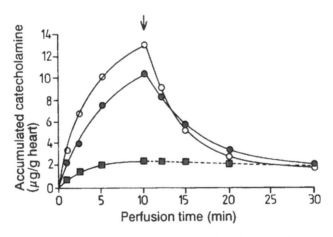

Figure 2.20 Uptake of noradrenaline (●, ■) and adrenaline (O) by rat isolated hearts perfused for different times with a low concentration (500 ng/ml, ■) or high concentration (5 μg/ml, ●). At the arrow, perfusion for various times with amine-free medium was commenced. Note that uptake at the low concentration is saturable and is not removed by washout; this is uptake$_1$, or neuronal uptake. High concentration uptake was the first demonstration of uptake$_2$ or extraneuronal uptake which was removed by washout to a level equivalent to uptake$_1$ saturation. Adapted from Iversen (1963, 1965).

sympathetic neurone and that the neurone is metabolically active. Net accumulation of noradrenaline appears to be impaired by inhibition of glycolysis and oxidative metabolism. Hypoxia and glucose deprivation cause a net loss of noradrenaline by enhancing the *efflux* of noradrenaline via the carrier process in the reverse direction. This has important implications in hypoxic and ischaemic tissues which are known to release noradrenaline in organs such as the heart (Trendelenburg 1991).

The accumulation of noradrenaline is Na^+- and Cl^--dependent and is enhanced by an increase in extracellular Na^+ (Paton 1976a). The carrier or transporter for noradrenaline has three binding sites, one for Na^+, one for Cl^-, and a third for the protonated substrate (noradrenaline). The binding of Na^+ increases the affinity of noradrenaline for the carrier. The carriers are mobile and able to cross the neuronal membrane when all binding sites are free. When it binds Na^+, it is immobile. Under resting conditions the external Na^+ levels are higher than the intracellular levels, being maintained by the Na^+,K^+-ATPase-driven outward Na^+ pump. This inward Na^+ gradient therefore results in there being more immobile (Na^+ bound) carriers on the outside of the membrane. These have a greater affinity for noradrenaline and are therefore available to carry noradrenaline into the neurone. Once the carriers have bound noradrenaline and Cl^- they are again mobile for inward transport of noradrenaline (Figure 2.21).

The dependence upon the Na^+ pump is illustrated by the fact that ouabain inhibits the accumulation of noradrenaline. The inhibition of Na^+,K^+-ATPase causes an

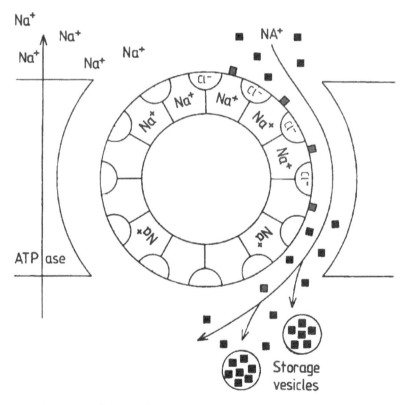

Figure 2.21 Carrier mechanism for neuronal uptake (uptake₁) of noradrenaline (NA). Noradrenaline is transported in the charged form (NA⁺, ■) when the carrier has bound sodium (Na⁺) and chloride (Cl⁻) and is then stored in vesicles.

increase in intracellular Na$^+$. This reduces the inward Na$^+$ gradient and the inward carrier process. Uptake will also be driven by the fact that once inside the neurone, noradrenaline is avidly bound to the storage vesicles or metabolized via MAO. A change in extracellular K$^+$ also affects the accumulation of noradrenaline, a reduction having an inhibitory effect. However, care must be taken in interpreting early data on factors such as K$^+$ which influence the net accumulation of noradrenaline since, as will be seen later, the efflux of noradrenaline is also affected (Paton 1976b).

The saturability of the accumulation of [^{3}H]noradrenaline is due to the secondary processes rather than saturation of the carrier. Once saturation of accumulation is reached in perfused hearts, the noradrenaline can only be washed out from extracellular spaces, the extraneuronal compartment and the vasculature. However, if vesicular storage and intraneuronal metabolism are inhibited by reserpine pretreatment and MAO inhibition, respectively, cytoplasmic noradrenaline levels rise. Noradrenaline can then be washed out by a reversal of the uptake process. This efflux of noradrenaline is via the same carrier as the uptake$_1$, which is favoured by a reversal of the Na$^+$ gradient. Thus, a rise in intracellular Na$^+$ (or fall in external Na$^+$) will enhance efflux under these circumstances. Ouabain, by inhibiting the Na$^+$,K$^+$-ATPase-driven outward pump, will raise intracellular Na$^+$ and promote efflux. This ouabain-induced efflux is susceptible to blockade by the uptake blockers such as desmethylimipramine (DMI), suggesting a common route via the same carrier. Raising the external K$^+$ also enhances efflux. However, in normal tissues pretreated with reserpine or MAO inhibitor, the depolarizing effect of K$^+$ causes *release* of noradrenaline by exocytosis which, unlike efflux, is Ca^{2+}-dependent (Figure 2.22).

The outward transport of noradrenaline is facilitated by the inward transport of a substrate because it makes carrier available on the inside of the membrane. This is probably the mechanism by which indirectly acting sympathomimetic amines release noradrenaline, that is, through outward transport by a reversal of the neuronal uptake process (Figure 2.22). Their action does not involve exocytosis and is Ca^{2+}-independent. Amines having the greatest affinity for efflux are tyramine and metaraminol, but O-methylated (eg normetanephrine) or N-substituted (eg isoprenaline) derivatives have low affinity. There is no evidence for stereospecificity. Substrates of uptake$_1$ are therefore good indirectly acting sympathomimetic amines, but only if they are also substrates for vesicular uptake. Without any accompanying mobilization of stored amine, the normally very low cytoplasmic noradrenaline levels would not sustain effective outward transport (Paton 1976b, Bönisch & Trendelenburg 1988, Trendelenburg 1991).

2.5.4.1 *Structural requirement of substrates for uptake$_1$*

The most commonly used methods for measuring uptake of sympathomimetic amines are from the tissue level after infusion with the test amine, from arteriovenous differences in perfused organs, or from uptake into synaptosomes. The use of reserpine- and MAO inhibitor-treated tissues avoids influence of metabolism and vesicular uptake. However, it does have the disadvantage of creating artificial conditions of raised cytosolic noradrenaline levels and the possible interference from the cellular toxicity of excessive reserpinization. Alternatively, the inhibition of uptake of [^{3}H]noradrenaline by structural analogues also provides a measure of their affinity for neuronal uptake.

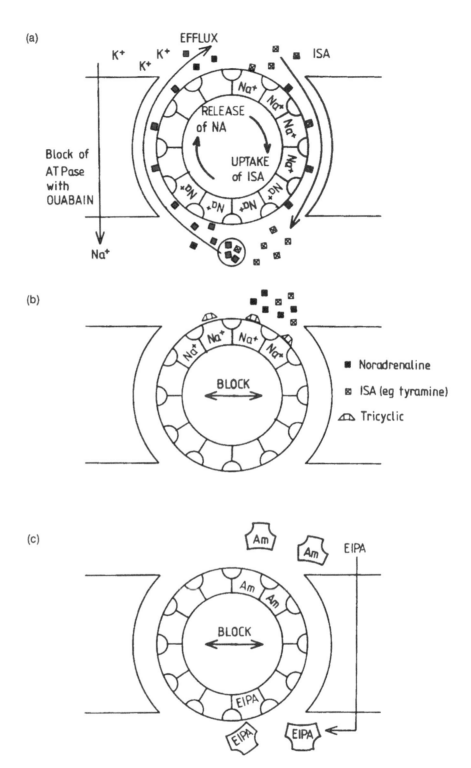

Figure 2.22 Efflux of noradrenaline (NA) and its release from noradrenergic neurones by indirectly acting sympathomimetic amines (ISAs) such as tyramine and guanethidine (a). Blockade of noradrenaline uptake and release by the neuronal uptake inhibitors such as tricyclic antidepressants (imipramine)(b) and by amiloride (Am) and its ethylisopropyl derivative (EIPA) (c). Key as in Figure 2.21 showing uptake mechanism.

More information on the uptake process can be derived from kinetic analysis by obtaining Lineweaver–Burke plots for the reciprocals of rate of uptake ($1/V$) against substrate concentration ($1/S$). These yield straight lines from which the V_{max} (maximum velocity) can be derived from the intercept on the $1/V$ axis and K_m (dissociation constant) derived from the slope (K_m/V_{max}). The Lineweaver–Burke plot obtained in the presence of competitive inhibitors of uptake, including the structural analogues of the substrate, intercepts the $1/V$ axis at the same point as does the control ($1/V_{max}$) and thus yields an altered K_m value. In the presence of non-competitive inhibitors of uptake, such as ouabain, the plot has the same intercept on the $1/S$ axis but a reduction in V_{max}. The K_m is unaltered (Figure 2.23) (Berti & Shore 1967).

The consensus of opinion is that uptake (like efflux) is not stereoselective. Where the (−)-isomer is apparently preferentially taken up, this is probably a reflection of the selectivity of the secondary processes of metabolism by MAO and vesicular uptake. Amines lacking the chiral centre, such as dopamine, are

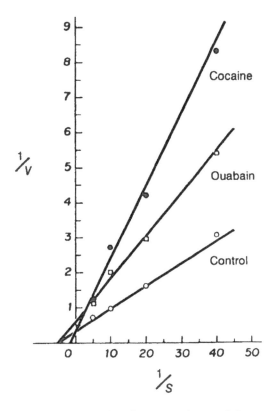

Figure 2.23 Neuronal uptake measured as the accumulation of the noradrenaline analogue, (−)-metaraminol (see Table 4.1), by rabbit heart slices. Uptake is presented as Lineweaver–Burke plots of net rate of uptake over 30 min ($1/V$) against substrate concentration ($1/S$). Inhibition by cocaine (3×10^{-6} M) is competitive as shown by the same intercept on the ordinate ($1/V$) as for the control regression line. Inhibition by ouabain (3×10^{-6} M) is non-competitive as shown by the different intercept on the $1/V$ axis but a common point of interaction on the abscissa. Reproduced with permission from Berti & Shore (1967).

therefore also good substrates for neuronal uptake. Adrenaline is less readily taken up than noradrenaline by a factor of two. In the amphibian heart, however, the selectivity is reversed with adrenaline being preferentially taken up. This reflects the fact that in the amphibian heart adrenaline is the neurotransmitter. Larger N-substituents further decrease affinity of catecholamines for neuronal uptake, with isoprenaline (N-isopropyl) not being accumulated by noradrenergic neurones. However, N-substitution has little effect upon the uptake of non-catechols. Phenylethylamines with phenolic hydroxyl groups have greater affinity for the uptake carrier site. The *m*-OH group appears to be more important since metaraminol, *m*-octopamine and *m*-tyramine are all actively accumulated. Phenylethylamines without ring substituents, such as amphetamines, have yielded confusing results. They appear to be taken up since their pharmacological effects are prevented by uptake inhibitors such as cocaine. However, neither cocaine nor ouabain appears to prevent their accumulation and it has therefore been questioned whether they are transported by the neuronal uptake process. One possibility is that the inhibitory effect of cocaine on the pharmacological response is on the efflux of noradrenaline induced by phenylethylamines. Although earlier studies showed that the uptake of (+)-amphetamine was not cocaine-sensitive or sodium-dependent, more recently its uptake into rat cultured phaeochromocytoma cells has been shown to have a cocaine- and sodium-dependent component. These cells have the advantage of there being no other cells for diffusion of the highly lipophilic (+)-amphetamine. Normally, the neuronal uptake via the carrier is probably dominated by diffusional entry (Bonisch & Trendelenburg 1988).

O-Methylation of catecholamines reduces affinity and [^{3}H]normetanephrine is not taken up. Substitution of methyl groups on the side-chain α-carbon appears to have little influence on affinity for neuronal uptake. The structures of these amines are shown in Table 4.1 in Chapter 4.

Finally, the adrenergic neurone blockers, bretylium, guanethidine, bethanidine and debrisoquine, are accumulated in sympathetic nerves by a Na$^+$-dependent mechanism. This is prevented by uptake inhibitors as is their noradrenaline-releasing activity. They are therefore considered to be transported into the neurone via the noradrenaline carrier mechanism (Figure 2.22) (Maxwell *et al.* 1976, Ross 1976).

2.5.4.2 *Uptake into dopaminergic neurones*

The uptake of dopamine into peripheral neurones appears to display differences from noradrenaline uptake into noradrenergic neurones. [^{3}H]Dopamine is particularly taken up by neurones in various brain regions including the nigrostriatal and mesolimbic systems. The uptake has a powerful concentrating effect (100:1), is saturable and Na$^+$-dependent but Ca^{2+}-independent. It is blocked by ouabain and is temperature-dependent. It therefore has similar characteristics to noradrenaline uptake, but noradrenaline shows less affinity ($K_m = 20$ μM) compared with that for dopamine (0.2–0.4 μM). The transport of dopamine and noradrenaline also has different susceptabilities to inhibitors. The tricyclic antidepressants (eg DMI) are much weaker inhibitors of dopamine uptake than noradrenaline (1000-fold). Selective inhibitors of dopamine uptake include benztropine and GBR 12909. Pretreatment of animals with 6-hydroxydopamine in the presence of GBR 12909 causes selective depletion of noradrenergic neurones (Bell & Sunn 1990), while

6-hydroxydopamine in the presence of DMI causes selective dopaminergectomy (Sunn *et al.* 1990).

2.5.4.3 *Inhibitors of neuronal uptake*

Members of this class of drug include substances with a diverse range of pharmacological classifications. They include the substrates for neuronal uptake described above which are structurally related to phenylethylamine, tricyclic antidepressants (eg DMI), cocaine, the irreversible α-adrenoceptor antagonist, phenoxybenzamine (Chapter 5), and adrenergic neurone blockers such as guanethidine and bethanidine (Chapter 6).

Phenylethylamine analogues have affinity for the noradrenaline carrier as shown by their ability to inhibit uptake of noradrenaline. A non-phenolic phenylethylamine such as amphetamine inhibits uptake but is not taken up itself by the neuronal carrier, since the accumulation is not sensitive to uptake inhibitors. It accumulates by diffusion due to its high lipophilicity. An anomolly however is that cocaine blocks the pharmacological effects of these amines. The site of the inhibitory action of cocaine and DMI is probably on the *efflux* of noradrenaline induced by these amines.

Tricyclic antidepressants including imipramine (Tofranil[R]), desmethylimipramine (DMI) and amitriptyline (Tryptizol[R]) are all potent inhibitors of noradrenaline transport. DMI is the most active and is frequently used as a tool drug in pharmacological studies requiring uptake inhibition. These agents are also inhibitors of the transport of 5-HT, to varying degrees. DMI selectively inhibits the uptake of noradrenaline by >100-fold, whereas amitriptyline is equally effective against both noradrenaline and 5-HT. The antidepressant properties are attributed in part to the inhibition of uptake of noradrenaline and 5-HT and consequent potentiation of central neurotransmission by these monoamines. Although the inhibition of uptake occurs promptly, the development of antidepressant activity takes several weeks. Thus, secondary consequences upon central adrenoceptors arising from elevated transmitter levels may also be involved. In addition, these compounds are α-adrenoceptor antagonists, particularly amitriptyline, which may contribute to their antidepressant activity. This α-blocking property does explain the peripheral side-effect of postural hypotension. It may also complicate experiments in which α-adrenoceptor-mediated responses are under examination and neuronal uptake is inhibited by DMI; for this reason cocaine is now the preferred inhibitor (Williamson & Broadley 1989).

Tricyclic antidepressants also block muscarinic receptors (Chapter 8) which explains other side effects of dry mouth, blurred vision, constipation and urinary retention. They are also involved in many clinically important drug interactions related to their pharmacology at the noradrenergic neurone. Since they prevent the transport of noradrenaline and other indirectly acting amines, they prevent the pharmacological actions of the latter group of drugs and thus their therapeutic efficacy (eg as nasal decongestants). Since the adrenergic neurone blockers, guanethidine and bethanidine, are also transported into the neurone via the same route, their antihypertensive activity is impaired by concurrent use of tricyclic antidepressants.

DMI blocks both the inward and outward transport of noradrenaline. It is bound preferentially to the carriers on the surface which, like noradrenaline, is dependent on Na^+ ions (Figure 2.22). The blockade is competitive. In spite of being highly lipophilic, DMI does not enter the neurone and the carrier remains immobile. The possibility exists that the larger bulk of these molecules than noradrenaline enables additional binding to ancillary hydrophobic binding sites on the carrier.

Cocaine is a local anaesthetic obtained from the leaves of the South American bush, *Erythroxylum coca*. Its general stimulant action on the CNS and the elevation of mood were recognized by the natives who have chewed the leaves for at least 1500 years. Following its isolation as the hydrochloride in 1855, it has become a major drug of abuse, more recently being smoked as the free base ('crack'). The central stimulant properties are attributed primarily to the inhibition of dopamine uptake in the brain, which allows increases in extracellular synaptic dopamine levels in certain brain areas. The elevated dopamine levels appear to activate both D_1 and D_2 receptors. This action, together with a minor inhibition of uptake of noradrenaline and 5-HT, are thought to result in a mixture of inhibition and excitation of neurotransmission in local circuits in the brain leading to the behavioural effects (Woolverton & Johnson 1992). The central stimulant property is associated with vasomotor stimulation of sympathetic nerves resulting in tachycardia and vasoconstriction. A peripheral component of the increase in heart rate and vasoconstriction also arises from the inhibition of noradrenaline uptake, allowing more noradrenaline released from sympathetic nerves to reach the β- and α-adrenoceptors, respectively. The uptake of circulating adrenaline is to a lesser extent also inhibited.

The inhibition of the uptake carrier by cocaine is competitive although the blockade of efflux is non-competitive. The inhibition is unrelated to its local anaesthetic activity and occurs at lower concentrations. Indeed, in isolated tissues a concentration of 2 μg/ml potentiates the effects of sympathetic nerve stimulation by inhibition of uptake, while an increase to 10 μg/ml causes blockade through inhibition of nerve conduction.

Also evident after cocaine administration is a sympathetic-mediated dilation of the pupil (mydriasis). Cocaine was once used extensively as a local anaesthetic for ophthalmological procedures; when applied to the eye its vasoconstrictor effect resulted in corneal blanching. However, because of shedding of the corneal epithelium and the potential for abuse, it is no longer used for this purpose. The raised sympathetic activity of the cardiovascular system in habitual cocaine users may ultimately result in coronary vasoconstriction leading to myocardial ischaemia and infarction and arrhythmias. The cardiac toxicity may be treated with β-blockers, although this would leave the vasoconstriction unopposed, resulting in hypertension. This could be prevented by use of the combined α- and β-blocker, labetalol (Trandate[R]). An alternative approach to treatment could be Ca^{2+}-channel blockers such as nifedipine (Adalat[R]) or nicardipine (Cardene[R]). Vasoconstriction of the nasal mucosa after repeated intranasal insufflation of cocaine ('snorting') results in disfiguring necrosis of nasal tissue, rhinorrhoea and perforation of the nasal septum.

Adrenergic neurone blocking agents including guanethidine, bethanidine and debrisoquine (Chapter 6) are substrates for neuronal uptake as indicated by the fact that their neurone blocking and indirect sympathomimetic actions are prevented by the neuronal uptake inhibitors. They are accumulated by sympathetic nerves and this is also antagonized by DMI. As a consequence, the neurone blocking agents are also inhibitors of subsequent uptake of noradrenaline and thus potentiate the actions of exogenous noradrenaline. At the same time they inhibit the effects of sympathetic nerve stimulation (neurone blockade) and indirectly acting sympathomimetic amines (Figure 2.24). The concurrent use of these neurone blocking

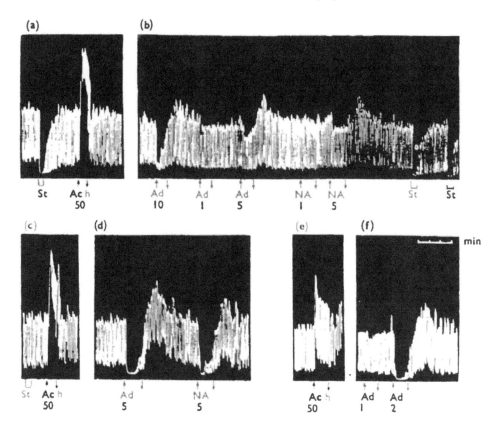

Figure 2.24 Blockade by guanethidine of the inhibitory responses of rabbit ileum to sympathetic nerve stimulation at 50/sec for 20 sec (St) and potentiation of responses to adrenaline (Ad) and noradrenaline (NA). Ach-induced contraction was unaffected. Responses were obtained before guanethidine in (a) and (b), after guanethidine (1 μg/ml) in (c) and (d), and after guanethidine (10 μg/ml) in (e) and (f). Numerals refer to the doses (in micrograms) added to a 70 ml organ bath. Reproduced with permission from Day & Rand (1961).

agents as antihypertensive drugs with other drugs acting at the noradrenergic neurone, such as the tricyclic antidepressants, MAO inhibitors and indirectly acting sympathomimetic amines is therefore contraindicated.

Phenoxybenzamine (Dibenyline, Dibenzyline[R]) is an irreversible α-adrenoceptor antagonist (Chapter 5). It blocks the neuronal carrier also by an irreversible non-competitive mechanism. The degree of inhibition is therefore dependent upon the time of exposure. The mechanism is presumed to be similar to that for its irreversible antagonism at α-adrenoceptors, whereby there is alkylation of the binding site after formation of an intermediate ethyleniminium ion. Unlike the other inhibitors of neuronal uptake described above, phenoxybenzamine is also an irreversible inhibitor of extraneuronal uptake. It is therefore a useful tool drug in experimental pharmacology for eliminating the influence of both uptake processes and α-adrenoceptors. It is only necessary to incubate isolated tissues with phenoxybenzamine for 30 min and then remove the free drug by washing out before commencing the experiment (Furchgott 1972).

Amiloride (Midamor[R]) is a K^{+}-sparing diuretic which is also an inhibitor of uptake$_1$, but via a completely different mechanism. Amiloride is an inhibitor of Na$^+$ transport in the late distal tubule of the kidney. It inhibits primarily the electrogenic entry of Na$^+$ into the tubular epithelium but additionally inhibits Na$^+$-H$^+$ exchange and Na$^+$,K$^+$-ATPase at higher concentrations. At the noradrenaline uptake carrier, amiloride and ethylisopropyl-amiloride (EIPA) compete with Na$^+$ for the Na$^+$ binding site (Figure 2.22). This prevents the Na$^+$-dependent uptake of noradrenaline. EIPA has high lipid solubility and readily diffuses into the neurone where the low Na$^+$ concentration permits more ready competition for Na$^+$ binding sites. This results in an uncompetitive inhibition characterized by a decrease in both V_{max} and K_m for the uptake process (Trendelenburg 1991).

Inhibitors of neuronal uptake potentiate the effect of noradrenaline and other substrates for neuronal uptake by inhibiting a major site of loss, namely uptake into the noradrenergic neurone (Figure 2.25). This permits a higher concentration of the catecholamine in the synaptic cleft in the proximity of the adrenoceptor and hence an enhanced response. The potentiation is very pronounced in tissues in which the sympathetic innervation is dense and the synaptic distance is relatively narrow such as the vas deferens and nictitating membrane. It is prevented by denervation of these organs. The potentiation is seen as a parallel leftwards shift of the dose–response curve for noradrenaline (Figure 2.26). This represents a supersensitivity of the presynaptic type (type I) (Kalsner 1974). There has been debate whether uptake inhibition is the only explanation for the cocaine-induced supersensitivity, since it can potentiate responses in nerve-free vascular preparations and in certain denervated preparations. This minor action is proposed to be a postsynaptic effect due to a sensitization at the α-adrenoceptor or alterations in the binding or permeability of Ca^{2+} (Triggle & Triggle 1976). Another possible mechanism may be related to MAO inhibitory properties of cocaine.

Uptake inhibitors do not potentiate amines such as isoprenaline or salbutamol which are not substrates for neuronal uptake (Figure 2.26). The responses to

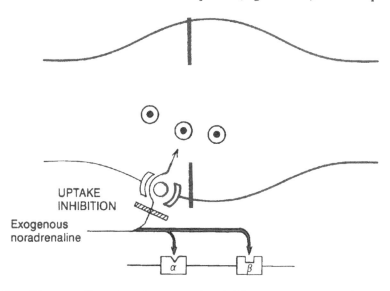

Figure 2.25 Diagram illustrating how uptake inhibition potentiates directly acting sympathomimetic amines that are substrates for neuronal uptake. The concentration of amine in the biophase that is available for receptor activation is raised.

88

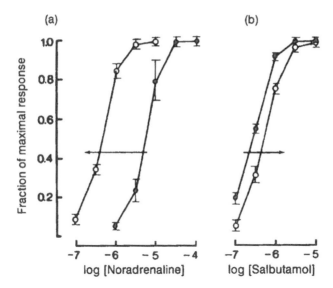

Figure 2.26 Effect of neuronal uptake inhibition upon the dose–response curves of (a) a directly acting sympathomimetic amine that is a substrate for neuronal uptake (noradrenaline) and (b) one that is not a substrate (salbutamol). Mean relaxation responses (n = 6) of the guinea-pig trachea are plotted in the absence (●) and presence of (○) cocaine (100 μM), which shifted the curve for noradrenaline only to the left. Reproduced with permission from Kenakin (1980)

sympathetic nerve stimulation are enhanced, although not as effectively as with exogenous noradrenaline, as will be explained in the next chapter. The effects of indirectly acting sympathomimetic amines which release neuronal noradrenaline are, however, inhibited. This has been attributed to inhibition of the carrier for the amine into the neurone. It also appears that the outward transport of noradrenaline will also be inhibited, which may be of relevance to those indirectly acting sympathomimetic amines that enter the neurone merely by diffusion (eg amphetamine).

The noradrenaline, dopamine and 5-HT carriers are identifiable protein components of the neuronal membrane. Their distribution may be determined *in vivo* and on synaptosomes by the use of radiolabelled antagonists having the desired selectivity for the transporter under study. Radiolabelling may also be used to examine loss of transporter after treatment with neurotoxins. There is little evidence of regulation of the carrier by chronic activation or inhibition. The separate carriers for noradrenaline, dopamine and 5-HT have now been cloned, but an adrenaline carrier on PNMT-containing brain neurones has not (Pacholczyk *et al.* 1991).

2.5.5 *Extraneuronal Uptake of Noradrenaline (Uptake$_2$)*

The early studies of Iversen (1965), in which noradrenaline was perfused through rat isolated hearts, first demonstrated the existence of an uptake process additional to neuronal uptake. These experiments showed that when the concentrations of noradrenaline or adrenaline were raised above 1 μg/ml, a second phase of uptake occurred which, unlike neuronal uptake, was not readily saturable or stereospecific.

It also did not depend upon the presence of a sympathetic innervation since it was unaffected by prior immunosympathectomy of the tissue. The accumulation of noradrenaline at these high concentrations was also readily removed from the tissue when perfused with amine-free solution, indicating poor retention and a lack of intracellular binding. Perfusion with amine-free solution reduced the noradrenaline level to that of neuronal uptake (Figure 2.20). This uptake process was termed uptake$_2$ or extraneuronal uptake. These early studies simply measured the levels of unchanged noradrenaline in the heart and did not account for the metabolism via COMT subsequent to uptake (Figure 2.9). This explains why the extraneuronal uptake process appeared to take place only at high concentrations of noradrenaline, since only then were measurable levels of unchanged noradrenaline accumulated. It is therefore necessary to measure extraneuronal uptake as *accumulation* plus *metabolism*, that is, from the levels of unchanged noradrenaline and its metabolites. In fact, extraneuronal uptake of noradrenaline does occur at physiological concentrations and it is saturable. To distinguish neuronal and extraneuronal uptake the latter is not blocked by cocaine or DMI, but is blocked by phenoxybenzamine, steroids and normetanephrine.

The extraneuronal uptake of noradrenaline is followed by its O-methylation via COMT to normetanephrine (NMN) (Figure 2.9) (Langer 1974). There is probably little extraneuronal uptake and O-methylation of noradrenaline during spontaneous transmitter release since the resting levels of [^{3}H]NMN formed after preloading the tissues with [^{3}H]noradrenaline are not reduced by the uptake$_2$ inhibitors, phenoxybenzamine and hydrocortisone. The NMN is therefore probably formed from COMT located intraneuronally. The formation of [^{3}H]NMN from noradrenaline released by nerve stimulation, however, is due to activity of postsynaptic COMT located in an extraneuronal uptake compartment.

There are also sites of uptake into the extracellular spaces and onto collagen which have to be accounted for when measuring accumulation of radiolabelled noradrenaline. These two sites are not blocked by either cocaine or steriods and therefore do not constitute the extraneuronal uptake sites proper. Uptake onto collagen appears to be a surface binding phenonenom which can be prevented by oxytetracyclin. That this binding is an important means of inactivating noradrenaline in tissues with a high level of connective tissue, for example the rabbit ear artery, is demonstrated by the fact that oxytetracyclin potentiates the responses to noradrenaline (Powis 1973).

The mechanism of uptake into the extraneuronal binding compartment probably involves a facilitated diffusion process. It is not blocked by ouabain, glucose deprivation or anoxia. It shows modest Na$^+$-dependence and therefore probably involves a Na$^+$-dependent carrier. Hyperpolarization of the tissue, for example by smooth muscle relaxation with a β-agonist, appears to enhance uptake$_2$ (Bryan-Lluka & Vuocolo 1992). There are structural requirements for substrates which differ from those already established for neuronal uptake. All amines taken up by neuronal uptake are also transported into the extraneuronal sites. Adrenaline has a higher rate of uptake than does noradrenaline. Isoprenaline, which is not taken up into neuronal sites, is readily taken up and O-methylated by the extraneuronal compartment. Thus, N-substitution, β-hydroxylation and O-methylation in the *meta*-position of catecholamines all increase affinity for uptake$_2$. Hydroxylation in the *meta*- and *para*-positions or methylation of the side chain α-carbon appear to decrease affinity.

[^{3}H]Isoprenaline ([^{3}H]ISO) is therefore a useful substrate for examining the characteristics of extraneuronal uptake. Measurements have been made both of influx [as accumulation of unchanged isoprenaline and its O-methylated metabolite, O-methylisoprenaline (OMI)] and of efflux from perfused hearts. Extraneuronal uptake has also been examined by incubating isolated atria, lungs, trachea and aorta with [^{3}H]ISO and measuring the tissue levels of [^{3}H]ISO and tissue and medium levels of [^{3}H]OMI (Figure 2.17). The level of uptake appears to be considerably less, with tissue:medium ratios reaching only two-fold compared with nearly 10-fold for perfused hearts. In perfused hearts it appears that extraneuronal uptake occurs into both the vascular smooth muscle and cardiac muscle.

Inhibition of COMT by dihydroxy-2-methylpropiophenone (U-0521) (Figure 2.18) results in the failure of [^{3}H]OMI to be detected in heart perfusates and incubated tissues. However, the accumulation of unchanged [^{3}H]ISO increases. This confirms that extraneuronal uptake and O-methylation are located in series. Inhibition of extraneuronal uptake by metanephrine should prevent the O-methylation of [^{3}H]ISO. Metanephrine and cortisone do not completely inhibit O-methylation which indicates the presence of two O-methylating compartments, only one of which is metanephrine/steroid-sensitive (Broadley & Paton 1990). Efflux studies have also indicated the possibility of two extraneuronal uptake compartments with different half-times and only one being blocked by corticosteroid. The possibility exists, however, that although uptake and efflux of catecholamines are blocked by metanephrine and corticosteroid, the efflux of the O-methylated metabolites are not blocked (Trendelenburg 1976). A third extracellular O-methylating site may also occur (Head *et al.* 1985, Broadley & Paton 1990).

2.5.5.1 *Pharmacology of extraneuronal uptake inhibitors*

The three groups of agents that inhibit extraneuronal uptake$_2$ have already been mentioned. These are the O-methylated catecholamines (normetanephrine, metanephrine and OMI), corticosteroids such as hydrocortisone, and phenoxybenzamine. In the presence of these inhibitors of uptake, the responses to catecholamines are potentiated. Since extraneuronal uptake represents a site of loss, its inhibition results in a raised level of the amine in the vicinity of the receptor and hence an elevated response. This may be seen as a leftwards shift of the dose–response curve to amines that are substrates for uptake$_2$ such as isoprenaline (Kaumann 1972). Similarly, inhibition of O-methylation by U-0521 potentiates the response to catecholamines (Figure 2.19). If extraneuronal uptake and O-methylation are arranged in series, then once the dose–response curve for isoprenaline has been displaced to the left by corticosterone, it should not be further displaced by inhibition of O-methylation. This appears to be the case with hydrocortisone (Kaumann 1972), although the present author has found further potentiation of isoprenaline in guinea-pig isolated atria by U-0521 when uptake is maximally inhibited by metanephrine. The effectiveness of inhibition of extraneuronal uptake in potentiating the responses to catecholamines also depends upon the relationship between the concentration of amine and its K_m value for uptake. If high concentrations are used, for example in the presence of receptor blockade, then uptake will be saturated. It will then no longer serve as a site of loss. Inhibition by metanephrine under these circumstances will have little effect upon sensitivity.

2.5.5.2 *The physiological role of extraneuronal uptake*

Extraneuronal uptake of noradrenaline gains importance when neuronal uptake is inhibited: the response is potentiated by hydrocortisone only in the presence of cocaine (Kaumann 1972). It is also of importance when high concentrations of catecholamines are used and neuronal uptake is saturated. Thus in tissues with a rich sympathetic innervation, neuronal uptake is the dominant site of loss; whereas in poorly or non-innervated tissues, extraneuronal uptake is the site of removal of circulating catecholamines. The extraneuronal uptake site is well developed in such tissues, including the lung and aorta. Extraneuronal uptake is therefore an important means of inactivation of certain therapeutically useful sympathomimetic amines which are not substrates for neuronal uptake. These include orciprenaline (Alupent[R]), terbutaline (Bricanyl[R]) and salbutamol (Ventolin[R]) (Table 4.1, Chapter 4).

There is no therapeutic application for the inhibitors of extraneuronal uptake, although the implication of this property of phenoxybenzamine and the steriods should be understood. This property of steroids may explain their facilitation of lipolysis, for example, since they have no direct effect of their own upon adipose tissue.

Adrenoceptors: Classification and Distribution

3.1 Introduction

Compounds with pharmacological actions resembling those of adrenaline or of sympathetic nerve stimulation were first described as sympathomimetic amines by Barger & Dale (1910). Dale (1906) recognized that adrenaline or the neurotransmitter released from sympathetic nerve endings exerted their effects by interacting with a 'receptor substance for adrenaline' and this term could replace what he had called the sympathetic myoneural junctions. The work of Dale (1906) had shown that the excitatory or motor responses of various organs to adrenaline or nerve stimulation were 'paralysed' by ergot alkaloids, whereas the inhibitory actions were unaffected. This was the first indication of the presence of different receptors for the excitatory and inhibitory effects of the sympathetic nervous system. This concept was directly opposed to that of two mediator substances, sympathin E and sympathin I, supported by Cannon & Rosenblueth (1933). The limitation of the two-mediator theory and its ultimate rejection have been discussed in Chapter 2.

To explain the opposing excitatory and inhibitory responses to sympathetic nerve stimulation, Ahlquist (1948) proposed the existence of two distinct types of 'adrenotropic' receptors; α- and β-adrenotropic receptors. He provided several arguments against the theory of two transmitters and like the present writer could find no satisfactory explanation for the anomolous behaviour of liver sympathin, which he considered to be a unique product. Ahlquist therefore supported the idea of a single neurotransmitter which he persisted in identifying as adrenaline rather than noradrenaline, based on the greater overall potency of the former in a range of test systems. However, the differences in activity of this common transmitter were attributed to two types of receptor. These are now known as adrenoceptors, and sympathomimetic amines are referred to as adrenoceptor agonists.

The basis of the subclassification was the order of potency upon a range of tissues, taken from four mammalian species, of the sympathomimetic amines, adrenaline, noradrenaline, methylnoradrenaline, methyladrenaline and isoprenaline. The tissue responses fell into two main potency orders. Those with the order shown above were designated α-adrenoceptor-mediated, while those with the order isoprenaline > adrenaline > methyladrenaline > methylnoradrenaline > noradrenaline were assigned to the β-type. α-Adrenoceptors were associated with most of the excitatory functions of the sympathetic nervous system including vasoconstriction and contractions of the uterus, nictitating membrane, ureter and *dilator pupillae*. β-Adrenoceptors were associated with the inhibitory functions, including vasodilatation and relaxation of the uterus and bronchial musculature. There were, however, two notable exceptions to this general rule: the adrenoceptor mediating the inhibitory response of the gastrointestinal tract was assigned to the α-type, whereas in the heart the adrenoceptor mediating excitatory increases in rate and force of contraction are of the β-type.

At about the same time, the first synthetic antagonists of sympathetic effects, the β-haloalkylamines, were described (Nickerson 1949). Their naturally occurring forerunners, the ergot alkaloids, were reversible competitive antagonists of the excitatory responses to adrenaline and sympathetic nerve stimulation. However, the β-haloalkylamines, including dibenamine, inactivated the receptor by alkylation in an irreversible manner (Chapter 5) and therefore produced virtually complete blockade of the contractile responses of smooth muscle without affecting the excitatory responses of the heart. This provided further support for the α- and β-adrenoceptor subclassification of Ahlquist (1948); responses blocked by these adrenergic blocking agents were attributed to α-adrenoceptors, while responses not susceptible to blockade were attributed to β-adrenoceptor activation. The relaxation of the intestine was not affected, but remained classified as an α-adrenoceptor-mediated response.

It was not until 1958 that the first agent to block β-adrenoceptors, dichloroisoprenaline (DCI), was described. This compound was shown to inhibit the adrenergic inhibitory responses of a range of smooth muscles (Powell & Slater 1958) and the stimulatory responses of the heart (Moran & Perkins 1958) without affecting contractile responses of smooth muscle. This, and the subsequent introduction of more potent β-blockers, further strengthened the α- and β-adrenoceptor subclassification. The responses of the gastrointestinal tract, however, remained an issue of contention. The inhibitory effects of noradrenaline and adrenaline in the rabbit isolated duodenum were found by Furchgott (1959) to be not fully blocked by either DCI or the potent α-blocking drugs, phentolamine or dibenamine. On this basis he proposed the existence of a third receptor, the delta receptor, in the intestine. However, this was soon shown to be an erroneous conclusion, when Ahlquist & Levy (1959) demonstrated that although individually DCI and an α-adrenoceptor blocking agent did not attenuate the inhibitory response of the dog small intestine to single doses of adrenaline, in combination they produced effective blockade. This was subsequently confirmed by Furchgott (1960) in the rabbit isolated intestine. Thus, both α- and β-adrenoceptors are present in the intestine but *both* are responsible for the inhibition of intestinal motility. This was therefore the first recognition of the heterogeneity of adrenoceptors mediating the responses of a tissue – the existence of more than one receptor subtype mediating a common response. This will be discussed in more detail later in this chapter.

The class of adrenoceptor involved in individual responses to sympathetic nerve stimulation or to exogenously applied sympathomimetic amines is therefore unrelated to the type of response produced. An excitatory or contractile response can be produced by either α- (smooth muscle contractions) or β-adrenoceptors (tachycardia). The basis of receptor classification for a given response is therefore the agonist potency orders in the first instance, confirmation then follows when appropriate selective antagonists become available. The approaches used in the classification of adrenoceptors and their further subclassification, for example into β_1, β_2, α_1 and α_2 subtypes, are described in the following section.

3.2 General Considerations in the Classification of Adrenoceptors

3.2.1 Radioligand Binding

The concept of a physiological or pharmacological response to sympathetic nerve stimulation and sympathomimetic amines being the result of an interaction between an agonist molecule and a receptor was developed with little knowledge of the nature of these receptors or that they even existed. The first step towards identifying adrenoceptors was the development of radioligand binding techniques (Williams & Lefkowitz 1978, Stadel & Lefkowitz 1991). These techniques demonstrated the presence in tissue homogenates of specific binding sites for agonists and antagonists which displayed the same characteristic of selectivity as obtained by measuring the functional response of the tissue. These binding sites and adrenoceptors are therefore regarded as synonomous. The description of radioligand binding to adrenoceptors that follows is equally applicable to the muscarinic receptors of the parasympathetic nervous system (Chapters 7–9).

The binding of radiolabelled antagonists (or, less commonly, agonists) incubated with membrane fractions of tissue homogenates is detected after separation of the membrane fraction from the free unbound radioligand, usually by filtration. The total binding is then determined by scintillation counting of tritiated ligands such as [^{3}H]dihydroalprenolol ([^{3}H]DHA) or in a gamma counter ([^{125}I]-labelled ligands such as [^{125}I]iodocyanopindolol)([^{125}I]CYP) (see Table 5.3). Antagonists are preferable to agonists as radioligands because they do not bind to the uptake or metabolism sites. Iodinated ligands have the advantage of an increased sensitivity to detection due to much higher specific activity, whereas certain of the tritiated ligands have the advantage of stereoisomeric purity and stability (Stiles *et al.* 1984). [^{3}H]DHA and [^{125}I]CYP are hydrophobic. They therefore readily enter intact cells which may be a disadvantage if it is intended to examine cell surface β-adrenoceptors in whole cells. The hydrophilic antagonist, CGP-12177, is used to measure cell surface receptors in intact cells such as lymphocytes. Comparison between binding of these agents permits examination of receptors that are internalized or sequestered into the cell membrane, for example, by prolonged exposure to agonists (see Chapter 14).

Non-specific binding to non-receptor components of the tissue fraction and to the materials used in the assay (eg filters and glassware) is measured after displacing the ligand from the receptor with a relatively high concentration of unlabelled agonist or antagonist with high affinity and specificity for the receptor under study. For example, for α-adrenoceptors phentolamine or prazosin are used, and for β-adrenoceptors isoprenaline is used. The receptor-specific binding is therefore always derived from the difference between total and non-specific binding calculated under identical conditions. Clearly the proportion of specific to non-specific binding should be as high as possible, and this is achieved by using ligands with high affinity for the receptor so that relatively low concentrations can be employed. To aid detection of low levels, the ligand must also have a high specific activity (μCi/mmol) or level of labelling with ^{14}C, ^{3}H or ^{125}I. Radioiodinated ligands tend to have higher specific activity than do tritiated ligands.

Graphs of total binding (pmol/mg protein) at increasing concentrations of ligand (M) are often quite linear with no apparent saturation as concentration increases. However, specific binding is saturable, the plateau indicating a finite number of binding sites (B_{max}) (Figure 3.1). From these saturation-binding curves, the affinity of the radioligand for its binding site (K_D) can be determined as the concentration required for 50% occupancy. It is preferable to convert them to a Scatchard (1949) plot of bound/free ligand versus bound ligand. This avoids the need to use large concentrations of ligand to achieve saturation where binding may become atypical and non-specific binding dominates. It also converts to a linear relationship for more

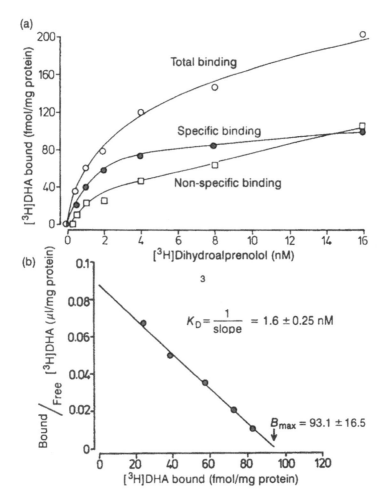

Figure 3.1 (a) Saturation curves for the binding of increasing concentrations of [³H]dihydroalprenolol ([³H]DHA) to β-adrenoceptors in guinea-pig heart ventricular membranes. Total binding (O), non-specific binding in the presence of unlabelled (−)-isoprenaline (200 μM) (□) and specific β-adrenoceptor binding (●), the difference between total and non-specific binding, are shown. (b) Scatchard analysis of the saturation curve for specific binding illustrated in (a). The affinity of binding, K_D, is given by 1/slope and the total number of binding sites, B_{max}, is given by the intercept on the [³H]DHA bound axis.

accurate curve fitting. The B_{max} (fmol/mg protein) is derived by extrapolation to the intercept on the bound ligand axis and the K_D obtained from the slope (where slope = $-1/K_D$) (Figure 3.1).

The binding of a radioligand (L) to its receptor (R) to form a complex (RL), like any drug–receptor interaction, is based upon the Law of Mass Action. The affinity of the interaction is given by the equilibrium dissociation constant (K_D), where

$$K_D = \frac{(R)(L)}{(RL)} \tag{3.1}$$

The total receptor population $R_t = (R) + (RL)$ where R_t is equivalent to B_{max}, (R) is the number of free unbound receptors and (RL) is the concentration of bound ligand.

Therefore:

$$K_D = \frac{(R_t - RL)(L)}{(RL)},$$ (3.2)

and rearrangement gives:

$$\frac{RL}{R_t} = \text{fractional receptor occupancy} = \frac{(L)}{K_D + (L)}.$$ (3.3)

This represents the saturation curve for binding from which the concentration (L) at 50% receptor occupancy $\{(RL)/(R_t) = 0.5\}$ is the K_D. The assumptions are that there is a bimolecular interaction between ligand and receptor and that measurements are made at equilibrium. Providing only a small fraction of the ligand is bound, then the concentration added can be used to approximate for 'free' ligand (L), however, the actual amounts bound can be calculated and subtracted from the concentration added.

Rearranging equation (3.3) gives:

$$\frac{(RL)}{(L)} = \frac{(R_t)}{K_D} - \frac{(RL)}{K_D}$$ (3.4)

yields the Scatchard (1949) plot of bound/free $\{(RL)/(L)\}$ against bound ligand (RL). Hence, B_{max} (R_t) is obtained from the intercept on the bound ligand axis and K_D is $-1/$slope (Titeler 1989).

The affinities of unlabelled agonists and antagonists for the receptor can be determined from their ability to displace the radioligand. Increasing concentrations of unlabelled compounds are incubated with a single low concentration (L) of the ligand close to its K_D to minimize non-specific binding. From the displacement curves of percentage inhibition of specific binding against concentration (Figure 3.2), the equilibrium dissociation constant (K_I) or affinity is determined by means of the Cheng & Prusoff (1973) equation:

$$K_I = \frac{IC_{50}}{1 + [(L)/K_D]},$$ (3.5)

where K_D is the dissociation constant of radioligand and IC_{50} is the molar concentration which inhibits binding by 50%. As evidence that the agonists or antagonists are interacting with a specific receptor related to that involved in the pharmacological response of the tissue, the order of affinities should be close to their potency order for the response. The potency orders of β-adrenoceptor agonists for their positive inotropic effect in isolated atria, their ability to displace [^{3}H]DHA binding from β-adrenoceptors and their ability to stimulate cardiac adenylyl cyclase (see Chapter 13) are all comparable (Figure 3.2). The displacement must also display stereoselectivity, with the R($-$)-isomers of β-agonists showing greater potency. It is also usual for radioligand displacement and pharmacological activity to occur over similar concentration ranges, however this is not essential since there may be considerable amplification of signal in the functional tissue response (Watanabe *et al.* 1982).

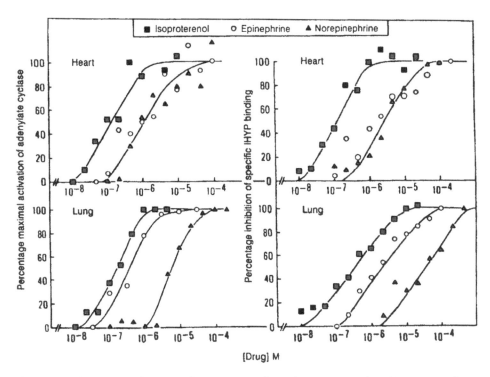

Figure 3.2 Comparison between the potency orders of (−)-isoprenaline (isoproterenol, ■), (−)-adrenaline (epinephrine, O) and (−)-noradrenaline (norepinephrine, ▲) as activators of adenylyl cyclase (left-hand panels) and inhibitors of β-adrenoceptor binding by radiolabelled [^{125}I]cyanopindol (IHYP) (right-hand panels). Membrane fractions were obtained from rat heart and lung which show different potency orders associated with their predominant receptor populations; β_1 in heart and β_2 in lungs. Reproduced with permission from Minneman *et al.* (1979).

Thus, radioligand binding is now routinely employed for the rapid screening of new chemical entities which have affinity for a wide range of receptor types. It is not immediately apparent, however, whether the functional response of such compounds will be agonistic or antagonistic. One difference between agonists and antagonists at α- or β-adrenoceptors in radioligand binding methods is how they are affected by guanosine 5'-triphosphate (GTP) or a stable non-hydrolysable derivative Gpp(NH)p (guanyl-5'yl-imidotriphosphate). In the presence of GTP, displacement of radioligand by agonists is inhibited to lower affinity binding; the displacement curves are shifted to the right and tend to be steepened. In contrast, antagonist displacement of the radioligand is unaffected. A further difference is that divalent cations such as Mg^{2+} displace competition curves for agonists to high affinity (ie to the left). Antagonist displacement curves are not affected. Thus, the receptor binding sites for agonists appear to be converted to high affinity form by Mg^{2+} but to low affinity form by GTP. The role of GTP and its binding proteins in receptor activation of second messenger signalling pathways is discussed in Chapter 13.

The relative proportion of the subtypes of a receptor (eg β_1 and β_2) present in a tissue can be determined from the inhibition of binding by unlabelled antagonists that are selective, of a radioligand that shows similar affinity for the two subtypes.

For example, atenolol, betaxolol and CGP 20712A displace selectively from β_1-adrenoceptors while ICI 118551 is selective for β_2-adrenoceptors. If both receptor subtypes are present, the displacement curves are biphasic or have a shallow slope. Steep and monophasic displacement curves are obtained if only one receptor type is present. These curves are now analysed by iterative nonlinear least-squares fitting by means of computer programs. Alternatively, they may be transformed into a linear form such as pseudo-Scatchard plots (Hofstee plots) of percentage inhibition of radioligand binding against percentage inhibition/concentration of displacing agent.

Radioligand binding techniques are also useful for determining changes in receptor characteristics in clinical disorders or after drug treatments. Changes in adrenoceptor numbers have been observed in a wide range of diseased states. Scatchard analysis of the radioligand binding to tissue homogenates from patients has revealed reductions in cardiac β-adrenoceptor numbers in the ischaemic heart of cardiac failure patients as a fall in B_{max} (Brodde 1991). Also, in asthmatic and cystic fibrosis patients there is a reported lowering of pulmonary β-adrenoceptors (Figure 3.3) (Sharma & Jeffery 1990). Chronic treatment with β-adrenoceptor agonists has also been shown to induce a fall in β-adrenoceptor number in the airways and lymphocytes, which may explain a possible tolerance to the bronchodilator effect of β-agonists (Conolly *et al*. 1982).

A reduction of B_{max} may also occur when receptor number are effectively reduced by incubation with an irreversible α-adrenoceptor antagonist such as phenoxybenzamine (see Chapter 5). In this case there is a parallel leftwards shift of the Scatchard plot with a reduction in B_{max}. This contrasts with the effect of a competitive antagonist, such as phentolamine, which displaces the Scatchard plot with a reduced slope indicating a change in affinity but with no change in intercept on the bound ligand axis (B_{max}) (Figure 3.4).

Whether binding sites identified from radioligand binding to membrane fractions are identical to the functional adrenoceptors mediating physiological responses of intact tissues is not certain. The affinity of antagonists for β-adrenoceptors measured by binding is often less than when measured from the functional responses. This is claimed to be because putative accessory site(s) are lost or damaged during the tissue homogenization and membrane fraction preparation (Kaumann 1987). Notwithstanding this limitation, radioligand binding techniques have proved extremely useful in the process of identifying the molecular structures of the α- and β-adrenoceptors.

Initially, radioligands that covalently bound to the receptor were employed to affinity label the receptor in whole cells or membranes. This enabled visualization of the receptor protein after separation by electrophoresis and showed that the β-adrenoceptor consisted of a single polypeptide chain of molecular mass in the range 52–65 kDa. Subsequent purification of adrenoceptors was achieved by affinity chromatography and high-performance liquid chromatography (HPLC). The next step towards receptor identification was preparation of antibodies to the purified receptor. These could be either antibodies directed against the receptor material or antibodies to a binding ligand (eg anti-alprenolol antibodies). These in turn could be made to produce anti-anti-alprenolol antibodies which bind to the receptor, as demonstrated by their ability to inhibit ligand binding. The availability of antibodies that recognized a specific receptor type has allowed application of immunoblotting techniques after electrophoretic separation of receptor protein. The IgG antibodies are readily radiolabelled with ^{125}I by iodination *in vitro*.

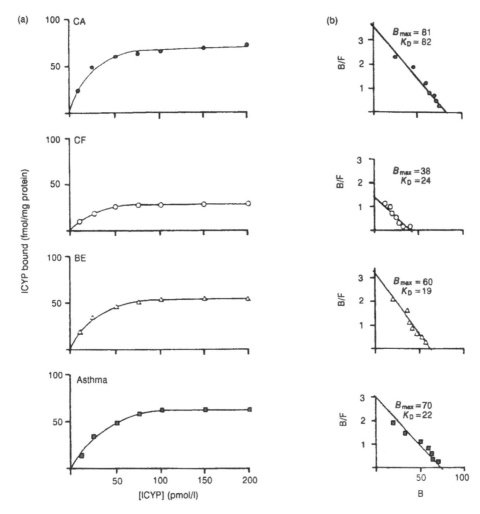

Figure 3.3 β-Adrenoceptor density (B_{max}) and binding affinity (K_D) of [^{125}I]cyanopindol (ICYP) in membrane preparations from the lungs of patients with carcinoma (CA), cystic fibrosis (CF), bronchiectasis (BE) or asthma. (a) Saturation curves for specific binding. (b) Scatchard plots of bound ligand (B, pmol) against the ratio of bound to free (B/F, fmol/mg protein/pmol/l). Units for B_{max} are fmol/mg protein, and for K_D are pmol/l. Reproduced with permission from Sharma & Jeffery (1990).

As will be seen in Chapter 13, the human adrenoceptor subtypes have now been cloned. Cloned cDNA derived from human gene libraries has been introduced into mammalian cell lines, such as Chinese hamster oocytes, or the bacterium *Escherichia coli*, which are normally without adrenoceptors. These cells then express the receptor that the cDNA encodes. Verification of the presence of the receptor is achieved by radioligand binding, which displays the characteristic of the natural receptor in terms of rank order of potency of displacing agents and stereoselectivity. It is interesting that β-adrenoceptors cloned in *E. coli* generally exhibit lower ligand binding affinities for agonists than are observed in mammalian cells. This is because mammalian cells contain a guanyl nucleotide binding protein (G_s) that forms

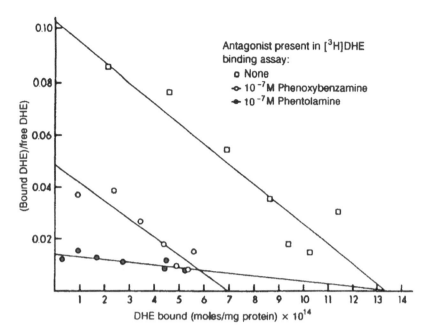

Figure 3.4 Comparison of the effects of competitive (phentolamine 10^{-7} M, ●) and irreversible (phenoxybenzamine 10^{-7} M, O) α-adrenoceptor antagonists on the Scatchard analysis of [³H]dihydroergocryptine ([³H]DHE) binding to α-adrenoceptors. Rabbit uterine membranes were incubated for 5 min in the absence (□) or presence (●,O) of antagonist at 25 °C and then used directly (without washing) for the [³H]DHE binding assay. Note the parallel displacement of the linear regression line by phenoxybenzamine with change in intercept on the concentration axis (B_{max}) indicating a reduction in receptor number. Phentolamine does not alter B_{max} but changes the slope. Reproduced with permission from Williams & Lefkowitz (1978).

a ternary complex with the receptor and agonist, and which is absent from *E. coli*. The agonist binds with greater affinity to this complex than to the receptor alone. However, if experiments are performed in which recombinant G protein is also expressed in the *E. coli*, at least for the a_s subunit of the protein, then the same ligand binding affinity is obtained as for mammalian cell membrane fractions. Radioligand binding is the only means of verifying the successful cloning of the receptor in *E. coli*. Normally, there is no linking of the receptor through the G_s protein to second messengers, therefore no biochemical response is measurable. In mammalian cells, radioligand binding and second messenger production can be measured, however they do suffer the disadvantages of requiring considerable effort and technical skill to maintain the cell lines and prevent non-expression. Mammalian cells also contain receptors and effector proteins which may interfere with the analysis of the products of transfected genes. Bacterial cell culture is, in contrast, simpler and more reliable and has advantages for the expression of chimaeric or mutant receptors. DNA chimaeras are composite DNA molecules in which a foreign segment of DNA is inserted into a vector molecule. When transfected into *E. coli*, the receptor is expressed with the amino acid sequence corresponding to the sequence of bases in the chimaeric DNA. This may be formed by combining portions of DNA from two receptor types (Strosberg 1992).

3.2.2 Characterization of Adrenoceptors from Pharmacological Responses

3.2.2.1 Quantifying the drug–receptor interaction

The dose–response curve for a tissue response to an agonist is the basis of all analysis of drug–receptor interaction in the characterization of adrenoceptors. The dose–response curve can be described from three parameters: the position with regard to the concentration of agonist, the slope and the maximum. The location parameter is the most widely used measure of the potency of agonists and is most often the EC_{50} (molar concentration for a 50% of maximum response). This can be a clumsy numerical value (eg 3.5×10^{-6} M) which is often simplified by converting to the negative logarithm ($-\log EC_{50} = pD_2 = 6.73$). This is then comparable with representations of dissociation constants and affinity values. Replicate concentration–response curves should be obtained from tissues taken from different subjects to provide several individual EC_{50} values. The concentration–response curve is plotted with concentration on a log scale and the log EC_{50} has been shown to be normally distributed (Fleming *et al.* 1972). Thus, *geometric* mean EC_{50} values must be quoted (antilog of the mean log EC_{50}) together with the 95% confidence limits or the $pD_2 \pm$ SEM. Statistical analyses are then made between logarithmic values. Another location parameter is the threshold concentration but this is less accurate and subject to error. It is also dependent upon the second parameter, the slope of the dose–response curve. The slope should comply with that predicted by the Law of Mass Action for the drug–receptor interaction.

The form of the Mass Action equation used for radioligand binding (equation 3.3) may be used:

$$\text{Fractional receptor occupancy} = \frac{RA}{R_t} = \frac{[A]}{K_A + [A]}, \tag{3.6}$$

where [A], the concentration of agonist, replaces the ligand concentration (L) and K_A is the equilibrium dissociation constant of the agonist (which is the reciprocal of the binding constant).

Ariëns (1954) introduced the term intrinsic activity (a) to explain the existence of compounds, termed partial agonists, which could not elicit the maximal tissue response even though there was presumed to be maximal receptor occupancy. This is a scaling factor which is a drug-related factor defining the ability of the combination of one drug molecule with one receptor to produce a response:

$$\frac{E_A}{E_{max}} = \text{functional tissue response} = \frac{a[A]}{K_A + [A]}. \tag{3.7}$$

According to this relationship, full agonists have an intrinsic activity = 1 and are able to produce the maximum possible response of the tissue via the receptor under study. Furthermore, antagonists would have zero intrinsic activity while partial agonists have values ranging between zero and unity; and the K_A is given by the EC_{50}.

The scheme proposed by Ariëns, however, implies a linear relationship between response and receptor occupancy. This is clearly not the case in many systems since, for example, a full agonist may produce a maximum response by occupying only a fraction of the receptors. This has been demonstrated by the use of irreversible antagonists such as phenoxybenzamine at a-adrenoceptors (see Chapter 5). After incubation of the tissue with this antagonist, a fraction of the receptors is effectively removed (R_t is reduced). This does not necessarily reduce the maximum response but merely displaces the dose–response curve to the right (Figure 3.5). This indicates the existence of 'spare receptors' or a 'receptor reserve' which are not occupied in the production of a maximum response. Only when all of the spare receptors are removed by the irreversible antagonist is the maximum response to the full agonist reduced (Figure 3.5). Thus, full agonists clearly do not all have the same scaling factor. From a knowledge of the K_A value for an agonist, derived either from binding data or from calculations involving the irreversible antagonist (see later), receptor occupancy may be determined from equation 3.6 and occupancy–response curves plotted. These are nonlinear and typically hyperbolic, a steep slope indicating efficient coupling (Figure 3.6). To explain this nonlinear relationship, Stephenson (1956) introduced the term stimulus (S). The response of a tissue was thus some undefined function (f) of the stimulus; two agonists generating the same stimulus would produce identical responses regardless of receptor occupancy:

$$\frac{E}{E_{max}} = f(S).\qquad(3.8)$$

It is this relationship that is responsible for the nonlinearity between occupancy and response. The coupling of adrenoceptor occupancy to the tissue response consists of a series of events, each with an amplifying capacity and a reserve capacity (Figure 3.7). As will be discussed in Chapter 13, the second messenger for β-adrenoceptor activation is cAMP, produced by activation of adenylyl cyclase (stimulus, S). This is nonlinearly related to β-adrenoceptor occupancy with approximately one-quarter of maximum adenylyl cyclase stimulation being sufficient for a maximum tissue response of the

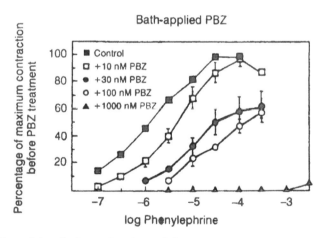

Figure 3.5 Effect of incubating isolated vas deferens with increasing concentrations of an irreversible α-adrenoceptor antagonist to show initial parallel shift of the agonist dose–response curve followed by depression of the maximum. Dose–response curves for the contractile responses to phenylephrine were obtained before (■) or after incubation with 10^{-8} (□), 3×10^{-8} (●) 10^{-7} (○) or 10^{-6} M (▲) phenoxybenzamine (PBZ) for 10 min. The antagonist was washed from the organ bath before constructing the agonist dose–response curve. Contractions to phenylephrine are plotted as a percentage of the maximum response before phenoxybenzamine treatment. Reproduced with permission from Minneman & Abel (1984).

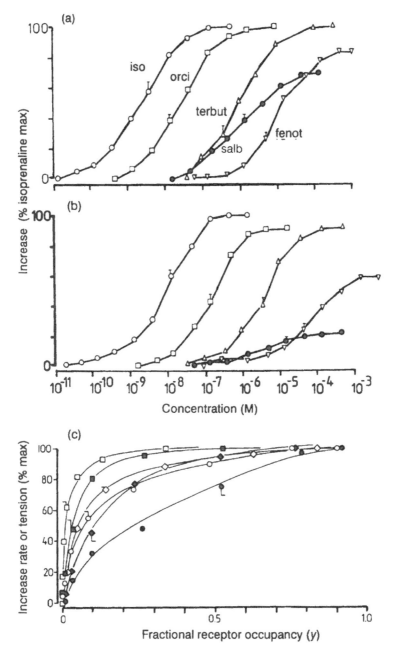

Figure 3.6 Concentration–response curves for (a) the rate and (b) the tension responses of guinea-pig isolated right and left atria to a range of β-adrenoceptor agonists. Agonists are isoprenaline (iso), orciprenaline (orci), terbutaline (terbut), salbutamol (salb) and fenoterol (fenot), the structures of which are shown in Table 4.5. Note similar potency orders and the partial agonist activity of salbutamol and fenoterol. (c) Occupancy–response curves for isoprenaline (◊, ◆), terbutaline (○,●) and fenoterol (□,■) on rate and tension, respectively. Fractional occupancy (y) was calculated from K_A values derived by use of receptor desensitization (Herepath & Broadley 1990a) from equation 3.6. Note that the partial agonist tension curve has become virtually linear.

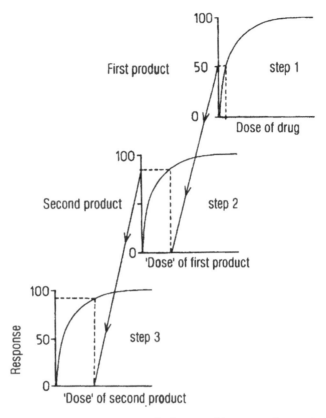

Figure 3.7 Diagrammatic representation of the amplification of signal from receptor occupancy (upper panel) through several intermediate steps to production of the response (lower panel). The arrowed line joining each step indicates the effect of a dose generating 50% of maximum first product. Since step 2 has a reserve capacity, only a fraction of the maximum first product is required to stimulate the second product formation. This process will be repeated at each step having reserve capacity, so that 50% receptor occupancy achieves >50% response.

heart (Kaumann 1981, Lemoine *et al.* 1989). A consequence of this amplification is that the EC_{50} occurs at a lower concentration than the K_D, thus EC_{50}s for full agonists are a poor estimate of the dissociation constant or affinity of an agonist.

The relationship between receptor occupancy (RA/R_t) and stimulus (S) is defined by the term efficacy (e):

$$S = e(RA/R_t).\tag{3.9}$$

From the Mass Action equation 3.6:

$$S = e\,\frac{[A]}{K_A + [A]},\tag{3.10}$$

and from equation 3.8:

$$\frac{E}{E_{max}} = f\left[\frac{e[A]}{K_A + [A]}\right].\tag{3.11}$$

According to Stephenson's definition, efficacy is both a drug- and tissue-related factor, since it depends upon R_t [rearranging equation 3.9; $e = (SR_t)/RA)$]. Furchgott (1966) therefore introduced the strictly receptor-related term, intrinsic efficacy (ε), to define the relationship between occupation of a single receptor and the stimulus:

$$e = \varepsilon.R_t. \tag{3.12}$$

According to current thinking on the occupancy model of drug–receptor interaction, agonist dose–response relationships are therefore described by the equation:

$$\frac{E}{E_{max}} = f\left[\frac{\varepsilon[R_t] \cdot [A]}{K_A + [A]}\right]. \tag{3.13}$$

The tissue-related factors are R_t and f (the function relating stimulus to response), while the drug-related factors are its dissociation constant, K_A, intrinsic efficacy and concentration (Ruffolo 1982, Kenakin 1984, 1987, 1993).

Compliance with the Law of Mass Action and this general equation may be determined by transformation of the dose–response data to a pseudo-Hill plot. This takes the form derived by Hill to determine the number of oxygen molecules binding to haemoglobin:

$$Hb(O_2)n \rightleftharpoons Hb + nO_2.$$

Thus, fractional saturation binding:

$$y = \frac{(pO_2)^n}{(K_A)^n + (pO_2)^n}. \tag{3.14}$$

For receptor occupany or tissue response this becomes:

$$\frac{R_A}{R_t} \quad \text{or} \quad \frac{E}{E_{max}} = \frac{[A]^n}{K_{A^a} + [A]^n}. \tag{3.15}$$

Rearranging according to Hill (1913) gives:

$$\frac{E}{E_{max} - E} = \left[\frac{[A]}{K_A}\right]^n. \tag{3.16}$$

Taking logarithms gives:

$$\log_{10}\left[\frac{E}{E_{max} - E}\right] = n\log_{10}[A] - n\log_{10}K_A. \tag{3.17}$$

A graph of $\log_{10}\{E/(E_{max} - E)\}$ against log concentration (a Hill plot) should yield a straight line with a slope (n, Hill coefficient) of unity. That is, a one-to-one binding of drug molecule with receptor. Deviation would indicate lack of compliance with the Mass Action equation and co-operativity of binding. A value greater than unity indicates that the binding of the drug to the receptor facilitates further binding to other sites; this is the situation with the binding of oxygen to haemoglobin. Values of less than one indicate negative co-operativity or the involvement of more than one receptor type with the response.

The characterization of the adrenoceptors involved in the responses to sympathomimetic amines is based upon the drug–receptor theory described above.

Adrenoceptors could be classified according to (1) the type of stimulus or response produced, (2) the linkage to second messengers or (3) their location. For example, we will see in Chapter 13 that β-adrenoceptors are coupled through a G protein to adenylyl cyclase, whereas α_1-adrenoceptors are linked to phosphoinositide turnover. However, this does not distinguish between the β_1- and β_2-adrenoceptor subtypes, which are both linked to adenylyl cyclase. Additionally, α_2-adrenoceptors were thought to lie postjunctionally on the effector site while α_2-adrenoceptors were located prejunctionally on nerve terminals, but it is now clear that α_2-adrenoceptors are also found postjunctionally. Thus, these methods of classifying adrenoceptors are unreliable. A pharmacological approach based upon the above equations derived from the Law of Mass Action is now the accepted method of receptor classification.

The agonist dose–response curve is the starting point which provides a measure of the agonist potency as the EC_{50}. However, more detail on the drug–receptor interaction may be obtained from the affinity and intrinsic efficacy of agonists. In spite of the development of equations relating dose and response, calculation of the affinity and intrinsic efficacy of an agonist is hampered by the unknown quantitative nature of the relationship that exists between stimulus and response. Since the stimulus (S) is linearly related to receptor occupancy (RA/R_t), receptor number (R_t) and intrinsic efficacy (ε) (equation 3.9), the nonlinear hyperbolic relationship between response and occupancy (Figure 3.6) must lie in the stimulus–response function (f). To eliminate the unknown tissue factors which control this relationship, assessment of drug–receptor interactions is made by means of null methods. That is, comparisons are made at equal responses either between agonists or for the same agonist in the absence and presence of an antagonist.

3.3 Methods of Classification of Adrenoceptors

Receptors mediating pharmacological responses are therefore classified by four methods: (1) agonist potency ratios or rank orders of potency; (2) activity of selective agonists; (3) comparisons of agonist affinities or intrinsic efficacies; and (4) competitive antagonist affinity.

3.3.1 Agonists in Adrenoceptor Classification

The *agonist potency ratio* is the ratio of equi-effective molar concentrations of one or more agonists relative to a standard agonist, usually the most potent in a series. From equation 3.10:

$$S = \frac{\varepsilon[R_t]\cdot[A]}{K_A + [A]}.$$

(3.18)

For a full agonist with a substantial receptor reserve, K_A significantly exceeds $[A]$ (see Figure 3.6), and the latter can be ignored in the denominator of equation 3.15. Thus, for two full agonists, A_1 and A_2, in a tissue:

$$S_1 = \frac{\varepsilon_1[R_t]\cdot[A_1]}{K_{A_1}} \quad \text{and} \quad S_2 = \frac{\varepsilon_2[R_t][A_2]}{K_{A_2}}.$$

(3.19)

The ratio of molar concentrations producing equal responses (eg EC_{50} values) when $S_1 = S_2$ is:

$$\frac{[A_1]}{[A_2]} = \frac{\varepsilon_2[R_t] \cdot K_{A_1}}{\varepsilon_1[R_t] \cdot K_{A_2}} = \frac{\varepsilon_2 \cdot K_{A_1}}{\varepsilon_1 \cdot K_{A_2}} \qquad\qquad (3.20)$$

The tissue is the same, therefore R_t is constant.

The potency ratio for two *full agonists* is therefore a measure of the intrinsic efficacy (ε) and affinity (K_A) of the drugs, independent of tissue factors. Clearly it is only relevant when using full agonists. Partial agonists can give misleading results when included for measurement of potency ratios at the EC_{50}. Partial agonists do have a place in receptor classification where they serve as an example of a *selective agonist*. An agonist of this type might be selective for a particular receptor subtype, and at that receptor produce full agonist activity. The finding that such an agonist fails to produce a maximum response in a particular tissue would therefore suggest the absence of that receptor subtype. An example would be the β_2-adrenoceptor selective agonist, salbutamol, which is a full agonist at β_2-adrenoceptors, but only a partial agonist in cardiac tissue where it normally acts on β_1-adrenoceptors (Figure 3.6) (Lumley & Broadley 1977). This approach, however, has led to erroneous conclusions in the case of prenalterol which was thought to be a β_1-selective agonist based on its activity in certain cardiac tissues (β_1) but not in vascular tissue (β_2). However, it failed to produce a response in the dog coronary artery, a tissue known to possess only β_1-adrenoceptors. It was concluded that prenalterol failed to exert responses in tissues where the stimulus–response coupling was inefficient because it is a weak partial agonist (Kenakin & Beek 1982a), not because of any selectivity for a particular receptor subtype.

The potency ratios of agonists relative to a standard, such as isoprenaline for β-adrenoceptors or noradrenaline for α-adrenoceptors, will allow them to be placed in *a rank order of potency*. This has become a most useful means of making the first characterization of receptor subtypes involved in a range of responses. Different rank orders for a simple range of endogenous and synthetic analogues in several tissues indicated that there was more than one adrenoceptor subtype. A rank order of potency adrenaline $\geqslant$ noradrenaline > phenylephrine > isoprenaline characterizes responses mediated via α-adrenoceptors, whereas the orders isoprenaline > noradrenaline > adrenaline > phenylephrine or isoprenaline > adrenaline > noradrenaline > phenylephrine characterize β-adrenoceptor-mediated responses of the β_1 and β_2 subtypes, respectively (Furchgott 1972). The same potency order in two tissues, however, does not necessarily mean that the same receptor is involved. The quantitative value of the potency ratio is a more definitive characteristic of the receptor subtype involved than just the potency order.

3.3.2 *Optimal Conditions for Characterizing Adrenoceptors*

To comply with drug–receptor theory, it is essential to measure the agonist response under equilibrium conditions. The concentration in the vicinity of the receptor (biophase) must not vary during the period of exposure and is the same as that which has been added to the tissue and is therefore measureable. These conditions are less readily achieved *in vivo* where additional factors such as reflexes and other feedback control systems operate. For example, increases in blood pressure are a poor measure of α-adrenoceptor-mediated vasoconstriction since they represent the combined effects on many vascular resistance beds and of any β-adrenoceptor-mediated cardiac effects. They are blunted by baroreceptor reflexes. Isolated tissues therefore allow for equilibrium conditions to be achieved and the response usually arises from changes in one predominant tissue type. In isolated ileum set up in an organ bath to record lengthwise contraction or relaxation, it is the responses of predominantly the outer longitudinal smooth muscle layer that are measured. Of course, interference from other tissue within the segment should be considered. For example, lung parenchymal strips have been used to study airway β-adrenoceptors, but the responses of blood vessels within this tissue may be a complicating factor (Clayton *et al.* 1981).

The optimal conditions for classifying the receptors in isolated tissues have been listed by Furchgott (1972) and are shown in Table 3.1.

When determining the rank orders or potency ratios of agonists from their EC_{50} values, it is essential to eliminate factors that may affect the potency of each agonist differently. For example, the stimulation of another receptor type present in the tissue may have an opposing action and thus interfere with the estimate of EC_{50}. The question of heterogenous adrenoceptor populations will be considered later in this chapter, but clearly if a tissue has both α- and β-adrenoceptors then characterization of the β-adrenoceptors must be performed with influence of α-adrenoceptor stimulation removed by inclusion of an α-antagonist throughout the experiment. Another way in which an agonist may produce an unwanted response is through the release of endogenous substances. Although histamine, Ach and 5-HT may be released from tissues, it is noradrenaline itself that is most commonly released by

Table 3.1 Optimal conditions for the pharmacological characterization of adrenoceptors in isolated tissues (Furchgott 1972).

1 The response of the tissue to an agonist should be due solely to a direct action on one receptor type. It should not be the resultant of actions at more than one receptor type or due to indirect activity (eg release of noradrenaline).

2 The altered sensitivity to an agonist in the presence of a competitive antagonist should be due solely to competition between the agonist and antagonist for the receptor. Altered sensitivity after treatment with an irreversible antagonist should be due solely to inactivation of the receptor.

3 The response to each dose of agonist should be measured at the peak effect.

4 The free concentrations of agonist and competitive antagonist should be known and maintained at a steady level at the time the response is measured. The concentration of irreversible antagonists in solution should be zero during measurement of the agonist response.

5 The concentration of agonist and competitive antagonist in the region of the receptor should be in equilibrium with that in the external solution. That is, the rate of removal (eg uptake, metabolism or binding) should be negligible compared with the rate of diffusion back into the external solution. In the case of an irreversible antagonist, the fraction of receptors not inactivated should remain constant throughout the period when responses are measured.

6 Proper controls should be performed to measure and correct for any time-dependent changes in sensitivity during the course of the experiment.

sympathomimetic amines. Those amines capable of releasing noradrenaline from the sympathetic neurone are known as indirectly acting sympathomimetic amines and these will be discussed in more detail in Chapter 4. However, for receptor character-ization it is the direct sympathomimetic activity at the receptor that must be measured. To eliminate indirect activity, the tissues are depleted of their noradrena-line stores by pretreatment of the animal with reserpine or 6-hydroxydopamine (6-OHDA) or by *in vitro* depletion of the tissue with 6-OHDA (see Chapter 6).

Further influences that must be removed are those that affect the concentration of the agonist in the synaptic cleft in the vicinity of the receptor. If agonist is continuously removed from the region of the receptor then equilibrium conditions cannot be achieved. The main processes of removal and inactivation are via neuronal and extraneuronal uptake and metabolism by COMT. MAO, being an intraneuronal enzyme, does not have a primary role in reducing the concentration of most amines near the receptor; its inhibition does not normally potentiate the responses to noradrenaline (see Chapter 2). If the agonists under study have different affinities for neuronal uptake then agonists with greater affinity will be removed from the biophase more effectively and their apparent potency reduced accordingly. Thus, noradrenaline is more effectively taken up than adrenaline and frequently appears less potent than adrenaline. A rank potency order of isoprenaline > adrenaline ⩾ noradrenaline (ratios 70 : 1.2 : 1) is obtained in the guinea-pig left atrium, which might suggest a β_2-adrenoceptor. However, in the presence of cocaine to inhibit neuronal uptake, the rank order becomes isoprenaline > noradrenaline > adrenaline (4 : 1 : 0.5), indicating that the β-adrenoceptor is in fact of the β_1-subtype. In the rabbit aorta, however, where innervation and neuronal uptake is minimal, the presence of cocaine does not influence potency order for the β-adrenoceptor-mediated relaxation response (Furchgott 1972). It is also important to realize that uptake does not influence potency if it is already saturated when high concentrations of agonist are used.

A particular form of agonist potency-ratio is the isomeric ratio. This can be obtained when the agonist exists in at least two chimeric forms and is the potency-ratio of two isomers (Patil *et al.* 1971). For the β-adrenoceptor agonist, isoprena-line, the $(-)/(+)$ ratio {R$(-)$/S$(+)$ or L/D} is greater at β_1- than at β_2-adrenoceptors (Morris & Kaumann 1984).

When receptor types in different tissues are under study, the conditions such as bathing medium and temperature should be identical, but it is surprising how often this is not the case. For example, it is common practice to set up isolated cardiac tissues at temperatures of 30–32 °C to achieve greater stability and minimize arrhythmias. More stable smooth muscle preparations however are set up nearer to body temperature (37.5 °C). Bath temperature affects responses mediated by receptor subtypes differently; β_1-adrenoceptor-mediated responses are enhanced at lower temperatures whereas β_2-adrenoceptor-mediated responses are little affected (Williams & Broadley 1982). Thus, if different temperatures are used, a greater potency of an agonist at cardiac β_1-adrenoceptors than at vascular β_2-adrenoceptors may be artifactual.

To summarize, characterization of the receptors mediating a pharmacological response should be performed in an isolated tissue with only a single receptor type being involved in the response. Of course, such knowledge only becomes available with time as further receptor subtypes are identified and more selective agonists are developed; receptor pharmacology advances by a series of circular progressions. The 'unwanted' receptor effects are removed by appropriate antagonists. Uptake and metabolism of sympathomimetic amines are prevented and indirect sympathomimetic activity is eliminated by using tissues from reserpine pretreated animals. A useful pretreatment of tissues is with phenoxybenzamine, which blocks α-adrenoceptors and uptake$_1$ and uptake$_2$. For study of a-adrenoceptors, propranolol, cocaine and metanephrine are useful for blockade of all β-adrenoceptor subtypes, uptake$_1$ and uptake$_2$, respectively. There may be a price to pay for such a diligent attempt to obtain optimum conditions – the inhibitors may also have undesirable actions. For example, metanephrine has some β-blocking activity (Kenakin 1980) and the commonly used uptake$_1$ blocker, DMI, is a weak α-blocker (Williamson & Broadley 1989). A consequence of this is that maximum potentiation of the response by uptake inhibition may not be obtained since at higher concentrations it is counteracted by an opposing block at the receptor.

3.3.3 Agoninst Affinity and Efficacy

Affinity and intrinsic efficacy values of agonists provide a more quantitative and thorough estimate of the drug–receptor interaction, which permits comparisons with dissociation constants derived from radioligand binding data and biochemical analysis of second messengers. However, calculations of the affinity of full agonists are not easily made because of the unknown factor relating stimulus and response. For partial agonists, where there is complete receptor occupancy at the maximum response, the EC_{50} is a close approximation of the dissociation constant (K_D) (equations 3.7 and 3.11). In this case the occupancy–response relationship (f) approaches linearity (Figure 3.6). The affinity of a full agonist, the reciprocal of the dissociation constant ($1/K_A$), may be calculated by a null method involving occlusion of a fraction of the receptors by use of an irreversible antagonist. The receptor reserve is eliminated and a maximum response can no longer be produced (Figure 3.5). For a-adrenoceptors, phenoxybenzamine has been used to irreversibly alkylate the receptors (see Chapter 5). Furchgott & Bursztyn (1967) derived a method in which equi-effective concentrations of the full agonist before (A) and after (A') irreversible occlusion of a fraction of the receptors are compared. The stimulus–response coupling and intrinsic efficacy are therefore common for each response and cancel. A condition of the method is therefore that the irreversible antagonist does not interfere with this coupling. Thus, from equation 3.13 for before and after receptor occlusion, respectively:

$$\frac{\varepsilon R_t [A]}{[A] + K_A} = \frac{\varepsilon (R_t . q)[A']}{[A'] + K_A},$$ (3.21)

where q is the fraction of the total receptor population (R_t) remaining after irreversible alkylation. R_t and ε cancel.

Rearranging:

$$\frac{1}{[A]} = \frac{1}{[A']} \cdot \frac{1}{q} - \frac{1}{K_A} \cdot \frac{(1-q)}{q}$$ (3.22)

The plot of the reciprocals of several equi-effective concentrations of agonist before ($1/[A]$) and after ($1/[A']$) receptor occlusion yields a straight line with a slope of

$1/q$ and intercept $\{(1-q)/(q.K_A)\}$ on the $1/A$ axis. The dissociation constant, K_A, can therefore be calculated from:

$$K_A = \frac{\text{slope} - 1}{\text{intercept}}. \tag{3.23}$$

Double reciprocal plots are notoriously prone to error and various refinements have been made to improve accuracy, for example, by suitable weighting of the values at the lower end of the dose−response curve. Kenakin (1987, 1993), who has made a thorough appraisal of these methods, recommends the plotting of $[A']/[A]$ against $[A]$ to minimize these errors, the K_A being derived as (intercept − 1)/slope.

Irreversible antagonists for β-adrenoceptors have become available in recent years. Examples include bromoacetylated alprenololmenthane (BAAM) and Ro 03-7894, both of which have been employed in the calculation of K_A values for isoprenaline at cardiac β_1-adrenoceptors (see Table 5.5) (Nicholson & Broadley 1978, Doggrell 1990). However, the use of both these antagonists has been questioned on the grounds that they also have postreceptor activity and may therefore contravene the basic assumptions of equation 3.22 (Krstew *et al.* 1984, Ng & Malta 1989).

An extention of this method of calculating agonist affinity is to remove effective receptors by desensitization. Prolonged incubation of a tissue with an agonist causes desensitization, the mechanism of which will be discussed more fully in Chapter 14. However, the consequence is that the receptors are progressively down-regulated or internalized until there is no longer a receptor reserve for a full agonist and the maximum response is depessed. If this occurs, the same equation (3.22) may again be applied and reciprocals of equi-active concentrations of the agonist before and after desensitization plotted to yield a straight line from which $K_A = (\text{slope} - 1)/$ intercept. This method has been applied to cardiac β_1-adrenoceptors and the K_A values for isoprenaline compare favourably with those derived by irreversible alkylation (Herepath & Broadley 1990a). The K_A values by both methods are consistently more by about one order of magnitude (a logarithmic unit of concentration) than the EC_{50} value (3.2 and 57 nM, respectively, for guinea-pig right atrial responses to isoprenaline), confirming that amplification of the stimulus signal occurs. The affinity values derived from receptor desensitization and from irreversible alkylation of the receptor also differ from those obtained with radioligand binding, the latter being ~10-fold less. It has been argued that the binding values are the correct values (Leff *et al.* 1985, Molenaar & Malta 1986); however, as pointed out by Kaumann (1987), binding values may be less because of a loss of accessory binding sites during membrane preparation.

A further method for calculating the affinity of a full agonist, based on the receptor occlusion method, has been by the use of *functional antagonism*. A functional (or physiological) antagonist is an agonist that exerts an opposing stimulus to that of the agonist under study. In the presence of the functional antagonist, the responses to the test agonist, including the maximum, are depressed. It has been argued that the effect of reducing the stimulus is equivalent to a reduction of receptors and the same principal can be applied as for irreversible antagonism (Buckner & Saini 1975). Similar K_A values have been obtained for the cardiac β_1-adrenoceptors when using an irreversible alkylating agent and functional antagonism by the muscarinic agonist, carbachol (Williams & Broadley 1983). The method of plotting the dose−response curve, as with most quantitative pharmacological analysis, can seriously influence the calculated values. Equiactive concentrations in the absence and presence of the functional antagonist should be obtained at equivalent *tissue states* and not at equivalent responses (eg changes in tension). The curve in the presence of functional antagonist should not be normalized (Kenakin 1984, 1987). Mackay (1982) has classified the types of functional antagonism according to the site of interaction of the two opposing agonists and the appearance of the dose–response curve in the presence of the functional antagonist. In type I, the functional antagonist changes the baseline of the tissue before depressing the maximum response. In type IIA, there is only an initial parallel shift of the curve before depression of the maximum with higher concentrations of functional antagonist. Type IIB shows no initial parallel displacement but a progressive depression of the maximum response. On theoretical grounds, the double reciprocal plot of $1/A$ versus $1/A'$ is more valid for type IIB interactions (Mackay 1982, Kenakin 1987). Thus, functional antagonism can yield quantitative measures of affinity of full agonists but the potential for error is great due to the complex nature of the interaction.

Having determined the affinity of a full agonist, the *intrinsic efficacy* relative to a reference agonist can be calculated. The efficacy is a dimensionless proportionality constant relating receptor occupancy and stimulus (equation 3.9):

$$S = e(RA/R_t) \text{ or } \varepsilon.R_t.P, \tag{3.24}$$

where P is the fraction of the total receptor population (R_t) occupied. Using the null approach for equal responses of two agonists, the stimuli are the same:

$$S_1 = \varepsilon_1.P_1.R_t \text{ and } S_2 = \varepsilon_2.P_2.R_t.$$

Since the total receptor population is the same:

$$\varepsilon_1/\varepsilon_2 = P_2/P_1, \tag{3.25}$$

the fractional occupancy $(P_1$ or $P_2)$ for each concentration of agonist is calculated from equation 3.6:

$$P = \frac{RA}{R_t} = \frac{[A]}{K_A + [A]}.$$

The value of K_A is determined by use of irreversible antagonism, functional antagonism or receptor desensitization. Although some authors use values derived from binding studies, these may not be strictly valid due to the discrepancies which occur between functional and binding data. The dose–response curves for each agonist are then replotted as the response against log fractional occupancy. The displacement between the curves along the log occupancy axis, converted to the antilog, is the relative efficacy between the two agonists (Furchgott 1966). Since R_t is constant, this is also a measure of relative intrinsic efficacy $(\varepsilon_1/\varepsilon_2)$. An example is shown in Figure 3.8 for isoprenaline and orciprenaline in guinea-pig atria.

Thus, agonist activity depends upon both the affinity and the efficacy. The determination of affinity from binding data may not be sufficient to predict the ability of drugs to induce a pharmacological response. The affinity and efficacy of a series of agonists may have independent chemical structural requirements. For example, in a series of α-adrenoceptor agonists producing vasoconstriction of the rat aorta, a progressive fall in affinity from naphthazoline through xylometazoline to phenylephrine was associated with an increase in relative efficacy of the latter agonist (Ruffolo *et al.* 1979). Similarly, at β_1-adrenoceptors of guinea-pig atria, the

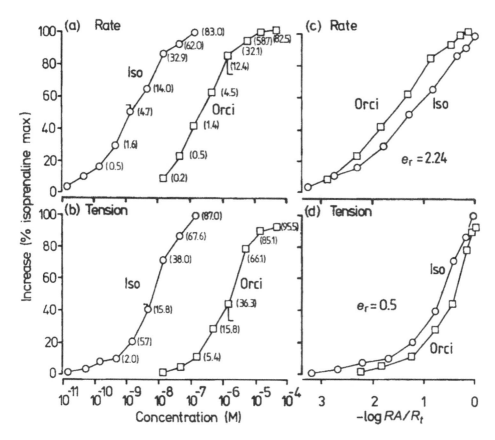

Figure 3.8 Concentration–response curves for isoprenaline (Iso, O) and orciprenaline (Orci, □) on the rate (a) and tension (b) of guinea-pig isolated right and left atria. Values in parenthesis are the percentage receptor occupancy calculated from equation 3.6 where RA/R_t (or P) = $[A]/(K_A + [A])$, and using affinity values (K_A) derived from irreversible antagonism (Broadley & Nicholson 1981). Responses plotted against $-\log RA/R_t$ are shown in (c) and (d), the antilog of the distance along the abcissa scale between the agonist curves yielded the relative efficacy values (e_r) shown.

order of affinities is isoprenaline > noradrenaline > adrenaline, whereas adrenaline is the most efficacious (McPherson *et al.* 1985). The contribution of affinity and efficacy to agonist activity is also important when comparing responses in different tissues. There are dangers in assuming that a lack of response to a particular agonist in one tissue reflects selectivity of that agonist or a different receptor type in that tissue. For example, prenalterol, which was first described as a selective β_1-agonist, produces a maximum response in rat left atria but no response in dog coronary artery. The response of the latter tissue to isoprenaline is, however, due to β_1-adrenoceptor stimulation and the lack of response to prenalterol is attributed to its low efficacy in a tissue with a poor stimulus–response coupling (Kenakin & Beek 1982a).

As an alternative to the efficacy term to describe the coupling between occupancy and response, Black & Leff (1983) introduced an operational model in which efficacy is quantified by a transducer ratio (τ), where:

$$\tau = [R_t]/K_E,$$ (3.26)

and $[R_t]$ is the total receptor population and K_E is the equilibrium dissociation constant for the interaction between drug–receptor complex and a transducer to form a ternary complex (eg drug–receptor–GTP binding protein complex). K_E is the receptor occupancy $[AR]$ for a half-maximal saturation of the transducer. Therefore an agonist with high efficacy (or τ) will have a high ratio between maximal occupancy (R_t) and K_E, the occupancy for a half-maximal effect (Figure 3.9) (Leff *et al.* 1990). This method allows for direct fitting of dose-response curves from which affinity (K_A) and efficacy (τ) may be derived by computation of these fitted curves. Thus, it differs from null methods in that it does not attempt to ignore the unknown relationship between stimulus and response. The operational model has also been applied to calculation of agonist affinity by use of irreversible antagonists and functional antagonism (Kenakin 1993).

3.3.4 *Antagonists in Adrenoceptor Classification*

Competitive reversible antagonists have proved to be by far the most accurate and reliable means of classifying receptors. The affinity of the antagonist may be measured more accurately than for agonists as the activity is not influenced by the stimulus-related factor, intrinsic efficacy. The calculation of affinity of an antagonist for its receptor is based on the equation of Gaddum (1937):

$$\frac{[RA']}{R_t} = \frac{[A']}{K_A(1 + [B]/K_B) + [A']}. \tag{3.27}$$

This represents the fractional occupancy of a concentration $[A']$ of agonist in the presence of antagonist concentration $[B]$. Comparison with equation 3.6, which represents the situation in the absence of antagonist, shows that the fractional

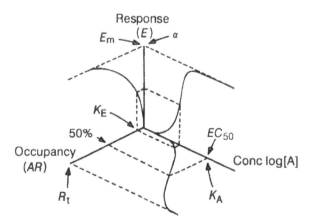

Figure 3.9 Diagrammatic representation of the operational model of agonist action. Shown are the three-dimensional relationships between (1) drug concentration (conc log [A]) and response (E), (2) receptor occupancy (AR) and drug concentration, and (3) occupancy and response. K_E is the receptor occupancy for half-maximal effect. A high ratio (τ) between maximum receptor occupancy (R_t) and K_E occurs with agonists with high efficacy. Thus, in this example, a greater concentration of agonist is required for 50% receptor occupancy (log K_A, affinity) than for a 50% response (log EC_{50}) due to the amplification factor (τ).

occupancy is less in the presence of antagonist by the factor $(1 + [B]/K_B)$ applied to the K_A. In the presence of antagonist, the agonist dose–response curve is shifted to the right in a parallel fashion and the degree of shift is quantified as the dose–ratio, usually at the 50% response level. This can be used to generate a quantitative measure of drug antagonism in the form of the pA_2 value (Arunlakshana & Schild 1959). For the same response obtained in the absence and presence of antagonist, equations 3.6 and 3.27 may be equated, since the same receptor occupancies are involved:

$$\frac{[A]}{K_A + [A]} = \frac{[A']}{K_A(1 + [B]/K_B) + [A']}. \tag{3.28}$$

Therefore the dose–ratio (DR) for equal responses in the absence and presence of antagonist:

$$DR = \frac{[A']}{[A]} = \frac{K_A(1 + [B]/K_B) + [A']}{K_A + [A]},$$

therefore:

$$DR = \frac{[B]}{K_B} + 1. \tag{3.29}$$

In logarithmic form:

$$\log(DR - 1) = \log[B] - \log K_B. \tag{3.30}$$

This is the equation for the Schild regression of $\log(DR - 1)$ versus log molar concentration of antagonist. It is usual practice to examine at least three concentrations of antagonist and to plot the $\log(DR - 1)$ versus molar concentration to produce a Schild plot. If the antagonism is competitive and is produced under equilibrium conditions, the Schild regression will be linear and have unity slope (Figure 3.10). The intercept on the concentration axis is $-\log K_B$, the negative log of the antagonist dissociation constant. Since this occurs at the zero value of log $(DR - 1)$, it represents the concentration required for a $DR = 2$, which by Schild's (1947) definition is the pA_2 value:

$$pA_2 = -\log K_B. \tag{3.31}$$

The pA_2 is therefore a quantitative measure of the antagonist affinity for the receptor, *providing* the Schild plot is linear and of unity slope. If these conditions do not apply, then the pA_2 is not the K_B of the antagonist. If it is known that the antagonism is competitive then the equation:

$$pA_2 = \log(DR - 1) - \log[B] \tag{3.32}$$

can be used for the shift of the dose–response curve by a single concentration $[B]$ of antagonist (Mackay, 1978). The pA_2 is independent of the concentration of antagonist and the mean pA_2 value of an antagonist may be calculated with its error ($\pm$SEM) for statistical comparison, for example, between agonists.

The pA_2 value can therefore be regarded as a constant for the antagonist at a particular receptor. The value is therefore independent of the agonist used, providing that the agonist acts solely at the specific receptor. It is also useful for characterizing

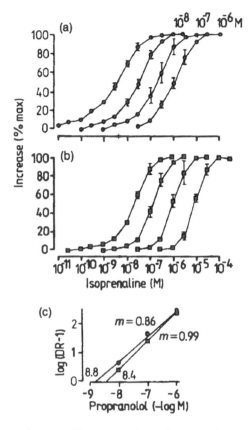

Figure 3.10 Illustration of competitive antagonism. Concentration–response curves for the increases in (a) rate (●) and (b) tension (■) by isoprenaline of guinea-pig isolated right and left atria, respectively, are displaced progressively rightwards in a parallel fashion in the presence of increasing concentrations (10^{-8} 10^{-7} 10^{-6} M) of the competitive β-adrenoceptor antagonist, propranolol. The displacement is expressed as a dose-ratio (DR), and a Schild plot of log (DR -1) versus propranolol concentration ($-$log M) produced (c). The pA_2 values for antagonism of rate (8.8, ●) and tension (8.4, ■) responses were calculated from the intercept and were not significantly different, indicating involvement of the same receptor. Compliance with receptor theory is indicated by slopes (m) of near unity for the regression lines.

the receptors in different tissues; these will yield similar pA_2 values if the receptors mediating their pharmacological responses are identical. There are, however, circumstances when Schild plots deviate from unity slope and yield differing pA_2 values from those expected of the receptor under study. Care must be taken not to conclude that a different receptor is involved. Examples of these potential errors are described in the following paragraphs.

3.3.4.1 Non-equilibrium conditions

A common problem with sympathomimetic amines is the influence of the uptake processes upon Schild plots. If the agonist is a substrate for saturable neuronal uptake, slopes of less than unity are often obtained leading to an overestimate of the pA_2 value because the intercept is displaced to the left (Figure 3.11). In the presence

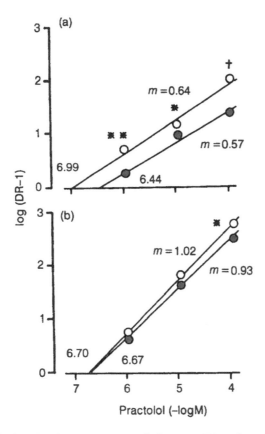

Figure 3.11 Schild plots for the antagonism of the rate (O) and tension (●) increases to (−)-noradrenaline of guinea-pig right and left atria, respectively. The dose-ratios (DR) of the shift of the concentration–response curves by increasing concentrations of practolol are plotted as log (DR − 1) versus antagonist log molar concentration (M). The method is as depicted in Figure 3.10. Dose-ratios were corrected for time-dependent changes in sensitivity from control experiments without the antagonist. (a) Untreated preparations, (b) preparations incubated throughout with the neuronal uptake inhibitor, cocaine (1.64×10^{-5} M). Note the low non-unity slopes of the regression lines in the untreated preparations (a) and the apparently different pA_2 values for rate (6.99) and tension (6.44). When neuronal uptake was inhibited, the regression lines have close to unity slope and the pA_2 values are similar.

of an uptake inhibitor (eg cocaine), the slope is raised to unity indicating the true competitive nature of the antagonism. In fact, observation of the dose–response curves alone shows parallel displacement of the curves with no reduction of the maxima – characteristic features of competitive antagonism. It is only after conversion to the Schild plot that the role of uptake is revealed. The key point is the saturability of the uptake process. In the absence of antagonist, the dose–response curve is produced by low concentrations of agonist and inhibition of uptake will produce a leftwards shift. However, as the curves are displaced to the right by the antagonist, higher concentrations of agonist are used and uptake becomes saturated. Thus, in the presence of higher concentrations of antagonist, the addition of an uptake inhibitor has no effect on the concentration of agonist in the biophase and the dose–response curve is unaffected by the presence of the uptake inhibitor. It is therefore the reference curve in the absence of antagonist that is displaced leftwards by the uptake inhibitor, causing an increase in dose-ratio for higher concentrations of antagonist. This raises the Schild plot at higher concentrations but not at low concentrations of antagonist and hence steepens to unity slope.

When the left and right atrial responses to noradrenaline were antagonized by a β-adrenoceptor antagonist, the Schild plots were separated, yielding pA_2 values of 6.44 and 6.99, respectively. The initial conclusion might be that different β-adrenoceptors mediate the positive inotropic and chronotropic responses to noradrenaline. However, the Schild plots had less than unity slope (0.57 and 0.64) and when repeated in the presence of cocaine, unity slopes were obtained and the pA_2 values were identical (Figure 3.11) (Lumley & Broadley 1975). The interpretation was that saturation of uptake occurred at different concentrations of noradrenaline for the left and right atria.

3.3.4.2 Heterogeneous receptor populations

If more than one receptor is involved in the response and the agonists used have different selectivities for these receptors, then the general rule that pA_2 values are constant for all agonists will fail. The pA_2 values will then differ according to the selectivity of the agonist used. This depends upon the relative proportions of the two receptors, the relative affinity of the antagonist for the two receptors and the stimulus produced by activation of each receptor. If two agonists with selectivities for each of the receptor types involved in the response are used, then it is possible to obtain separated linear Schild plots with slopes of unity. These will be described later for specific examples such as the mixed β_1 and β_2-adrenoceptor populations of the guinea-pig trachea (see Figure 3.13). Each Schild plot represents the dominant antagonism of one receptor subtype by the antagonist. It may be possible to identify biphasic Schild plots at low concentrations of antagonist (dose-ratios of <20) where a nonlinear portion of slope less than unity links two linear sections. If the Schild regressions for two selective agonists in combination with a selective antagonist are superimposed, a single receptor (homogeneous) population mediating the response is indicated. An example of this situation is the β-adrenoceptor mediating relaxation of the guinea-pig lung parenchyma (see Figure 3.13).

Another observation that arises from examination of antagonists is that of a limit to the shift of dose–response curves as the concentration of antagonist is increased. According to receptor theory, for each 10-fold increase in concentration, $(DR - 1)$ should correspondingly increase by 10-fold (equation 3.30). In the presence of high concentrations of antagonist, high concentrations of agonist are also required to produce a response. The potential for non-specific effects of both agonist and antagonist is then enhanced. For example, high concentrations of histamine, as the hydrochloride, exert negative chronotropy in cardiac preparations by a pH effect (Black *et al.* 1981). Thus, a histamine receptor antagonist (eg cimetidine) does not block this component of the response and a limit to the shift of the curve may occur. Non-specific effects of the agonist can dominate the responses mediated via adrenoceptors so that the antagonist has no apparent blocking activity. For example, in the guinea-pig ileum certain β-adrenoceptor agonists have membrane stabilizing activity sufficient to induce relaxation that is unaffected by β-adrenoceptor blockade (Grassby & Broadley 1987).

It is possible that a halt in the progressive displacement of dose–response curves by an antagonist arises because the agonist is no longer mediating its response via receptors blocked by the antagonist. Such a halt in the shift of the dose–response curves occurred with the antagonism of the inhibitory effects of $(-)$-isoprenaline in guinea-pig ileum by the β-adrenoceptor antagonist, nadolol (Bond & Clarke 1988). This was attributed to isoprenaline stimulating an atypical β-adrenoceptor at this point; the receptor was later defined as the β_3-adrenoceptor (see later). Similarly, in rat left atrium, the Schild plot for the antagonism of the inotropic response to adrenaline by the β-antagonist, pindolol, comes to a halt, but can then be further displaced by prazosin, indicating that the response shifts from β- to α-adrenoceptor-mediated (Williamson & Broadley 1989). The displacement of the dose–response curves for noradrenaline by pindolol and timolol also came to a halt but, unlike with adrenaline, prazosin did not cause further displacement. This suggested that noradrenaline was acting on neither α- nor β-adrenoceptors at this point.

To summarize, the type of adrenoceptor mediating the response of a tissue and the possibility of the involvement of more than one receptor can be classified from the rank order of potency of a series of agonists or their potency ratios relative to a standard. Calculations of affinities and relative intrinsic efficacies may provide more insight into the structural requirements of the agonist for producing the response. Antagonists provide a more reliable and accurate means of receptor classification; examination of complete dose–response curves for the agonist in the absence and presence of increasing concentrations of antagonist generates a Schild regression. Linearity and unity slope of the regression line confirms that the assay has been performed under controlled equilibrium conditions and that the antagonism is competitive. Deviations of the Schild plot when different selective agonists are used may indicate the involvement of more than one receptor type. This chapter will now consider the classification of adrenoceptors and their distribution throughout the body.

3.4 Adrenoceptor Subtypes and their Distribution

Ahlquist's (1948) subclassification of adrenoceptors into the α- and β-subtypes was based upon potency orders of a simple range of five catecholamine agonists and was soon supported by the observation that only those responses attributed to α-adrenoceptor stimulation were blocked by the β-haloalkylamine antagonists. The gastrointestinal tract was the first tissue in which the response (inhibition of myogenic activity and relaxation) was attributed to a mixed or heterogeneous receptor population; both α- and β-adrenoceptors mediate relaxation in this tissue. Although this was regarded as exceptional, it is now clear that most tissues have a mixed population of functional adrenoceptors of the two major types (α and β) and the ever increasing list of their subtypes. Where the two major adrenoceptor subtypes occur in a single tissue they may mediate the same response or opposing responses. For example, in the ileum the relaxation is due to both α- and β-adrenoceptors. In vascular smooth muscle, however, the relaxation response (vasodilatation) is due to β-adrenoceptor stimulation whereas the contraction (vasoconstriction) is α-adrenoceptor-mediated. Similarly, uterine smooth muscle contracts to α-adrenoceptor stimulation and relaxes in response to β-adrenoceptor agonists. In the majority of tissues, one response predominates, usually the response which occurs when sympathetic nerves discharge onto it or accompanying the release of adrenaline. The heart, for example, responds to sympathetic nerve stimulation and adrenaline by increases in rate and force of contraction mediated via adrenoceptors of the β-major type. A minor population of α-adrenoceptors has also been identified which also mediate an increase in force of cardiac contraction but these are only detected when β-adrenoceptors are blocked or by the use of highly selective α-adrenoceptor agonists (Benfey 1987). Thus, the ability of an agonist to produce a particular response in a tissue will depend upon the predominant receptor type mediating the response and both the selectivity and affinity of the agonist for that receptor. Isoprenaline, being a selective β-agonist, would be expected to cause vasodilatation and a fall in blood pressure, while phenylephrine, being an α-adrenoceptor agonist, would induce vasoconstriction and a rise in blood pressure. The major and minor adrenoceptors mediating responses of the various organs of the body are shown in Table 3.2.

3.5   β-Adrenoceptor Subclassification

Lands and colleagues (1967) compared the relative potencies of a series of sympathomimetic amines upon lipolysis of rat adipose tissue, bronchodilation of the guinea-pig perfused lung, cardiac stimulation of the rabbit perfused heart and vasodepression in the anaesthetized dog. Their data showed a strong correlation of the relative potencies between lipolysis and cardiac stimulation and between bronchodilation and vasodepression, but not between other combinations. They concluded that lipolysis and cardiac stimulation are mediated via β_1-adrenoceptors and vasodepression and bronchodilation are mediated via β_2-adrenoceptors. This conclusion, apart from lipolysis, has in general terms been proven correct and supported by the susceptibilities to selective β-adrenoceptor antagonists. However, the conclusion was reached under conditions which broke many of the rules already described for satisfactory receptor classification. For example, the vascular β-adrenoceptor was

Table 3.2 Adrenoceptors mediating responses of various organs

Organ/location	Response	Predominant receptor	Minor receptor involved[a]
Eye			
Iris			
Radial muscle	Contraction (mydriasis)	α_1	
Circular muscle		–	
Ciliary muscle	Relaxation	β_2	
Nictitating membrane	Contraction	α_1	Relaxation β_2
Heart			
Sinuatrial (SA) node	Tachycardia	β_1	$\alpha_1\beta_2$
Atria/ventricles	Increased contractility	β_1	$\alpha_1\beta_2$
Conducting tissue	Increased conduction	β_1	
Blood vessels			
Arterioles			
coronary, renal	Vasoconstriction	α_1	α_2
	Vasodilatation	β_2	
skin, salivary glands	Vasoconstriction	α_1	α_2
visceral	Vasoconstriction	α_1	Vasodilatation β_2
skeletal muscle, liver	Vasodilatation	β_2	
Veins	Constriction	α_1	
	Dilatation	β_2	

Table 3.2 (*Continued*)

Organ/location	Response	Predominant receptor	Minor receptor involved[a]
Lungs			
Trachea	Relaxation	β_2	β_1
	Contraction	–	α
Bronchioles	Relaxation	β_2	
Mast cells	Inhibition of histamine release	β_2	
Mucus secretion	Stimulation	β_2	
	Inhibition	α	
Ciliary beating	Increased	β_2	
Gastrointestinal tract			
Motility and tone	Decreased	α_1/β_1	β_3
Sphincters	Contract	α_1	
Secretions	Inhibition	α_2	Promotion of acid β_3
Spleen capsule	Contraction	α_1	Relaxation β_2
Anococcygeus muscle	Contraction	α_1	
Salivary glands			
Acinar cell	Electrolyte/water secretion	α_1	
	Amylase secretion	β_1	
Kidney			
Renin release	Increase	β_1	
	Decrease	α_2	
Sodium reabsorption		α_1	
Urinary bladder			
Trigone and sphincter	Contraction	α_1	
Sex organs			
Vas deferens	Contraction	α_1	Relaxation β_2
Uterus	Contraction	α_1	
(Longitudinal muscle)	Relaxation	β_2	
Skin			
Sweat glands	Secretion	α_1	
Piloerector muscles	Contraction	α_1	
Skeletal muscle			
Tetanic contractions	Reduced	β_2	
Tremor	Increased	β_2	
K$^+$ uptake	Increased (hypokalaemia)	β_2	
Metabolic			
Glycogenolysis (liver)	Hyperglycaemia	β_2	α_1
Lipolysis		β_3	
Insulin secretion	Decreased	α_2	
	Increased	β_3	
Plasma K$^+$	Hypokalaemia	β_2 (Skeletal muscle)	
	Hyperkalaemia	α_1 (liver)	
Platelets			
Aggregation[c]	Increased	α_2	
	Inhibited	β_2	
Lymphocytes	Inhibit immune function	β_2	

Notes: [a] Response is the same as that for the predominant receptor except where indicated.
[b] Pig coronary arteries β_1.
[c] Proaggregatory in certain species, eg rabbit. No inhibitory effect in rat platelets.

characterized from use of *in vivo* data in the anaesthetized dog and no attempts were apparently made to inhibit uptake processes throughout. Inspection of the data in fact shows a potency order of isoprenaline > adrenaline > noradrenaline for both cardiac (β_1) and lung (β_2) responses which inhibitors of uptake would presumably have converted to isoprenaline > noradrenaline > adrenaline for the heart. Notwithstanding these criticisms of the study of Lands and colleagues, it did provide the impetus for the further subclassification of β-adrenoceptors into two subclasses – β_1 and β_2. This in turn soon led to the development of selective agonists and antagonists for these two receptor subtypes. A third subtype, the β_3-adrenoceptor, has recently been identified initially in adipocytes, based upon the relative insensitivity to commonly used antagonists (Bylund *et al.* 1994).

The β-adrenoceptor subtypes mediating the responses of the major organs are shown in Table 3.2. In general, smooth muscle relaxation is mediated via adrenoceptors of the β_2-subtype. However, there are certain exceptions to this rule and it is worth considering the receptor distribution of specific tissues or organs determined from their functional responses and by radioligand binding.

3.5.1 *Heart*

Cardiac activity and initiation of the heart beat does not depend upon the presence of the sympathetic innervation, but stimulation of cardiac β-adrenoceptors by noradrenaline or adrenaline exerts a modulating effect. The force of both atrial and ventricular cardiac muscle contraction is increased. The maximum velocity of force or tension development (dT/dt_{max}) is increased, which *in vivo* is represented by an increase in the rate of rise of ventricular pressure (dP/dt_{max}). This is independent of any changes in pre- or after-load caused by simultaneous increases in venous return or aortic blood pressure, respectively. There is also a characteristic shortening of the relaxation time and abbreviation of the overall contraction (Figure 3.12). This is known as the relaxant effect of catecholamines. It is most readily observed at high heart rates both *in vivo* and *in vitro* and its physiological importance is to allow more time for complete ventricular relaxation to occur and thus permit adequate ventricular filling during β-adrenoceptor-mediated increases in heart rate (Scholz 1980). The relaxant effect can be demonstrated independently of the positive inotropic effect by use of K^+ depolarized cardiac muscle, which is relaxed by β-adrenoceptor agonists and is susceptible to β-adrenoceptor antagonists (Kaumann 1977). The mechanism is thought to be due to enhanced sequestering of Ca^{2+} by the SR (Tsien 1977). Furthermore, like other β-adrenoceptor-mediated responses, it is probably due to stimulation of adenylyl cyclase and intracellular accumulation of cAMP (Kaumann 1977).

The rate of discharge of the sinuatrial (SA) node and other pacemaker tissue [atrioventricular (AV) node] is enhanced, leading to an elevated rhythmicity of the heart. Indeed, normally quiescent isolated ventricular preparations become spontaneously beating in the presence of β-adrenoceptor agonists. Atrioventricular conduction is accelerated and the refractory period of cardiac and conducting tissue is reduced.

These changes can be associated with the opening and closing of ion channels across the cell membrane, which are in turn reflected by the electrophysiological properties of the heart. The effect of β-adrenoceptor agonists upon the electrophysiological properties of the heart can be described by their action on the various phases of the cardiac action potential (Noble 1979, 1984) (Figure 3.12). The onset of cardiac contraction is associated with a rapid depolarization (Phase 0) generated by the

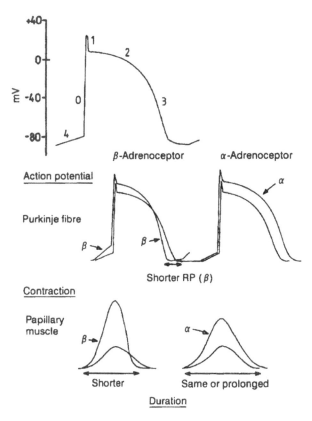

Figure 3.12 Effects of α- and β-adrenoceptor agonists on the cardiac action potential of Purkinje fibres and the single contraction cycle of isolated papillary muscles. The phases of the action potential are labelled 0–4, the ionic movements during each phase being described in the text. Note that β-adrenoceptor stimulation shortens the action potential plateau in Purkinje fibres but may lengthen the plateau in papillary muscles. There is an increase in the rate of rise of the spontaneous phase 4 discharge in spontaneously active cardiac cells such as the SA node and Purkinje fibres and the refractory period is shorter. Refractory period (RP) is shortened. The rate of rise of muscle contraction, relaxation rate and peak height of muscle contraction are increased, so that contraction duration is shorter. α-Adrenoceptor stimulation increases the height of muscle contraction, but rate of contraction and relaxation are little affected so that contraction duration is unaltered or even prolonged.

influx of Na^+ through voltage-dependent channels (i_{Na^+}). These channels are rapidly inactivated and close during the plateau phase; they are not reactivated until the membrane is repolarized. Phase 1 is a rapid repolarization to the plateau of depolarization (zero potential) of the action potential. This is due, among other factors, to inactivation of the i_{Na^+} and a transient outward K^+ current. The plateau (Phase 2) is due to the influx of Ca^{2+} through voltage-dependent Ca^{2+} channels primarily of the L-type, although T-type channels are also involved. This Ca^{2+} current (i_{Ca}) is inactivated slowly during the plateau. Phase 3 is the repolarization which terminates the plateau and is due to an outward current carried by the efflux of K^+ (i_K). This channel opens at ~40 mV, the current reaching maximum values towards the end of the plateau and closing at values more negative than −50 mV. Phase 4 occurs only in spontaneously discharging cardiac cells, including nodal tissue and the His-Purkinje

125

conducting system. In other atrial and ventricular cardiac muscle, the resting diastolic potential remains constant during Phase 4 and is only depolarized by an external stimulus or the arrival of a normal propagated impulse. In spontaneously-firing cells or pacemaker cells, there is a spontaneous depolarization due to the pacemaker current. The usual explanation for this spontaneous depolarization is a slow loss of K^+ permeability leading to a decay of the outward K^+ current (i_K). While this may contribute to the depolarization, it is now considered to involve activation of an inward current carried by K^+ and Na^+ (i_f) and activated by the diastolic hyperpolarization. The firing rate of the pacemaker depends upon the slope of Phase 4.

The rate of development of Phase 4 depolarization of the SA node is increased by β-adrenoceptor stimulation through increased rate of activation of the pacemaker current. Precisely which ion channels are involved in this process is not known. β-Adrenoceptor stimulation increases the outward K^+ current (i_K). However, this alone would paradoxically have the effect of slowing pacemaker activity since it would reverse the decay of the outward K^+ current. Thus, increased pacemaker activity appears to be due to activation of the inward pacemaker current (i_f). The role of the activation of the i_K current is to accelerate repolarization (Phase 3) and shorten the action potential duration. This will allow the heart to speed up by keeping action potential duration within the shorter cycle length (DiFrancesco 1993). The normally quiescent Purkinje fibres can display spontaneous activity when exposed to β-adrenoceptor agonists, an effect probably only mediated via β_1-adrenoceptors (Masini *et al.* 1991). Catecholamines also cause disturbances of rhythm due to induction of delayed after-depolarizations. These are secondary depolarizations that take place after repolarization has occurred and which precede the Phase 0 spontaneous depolarizations. When these reach the threshold level, they trigger a single premature depolarization and contraction. If these are repetitive they cause tachyarrhythmias or multiple extrasystoles.

Ischaemia of the myocardium caused by inadequate coronary circulation in angina or myocardial infarction promotes the local release of catecholamines (Schömig 1988). During the early phase of myocardial ischaemia there are also disturbances of membrane transport of ions, possibly because of reduced energy supply. There is net accumulation of intracellular Na^+ and Cl^-, but loss of K^+. The increase in extracellular K^+ reduces action potential duration and amplitude which facilitate the appearance of cardiac arrhythmias (Hearse & Dennis 1982). Elevated extracellular K^+ also causes partial membrane depolarization sufficient to inactivate the fast inward Na^+ current and raise the threshold for the slow inward Ca^{1+} current. This favours slow conduction and ectopic activity. The increased levels of free Ca^{2+} in the ischaemic myocardial cells that follows long periods of ischaemia and the intense catecholamine release leads to cellular damage known as Ca^{2+} overload (Parratt 1982).

β-Adrenoceptor agonists appear to have conflicting effects upon the action potential. They certainly raise the plateau through increasing the probability of L-type Ca^{2+} channel opening during depolarization (Figure 3.12). However, β-adrenoceptor agonists appear to have a dual effect upon the action potential duration (Kaumann 1991). In atrial muscle there is prolongation (Pappano 1971) which is probably related to the increased Ca^{2+} current in cardiac muscle during depolarization (Reuter 1983, Kameyama *et al.* 1985). Both β_1- and β_2-adrenoceptors mediate the increase in Ca^{2+} influx during the plateau phase. The overall duration of the action potential in Purkinje fibres is shortened by isoprenaline, an effect probably due to activation of an outward K^+ channel (Walsh *et al.* 1989, Yazawa & Kameyama

1990). This effect is probably related to the shortening of the refractory period produced by β-adrenoceptor stimulation (Szekeres & Papp 1980). A further channel that appears to be activated by isoprenaline is that carrying the inward Na$^+$ current (i_{Na}) which has also been implicated in catecholamine-induced arrhythmias (Egan *et al.* 1987).

3.5.1.1   Heterogeneous β-adrenoceptor populations

The possibility of a mixed population of functional β_1- and β_2-adrenoceptors in the heart was first intimated by the preferential reduction of the rate responses of isolated cat hearts to noradrenaline (selective for β_1-adrenoceptors) by practolol and the preferential reduction of the rate response to salbutamol (β_2-selective) by H35/25 (Carlsson *et al.* 1972). The conclusion that the positive chronotropic response of the cat heart was mediated via a heterogeneous population of β_1- and β_2-adrenoceptors was further substantiated in anaesthetized cats (Carlsson *et al.* 1977). In contrast, Carlsson's group observed that the positive inotropic responses were mediated through a homogeneous or predominantly β_1-adrenoceptor population in the myocardium.

Subsequently, more thorough analysis of the role of β_1- and β_2-adrenoceptors in the rate and force responses of cardiac tissues have been made by determination of pA_2 values for the antagonism of selective agonists by selective antagonists (Broadley 1982). In the cat right atria, atenolol (β_1-selective) had a higher pA_2 value when noradrenaline was the agonist (β_1-selective) than when fenoterol (β_2-selective) was used (O'Donnell & Wanstall 1979a) (Figure 3.13). A similar differential blockade of the cat right atrial

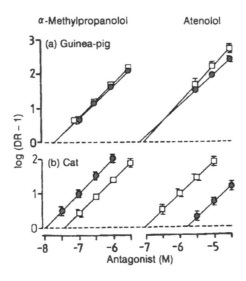

Figure 3.13   Schild plots for the antagonism of a β_1-selective agonist (noradrenaline, □) and a β_2-selective agonist (fenoterol, ●) by the β_2-selective antagonist, α-methylpropranolol (left) and the β_1-selective antagonist, atenolol (right). Increase in rate of contraction responses of (a) guinea-pig and (b) cat right atria were recorded in the absence of and presence of increasing concentrations of antagonist (M). Displacement of the concentration–response curves was expressed as the dose-ratio (DR) and Schild plots of log (DR − 1) against molar concentration of antagonist(log M) were constructed. Note the separation of regression lines for the two agonists in cat atria indicating a mixed population of β_1- and β_2-adrenoceptors. Redrawn from O'Donnell & Wanstall (1979a) with permission.

Table 3.3 pA_2 values and Schild plot slopes for antagonism of the cardiac responses to fenoterol (FEN) (β_2-selective) and noradrenaline (NA) (β_1-selective) by practolol (β_1-selective) and ICI 118,551 (β_2-selective)

		Practolol		ICI 118,551	
		pA_2	Slope	pA_2	Slope
Right atrial rate	FEN	6.36 ± 0.13	1.07	6.43 ± 0.13	0.92
	NA	6.25 ± 0.06	0.86	6.51 ± 0.08	1.11
Left atrial tension	FEN	6.59 ± 0.05	1.12	6.64 ± 0.12	1.23
	NA	6.34 ± 0.05	0.86	6.30 ± 0.07	1.10
Papillary muscle tensions	FEN	6.43 ± 0.11	1.14	6.56 ± 0.13	1.05
	NA	6.10 ± 0.05	1.17	6.64 ± 0.10	1.50

rate response to noradrenaline and salbutamol by practolol was observed by Vlietstra & Blinks (1976) who also showed that the left atrial inotropic responses were not blocked differentially. This evidence confirmed that a mixed β-adrenoceptor population in the SA node mediates the increases in rate of contraction of the cat heart while a homogeneous population mediates the myocardial positive inotropic response. In the guinea-pig, in contrast, similar pA_2 values are obtained for selective antagonists irrespective of the agonist used. The rate increases of the right atria and tension increases of the left atria and papillary muscles of the guinea-pig to noradrenaline and fenoterol are all similarly blocked by the β_2-selective antagonist, ICI 118,551 (Broadley & Hawthorn 1983) (Table 3.3). The responses of the various regions of the guinea-pig heart therefore appear to be mediated via a homogeneous population of β_1-adrenoceptors. A similar conclusion was reached for the rat and rabbit heart (Wilson & Lincoln 1984). However, the generation of a positive chronotropic response in guinea-pig right atria by the highly selective β_2-agonist, procaterol, has led to the suggestion that a small population of β_2-adrenoceptors may be involved in the guinea-pig right (Molenaar & Summers 1987) and left atria (Johansson & Persson 1983).

Radioligand binding data confirms that the guinea-pig heart contains a mixed population of β_1- and β_2-adrenoceptors. Competition for the binding of [^{3}H]dihydroalprenolol by the antagonists practolol (β_1-selective) and ICI 118,551 (β_2-selective) has revealed biphasic displacement curves. When these are plotted as Hofstee (or pseudo-Scatchard) plots of percentage binding against percentage binding/antagonist concentration, two straight lines of different slope are obtained (Figure 3.14). The proportion of β_1-adrenoceptors varies between different studies, but lies in the range 50–75% (Hedberg *et al.* 1980, Broadley & Hawthorn 1983, Molenaar & Summers 1987). Autoradiographic studies have also been performed whereby the receptors are localized by incubating cardiac slices with radiolabelled antagonists and then storing them in close contact with photographic film. The binding of radioligand shows up as bright dots, the locations of which are identified by comparison with the appearance of an adjacent stained section seen under the light microscope. The use of selective antagonists to displace binding has also revealed mixed populations of β-adrenoceptors in the guinea-pig atria (Summers *et al.* 1991). While there is no doubt that a substantial population of β_2-adrenoceptors exists in the guinea-pig heart, the functional relevance in mediating inotropic or chronotropic responses appears to be minimal. These receptors may be 'silent' in that

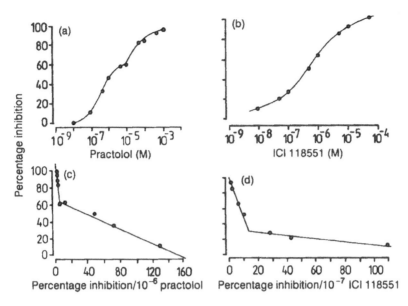

Figure 3.14 Displacement of specifically bound [³H]dihydroalprenolol from cardiac ventricular membranes by the β-adrenoceptor-selective antagonist, practolol (a), and the β_2-selective antagonist, ICI 118,551 (b). Upper graphs show percentage inhibition of binding against antagonist concentration. Lower graphs show Hofstee plots of percentage inhibition against percentage inhibition/antagonist concentration for practolol (c) and ICI 118,551 (d). Note the biphasic Hofstee plots indicative of high and low affinity displacement (β_1- and β_2-adrenoceptors). These are representative graphs of a single experiment, each point being the mean of duplicate determinations.

they are not coupled to the subsequent biochemical events that initiate the tissue response, that is, not coupled to the regulatory nucleotide binding protein G_s and adenylyl cyclase (see Chapter 13). Additionally, they may mediate other physiological or metabolic responses such as glycogenolysis. Binding data would also include β_2-adrenoceptors from vascular tissues (Broadley 1982).

Human myocardium has now been extensively examined to determine the distribution of β_1- and β_2-adrenoceptors. Radioligand binding (Brodde 1987) and autoradiography (Buxton *et al.* 1987) have revealed mixed populations of β-adrenoceptors in human atrial and ventricular tissue. In human isolated paced right and left atria and left papillary muscles, isoprenaline and adrenaline increase force of contraction via both β-adrenoceptor subtypes, since the concentration–response curves are displaced by both the selective β_1-antagonist, CGP 20712A, and the selective β_2-antagonist, ICI 118,551 (see Table 5.3). However, noradrenaline exerts its effects by stimulating only β_1-adrenoceptors, since they are not blocked by ICI 118,551 (Motomura *et al.* 1990) (Figure 3.15). *In vivo*, exercise-induced tachycardia, which is also due mainly to the release of noradrenaline from sympathetic neurones, is also unaffected by ICI 118,551, but is blocked by selective β_1-antagonists such as bisoprolol. It is therefore concluded that under normal physiological conditions, cardiac contractility and rate are regulated only by β_1-adrenoceptors but under stress situations, when adrenaline is also released from the adrenal medullae, stimulation of cardiac β_2-adrenoceptors can also contribute to additional increases in force and rate of contraction (Motomura *et al.* 1990, Kaumann 1991).

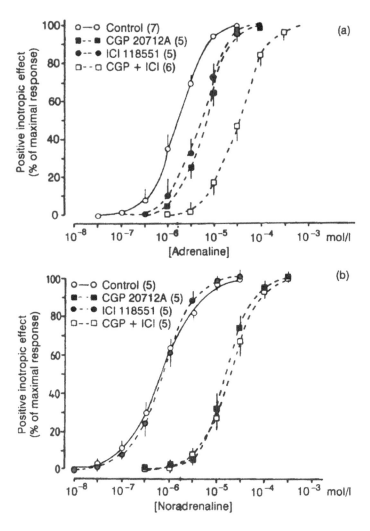

Figure 3.15 Effects of the β_1-adrenoceptor-selective antagonist, CGP 20712A (■, 3×10^{-7}M) and β_2-selective antagonist, ICI 118,551 (●, 3×10^{-8} M), either alone or in combination (□) upon dose–response curves for the positive inotropic responses of human isolated electrically driven left atria to (a) adrenaline or (b) noradrenaline. The dose–response curves in the absence of antagonists are indicated by open circles (O). Note that adrenaline is blocked by both antagonists and therefore acts through both receptor subtypes, whereas noradrenaline is only blocked by CGP 20712A and therefore acts only on β_1-adrenoceptors. Reproduced with permission from Motomura *et al.* (1990).

3.5.1.2 *Cardiac β-adrenoceptors in disease and after drug treatment*

The density of cardiac β-adrenoceptors has been found to alter after drug treatment and in disease states. In patients with advanced heart failure due to idiopathic dilated cardiomyopathy and end-stage ischaemic cardiomyopathy, the density of β-adrenoceptors determined by radioligand binding has been shown to fall. This is coupled with a loss of adenylyl cyclase stimulating activity and reduced inotropic responsiveness (Feldman & Bristow 1990). These changes are thought to be due to the chronic release of endogenous catecholamines from the failing myocardium to provide compensatory inotropic support, which exerts a down-regulatory effect upon the β-adrenoceptors. There is a selective reduction of β_1-adrenoceptor density and sensitivity in tissues from these patients, which has been attributed to a selective susceptibility of this adrenoceptor subtype to down-regulation. However, it now appears that both β_1- and β_2-adrenoceptors are capable of down-regulation. Chronic treatment with the selective β_1-partial agonist, xamoterol, attenuated all β_1-adrenoceptor-mediated effects but not β_2-mediated effects, whereas chronic treatment with the β_2-adrenoceptor agonist, procaterol, reduced only β_2-adrenoceptor-mediated responses. Thus, long-term treatment with subtype-selective β-agonists causes subtype-selective desensitization. The selective loss of β_1-adrenoceptors in cardiac myopathy may arise because noradrenaline is the predominant catecholamine released and this preferentially stimulates the β_1-adrenoceptors, sparing the β_2-adrenoceptors from down-regulation (Brodde 1991). Thus, in congestive cardiomyopathy, β_2-adrenoceptors appear to take on more importance, which may explain the beneficial effect of selective β_2-adrenoceptor agonists in severe cardiac failure (Brodde *et al.* 1986). In heart failure associated with mitral valve disease, there appears to be a gradual decrease of both β-adrenoceptor subtypes in atria and left ventricles, which is related to the severity of the disease (Brodde 1991).

The coupling of the β_2-adrenoceptor to adenylyl cyclase is more efficient than for the β_1-adrenoceptor, but in heart failure the human ventricular β_2-adrenoceptor appears to be partially uncoupled from adenylyl cyclase (Kaumann 1991). Thus, although β_2-adrenoceptor density does not always appear to be reduced, there may be a slight loss of β_2-adrenoceptor sensitivity due to uncoupling. This may explain why the responses to the selective β_2-adrenoceptor partial agonist, dopexamine, are impaired in isolated ventricular strips from patients with heart failure (Böhm *et al.* 1989a).

Reduced positive inotropic and chronotropic activity from β-adrenoceptor stimulation has also been demonstrated *in vivo* in patients with chronic heart failure. The heart rate response to exercise is regarded as a β_1-adrenoceptor-mediated response to neuronally released noradrenaline and this is reduced in patients with heart failure. Isoprenaline infusions stimulate both β_1- and β_2-adrenoceptors and the heart rate responses are also reduced in heart failure.

A recent observation of relevance to the cardiac β-adrenoceptor is that in heart failure an increase in the G_i nucleotide binding protein has been found. This is linked negatively to adenylyl cyclase, thus inhibiting the accumulation of cAMP and opposing the effect of β-adrenoceptor stimulation, which acts through the G_s regulatory protein. This may explain the uncoupling of the β_2-adrenoceptor-mediated effects (Feldman & Bristow 1990). However, negative inotropic responses mediated through G_i, such as those to adenosine, do not appear to be enhanced in atrial or ventricular preparations from patients with cardiomyopathy (Böhm *et al.* 1990).

Hypertension has also been found to affect cardiac adrenoceptor sensitivity. Isolated atria from hypertensive rats have been shown to produce reduced maximum inotropic and chronotropic responses mediated via both α- and β-adrenoceptors (Kunos *et al.* 1978). A decrease in myocardial β-adrenoceptors (Yamada *et al.* 1984) and reduced sensitivity of adenylyl cyclase to stimulation by isoprenaline (Bhalla *et al.* 1978) has been observed in spontaneously hypertensive rats (SHR). The reduced β_1-adrenoceptor density in SHR is associated with an elevation of β_2-adrenoceptors (Brodde & Michel 1992). In humans, myocardial tissue from hypertensive patients is not readily accessible. However, vascular β_2-adrenoceptor sensitivity has been measured, together with lymphocyte binding, which is a valid index of vascular β-adrenoceptor responsiveness. Vascular β-adrenoceptor sensitivity, however, may not reflect cardiac β-adrenoceptor responsiveness (Michel *et al*. 1986). Vascular β-adrenoceptor sensitivity and lymphocyte binding sites are generally reduced in hypertensive subjects (Borkowski 1988, Feldman 1991). One possible explanation is a down-regulation due to the elevated levels of adrenaline that are often found in patients with essential hypertension (Goldstein 1980, Kjeldsen *et al*. 1983), although renal β_2-adrenoceptors are not desensitized (Brodde & Michel 1992). A possible reduction of G protein coupling of the receptor to vasodilatation (Asano *et al*. 1988) has not been confirmed from direct measurement of vascular G protein (Brodde & Michel 1992).

Aging also influences the sensitivity of the heart to sympathetic stimulation. The positive chronotropic response to levels of exercise that increase plasma catecholamines is blunted in aged humans (Petrofsky & Lind 1975). Whether this is because of reduced β-adrenoceptor responsiveness due to catecholamine levels that are known to rise with age (Ziegler *et al*. 1976) is uncertain. The reduced β-adrenoceptor function of the heart and vasculature with age is not due to reduced receptor number but probably to an impaired coupling to the G protein resulting in reduced formation of cAMP (see Chapter 13) (Folkow & Svanborg 1993). Both the hypertension- and age-related reduction in vascular β-adrenoceptor sensitivity are restored by dietry sodium restriction (Feldman 1992).

Chronic treatment of animals and patients with β-adrenoceptor agonists causes desensitization of cardiac β-adrenoceptor-mediated responses and a down-regulation of β-adrenoceptor binding sites (Harden 1983). Furthermore, chronic treatment with β-adrenoceptor antagonists leads to an increase in sensitivity of isolated cardiac tissue taken from pretreated animals (Kunos *et al*. 1978, Chess-Williams & Broadley 1984). In man, there is a rebound increase in the heart rate response to isoprenaline infusion after cessation of propranolol treatment administered for several weeks (Ross *et al*. 1981). The mechanisms of these changes will be discussed more fully in Chapter 14. However, a recent unexpected finding has been that when patients are treated with selective β_1-adrenoceptor antagonists, it is the sensitivity to β_2-adrenoceptor stimulation that is enhanced, not β_1-adrenoceptors (Kaumann 1991). β_2-Adrenoceptor binding sites, however, are unchanged. Since the β_1-adrenoceptor density of atria from β_1-adrenoceptor antagonist treated patients measured from binding data is enhanced (Michel *et al*. 1988), the apparently unchanged β_1-adrenoceptor sensitivity must be associated with an increased coupling between β_1-adrenoceptor and the subsequent events. This further illustrates the dissociation between functional tissue responsiveness and radioligand binding data. The precise mechanism for the apparent increase in β_2-adrenoceptor function after chronic β_1-antagonist treatment remains a mystery. However, possible cross-talk between receptors at the level of the regulatory G_s proteins leading to heterologous desensitization of different receptor types is under investigation.

Finally, the heart may also contain a population of atypical β-adrenoceptors. This proposal is based upon the observation that partial agonists, usually related to pindolol, exert stimulant effects upon isolated cardiac preparations at concentrations in excess of their blocking activity of classical β-adrenoceptor agonists (Kaumann 1989). Whether the stimulant responses to this group of partial agonists are due to activation of an atypical β-adrenoceptor remains to be established. A β_3-adrenoceptor has been described for pharmacological responses of the ileum, airways and adipocytes and will be discussed in more detail later. Whether the atypical cardiac β-adrenoceptor is the same as this β_3-adrenoceptor awaits the results of further study.

3.5.2 *Lungs*

The airway smooth muscle is relaxed by β-adrenoceptor stimulation causing bronchodilatation. This was first characterized as a β_2-adrenoceptor-mediated response from the antagonism of histamine-induced bronchoconstriction of guinea-pig isolated perfused lungs by Lands *et al.* (1967). However, more thorough analysis of the responses of isolated tissues of preparations from different parts of the airways has also revealed heterogeneous receptor populations. The relaxation of the guinea-pig trachea was first shown to be mediated by a heterogeneous population of β_1- and β_2-adrenoceptors by Furchgott (1976). Subsequently, several groups have shown that the pA_2 values for the selective β_2-antagonist, H35/25, differs depending upon whether noradrenaline (β_1-selective) or fenoterol (β_2-selective) is the agonist (O'Donnell & Wanstall 1979b, Zaagsma *et al.* 1983). Separated Schild plots of unity slope for the two agonists illustrates this clearly in Figure 3.13. Similarly, in the cat trachea a mixed population of β_1- and β_2-adrenoceptors can mediate relaxation. In contrast, in the more peripheral airways of the guinea-pig, as measured in lung parenchymal strips, the relaxation is mediated via a homogeneous population of β_2-adrenoceptors (Zaagsma *et al.* 1983). This was also the situation in the human tracheal smooth muscle, main bronchus and intrapulmonary smooth muscle obtained from patients undergoing pneumocotomy (Zaagsma *et al.* 1983).

Radioligand binding studies with homogenates from human, guinea-pig and rat lung have, however, found a mixed population with ~20–30% of β_1-adrenoceptor binding sites (Engel 1981). These β_1-adrenoceptors would therefore appear to serve other functions than relaxation of airway smooth muscle. β-Adrenoceptors are also located on airway epithelium where they may control beat frequency of the cilia; β_2-adrenoceptor agonists increase their beat frequency and may facilitate mucociliary clearance (O'Donnell 1991). The airway epithelium acts as a site of loss or metabolism (eg extraneuronal uptake) of catecholamines. It may also release an epithelium-derived inhibitory factor (EpDIF) in response to β_2-adrenoceptor stimulation so that epithelium removal inhibits the relaxant effect of isoprenaline in many isolated tissues (Goldie *et al.* 1990). Pulmonary mast cells are a further location of β-adrenoceptors since their ability to release inflammatory mediators such as histamine and leukotrienes is inhibited *in vitro* by β-agonists such as isoprenaline. Whether this property occurs *in vivo* is unclear. Certainly, the initial bronchoconstriction of asthmatic patients induced by allergen challenge is inhibited by β-agonists. However, the late-onset reaction to allergen challenge is not diminished. This represents the secondary inflammatory response to released mediators and indicates that β-agonists may not necessarily suppress the inflammatory response. More recently, a long-acting β-adrenoceptor agonist, salmeterol, has been introduced which is claimed to have anti-inflammatory properties and attenuates both the early- and late-phase bronchoconstriction to allergens. However, this has been the subject of some dispute and it is now agreed that the use of salmeterol in asthma should still be accompanied by anti-inflammatory therapy with an inhaled steroid such as beclomethasone (Becotide[R]) (Morley 1991).

A further site of anti-inflammatory activity through β_2-adrenoceptor stimulation is the bronchial vascular endothelium. The postcapillary venules of the circulation supplying the tracheobronchial airways increase in permeability in response to inflammatory mediators including histamine, leukotrienes and PAF. This results in exudation of plasma and associated plasma proteins into the airway tissue and lung lumen. This exudate contributes to the bronchial oedema that accompanies the inflammatory response. There is evidence that the endothelial cells have contractile properties and that the increased microvascular leakage is induced by opening of intercellular junctions. β_2-Adrenoceptor agonists attenuate the increased microvascular leakage induced by bradykinin, a mediator likely to be generated from the plasma exudate. This effect may be produced by β-adrenoceptor-mediated relaxation of the vascular endothelial cells (O'Donnell 1991).

β_2-Adrenoceptors also appear to be involved in improving mucociliary clearance effected by the ciliated epithelium of the airways. This is a particularly important aspect of the therapeutic efficacy of β_2-adrenoceptor agonists in infective lung disease. The inhibitory effect of pyocyanin, a bacterial toxin from *Pseudomonas aeruginosa* which occurs in airways infections, on nasal epithelial ciliary beating has been shown to be inhibited by salmeterol (Kanthakumar *et al.* 1994).

β-Adrenoceptor abnormalities have been implicated in several pathological conditions of the airways including cystic fibrosis and asthma. Cystic fibrosis (CF, mucoviscidosis) is a genetic disorder of exocrine secretion from the salivary and sweat glands and the pancreas. Salivary and sweat glands from patients with CF display reduced secretory responses to stimulation with β-adrenoceptor agonists. Cultured airway epithelial cells from patients with CF show diminished Cl^- secretion in response to stimulation by β-adrenoceptor agonists. In mucosal strips, the secretion of mucus in response to isoprenaline is also reduced. These changes are associated with a fall in density (B_{max}) of β-adrenoceptors assessed by radioligand binding in lung membranes from patients undergoing heart–lung transplant. Autoradiography showed that these changes were most apparent in the alveolar wall and bronchial epithelium. Whether these changes are the cause or result of airway inflammation remains to be established (Sharma & Jeffery 1990). Asthma has also been associated with a reduced β-adrenoceptor responsiveness (Szentivanyi 1968). Although not now regarded as a cause of the symptoms of asthma, changes in β-adrenoceptors have been found. A decrease in the sensitivity to β-adrenoceptor-mediated relaxation of airway smooth muscle from severely asthmatic patients has been observed. This is related to the severity of the disease and is independent of any down-regulation due to previous β-adrenoceptor agonist therapy. Lymphocytes from asthmatic patients have been found to have fewer β_2-adrenoceptor binding sites and to be less responsive to β-agonists in stimulating cAMP production. Allergen challenge has also been shown to reduce lymphocyte β-adrenoceptor function in asthmatics, suggesting that the reduced β-adrenoceptor function is secondary to the disease and not the primary cause. Lung tissue has also been shown to have reduced β-adrenoceptor binding sites (Lulich *et al.* 1988). It is generally regarded that binding sites do not alter but that the reduced β-adrenoceptor function arises from uncoupling between β-adrenoceptor and adenylyl cyclase (Goldie *et al.* 1990).

3.5.3 *Lymphocytes*

Related to the presence of β-adrenoceptors on mast cells and their involvement in the inflammatory response is their location on lymphocytes. Lymphocytes contain a homogeneous population of β_2-adrenoceptors and, because of the ready availability of blood, they have been widely used to monitor β-adrenoceptor function in man during drug treatments and in diseased states. For example, lymphocyte β_2-adrenoceptor density, measured from $(-)H[^{125}I]$iodocyanopindolol binding, correlated well with the β_2-adrenoceptor density of right atrial appendage from patients undergoing coronary artery bypass grafting (Michel *et al.* 1986). Long-term administration of the non-selective β-adrenoceptor antagonist, propranolol, causes an increase in lymphocyte β-adrenoceptor density, which is matched by a raised sensitivity of isolated cardiac tissue to β-adrenoceptor agonists. However, chronic treatment with a β_1-selective antagonist, such as atenolol, does not cause changes in lymphocyte β_2-adrenoceptor density. One non-selective antagonist, tertatolol, behaves in an atypical manner in that it produced an unexpected decrease in lymphocyte β_2-adrenoceptor binding sites after chronic treatment (De Blasi *et al.* 1986). Prolonged treatment with selective β_2-agonists, such as terbutaline, induce a down-regulation of lymphocyte β_2-adrenoceptors. The value of lymphocyte β-adrenoceptor measurements in predicting airway β-adrenoceptor characteristics has been questioned since, although chronic treatment with β-agonists has been shown to reduce lymphocyte β-adrenoceptor density, no change in lung tissue was found (Hauck *et al.* 1990).

The function of the lymphocyte β-adrenoceptor is uncertain but, in view of the role of lymphocytes in the immunological response to foreign antigenic protein, this may indicate a connection between the sympathetic autonomic nervous system and the immune response. Sympathetic fibres innervate the spleen and lymph nodes, the nerve terminals lying in close proximity to the lymphocytes. The various types of lymphocytes – B cells, which mature in the bone marrow, and T cells, which mature in the thymus gland – all have β_2-adrenoceptors. The B-lymphocytes are involved in the primary immune response and react to antigen by producing specific soluble antibodies (IgE, IgG). The T cells respond to antigen on subsequent exposure by releasing lymphokines which attract phagocytes, induce macrophages to destroy antigens and stimulate B-lymphocytes to produce antibodies. T_{helper} cells activate B cells, the lymphokine responsible being interleukin 2 (IL-2). Lymphocyte function can be measured by the secretion of IL-2, their proliferation or inositol phosphate generation induced by mitogen (eg concanavalin A), antibody formation by B cells, or the cytolytic activity of T cells. β_2-Adrenoceptor agonists inhibit these functions of lymphocytes and therefore circulating adrenaline would appear to exert a modulatory effect upon the immune response (Brodde *et al.* 1991).

3.5.4 *Adipocytes*

Lipolysis, the breakdown of triglycerides to free fatty acids (FFA) and glycerol by the action of hormone-sensitive lipase, occurs in adipocytes of white (WAT) and brown adipose tissue (BAT). This enzyme catalyses the first rate-limiting step and second step of the degradation to diacylglycerol and 2-monoacylglycerol, respectively; a separate monoacylglycerol lipase causes the final release of glycerol. WAT is the fat storage tissue making up ~10% of body weight of the non-obese subject. It is mobilized during fasting and by exercise. Lipolysis is also stimulated by circulating adrenaline but release of FFA is inhibited by the accompanying vasoconstriction. This latter effect is probably due to limited availability of albumin, which is necessary for the binding of FFA (Vernon & Clegg 1985). Hormone-sensitive lipase is stimulated by those hormones that increase the levels of cAMP including noradrenaline, glucagon and ACTH. Activation is thought to be due to phosphorylation of the lipase by a cAMP-dependent protein kinase. This stimulation of lipolysis is inhibited by insulin through dephosphorylation of the enzyme, although insulin does not affect basal or non-stimulated lipolysis (Belfrage 1985).

WAT receives a sparse sympathetic innervation, the nerve terminals appearing to supply mainly the vasculature. Nerve stimulation therefore causes predominantly vasoconstriction by the action of noradrenaline on α-adrenoceptors which reduces blood flow through the tissue. β-Adrenoceptor-mediated vasodilatation also occurs but these receptors lie distant from the innervation and are probably activated by circulating adrenaline, as do the β-adrenoceptors mediating lipolysis (Fredholm 1985). Adrenaline and noradrenaline cause lipolysis in humans, as shown by an elevation of plasma glycerol (Hjemdahl & Linde 1983). Circulating adrenaline can, however, have an opposing antilipolytic effect through stimulation of α_{2A}-adrenoceptors (Tarkovács *et al.* 1994).

BAT is restricted to small interscapular and axillary fat pads. It is physiologically active during arousal from hibernation, in the newborn and during acclimatization to cold in adult humans. It is therefore not a fat storage tissue but has the function of thermogenesis – the generation of heat following lipolysis. The production of heat by BAT occurs because of the uncoupling of oxidative phosphorylation. The normal oxidation of fatty acids in mitochondria leads to oxidative phosphorylation of ADP to ATP. However, in BAT the fatty acids produced by lipolysis are uncouplers of oxidative phosphorylation so that the energy is converted to heat rather than conversion of ADP to ATP. This is probably an oversimplification and the precise mechanism for uncoupling remains uncertain. The normal activity of the respiratory chain pumps protons (H^+) from mitochondria and these may re-enter through the ATP synthetase pathway to generate ATP. However, in BAT mitochondria, an ion channel may allow outward movement of OH^- to neutralize the protons and thus inhibit ATP formation. The protein regulating this ion channel or uncoupler pathway has been termed thermogenin and it is suggested that fatty acids or acyl CoA activate this channel (Figure 3.16) (Cannon & Nedergaard 1985).

BAT is activated by cold exposure, which increases the metabolic rate (non-shivering thermogenenis, NST). Thermogenesis may also be induced by hyperphagia (diet-induced thermogenesis) in which animals are fed palatable and flavourful foods, the so-called cafeteria diet. Certain strains of animals can overeat such a diet without becoming obese because energy expenditure through thermogenesis rises to compensate for the increased energy intake. Hyperphagic or cold-adapted rats show increased BAT growth and activity. The sympathetic activity to BAT increases and there is an elevated thermogenic response to noradrenaline (Himms-Hagen 1985). This activation can be prevented by sympathetic denervation. There is good evidence that the adipocytes of BAT receive a rich innervation compared with WAT. Endotoxin (*E. coli* lipopolysaccharide) also induces fever associated with BAT thermogenesis. This response is blocked by β-adrenoceptor antagonist and by denervation of the interscapular fat pad and is therefore due to enhanced sympathetic activity. The thermogenic response to sympathetic stimulation is enhanced by thyroid hormone, an effect due to the raised sensitivity to noradrenaline rather than to a direct thermogenic response. The levels of thyroid hormone are elevated in cold-acclimatized and cafeteria-fed rats.

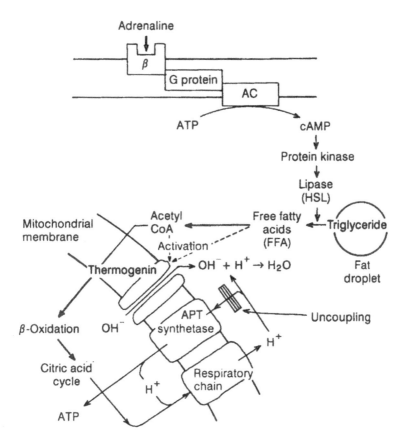

Figure 3.16 Diagrammatic depiction of the putative mechanisms for β-adrenoceptor-mediated lipolysis and thermogenesis in brown adipose tissue (BAT). Adrenaline stimulates triglyceride breakdown in the fat tissue via the $β_3$-adrenoceptor, adenylyl cyclase (AC) and hormone sensitive lipase (HSL). The free fatty acids (FFA) undergo β-oxidation in the mitochondria and ultimately H^+ passes out via the respiratory chain. Normally these would re-enter through ATP synthetase. In BAT, however, there is *uncoupling* of oxidative phosphorylation probably through combination of the H^+ ions with OH^- which are extruded from the mitochondria by the regulatory protein, thermogenin, as a result of its activation by FFA and acetyl coenzyme A (acetyl CoA).

Sympathomimetic amines induce thermogenesis. This response may be measured as an increase in metabolic rate, an increase in oxygen comsumption of the whole animal or isolated adipocytes, or an increase in BAT temperature. The increased rate of lipolysis results in loss of body weight entirely due to reduction of body lipid content. This is especially true of genetically obese rodents since there is little effect on body weight of lean mice.

Early studies showed that the sympathomimetic compound, phenylpropanolamine (norephedrine), could be used in the treatment of obesity to produce weight loss (Hirsch 1939). This compound has consequently been available in the USA as an OTC weight-reducing remedy for many years. In spite of some controversy over its efficacy, the FDA has approved the continued sale of phenylpropranolamine-containing products for weight loss. Its combination with caffeine, however, is no longer used. Unlike amphetamine, which is closely related structurally (see Table 4.1), phenylpropranolamine does not appear to exert significant appetite supression and weight loss by a central anorexic mechanism. It is virtually free of CNS stimulant activity and therefore should not be regarded as a look-alike, amphetamine-type, stimulant drug of abuse, although it may appear on the streets as such. More recent studies showed that ephedrine reduced body lipid (but not protein) in various obese mice (Massoudi & Miller 1977). Also, phenylpropranolamine has been shown to increase thermogenic energy expenditure in normal and genetically obese mice. This was associated with decreased body weight gain, decreased total body fat and reduced food intake (Arch *et al.* 1982). Phenylpropanolamine stimulates BAT thermogenesis in rats, an effect that is abolished by pretreatment with reserpine to deplete endogenous catecholamines. This suggests that the effect is due to release of noradrenaline from noradrenergic neurones (Lasagna 1988).

3.5.4.1 β-Adrenoceptor subtype – β₃-adrenoceptors

The potential for sympathomimetic amines as anti-obesity agents is therefore apparent. Characterization of the adrenoceptor subtype with a view to developing selective compounds became a viable target in the pharmaceutical industry. Lands *et al.* (1967) had classified β-adrenoceptors mediating lipolysis in rat WAT as the typical β_1-subtype and this has been supported by radioligand binding techniques (Bojanic & Nahorski 1983). Similarly, binding studies using BAT indicate a β_1-adrenoceptor subtype (Bukowiecki *et al.* 1980). However, studies on the lipolytic response with selective agonists and antagonists have failed to support this subclassification. The pA_2 values of antagonists for inhibition of agonist-induced lipolysis are generally low compared with other β_1- or β_2-adrenoceptor-mediated responses. This led to the suggestion that rat white adipocytes possess both β_1-adrenoceptors, which are detected by conventional ligand binding but are not coupled to lipolysis, and atypical β-adrenoceptors, which mediate lipolysis but are not detected by ligand binding (Bojanic *et al.* 1985). Final support for an atypical β-adrenoceptor came from the description of a series of novel agonists from the Beecham laboratories (eg BRL 37344 and BRL 28410) with selective lipolytic activity in rat white and brown adipocytes (Arch *et al.* 1984).

Subsequently, other selective agonists have been introduced by Hoffman-La Roche (Ro 16-8714 and Ro 40-2148), Eli Lilly (LY 79771 and LY 104119) and ICI (ICI 198,157 and ICI D7114) which, apart from differences in potency, appear to have common effects upon body weight, body composition and metabolic rate (see Table 4.5, Chapter 4). They reduce body weight or body weight gain and body lipid content in obese rats and mice and increase metabolic rate in obese and lean rodents and man. The reduction in weight gain of obese rats appears to be potentiated by exercise (Santti *et al.* 1994). These agents behave as selective stimulants of thermogenesis compared with their effects on heart rate (β_1) or vasodepression (β_2). Furthermore, the low potency of propranolol in blocking thermogenic responses supports the proposal that they are mediated via the atypical β-adrenoceptor located in BAT. The possibility exists that skeletal muscle is also an important site of thermogenesis induced by these agents (Arch *et al.* 1991).

The atypical β-adrenoceptor of brown adipocytes has been designated as being of the β_3-subtype. The pharmacological identification of this subtype has been supported by the cloning of a β_3-adrenoceptor by use of a homologous gene isolated from the genes for the human β_1- and β_2-adrenoceptor in the human cDNA library. When the gene coding for the β_3-adrenoceptor was introduced into Chinese hamster ovary (CHO) cells, which normally contain no β-adrenoceptors, they were shown to have ligand binding sites and to generate cAMP in a manner similar to that for lipolysis in rodent adipocytes. The β_3-adrenoceptor contains 402 amino acids and the sequence was translated from the nucleotide sequence of the gene. The extent to which this sequence coincided with that determined for the human β_1- and β_2-adrenoceptors (ie homology) was found to be 50.7 and 44.5%, respectively. Thus, clear differences existed, but the features common to other receptors coupled to guanine nucleotide binding proteins (so-called R_7G receptors) could be identified. That is, there are seven well-conserved hydrophobic segments probably arranged in transmembrane α-helices with three intracellular and three extracellular loops and an intracellular C-terminal tail (see Chapter 13). Also, the β_1- and β_2-adrenoceptors contain several Ser- and Thr-rich regions in the third intracellular loop and in the carboxy-terminal region which are the target sites for phosphorylation by protein kinase A (PKA) and β-adrenoceptor kinase (βARK). Phosphorylation of these sites is required for desensitization of the β-adrenoceptor (see Chapter 14). The β_3-adrenoceptor, however, contains no PKA sites and few Ser residues, suggesting that its regulation by desensitization may differ from that of β_1- and β_2-adrenoceptors (Emorine *et al.* 1992). The β_3-adrenoceptor identified in human adipose tissue is identical to that of mice and rats, being encoded by a common gene. Its expression in human adipose tissue decreases with age in parallel with a morphological regression of brown adipocytes (Emorine *et al.* 1994).

3.5.5   *Gastrointestinal Tract Smooth Muscle*

The relaxation and slowing of myogenic activity of the ileum by sympathetic nerve stimulation was extensively described by Finkleman (1930). The response of the rabbit isolated intestine to adrenaline was attributed to the combined activation of both α- and β-adrenoceptors (Furchgott 1960). Subsequent studies have shown that the effect of sympathomimetic amines on longitudinal contraction of the ileum involves a complex interaction with α- and β-adrenoceptors at both pre- and postjunctional sites. The dominant postjunctional response is the β-adrenoceptor-mediated direct relaxation of longitudinal smooth muscle seen equally in ileum contracted by carbachol and by potassium-induced depolarization and by transmural stimulation of parasympathetic nerve endings (Paton preparation, 1957). In the latter tissue, an additional prejunctional α_2-adrenoceptor mediated stimulation abolishes the twitch due to inhibition of transmitter release. The partial relaxation of the potassium- and carbachol-contracted ileum is due to a direct effect on postjunctional α-adrenoceptors. Finally, minor excitatory effects have been identified as a direct α-adrenoceptor-mediated contractile response to noradrenaline and as a β-adrenoceptor-mediated potentiation of transmural stimulation by isoprenaline, possibly due to facilitation of parasympathetic transmitter release (Figure 3.17) (Broadley & Grassby 1985).

   The β-adrenoceptors mediating the direct relaxation of the guinea-pig ileum have been characterized as a homogeneous population of the β_1-subtype. The Schild plots for the antagonism by practolol (β_1-selective) of the relaxation of K^+-contracted ileum by noradrenaline (β_1-selective) and fenoterol (β_2-selective) were virtually

(a)   α-Adrenoceptor responses

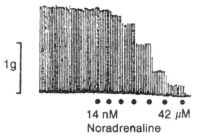

(b)   β-Adrenoceptor responses

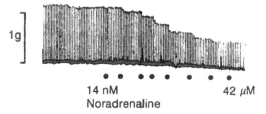

(c)   β-Adrenoceptor responses

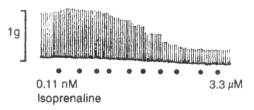

Figure 3.17 Typical concentration-dependent inhibitory responses of the transmurally-stimulated guinea-pig ileum (Paton preparation, 1957) mediated via α-adrenoceptors (a) and β-adrenoceptors (b, c). α-Adrenoceptor-mediated responses to noradrenaline were obtained in the presence of propranolol (10^{-6} M). β-Adrenoceptor-mediated responses to noradrenaline (b) and isoprenaline (c) were obtained in the presence of phentolamine (5×10^{-6} M), the inhibition being on postjunctional smooth muscle receptors. Note the increase in twitch height with low concentrations of isoprenaline which are probably due to activation of prejunctional β-adrenoceptors facilitating transmitter (Ach) release. Reproduced with permission from Broadley & Grassby (1985).

superimposed. Similarly, the antagonism of these two selective agonists by the β_2-selective antagonist, ICI 118,551, was identical. The direct relaxation is therefore mediated primarily via β_1-adrenoceptors with a small contribution from α-adrenoceptors (Grassby & Broadley 1984).

More recently, however, a further subtype of β-adrenoceptor has been identified which mediates relaxation of the gastrointestinal tract. The relaxation of histamine-contracted guinea-pig ileum by isoprenaline and noradrenaline was shown to be blocked by propranolol and nadolol but a limit to the displacement of the dose–response curves occurred. Much higher concentrations of nadolol produced further displacement. The initial blockade by nadolol was due to the antagonism of β_1-adrenoceptor-mediated relaxation, but at higher doses these agonists appear to stimulate an atypical β-adrenoceptor for which nadolol has a low affinity. Agonists already shown to be selective for the atypical or β_3-adrenoceptor of adipocytes, such as BRL 37344, also showed selectivity for this atypical β-adrenoceptor of guinea-pig ileum (Bond & Clarke 1988). Similar relaxation responses mediated via β_3- or atypical β-adrenoceptors have been observed in guinea-pig stomach (Coleman *et al.* 1987), rat small intestine (Van der Vliet *et al.* 1990) and rat oesophagus (Ford *et al.* 1992). In the rat intestine, the selective β_2-adrenoceptor agonists, BRL 37344 and ICI D7114 (now ZD 7114), become partial agonist and antagonist, respectively (Growcott *et al.* 1993). A more selective antagonist of the β_3-adrenoceptor of the gastrointestinal tract is cyanopindolol (Ford *et al.* 1992).

β-Adrenoceptor agonists, such as isoprenaline, promote acid secretion from the rat stomach. This response is antagonized by propranolol but not by selective β-adrenoceptor antagonists, suggesting that they are also mediated via an atypical β-adrenoceptor. This is supported by the fact that the response is mimicked by the selective β_3-adrenoceptor agonist, BRL 37344 (Canfield & Paraskeva 1992). This contrasts with the antisecretory effect of noradrenaline on mucosal cells of rat jejunum which is a postjunctional α_2-adrenoceptor-mediated effect (Williams *et al.* 1990).

3.5.6 *Uterus*

The uterus displays slow spontaneous contractions, the level of which varies with the stage of oestrus. The level of spontaneous activity and responsiveness to autonomic nerve stimulation, hormones and pharmacological agents increases at puberty. Catecholamines cause inhibition of uterine contraction via β-adrenoceptors and in some species (human, rabbit and rat) stimulation is induced via α-adrenoceptors. Whether an α- or β-adrenoceptor-mediated effect dominates depends upon the hormonal influences of the steroid hormones. Thus, the response can vary with the stage of the ovulatory cycle and whether the subject is pregnant. In the pregnant human uterus, the raised progesterone levels suppress spontaneous activity and sympathetic nerve stimulation causes β-adrenoceptor-mediated relaxation.

In the rabbit uterus, oestrogen domination favours the α-mediated contractile response to hypogastric nerve stimulation while, progesterone treatment favours the β-adrenoceptor-mediated relaxation and decreases the number of α-adrenoceptor binding sites. The contractile response is mediated solely via α_1-adrenoceptors, yet α_2-binding sites are also found. Under the influence of oestrogen, α-binding sites increase, with a selective effect on the α_2-subtype (Hoffman 1987, Ruffolo *et al.* 1993). The rat has a short, 4-day ovulatory cycle in which four phases can be distinguished by microscopical examination of a vaginal smear – dioestrus (mucus and leucocytes); proestrus (nucleated epithelial cells); oestrus (cornified cells); and metoestrus (degenerating cornified cells). Dioestrus is associated with progesterone domination, high spontaneous activity and relaxation responses to catecholamines. At proestrus, oestrogen levels rise, spontaneous motility is reduced, the levels of adrenaline rise and a contractile response to catecholamines occurs. From morphological evaluation, the innervation appears to be most dense in oestrus and least dense in dioestrus or after progesterone treatment and this correlates with the noradrenaline content of the tissue. In the cat, the situation is opposite; the non-pregnant or oestrogen-dominated cat uterus reveals a β-adrenoceptor-mediated relaxation, while in pregnancy or progesterone domination an α-adrenoceptor-mediated contraction occurs (Marshall 1973).

Examination of the β-adrenoceptor subpopulation of the rat uterus has revealed that after oestrogen treatment of rats both β_1- and β_2-binding sites exist and the relaxation of electrically-stimulated preparations by agonists is due to both receptor subtypes. In the progesterone pretreated rat, however, relaxation and binding is due to a homogeneous population of β_2-adrenoceptors. No change, however, has been observed throughout the natural oestrus cycle of the rat, relaxations of the K^+-contracted uterus being due to only the β_2-subtype (Johansson *et al.* 1980, Piercy 1987).

3.5.7 Liver

Adrenaline accelerates both gluconeogenesis and glycogenolysis in the liver to elevate blood glucose levels (see Figure 13.7). Hepatocytes of the liver undergo glycogenolysis primarily by the stimulation of β_2-adrenoceptors but this organ is one of the few where α_1-adrenoceptors produce the same response. In rat hepatocytes, the proportion of α_1-adrenoceptor-mediated glycogenolysis increases with age (Hoffman 1987).

3.6 α-Adrenoceptor Subclassification

3.6.1 α_1 and α_2 Subtypes

The subclassification of α-adrenoceptors was first made on the basis of their anatomical location (Langer 1977). The observation leading to this subdivision was the potentiation of radiolabelled noradrenaline release from sympathetic nerves by low concentrations of the α-adrenoceptor antagonist, phenoxybenzamine (Starke *et al.* 1971). This was attributed to the presence of prejunctional receptors, termed α_2-adrenoceptors, stimulation of which causes inhibition of transmitter release (see Chapter 6). Thus, their blockade exerts a facilitatory effect upon transmitter release. Postjunctional α-adrenoceptors mediating excitatory responses such as smooth muscle contraction were designated α_1-adrenoceptors. The distribution of α-adrenoceptor subtypes in various organs is shown in Table 3.4.

The potency order of agonists for the prejunctional inhibitory α_2-adrenoceptor was adrenaline > oxymetazoline > tramazoline > α-methylnoradrenaline > clonidine > noradrenaline > naphazoline > phenylephrine > methoxamine. The differences in agonist selectivity for the inhibition of electrically-stimulated [^{3}H]noradrenaline release from pulmonary artery segments *in vitro* (α_2) and for contractile activity on pulmonary artery strips (α_1) is illustrated by the ratios of EC_{20} values for these responses. Methoxamine and phenylephrine are clearly selective for the α_1-adrenoceptor-mediated contraction while, clonidine and α-methylnoradrenaline are

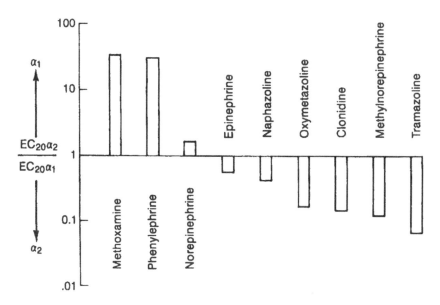

Figure 3.18 Agonist selectivities for α_1- and α_2-adrenoceptors. Selectivity is expressed as the ratio of EC_{20} values for the α_1- and α_2-adrenoceptor-mediated responses ($EC_{20}\alpha_2/EC_{20}\alpha_1$). $EC_{20}\alpha_1$ is the concentration of agonist causing 20% of maximal contraction of isolated pulmonary artery strips and $EC_{20}\alpha_2$ is the concentration of agonist causing 20% of maximal inhibition of electrically stimulated [^{3}H]noradrenaline release from pulmonary artery segments. α_1-Adrenoceptor-selective agonists have a ratio greater than unity while α_2-selective agonists have a ratio less than unity. Reproduced with permission from Berthelsen & Pettinger (1977).

selective for prejunctional α_2-adrenoceptors (Figure 3.18) (Berthelsen & Pettinger 1977). Similar conclusions were reached for the increase in heart rate of the pithed rat when the sympathetic outflow in the cardiac nerves from the spinal cord was stimulated via the pithing rod; stimulation of prejunctional α_2-adrenoceptors facilitates the tachycardia. In contrast, the pressor effects of α-agonists were used as a measure of postjunctional α_1-adrenoceptor activity. Support for this subclassification was provided with the development of more selective agonists and antagonists. Greater selectivity of agonism between the α_1- and α_2-adrenoceptor subtypes has been achieved with compounds such as guanabenz, xylazine, B-HT 920, B-HT 933 (azepexole) and UK-14,304 which are α_2-selective and amidephrine, naphazoline, oxymetazoline, St 587 and cirazoline which are α_1-selective (De Marinis *et al.* 1987) (see Table 4.3). Antagonists which display selectivity between α_1- and α_2-adrenoceptors include prazosin, doxazosin, WB 4101 and corynanthine (α_1) and yohimbine, idazoxan and rauwolscine (α_2) (see Table 5.2).

Other sites for α_2-adrenoceptor-mediated responses were soon identified from similar potency orders, in which α-methylnoradrenaline or clonidine displayed high potency. These included inhibition of melanocyte stimulating hormone-induced darkening of frog skin, inhibition of renin release from the kidney and inhibition of sympathetic outflow, as assessed by a fall in blood pressure or heart rate, when injected

143

Table 3.4 Locations of and responses mediated by α-adrenoceptor subtypes

Organ/location	Response	α_1 or α_2	Subtype Functional	Subtype Binding
Prejunctional on autonomic nerves	Inhibit noradrenaline release	α_2	Prostatic vas deferens α_{2A} Atria α_{2B} Cardio-acceleration α_{2D} Ileum α_{2A}	α_{2A} α_{2B} (Rat) α_{2C} (Opossum)
	Inhibits Ach release	α_2		
Melanocyte stimulating hormone	Inhibition of skin darkening (frog)	α_2		
Platelets	Aggregation	α_2		
Kidney				
renin release	Inhibition	α_2		
tubular epithelial cells	Na$^+$ reabsorption (antinatriuresis)	α_1		
	K$^+$, Na$^+$ and water excretion	α_2		
renal artery	Vasoconstriction	α_1	α_{1A}	
Blood vessels				
mesentery	Vasoconstriction	α_1	Rat	α_{1A}
aorta (rabbit and rat)	Vasoconstriction	α_1	Rat	α_{1D}
saphenous vein (human)	Vasoconstriction	α_1		
rabbit ear vein	Vasoconstriction	α_2	α_{2A}	
rat portal vein	Vasoconstriction	α_2		
rat tail	Contraction	α_1	α_{1A}	
	Vasoconstriction	α_2		
endothelium	EDRF release (vasodilatation)	α_2		

144

Vas deferens	Contraction	α_1	Epididymal Human	α_{1A} α_{1A}		
Prostate	Contraction	α_1	Rat Human	α_{1A} $\alpha_{1A}(\alpha_{1C})$		Bovine α_{1B}
Urinary bladder	Contraction	α_1				α_2 in base
Uterus	Contraction	α_1				$\alpha_2\uparrow$ after oestrogen
Anococcygeus muscle	Contraction	α_1				
Spleen capsule	Contraction	α_1		α_{1B}		
Heart	Increased contractility	α_1		α_{1B}		α_{1B} and α_{1A}
Gastrointestinal tract						
mucus cells	Inhibition of secretion	α_2				
fluid absorption	Increased } antidiarrhoel	α_2				
motility	Reduced }	α_2				
salivary glands	Secretory (electrolyte/water)	α_1				Prejunctional α_{2D}-like Postjunctional α_{1A}
Liver						Rat α_{1B}
Lung	Bronchoconstriction	α_1				α_{2B}
Adipose tissue	Inhibition of lipolysis	α_2		α_{2A}		α_{2A}
Pancreatic β cells	Inhibition of insulin Release (hypoinsulinaemia)	α_2	Hyperglycaemia			α_{2A}

145

into regions of the brain including the cisterna magna, fourth ventricle and nucleus tractus solitarii. Central receptors inhibitory towards salivation, ACTH release and heat conserving mechanisms may also be of the α_2-subtype. Metabolic responses involving peripheral α_2-adrenoceptors include inhibition of isoprenaline-induced lipolysis in isolated epididymal adipocytes and inhibition of insulin release from the β cells of the pancreas leading to hyperglycaemia and hypoinsulinaemia (Berthelsen & Pettinger 1977).

The subclassification of α-adrenoceptors into the α_1- and α_2-subtypes was therefore clearly unrelated to their location, with α_2-adrenoceptors not, as originally thought, restricted to the prejunctional sites where they inhibit transmitter release. Other examples of inhibitory postjunctional α_2-adrenoceptors were also identified. However, a functional basis for the subclassification is also not valid since it soon became apparent that α_2-adrenoceptors could also mediate postjunctional excitatory responses, including vasoconstriction. The α-adrenoceptor-mediated responses of the vasculature will be considered in detail later. The stimulation of platelet aggregation by adrenaline was also identified as an α_2-adrenoceptor-mediated response (Grant & Scrutton 1980).

3.6.2 Subtypes of the α_1-adrenoceptor

There have been several attempts to subdivide the α_1-adrenoceptor (reviewed by Ruffolo *et al.* 1991). A nearly 100-fold range of potency of prazosin against a series of α_1-adrenoceptor-mediated responses was suggested to indicate two subtypes of α_1-adrenoceptor. The subtype with high affinity was suggested to be directly innervated, while the lower sensitivity receptor was more distant from the nerve terminals (Medgett & Langer 1984). This led to the proposal of two subtypes, α_{1H} and α_{1L}, with high and low affinity for prazosin, respectively (Flavahan & Vanhoutte 1987). A third subtype in this subclassification is the α_{1N}-adrenoceptor which has low affinity for prazosin but has high affinity for the antagonist, HV723. More recently, this subclassification has been extended by using the antagonist, chloroethylclonidine (CEC), which selectively blocked the α_{1H}-adrenoceptor-mediated contractile response to noradrenaline of rabbit aorta (Muramatsu *et al.* 1990). The contractions of the vas deferens and rat portal vein are not, however, blocked by CEC and are therefore α_{1L}-adrenoceptor-mediated (Ohmura *et al.* 1992, Sayet *et al.* 1993). An alternative classification was proposed on the basis that α_1-adrenoceptor agonists belong to two structural groups – phenylethylamines or imidazolines (Tables 4.1 and 4.3). Phenylethylamines are generally α_1-selective while imidazolines interact with both α-adrenoceptor subtypes. It is thought that imidazolines and phenylethylamines interact with the α_1-adrenoceptor in different manners. Evidence pointing to this is the fact that cross-desensitization occurs between different imidazolines for the contractile response of the vas deferens but not between imidazolines and phenylethylamines (Ruffolo *et al.* 1977). This difference was suggested to indicate subtypes of the α_1-adrenoceptor (α_{1a} and α_{1b}) in the anococcygeus muscle of the rat or mouse. Both were sensitive to prazosin blockade, but imidazolines are supposed not to stimulate the α_{1b}-subtype (McGrath 1982). However, this idea is not supported by antagonist studies which showed no difference in their ability to block phenylethylamines (amidephrine) and imidazolines (oxymetazoline) (McGrath 1984).

Radioligand binding studies could not distinguish two subtypes with prazosin binding. However, a subdivision of the α_1-adrenoceptor was suggested by the fact that WB 4101 and phentolamine displayed two-site inhibition curves for [^{3}H]prazosin binding to rat cerebral cortex membranes (Morrow & Creese 1986). Subsequently, the irreversible alkylating antagonist, chloroethylclonidine (CEC), was shown to inactivate only half of the α_1-adrenoceptor binding sites. Membranes from rat liver and spleen were found to contain predominantly CEC-sensitive binding sites while the vas deferens was virtually CEC insensitive. The selectivity of CEC for blocking functional contractile responses of the spleen capsule but not of the vas deferens was also demonstrated. The competitive antagonist WB 4101 (see Table 5.2), which had been shown earlier to distinguish two subtypes of [^{3}H]prazosin binding sites (Morrow & Creese 1986), and its congener, benoxathian, were found to be substantially more effective in CEC-insensitive tissues. Furthermore, WB 4101 was more potent at inhibiting noradrenaline-induced contractions of the vas deferens (Han *et al.* 1987). The two subtypes of α_1-adrenoceptor are termed α_{1A} and α_{1B}. The α_{1A}-subtype has a high affinity for WB 4101 and is resistant to inactivation by CEC ($\geqslant 100$ μM), while α_{1B}-adrenoceptors have a 20-fold lower affinity for WB 4101 and are inactivated by CEC at concentrations of ~1 μM.

Subsequently, two further competitive antagonists have been described with selectivity for α_{1A}-adrenoceptors. These are 5-methylurapidil (see Table 5.2) and (+)-niguldipine which have 70- and 200-fold higher affinities, respectively, for the α_{1A}- than the α_{1B}-adrenoceptor. At present the selectivity of agonists for these two receptor subtypes is uncertain, however, (−)-dobutamine is thought to be α_{1A}-selective. The α_{1A}-subtype is also resistant to the photoaffinity of azidoprazosin, but there is no difference between the two subtypes when it is used as a reversible ligand (Minneman *et al.* 1991, Bylund 1992). It appears that the two subtypes are linked to the same second messenger, increasing levels of inositol (1,4,5) trisphosphate [Ins(1,4,5)P$_3$] and diacylglycerol, which release intracellular stored Ca^{2+} and activate protein kinase C, respectively. The two receptor subtypes may differ in their requirements for influx of extracellular Ca^{2+} in the contractile response. The contraction of the rat spleen (α_{1B}) was unaffected by the dihydropyridine Ca^{2+} channel antagonist, nifedipine, and therefore assumed to be independent of extracellular Ca^{2+}. In contrast, the contraction of the rat vas deferens to noradrenaline (α_{1A}) was blocked and therefore dependent upon the influx of Ca^{2+} through dihydropyridine Ca^{2+} channels (Minneman *et al.* 1991).

The existence of subtypes of α_1-adrenoceptors has now been confirmed by molecular cloning, with currently four subtypes described. The relationship of these cloned receptors (indicated by a lower case suffix) to the native receptors defined in functional tests, however, remains controversial (Bylund *et al.* 1994). The α_{1b}-adrenoceptor was the first to be cloned, being isolated from a hampster smooth muscle cell line (DD$_1$,MF2 cells). The amino acid sequence and structure of the receptor encoded by the cloned cDNA have been deduced. They are consistent with the seven membrane spanning model of G protein-coupled receptors (R$_7$G protein) based on the known structure of bacteriorhodopsin, that has been identified by X-ray crystallography. These structural features will be described in more detail in Chapter 13. The cloned human and hampster α_{1b}-receptors appear to be identical to the native α_{1B}-adrenoceptor (Bylund *et al.* 1994, Ford *et al.* 1994).

An α_{2A}-adrenoceptor with similar properties to the native receptor was cloned

147

from a rat cerebral cortex cDNA library. This, however, displayed certain differences from the naturally expressed α_{1A}-adrenoceptor. For example, in tissues and cell lines, oxymetazoline selectively displaces radioligand binding from α_{1A}- compared with α_{1B}-adrenoceptors. In contrast, the affinity of oxymetazoline is greater for the cloned α_{1b}- than for the cloned α_{1A}-receptor. Furthermore, the cloned α_{1A}-adrenoceptor has low affinity for the more selective α_{1A}-adrenoceptor antagonists, 5-methylurapidil and (+)-niguldipine. Consequently, this clone has been redefined as the α_{1d}- or $\alpha_{1a/d}$-adrenoceptor (Bylund *et al.* 1994). There is little evidence of a functional α_{1D}-adrenoceptor although it has recently been suggested to exist in the rat vasculature (Kenny *et al.* 1995). In rat aorta, (−)-discretamine from the Chinese plant, *Fissistigma glaucescens* (Ko *et al.* 1994), and BMY 7378 (see Table 5.2) (Kenny *et al.* 1995) have been shown to be selective antagonists of α_{1D}-adrenoceptors.

A further subtype of α_1-adrenoceptor was cloned from a bovine brain cDNA library and named the α_{1c} subtype. The pharmacological profile is distinct from either the α_{1B}- or α_{1D}-adrenoceptors, having a relatively high affinity for CEC, 5-methylurapidil and WB 4101. Recently, mRNA for this α_{1c}-adrenoceptor has been found in the human prostate and the contractile response has been attributed to receptors having the same pharmacological characteristics (Couldwell *et al.* 1993). However, the cloned α_{1c}-adrenoceptor is now regarded as being equivalent to the native functional α_{1A}-adrenoceptor (Ford *et al.* 1994). Contractions of the human prostate are therefore mediated via α_{1A}-adrenoceptors (Marshall *et al.* 1995). Thus, there are currently functional α_{1A}-, α_{1B}- and α_{1D}-adrenoceptors and cloned α_{1b}-, α_{1c}- and α_{1d}-adrenoceptors. The four subtypes of α_1-adrenoceptor have similar binding affinities with [^{3}H]prazosin, producing monophasic saturation curves. Thus, while the α_{1A} and α_{1B} subclassification does not explain the discrepancies of the variable pA_2 values for prazosin obtained in functional studies, it has now gained acceptance in both functional and radioligand binding studies. The relative merits of each subclassification system are discussed by Docherty (1989), Wilson *et al.* (1991), Bylund *et al.* (1994) and Ford *et al.* (1994). The structures of the various antagonists are shown in Table 5.2.

3.6.3 Subtypes of the α_2-adrenoceptor

The possibility that the α_2-adrenoceptor existed in more than one subtype was indicated by radioligand binding experiments (Ruffolo *et al.* 1991, Bylund 1992, Bylund *et al.* 1994). The ability of several α-adrenoceptor ligands, including oxymetazoline, ARC 239 and prazosin, to displace [^{3}H]yohimbine binding from membranes of neonatal rat lung, rat cerebral cortex and rat kidney (α_{2B}) was found to differ significantly from that observed with human platelets (α_{2A}). Oxymetazoline (in fact an α_1-selective agonist) shows a 100-fold selectivity for the α_{2A}-receptor, whereas ARC 239 (100-fold) and spiroxatrine (40-fold) show selectivity for the α_{2B}-receptor. Prazosin shows 60-fold selectivity for the α_{2B}- over the α_{2A}-receptor, but the fact that it is primarily an α_1-selective antagonist should not be forgotten.

A third subtype (α_{2C}) has been identified by radioligand binding in the opossum kidney and in an opossum kidney (OK) cell line. The selectivity of antagonists is, however, similar to that of the α_{2B}-adrenoceptor, but it may be distinguished by means of the drugs, WB 4101 and BAM 1303, which are selective for the α_{2C}-subtype. A fourth subtype (α_{2D}) has been described in the bovine pineal gland, which is characterized by a lower affinity for rauwolscine compared with the other three subtypes (Bylund 1992). This may be similar to the α_2-adrenoceptor binding sites located in the rat submaxillary salivary gland which differ from the α_{2A}- and α_{2B}-subtypes (Michel *et al.* 1989, Limberger *et al.* 1992). These probably function as prejunctional autoreceptors.

The three primary subtypes of α_2-adrenoceptor are linked to adenylyl cyclase and their stimulation results in inhibition of cAMP production. The subtype classification has been confirmed from the ability of selective antagonists to inhibit the cAMP response. No agonist selectivity has been identified for the three subtypes, with noradrenaline and adrenaline having similar affinities for each subtype. The relevance of this receptor subtyping to the pharmacological responses mediated via α_2-adrenoceptors in other tissues, such as vasoconstriction and prejunctional inhibition of transmitter release, remains to be established. For example, it has been suggested that pre- and postjunctional α_2-adrenoceptors may be differentiated by use of SKF 104078 in functional tests, this agent being selective for postjunctional sites. However, binding data has shown that SKF 104078 does not distinguish between the α_{2A}, α_{2B} or α_{2C}-subtypes (Bylund *et al.* 1991). Furthermore, the 5-HT$_{1A}$-receptor ligand, 8-OH-DPAT [8-hydroxy-2-(N-dipropylamino)tetralin], showed a greater potency as an antagonist of prejunctional α_2-adrenoceptors than the postjunctional receptors mediating contraction of the human saphenous vein. This, and the higher potency of ARC 239 (see Table 5.2) at prejunctional receptors of the right atrium, indicate that atrial prejunctional receptors are of the α_{2B}-type, while those of the saphenous vein are α_{2A}-adrenoceptors (Borton *et al.* 1991).

The prejunctional α_2-adrenoceptors of the vas deferens and submandibular salivary gland, however, have recently been classified as the α_{2A}-subtype (Smith *et al.* 1992). The opposing classification of atrial and salivary gland prejunctional receptor as the α_{2D}-subtype has also been made (Limberger *et al.* 1992). The situation at the time of writing therefore remains confusing (Ruffolo *et al.* 1993, MacKinnon *et al.* 1994).

The existence of multiple sub-types of α_2-adrenoceptors has been confirmed by the molecular cloning of the three major subtypes. The human platelet α_2-adrenoceptor was purified and sequenced; then enzymatically cleaved peptide segments were used to obtain oligonucleotide probes for screening the human genomic DNA library. The deduced amino acid sequence was consistent with the characteristic arrangements for G protein-linked membrane receptors with seven hydrophobic transmembrane domains. The gene coding for the human platelet receptor was located to chromosome 10 and the cloned receptor was described as α_{2-C10}. The amino acid sequence of the rat cloned α_2-adrenoceptor had 89% similarity with the platelet receptor. Both have pharmacological properties close to that of the naturally expressed human platelet α_2-adrenoceptor and are therefore probably similar to the α_{2A}-adrenoceptor. A cloned receptor has been derived from the human kidney cDNA library having characteristics of the α_{2C}-subtype found in opossum kidney cell line. The gene corresponding to this cDNA is located on human chromosome 4 and the cloned receptor is described as α_{2-C4}. The α_{2B}-receptor has been cloned from the rat and human and pharmacologically resembles that of the neonatal rat lung. The gene for this receptor is found on chromosome 2 and is therefore referred to as the α_{2-C2}-receptor. This receptor differs from other G protein-linked receptors in that it is not glycosylated on the extracellular amino terminal strand of the peptide (Bylund 1992).

3.6.4 *Imidazoline Binding Sites*

From the earlier discussions it is clear that α-adrenoceptor agonists of the imidazoline and phenylethylamine types can interact with α_1- and α_2-adrenoceptors in different ways. However, there is also the possibility that imidazolines can bind to a separate *imidazoline* receptor. The α_2-adrenoceptor agonist, clonidine (see Chapter 6), lowers blood pressure when injected into the medulla, a response shared by other imidazolines such as cirazoline, which are not α_2-agonists (Bousquet *et al.* 1984). The presence of imidazoline binding sites in the brain has been identified using [^{3}H]clonidine as the binding ligand. This binding is poorly displaced by the α_2-antagonists, yohimbine and rauwolscine (Ernsberger *et al.* 1987) and the endogenous catecholamines, which distinguishes it from the α_2-adrenoceptor. Two types of imidazoline receptor have been identified, an I_1 receptor is preferentially labelled by [^{3}H]clonidine and an I_2 receptor by [^{3}H]idazoxan. A high density of I_1-receptors is found in the brain and may explain the antihypertensive effect of clonidine when administered centrally (Hieble & Kolpak 1993). This imidazoline binding site has also been identified on platelets and rat liver membranes by using [^{3}H]idazoxan, the binding of which is displaced by imidazoline (naphazoline, UK 14,304) and guanidine (guanabenz, guanoxan) -type α_2-adrenoceptor agents (see Table 4.3).

A naturally occurring substance that interacts with these imidazoline receptors has been isolated and purified from calf brain and human blood, and is known as clonidine displacing substance (CDS) (Atlas & Burstein 1984). CDS has biological activity similar to that of clonidine including inhibition of adrenaline-induced platelet aggregation, relaxation of precontracted aortic rings, inhibition of electrically stimulated rat vas deferens, and a change in arterial blood pressure when injected centrally (Synetos *et al.* 1991). It may therefore be an endogenous material involved in both the central and peripheral control of blood pressure by interacting with imidazoline receptors. These receptors appear to be distinct from the α_2-adrenoceptor and clonidine may exert its antihypertensive activity by interacting with both receptor subtypes.

Further roles for imidazoline receptors are now emerging. Imidazolines, such as efaroxan (see Table 5.2), enhance the rate of insulin secretion from the islets of Langerhans, possibly through I_2 receptors (Chan *et al.* 1994). Clonidine and the selective I_1 imidazoline agonist, moxonidine (see Table 4.3), inhibit gastric acid secretion and reduce ethanol-induced gastric mucosal injury (Glavin & Smyth 1995). Moxonidine also induces sodium and water excretion when given centrally (Penner & Smyth 1994). However, peripheral administration of the α_2-agonists, rilmenidine (I_1) and guanabenz (I_2), increased blood pressure and urine excretion via α_2-adrenoceptors (Evans & Anderson 1995).

The classification of α-adrenoceptors mediating the responses of several organs and systems will be considered in the following sections.

3.6.5 Vascular α-adrenoceptors

In vivo, the pressor responses of the pithed rat to noradrenaline were first shown to be mediated by more than one postjunctional α-adrenoceptor, since only a portion of the response was antagonized by the selective α_1-adrenoceptor antagonist, prazosin. The response to phenylephrine (α_1-selective), however, was competitively antagonized by prazosin (Figure 3.19) (Drew & Whiting 1979). Subsequent studies have demonstrated the antagonism of the pressor responses to α_2-adrenoceptor agonists, such as B-HT 933, by the α_2-adrenoceptor antagonist, yohimbine, but not by prazosin (Timmermans & Van Zwieten 1980).

In isolated blood vessels, the existence of excitatory postjunctional α_2-adrenoceptors was more difficult to identify. It is most likely that only arteriolar constriction accounts for the changes in blood pressure observed *in vivo* and therefore the postjunctional α_2-adrenoceptors might occur only in resistance vessels (arterioles) and not in isolated arteries. It has become clear that although the α_1-adrenoceptor is the predominant subtype mediating vasoconstriction, there are many different vessel types that also have postjunctional α_2-adrenoceptors. For example, the femoral and saphenous veins of the dog and saphenous vein of the rabbit have a mixed population of α_1- and α_2-adrenoceptors. In the isolated saphenous vein of humans (easily obtained from the legs of patients undergoing surgery for removal of varicose veins) the constrictor response to noradrenaline is mediated predominantly via α_2-adrenoceptors. Thus, α_2-adrenoceptors also play an important role in the regulation of venous function, their activation producing a reduction in venous capacitance (Ruffolo *et al.* 1993). Forearm blood flow in humans shows significant vasoconstriction in response to both methoxamine and clonidine, indicating the presence of both α_1- and α_2-adrenoceptors. The aortae of the rat, rabbit and guinea-

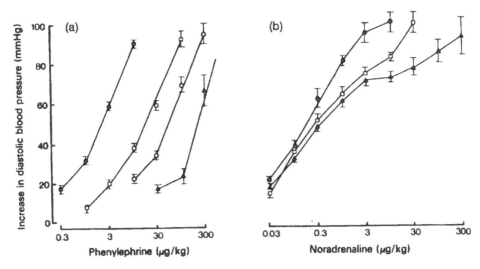

Figure 3.19 Effects of intravenous injections of the α_1-selective antagonist, prazosin (10,□; 30,○; and 100 μg/kg, ▲), on the increase in diastolic blood pressure of pithed rats to intravenous phenylephrine (a) and noradrenaline (b). Control responses in the absence of prazosin are indicated by closed circles (●). The lack of displacement of lower doses of noradrenaline indicates that it is not stimulating α_1-adrenoceptors, but probably α_2-adrenoceptors. Reproduced with permission from Drew & Whiting (1979).

pig, in contrast, contain homogeneous populations of α_1-adrenoceptors. Similarly, in the renal vascular bed α_1-adrenoceptors appear to almost exclusively mediate the vasoconstriction. However, a few vascular tissues have only α_2-adrenoceptors; these include cerebral vascular beds of the cat and dog but not humans, the digital artery of humans and the ear vein of the rabbit (Langer & Hicks 1984).

The sensitivity of blood vessels to α_1-adrenoceptor stimulation appears to vary along their length, between species and with age. Abrupt changes in sensitivity have been reported to occur at branches of the arteries. The changes in sensitivity are associated with differences in agonist affinity and receptor reserve. An increase in receptor density or in efficiency of coupling between receptor occupancy and response should increase the potency of an agonist. Thus, in tissues with low density or coupling efficiency there is generally a low receptor reserve and agonists tend to be partial agonists. With more efficient coupling, there is an increase in potency of full agonists, and partial agonists become full agonists. However, there are also examples of blood vessels where differences in potency of noradrenaline are not due to changes in receptor reserve. In these cases, the affinity differs and agonist affinities can vary by over two log units of concentration in the aortae of different species, with the rat displaying the highest. Affinity of the antagonist, prazosin, however, does not differ. This has given rise to the 'variable affinity theory' for α_1-adrenoceptor agonists to explain differences between tissue and species rather than different receptor subtypes (Bevan *et al.* 1988). The α_1-adrenoceptors of the rat aorta, however, have been proposed to be atypical α_1-adrenoceptors on the basis of their high affinity for clonidine and yohimbine (Flavahan & Vanhoutte 1987). It is now clear that there are at least two subtypes of α_1-adrenoceptor as discussed earlier (α_{1A} and α_{1B}). The rat aorta was initially shown to contain predominantly α_{1B}-adrenoceptors (Han *et al.* 1990), but more recently the contraction has been shown to be mediated via α_{1D}-adrenoceptors (Kenny *et al.* 1995).

Smooth muscle contraction mediated via α_{1B}-adrenoceptors is thought to be independent of Ca^{2+} influx through dihydropyridine-sensitive channels, since they are not blocked by the Ca^{2+} channel antagonist, nifedipine. The contractions of the rat aorta to noradrenaline are resistant to nifedipine but are blocked by CEC. However, (−)-dobutamine, which is thought to be an α_{1A}-selective agonist, is blocked by nifedipine. A possible explanation for this discrepancy is that nifedipine does not block contractions when there is a high receptor reserve for Ca^{2+} influx. However, it will block partial agonists or full agonists such as noradrenaline when a portion of the α_1-adrenoceptors has been removed by irreversible antagonists (Nichols & Ruffolo 1991). Vasoconstriction of the rat renal circulation induced by sympathetic nerve stimulation or noradrenaline is clearly mediated via α_{1A}-adrenoceptors, linked to dihydropyridine-sensitive Ca^{2+} channels (Blue *et al.* 1992). In the mesenteric artery and portal vein, a mixture of the α_{1A}- and α_{1B}-subtypes are involved (Han *et al.* 1990).

Postjunctional α_2-adrenoceptors appear in arteries of young but not older animals, whereas α_1-adrenoceptors are present in both old and young. Thus, most vessels may originally have both subtypes, but with development the α_2-adrenoceptor is only retained in specific vascular beds.

The reason why postjunctional α_2-adrenoceptors appear in mainly venous and cutaneous beds may be related to their role in temperature regulation. The α_2-adrenoceptors are more dependent upon temperature changes and become more sensitive with cooling. Thus, the superficial vessels close down more readily in the cold and restrict blood flow. This is achieved by a reflex sympathetic cutaneous vasoconstriction which redirects blood to deeper veins which correspondingly undergo cold-induced dilatation as a result of the colder blood coming from the skin (Flavahan & Vanhoutte 1987). The increase in α_2-adrenoceptor sensitivity is associated with an increase in agonist affinity and has been demonstrated by this author for several α_2-adrenoceptor-mediated responses including prejunctional α_2-adrenoceptors (Broadley & Al-Attar 1990). The α_1-adrenoceptor-mediated responses are not potentiated but in fact may be reduced when agonists with a low receptor reserve are examined. With noradrenaline, a high receptor reserve protects the depression of the α_1-adrenoceptor-mediated vasoconstriction in the cutaneous vessels by cooling. The α_2-adrenoceptor-mediated vasoconstriction, however, is enhanced, so reducing the superficial blood flow.

The enhanced sensitivity of the cutaneous circulation to cold in Raynaud's syndrome causes vasoconstriction and reduced blood flow in the extremities (Roath 1989). It is attributed to an enhanced α-adrenoceptor sensitivity. A reduced number of calcitonin gene-related peptide (CGRP)-releasing neurones has also been implicated; CGRP being a vasodilator of potential use in its treatment (Bunker *et al.* 1990). The raised α-adrenoceptor-mediated vasoconstriction may arise from an increase in receptor reserve for the α_1-response, so that the vasoconstriction is not opposed by the direct vasodilatation induced by cooling (Flavahan & Vanhoutte, 1987).

The vascular α_1- and α_2-adrenoceptors described above mediate smooth muscle contraction directly. The second messengers involved in this response and the role of Ca^{2+} will be discussed in Chapter 14. However, α-adrenoceptors are also located in the vascular endothelium where stimulation induces release of endothelium derived relaxant factor (EDRF). In isolated blood vessels with an intact endothelium and contracted with the thromboxane analogue, U46619, noradrenaline causes relaxation; whereas after removal of endothelium, noradrenaline produces a contraction (Figure 3.20) (Cocks & Angus 1983). In the presence of an intact endothelium, the normal vasoconstrictor response to noradrenaline is blunted by the simultaneous release from the endothelium of the relaxant factor. Its removal would potentiate the constrictor response. EDRF has now been identified as nitric oxide (NO), produced from L-arginine and molecular oxygen by the action of nitric oxide synthase (NOS) (Moncada *et al.* 1991). Nitric oxide probably causes smooth muscle relaxation by increasing intracellular levels of cyclic GMP (cGMP) through activation of guanylyl cyclase by binding to the haem group of this enzyme (Ignarro 1990). The effect can be blocked by various guanidino-substituted analogues of L-arginine including N^G-monomethyl-L-arginine (L-NMMA), N^G-nitro-L-arginine methyl ester (L-NAME) and nitro-L-arginine (NOARG). This inhibition of NO synthesis is prevented by L-arginine but not the inactive D-arginine. The endothelial receptors are of the α_2-subtype (Vanhoutte *et al.* 1986).

3.6.5.1 *Vascular α-adrenoceptors in pathological states*

Vascular α-adrenoceptor characteristics have been shown to alter during various pathological states. In animal models of hypertension, increased pressor responsiveness to α-adrenoceptor agonists have been reported (Yamaguchi & Kopin 1980). Increased sensitivity to noradrenaline of resistance vessels in the fingers of hypertensive patients has also been observed but this is generally regarded to be

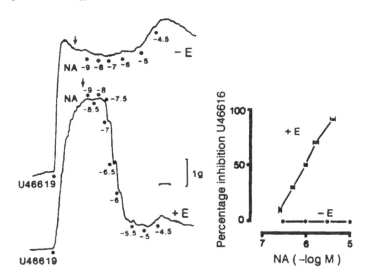

Figure 3.20 Demonstration of the dependence on vascular endothelium for the relaxation of pig circumflex coronary artery rings by noradrenaline. Rings were constricted with the thromboxane mimetic, U46619, and then noradrenaline (NA) added cumulatively. In the presence of endothelium (+E) there is concentration-related relaxation whereas in the absence of endothelium (−E) there is only further constriction at higher concentrations. Experiments were performed in the presence of propranolol to block β-adrenoceptors and prazosin to block α_1-adrenoceptors. The endothelium-dependent relaxation is therefore due to α_2-adrenoceptor stimulation of EDRF (nitric oxide) release. Reproduced with permission from Cocks & Angus (1983).

non-selective in nature since it also occurs with 5-HT and angiotensin. These changes are therefore probably a result of structural changes in the vasculature (Folkow 1982). An increase in coupling to inositol phosphate turnover (see Chapter 13) may explain the increase in sensitivity (Brodde & Michel 1992). The raised sensitivity to α-adrenoceptor agonists, combined with the fall in β-adrenoceptor responsiveness of the vasculature, may be contributing factors in the development of hypertension (Marin 1993). Stimulation of vascular α_2-adrenoceptors may take on more importance in hypertensive states, in particular where levels of circulating adrenaline are raised (Ruffolo *et al.* 1993).

α_1-adrenoceptor function of the vasculature appears to be well preserved with aging, but α_2-adrenoceptor-mediated responses are attenuated both pre- and postsynaptically. As a consequence there is reduced negative feedback on noradrenaline release resulting in increased transmitter output. This, and an impaired neuronal uptake, probably explains the raised plasma noradrenaline levels with age (Folkow & Svanborg 1993). Senescent rats display reduced α-adrenoceptor responsiveness, but this is probably a generalized phenomenon for all vasoconstrictors since responses to 5-HT are also reduced (Docherty 1986).

Vascular sympathetic responses may be altered in disease states in which autonomic nerve function deteriorates. Sympathetic autonomic failure can be associated with diabetes mellitus and with certain primary neurological diseases including Parkinson's disease, pure autonomic failure (idiopathic postural hypotension) and multiple system atrophy (Shy Drager syndrome). Autonomic disfunction may also occur in Alzheimer's disease. In patients with sympathetic failure, there is an increased vascular responsiveness to the pressor effects of α-adrenoceptor agonists and a raised density of α_2-adrenoceptor binding sites on platelets. These changes are due to sympathetic denervation equivalent to that seen after drug-induced depletion of noradrenaline stores (see Chapter 6). Thus, the baroreceptor reflex to postural changes is impaired in these patients which results in postural hypotension on standing. The raised vascular sensitivity to circulating catecholamines, however, compensates when the patient is supine and this leads to a hypertensive state (Davies 1991).

3.6.6 Vas Deferens

Contraction of the longitudinal muscle of the vas deferens has been widely used as a typical α-adrenoceptor-mediated response to characterize the innervation of the receptor and its coupling to contractile events. It is particularly useful in view of the availability of fresh human tissue subsequent to vasectomy. Stimulation of the intramural portion of the sympathetic hypogastric nerve by field stimulation or of the extrinsic nerve produces a contraction of the rat vas deferens which is biphasic when a single stimulus is applied (Figure 3.21). The initial component is resistant to blockade by α_1-adrenoceptor antagonists such as prazosin, corynanthine and WB 4101, and to noradrenaline depletion

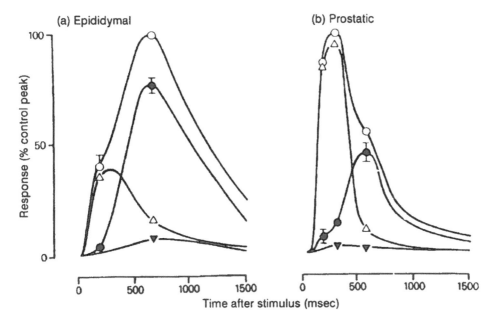

Figure 3.21 Mean time courses for the isometric contractions of rat isolated vas deferens to single supramaximal electrical stimulation (0.5 msec). The effects of the L-type Ca^{2+} channel blocker, nifedipine (●), and the α_1-adrenoceptor antagonist, corynanthine (△), alone or in combination (▼) are shown on the (a) epididymal and (b) prostatic portions of the vas deferens. Note the biphasic nature of the twitch. The first component is blocked by nifedipine and dominates in the prostatic portion. The second component is α_1-adrenoceptor-mediated and dominates in the epididymal portion. Reproduced with permission from Brown *et al.* (1983).

by reserpine and 6-OHDA pretreatment. It is therefore regarded as non-adrenergic and is attributed to the release of a co-transmitter, most likely to be ATP (see Chapter 12). This component of the response is blocked by the purine antagonist, suramin, and by receptor desensitization with α,β-methylene ATP. It is also blocked by the Ca^{2+} channel antagonist, nifedipine, and is therefore due to excitation of calcium channels and influx of Ca^{2+}. The second component of the response is blocked by antiadrenergic drugs and is therefore due to noradrenaline release onto α_1-adrenoceptors. It is not blocked by nifedipine, although the contraction to exogenously added noradrenaline *is* abolished by nifedipine. Similarly, with trains of pulses, or after prolongation of the action of noradrenaline by inhibition of uptake with cocaine, there is a third adrenergic phase which is nifedipine-sensitive. An explanation for this differential effect of nifedipine is that the rapid action of noradrenaline released by the first pulse may act on a different subtype of α_1-adrenoceptor to that responsible for the sustained effect of exogenously added noradrenaline. This different subtype would have an atypical coupling to the Ca^{2+} channel.

The noradrenergic component of the response to a single pulse predominates in the epididymal end of the rat vas deferens whereas the non-adrenergic component dominates in the prostatic portion. The epididymal end of the vas deferens yields larger contractions in response to α-adrenoceptor agonists and the response is associated with the existence of spare receptors. The epididymal end is therefore selected for evaluation of noradrenergic α_1-adrenoceptor-mediated responses. In contrast, the prostatic end is preferred for examination of prejunctional α_2-adrenoceptor-mediated inhibition of field stimulation-induced contractions. The twitch height is inhibited by the α_2-adrenoceptor agonists, clonidine and xylazine. This occurs for both the noradrenergic component obtained in epididymal vas deferens in the presence of nifedipine and for the non-adrenergic component obtained in prostatic vas deferens. This response is antagonized by the α_2-antagonists, idazoxan and yohimbine. Thus, prejunctional α_2-adrenoceptors inhibit release of both co-transmitters (Brown *et al.* 1983). These prejunctional receptors are probably involved in autoinhibition since the noradrenergic response to a second pulse is inhibited at frequencies of 0.5 Hz and this inhibition is prevented by the α_2-antagonist, yohimbine. The autoinhibition of the non-adrenergic contraction, however, requires higher frequencies. The autoinhibition of the adrenergic component, measured on the second of twin pulses, has recently been disputed, however, on the grounds that idazoxan, a highly selective α_2-antagonist, failed to restore the response (Mallard *et al.* 1992). The subtype of prejunctional α_2-adrenoceptor in the guinea-pig and rat vas deferens is still a matter of debate, but at present it appears to be of the α_{2A}-subtype since from functional studies it is poorly inhibited by urapidil (Connaughton & Docherty 1990).

The contractile response of the rat vas deferens to noradrenaline is mediated via α_1-adrenoceptors of the α_{1A}-subtype (Han *et al.* 1987), since it is unaffected by chloroethylclonidine (CEC) but sensitive to WB 4101. The receptor is probably coupled to the dihydropyridine-sensitive L-type $Ca2^+$ channel and therefore the response is mediated via influx of extracellular Ca^{2+} since the response to noradrenaline is blocked by nifedipine. The initial nifedipine-resistant transient peak of contraction due to endogenously released noradrenaline appears to be mediated via α_{1B}-adrenoceptors since it is blocked by CEC (Spriggs *et al.* 1991).

The responses of the vas deferens to added noradrenaline and methoxamine appear to differ qualitatively since the latter induces rhythmic contractions. Although methoxamine causes prostaglandin release from the vas, it is not thought to be responsible for the rhythmic contractions since they are not blocked by indomethacin at cyclo-oxygenase-inhibiting concentrations (Patra *et al.* 1990). In the human vas deferens, noradrenaline also induces rhythmic activity. However the human vas differs from that of rodents in that the innervation is mainly adrenergic (Amobi & Smith 1993).

3.6.7 *Heart*

Stimulation of sympathetic nerves to the heart induces positive inotropy and chronotropy through the release of noradrenaline primarily onto β-adrenoceptors. However, α-adrenoceptors have been identified in cardiac muscle by both radioligand binding and measurement of functional responses. The binding sites have been characterized as belonging to the α_1-subtype from the binding of [^{3}H]prazosin. α_2-Adrenoceptors are absent on the myocardial cells. The density of binding sites is dependent on the species and the region of the heart under study. The α_1-adrenoceptor binding sites are equally distributed between left and right sides of the heart, but in most species the density is greater in the ventricular than the atrial myocardium. The density of α_1-adrenoceptor binding sites is greater in the rat than in most other species, and this is reflected in a more substantial positive inotropic response to the α-agonists, phenylephrine and methoxamine, of isolated cardiac tissues from this species. Phenylephrine exerts positive inotropy via both β- and α-adrenoceptors in isolated cardiac tissues, but which receptor type is more important in a particular tissue or species depends upon the ratio of β to α densities (Chess-Williams *et al.* 1990).

The α-adrenoceptor mediating positive inotropy of the myocardium appears to be typically of the α_1-subtype. The increases in force of contraction by selective agonists such as phenylephrine are antagonized by prazosin with a pA_2 value (9.05) comparable with those for other α_1-adrenoceptor-mediated responses of the rat anococcygeus muscle (8.2–9.4) and aorta (8.83–10.6) (Williamson & Broadley 1987). More recently, the cardiac α_1-adrenoceptor mediating increases in force of contraction of rat isolated ventricular muscle by adrenaline (in the presence of propranolol) has been further subclassified as the α_{1B}-subtype. This was based on its lower susceptibility to blockade by 5-methylurapidil and WB 4101 than in the vas deferens, and effective blockade of the responses of rabbit papillary muscle by CEC. It was confirmed by the poor displacement of [^{3}H]prazosin binding by 5-methylura-pidil from rat cardiac membranes (Hanft & Gross 1989, Endoh *et al.* 1992). A minor population of α_{1B}-adrenoceptors may exert positive inotropy when α_{1A}-adrenoceptors are inhibited (Michel *et al.* 1994). The adult mouse ventricle appears to differ from other species in that phenylephrine in the presence of β-adrenoceptor blockade exerts negative inotropy mediated via α_1-adrenoceptors (probably α_{1A}) (Tanaka *et al.* 1995).

Increases in rate of contraction mediated via α-adrenoceptors have been observed in only a few species, such as the rat, where an α_1-adrenoceptor is involved. This suggests that the sinuatrial node is poorly endowed with coupled α-adrenoceptors (Broadley 1982). Both increases and decreases in rate of firing of dog Purkinje fibres have been noted and these are attributed to α_{1A}- and α_{1B}-adrenoceptors, respectively (Anyukhovsky & Rosen 1991).

There has been considerable debate with regard to the action of the endogenous catecholamines, noradrenaline and adrenaline, upon cardiac α-adrenoceptors. Normally they exert positive inotropic activity entirely through β-adrenoceptor stimulation. However, in the presence of β-blockade, adrenaline has been shown to produce positive inotropy via α-adrenoceptors in isolated tissues from several species including humans. This response is susceptible to blockade by phentolamine and prazosin. With noradrenaline, there are several reports which fail to reveal an α-adrenoceptor-mediated inotropic response of rat, rabbit and human cardiac muscle in the presence of β-blockade (Schümann 1983, Williamson & Broadley 1989). One group has observed a prazosin-sensitive response to noradrenaline by analysing the various components of each contraction cycle (Skomedal *et al.* 1985), but it remains open to dispute whether noradrenaline stimulates cardiac α-adrenoceptors.

The effects of selective α-adrenoceptor agonists on the characteristics of each contraction cycle differ from those of β-adrenoceptor stimulation (Figure 3.12) (Broadley 1982). Although both increase the height of individual contractions, α-stimulation does not shorten the time to peak tension. This contrasts with β-adrenoceptor stimulation which does shorten time to peak tension (positive klinotropy) and increase the rate of rise of tension (dT/dt). The rate of onset of the inotropic response (time to peak tension developed) is also slower for α- than β-adrenoceptor-mediated responses. The rate of relaxation is increased by β-adrenoceptor stimulation (lisotropy) but is generally regarded as being unaltered or slowed by α-adrenoceptor stimulation, although an increased relaxation rate has been demonstrated in rat papillary muscles under isotonic conditions (El Amrani *et al.* 1989). Thus, the overall duration of the contraction is shortened by β-adrenoceptor stimulation and prolonged by α-adrenoceptor stimulation. At high rates of contraction associated with β-stimulation, the improved relaxation time allows more time for complete ventricular relaxation and adequate ventricular filling. Such an action does not occur with α-adrenoceptor stimulation.

The electrophysiological effects of cardiac α-adrenoceptor stimulation also differ from the β-mediated effects. Phenylephrine prolongs refractory period of atrial muscle through α-stimulation. This contrasts with the β-adrenoceptor-mediated shortening of refractory period. The action potential duration is prolonged by α-adrenoceptor stimulation, as is the plateau (Figure 3.12). This effect has been observed with adrenaline and noradrenaline in a wide range of cardiac tissues from various species and appears to be due to a decrease in outward K^+ currents. Stimulation of α-adrenoceptors suppresses the delayed rectifier K^+ current but has no effect on voltage-activated Ca^{2+} currents. There may be a transient increase in $[Ca^{2+}]_i$ by Na^+/Ca^{2+} exchange and Ca^{2+} release from the sarcoplasmic reticulum.

The positive inotropic effect may also arise partly from increased myofibrillar Ca^{2+} sensitivity directly and by raised intracellular pH. The electrophysiological changes appear to be mediated via both α_{1A}- and α_{1B}-adrenoceptors. Finally, it appears that myocardial α-adrenoceptor stimulation may provide a negative feedback inhibition of the β-adrenoceptor-mediated stimulation (Benfey 1993).

It is worth noting that various factors have been identified which can modify cardiac α-adrenoceptor sensitivity and density. Early reports on isolated cardiac tissues showed that α-adrenoceptor-mediated responses to phenylephrine were revealed when preparations were paced at lower frequencies (0.5 Hz) and when the bath temperature was lowered to 25 °C (Broadley 1980, 1982). Indeed it had been claimed that lowering bath temperature caused a conversion of β- to α-adrenoceptor

(Kunos & Nickerson 1977) because the response to phenylephrine was antagonized by phentolamine to a greater extent at 17 than at 31 °C. This hypothesis is, however, no longer accepted since many authors have failed to confirm the antagonism at lower temperatures (eg Martinez & McNeill 1977, Benfey 1979).

3.6.7.1 Cardiac α-adrenoceptor function in diseased states

Hormonal and disease states have been shown to affect cardiac α-adrenoceptor sensitivity. Increases in myocardial α_1-adrenoceptor density, but not binding affinity, have been demonstrated in the hearts of several species following coronary artery occlusion or hypoxia in the anaesthetized animal. This appears to be due to accumulation of sarcolemmal long-chain acylcarnitines since the carnitine acyltransferase inhibitor, POCA (sodium 2-[5-(4-chlorophenyl)pentyl]-oxiran-2-carboxylate), abolishes their accumulation and the increase in α_1-adrenoceptor density in dog hypoxic ventricular myocytes. Incubation of normoxic myocytes with palmitoyl carnitine increased α_1-adrenoceptor density. The increase in receptor density induced by coronary occlusion and hypoxia is associated with an increase in sensitivity of the heart to α_1-adrenoceptor stimulation. For example, in the anaesthetized cat, an enhanced response to efferent sympathetic nerve stimulation via the left stellate nerve measured as the increase in idioventricular rate of beating was observed. Whereas the response was blocked by propranolol (β-adrenoceptor-mediated) before occlusion, after reperfusion it was susceptible to α-adrenoceptor blockade by phentolamine. Furthermore, after reperfusion, the α_1-agonist methoxamine was effective at increasing idioventricular rate (Sheridan *et al.* 1980). In guinea-pig isolated perfused hearts, methoxamine became arrhythmogenic during global ischaemia (low flow) and reperfusion. The ischaemia- and reperfusion-induced arrhythmias were prevented by catecholamine depletion with 6-OHDA, indicating a role for endogenous catecholamines. Methoxamine was then arrhythmogenic, an effect blocked by several α-adrenoceptor antagonists (Culling *et al.* 1987). Similar observations have been made by others and it is therefore suggested that α_1-adrenoceptors become important after myocardial ischaemia and reperfusion for the generation of cardiac arrhythmias. Non-selective or α_1-adrenoceptor-selective antagonists have antiarrhythmic properties. Some of this activity, however, may arise from direct electrophysiological effects not related to α-blockade, such as prolongation of action potential duration or slowing of heart rate when used at high dose levels. It is also possible that blockade of coronary vascular α-adrenoceptors causes vasodilatation and improved regional coronary blood flow to the ischaemic myocardium (Benfey 1990). Thus, α-adrenoceptor antagonists may have potential in the treatment of ischaemia- and reperfusion-induced cardiac arrhythmias. However, concommitant blockade of peripheral α_1-adrenoceptors will lower blood pressure and induce detrimental reflex tachycardia. To eliminate this disadvantage, selective antagonists for the myocardial α_1-adrenoceptor need to be found. This will depend upon identifying a different subtype of α_1-adrenoceptor in the myocardium as compared with the vasculature.

When ischaemic heart is reperfused, contractile function may remain impaired, an effect known as myocardial stunning. This may be prevented by moderate α_1-adrenoceptor stimulation through enhanced release of adenosine. However, α-agonists have not been used in myocardial stunning in man (Bolli 1992).

Whether α_1-adrenoceptor density is altered in severe human heart failure is uncertain. There are reports showing an increase (Steinfath *et al.* 1992), no change

(Böhm *et al*. 1988, Bristow *et al*. 1988), or down-regulation (Limas *et al*. 1989) of α_1-adrenoceptors. Chronic α-adrenoceptor stimulation causes cardiac hypertrophy, which may be partially inhibited by α-blockade (Benfey 1993).

Hypertension has been reported to cause decreased, increased or unchanged numbers of α_1-adrenoceptors in rat hearts, depending on the model of hypertension. In the human heart, there is little information on any changes to the small α_1-adrenoceptor population (Brodde & Michel 1992). A decline in α_1-adrenoceptor binding sites has been reported in aging rats (Gascón *et al*. 1993).

Hyperthyroid patients and animals display marked changes in cardiac rate and force of contraction. Similarities between hyperthyroidism and overactivity of the sympathetic nervous system or catecholamines have been noted and, despite much research, it remains controversial whether hyperthyroidism enhances the cardiac effect of sympathomimetic amines. Cardiac responses mediated via β-adrenoceptors predominate in the hearts of euthyroid rats or those made hyperthyroid by daily injections of triiodothyronine (T3) or thyroxine (T4) for several days. However, after thyroidectomy, either by surgery or with antithyroid drugs, such as propylthiouracil or methimazole, the balance shifts from β- to α-adrenoceptor-mediated responses. This change in sensitivity is associated with an increase in the number, but not affinity, of cardiac α-adrenoceptors. Furthermore, the ability of β-agonists to stimulate adenylyl cyclase is usually reduced in the hypothyroid states. At one time it was suggested that there was an interconversion of β-adrenoceptors to those of the α-type (Kunos *et al*. 1974). However, this has not been supported from pharmacological data, and with the amino acid sequences of receptors now known and genetically predetermined, it seems unlikely that short-term changes in thyroid function could alter this fundamental structure of the receptor. These observations on thyroid-induced changes have been confirmed in isolated myocardial cells in culture exposed to T3, where β-adrenoceptor binding sites increased and α-binding sites decreased, with no change in affinity (Kupfer *et al*. 1986). Thyroid status also appears to affect the synthesis and turnover of catecholamines. Hyperthyroidism is associated with decreased catecholamine synthesis and turnover rate in the heart (although actual levels may not change), and plasma noradrenaline levels are reduced. However, thyroid hormone does not induce supersensitivity of the β-effects of catecholamines by inhibition of their uptake, nor does it cause a decentralization type of supersensitivity (see Chapter 14) (Ishac *et al*. 1983, McNeill 1985).

Diabetes mellitus is associated with a higher than usual incidence and mortality from heart disease and with autonomic dysfunction. It arises from a total or partial failure of the β-cells of the pancreatic islets of Langerhans to synthesize or release insulin (Type I or insulin-dependent diabetes), or to a combination of a reduced secretion of insulin and decreased responsiveness to the hormone (Type II or non-insulin-dependent diabetes). Diabetes is characterized by chronic hyperglycaemia which is corrected by insulin (Type I) or oral hypoglycaemic drugs. Experimental diabetes may be induced in animals by pretreatment with streptozotocin or alloxan which destroy the insulin-secreting β-cells of the pancreas. In diabetic patients and animals, cardiac function is depressed and there is autonomic neuropathy (degeneration of sympathetic neurones) together with atherosclerosis. These changes can be partially reversed or prevented by appropriate insulin treatment. Of particular relevance to cardiac adrenoceptor function is the observation that diabetic rats display a reduced number of cardiac α-adrenoceptor [^{3}H]prazosin binding sites, yet the sensitivity of cardiac tissues to α-agonists is not always affected. The changes appear to be small and dependent upon the duration of the diabetes. Increases in α-adrenoceptor sensitivity have been observed (eg Heijnis & Van Zwieten 1992) but they occur at different time intervals (McNeill 1985). The cause of these changes is still uncertain. There appear to be several interacting factors involved in diabetes, all of which are capable of influencing adrenoceptor status of the heart and vasculature. For example, the diabetes-induced sympathetic neuropathy could result in reduced output of noradrenaline which will affect adrenoceptor function by allowing receptor up-regulation. Diabetes is also known to result in a state of hypothyroidism in man and experimental animals, with T3 and T4 levels being depressed. As noted above, this state will increase α-adrenoceptor function and binding. In the alloxan-treated rabbit, no hypothyroid state occurred and there was no change in α-adrenoceptor sensitivity. Furthermore, β-adrenoceptor sensitivity was reduced, indicating that changes in both receptor type are independent of the hypothyroid state (Grassby & McNeill 1988).

3.6.8 *Platelets*

Adrenaline induces aggregation of human platelets and release of 5-HT. This response is also produced by the α_2-adrenoceptor agonists, UK 14,304 and clonidine, the latter being a partial agonist, and is blocked by yohimbine (Figure 3.22). In other mammalian species, such as the rabbit, there is no direct aggregatory

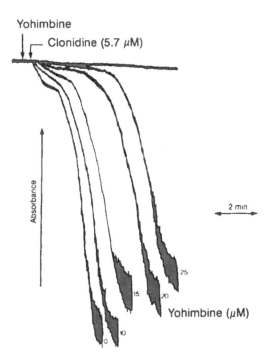

Figure 3.22 Aggregation of platelet-rich plasma prepared from a human donor induced by the α_2-adrenoceptor agonist, clonidine (5.7 μM). Aggregation is measured as a fall in light absorbance in the aggregometer. The response is delayed and inhibited by increasing concentrations of yohimbine and therefore due to activation of α_2-adrenoceptors. Reproduced with permission from Grant & Scrutton (1980).

161

response, but a proaggregatory response may be measured in rabbit platelet-rich plasma. In the presence of adrenaline or UK 14,304, the addition of calcium ionophore A-23187 or ADP, at concentrations which have no direct aggregatory response, can produce an α_2-adrenoceptor-mediated aggregation. An increase in cytosolic Ca^{2+} concentration is therefore a prerequisite for expression of the aggregatory response to α_2-agonists in the rabbit. In the human platelet, but not rabbit, there is also a small α_1-adrenoceptor-mediated component to the response to adrenaline (Grant & Scrutton 1980). Binding experiments confirm the existence of α_2-adrenoceptors on human platelets and these are of the α_{2A}-subtype (Bylund 1992).

3.6.9 Lung α-adrenoceptors

α_1-Adrenoceptors have been identified in human lungs from radioligand binding with [^{3}H]prazosin and there is an increase in patients with obstructive airway disease. Their precise location to bronchiolar smooth muscle is still uncertain since α_1-adrenoceptors are found on pulmonary blood vessels and secretory cells. Inhalation of α_1-adrenoceptor agonists causes increased airway resistance in asthmatics after β-blockade. While this response may be due to constriction of the airway smooth muscle, increased mucus secretion and release of constrictor mediators from mast cells may also be involved. Inhaled α-adrenoceptor antagonists such as prazosin do not appear to influence airway calibre and are generally ineffective in the treatment of bronchial asthma (see also Chapter 5). In spite of this apparent lack of a functional role for α-adrenoceptors in the airways, in isolated lung tissues from various species, including man and rabbits, contractile responses may be induced by α_1-adrenoceptor agonists, particularly in the presence of β-blockade. The α_2-adrenoceptor agonist, clonidine, has been shown to improve airway function when inhaled. This effect is attributed to activation of prejunctional α_2-adrenoceptors which inhibits parasympathetic bronchoconstriction mediated via release of Ach, and secondly to inhibition of inflammatory mediator release (Goldie *et al.* 1990).

3.7 Physiological Significance of Adrenoceptor Subtypes

What is the physiological reason for having these multiple subtypes of adrenoceptors? There are only two major transmitters: neuronally released noradrenaline and adrenomedullary adrenaline. Thus, one reason for having subtypes could be for preferential actions by adrenaline or noradrenaline. This may be the case for the β-adrenoceptor subtypes. It has been suggested that β_1-adrenoceptors are stimulated primarily by noradrenaline (once called β_T, for transmitter), whereas β_2-adrenoceptors are primarily stimulated by circulating adrenaline (β_H, for hormone) (Ariëns & Simonis 1983). Evidence to support this proposal includes the fact that neuronal uptake blockers, such as desmethylimipramine, potentiate β_1- but not β_2-adrenoceptor-mediated responses to noradrenaline, because the neuronal site of removal of noradrenaline is nearer to the β_1-adrenoceptor. Denervation by surgery or depletion of catecholamines with reserpine or 6-OHDA potentiates β_1- but not β_2-adrenoceptor-mediated responses (Broadley *et al.* 1986). The tissues that have predominantly β_2-adrenoceptors are known to have no or sparse innervation. Thus, the β_2-adrenoceptors are considered to be located extrasynaptically. The greater potency of noradrenaline compared with adrenaline for β_1-adrenoceptors argues for their role as neuronal receptors. In amphibians, where adrenaline is the neurotransmitter, β_2-adrenoceptors appear to serve as the neuronal receptors. Prejunctional β-adrenoceptors are of the β_2-subtype and although not extrasynaptic, are probably stimulated preferentially by circulating adrenaline, which is therefore a mechanism whereby the circulating hormone can modulate sympathetic neurotransmission (Rand *et al.* 1980). β_3-Adrenoceptors of adipocytes in BAT, at least, appear to be analogous to the β_1-adrenoceptor in being innervated (Grassby *et al.* 1987).

The situation with α-adrenoceptors may be similar since postjunctional α_2-adrenoceptors mediating responses such as vasoconstriction have been suggested to be located outside the synapse, whereas the α_1-adrenoceptors lie within the synapse. This was suggested by the fact that prazosin was more effective in blocking the responses to noradrenaline released by nerve stimulation than by exogenously added noradrenaline. Furthermore, in rats that have been adrenalectomized to remove the influence of released adrenaline which does stimulate α_2-adrenoceptors, the ganglion stimulant, DMPP (1,1-dimethyl-4-phenylpiperazine iodide), produces a pressor response that is due solely to α_1-adrenoceptor stimulation (Wilffert *et al.* 1982). The interpretation of these results is that exogenous noradrenaline stimulates both α_1- and α_2-adrenoceptor-mediated vasoconstriction, whereas transmitter released by nerve activity only stimulates α_1-adrenoceptors. The latter receptor type possibly lies at the adventitial–medial border where noradrenergic nerves terminate. The α_2-adrenoceptors probably lie near the intima and are stimulated primarily by circulating adrenaline (Langer & Hicks 1984). However, a few exceptions have been identified, such as the dog saphenous vein, where the effects of nerve stimulation and the indirectly acting sympathomimetic amine, tyramine, are blocked more effectively by the α_2-antagonist, rauwolscine, than by prazosin. Thus, innervation of this cutaneous vessel appears to involve both α_1- and α_2-adrenoceptors, as it does in human cutaneous veins. An apparent lack of antagonism by α_2-antagonists against nerve-mediated responses may be due to simultaneous blockade of the prejunctional α_2-adrenoceptors. Blockade of these inhibitory receptors would enhance transmitter release, thereby overcoming the inhibition of postjunctional effects (Flavahan & Vanhoutte 1987). Thus, the distinction between synaptic α_1- and extrasynaptic α_2-adrenoceptors remains to be firmly established. The more recent further subdivision of α_1-adrenoceptors into α_{1A}- and α_{1B}-subtypes has raised the possibility that the α_{1A}-adrenoceptor is innervated in the kidney (Blue *et al.* 1992). Other studies in the vas deferens suggest that it is the α_{1B}-subtype that is located in the synaptic cleft and stimulated only by endogenously released noradrenaline. α_{1A}-Adrenoceptors may, however, be located extrasynaptically and stimulated by exogenously added noradrenaline or that which leaks out from the cleft during trains of impulses (Spriggs et al. 1991).

4

Sympathomimetic Amines: Actions and Uses

4.1 Introduction

The previous chapter was receptor-orientated and dealt with the adrenoceptors of the sympathetic nervous system, their classification, distribution and modulation. This chapter is drug-orientated and covers the pharmacological actions and uses of agonists at these receptors: the sympathomimetic amines. The activity of a sympathomimetic at an organ or tissue is dependent upon the receptor type or types that are present and upon the selectivity and potency of the agonist for the receptor(s). These drug-related factors are determined by the chemical structure, and structure–activity relationships (SARs) will be briefly considered. The effect of structure upon the uptake (extraneuronal and neuronal) and metabolism (MAO and COMT) of sympathomimetic amines was dealt with in Chapter 2. This is relevant to the duration of action and effectiveness of the drugs in clinical use and will be considered where appropriate.

The structure of a sympathomimetic amine dictates whether it will have direct activity upon the adrenoceptor, indirect sympathomimetic activity through the release of endogenous noradrenaline, or a mixture of both types of activity. This subclassification will be considered first.

4.2 Indirectly Acting Sympathomimetic Amines (ISAs)

The first indication that sympathomimetic amines could exert their effects by different pharmacological mechanisms was the observation that cocaine potentiated the effects of adrenaline but antagonized the effects of tyramine (Tainter & Chang 1927). A similar finding was that after chronic sympathetic denervation, responses to adrenaline were potentiated whereas tyramine was antagonized (Burn & Tainter 1931). This led to the first classification of sympathomimetic amines into directly acting, neurosympathomimetic amines (later to be termed 'indirectly acting') and an intermediate group which has both direct and indirect activity (Fleckenstein & Burn 1953). Support for the indirect mechanism of action of tyramine and related drugs came from the use of reserpine to deplete noradrenaline stores (Chapter 6). After depletion of noradrenaline by reserpine pretreatment, tyramine became ineffective (Burn & Rand 1958).

In general, an amine will fall into the direct, indirect or intermediate group by each of these three procedures – denervation, cocaine- and reserpine-pretreatment. Thus, a purely indirectly acting amine will be inhibited by these procedures. Certain problems in classification, however, may be encountered. For example, cocaine only potentiates directly acting amines that are substrates for neuronal uptake; isoprenaline is neither inhibited nor potentiated by cocaine. The responses to amines with mixed activity may be unaffected by these procedures because the potentiation of the direct activity is balanced by the inhibition of the indirect activity. A complication of the use of reserpine and denervation is that with extended treatment beyond 3 days a postjunctional supersensitivity may develop (Chapter 14) due to the up-regulation of α- and β-adrenoceptors. This will enhance the direct activity independent of whether the amine is taken up. A second problem with reserpine is that of inadequate depletion so that a small indirect response remains which is mistaken for direct activity. Notwithstanding these limitations, reserpine pretreatment is generally regarded as the most reliable method for classifying sympathomimetic amines into direct, indirect or mixed action types (Trendelenburg 1972).

The structural requirements of sympathomimetic amines for direct and indirect activity are discussed below. In general, as direct activity is reduced with loss of affinity for adrenoceptors, so indirect activity becomes apparent. To induce indirect activity, the amine must be able to enter the neurone and thus have affinity for the uptake process.

4.2.1 Mechanisms of Indirectly Acting Sympathomimetic Amines

Indirectly acting sympathomimetic amines (ISAs) therefore exert their pharmacological response by release from sympathetic neurones of noradrenaline in active form so that it stimulates the α- and β-adrenoceptors. Since noradrenaline has selectivity for α- and β_1-adrenoceptors, it might be argued that ISAs have weak β_2-adrenoceptor-mediated pharmacological responses. This is not always the case since, for example, ephedrine has β_2-mediated bronchodilator activity. This can be attributed to the fact that this amine has mixed activity, its direct action being responsible for the bronchodilatation. A further aspect of the minimal β_2-adrenoceptor-mediated responses of ISAs is that in general β_2-adrenoceptors are not innervated by sympathetic nerves and are therefore present usually in non-innervated tissues (Chapter 3). Thus, in lung parenchymal strips which receive no sympathetic innervation, tyramine exerts virtually no relaxation response. In the same tissue, β-phenylethylamine, however, caused a contraction which was not α-adrenoceptor-mediated as judged by the failure of phentolamine to block it. It was therefore attributed to stimulation of a novel phenylethylamine receptor. Tyramine also has minimal relaxant effects in the uterus, although the uterus is clearly innervated by the hypogastric sympathetic nerve. The β_2-adrenoceptors are therefore not under the influence of noradrenaline released by the ISA (Hawthorn & Broadley 1984, Hawthorn *et al.* 1985).

To induce release of noradrenaline from sympathetic neurones, the ISA must enter the neurone via the Na^+-dependent uptake carrier (Figure 2.22, Chapter 2). The affinity of the amine for the carrier is facilitated by the binding of Na^+ and when both are bound, the carrier is mobile to transport the amine into the neurone. Thus, the inward transport of a sympathomimetic amine such as tyramine is inhibited by a decrease in external Na^+ or a rise in internal Na^+ by exposure to veratridine or by blocking the Na^+,K^+-ATPase-driven outward Na^+ pump with ouabain. This inward transport of the ISA via the carrier makes carrier available on the inside of the neuronal membrane. Those amines with high affinity for uptake$_1$ are the most effective releasers of [^3H]noradrenaline. The increased levels of carrier at the inside of the membrane can then transport noradrenaline outwards by facilitated exchange diffusion. Noradrenaline in pharmacologically active form is released; the levels of the major deaminated metabolite, DOPEG, are relatively low. The release is not directly from vesicles via exocytosis as occurs with a propagated action potential induced by nerve stimulation or by nicotine, but is from the cytoplasmic pool of noradrenaline. Confirmation of this is the fact that tetrodotoxin (TT_X) does not abolish the positive inotropic and chronotropic responses to tyramine, or the release of noradrenaline, in guinea-pig atria. It does, however, block the effects of low concentrations of nicotine and of electrical stimulation. The release of noradrenaline also contrasts with that of a propagated action potential in that it does not involve membrane depolarization and does not require extracellular Ca^{2+} (Sarantos-Laska *et al.* 1981, Brasch 1991). That exocytosis does not occur is also illustrated by the fact that the ISA does not release DBH. It is, however, necessary for the ISA to release noradrenaline from the vesicles since the available levels of noradrenaline in the cytoplasmic pool are relatively low. This is because although noradrenaline leaks out of the vesicles, it is avidly retaken up by the vesicle and metabolized via MAO. Once taken into the neurone, the ISA prevents the reuptake of leaked noradrenaline into the vesicles, possible by saturation of the vesicular uptake sites themselves. Since there is little deaminated metabolite released by the ISA, the release process must be of the cytoplasmic pool of noradrenaline derived from vesicles located close to the neuronal membrane.

It is thought that facilitated exchange diffusion is not the sole mechanism for release of noradrenaline by ISAs. Neuronal reuptake of noradrenaline after its release may be prevented by the amine if it has high affinity for uptake$_1$ (see Chapter 2). The co-transport of Na^+ and Cl^- into the neurone increases the affinity (reduced K_m) for uptake$_1$ and this also controls the release of noradrenaline. It is not certain whether inhibition of MAO by the ISA is involved in its releasing activity. It might be argued that if those amines that are substrates for MAO (eg tyramine, phenylephrine) could inhibit MAO by saturation, this would increase the availability of axoplasmic noradrenaline.

The mechanisms of ISA activity have been studied in tissues with intact nerve endings, but the levels of cytoplasmic noradrenaline are relatively low so that release of noradrenaline by tyramine is relatively weak. The availability of cytoplasmic noradrenaline can be increased by pretreating with an irreversible inhibitor of MAO, such as pargyline, or with reserpine and a MAO inhibitor. Reserpine impairs the reuptake into the vesicles of leaked noradrenaline. The tissues are then loaded with [^{3}H]noradrenaline which accumulates in the cytoplasm and its release by tyramine can be monitored in Ca^{2+}-free medium to inhibit any exocytosis. Reserpinization alone would of course inhibit indirect sympathomimetic activity by depletion of noradrenaline from vesicles, thus making less available for release. The major difference between the release of noradrenaline by reserpine and an ISA is that the noradrenaline is deaminated to DOPEG before release by reserpine and is therefore inactive. Also, reserpine enters the neurone by passive diffusion, not by the uptake$_1$ carrier. Thus, carrier is not provided to the inside of the neuronal membrane for outward facilitated exchange diffusion of noradrenaline. Certain drugs have both properties; these include the adrenergic neurone blocker, guanethidine (Chapter 6). This causes both an initial sympathomimetic response and with continual usage it can deplete noradrenaline (Bönisch & Trendelenburg 1988).

4.2.2 Tachyphylaxis to ISAs

A characteristic feature of ISAs is that with repeated dosage the responses gradually decline (Figure 4.1). This is probably due to a progressive exhaustion of superficial storage vesicles rather than of total noradrenaline. The noradrenaline is probably replaced in the vesicles by the ISA which, being less potent, exerts weaker effects. This tachyphylaxis is easily reversed by exposure to exogenous noradrenaline which readily reloads these stores with noradrenaline.

The tachyphylaxis to ISAs that are immune to breakdown by MAO (α-methylated amines such as (+)-amphetamine and α-methyltyramine) is rapid and slowly reversible. The rate of decline of the responses to tyramine and phenylethylamine, however, is slower and easily reversed. After inhibition of MAO, however, they behave like their α-methylated derivatives and display rapid tachyphylaxis. Support for the idea of a gradual loss of a vital pool of noradrenaline comes from the observation of cross-tachyphylaxis whereby the development of tachyphylaxis to one amine (α-methyltyramine) is associated with the reduction of the response to a

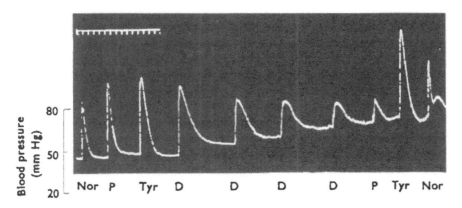

Figure 4.1 Tachyphylaxis of the pressor response of the pithed rat (280 g) to indirectly acting sympathomimetic amines. Tachyphylaxis to repeated intravenous doses of dexamphetamine (D, 25 μg) was established. There was cross-tachyphylaxis with the indirectly acting amine, phenylethylamine (P, 25 μg) but not with tyramine (Tyr, 25 μg) or the directly acting noradrenaline (Nor, 50 ng). Reproduced with permission from Day (1967).

second amine (tyramine). However, there are examples of lack of cross-tachyphylaxis. For example, the development of tachyphylaxis to the pressor effect of (+)-amphetamine is crossed to phenylephrine but not to tyramine (Figure 4.1). A possible explanation was that tyramine is converted to the more potent directly acting agonist, octopamine, by dopamine-β-hydroxylase (Figure 2.7, Chapter 2). However, the DBH inhibitor, disulphuram, failed to consistently promote cross-tachyphylaxis or selectively reduce the response to tyramine. Another possibility was that during the development of tachyphylaxis to (+)-amphetamine, this amine caused inhibition of MAO thereby potentiating the response to tyramine to apparently offset any tachyphylaxis. This theory is also discounted since inhibition of MAO in fact enhances the rate of tachyphylaxis to tyramine. It would appear that the lack of cross-tachyphylaxis is indicative of separate mechanisms for the release of noradrenaline by (+)-amphetamine and phenylethylamine on the one hand and by tyramine and α-methyltyramine on the other (Day 1967). This difference may be related to the high lipophilicity of (+)-amphetamine and the possibility of its diffusion into the neurone independently of the carrier mechanism described above.

4.3 Structure-activity Relationships (SARs) of Sympathomimetic Amines

It is beyond the scope of this book to undertake a complete analysis of all of the structural modifications that have been attempted and therefore only the more obvious structural trends are described. The chemical structure of a sympathomimetic amine may influence its activity in several ways: it will affect the proportion of direct or indirect activity, its metabolism by MAO and COMT, extraneuronal and neuronal uptake, lipophilicity and hence CNS activity, and finally the potency at α- and β-adrenoceptors. The effects upon metabolism and uptake have been considered in Chapter 2.

The structural requirements for direct sympathomimetic activity provide useful information on the drug–receptor interactions with the different adrenoceptor subtypes. The directly acting agonists have provided the basis of the receptor subclassifications described in Chapter 3. The structural requirements for direct activity at the various receptor subtypes will therefore be considered briefly.

The usefulness of pharmacological data is limited according to whether factors other than receptor activation have been eliminated, for example, uptake and metabolism differences between agonists will indirectly influence the magnitude of their responses (Chapter 2). Early studies did not consider these factors. Furthermore, SAR studies *in vivo* are limited by the possible modification of the responses via reflexes and other regulatory mechanisms. For example, blood pressure may be affected by the activation of multiple receptor types in both the vasculature (α_1, α_2, β_2) and myocardium (β_1), by haemodynamic control of the heart by pre-load and after-load changes and by other mechanisms such as release of renin from the kidney. While it is of relevance to determine how the structure of a sympathomimetic amine affects its overall activity in the intact animal or human, it does not provide any information on how the chemical structure affects the activity at a specific target receptor or organ. Radioligand binding will only yield measures of affinity and not of functional pharmacological activity and is therefore not relevant to SAR studies in this context.

The first SAR study was made by Barger & Dale (1910) and later extended by Chen *et al.* (1929). These studies examined the activity of phenylethylamines in both intact animals and isolated tissues. They concluded that the catechol nucleus was not essential but that 3,4-dihydroxy substitution on the phenyl ring increased activity substantially. The primary and secondary amines are most potent, and optimum activity occurred when the amine was separated from the phenyl ring by two carbon atoms. The β-hydroxyl group only increased activity in catecholamines. These fundamental structural requirements are still valid today but they do not distinguish between α- and β-adrenoceptor activity and generally describe the direct activity.

The requirements for indirect sympathomimetic amine activity are that the amine should have affinity for the neuronal uptake carrier and have the ability to release noradrenaline from the intraneuronal storage vesicles. The structural requirements are therefore relatively more stringent than for direct receptor activity. The phenylephrine structure is necessary and generally loss of the aromatic hydroxyl groups favours indirect activity, while it is impaired by N-substitution.

With the introduction of newer classes of sympathomimetic amines, they may conveniently be divided into two principal groups: the phenylethylamines and imidazolines.

4.3.1 Phenylethylamines

Structural requirements for the activity of phenylethylamines have been extensively studied by substitution on the aromatic ring, the α- and β-carbon atoms of the ethylamine side-chain and the amino group. The structures of a range of phenyl-ethylamines are shown in Table 4.1.

4.3.1.1 Substitution on the aromatic ring

Greatest direct activity is achieved with hydroxyl groups in the 3 and 4 positions; these are the catecholamines. When one or both of these groups are lost, there is a substantial reduction in potency and ISA is favoured. In the monohydroxy

170

Table 4.1 Structures of phenylethylamine adrenoceptor agonists

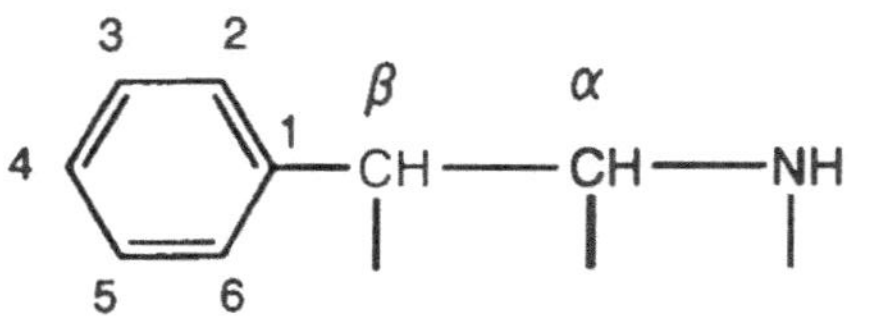

Agonist			β	α	NH	Mode of action[a]	Selectivity
β-Phenylethylamine			H	H	H	I	
Noradrenaline (norepinephrine)	3-OH	4-OH	OH	H	H	D	$\alpha\beta_1$
α-Methylnoradrenaline (cobefrine)	3-OH	4-OH	OH	CH_3	H	D	α_2
Adrenaline (epinephrine)	3-OH	4-OH	OH	H	CH_3	D	$\alpha\beta$
Dopamine	3-OH	4-OH	H	H	H	M	$\alpha\beta\,D_1,D_2$
Epinine	3-OH	4-OH	H	H	CH_3	M	$\beta_2\,D_1,D_2$
Phenylephrine (m-sympatol)	3-OH		OH	H	CH_3	D	α_1
Synephrine (p-sympatol)		4-OH	OH	H	CH_3	D	α/β
Metaraminol	3-OH		OH	CH_3	H	M	
Octopamine		4-OH	OH	H	H	M	α
p-Tyramine		4-OH	H	H	H	I	
m-Tyramine	3-OH		H	H	H	I	
Ephedrine			OH	CH_3	CH_3	M	
Phenylpropanolamine (norephedrine)			OH	CH_3	H	I	
Amphetamine			H	CH_3	H	I	
Mephentermine			H	$(CH_3)_2$	CH_3	I	
Phentermine			H	$(CH_3)_2$	H		Anorectic
Fenfluramine	3-CF_3		H	CH_3	C_2H_5		Anorectic
Diethylpropion			=O	CH_3	$(C_2H_5)_2$		Anorectic
Methoxamine	2-OCH_3	5-OCH_3	OH	CH_3	H	D	α_1
Amidephrine	5-$NHSO_2CH_3$	4-OH	OH	H	CH_3	D	α_1
Isoprenaline (isoproterenol)	3-OH	4-OH	OH	H	$CH(CH_3)_2$	D	β
Orciprenaline (metaproterenol)	3-OH	5-OH	OH	H	$CH(CH_3)_2$	D	β_2
Soterenol	3-$NHSO_2CH_3$	4-OH	OH	H	$CH(CH_3)_2$	D	β

Notes: [a] Mode of action: D, direct; I, indirect; M, mixed.
 Further β-adrenoceptor agonists are shown in Table 4.5. D_1 and D_2 dopamine receptor activity is direct.

171

substituted amines it is the *meta*-hydroxyl group that confers greatest activity, so that phenylephrine (*m*-sympatol) is more potent than synephrine (*p*-sympatol) and displays high selectivity for α_1-adrenoceptors. The *meta*-hydroxyl group probably forms a hydrogen bond with the hydroxy group on Ser-204 of the β-adrenoceptor (see Chapter 13). The loss of activity is due to a loss of affinity rather than loss of efficacy.

Loss of hydroxyl substitution or replacement with alkyl groups results in the amine being less polar, and the greater lipophilicity results in improved penetration of the blood–brain barrier. Thus, ephedrine and amphetamine are potent CNS stimulants. Loss of the hydroxyl groups in both the 3 and 4 positions confers immunity from metabolism by COMT; only catecholamines are O-methylated (Chapter 2).

Other substituents on the aromatic ring affect selectivity for α_1-, α_2- and β-adrenoceptors. Hydroxy groups in the 3 and 5 positions in compounds with large N-substituents (resorcinols) show selectivity for the β_2-adrenoceptors; examples include the bronchodilators, orciprenaline (metaproterenol) and terbutaline (O'Donnell & Wanstall 1974). A variety of substituents can replace the *meta*-hydroxyl group in amines carrying the isoprenaline-type side-chain with similar or improved β-agonist activity. For example, replacement of the *m*-OH in *t*-butyl-noradrenaline with an alkanesulphonamide group (CH_3SO_2NH-) yields soterenol, which although slightly weaker has three-fold selectivity for tracheal β_2-adrenoceptors. *m*-Substitution with other amino groups such as NH_2CONH-, $(CH_3)_2N$- and CH_3HN- also confer β_2-selectivity (Kaiser *et al.* 1974). The saligenin derivatives carry a -CH_2OH group in the *meta* position, the *t*-butylnoradrenaline derivative, salbutamol, being approximately 2000-fold selective for β_2-adrenoceptors (Brittain *et al.* 1970).

Fluoro-substitution of noradrenaline in ring positions 2, 5 or 6 affects selectivity for α_1-, α_2- and β-adrenoceptors. 6-Fluoronoradrenaline has been found to be an α_2-selective agonist; compared with noradrenaline it is three-fold more potent at α_2-, nine-fold weaker at α_1- and 100-fold weaker at β-adrenoceptors. A methoxy group in the 3 position of noradrenaline and adrenaline yields the products of COMT metabolism, normetanephrine and metanephrine, which are relatively inactive at β-adrenoceptors but retain some α-activity. They do have β-blocking activity and, as substrates of uptake$_2$, function as inhibitors of this process (Chapter 2). Methoxy substitution in the 2 and 5 positions of the aromatic ring, with the side-chain of 2-methyl noradrenaline, yields methoxamine. This has high selectivity as an α_1-adrenoceptor agonist and modest β_2-blocking activity. Finally, replacement of the catechol hydroxyl groups of noradrenaline, adrenaline and isoprenaline with chloride groups led to the first recognition of β-adrenoceptor antagonism. Dichloroisoprenaline (DCI) has a pD_2 value for relaxation of calf tracheal muscle of 6.0 compared with a value of 7.5 for isoprenaline. The pA_2 value for DCI as an antagonist was 4.9 (Ariëns 1960a). Thus, although DCI retains substantial intrinsic sympathomimetic activity, it behaves as a true partial agonist with additional antagonist properties (Powell & Slater 1958).

4.3.1.2 *Substitution on the amino group*

Increase in the size of the N-substitution increases selectivity for β-adrenoceptors over α-adrenoceptors. Thus, of the catecholamines, noradrenaline has greatest α-

selectivity, whereas β-selectivity is optimum with N-isopropyl or N-butyl substitution. Potency at the α-adrenoceptor is, however, optimum with adrenaline (N-methyl), the pD_2 values for noradrenaline, adrenaline and isoprenaline at α-adrenoceptors of the rat vas deferens being 5.1, 5.6 and 4.5, respectively (Ariëns 1960b). Phenylephrine (N-methyl substitution) is also almost entirely selective for α-adrenoceptors. The reduced α-adrenoceptor activity with increased N-substitution is due mainly to loss of efficacy rather than of affinity. The interaction of the amino group with the receptor involves both ion–ion interaction with a receptor anionic group and hydrogen bonding. This appears to be less important for the β-adrenoceptor. The presence of a methyl group would increase the charge distribution on the nitrogen and the potential for ion–ion interaction, although the nitrogen will be charged at physiological pH in a high proportion of molecules whether it is a primary (noradrenaline) or secondary amine (adrenaline). The methyl group will also impart hydrophobicity to the molecule which will favour interactions with hydrophobic pockets of the receptor. Further increases in bulk of the substitution will counteract these effects by increasing the steric interference. The presence of an amine group is essential for activity since its substitution by isosteric C or O results in inactivity. An exception to the general rule of reduced α-activity with increased size of N-substitution is dobutamine, the ($-$)-enantiomer of which is a potent α-adrenoceptor agonist with only weak β-agonist activity (see Table 4.5B). The ($+$)-enantiomer, however, has the reverse profile, namely very weak α_1-agonist activity but potent β_1- and β_2-stimulant properties. As a consequence, the racemate is a selective inotropic agent because it stimulates cardiac force of contraction via β_1- and α-adrenoceptors but only weakly affects rate because of a lack of α-adrenoceptors involved in positive chronotropic responses. Vasoconstrictor activity of the ($-$)-enantiomer is negated by the β_2-adrenoceptor-mediated vasodilatation of the ($+$)-isomer (Hieble & Ruffolo 1987).

4.3.1.3 Substitution on the α-carbon

Introduction of an α-methyl group in noradrenaline produces α-methylnoradrenaline (cobefrine) which is generally less potent. This is associated with a loss of affinity for α_1-adrenoceptors of rabbit aorta as determined by receptor occlusion with phenoxybenzamine (Chapter 3) (Besse & Furchgott 1976). This substitution creates two asymmetric centres and four stereoisomers are possible, of which only 1R, 2S-($-$)-erythro-α-methylnoradrenaline is a potent α_1-adrenoceptor agonist. 1S, 2R-($+$)-erythro-α-methylnoradrenaline has indirect sympathomimetic activity but in vas deferens from reserpine-pretreated rats is devoid of agonist activity at α_1-adrenoceptors (Patil & Jacobowitz 1968). Of the four stereoisomers of ephedrine, only 1R, 2S-($-$)-erythro-ephedrine has direct activity at α_1-adrenoceptors, the remaining isomers have indirect activity and some weak direct activity at β-adrenoceptors. This direct activity is attributed to the presence of the β-OH group (Figure 4.2).

   The α-methyl substitution in α-methylnoradrenaline also favours selectivity for α_2-adrenoceptors. At β-adrenoceptors, an α-methyl substitution increases selectivity for the β_2-subtype. This was clearly shown in the original $\beta_1 : \beta_2$ subclassification study of Lands *et al.* (1967), where the increase in size of α-substitution in isoprenaline resulted in loss of both β_1- and β_2-adrenoceptor activity, but the loss of cardiac β_1-activity was greater. The α-ethyl derivative is isoetharine (Broncometer[R]), a useful bronchodilator (Tables 4.2 and 4.5).

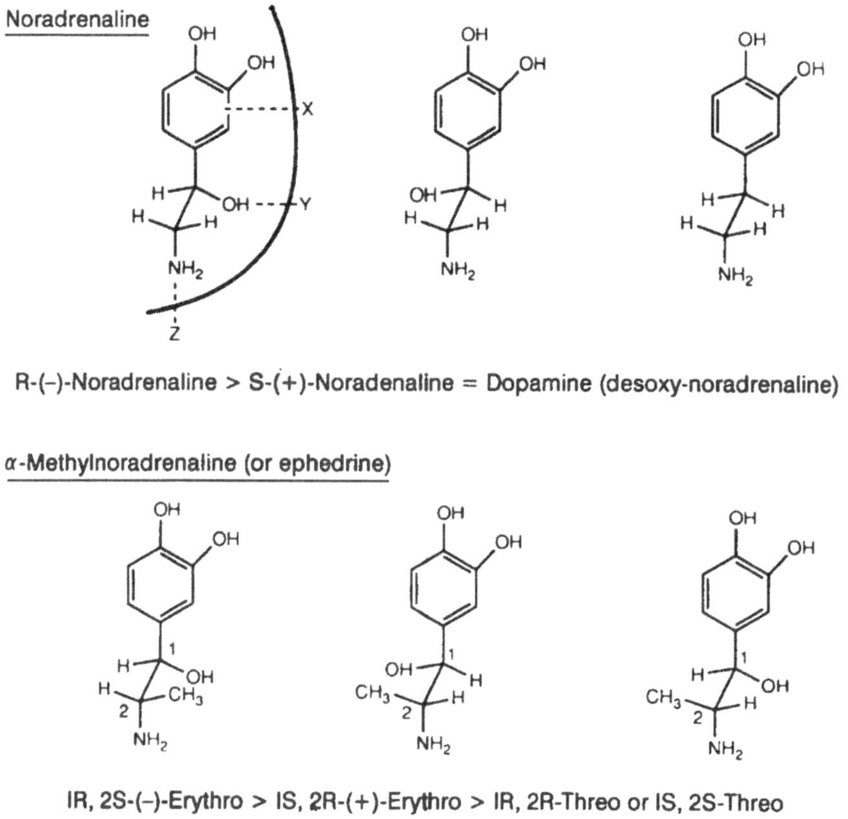

Noradrenaline

R-(–)-Noradrenaline > S-(+)-Noradenaline = Dopamine (desoxy-noradrenaline)

α-Methylnoradrenaline (or ephedrine)

IR, 2S-(–)-Erythro > IS, 2R-(+)-Erythro > IR, 2R-Threo or IS, 2S-Threo

Figure 4.2 Steric configuration of noradrenaline and α-methylnoradrenaline. According to the Easson–Stedman hypothesis, the R-(–)-enantiomer has optimum configuration for a three-point binding to the receptor (X, Y and Z). The potency order for ephedrine isomers is identical to that of α-methylnoradrenaline; ephedrine lacks the catechol hydroxyl groups and has an N-methyl group.

In non-catecholamines, where COMT is not a route of metabolism, introduction of an α-methyl group additionally imparts immunity to MAO. Thus, ephedrine and amphetamine have a much prolonged duration of action. Their resistance to MAO also improves the degree of indirect activity since they persist for longer in the adrenergic neurone.

4.3.1.4 Substitution on the β-carbon

The β-hydroxyl group is essential for direct sympathomimetic activity of the phenylethylamines. It provides a chiral centre and for optimum activity must be in an R absolute configuration (Figure 4.2). This stereospecificity of directly acting catecholamines is greater for β- than for α-adrenoceptor-mediated responses and greater for β_1- than for β_2-adrenoceptors. Easson & Stedman (1933) provided an early description of the stereospecificity of binding of phenylethylamines to the receptor according to a three-point interaction involving the β-OH, catechol-OH groups and amine group. These requirements are only met with the R(–)-enanti-

174

Table 4.2 Effects of α-methyl substitution of catecholamines on $\beta_1 : \beta_2$-adrenoceptor sensitivity

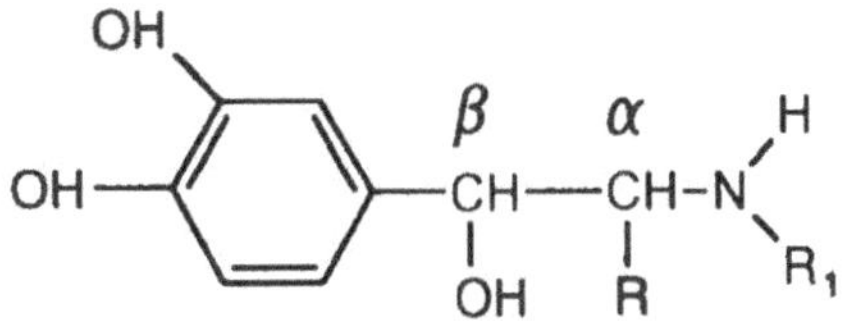

Activity relative to isoprenaline (1000)

	R	R₁	Lipolytic[a] (β_3)	Cardiac[b] (β_1)	Bronchodilator[c] (β_2)	Vasodepressor[d] (β_2)
(−)-Isoprenaline	H	CH(CH₃)₂	1000	1000	1000	1000
	CH₃	CH(CH₃)₂	28	12	112	133
(±)-Isoetharine	C₂H₅	CH(CH₃)₂	4	3	115	67
	H	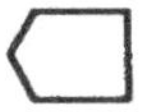 ⬠	214	208	350	200
	CH₃	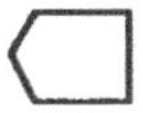 ⬠	68	16	166	61
	C₂H₅	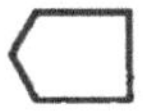 ⬠	6	7	206	80

Notes: [a] Minced testicular adipose tissue.
[b] Rabbit isolated perfused heart.
[c] Guinea-pig perfused lung.
[d] Fall in blood pressure of anaesthetized dog.
Note the greater loss of cardiac β_1 potency relative to isoprenaline with α-methyl substitution compared with loss of β_2 bronchodilator potency. These compounds become β_2-selective. Data from Lands *et al.* (1967).

omers which are therefore more active. When these conditions are met, the molecule has a hydrophobic side and a hydrophilic side, the β-OH being on the same (hydrophilic) side as the *meta*-phenolic OH group of the aromatic ring. In the S(+)-enantiomer, the β-OH group is on the opposite side of the molecule to the *m*-OH and activity is less, but of similar magnitude to that of the desoxy derivative (eg dopamine compared with noradrenaline). Thus, the Easson–Steadman hypothesis has been found to be valid for interactions of phenylethylamines with α_1-, α_2-, β_1- and β_2-adrenoceptors. The β-OH is essential for the binding and functional affinity of agonists but does not contribute to efficacy, of α_1-adrenoceptor activation at least (Ruffolo *et al.* 1983).

The preferred conformation of R(−)-noradrenaline is the extended *trans* conformation shown in Figure 4.2, in which the amino and phenyl groups are at a dihedral angle of 180 °C to one another (Timmermans *et al.* 1990).

4.3.2 *Imidazolines*

The group of compounds known as the imidazolines have an imidazoline ring usually linked by either a single carbon or nitrogen bridge to a phenyl ring. These compounds have little or no β-adrenoceptor activity and, because they are distinct from the phenylethylamines, they have no indirect activity. Their selectivities are therefore between the subtypes of α-adrenoceptors. The structures and selectivities of a range of imidazoline α-adrenoceptor agonists are shown in Table 4.3.

Tolazoline (formerly available as Priscol^R) has the basic structure of an imidazoline ring linked to a benzene ring by a carbon bridge and is a weak partial agonist at α_1-adrenoceptors and non-selective antagonist at both α_1- and α_2-adrenoceptors (pD_2 6.59 and maximum relative to a full agonist of 0.16 on rat aorta). Substitution of a phenyl hydroxyl group increases agonist activity at α_1-adrenoceptors, with the 3-OH (*meta*) substitution giving greater potency than the 4-OH (*para*) but 3,4-dihydroxy substitution giving optimum potency and full agonist activity. Thus, hydroxyl substitution of imidazolines appears to have comparable effects upon potency as with the phenylethylamines. However, in the case of the imidazolines, the increase in potency is associated with increase in efficacy rather than affinity. Methoxysubstitution of the benzene ring also affects potency and selectivity between α_1- and α_2-adrenoceptors. In contrast to the hydroxy substitution, the 3,4-methoxy derivative of tolazoline is a weaker α_1-agonist than tolazoline, but has α_2-adrenoceptor antagonist activity. Dimethoxy substitution in the 2,5- or 3,5-positions, however, produces potent highly selective α_1-agonists. This is analagous to the phenylethylamines, where methoxamine is a 2,5-methoxy derivative (Table 4.1). The 2,3-dimethoxy derivative of tolazoline is a potent α_2-adrenoceptor agonist.

The nature of the bridge between the phenyl and imidazoline rings does not appear to affect activity; substitution of the methylene group in 3,4-dihydroxytolazoline with a nitrogen does not affect potency. However, increasing the chain length substantially reduces activity. No clear trend emerges with regard to selectivity between α_1- and α_2-adrenoceptors since imidazolines with either nitrogen or carbon bridges have activity at both receptor subtypes. Although clonidine is the prototype α_2-selective agonist of the imidazoline group, and has the nitrogen bridge, structural variations of this compound yield both α_1- (St 587) and α_2-adrenoceptor selectivity (UK 14,304, xylazine).

Substitution on the methylene bridge produces asymmetry and the opportunity to examine the effects of the optical isomers. Both enantiomers of hydroxyl substituted imidazolines are *less* potent α_1-adrenoceptor agonists than is the desoxy (non-substituted) compound. This is in direct contrast to the situation with phenylethylamines where the order of potency is $R(-) > S(+) = $ desoxy (Figure 4.2). Thus, the Easson–Stedman hypothesis of a three-point attachment of phenylethylamines to the adrenoceptor does not apply to the imidazolines. Their interaction with the α_1-adrenoceptor appears to involve only a two-point attachment. Interestingly, the clonidine derivative, chloroethylclonidine (CEC), has been used to characterize α_{1B}-adrenoceptors at which it is an irreversible antagonist. However, it is also an irreversible agonist at prejunctional α_2-adrenoceptors (Bültmann & Starke 1993).

Aminotetralines are also selective directly acting α_1-adrenoceptor agonists (eg SK&F 89748). Two enantiomers of this compound exist and, in common with the imidazolines, the stereoselectivity is poor. Other similarities with the imidazolines with regard to substitution patterns suggest that the aminotetralines also interact with the α_1-adrenoceptor by a two-point attachment (De Marinis *et al.* 1987).

Table 4.3 Imidazolines and related α-adrenoceptor agonists

Agonist	Structure	Selectivity	(Imidazole binding)
Tolazoline	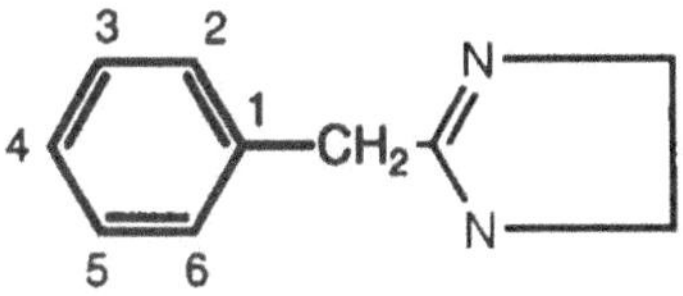	α_1 (Weak partial agonist)	
3,4-Dihydroxy-tolazoline		α_1 Full agonist	
2,5- and 3,5-Meth-oxytolazoline		α_1 Full aconist	
Naphazoline	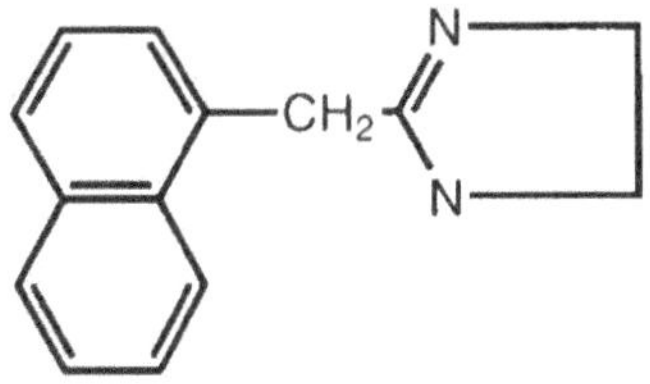	α_1	
Cirazoline	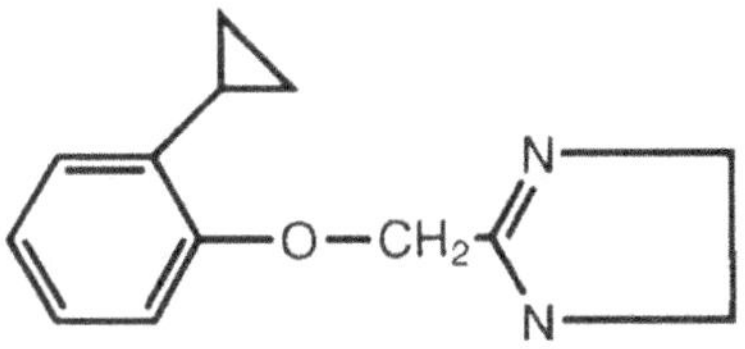	α_1	
Oxymetazoline	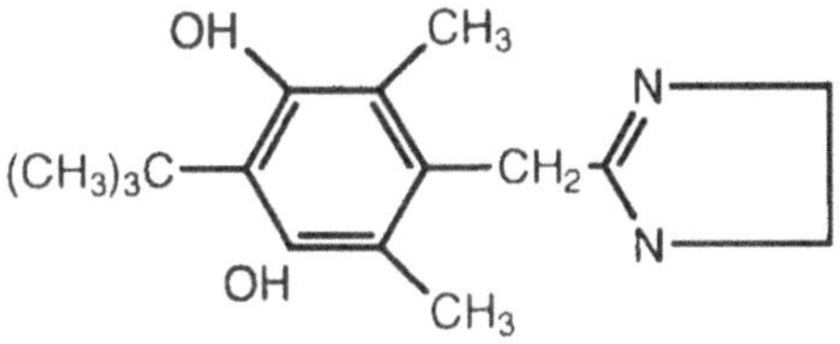	α_1	
Xylometazoline	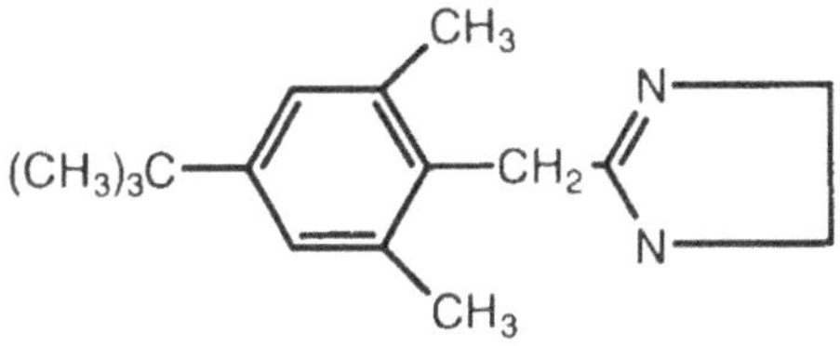	α_1	
Lofexidine (Lofetensin[R]) (Britlofex[R])	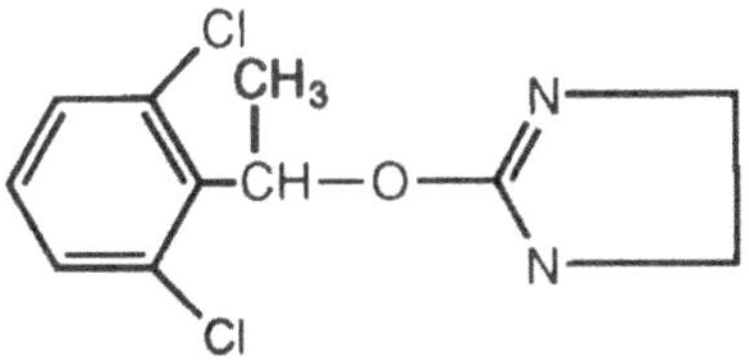	α_2	
Detomidine (Domosedan[R])	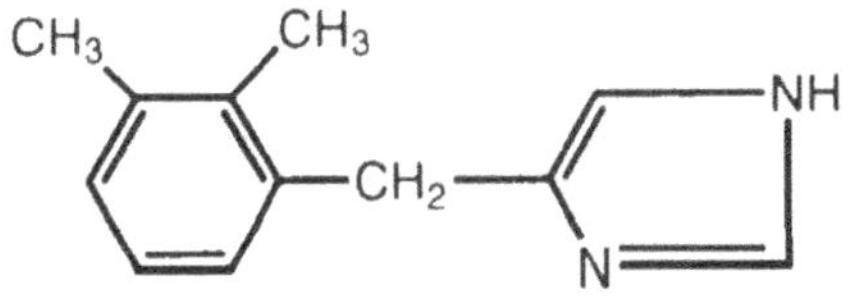	α_2(Sedative)	

(continued)

Table 4.3 *Continued*

Agonist	Structure	Selectivity	(Imidazole binding)
Dexmedetomidine (Domitor^R)		α_2	
Clonidine (Catapres^R)		α_2	(I$_1$)
UK 14,304		Highly α_2	
Moxonidine		α_2	(I$_1$)
Rilmenidine (Hyperium^{Fr})		α_2	(I$_1$)
Xylazine		α_2	
St 587		α_1	

(continued)

Table 4.3 *Continued*

Agonist	Structure	Selectivity	(Imidazole binding)
Sgd 101/75 (Indanidine)	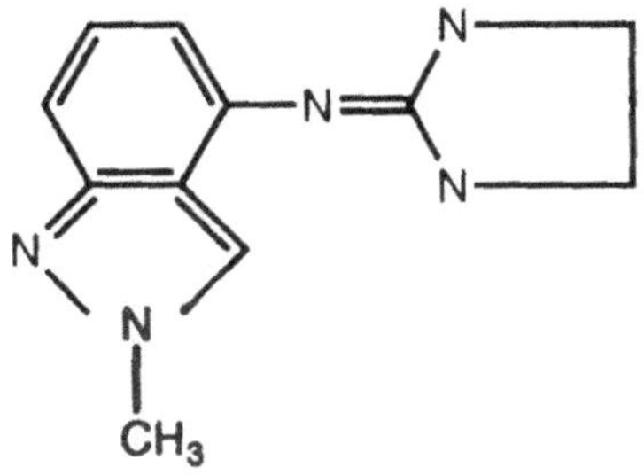	α_1	
B-HT 920 (Alefexole) B-HT 933 (Azepexole) B-HT 958	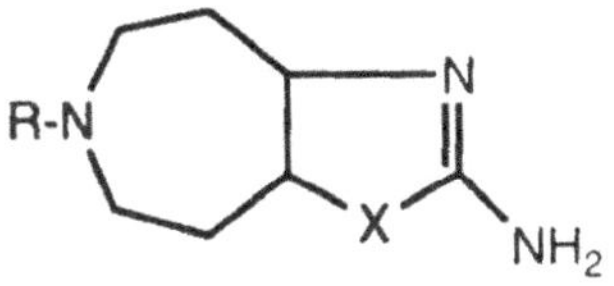 R CH$_2$=CH-CH$_2$- C$_2$H$_5$- CI-C$_6$H$_4$-CH$_2$- X S O S	α_2	
Guanabenz	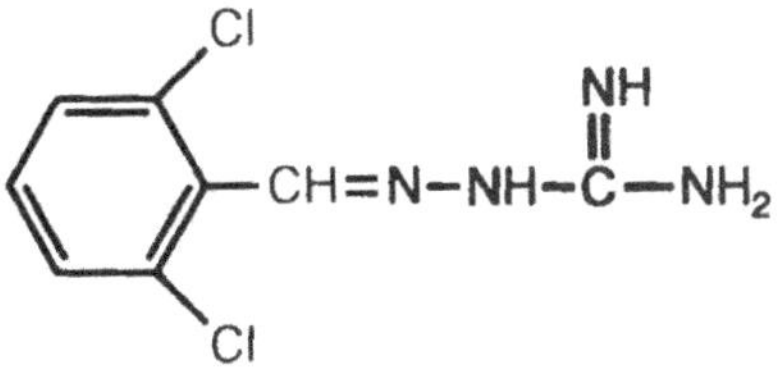	α_2	(I$_1$)
Guanfacin	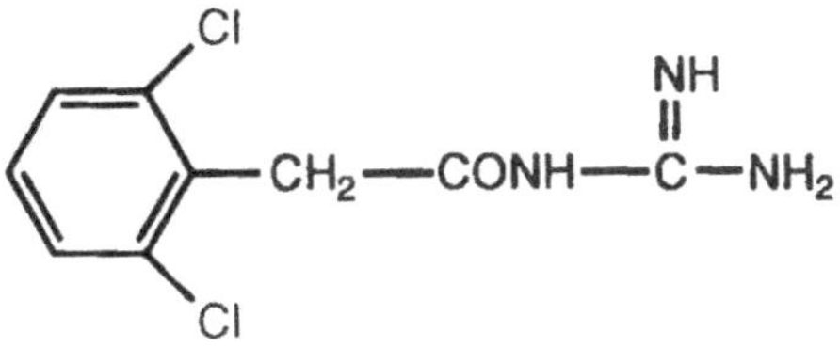	α_2	
Aminotetralines SK&F 89748	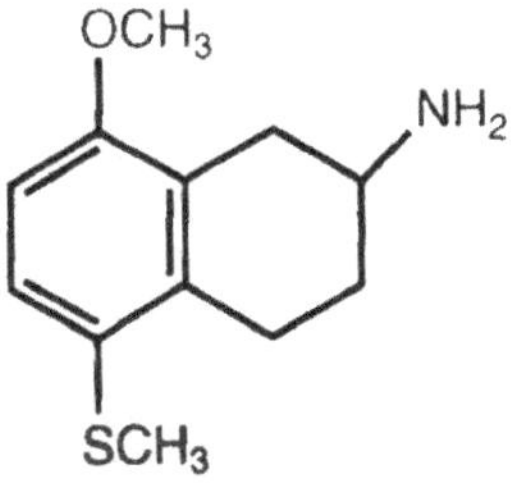	α_1	
8-OH-DPAT (8-Hydroxy-2-dipropyl-aminotetralin)	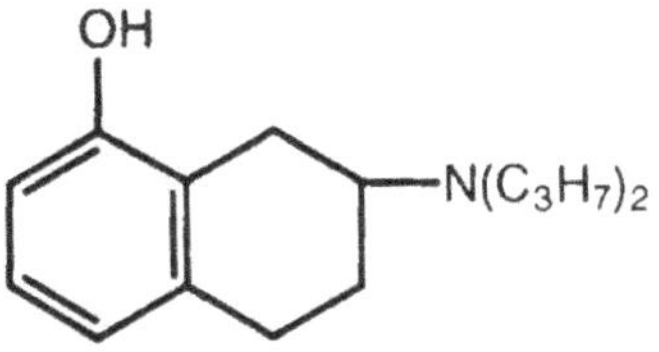	α_2-Antagonist	

4.4 Pharmacological Properties of Sympathomimetic Amines

The pharmacological responses to sympathomimetic amines are generally the same as are the physiological responses to sympathetic nerve stimulation (Chapter 1, Table 1.1). However, pharmacological responses may be obtained in non- or sparsely innervated tissues. Furthermore, multiple receptor types are present and these often exert opposing responses. The response profile of any sympathomimetic amine will therefore be dependent upon its selectivity for the dominant receptor type. An overview of the pharmacological properties of a non-selective sympathomimetic amine such as adrenaline is therefore provided before consideration of the actions and uses of specific groups of agents.

4.4.1 Cardiovascular Responses

Adrenaline, administered intravenously, increases systolic blood pressure due to α-adrenoceptor-mediated vasoconstriction in skin, mucosal and visceral (mesenteric) arterioles and to increased cardiac output due to the β_1-adrenoceptor-mediated positive inotropic and chronotropic responses. Venules are also constricted. The increase in diastolic blood pressure is smaller or it may even fall due to opposing β_2-adrenoceptor-mediated vasodilatation in skeletal muscle resistance vessels. Indeed, at low doses only a fall in pressure may be observed. In some cases biphasic responses may occur showing an intermediate fall in pressure separating an initial increase (cardiac output) and more sustained secondary rise in blood pressure (α-vasoconstriction) (Figure 4.3). Thus, blood flow to the skin is impaired while that to skeletal muscle is improved. The skin blanches after local application, as does the conjunctiva of the eye. Blood flow to mucous membranes, such as the nasopharyngeal mucosa, is reduced, the thickness of which is also diminished. After local application of adrenaline, only a small proportion is absorbed because of the reduced blood flow. Blood flow in the kidneys is impaired through α_{1A}-adrenoceptor-mediated vasoconstriction at doses of adrenaline that have little effect on mean arterial blood pressure. Excretion of Na^+, K^+ and Cl^- in the urine is lowered but effects on urine output are variable. Blood flow to the liver increases together with a decrease in splanchnic vascular resistance. This is accompanied by an increase in hepatic glucose output and oxygen consumption. Coronary blood flow increases even at doses that have no effect upon arterial blood pressure. This is primarily because of metabolic autoregulation due to increased myocardial activity which creates a relative hypoxia, and consequent release of vasodilator mediators such as adenosine. A small direct β_2-adrenoceptor-mediated coronary vasodilatation also contributes to the increase in blood flow. Opposing these effects are a reduced blood flow due to extravascular compression from the raised cardiac contractility and a direct α-adrenoceptor-mediated vasoconstriction (see coronary circulation, Chapter 1). Venous constriction also occurs and contributes to the raised blood pressure. Precapillary sphincters also contract. Pulmonary venous and arterial pressures increase, and large doses of adrenaline cause death by pulmonary oedema through the elevated pulmonary capillary filtration pressure.

   The *cardiac effects* of adrenaline consist of β_1-adrenoceptor-mediated increases in rate and force of contraction.There is an increase in cardiac output and oxygen consumption. The electrocardiogram (ECG) shows T-wave reduction and even inversion at higher doses indicative of myocardial ischaemia. With higher doses of adrenaline, premature ventricular systole may occur and there is a risk of ventricular arrythmias. The enhanced cardiac automaticity and precipitation of arrhythmias is antagonized by β-adrenoceptor antagonists. Certain general anaesthetics (halogenated hydrocarbons such as halothane, but not cyclopropane and isoflurane) may sensitize the heart to the arrythmogenic effects of exogenously administered or endogenous adrenaline. Endogenous adrenaline may be released during surgery because of inadequate ventilation or build up of arterial pCO_2. These arrythmias may be prevented during anaesthesia by the use of α- rather than β-adrenoceptor antagonists. This action is probably because they prevent the rise in blood pressure which may be involved in the sensitization of the myocardium to these arrhythmias. The rise in blood pressure also triggers reflex vagal slowing of the heart (Figure 4.3). The cardiac effects of α- and β-adrenoceptor stimulation are described in more detail in the relevant section of Chapter 3.

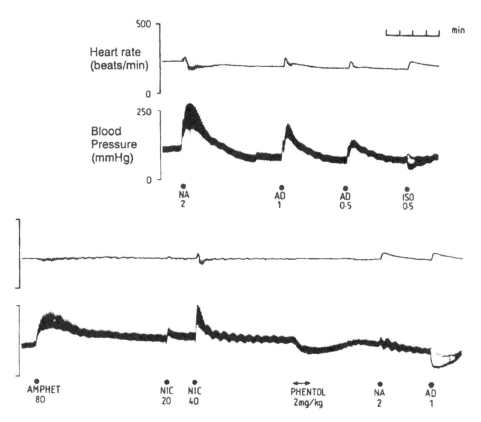

Figure 4.3 Effects of sympathomimetic amines on the heart rate and blood pressure of a cat (3.4 kg) anaesthetized with chloralose (80 mg/kg iv) after induction with halothane: nitrous oxide. Noradrenaline (NA), adrenaline (AD), isoprenaline (ISO), amphetamine (AMPHET) and nicotine (NIC) were administered as single intravenous bolus doses (micrograms/ kilogram). The α-adrenoceptor antagonist, phentolamine, given as an infusion, converted the pressor responses of NA and AD to a substantially reduced pressor response and a depressor response, respectively. It also abolished the reflex bradycardia seen with NA.

Excessively high intravenous doses of adrenaline (0.5–1.5 mg) cause ventricular tachycardia or fibrillation, cerebral haemorrhage due to the high blood pressure and pulmonary oedema.

Noradrenaline, having weaker β_2-adrenoceptor activity, causes both systolic and diastolic pressures to increase. This rise in pressure induces baroreceptor-mediated vagal slowing of the heart which outweighs the direct β_1-adrenoceptor-mediated cardiac stimulation (Figure 4.3). Thus, cardiac output may fall and stroke volume increases. Blood flow in most vascular beds (hepatic, renal and skeletal muscle) is reduced due to arteriolar and venous constriction. α-Adrenoceptor blockade does not therefore expose a vasodilator component but merely reduces the pressor response (Figure 4.3).

Isoprenaline has predominant β-adrenoceptor activity and reduces diastolic blood pressure due to a fall in vascular resistance primarily in skeletal muscle vascular beds but also in renal and mesenteric beds. Systolic blood pressure may remain unchanged or even rise because of the substantial increases in rate and force of cardiac contractions mediated via β_1-adrenoceptors (Figure 4.3, see also Figure 3.6). These may lead to palpitations and arrhythmias. Large doses have been shown to cause cardiac necrosis after infusions into experimental animals, similar to the lesions seen in myocardial infarction and in patients with phaeochromocytoma.

4.4.2   Other Smooth Muscle

Adrenaline, isoprenaline and other β-adrenoceptor agonists are bronchodilators. They relieve the bronchospasm of asthma and that induced by spasmogens through a physiological antagonism. This improves the airway exchange of CO_2 and O_2 to restore to normal the lowered pO_2 and pCO_2 that occurs in severe asthma attacks. There may be little effect in the absence of bronchoconstriction when smooth muscle tone is low. Gastrointestinal smooth muscle (α and β_1) is relaxed by adrenaline and isoprenaline, as is the uterus, in particular during the last month of pregnancy (β_2). The detrusor muscle of the bladder is relaxed (β) while the trigone and sphincter muscles contract (α) leading to urinary retention.

4.4.3   Metabolic and Other Effects

Adrenaline causes hyperglycaemia by stimulation of glycogenolysis in the liver (β_2) (see Figure 13.7) associated with a predominant inhibition of insulin release through α_2-adrenoceptors. Lipolysis is stimulated (β_3) to increase levels of FFA in the plasma and to exert a thermogenic effect which raises metabolic rate. On the eye, instillation of adrenaline inconsistently causes mydriasis (pupil dilatation) through contraction of the radial muscles of the iris. Sympathomimetic amines that cross the cornea cause mydriasis and reduce intraocular pressure. This action is probably due to vasoconstriction (α) reducing the production of aqueous humour from the ciliary body and improved drainage (Grant 1969).

In skeletal muscle, adrenaline facilitates transmitter release from somatic nerves through α-adrenoceptors, an effect probably due to increased Ca^{2+} influx into the nerve terminal. This contrasts with the effect of prejunctional α-adrenoceptors on autonomic nerves, activation of which causes *inhibition* of transmitter release.

Adrenaline has little effect upon normally contracting skeletal muscle but restores the contractions of muscles that have been fatigued by repetitive stimulation. This transmission failure is probably the result of hypoxia and accumulation of extracellular K^+ due to depolarization of the nerve terminal. This property of adrenaline may explain the modest improvement in muscle strength of myasthenia gravis patients. It is also responsible for the reversal of neuromuscular paralysis by non-depolarizing (competitive) blockers such as tubocurarine (Bowman 1981).

The β_2-adrenoceptor stimulating properties of adrenaline also cause shortening of the active state of the red, slow-contracting skeletal muscle fibres which are rich in myoglobin. These are often termed intermediate fibres and are dominant in such muscles as the soleus muscle of mammalian species. They are involved in low-speed sustained activities which maintain posture. The active state of skeletal muscle is the tension developed by the contractile component of the muscle due to myofilament cross-bridging (Chapter 1). This shortening of the contractile elements stretches the series elastic components of the muscle before the whole muscle can contract. This takes time, and with single muscle twitches, the cross-bridging between myofilaments is too brief to allow full development of whole muscle contraction. Only with repeated stimulation during tetanic pulses can maximum contractions be achieved. The active state development is probably incomplete because the short duration of the action potential does not permit optimum levels of activation of contractile filaments by intracellular Ca^{2+}. Thus, increased sequestering of Ca^{2+} by the sarcoplasmic reticulum will reduce levels of intracellular Ca^{2+} and shorten the active state further. This appears to be the action of β_2-adrenoceptor stimulation. At physiological rates of nerve stimulation, the tetanic contractions are normally fused to maintain steady muscle tone. However, shortening of the active state by β_2-adrenoceptor stimulation leads to incomplete fusion and reduced tension of tetanic contractions (Waldeck 1976) (Figure 4.4). This property, together with β-adrenoceptor-mediated enhancement of muscle spindle discharge, is thought to explain the development of muscle tremor often experienced by patients using β_2-agonists as bronchodilators. In bundles of fibres from slow- and fast-twitch muscles, terbutaline has recently been shown to induce a contrasting *potentiation* of twitch and tetanic contractions (Cairns & Dulhunty 1993).

Skeletal muscle is also a site of potassium homeostasis by the sympathoadrenal system. Adrenaline injections produce rapid and transient increases in plasma potassium. This is mimicked in several species by selective α_1-adrenoceptor agonists such as phenylephrine and amidephrine. This hyperkalaemia is blocked by prazosin and by apamin, a toxin from bee venom which blocks Ca^{2+}-channel-activated K^+ channels and the efflux of K^+ from isolated liver slices. The rise in plasma K^+ is therefore due to K^+ release from the liver through activation of α_1-adrenoceptors (Reverte *et al.* 1991). The response to adrenaline is followed by a more prolonged fall in plasma K^+ due to its β-adrenoceptor activity. The hypokalaemia is due to stimulation of the Na^+, K^+-ATPase-driven pump coupled to β_2-adrenoceptors. This increases the ability of the Na^+,K^+-pump to push Na^+ out of the cell and facilitate intracellular accumulation of K^+, thereby lowering plasma levels. This property of β_2-agonists has implications in the treatment of hyperkalaemic familial periodic paralysis, which is characterized by muscle weakness. Thus, hypokalaemia due to elevated catecholamines may also occur in myocardial infarction and during stress, for example on admission of patients to hospital. The hypokalaemia may precipitate arrhythmias. Hypokalaemia may also accompany the use of β_2-adrenoceptor

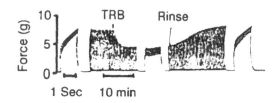

Figure 4.4 Depression by the β_2-adrenoceptor agonist, terbutaline (TRB) of subtetanic contractions of the guinea-pig isolated soleus muscle (skeletal muscle). The muscle was stimulated electrically with pulses (0.5 msec) delivered every 22 sec for 1.5 sec at 14 Hz. The partially fused contractions are shown at fast and slow moving chart speeds before and after addition of terbutaline (20 μg). Reproduced with permission from Waldeck (1976).

agonists such as salbutamol in the management of asthma and has been linked to an increased incidence of tachyarrhythmias (Du Plooy *et al.* 1994). Furthermore, since cardiac glycosides also owe their cardiac activity to an interaction with the Na^+,K^+-pump in cardiac muscle, the hypokalaemia sensitizes the heart to the effects of glycosides, increasing their potential for toxicity (Bowman 1981).

4.5   Therapeutic Uses of Sympathomimetic Amines

The general properties of sympathomimetic amines have been illustrated with adrenaline. Selective activity is targeted to specific receptor subtypes and organs for the treatment of a wide range of disease states. It is therefore convenient to consider those sympathomimetic amines in clinical use according to the therapeutic applications of each pharmacological property.

4.5.1   *Vasoconstrictor Activity*

The vasoconstrictor activity currently utilized in therapeutics is α_1-adrenoceptor-mediated constriction of arterioles and veins. This property is exploited in the treatment of nasal and bronchial congestion, in hypotensive states, as an adjunct to local anaesthesia and for minor haemostasis (Table 4.4).

4.5.1.1   *Decongestants*

α-Adrenoceptor agonists of the direct or indirect type are widely used in the symptomatic relief of nasal congestion associated with upper respiratory tract infection such as the common cold, which leads to non-allergic or vasomotor rhinitis. They are also used in allergic rhinitis, for example, hay fever. In these conditions, the mucous membranes become engorged and swollen and secretion of mucus is greatly increased. This reduces airflow through the nasal passages. The α-adrenoceptor agonists decrease the volume of the mucosa by vasoconstriction of the mucosal blood vessels and an action on erectile tissue. They also reduce mucus secretion, possibly as a consequence of the reduced blood flow, and also due to direct α-adrenoceptor-mediated inhibition of glandular activity together with inhibition of parasympathetic activation of secretion via prejunctional α-adrenoceptors (presumably α_2). These actions reduce the resistance to airflow and improve

184

Table 4.4 Sympathomimetic amines used as vasoconstrictors

Sympathomimetic	Combined with	Indication	Product	Formulation
Decongestants				
Ephedrine	—	Nasal congestion	BNF	Drops
	Chlorpheniramine		Haymine[R]	Tab
Pseudoephedrine	—	Nasal/sinus congestion	Sudafed[US/UK]	SR cap
			Novafed[US]	Liquid
	Triprolidine (antihist)/ dextromethorphan (antituss)	Cough/congestion	Actifed[UK] Afrin[US]	Tablet/ liquid
	Brompheniramine (antihist)/ guaiphenesin (expectorant)	Cough/congestion	Dimotane[UK] Dimetane[US]	Liquid
	Codeine	Cough/congestion	Tussar[US]	Liquid
	Terfenadine (antihist)	rhinitis	Seldane-D[US]	SR tab
Phenyl-propanolamine	Chlorpheniramine (antihist)	Congestion	Contac 400[UK] Triaminic[US]	SR cap Tab
	Pheniramine (antihist)		Triominic[UK]	
	Diphenylpyraline (antihist)	Congestion/ rhinitis	Eskornade[UK] Ornade[US]	SR cap
	Codeine (analgesic) Guaiphenesis (expectorant)	Cough/congestion	Naldecon[US]	Liquid
	Dextromethorphan/ guaiphenesin	Cough/congestion	Robitussin[UK/US]	Liquid
	Paracetamol	Catarrh	Mucron[UK]	Tab
		Nasal congestion	Sinutab[UK] Triogesic	
Phenylephrine	Chlorpheniramine/ methscopolamine (antimusc)	Rhinitis/ congestion	Extendryl[US]	Tab/ liquid
	—	Congestion	Neosynephrine[US]	Spray/ drops
	Promethazine (antihist)	Rhinitis/congestion	Phenergan[US]	
	Brompheniramine (antihist)/ Phenylpropanolamine	Catarrh/rhinitis	Dimotapp[UK]	SR tab
	—	Nasal congestion	Fenox[UK]	Spray/ drops
	Thenyldiamine (antihist)	Hay fever	Hayphrin	Spray
Xylometazoline	—	Nasal congestion	Otrivin(e)[UK/US]	Spray/ drops
	Sodium cromoglycate	Allergic rhinitis	Resiston	Spray
Tramazoline	Dexamethasone (steroid)/ Neomycin (antibiotic)	Allergic rhinitis	Dexa-rhinaspray[UK]	Spray
Oxymetazoline	—	Nasal congestion	Afrazine[UK] Iliadin[UK]	Spray Drops
Naphazoline	Antazoline	Nasal congestion/ rhinitis	Antistine-Privine (discontinued)	Drops/ spray
Haemorrhoidal preparations				
Phenylephrine	Betamethasone (anti-inflam)/ Lignocaine (local anaesthetic)		Betnovate Rectal	Ointment
Ephedrine	Allantoin/ Lignocaine		Anodesyn[UK]	Ointment/ suppositories
Pressor agents				
Metaraminol	—	—	Aramine	Injection
Noradrenaline	—	—	Levophed	Injection
Methoxamine	—	—	Vasoxine	Injection

Notes: Antihist, antihistamine; antituss, antitussive; antimusc, antimuscarinic cholinergic; SR sustained release; tab, tablet; cap, capsule; anti-inflam, anti-inflammatory.

ventilation. The drainage of the paranasal sinus is enhanced, the postnasal space is decongested and the eustachian tube opens.

A major limitation in the use of sympathomimetic decongestants is a rebound congestion that occurs with prolonged use or when the treatment is stopped. This may be due to desensitization of the α-adrenoceptors or to damage of the mucosa caused by the prolonged constriction of mucosal vasculature. Selective α_1-adrenoceptor agonists appear to produce less damage (DeBernardis *et al.* 1987).

The nasal decongestants are used topically by instillation into the nose as drops or by spray, the latter reaching a greater area. Oral administration is also used to relieve congestion of the nasal passages and of the lower airways in cough and cold remedies. The effectiveness in bronchial congestion is less convincing. Local application has the advantage of minimizing systemic effects such as an increase in blood pressure and CNS stimulant activity. The potential CNS stimulation, improved cardiovascular activity and effects on hostility and competitiveness have led to their misuse in sports. Indeed, there have been several deaths when normal doses of amphetamines have been used under conditions of maximal physical activity. As a consequence, the International Olympic Committee (IOC) Medical Commission considers sympathomimetic amines as stimulants and as such they are banned. This includes all phenylethylamines used as decongestants, however imidazole preparations, such as oxymetazoline, are acceptable for topical use (Badewitz-Dodd 1991).

Derivatives of amphetamine have become available through illicit sources as 'designer drugs'. These include methylenedioxymethamphetamine (MDMA), better known as 'ecstacy', which is an indirectly acting sympathomimetic amine with CNS stimulant and hallucinogenic properties. The hypertensive and cardiac stimulant effects have probably contributed to deaths resulting from its use (Dowling *et al.* 1987).

Sympathomimetic amine decongestants should be used with care in hypertensive, diabetic and hyperthyroid subjects and men with prostatic enlargement. They are contraindicated in patients being treated for depression with MAO inhibitors (see Chapter 2). Oral administration is more likely to induce systemic complications but the risks of rebound congestion are reduced. Sympathomimetic amines are often combined with antihistamines and anticholinergics in decongestant formulations and with analgesics in OTC cough and cold remedies. The most popular are phenylpropanolamine, pseudoephedrine and ephedrine as general purpose decongestants, while oxymetazoline and xylometazoline are used for rhinitis. Naphazoline is no longer recommended. Table 4.4 shows those sympathomimetic amines in common use.

4.5.1.2 *Local vasoconstrictor properties*

Classically, adrenaline solutions have been used for haemostatis of minor cuts and abrasions. A 1:1000 solution of adrenaline is the only agent authorized for use in the boxing ring to reduce bleeding from fighters cuts. The vasoconstrictor action will only slow bleeding time and allow clotting from minor blood vessels. This activity is used during surgical procedures to shrink the mucosa of the oral and nasopharyngeal regions and to improve visibility in the surgical field by limiting haemorrhage.

Sympathomimetic amines with α-vasoconstrictor activity are often included with local anaesthetics. They reduce local blood flow, preventing the local anaesthetic from being dispersed from its site of application or injection and thereby prolong its local action, for example, in dental procedures. Care must be taken not to produce severe reduction of blood flow, since the ischaemia may cause tissue necrosis. Adrenaline is frequently combined with lignocaine (lidocaine, xylocaine) and procaine. It must not be combined with cocaine which has constrictor properties in its own right and potentiates the effects of adrenaline through inhibition of uptake$_1$. This property can also be utilized in ointment and suppository preparations for the treatment of haemorrhoids (piles) which are varicose veins of the anal region. The α-vasoconstriction localizes the anaesthetic to relieve the pain and pruritis associated with inflammation and may also constrict the swollen veins directly to aid their retraction. Phenylephrine and ephedrine have been used for this purpose (Table 4.4).

4.5.1.3 Hypotensive states

Hypotension is a characteristic feature of cardiovascular shock. It is caused by loss of fluid or dehydration (hypovolaemia), cardiac failure (myocardial infarction), obstructed cardiac output (eg pulmonary embolism), peripheral sepsis (septic or endotoxic shock) or anaphylaxis (acute allergic hypersensitivity). There is a rapid life-threatening failure to adequately perfuse the vital organs and supply nutrient and oxygen. The treatment is aimed at correcting the underlying pathogenesis of the shock, together with restoring blood flow to the organs. The fall in blood pressure activates reflex sympathetic vasoconstriction which initially serves to maintain blood pressure but is a major contribution to the impaired organ blood flow. Thus, although an α_1-adrenoceptor vasoconstrictor, such as noradrenaline, metaraminol, methoxamine or phenylephrine, would raise blood pressure, it would further compromise the reduced blood flow and place an additional burden on the heart.

The treatment of shock is complex and depends upon the causative factors. Fluid loss caused by haemorrhage, burns or dehydration is treated with volume expanders such as saline or dextran. Cardiogenic shock arising from myocardial infarction requires inotropic support (see later). Septic shock is caused by endotoxins (bacterial lipopolysaccharide) from Gram-negative bacteria which injure tissues and release inflammatory products. These include the cytokines which promote vasodilatation and increase capillary permeability and thus contribute to the fluid loss. Treatment includes use of antibiotics, maintenance of cardiac output and fluid replacement. Endotoxin and cytokines have been shown to induce nitric oxide (NO) formation from phagocytic cells and endothelial cells, the vasodilator properties of which contribute to the hypotension and reduces vasoconstrictor sensitivity to α-agonists. The NO production and hypotension are reversed by NO-synthase inhibitors such as L-NAME, which therefore have potential in the treatment of endotoxic shock (Thiemermann & Vane 1990). In all of these types of shock, α-adrenoceptor agonists are more likely to have a detrimental effect. Only in rare cases is the fall in blood pressure so severe that their use is indicated to raise blood pressure sufficiently to maintain perfusion of the brain (Higgins & Chernow 1987).

α_1-Adrenoceptor agonists do, however, have a place in the treatment of anaphylactic shock and in hypotension associated with sympathetic nervous failure such as after injury and spinal anaesthesia or overdose with antihypertensive drugs. In anaphylactic shock, due to hypersensitivity reactions to a food or drug allergy or a bee sting, adrenaline is the treatment of choice. It relieves the oedema which causes swelling of eyes, lips, tongue and glottis; the latter threatening to cause suffocation. It also restores the blood pressure and, through its β_2-adrenoceptor stimulant activity, will improve airway ventilation and impair release of inflammatory mediators from mast cells (see Chapter 3).

4.5.2 α-Adrenoceptor-mediated Ophthalmic Uses

The mydriatic effects of α-adrenoceptor agonists such as adrenaline and phenylephrine may be used to dilate the pupil to enable examination of the fundus. However, unlike atropinic drugs (see Chapter 9), this action does not paralyse the light reflex and thus pupil constriction will occur on shining a light into the eye. They do not cause cycloplegia (paralysis of lens accommodation for near vision). The contraction of the radial muscles of the iris does, however, cause the iris to narrow the drainage angle to the canal of Schlemm (Figure 1.7, Chapter 1). Thus, there is a risk of precipitating narrow-angle glaucoma in susceptible individuals. Paradoxically, adrenaline does have a use in the treatment of glaucoma of the open-angle type. In this condition there is an increase in intraocular pressure not associated with narrowing of the drainage angle to the canal of Schlemm but with impaired absorption of the aqueous humour, probably because of reduced diffusion through the trabecular network to the canal of Schlemm. It may be diagnosed by application of glucocorticoids which increase intraocular pressure in susceptible individuals. The beneficial effect of adrenaline in open-angle glaucoma is attributed to inhibition of secretion of aqueous humour from the ciliary body due to α-adrenoceptor-mediated vasoconstriction in the afferent arteriole. There is also facilitated absorption of aqueous humour probably due to a β-adrenoceptor vasodilator action in the episcleral veins that drain the canal of Schlemm.

A recent improvement upon adrenaline has been the introduction of dipivalyl adrenaline (Propine[R]). This is a prodrug of adrenaline formed by the addition of two pivalyl side-chains to the catechol hydroxyl groups of the parent compound (Table 4.5). It is >100 times more lipophilic than adrenaline, leading to substantially enhanced penetration of the cornea. It is more effective at lowering intraocular pressure than adrenaline and has fewer side-effects (Kohn *et al.* 1979). Phenylephrine is also a mydriatic used for this purpose.

A cosmetic use of α-adrenoceptor agonists by topical application to the eye is in preparations that blanch the conjunctiva by vasoconstriction to relieve 'blood-shot eyes'. Naphazoline is retained for use in preparations of this type (Eye Dew[R], Murine[R], Optrex Clearine[R]).

4.5.3 β-Adrenoceptor Bronchodilatation

β-Adrenoceptor agonists are recommended as the first-line treatment for mild bronchial asthma, according to the International Consensus Report on Diagnosis and Management of Asthma supported by the British Thoracic Society (1992). They owe their primary beneficial effect to relaxation of bronchial smooth muscle which relieves the bronchospasm that characterizes the asthma attack. Asthma is usually due to an allergy to inhaled antigenic material in the inspired air, such as house dust mites, *Dermatophagiodes pteronyssinus*, which are distributed in bedding, carpets and furniture and live off human skin scales. Other common antigens are grass pollens and the dander from animal fur, cats being the worst offenders. These patients are regarded as atopic and carry the appropriate antibody (IgE) in their plasma. They usually demonstrate other features of atopy, including a positive response to skin-prick testing with allergens. They frequently have a personal and family history of other forms of hypersensitivity such as eczema, hay fever and rhinitis. This is allergic or extrinsic asthma. Atopic patients may not always display symptoms of asthma. Where no obvious external trigger factor other than respiratory infection occurs, it is referred to as intrinsic asthma. Exacerbations of the bronchospasm (asthma attacks) may also be precipitated, in sensitive individuals, by exercise or breathing cold air.

Asthma is characterized by bronchoconstriction and airflow obstruction, inflammation of the airways and a hyperreactivity to a variety of bronchoconstrictor stimuli. As a result of allergic challenge, inflammatory mediators are released from mast cells, neutrophils, eosinophils and macrophages. Preformed histamine, released by mast cell degranulation, is responsible for the early onset or immediate bronchospasm. A so-called late response occurs from ~6 hours later and is thought to be a manifestation of the inflammatory response arising from the influx of inflammatory cells, mainly eosinophils, into the airways. These cells release mediators, such as leukotrienes, which cause bronchoconstriction and mucosal swelling. The latter response is due to congestion of blood vessels and plasma exudation. Other pathophysiological features include an increase in mucus secretion which may cause plugging in fatal asthma attacks, and shedding of the epithelium. The epithelial damage may be caused by compromised binding of superficial epithelial cells to the basal lamina or to toxic damage from inflammatory cell products such as oxygen free radicals, major basic protein or eosinophil cationic protein. The loss of epithelium may in part be responsible for the bronchial hyperreactivity which is associated with the late response (Jeffery 1992).

Many intermediates have now been identified in the cascade of events linking allergen challenge to the sequelae of asthma. The initial immunological challenge causes migration of T-lymphocytes to the lungs and their activation to release a range of cytokines. The cytokines, including interleukines IL-3, IL-4, IL-5 and granulocyte/macrophage colony stimulating factor (GM-CSF), induce the expression of IgE receptors on other inflammatory cells (eosinophils and macrophages). Interleukins also activate B-cells to produce IgE antibodies. These interact with the primed inflammatory cells which release inflammatory mediators including prostaglandins, thromboxanes, leukotrienes and platelet activating factor (PAF) (Corrigan & Kay 1992).

β-Adrenoceptor agonists inhibit the early response to allergen challenge. The action is due to relaxation of airway smooth muscle through β_2-adrenoceptors, although enhanced mucociliary clearance, decreased vascular permeability and inhibition of mediator release from mast cells may also contribute (see Chapter 3). The clinical relevance of the latter effect is not currently considered to be important. The selective β_2-adrenoceptor agonists have now succeeded the non-selective agonists, adrenaline and isoprenaline.

An increase in asthma mortality in the 1960s was associated with a rise in the OTC sales of isoprenaline. A subsequent spate of asthma deaths has occurred since the late 1970s, with a cluster occurring in New Zealand and linked to the introduction of the β_2-agonist, fenoterol. Fenoterol is used at particularly high doses and its selectivity for β_2-adrenoceptors is only marginal. It may be that adverse effects of the less selective agonists are due to cardiac β_1-adrenoceptor stimulation. Other

189

unwanted effects include β_2-adrenoceptor-mediated hypokalaemia due to increased K^+ uptake into skeletal muscle, which may increase susceptibility to cardiac arrhythmias and skeletal muscle tremor (see earlier). These systemic adverse actions are minimized by use of inhalation therapy by metered dose inhaler (MDI) or nebulizer. Further possible reasons why inappropriate use of β-agonists may in fact make asthma worse or increase mortality are the further induction of hyperreactivity, increased access of allergen to the lungs as a result of the bronchodilatation and β-agonist-induced increases in airway mucus secretion of a particularly viscous type, and tolerance (Barnes & Chung 1992). Inhibition of mast cell degranulation by β_2-adrenoceptor agonists will also prevent heparin release, which normally exerts inhibitory effects on smooth muscle cell proliferation (Page 1991). Thus, proliferation of airways smooth muscle could be promoted, leading to wall thickening and increased airway resistance. However, recent findings suggest that salbutamol in fact has antiproliferative properties (Tomlinson *et al.* 1994).

The tolerance, down-regulation or desensitization of β-adrenoceptors has been demonstrated in the lymphocytes of asthmatic patients chronically treated with β-agonists. A loss of bronchodilator activity is less convincingly demonstrated, as is the loss in the ability to protect from histamine-induced bronchospasm. However, there is often a progressive increase in use of β_2-agonists by individual patients, which is regarded as a warning of deterioration of the disease and indicates the need to introduce regular anti-inflammatory therapy. The down-regulation of β-adrenoceptors may be less evident in asthmatic patients because it is prevented by the co-administration of corticosteriods (Barnes & Chung 1992, Tattersfield 1992).

β-Adrenoceptor agonists for the treatment of asthma are of two types: short-acting, including salbutamol (Ventolin[R]) and terbutaline (Bricanyl[R]), and long-acting, such as salmeterol (Serevent[R]) and formoterol (Table 4.5). The short-acting agents have a duration of action of 4–6 hr and are the medication of choice for the treatment of exacerbations of asthma or for exercise-induced asthma. Ventolin[R] had the highest worldwide pharmaceutical sales in 1992 of all drugs acting on the autonomic nervous system, being ranked sixteenth of all products (*Script* No. 1825 p. 23, 1993). These drugs may be used alone in mild intermittent asthma to relieve symptoms if the patient has a pretreatment baseline peak expiratory flow (PEF) >80% of the predicted value and experiences only one or two episodes per week. If the β_2-agonist is required more than three times a week, the patient should be moved to the next step of care which includes an inhaled anti-inflammatory agent such as corticosteriods or, in the case of children, cromoglycate or nedocromil. Short-acting inhaled β_2-agonists are always used on a when-required basis ≤3–4 times daily for escape relief of exacerbations.

The long-acting β_2-agonists, salmeterol and formoterol, are more lipid soluble than is salbutamol and show higher non-specific binding to the cell membrane, which may partly explain their longer action. The large non-polar aralkyloxyalkyl substituent on the amino group of salmeterol has been proposed to participate in binding to an exo-site adjacent to the β_2-adrenoceptor. The relaxation of isolated tracheal preparations by salmeterol and another long-acting agonist, quinprenalin (quinterenol), persists for ≤12 hr despite continuous washout. However, these persistent responses are rapidly and fully reversed by the β-adrenoceptor antagonists, propranolol and sotalol, indicating that the effect is not due to irreversible binding to the β-adrenoceptor (Figure 4.5). The persistence and possible binding of salmeterol to an exo-site is illustrated by the fact that the inhibition is then re-established after removal of the sotalol (Ball *et al.* 1991). The bronchodilator actions of salmeterol and formoterol and their protection from histamine-induced bronchospasm are maintained for at least 12 hr. It has been suggested that salmeterol also has anti-inflammatory actions; indeed all β_2-adrenoceptor agonists are able to inhibit mediator release from mast cells. Salmeterol produces a prolonged inhibition of antigen-induced mediator release from sensitized human lung fragments (Butchers *et al.* 1991) and guinea-pig mast cells (Gentilini *et al.* 1994). Whether it has additional anti-inflammatory properties, however, and the clinical relevance of any such effect, is questionable. The early bronchoconstriction and microvascular leakage in response to inflammatory mediators released by antigen challenge are

Table 4.5 Structures and uses of β-adrenoceptor agonists

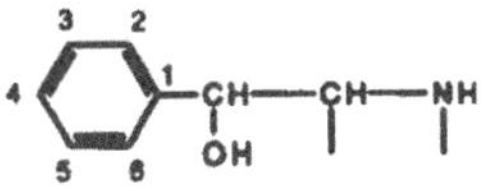

Drug				Activity	Product

(A) Bronchodilators

Drug				Activity	Product
Isoprenaline	3-OH 4-OH	H	CH(CH$_3$)$_2$	β_1/β_2	Isuprel Medihaler iso
Orciprenaline	3-OH, 5-OH	H	CH(CH$_3$)$_2$	Slight β_2	Alupent
Terbutaline	3-OH, 5-OH	H	C(CH$_3$)$_3$	β_2	Bricanyl (oral inh)
Fenoterol	3-OH 4-OH	H	A	β_2	Berotec (inh)
Isoetharine	3-OH 4-OH	C$_2$H$_5$	CH(CH$_3$)$_2$	β_2	Bronkometer[US]
Rimiterol	3-OH 4-OH	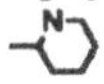		β_2	Pulmadil (inh)
Pirbuterol	B	H	C(CH$_3$)$_3$)	β_2	Exiril[UK]Maxair[US]
Tulobuterol	2 - CI	H	C(CH$_3$)$_3$	β_2	Brelomax [UK] (oral)
Reproterol	3-OH, 5-OH	H	C	β_2	Bronchodil[US]
Salbutamol (albuterol)	3-CH$_2$OH, 4-OH	H	C(CH$_3$)$_3$	β_2	Ventolin (inh) Volmax (oral)
Salmefamol	3-CH$_2$OH, 4-OH	H	D	β_2	
Broxaterol	E	H	C(CH$_3$)$_3$	β_2	Summair' (oral)
Bambuterol	3-&5-OCON(CH$_3$)$_2$	H	C(CH$_3$)$_3$	β_2	Ester prodrug of terbutaline
Ibuterol	3-&5-OCOCH(CH$_3$)$_2$	H	C(CH$_3$)$_3$	β_2	Ester prodrug of terbutaline
Salmeterol	3-CH$_2$OH, 4-OH	H	F	Long-acting β_2	Serevent (inh)
Formoterol	3-NHCHO, 4-OH	H	D	Long-acting β_2	

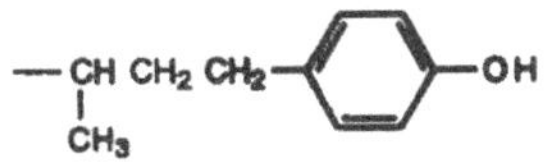

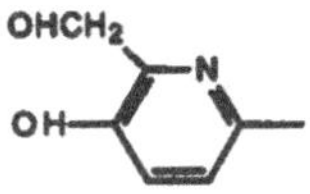

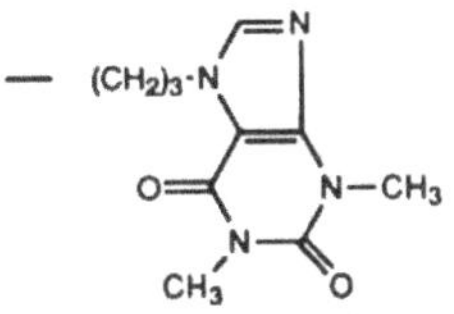

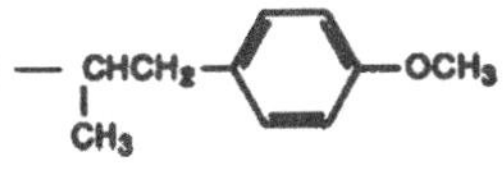

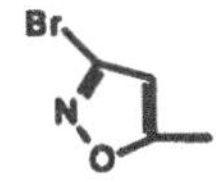

Table 4.5 *Continued*

Drug		Activity
(B) Positive inotropes		
Dobutamine	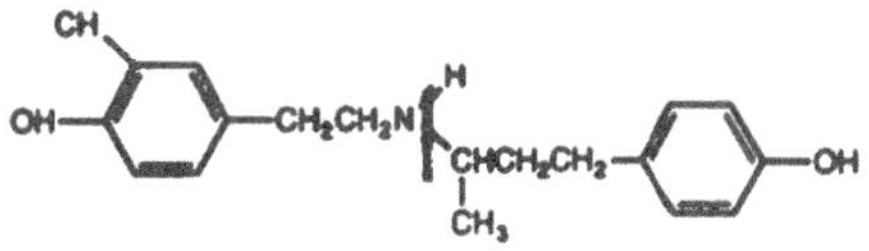	$\alpha_1\beta_1\beta_2$
Denopamine	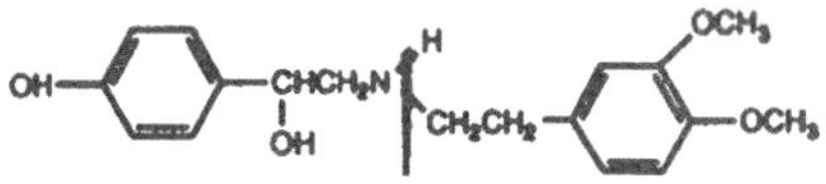	β_1
Dopexamine	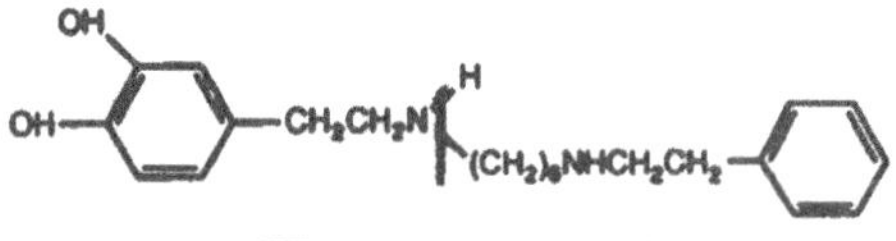	β_2 / D_1
(−)-Prenalterol (H133/80)	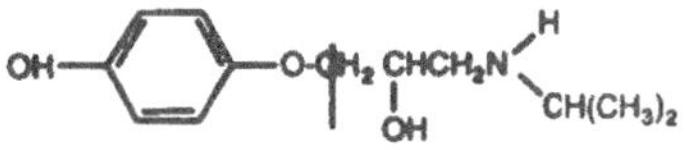	β_1 / β_2
Xamoterol	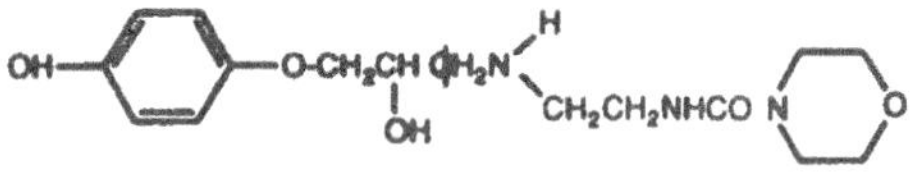	β_1 (partial agonist)
Oxyfedrine	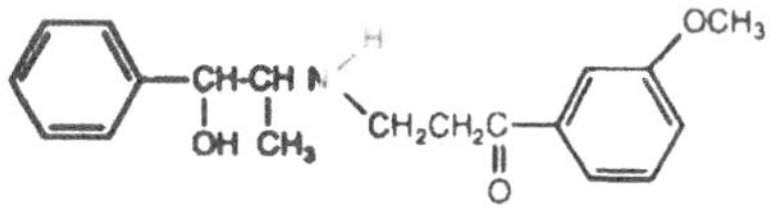	β_1 / β_2
Thyronamine	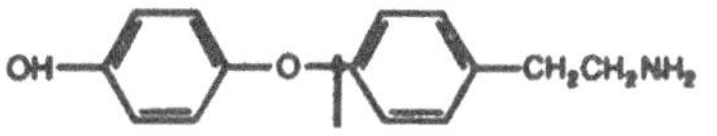	Mainly indirect

(C) β_2-Adrenoceptor-selective agonists

	R_1	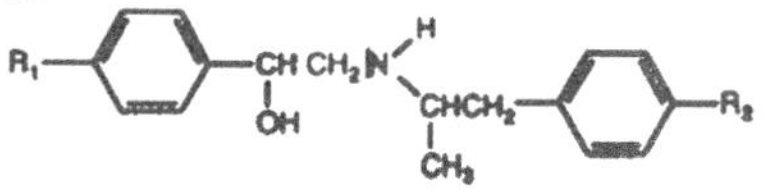 R_2	
BRL 26830	H	-COOCH$_3$	
BRL 28410	H	-COOH	Active metabolite
BRL 35135	CI	-OCH$_2$COOCH$_3$	
BRL 37344	CI	-OCH$_2$COOH	Active metabolite

	R_1	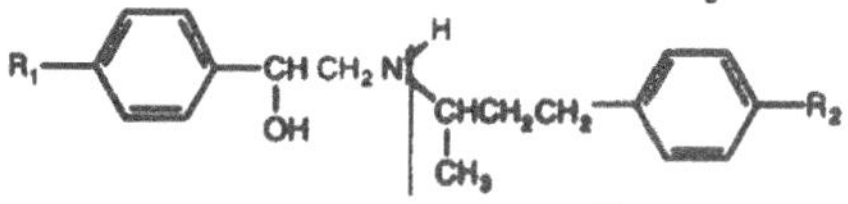 R_2	
LY 79771	H	-OH	
LY 104119	H	-CONH$_2$	
LY 99134	OH	-OH	Ractopamine

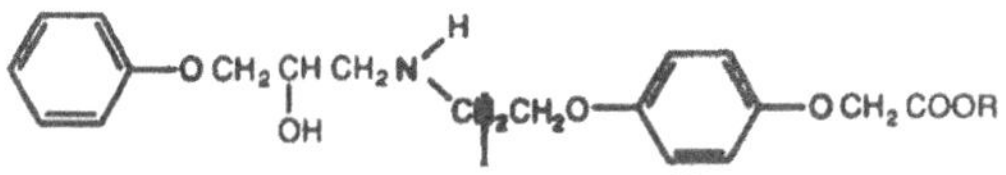

	R	
ICI 198, 157	-CH$_3$	
ICI 201, 651	-H	Active metabolite
ICI D7114	-NH CH$_2$ CH$_2$ OCH$_3$	(partial agonist)

(continued)

Table 4.5 *Continued*

Drug		Activity and use

(D) Others

Drug	Activity and use
Clenbuterol	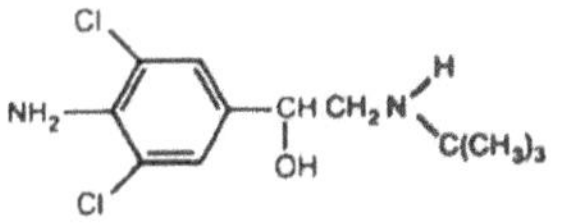 'β_2' Anabolic
Cimaterol	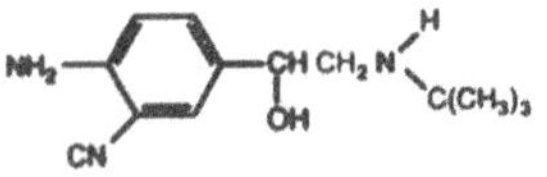 'β_2' Anabolic
BRL 46104	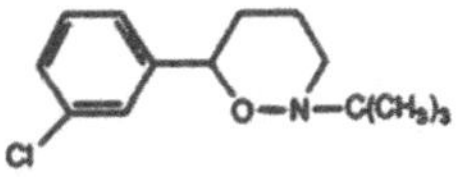 Anabolic
Isoxuprine (Duvadilan^R)	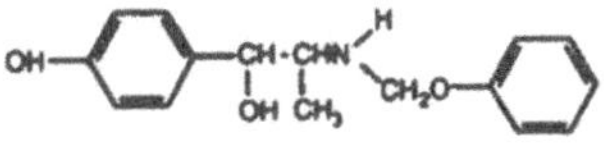 β_2 Uterine relaxant
Ritodrine (Yutopar^R)	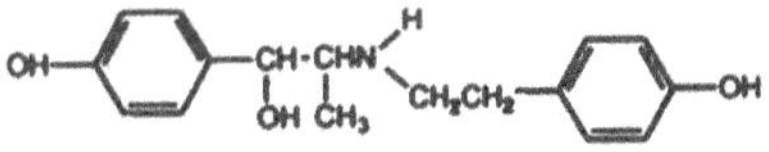 β_2 Uterine relaxant
Procaterol (OPC 2009)	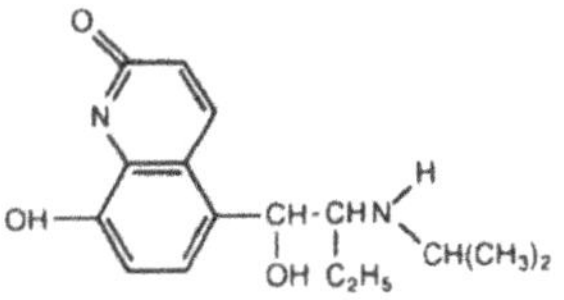 Highly selective β_2
Quinterenol (quinprenaline)	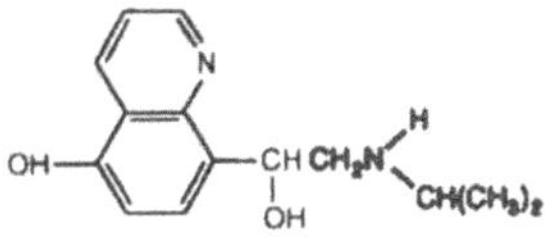 Long-acting β_2
Bitolterol	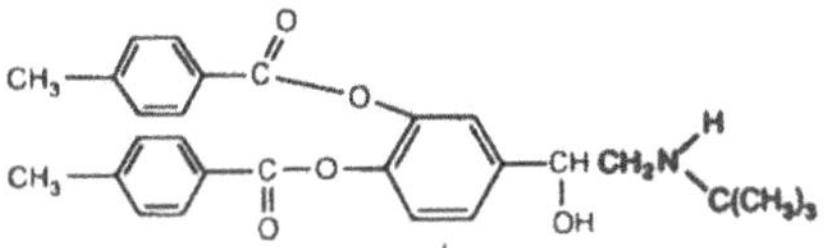 Inactive prodrug
	lung esterases
Colterol	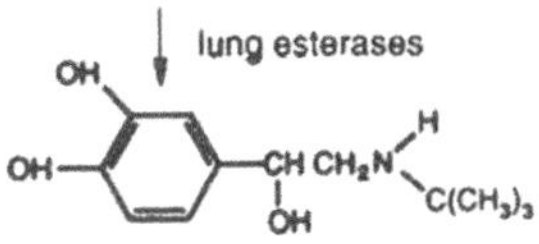 β_2-Selective product of bitolterol
Tazolol	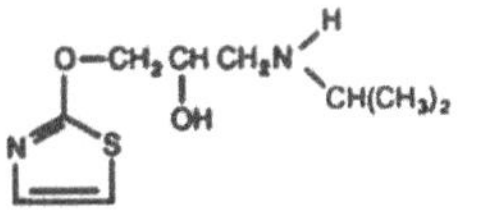 β_1-Selective partial agonist
Dipivalyl adrenaline (Propine^R)	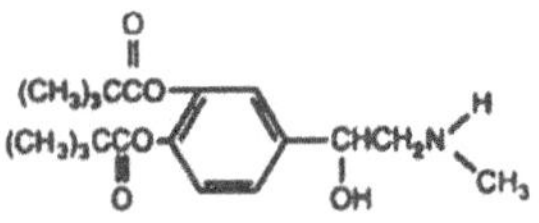 Adrenaline prodrug (glaucoma)
Trimetoquinol	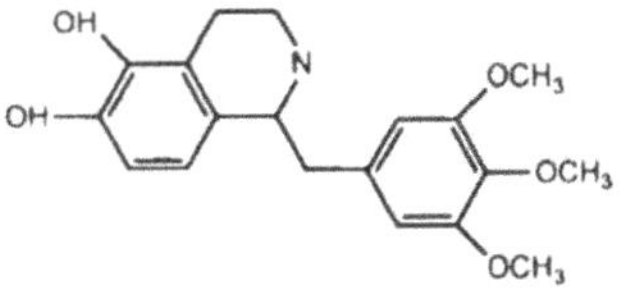 β_2-Agonist with thromboxane antagonist anti-platelet activity
RO363	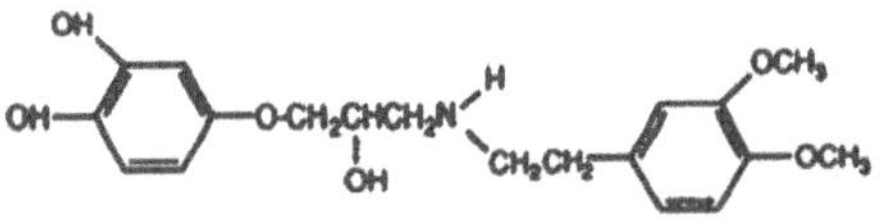 β-Selective agonist

193

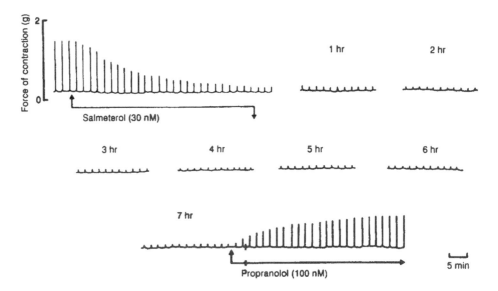

Figure 4.5 Prolonged inhibition of contractions of guinea-pig electrically stimulated superfused tracheal strip by the β_2-adrenoceptor agonist salmeterol (30 nM). The salmeterol was present during the indicated time and despite continuous washout by superfusion for 7 hr, the contractions remain inhibited, but were restored by propranolol. Reproduced with permission from Ball *et al.* (1991).

inhibited by both short- and long-acting β_2-agonists. Short-acting β_2-agonists have no effect against the late response or the associated hyperreactivity. Salmeterol, however, does attenuate the late response when given before allergen challenge. This may be a result of the persistent bronchodilator activity rather than to an anti-inflammatory action (Barnes & Chung 1992). The long-acting β_2-agonists are not used on an as-needed basis for acute symptomatic relief. They are used twice daily when corticosteriods fail to provide satisfactory control and are especially useful in nocturnal asthma.

There is concern that the regular use of short-acting β_2-agonists rather than their on-demand use may lead to more rapid development of tolerance and diminished control of the symptoms. This may be particularly so with the long-acting β-agonists and, although trials to date have failed to show loss of control of the symptoms with regular use of salmeterol, further evaluation of this potential problem is required. In spite of the uncertainties about the role of β-agonists on the long-term prognosis of asthma, they still form the basis of treatment for on-demand relief of the respiratory distress.

4.5.4 β_2-Adrenoceptor Uterine Relaxation

Relaxation of the uterine smooth muscle by β_2-adrenoceptor stimulation may be utilized to suppress preterm labour in pregnant women (tocolytic activity). Usually, bedrest and hydration are successful. However, if this approach fails, uterine relaxants may be employed. They are generally used between 20 and 34–36 weeks of gestation; beyond this time the foetus has excellent chances of normal development with modern neonatal care. Ritodrine (Yutopar[R]) is the only β_2-agonist approved for use in the USA while, additionally, terbutaline (Bricanyl[R]), salbutamol (Ventolin[R]) and isoxuprine (Duvadilan[R]) are used in the UK for this purpose, initially by intravenous infusion and then by oral maintenance doses. Although β_2-selective, they may cause tachycardia in both mother and foetus. They also increase blood sugar so that extreme caution should be exercised in diabetic women. The secretion of renin is enhanced, which through the renin–angiotensin–aldosterone system will promote Na^+ and water retention. This, with excessive hydration, may result in pulmonary oedema, so that total water intake at this stage of treatment should be restricted to 2 l in 24 hr. Recent deaths in the UK from pulmonary oedema have resulted in a tightening-up of prescribing information to emphasize that intravenous ritodrine should be administered with dextrose rather than saline. β_2-Agonists are contraindicated in threatened abortion, antepartum haemorrhage, toxaemia of pregnancy, cord compression, and where prolonging the pregnancy would be hazardous.

4.5.5 β-Adrenoceptor Inotropes

Stimulation of the force of cardiac contractions via cardiac β-adrenoceptors has been a target of drug discovery for the treatment of cardiac failure for the past decade. The potent full agonists at cardiac β-adrenoceptors, such as isoprenaline, are inappropriate since they also stimulate cardiac rate and thereby increase metabolic demands of an already compromised heart. Modest cardiac stimulation by means of a partial agonist may provide some inotropic support to the failing heart. The major problem with this approach, however, is that in chronic cardiac failure β-adrenoceptors are down-regulated (see Chapter 3), thus reducing the effectiveness of β-agonists. As discussed earlier, the β_2-adrenoceptor population of the human heart appears to be spared from this down-regulation and it has been argued that selective β_2-agonists may be of more value. A further complication is that chronic long-term use of β-agonists will lead to desensitization of β-adrenoceptors with consequent tolerance to their cardiotonic action. However, a range of orally active partial agonists has been introduced over the past few years as potential positive inotropic agents for treating congestive cardiac failure (Table 4.5). It is argued that the degree of down-regulation by partial agonists may be less than with full agonists.

Prenalterol (H 133/80, (−)isomer) was introduced as a cardioselective β_1-adrenoceptor partial agonist which displayed selective inotropic activity. Its selectivity is not now attributed to preferential binding at β_1-adrenoceptors but to a greater efficacy; it displays tissue selectivity (Chapter 3, Kenakin & Beek 1982a). It does not produce a fall in arterial blood pressure, probably because of the lack of efficacy at vascular β_2-adrenoceptors. This may be an advantage in not inducing reflex tachycardia, but a possible disadvantage is that it does not reduce after-load on the heart. The reduced tachycardia may be due to either reflex cardiac slowing because of modest increases in blood pressure or to β-adrenoceptor blockade when there is high sympathetic drive (Kenakin & Beek 1982b). Prenalterol was introduced for clinical use in the UK but has since been withdrawn. In isolated cardiac tissue from heart failure patients its inotropic activity is weak.

Xamoterol (Corwin[R]) is an orally active partial agonist with true selectivity for β_1-adrenoceptors, the maximum response in isolated tissues being 43–55% that of isoprenaline. It increases left ventricular contractility ($+dP/dt_{max}$) in healthy volunteers. There is a small increase in heart rate at rest when sympathetic tone is low, but it reduces heart rate on exercise due to β_1-adrenoceptor blockade. This latter effect may be detrimental in severe heart failure patients that depend upon a high level of sympathetic tone to maintain cardiac function. Thus, in mild chronic heart failure there is significant haemodynamic improvement and it is well tolerated, but in some patients with severe heart failure due to dilated cardiomyopathy there may be dramatic deterioration. It may not be an effective positive inotrope where β_1-adrenoceptors are down-regulated (Furlong & Brogden 1988).

Denopamine (TA-064) is also a β_1-adrenoceptor partial agonist, but has a maximum inotropic effect in isolated cardiac tissues nearer to that of isoprenaline (75–80%). It binds to cardiac α-adrenoceptors but the inotropic effect is exclusively via β-adrenoceptors (Kohi *et al.* 1993). Denopamine has been used in the long-term treatment of patients with heart failure but its actions are again considerably reduced in isolated tissue from heart failure patients, presumably due to the down-regulated status of β_1-adrenoceptors (Brodde 1991).

Pirbuterol (Maxair[R]) and *salbutamol* (Ventolin[R]) have received trials for the treatment of cardiac failure. The rationale for the use of these selective β_2-adrenoceptor agonists, which are better known for their application in asthma (Table 4.5), is that they lower blood pressure by β_2-adrenoceptor-mediated vasodilatation and thus reduce after-load on the failing heart. The success in treating heart failure with pirbuterol is not clear but a reduced effectiveness has been linked to down-regulation of lymphocyte β_2-adrenoceptors (Colucci *et al.* 1981a).

Thus, these inotropic partial agonists, whether β_1- or β_2-selective, are of limited value in the long-term treatment of chronic heart failure because of the possibility of receptor down-regulation. However, β-adrenoceptor agonists do have application in acute cardiac failure, for example, associated with myocardial infarction and cardiogenic shock, where they may be used over a short period without risk of down-regulation. Dopamine was the prototype of this class of drug, which also includes dobutamine and dopexamine.

Dopamine (Intropin[R]) exerts positive inotropic and chronotropic effects through mixed indirect and direct sympathomimetic activity. Its cardiac stimulant activity is mainly via β_1-adrenoceptors. Additionally, the haemodynamic properties of dopamine owe much to the stimulation of peripheral dopamine receptors in the vasculature and kidney; these properties will be described later (Chapter 12). Dopamine is administered by intravenous infusion to provide brief inotropic support in severe heart failure. Total peripheral resistance is usually unchanged with low to moderate doses, since a dopaminergic vasodilatation in the kidney and mesentery predominates over an α_1-adrenoceptor-mediated vasoconstriction elsewhere. However, at higher doses, the latter effect comes into play and may be significant in the shock patient where substantial reflex vasoconstriction may already exist. This will place further load on the already weakened heart. Tachycardia and the consequent risk of precipitating arrhythmias are a further potential disadvantage. Vasoconstriction may cause gangrene of fingers or toes during prolonged infusions. Dopamine is rapidly metabolized via COMT and MAO (it should be avoided in patients receiving MAO inhibitors) and any adverse effects soon disappear if the infusion is stopped.

Dopexamine (Dopacard[R]) was developed to overcome the disadvantages of dopamine by eliminating α_1-adrenoceptor vasoconstrictor activity and β_1-adrenoceptor-mediated cardiac stimulation. This sympathomimetic amine has 10-fold selectivity for

β_2-adrenoceptors, combined with agonist activity at dopamine D_1-receptors. The latter property increases renal blood flow and lowers blood pressure through direct vasodilatation, thereby reducing after-load. The β_2-adrenoceptor agonist activity is also aimed at reducing after-load through vasodilatation, but there is also inotropic support through stimulation of cardiac β_2-adrenoceptors. Dopexamine is also an inhibitor of neuronal uptake (Chapter 2), which may have the minor effect of enhancing the cardiac effects of circulating endogenous catecholamines. It has no indirect sympathomimetic activity and is not arrhythmogenic (Smith & O'Connor 1988).

Dopexamine is not orally active and must be administered by intravenous infusion; it is rapidly metabolized and has a short plasma half-life of ~3 min. It is used in the acute management of low cardiac output states. In isolated cardiac tissue from patients with cardiac myopathy, the positive inotropic activity of dopexamine is diminished, indicating that its long-term use in congestive cardiac failure may be limited (Brodde 1991).

The down-regulation of myocardial β-adrenoceptors in congestive cardiac failure and during the chronic use of β-agonists may be overcome by the use of measures that protect β-adrenoceptors. There is interest in the potential use of β-adrenoceptor antagonists in the management of heart failure; whether non-selective or β_1-adrenoceptor-selective antagonists are used would depend upon the status of β-adrenoceptor down-regulation. For example, if β_2-adrenoceptors were preserved, a β_1-selective antagonist would be sufficient and indeed this may cause an increase in β_2-adrenoceptor function (see Chapter 3, heart). An alternative approach to minimizing receptor down-regulation might be to reduce sympathetic tone to the heart. The success of the ACE inhibitors (eg captopril) in the treatment of congestive cardiac failure may be partly due to this, since by reducing production of angiotensin they lower the output of noradrenaline (Chapter 6) and the need for compensatory sympathetic reflexes. Clonidine may have a similar effect by reducing noradrenaline output from sympathetic nerves (Brodde 1991).

Dobutamine (Dobutrex[R]) is used as the racemic mixture, the pharmacological profiles of the two enantiomers being quite different. The $(-)$-isomer is a potent α_1-adrenoceptor agonist with little activity on β-adrenoceptors. The $(+)$-isomer, however, has α-adrenoceptor antagonist activity and directly acting full agonist activity at β_1- and β_2-adrenoceptors. Unlike dopamine and dopexamine, it does not have dopaminergic effects. *In vivo*, racemic dobutamine displays inotropic selectivity in both animals and humans. *In vitro*, this selective stimulation of force, compared with rate of contraction, is less easy to demonstrate. In guinea-pig tissues, which lack α-adrenoceptors, there is no inotropic selectivity (Lumley *et al.* 1977), whereas in rat tissues there is inotropic selectivity which is significantly reduced by the α-adrenoceptor antagonist, phentolamine (Kenakin 1981). Thus, the beneficial effects of dobutamine in acute cardiac failure appear to be due to the stimulation of force of contraction by the $(-)$-isomer via α_1-adrenoceptors and the $(+)$-isomer through β_1- and, to a lesser extent, β_2-adrenoceptors. The stimulation of cardiac rate is less because it is only through the $(+)$-isomer. Haemodynamic factors also probably play a part *in vivo*. The α_1-vasoconstriction by the $(-)$-isomer would produce both pre- and after-load effects on myocardial contractility. Furthermore, a modest rise in blood pressure would induce reflex bradycardia to oppose the direct β_1-adrenoceptor-mediated increases in heart rate. The α_1-adrenoceptor-mediated vasoconstriction would be offset by β_2-vasodilatation (Hieble & Ruffolo 1987, Docherty 1989). Dobutamine is used by short-term infusion to provide inotropic

197

support in patients with acute heart failure after cardiac surgery and myocardial infarction. It is readily metabolized by O-methylation and has a half-life of ~2 min.

4.5.6 β-Agonists Affecting Thermogenesis and Glucose Metabolism

Amphetamine was found to promote weight loss when used in patients suffering from narcolepsy (prolonged bouts of sleepiness). Dexamphetamine (Dexedrine[R]) has been used in the treatment of obesity, the weight loss being due to suppression of appetite rather than an effect on energy expenditure. Its effect is short-lived since tolerance develops and there are risks of hypertension and sleep disturbances. Amphetamine and related agents [phentermine (Ionamin, Duramine, Fastin[US]), diethylpropion (Tenuate Dospan[R]), phendimetrazine (Plegine[US]), benzphetamine (Didrex[US]), methamphetamine (Desoxyn[US])] have psychostimulant properties and have potential for abuse and habituation. They have therefore been largely replaced by non-stimulant derivatives such as fenfluramine (Ponderax[R]) and its dextro isomer, dexfenfluramine (Adifex[R]). Whereas amphetamines are indirectly acting sympathomimetic amines releasing noradrenaline from nerve endings, fenfluramine releases 5-HT and is sedative rather than stimulant; in fact depression is a potential adverse effect. Mazindol (Teronac[UK], Mazanor[US]) is a tricyclic antidepressant, structurally unrelated to amphetamine, which is an inhibitor of neuronal uptake. It supresses appetite and will interact with antihypertensives (eg neurone blockers).

Sympathomimetic amines such as phenylpropanolamine and ephedrine do not appear to exert significant appetite suppression through a central action. They promote weight loss primarily by their peripheral action on adipose tissue. As discussed in the previous chapter (adipocytes, Chapter 3), they promote weight loss by stimulating thermogenesis in BAT through the release of noradrenaline from noradrenergic neurones. The noradrenaline interacts with β-adrenoceptors on adipocytes mediating lipolysis, these adrenoceptors having now been classified as the β_3-subtype. With the identification of a novel receptor type involved in thermogenesis came the opportunity for developing selective agonists for potential use as anti-obesity agents. Beechams were the first to develop such compounds, with BRL 35135 and BRL 26830 being shown to have marked effects on metabolic rate and weight loss in animals. They are methyl esters which, although exerting activity in isolated adipocytes, do not appear in the circulation after oral administration. They are converted to the corresponding acids, BRL 37344 and BRL 28410, respectively, which are therefore responsible for the *in vivo* effects of the parent compounds. These have been followed by selective β-adrenoceptor agonists from other companies (Table 4.5). All appear to be selective stimulants of thermogenesis and increase metabolic rate in human subjects, and some have reached advanced stages of clinical trial. However, their effectiveness at promoting weight loss in obese humans has not matched that in animal models. Their primary action appears to be via stimulation of β_3-adrenoceptors; undesirable side-effects at β_1- and β_2-adrenoceptors, including tachycardia and tremor, may occur but are limited.

Glucose metabolism is also affected by selective β_3-adrenoceptor agonists, which therefore have potential application in the treatment of type 2 (maturity onset, non-insulin-dependent) diabetes mellitus. This condition occurs typically in older people who are often obese; the overeating having stimulated insulin secretion which in turn causes down-regulation of insulin receptors. The anti-obesity and antidiabetic properties of β_3-agonists are therefore a convenient combination. β-Adrenoceptor stimulation induces insulin release from the β-cells of the pancreas, thereby increasing glucose uptake by cells and lowering blood sugar levels. Opposing this action is stimulation of glycogenolysis in the liver (Table 3.2). This latter effect prevents β-agonists from producing severe hypoglycaemia. BRL 35135 has been shown to improve glucose tolerance in normal rats and mice, that is, the rise in blood glucose concentration after a subcutaneous dose of glucose was abbreviated. The release of insulin is mediated via β_3-adrenoceptors, whereas the glycogenolytic effects are probably through β_2-adrenoceptors. As a consequence, the β_2-selective agonists, fenoterol, salbutamol and terbutaline, impair glucose tolerance and the β_1-selective agonist, denopamine, has no effect. Repeated administration of the β_3-agonist also improves sensitivity to insulin, possibly by increasing insulin receptor numbers. Thus, β_3-agonists appear to have potential in the treatment of type 2 diabetics who retain the capacity to secrete insulin, but who are resistant to its action (Arch *et al.* 1991).

4.5.7   Increase in Skeletal Muscle Mass

Certain β-adrenoceptor agonists have been shown to increase body protein content and muscle mass. This property is particularly useful in anti-obesity agents since the fat loss is not accompanied by muscle wasting. There has consequently been considerable interest in such compounds as feed additives for meat-producing animals to obtain leaner and bigger carcasses. There are also clinical applications to increase muscle growth and strength in wasting conditions, such as trauma from burns, infection, limb immobilization due to fracture, and in cancer cachexia.

Two of the earliest β-agonists shown to increase muscle mass and protein accumulation were clenbuterol and cimaterol (Table 4.5), which are recognized as β_2-selective agonists. Clenbuterol was introduced in the early 1970s for the treatment of asthma and is still available in certain European countries. It exerts a potent anabolic effect on skeletal muscle after only 2 days of treatment, which is attributed to increased protein synthesis, and the effect persists for ~10 days after treatment stops. There has been considerable debate concerning the nature of the receptor involved in the anabolic effect. Although the β_3-selective agonist, BRL 35135, increases the weight of some muscles of the rat (eg soleus), BRL 26830 is ineffective and therefore a β_3-adrenoceptor does not appear to be involved. Most studies have failed to demonstrate an increase in muscle mass with other selective β_2-adrenoceptor agonists, including salbutamol and terbutaline. However, when administered to rats by continuous intravenous infusion via osmotic minipumps, salbutamol did increase muscle mass and the effect was blocked by the selective β_2-antagonist, ICI 118,551 (Choo *et al.* 1992). Similarly, in human volunteers, a sustained-release preparation of salbutamol has been shown to increase the strength by ~10% in certain muscles. Whether this was due to an increase in muscle mass, the dimensions of which did not alter, is unlikely; it may have arisen from a change in the contractile properties of the muscle (Martineau *et al.* 1992). There are also compounds that have no apparent β_2-adrenoceptor stimulant activity, such as ractopamine and BRL 46101 (Table 4.5), but which nevertheless stimulate skeletal muscle growth (Arch *et al.* 1991). Finally, the anabolic effects of clenbuterol are poorly antagonized by propranolol, and this weak antagonism is equally found in the

L- and D-isomers, indicating that it is not due to typical β-adrenoceptor blockade. While further work is still necessary, it appears that the β-adrenoceptor mediating the increase in muscle mass by clenbuterol may be of an atypical β_2-adrenoceptor type.

The anabolic activity of clenbuterol, originally used as an additive to animal feeds to promote lean meat production and now recognized as having therapeutic potential in wasting conditions, soon became identified by athletes for having performance-enhancing potential. Its application both in increasing muscle strength for explosive events such as weight-lifting and sprinting, and for increasing muscle mass in body building, soon became apparent. At the Barcelona Olympic Games, two British weight-lifters were disqualified from competition when they were found to have been using clenbuterol. The defence has been that they were using this drug for mild asthma. The IOC has confirmed that clenbuterol is a banned substance. However, this has raised the issue of whether all β_2-agonists should be banned, which would severely disadvantage asthmatics from competitive sport. At present there is no justification for such a ban, since administration of short-acting β_2-agonists by inhalation would not produce the sustained elevation of plasma levels that appear to be necessary to produce significantly enhanced muscle mass. Only bitolterol, orciprenaline, rimiterol, salbutamol and terbutaline in aerosol form are permitted by the IOC Medical Commission (Badewitz-Dodd 1991). There is good reason to prohibit the use of regularly administered oral clenbuterol and related compounds, since they cannot be considered the drugs of choice for the relief of asthma. They could, however, be claimed to give an unfair advantage in competitive sports. In non-competitive activities, such as body building, it could be argued that the anabolic activity of β-agonists including clenbuterol is a safer alternative to the much-abused anabolic steroids. Pharmacological principles have been applied by body builders in their use of clenbuterol to determine suitable doses. It is recommended that body temperature is measured, which monitors thermogenic activity. The number of daily tablets can be increased until the maximum temperature is reached (~99.2 °F) and the duration of each dose (~12 hr) can be monitored (Duchaine 1992).

5

Adrenoceptor Antagonists

5.1 Introduction

Drugs whose occupancy of the adrenoceptor results in a tissue response are agonists and these sympathomimetic amines were described in Chapter 4. Drugs which occupy adrenoceptors but generate no or negligible responses are the adrenoceptor antagonists. The occupancy of the receptor by these substances prevents the access to the receptor of the neurotransmitter, noradrenaline, and circulating adrenaline and exerts inhibitory effects upon sympathetic function. They also antagonize the effect of exogenously administered sympathomimetic amines and their pharmacological properties are the subject of this chapter. Most of these drugs are competitive in that they compete with the agonist for receptor binding and may be displaced or overcome by excessive levels of agonist. The characteristics of the blockade have been discussed in Chapter 3, together with the methods of calculating the affinities of competitive antagonists by Schild analysis. An important feature of competitive blockade is the parallel displacement of agonist dose–response curves in the presence of the antagonist, with no depression of maximum, indicating the surmountable nature of the antagonism. A few irreversible antagonists that bind covalently to adrenoceptors are also described.

The ever-increasing number of adrenoceptor subtypes to be described has been largely a result of antagonists showing selectivity for responses of some tissues compared with others. Thus, antagonists are conveniently subdivided primarily according to whether they are selective for α- or β-adrenoceptors. Antagonists displaying selectivity for adrenoceptors in the secondary subdivisions of α_1, α_2, β_1 and β_2 are also readily available and have well-defined therapeutic application or potential. The tertiary subdivision into the α_{1A}, α_{1B}, α_{2A} and α_{2B} types was made on the basis of selective antagonism or selective displacement of radioligand binding in different tissues, the uses and pharmacology of such drugs in the context of this receptor subdivision remains of academic importance only. Antagonists of this type, including chloroethylclonidine (α_{1B}), 5-methylurapidil (α_{1A}), (+)-niguldipine (α_{1A}), ARC 239 (α_{2B}), spiroxatrine (α_{2B}) and WB 4101 (α_{1A} and α_{2C}) were considered in Chapter 3 and will not be described further in this chapter.

5.2 α-Adrenoceptor Antagonists

Historically, the first α-adrenoceptor antagonist to be described was the mixture of alkaloids extracted from ergot which is a product of *Claviceps purpurea*, a fungal contaminant of rye grain. This mixture was first isolated in crystalline form and its pharmacological properties described in 1906 by Dale. The complex was termed ergotoxin and the antagonism of the pressor response to adrenaline, converting it to a depressor response, was the first description of the α-adrenoceptor blocking properties. However, it was to be another 40 years before the concept of α-adrenoceptors was proposed and the first α-adrenoceptor antagonists were introduced.

5.2.1 Non-selective α-Adrenoceptor Antagonists

5.2.1.1 Ergot alkaloids

The ergotoxin first extracted by Dale was in fact a mixture of ergocornine, ergocristine, α-ergocryptine and β-ergocryptine. Their structures are shown in Table 5.1, together with other ergot alkaloids including ergotamine and ergonovine (ergometrine). These alkaloids are amine derivatives of D-lysergic acid, to which they are converted when hydrolysed. The other products of hydrolysis are either amines, in the case of ergometrine (amine alkaloids), or amino acids, of which one is proline and the other is either phenylalanine, leucine, isoleucine or valine (amino acid alkaloids). The D-isomers are inactive. The amino acid alkaloids are active competitive α-adrenoceptor antagonists with a long duration of action. The amine alkaloids, which lack the amino acid side-chain, such as ergometrine, do not have α-adrenoceptor blocking properties. However, these compounds have a direct smooth muscle stimulant action on the uterus and blood vessels. This property is shared by the amino acid alkaloids and, although once thought to be independent of receptor-mediated events, it is now believed to involve activation of α- and 5-hydroxytryptamine (5-HT) receptors. The contractions of the uterus induced by ergometrine are antagonized by the 5-HT antagonist, cyproheptadine (Hashimoto *et al.* 1977).

Dihydrogenation at positions 9 and 10 of the lysergic acid nucleus increases α-adrenoceptor blocking activity and reduces the smooth muscle stimulating properties. Dihydroergocryptine (DHE) is non-selective between α_1- and α_2-adrenoceptor subtypes. It has, however, been employed in early radioligand binding studies of α_1-adrenoceptors, with prazosin used as the unlabelled ligand (Figure 3.4) (Bylund 1987). The vasoconstrictor activity is not completely eliminated by dihydrogenation, and dihydroergotamine retains partial agonist activity preferentially on capacitance vessels (veins).

Ergotamine produces a rise in blood pressure through arteriolar vasoconstriction when administered intravenously. This opposes the expected fall in blood pressure arising from α-adrenoceptor blockade of vasomotor tone. There is a reduction in blood flow through various organs, including the extremities, with the ergot alkaloids

Table 5.1 Structures of ergot alkaloids and related compounds

	R	
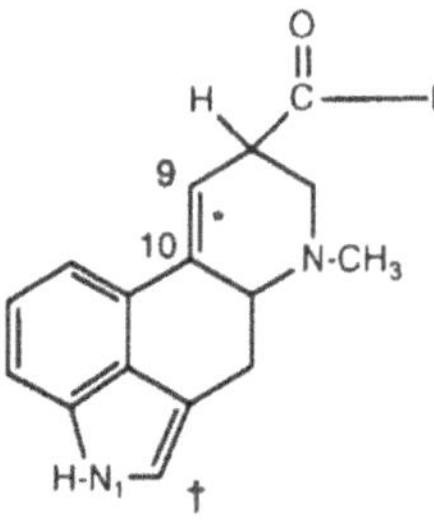	—OH	d-Lysergic acid
	—N(CH$_2$CH$_3$)$_2$	Amine alkaloids d-Lysergic acid Diethylamine (LSD)
	—NH-CHCH$_2$OH $\vert$ CH$_3$	Ergometrine (ergonovine)
	NH-CH $\genfrac{}{}{0pt}{}{C_2H_5}{CH_2OH}$	Methysergide (-CH$_3$ at N$_1$)
	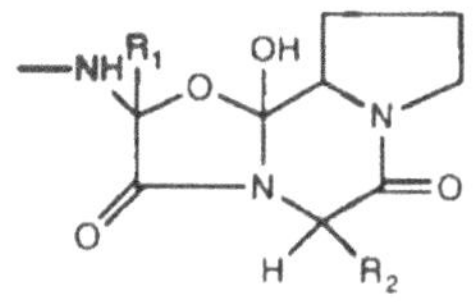 —NH structure	Amino acid alkaloids

R$_1$	R$_2$	
-CH$_3$	-CH$_2$-C$_6$H$_5$	Ergotamine
		Ergotoxine complex
-CH(CH$_3$)$_2$	-CH(CH$_3$)$_2$	Ergocornine
"	-CH$_2$C$_6$H$_5$	Ergocristine
"	-CH$_2$CH(CH$_3$)$_2$	α-Ergocryptine
"	-CH(CH$_3$)CH$_2$CH$_3$	β-Ergocryptine
"	-CH$_2$CH(CH$_3$)$_2$	Bromocriptine (Br at †) (Parlodel[R])

Note: * Dihydro derivatives produced by hydrogenation at positions 9 and 10.

and also damage to the capillary endothelium. Gangrene is a toxic result of their chronic intake. This was a characteristic feature of ergotism due to chronic ingestion of ergot-infested rye during the Middle Ages, when epidemics of gangrene of feet and hands occurred, and the extremities wasted away, accompanied by severe burning sensations. The disease was known as St Anthony's fire because it was relieved by pilgrimage to the shrine of St Anthony, probably because the sufferer moved to an area of contamination-free rye grain. In therapeutic use, however, providing appropriate contraindications are adhered to, including peripheral vascular disease and sepsis, then gangrene is not a problem.

Uses of ergot alkaloids Because of the multiple toxic effects and actions at other receptor sites, the ergot alkaloids have no therapeutic uses derived from their α-adrenoceptor blocking properties. Their therapeutic uses in the treatment of migraine and in obstetrics are due to other pharmacological properties.

In *cerebrovascular insufficiency*, the use of dihydroergotoxine may be in part due to α-adrenoceptor blockade. This is a mixture of the dihydro derivatives of ergocristine, ergocornine, α-ergocryptine and β-ergocryptine in the ratio 3 : 3 : 2 : 1, also known as co-dergocrine (Hydergine^R). The overall vasoconstrictor activity of the ergot alkaloids is replaced in co-dergocrine by vasodilatation, particularly of skin vessels, causing flushing. It has been found to be of benefit in treating elderly patients with moderate dementia; whether cerebral vasodilatation alone can account for this is unlikely. 'Steal' of blood flow away from ischaemic areas of the brain to already adequately perfused regions may occur because the ischaemic regions are probably fully dilated by local autoregulation. Co-dergocrine was believed to lower blood pressure through α-adrenoceptor blockade, however, more recently agonist activity at prejunctional dopamine D_2 receptors has also been identified. Stimulation of these receptors inhibits neurotransmitter release from sympathetic nerve endings, resulting in blockade of sympathetic tone to the heart and vasculature. The inhibition of pressor responses to periarterial nerve stimulation of the perfused mesenteric vascular bed by co-dergocrine is partially antagonized by the D_2 antagonist, domperidone (Lefebvre & Devreese 1991) (see Chapter 12).

In *obstetrics*, ergometrine or methylergometrine are used to contract the uterus to reduce the extent of postpartum haemorrhage. It is given after the delivery of the placenta by intramuscular injection. However, there is no strong evidence that it does reduce bleeding with normal vaginal delivery and its routine use may be unecessary, especially as it may cause side effects and increase uterine pain. Ergometrine may be used orally to stimulate the uterus in cases of delayed uterine involution.

Migraine headache is a major application of the ergot alkaloids. In 'classic' migraine, initial visual disturbances known as the aura precedes the headache by 10–30 min. The headache is throbbing, the pain often originating in the temple or forehead and lasting 2–3 hr but occasionally up to 12 hr. In 'common' migraine there is no aura, but both forms are associated with nausea and vomiting. The precise pathophysiology has still to be fully understood. The main hypothesis has been that large cerebral blood vessels are constricted by 5-HT possibly released from platelets and probably mediated via 5-HT$_2$ receptors. The consequent cerebral ischaemia may cause the aura. This is followed by an increased blood flow in cranial vassels particularly of the scalp and dura. A related hypothesis to the latter component is that arteriovenous shunts (anastomoses) dilate, allowing increased blood flow which in turn permits blood to bypass the cerebral tissues. These cause ischaemia which may produce the aura and engorgement in the cranial arteries in the cephalic region. The increased amplitude of pulsations in the cranial arteries, including the meningeal branches of the carotid artery, are thought to be the source of the pain. 5-HT released from platelets or from serotonergic nerves possibly arising from the midbrain raphe system may also act on sensory pain fibres via 5-HT$_3$ receptors.

Ergotamine and methysergide have selective vasoconstrictor activity on the carotid vascular bed and the arteriovenous shunts in the cephalic region. The resulting reduction in blood flow in these regions is believed to be a component of their antimigraine activities. Methysergide appears to behave as a 5-HT$_1$-like receptor agonist. The vasoconstrictor activities of ergotamine on the cephalic arteries and arteriovenous shunts is, however, not through 5-HT receptors (Saxena & Ferrari 1989). Dihydroergotamine is a 5-HT antagonist in rat aorta which is described as being competitive irreversible (Doggrell 1992). Thus, 5-HT$_1$-like agonists [eg methysergide (Deseril^R), sumatriptan (Imigran^R)] would appear to relieve migraine through constriction of the dilated arteriovenous shunts or cranial arteries, 5-HT$_2$ antagonism [methysergide, pizotifen (Sanomigran^R)] would block the 5-HT-mediated constriction of cerebral vessels, and 5-HT$_3$ antagonism [ondansetron (Zofran^R)] would block the sensory nerve pathways involved in pain and possibly the nausea. In view of the structural similarity of ergot alkaloids to agents active at 5-HT receptors, such as methysergide and LSD, their antimigraine properties may well be due to 5-HT-related activity. Whatever the precise mechanism of action of ergot alkaloids in the treatment of migraine, their α-adrenoceptor blocking properties do not appear to be involved since other α-antagonists do not have such an action.

Ergotamine (Lingraine^R tablets, Ergostat^R sublingual, Medihaler-Ergotamine^R metered dose aerosol) and dihydroergotamine (Dihydergot^R injection) are used for the treatment of acute migraine attacks. Ergotamine is sometimes combined with caffeine (Cafergot^R) to aid its absorption.

Toxicity and side-effects of ergot alkaloids The vasoconstrictor properties are a potential cause of side-effects of the ergot alkaloids and although these effects are less in the dihydrogenated derivatives, they may still be troublesome. They are contraindicated in patients with hypertension, peripheral vascular disorders and coronary artery disease since they may precipitate coronary vasospasm and angina attacks. They should not be combined with β-blockers, which are used in the prophylaxis of migraine, since they would further induce peripheral vasoconstriction. They should also not be co-administered with erythromycin which may enhance their effects. Other sites of toxicity in chronic ergotism are nausea and vomiting due to stimulation of the medullary vomiting centre, and muscular pain and weakness.

5.2.1.2 β-Haloalkylamines

This group was the first of the synthetic α-adrenoceptor antagonists. The pharmacological properties of dibenamine were first described by Nickerson & Goodman (1947) who showed that the blockade was slow in onset and very prolonged (Table 5.2). Subsequently, phenoxybenzamine (Dibenyline^{UK}, Dibenzyline^{US}) remains the only member of this class of antagonists to be used therapeutically. The mechanism of blockade by these antagonists is by irreversible covalent binding to α_1- and α_2-adrenoceptors. Phenoxybenzamine has slight selectivity for α_1-adrenoceptors. They are converted to a cyclized ethyleniminium (aziridium) ion by release of the halogen. The initial blockade is competitive in that it may be prevented by an excess of agonist. The initial competitive interaction with the receptor is of the haloalkylamine in the form of the ethyleniminium ion rather than the parent compound, being held in proximity to an anionic receptor site by the charged ring nitrogen (Figure 5.1). The binding then progresses to the irreversible phase when the anionic receptor site is alkylated. This effectively removes a proportion of the receptors.

The blockade is characterized by an initial parallel displacement of the concentration–response curves, followed by depression of the maximum response. The depressed maximum arises when all spare receptors have been inactivated (Figure 3.5, Chapter 3). Receptor protection experiments indicate the irreversible interaction with the α-adrenoceptor; in these experiments the alkylating agent and an

Table 5.2 α-Adrenoceptor antagonists

<u>β-Haloalkylamines</u> (irreversible)

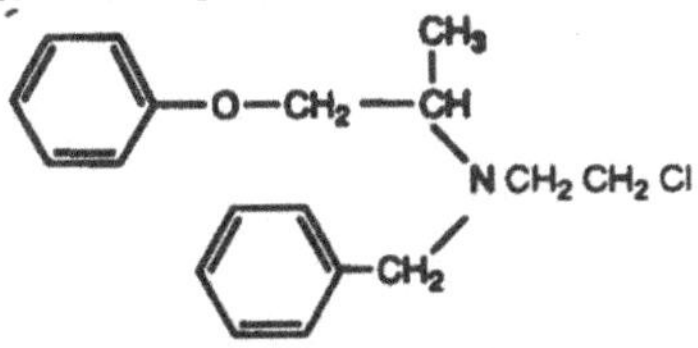

Phenoxybenzamine (α_1/α_2)
(Dibenyline[UK], Dibenzyline[US])

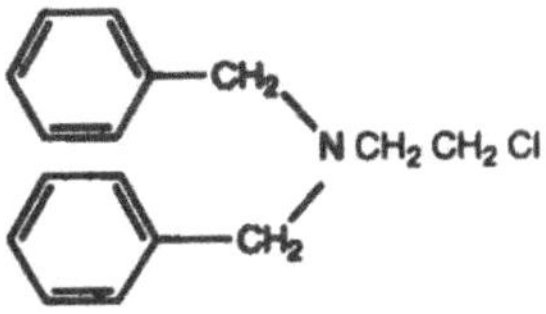

Dibenamine (α_1/α_2)

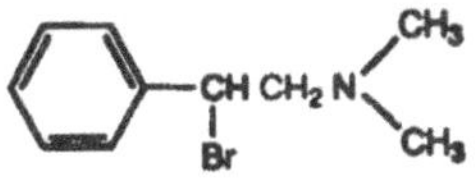

N,N-Dimethyl-2-bromo-2-phenylethylamine (DMPEA)

<u>Benextramine</u> (irreversible)

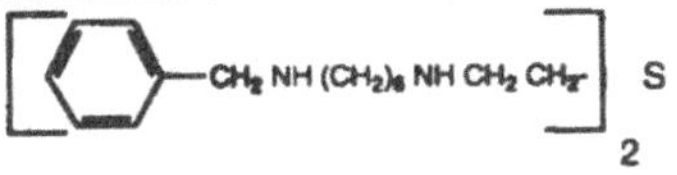

<u>Imidazolines</u>

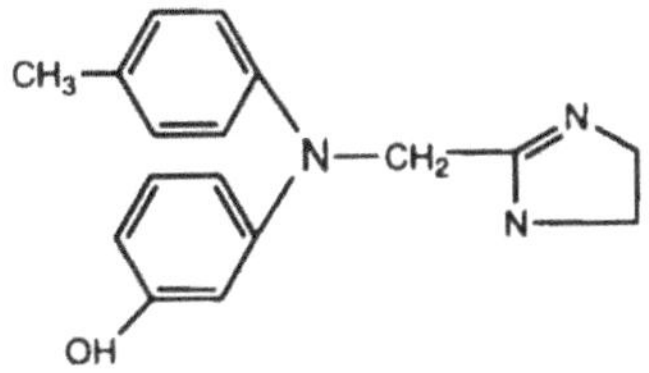

Phentolamine (Rogitine, Regitine[R])

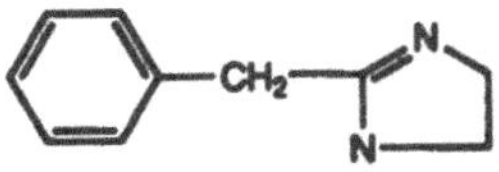

Tolazoline (Priscol[R])
(discontinued)

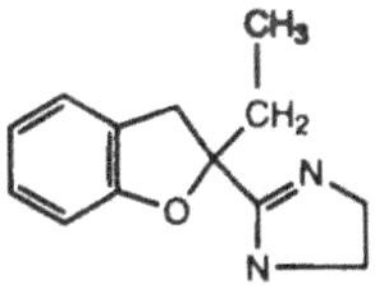

Efaroxan (α_2 and I_1 binding)

<u>Phenoxyalkylamines</u>

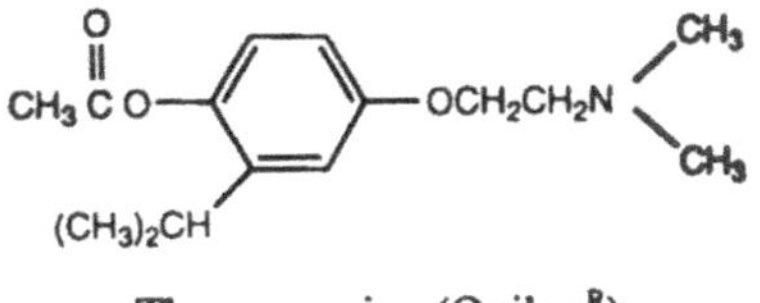

Thymoxamine (Opilon[R])

(Continued)

Table 5.2 *Continued*

Phenylpiperazines

Urapidil

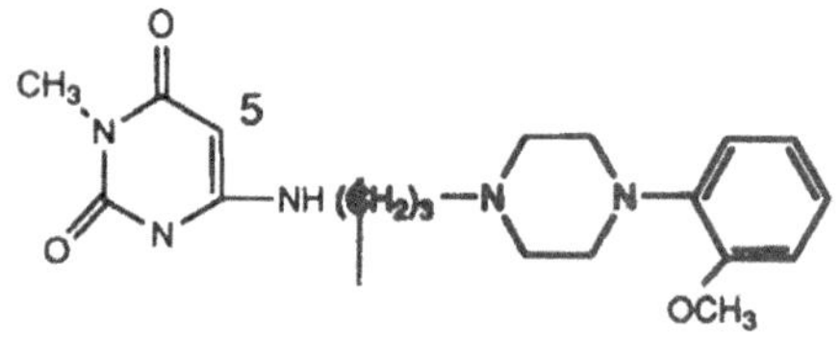

α_1

5-Methylurapidil

CH$_3$ group at position 5

ARC 239

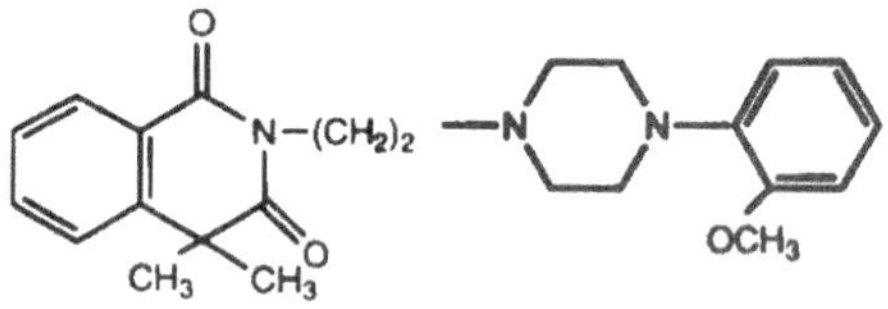

α_{2B}

BMY 7378

α_{1D}

(also a 5-HT$_{1A}$ agonist)

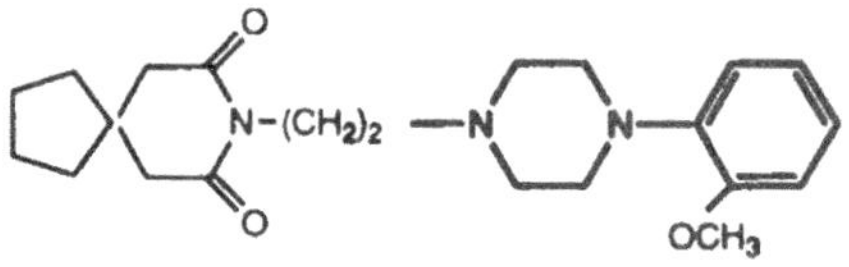

Quinazolines

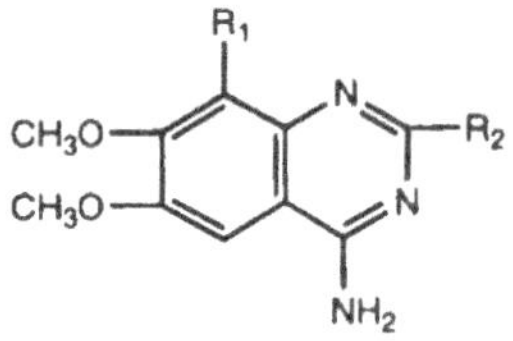

	R$_1$	R$_2$	
Prazosin (Hypovase^R, Minipress^R)	H	Ditto	
Doxazosin (Cardura^R)	H	Ditto	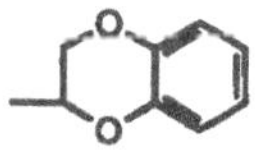
Terazosin (Hytrin^R)	H	Ditto	
Trimazosin (Cardovar^R)	-OCH$_3$	Ditto	-OCH$_2$C(CH$_3$)$_2$ OH
Bunazosin (Detantol^{JAP})	H	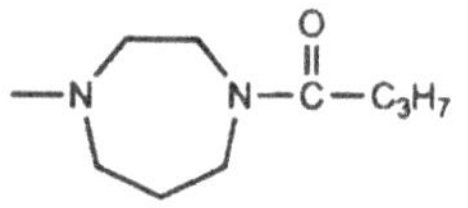	
Alfuzosin (Xatral^R)	H	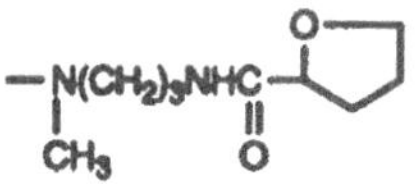	

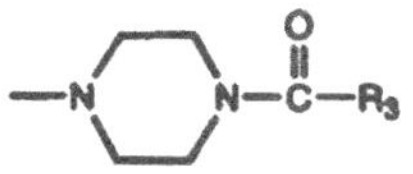

(Continued)

Table 5.2 *Continued*

Indoles and indolealkylamines
Indoramin (Baratol^R, Doralese^R)

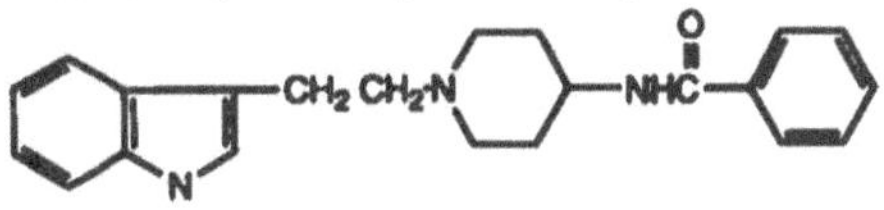

α_1

Yohimbane alkaloids

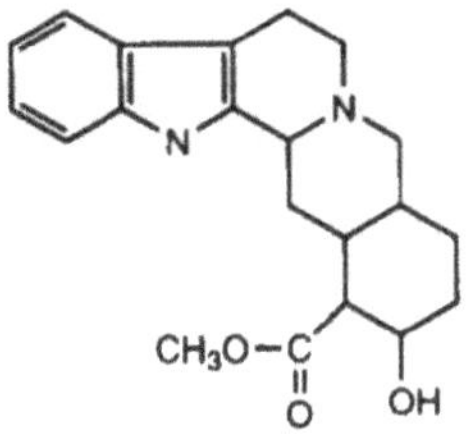

Yohimbine (α_2)
Rauwolscine (α_2) } diastereoisomers
Corynanthine (α_1)

Benzodioxans

Idazoxan

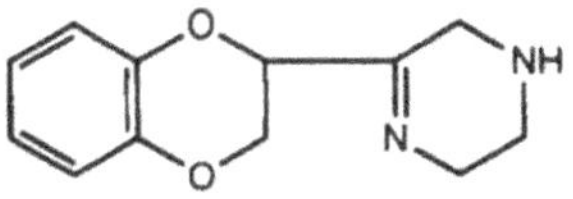

α_2

Imiloxan

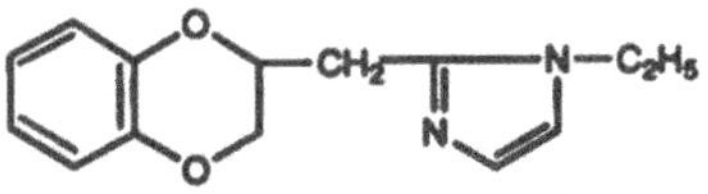

α_2

WB 4101

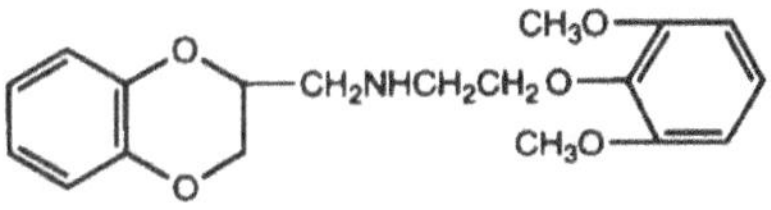

α_{1B}

Spiroxatrine

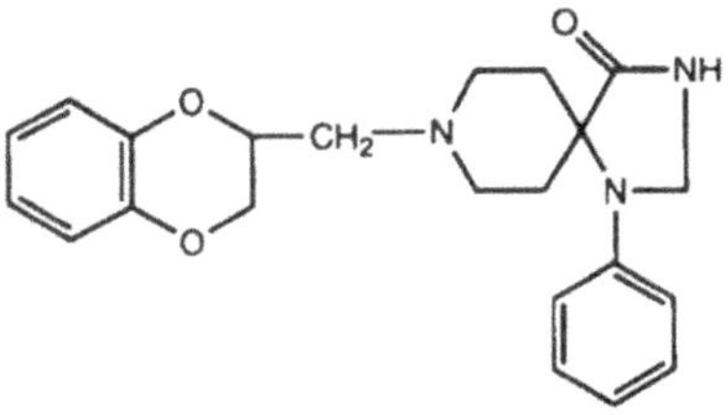

α_{2B}

Abanoquil
(UK-52,046)

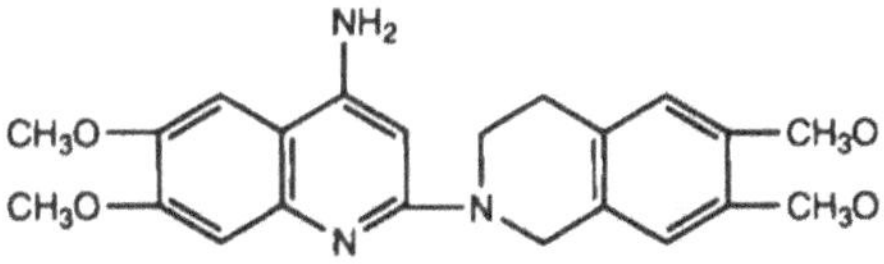

α_1 selective with
minimal reflex
tachycardia or
postural hypotension

RS-15385-197

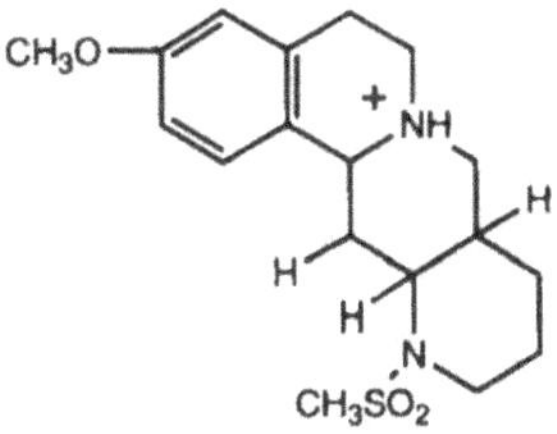

Cl$^-$

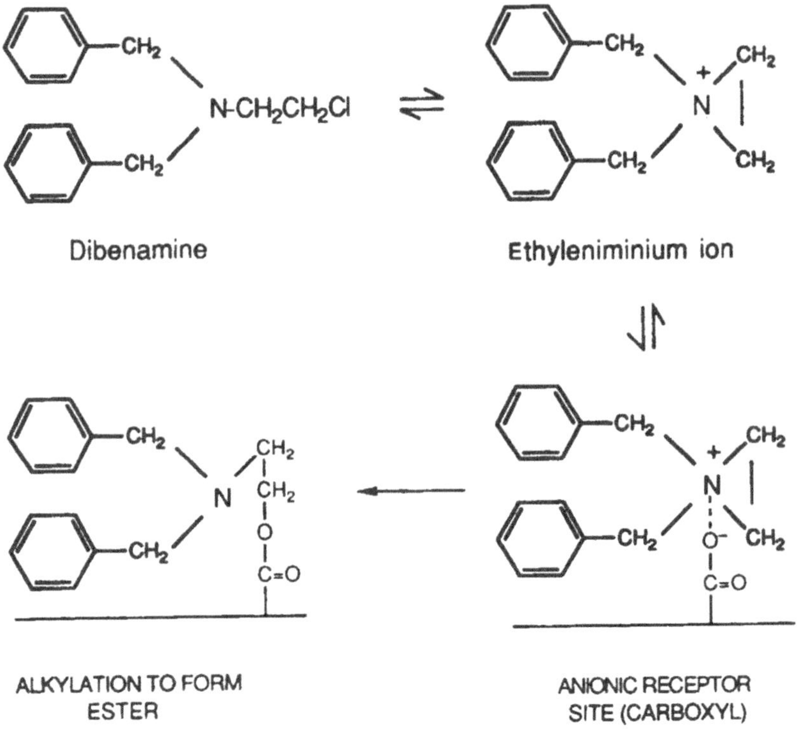

Dibenamine Ethyleniminium ion

ALKYLATION TO FORM ANIONIC RECEPTOR
ESTER SITE (CARBOXYL)

Figure 5.1 Interaction of the β-haloalkylamine, dibenamine, with the anionic site on the α-adrenoceptor.

agonist or antagonist with competitive activity are co-incubated with the tissue or membrane fraction (see Figure 9.4 for muscarinic receptor). They protect the receptor from inactivation and thus prevent blockade of the response or reduction of radioligand binding sites (see Figure 3.4). The degree of protection can, however, be low and unpredictable. Furthermore, pharmacological protection may be produced with little protection against the reduction of binding sites. This suggests that the sites of interaction of noradrenaline and dibenamine may not be identical, but that dibenamine alkylates an adjacent site on or near the α-adrenoceptor. The irreversible phase of antagonism is also prevented by incubating with sodium thiosulphate throughout the exposure to dibenamine. This behaves as a nucleophile that interacts with the highly reactive electrophilic ethyleniminium ion and impedes its subsequent attachment to the receptor. When added *after* the irreversible blockade is established, sodium thiosulphate is ineffective (Nickerson & Gump 1949).

Belleau (1960) made one of the first attempts to speculate on the nature of the sites that dibenamine alkylates. He proposed that although sulphydryl, carboxyl and phosphate groups were possibilities based on chemical considerations, the latter two were the most likely nucleophilic species for attack.

The blockade by dibenamine and phenoxybenzamine slowly reverses and in isolated tissues the response recovers by ~12–30% after 18 hr. There is a reasonable correlation between the ease of formation of the ethyleniminium ion, the time of onset, and the reversibility of antagonism by the haloalkylamines (Triggle 1965). Reversal is due in part to the hydrolysis of the ester formed by alkylation of the anionic receptor site. A group of N,N-dimethyl-2-halogeno-2-phenylethylamines (DMPEA, Table 5.2) cyclize to the ethyleniminium ion rapidly and the duration of their α-adrenoceptor blockade is only a matter of a few hours. There is some evidence of a two site interaction of the dibenamine class of antagonist with the receptor. Binding to one site is slowly reversible while binding to the other site is more rapidly reversed and may be the sole binding site for a DMPEA-type antagonist. Thus, the dibenamine class of antagonist may not simply alkylate the noradrenaline recognition site on the receptor, but may interact with one or more additional sites. The evidence with these antagonists is that the short-duration interaction is with a carboxylate ion which may be the same as the noradrenaline recognition site (Triggle & Triggle 1976).

Recovery from irreversible α-adrenoceptor antagonism *in vivo* is also due to the *de novo* synthesis of receptor protein. The reappearance of α_1-adrenoceptor binding sites after a single intravenous dose of phenoxybenzamine to rabbits takes 5–8 days to reach 50%, whereas α_2-adrenoceptor binding sites recover more quickly (2–3 days to achieve 50% recovery). The pressor response to phenylephrine, however, was completely restored at 8 days even though receptor numbers were still reduced; presumably the receptor reserve had not fully developed (Hamilton *et al.* 1982).

β-Haloalkylamines do not bind exclusively to the α-adrenoceptor; they also irreversible alkylate other receptors [histamine, 5-HT, Ach (Chapter 9, Figure 9.4)] and both neuronal and extraneuronal noradrenaline uptake sites (see Chapter 2).

Pharmacology and therapeutic uses   Phenoxybenzamine causes a progressive decrease in peripheral resistance, but the fall in supine blood pressure is relatively modest. In the standing position, however, there is normally a high degree of sympathetic vasomotor tone causing noradrenaline release onto vascular α_1-adrenoceptors (see Chapter 1). This is blocked by phenoxybenzamine with a consequent fall in blood pressure. There is tachycardia and increased cardiac output due to reflex sympathetic activity arising from the fall in blood pressure. Also contributing to the increased heart rate is an elevated sympathetic output to the heart due to blockade of prejunctional α_2-adrenoceptors. These normally exert negative feedback on transmitter release but their blockade results in enhanced noradrenaline release onto cardiac β-adrenoceptors, which of course are not blocked by phenoxybenzamine. Pressor responses to α-adrenoceptor agonists are blocked, and for those with vasodilator β-adrenoceptor activity (eg adrenaline), a depressor response is revealed. All other α-adrenoceptor-mediated responses are predictably blocked including most smooth muscle contractile effects. For example, the contraction of the vas deferens in response to sympathetic nerve stimulation is impaired resulting in ejaculation failure.

The principle use of phenoxybenzamine is in the management of phaeochromocytoma (tumour of the adrenal medulla). The increased mass of chromaffin cells secrete large amounts of adrenaline into the circulation causing secondary hypertension due to stimulation of vascular α_1-adrenoceptors, which may be paroxysmal (episodes of raised blood pressure) or sustained. Surgical removal of the tumour is the treatment of choice, however phenoxybenzamine does have a place in the management of the disease: (1) in the presurgical phase, (2) during surgery to prevent α-adrenoceptor stimulation by the excessive release of adrenaline caused by manipulation of the adrenal gland, and (3) as the main treatment in patients who are poor surgical risks or who have malignant phaeochromocytoma. The tachycardia associated with the disease may be antagonized by β-adrenoceptor antagonists, but only after administration of

211

phenoxybenzamine, otherwise the α-adrenoceptor-mediated pressor effects would be enhanced. Postural hypotension (Chapter 1) is the major adverse effect of phenoxybenzamine.

Benextramine is also an irreversible α-adrenoceptor antagonist displaying no selectivity for α_1- or α_2-adrenoceptors (Table 5.2). Analogues of this agent which lack the electrophilic centre are competitive antagonists.

5.2.1.3 Imidazolines

Phentolamine (Rogitine[UK], Regitine[US]) and tolazoline (Priscoline, no longer available) are competitive α-adrenoceptor antagonists with no selectivity for the α_1- or α_2-subtypes (Table 5.2). They are structurally related to the imidazoline α-adrenoceptor agonists (Table 4.3) and tolazoline has partial agonist activity. They have poor specificity for α-adrenoceptors and display a wide range of activity at other sites. For example, phentolamine blocks 5-HT receptors (Doggrell 1992) and stimulates histamine release from mast cells. They both stimulate the gut to cause diarrhoea and also stimulate the secretion of saliva, sweat and lacrymal secretions. These effects are blocked by atropine and are therefore mediated via muscarinic receptors, possibly through inhibition of cholinesterase. They also enhance gastric acid secretion, an effect similar to that of histamine to which they are closely related structurally. Another histamine-like effect is vasodilatation, which occurs independently of blockade of α-adrenoceptors and which contributes to the fall in blood pressure. Generally, the side effects are less pronounced with phentolamine than with tolazoline, the latter having become largely obsolete.

Phentolamine is used in the short-term management of hypertensive crises that occur in phaeochromocytoma. In common with all non-selective α-adrenoceptor antagonists, phentolamine has been notably ineffective, however, in the control of primary (essential) hypertension. Although the blockade of postjunctional vascular α_1-adrenoceptors will inhibit sympathetic vasomotor tone and favour a lower blood pressure, the blockade of prejunctional α_2-adrenoceptors will enhance neurotransmitter release. This will have a two-fold effect. (1) The increase in baroreceptor-mediated reflex sympathetic activity to the heart arising from the fall in blood pressure will lead to unopposed tachycardia via β-adrenoceptors, which will tend to restore blood pressure. (2) The enhanced release of noradrenaline from sympathetic neurones will compete with the antagonist at the postjunctional α-adrenoceptors thus reducing its effectiveness. A consequence of this is that these non-selective antagonists are more effective against the responses to exogenous noradrenaline than those to sympathetic nerve stimulation. Thus, side-effects such as nasal congestion, reflex tachycardia and ejaculation failure and low effectiveness make the non-selective antagonists unsuitable for use in essential hypertension. Phentolamine may be used to counteract hypertension associated with MAO inhibitor/sympathomimetic amine interactions (ie the 'cheese effect', Chapter 2) or to prevent tissue necrosis when adrenaline injections are inadvertently made outside the vein.

5.2.2 Selective α_1-Adrenoceptor Antagonists

5.2.2.1 Prazosin

Prazosin is a quinazoline and was the first selective α_1-adrenoceptor antagonist to be introduced (Table 5.2). It is a potent antagonist with pA_2 values ranging from 9.4 to

11.25 in isolated tissues where the responses are mediated via α_1-adrenoceptors. In contrast, the pA_2 values for α_2-adrenoceptor-mediated responses such as inhibition of contractile responses of the rat vas deferens and guinea-pig ileum to sympathetic and cholinergic nerve stimulation, respectively, is of the order of 5.5–6.5 (Docherty 1989). Its selectivity for α_1-adrenoceptors is therefore 1000-fold. Prazosin binds directly to the α_1-adrenoceptor, as illustrated by the specificity and saturability of binding of radiolabelled [³H]prazosin, and its displacement by unlabelled selective agonists (methoxamine) and antagonists (corynanthine). Furthermore, the irreversible blockade of α-adrenoceptors by phenoxybenzamine may be prevented by receptor protection with prazosin.

Prazosin antagonizes postjunctional α_1-adrenoceptors and therefore inhibits the effects of both sympathetic nerve stimulation and exogenously added noradrenaline equally. The major effects of prazosin are upon the cardiovascular system where blockade of postjunctional α_1-adrenoceptors of arterioles and veins causes vasodilatation. Peripheral resistance falls and venous return to the heart is impaired; this results in a hypotensive effect. Unlike the non-selective α-adrenoceptor antagonists (phentolamine) and vasodilators such as hydralazine, there is less likelihood of reflex tachycardia arising from the fall in blood pressure via baroreceptor mechanisms (see Chapter 1). The reason for this is two-fold. First, the prejunctional α_2-adrenoceptors, which remain unblocked, allow endogenous catecholamines to exert normal or enhanced inhibitory tone on noradrenaline release from sympathetic nerves. Thus sympathetic nerve activity to the heart may be reduced. In fact, plasma levels of noradrenaline are elevated due to the reflex sympathetic activity induced by the fall in blood pressure. Second, prazosin, along with other α_1-adrenoceptor antagonists including urapidil, alfuzosin and doxazosin, appears to have an effect via central α_1-adrenoceptors causing inhibition of central sympathetic tone (Ramage 1986).

Prazosin may cause increases in plasma renin and angiotensin II in response to the fall in blood pressure due to a modest reflex sympathetic activity to the kidneys. This may be more noticable with non-selective antagonists such as phentolamine where α_2-adrenoceptors mediating *inhibition* of renin release would be blocked, thus allowing the β-adrenoceptor-mediated facilitation of renin release to dominate.

Therapeutic doses of prazosin have a favourable profile of activity upon plasma lipids. Epidemiological studies have indicated that high concentrations of low density lipoprotein (LDL) cholesterol are a positive risk factor for coronary artery disease, while high plasma levels of high density lipoprotein (HDL) serve a protective role. While some antihypertensives such as atenolol and thiazide diuretics have a detrimental effect upon plasma lipids, during prazosin treatment there is an increase in HDL and fall in plasma total cholesterol and triglycerides.

Therapeutic uses of prazosin The effective blood pressure lowering activity of prazosin makes it a useful drug for the treatment of mild to moderate (Stages 1 or 2) essential hypertension (diastolic 90–99 and 100–109 mmHg, respectively). Prazosin and its newer congeners are now recognized as effective agents for the first-line treatment of hypertension. It lowers blood pressure in both the standing and recumbent positions. It may be used alone or more often as a second drug in patients not adequately controlled by a diuretic alone. The effects of prazosin are additive to those of diuretics, methyldopa (see Chapter 6), β-adrenoceptor antagonists and direct vasodilators (eg hydralazine). There does not appear to be tolerance to the antihypertensive activity.

The vasodilatation in veins and consequent reduction of venous pressure and preload on the heart is the basis for the use of prazosin in congestive cardiac failure (CHF). The short-term effect is to increase cardiac output and reduce the pulmonary congestion associated with heart failure. In contrast to the antihypertensive activity, however, tolerance develops to the vasodilator activity in CHF patients after a few months of treatment. Responsiveness may be re-established by increasing the dose of prazosin or by temporarily stopping the drug. The mechanism for this tolerance may be the accumulation of salt and water in the vascular walls which reduces vasodilator effectiveness. Increased diuretic dosage, particularly with spironolactone, has been found to reverse the tendency towards fluid retention in these patients. Alternative explanations include a progressively increasing sympathetic activity and stimulation of the renin–angiotensin system which exerts opposing vasoconstrictor activity. Prazosin does not appear to prolong the life of CHF patients (Stanaszek *et al.* 1983).

Prazosin may be useful in the treatment of patients with mitral valve dysfunction or aortic stenosis; the beneficial effects observed in limited trials suggest that the improvement in cardiac performance may be due to the reduction in after-load.

The effectiveness of prazosin in the treatment of Raynaud's syndrome (see Chapter 3) is equivocal. Some degree of improvement in subjective symptoms can be achieved in ~50% of patients (Stanaszek *et al.* 1983). On theoretical grounds, prazosin should be effective in the treatment of coronary vasospasm in patients with variant angina (Prinzmatal's angina) (see also under β-blockers), however no clear benefit has been demonstrated.

Finally, prazosin is used in the treatment of urinary obstruction due to benign prostatic hypertrophy. In this disorder, increased sympathetic tone to the prostatic and urethral smooth muscle exacerbates the degree of obstruction of the urethra. Prazosin relaxes the trigone muscle of the bladder and the urethra and thus eases bladder emptying in patients with prostatic obstruction or parasympathetic decentralization due to spinal injury (Andersson 1988). It has been argued that if the α_1-adrenoceptor subtype in the prostate is different from that of the vasculature then selective antagonists for prostatic α_1-adrenoceptors should have improved efficacy in the treatment of benign prostatic hypertrophy. Initial studies showed that contractions of rat prostatic smooth muscle were mediated via α_{1A}-, whereas in humans they were of the α_{1C}-subtype (Couldwell *et al.* 1993). More recently, however, the contractions of human prostate also appear to be mediated via the α_{1A}-adrenoceptor, which is pharmacologically regarded as being the same as the cloned α_{1C}-adrenoceptor (Marshall *et al.* 1995).

Side-effects of prazosin The most dramatic side-effect likely to occur with prazosin is the so-called 'first dose phenomenon'. The blood pressure falls rapidly and there may be postural hypotension and fainting after taking the initial dose. It is due to the rapid vasodilatation without compensatory tachycardia – indeed it is accompanied by bradycardia. This effect of the first dose may be minimized by limiting the initial dose to 0.5 mg at bedtime, increasing the dose to 0.5 mg twice daily for 3–7 days, then 1 mg two or three times daily and adding further antihypertensives cautiously. If these precautions are taken, orthostatic hypotension is a problem in only a few patients. Other side-effects are associated with α_1-adrenoceptor blockade and include nasal congestion and urinary urgency, while non-specific adverse effects include dry mouth, nausia, headache and dizziness. These side-effects do not, however, limit treatment with prazosin.

Since a minor population of α_1-adrenoceptors may be present in the human airways, which mediate bronchoconstriction (Table 3.2) (Goldie *et al.* 1990), prazosin might be expected to cause bronchodilatation. Several studies have demonstrated mild bronchodilator activity in subjects with bronchial hyperreactivity and airway obstruction. It does not appear to protect against cold-induced bronchospasm in asthmatics, but inhaled prazosin imparts partial protection against exercise-induced constriction. Thus, although any clinically useful bronchodilatation is marginal, prazosin will not aggravate bronchospasm and may therefore be used in asthmatics for treatment of hypertension (Stanaszek *et al.* 1983).

5.2.2.2 *Doxazosin, terazosin and trimazosin*

These are congeners of prazosin (Table 5.2) which have essentially the same profile of pharmacological activity as does the parent compound. Doxazosin has approximately half the potency of prazosin at postjunctional α_1-adrenoceptors, but it showed 400 times greater affinity for α_1- than α_2-adrenoceptors from radioligand binding studies. In common with prazosin, doxazosin has no direct vasodilator effects. It lowers blood pressure through arteriolar and venous vasodilatation without compensatory tachycardia and has a favourable effect on plasma lipids, increasing HDL cholesterol in addition to the HDL/total cholesterol ratio. Unlike prazosin, which has a relatively short onset of antihypertensive effect and a short plasma half-life (2.5 hr), doxazosin has a slower onset and much longer elimination half-life (22 hr). This permits a once-daily administration in the treatment of mild to moderate hypertension (Young & Brogden 1988).

Terazosin has essentially the same activity as doxazosin. It has an affinity for α_1-adrenoceptors of about one-third that of prazosin, and its selectivity for α_1-adrenoceptors is illustrated by the fact that more than 100-fold higher concentration is required to displace [^{3}H]rauwolscine from α_2-adrenoceptor binding sites of rat brain than is needed to displace [^{3}H]-prazosin from rat liver α_1-adrenoceptors. Terazosin is more water soluble than prazosin and its absorption from the gastrointestinal tract is more complete and predictable than with prazosin, thus enabling better titration of the dose. Its elimination half-life is ~12 hr, the effect lasting over 18 hr. It is therefore normally administered once daily for the treatment of hypertension, starting at a dose of 1 mg daily given at night to avoid the first dose effect. The daily dose is then increased gradually as the blood pressure stabilizes to a maximum dose of 20 mg daily (Titmarsh & Monk 1987).

Trimazosin also displays α_1-adrenoceptor selectivity with pharmacological properties and side-effects similar to prazosin. It exerts hypotensive effects by similar mechanisms but additionally has direct vasodilator activity. It has a slow onset of action, possibly because of conversion to the biologically active 1-hydroxy metabolite. The plasma half-life is equivalent to that of prazosin but it is regarded as having a longer duration of action similar to that of terazosin (Van Zwieten 1988).

5.2.2.3 *Indoramin*

This is a selective α_1-adrenoceptor antagonist that lowers blood pressure without inducing reflex tachycardia (Table 5.2). The bioavailability of indoramin is poor and it probably undergoes substantial first-pass metabolism in the liver. The elimination half-life is ~5 hr so that two or three times daily dosage is necessary. As with other α_1-adrenoceptor antagonists, the dose should be increased gradually. Adverse drug reactions include drowsiness, dry mouth, nasal congestion, weight gain and ejaculation failure. Indoramin reduces the occurrence of vasospasm of the digits in patients suffering from Raynaud's syndrome. It is used under the trade name, Doralese, to increase urine flow obstructed by prostatic hypertrophy.

5.2.2.4 *Abanoquil*

Abanoquil is a relatively new α_1-selective antagonist that produces minimal reflex tachycardia or postural hypotension. This agent has potential in the treatment of cardiac arrhythmias associated with myocardial ischaemia and reperfusion. In common with other α_1-adrenoceptor antagonists, it prevents reperfusion arrhythmias and prolongs action potential duration in ischaemic perfused hearts (Chapter 3, cardiac α-adrenoceptors) (Flores & Sheridan 1989).

5.2.3 *Selective α_2-Adrenoceptor Antagonists*

As yet, there does not appear to be a clear therapeutic application for α_2-adrenoceptor antagonism. However, agents with selectivity for α_2-adrenoceptors have been known for some time and these have proved to be useful pharmacological tools.

5.2.3.1 *Yohimbine*

This is a naturally occurring indolealkylamine alkaloid obtained from the bark of the *Pausinystalia yohimbe* tree and from *Rauwolfia* root (also the source of reserpine to which it is structurally related). Yohimbine, corynanthine and rauwolscine have the same structural formula but are diastereoisomers (Table 5.2). Of these, corynanthine is α_1-selective while rauwolscine and yohimbine are selective α_2-adrenoceptor antagonists. They are competitive antagonists with pA_2 values at α_1-adrenoceptors of 6.5–7.4, 5–7 and 5–7 for corynanthine, rauwolscine and yohimbine, respectively, while at α_2-adrenoceptors their respective pA_2 values are 4–6, 7.5–8.5 and 7–9, respectively (Wilson *et al.* 1991). The rigid structure prevents conformational changes to the molecule and corynanthine has been used as a model for the confirmational requirements for interaction with the α_1-adrenoceptor (DeMarinis *et al.* 1987).

Yohimbine is the only α_2-adrenoceptor antagonist approved for use in humans and formerly was used extensively as an aphrodisiac to treat male impotence; the product Potensan Forte which contains yohimbine HCl with strychnine, methyltestosterone and the stimulant, pemoline, is no longer available in the UK. The aphrodisiac property has not been clearly demonstrated but there is evidence for enhanced sexual activity in male rats (Clark *et al.* 1984) and of some benefit in patients with psychogenic impotence (Reid *et al.* 1987). The basis for this activity is uncertain but vasodilatation due to α_2-adrenoceptor blockade causing congestion of blood in the genital erectile tissue, combined with a CNS stimulant action, may contribute to the effect. It is also a local anaesthetic, an antagonist of 5-HT receptors, and releases antidiuretic hormone from the neurohypophysis.

Yohimbine readily enters the CNS to cause central effects including tremors and enhanced motor activity. It may also have antidepressant properties (Puech *et al.* 1979) and induces panic attacks in susceptible patients (Woods *et al.* 1988). Blockade of central α_2-adrenoceptors may induce a rise in blood pressure. However, interest has recently developed in the possible uses of a_2-adrenoceptor antagonists as antihypertensive agents. The peripheral effects of yohimbine favour a fall in blood pressure, with vasodilatation being produced in the human forearm (Bolli *et al.* 1983). The selective a_2-adrenoceptor antagonist, SK&F 86466 (6-chloro-N-methyl-2,3,4,5-tetrahydro-1-H-3-benzazepine), has been reported to effectively lower blood pressure in hypertensive rats, particularly those with high plasma noradrenaline levels (Roesler *et al.* 1986). Thus, postjunctional a_2-adrenoceptors appear to contribute to vascular tone and blood pressure and their blockade can exert an antihypertensive effect. This observation casts doubt on the role of prejunctional a_2-adrenoceptors in blood pressure maintenance since their blockade would oppose the block of postjunctional adrenoceptors and tend to raise blood pressure. However, this does assume that pre- and postjunctional a_2-adrenoceptors are the same and, as discussed in Chapter 3, this may not be the case. These findings also suggest that blockade of prejunctional a_2-adrenoceptors may not explain the lack of effectiveness of the non-selective antagonists, phentolamine and phenoxybenzamine, in hypertension. As pointed out earlier, phenoxybenzamine is in fact quite selective for a_1-adrenoceptors and at therapeutic doses probably does not block prejunctional a_2-adrenoceptors (Hieble & Ruffolo 1987). Antagonists selective for postjunctional a_2-adrenoceptors but without CNS activity therefore remain possible targets as antihypertensive agents.

A further potential use of a_2-adrenoceptor antagonists is to promote weight loss. This is based on the fact that a_2-adrenoceptor stimulation inhibits lipolysis, thus a_2-blockade should promote weight loss. In addition, yohimbine has been shown to reduce food intake (Ruffolo *et al.* 1993).

5.2.3.2 Idazoxan

One of the first synthetic selective a_2-adrenoceptor antagonists to become available was idazoxan which has been widely used as a research tool. Its pA_2 values for prejunctional a_2-adrenoceptors mediating inhibition of contractile responses to autonomic nerve stimulation range from 7.3 to 8.7, while at postjunctional a_1-adrenoceptors the pA_2 values are 5.0 to 6.3 (Doxey *et al.* 1983). Being an imidazoline, radiolabelled [^{3}H]idazoxan may bind not only to a_2-adrenoceptors but also to imidazoline binding sites (see Chapter 3) (Hamilton *et al.* 1988). These binding sites do not, as yet, appear to have a functional role in the periphery, although the binding of clonidine to central imidazoline receptors may contribute to its antihypertensive activity. [^{3}H]Idazoxan and [^{3}H]yohimbine appear to bind to different sites, the latter presumably not binding to the imidazoline receptor. Therefore idazoxan remains a useful antagonist for the study of functional peripheral a_2-adrenoceptor-mediated responses; unlike yohimbine it does not suffer the disadvantage of interacting with 5-HT receptors.

Whether idazoxan or other a_2-adrenoceptor antagonists have potential clinical applications remains to be seen. A likely use was in the treatment of maturity-onset type II diabetes. Since the a_2-agonist, clonidine, inhibits insulin secretion via stimulation of postjunctional a_2-adrenoceptors of the pancreatic β cells, selective antagonists might be expected to increase insulin secretion, providing there was sufficient tonic inhibition of insulin release from endogenous catecholamines (Table 3.4). The non-selective antagonist, phentolamine, increases basal and glucose-stimulated insulin levels in man (Robertson & Porte 1973). Idazoxan, however, failed to affect insulin levels in man in response to glucose, but facilitated the response to adrenaline (Struthers *et al.* 1985). It is also possible that concommitant blockade of prejunctional α_2-adrenoceptors on the sympathetic nerve will enhance noradrenaline release; if the postjunctional α_2-adrenoceptors are blocked the noradrenaline could activate β-adrenoceptors which promote secretion of insulin (Duval *et al.* 1991).

SL 840418 is a potent and selective α_2-adrenoceptor antagonist that does appear to stimulate basal insulin levels in man and produce hypoglycaemia (Bergougnan *et al.* 1990). A recently described α_2-adrenoceptor antagonist, RS-15385-197, appears to be the most selective agent at the time of writing. It is orally active and readily penetrates the brain. However, it has low binding activity at 5-HT receptors and imidazoline binding sites (Brown *et al.* 1993).

5.2.4 *Miscellaneous Agents with α-Adrenoceptor Antagonistic Properties*

Several other α-adrenoceptor antagonists have been used more widely in the past. These include dibenzazepines, such as azapetine, and benzodioxanes, such as piperoxan, the latter having α_2-selectivity. Thymoxamine (Opilon^R) is a phenoxyethylamine which remains in use as a peripheral vasodilator for the management of Raynaud's disease. It is said to be selective for α_1-adrenoceptors. Certain drugs with predominant pharmacological activity at other sites also have α-adrenoceptor blocking properties. Examples include the tricyclic antidepressants, such as amitriptyline and imipramine, where hypotensive side-effects are attributed to α-adrenoceptor blockade (U'Pritchard *et al.* 1978). The antipsychotic or major tranquillizer, phenothiazine, chlorpromazine (Largactil^R), has α-adrenoceptor antagonistic properties but also blocks 5-HT, histamine, dopamine and muscarinic receptors. How far each of these properties, or a combination of them, explains the central sedative effects and its use in schizophrenia are still uncertain. However, like other α-blockers, it does lower blood pressure and induce reflex tachycardia, although tolerance to the hypotension develops. Chlorpromazine is also an inhibitor of neuronal uptake and a direct vasodilator and has direct negative inotropic activity.

The potassium-sparing diuretic, amiloride (Midamor^R), has been shown to have α_1- and α_2-adrenoceptor blocking activity (Haussinger *et al.* 1987). Whether this property contributes to the diuresis is unclear but α_1-adrenoceptor stimulation has been found to enhance tubular sodium and water reabsorption so that α_1-blockade by amiloride may promote their loss (Smyth *et al.* 1985) (Table 3.4). However, α_2-adrenoceptors mediate an increase in renal Na$^+$, K$^+$ and water loss so that the α_2-blockade would appear to negate the α_1-mediated effects (Stanton *et al.* 1987).

Agents known primarily for their 5-HT receptor antagonist activity also have α-adrenoceptor blocking properties. These include ketanserin which is a selective 5-HT$_2$ antagonist but with substantial α_1-adrenoceptor blocking activity that probably accounts for much of its antihypertensive action (Zabludowski *et al.* 1985). Urapidil is also a 5-HT$_{1A}$ receptor antagonist having α-adrenoceptor blocking properties which may account in part for its blood pressure lowering activity (Fozard & Mir 1987). Urapidil (Table 5.2) is selective for α_1-adrenoceptors; it does not cause reflex tachycardia probably because of blockade of central α_1-adrenoceptors, a property shared with prazosin and its congeners. The 5-methyl derivative of urapidil is selective for α_{1A}-adrenoceptors (Hanft & Gross 1989)(see Chapter 3).

5.3 *β-Adrenoceptor Antagonists (β-Blockers)*

Antagonists of β-adrenoceptors have wide therapeutic application in the management of hypertension, myocardial infarction, angina and cardiac arrhythmias, in addition to other non-cardiovascular uses. In view of the large numbers of patients suffering from these disorders who may require lifetime treatment and the consequent market potential, there has been considerable activity in the pharmaceutical industry to develop improved β-blockers, with most of the major companies now having a product on the market.

The first agent to be identified with β-adrenoceptor blocking properties was the dichloro derivative of isoprenaline, dichloroisoprenaline (DCI) (Powell & Slater 1958). This compound is a partial agonist with profound stimulant activity at β-adrenoceptors and was therefore unsuitable for clinical use. The identification of this property led to the search for similar compounds at ICI Pharmaceuticals under the guidance of Sir James Black. The therapeutic target was antagonism of the cardiac β-adrenoceptor to protect the heart from sympathetic stimulation and the consequent raised oxygen demands and thus the treatment of angina. The first compound in the series was pronethalol (Table 5.3) which was withdrawn from clinical trial when it was found to cause tumours of the lymphoid tissue (thymus) in mice. However, it was soon followed by propranolol from the same company and this became the first compound to reach widespread clinical use (Black & Stephenson 1962). Initial indications were for angina and arrhythmias, but it was soon noted by Pritchard & Gillam (1964) that propranolol also lowered blood pressure (Scriabine 1979). The potential for use in hypertension was therefore realized. Since then many compounds have been introduced which all display the characteristics of competitive blockade of β-adrenoceptor-mediated responses.

5.3.1 *Structural Requirements*

The structures of the common β-adrenoceptor antagonists are shown in Table 5.3. Unlike the α-adrenoceptor antagonists, the β-adrenoceptor antagonists show a very close structural similarity to their agonist counterparts. The earlier β-adrenoceptor antagonists including DCI, butoxamine, pronethalol and sotalol were N-alkylphenylethanolamines based upon the agonist structure. Subsequently, the more effective agents are N-alkylphenoxypropanolamines. All of the antagonists have virtually the same side-chain as does the selective β-adrenoceptor agonist, isoprenaline, with an isopropyl or tertiary butyl substituent on the amino nitrogen. N,N-Disubstituted derivatives have reduced activity, but a high level of activity remains when a phenyl, hydroxyphenyl or methoxyphenyl group is added to the isopropyl or *t*-butyl substituent.

There is a consistent high degree of stereoselectivity with the (−) or laevorotatory isomer showing the greater activity. In general, the stereoselectivity is greater at β_1-adrenoceptors than at β_2-adrenoceptors, as indicated by isomeric ratios for displacement of $[^3H]$-(−)-propranolol binding (Morris & Kaumann 1984). The major differences between β-adrenoceptor antagonists lie in the nature of the aryl substitution on the aromatic ring. In the ethanolamines, a *para*-substituted methanesulfonamido-, carbomethoxy- or nitro-group produces compounds with activity significantly more than anticipated from the simple partitioning of this

Table 5.3 β-Adrenoceptor antagonists

Generic name	Proprietary name (UK)	Structure	Selectivity	ISA[a]	Membrane stabilization	Lipid solubility[f]
(A) β-Blockers in use or trial						
Propranolol	Inderal		β_1/β_2	–	+++	+++
Acebutolol	Sectral		β_1	+	+	+
Atenolol	Tenormin		β_1	–	–	–

(continued)

Drug	Structure			
Betaxolol Kerlone Betoptic (G)	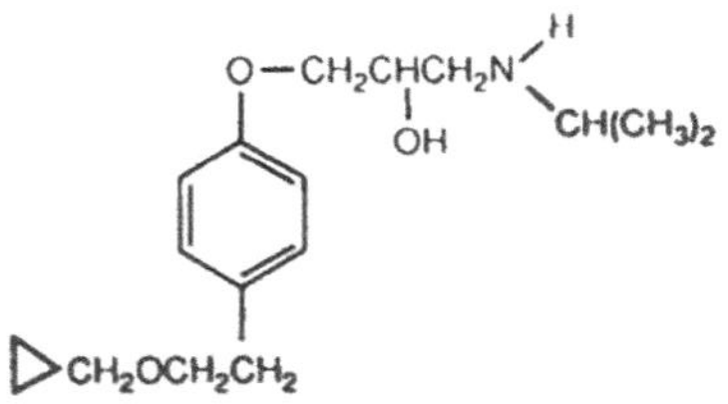 $O-CH_2CHCH_2N-CH(CH_3)_2$, OH, H; $CH_2OCH_2CH_2$ (cyclopropyl)	β_1	−	+/−	++
Bisoprolol Emcor	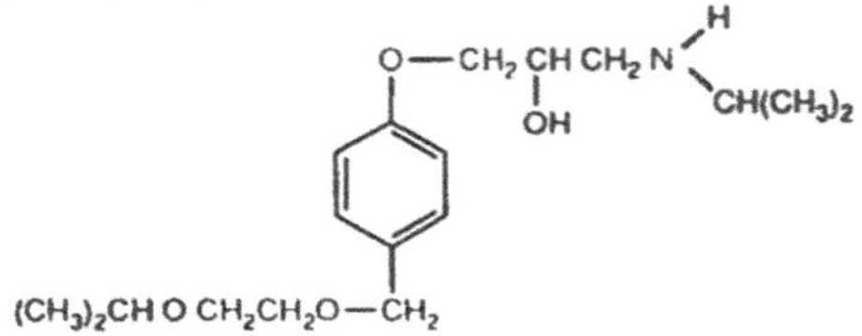 $(CH_3)_2CHO\,CH_2CH_2O-CH_2$ — $O-CH_2\,CH\,CH_2\,N-CH(CH_3)_2$, OH, H	β_1	−	−	−
Bopindolol Sandonorm	$O-CH-CH_2\,N-C(CH_3)_3$, $O-C=O$ — C_6H_5 (2), H	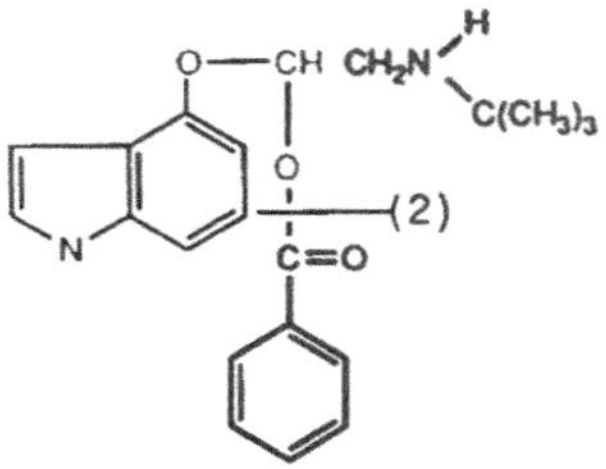 β_1/β_2	+	+	+
Bucindolol	$O-CH_2CHCH_2-N-CH_2-C-CH_2$, OH, H, CH_3, CH_3; $N\equiv C$; indole NH	β_1/β_2	+	+ (Vasodilator)	

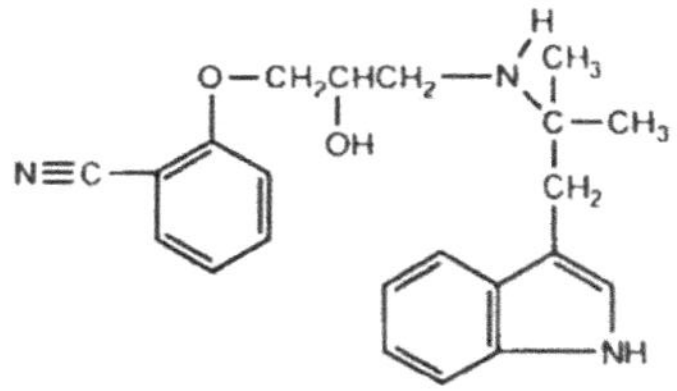

β_1/β_2 (Vasodilator)

(continued)

Table 5.3 Continued

Generic name	Proprietary name (UK)	Structure	Selectivity	ISA[a]	Membrane stabilization	Lipid solubility[b]
Bupranolol		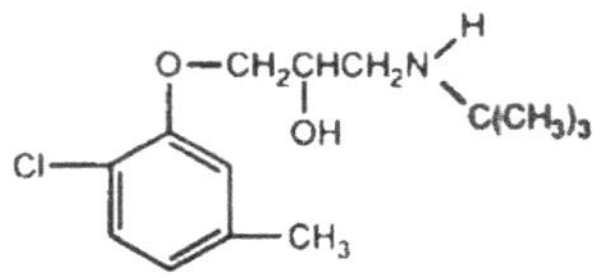 O—CH$_2$CHCH$_2$N—H—C(CH$_3$)$_3$, OH; CH$_3$; Cl	β_1/β_2	−		
Carazolol		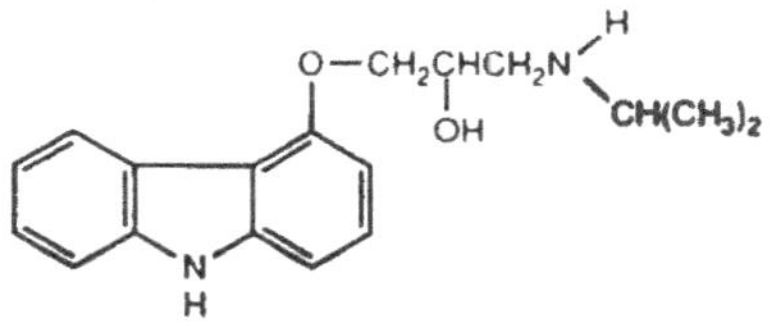 O—CH$_2$CHCH$_2$N—H—CH(CH$_3$)$_2$, OH				
Carteolol (OPC-1085)	Cartrol Teoptic (G)	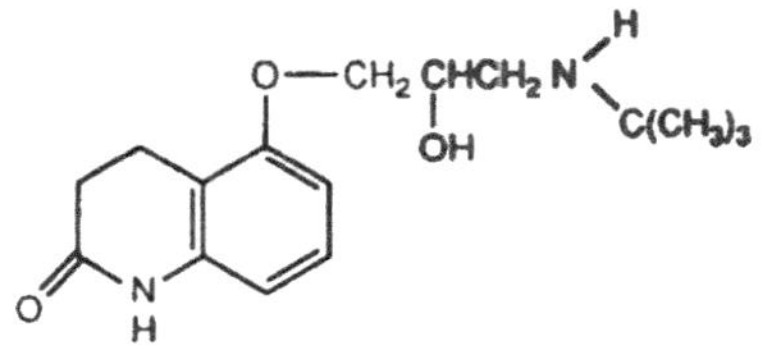 O—CH$_2$CHCH$_2$N—H—C(CH$_3$)$_3$, OH	β_1/β_2	+	−	−
Celiprolol	Celectol	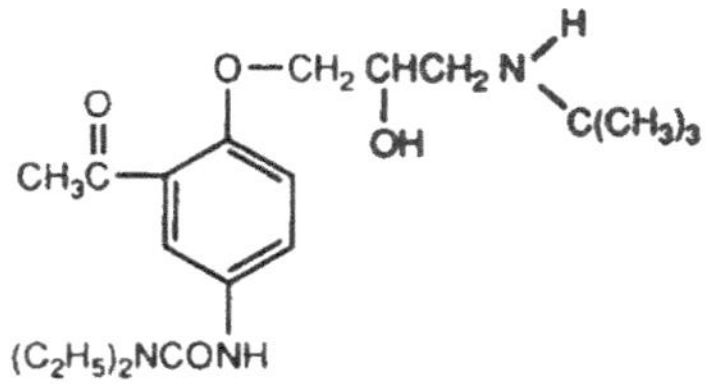 O—CH$_2$CHCH$_2$N—H—C(CH$_3$)$_3$, OH; CH$_3$C=O; (C$_2$H$_5$)$_2$NCONH	β_1	+ (β_2?)	−	−

(continued)

Drug	Structure				
Cicloprolol		β_1	+	?	?
Esmolol Brevibloc		β_1	–	–	?
Levobunolol Betogon (G)		β_1/β_2 High potency	–	–	?
Metipranolol (Trimepranol) Betanol Minims metipranolol (G)		β_1/β_2	–	–	++

(continued)

Table 5.3 Continued

Generic name	Proprietary name (UK)	Structure	Selectivity	ISA[a]	Membrane stabilization	Lipid solubility[a]
Metoprolol	Lopressor Betaloc	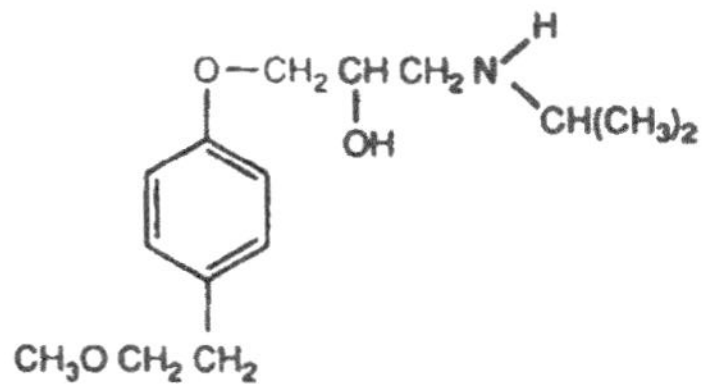	β_1	–	–	+
Nadolol	Corgard	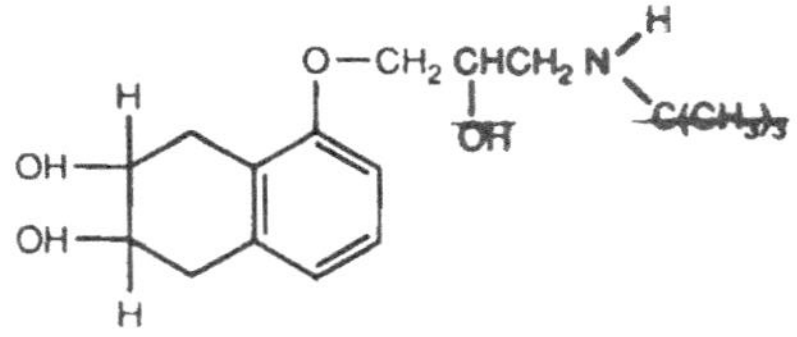	β_1/β_2	–	–	–
Oxprenolol	Trasicor	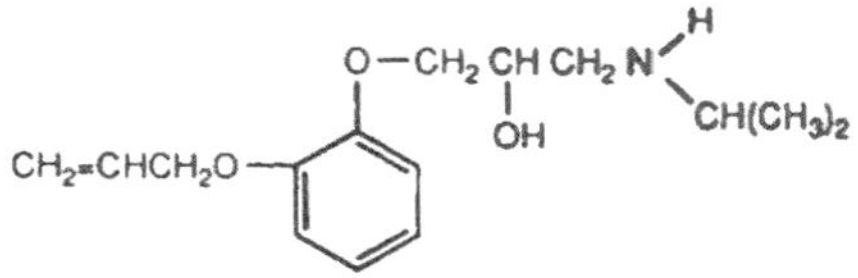	β_1/β_2	++	±	++
Penbutolol	Levatol	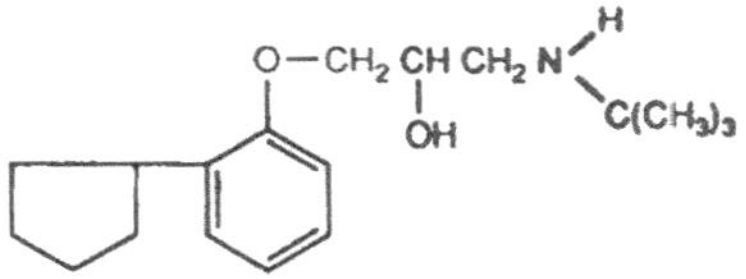	β_1/β_2	±	–	+

(continued)

Drug	Trade name	Structure		β selectivity		
Pindolol	Visken	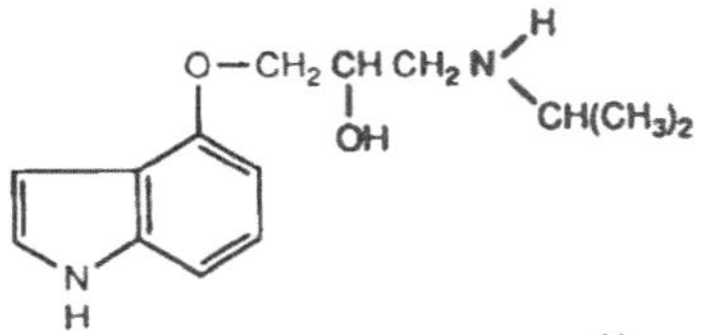 O–CH$_2$ CH CH$_2$ N–CH(CH$_3$)$_2$ / OH / H (indole ring)	β_1/β_2	+++	−	−/+
Sotalol MJ 1999	Beta-cardone Sotacor	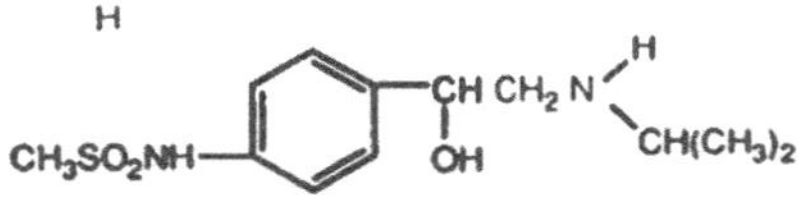 CH$_3$SO$_2$NH– ring –CH CH$_2$ N–CH(CH$_3$)$_2$ / OH / H	β_1/β_2	−	−	−
Tertatolol		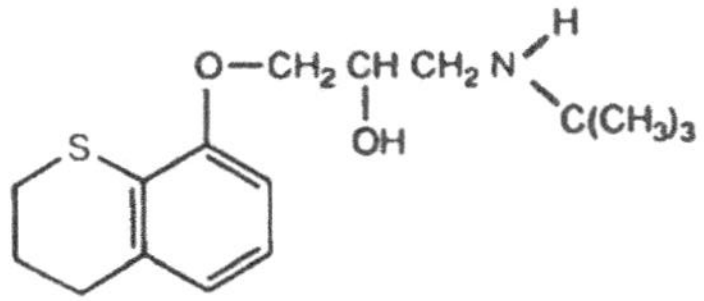 O–CH$_2$ CH CH$_2$ N–C(CH$_3$)$_3$ / OH / H (thiochroman ring, S)	β_1/β_2	−	−	?
Timolol	Blocadren Betim Timoptol (G)	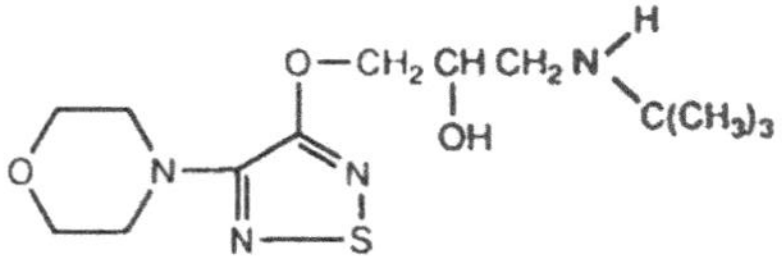 O–CH$_2$ CH CH$_2$ N–C(CH$_3$)$_3$ / OH / H (thiadiazole, N–S, N–N, morpholine ring O)	β_1/β_2	−	−	++

(B) Obsolete or experimental β-blockers

Drug	Structure		β selectivity		
Alprenolol (H 56/28)	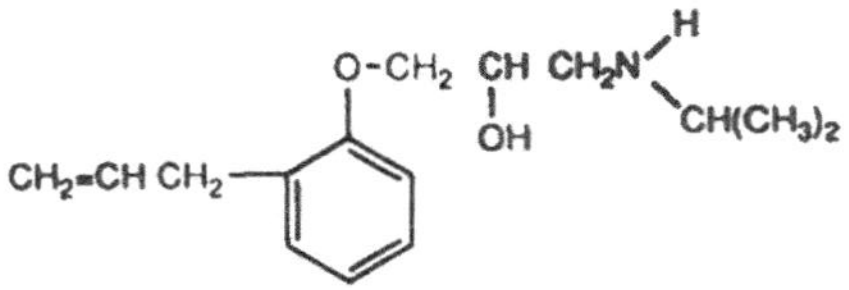 O–CH$_2$ CH CH$_2$N–CH(CH$_3$)$_2$ / OH / H; CH$_2$=CH CH$_2$ ring	β_1/β_2	+	+	++
Dihydroalprenolol (DHA) (or [^{3}H]DHA)[a]	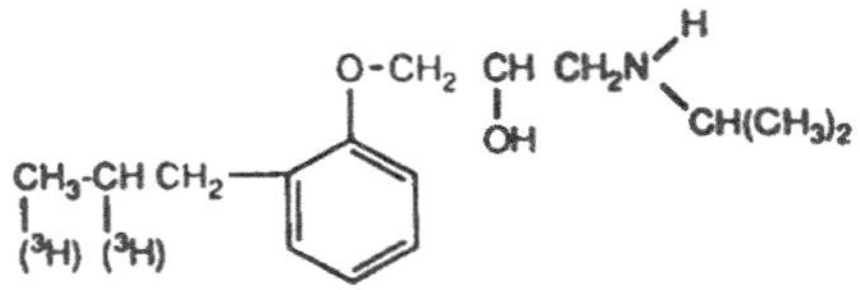 O–CH$_2$ CH CH$_2$N–CH(CH$_3$)$_2$ / OH / H; CH$_3$-CH CH$_2$ (^{3}H)(^{3}H) ring	β_1/β_2	?	+	+

(continued)

Table 5.3 Continued

Generic name	Proprietary name (UK)	Structure	Selectivity	ISA[a]	Membrane stabilization	Lipid solubility[b]
[¹²⁵I]Iodohydroxybenzylpindolol[d]		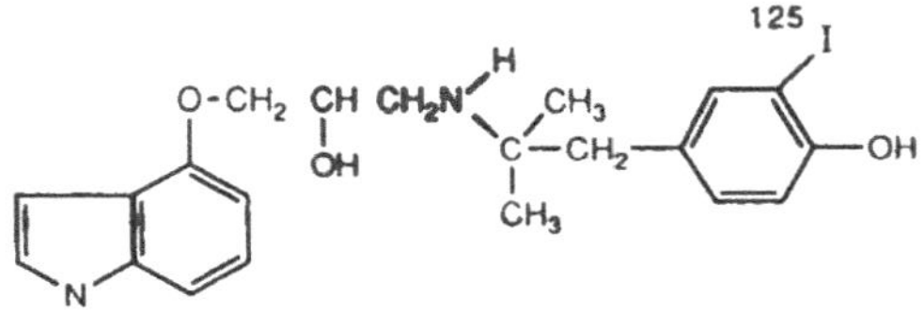	β_1/β_2	+	+ Radioligand with high specific activity	+
[¹²⁵I]Iodocyanopindolol ([¹²⁵I]CYP)[d,e]		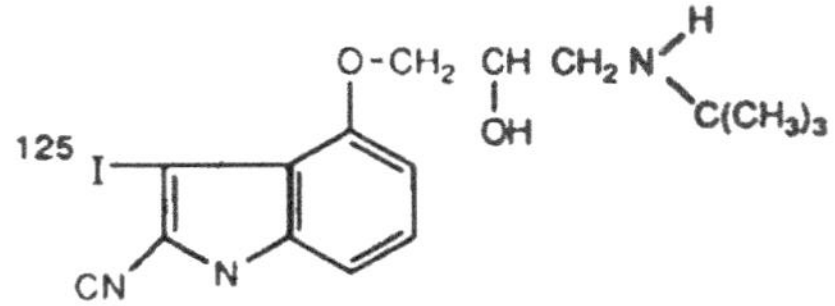	β_1/β_2	+	+ Radioligand with high specific activity	+
CGP-12177 (or [³H]CGP-12177)[d]		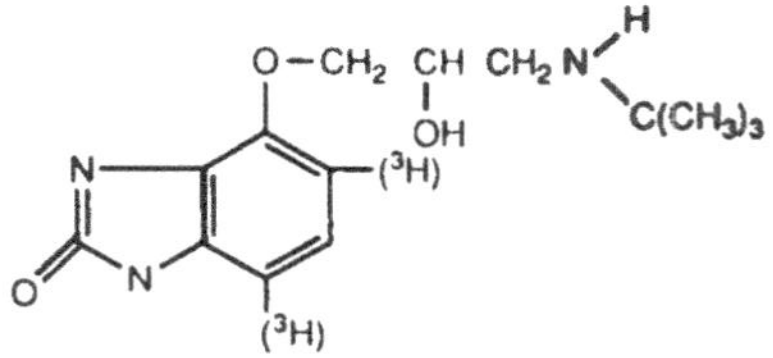	β_1/β_2		"	−

(continued)

Drug	Structure	Receptor			
CGP-20712A		β_1	$-$	$-$	?
Pronethalol (Alderlin) (withdrawn)	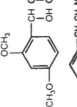	β_1/β_2	$+$	$+$	?
Practolol (Eraldin) (withdrawn)	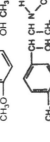	β_1	$++$	$+$	$-$
Butoxamine	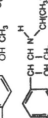	β_2			
H35/25	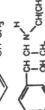	β_2			

(continued)

Table 5.3 Continued

Generic name	Proprietary name (UK)	Structure	Selectivity	ISA[a]	Membrane stabilization	Lipid solubility[b]
ICI 118,551		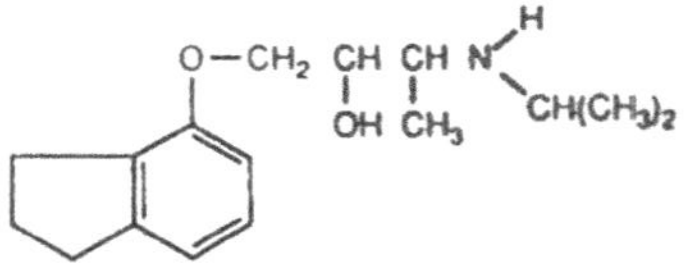	β_2	–	?	?
(C) β-Blockers with additional properties						
Bevantolol		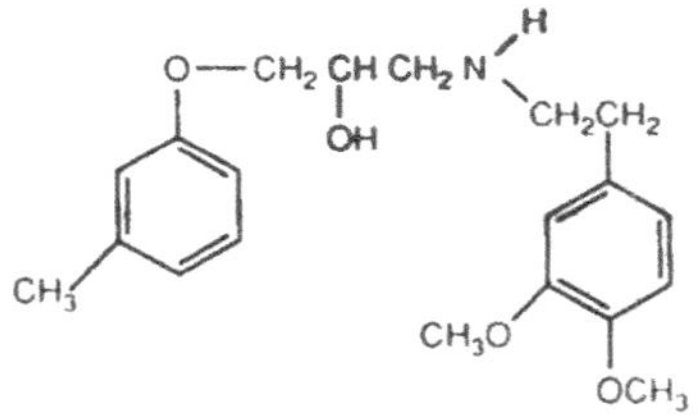	β_1/α_2	–	–	++
Carvedilol		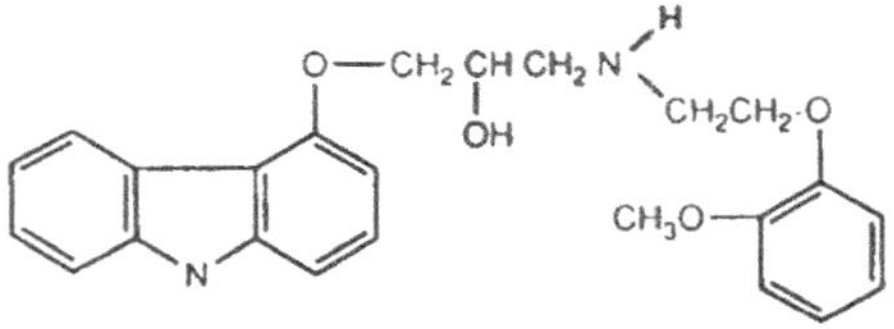	$\beta_1/\beta_2/\alpha_1$	–	–	++
Labetalol[f]	Trandate Normodyne[US]	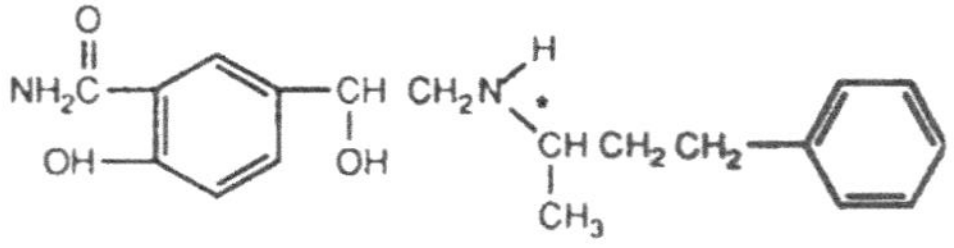	β_1/β_2	–	+	++

(continued)

(RR/SR/SS/RS mixture[e])

Dilevalol (Unicard)
RR isomer of labetalol

α_1 block in SR isomer
β_1/β_2 blockade, β_2 ISA

Tienoxolol

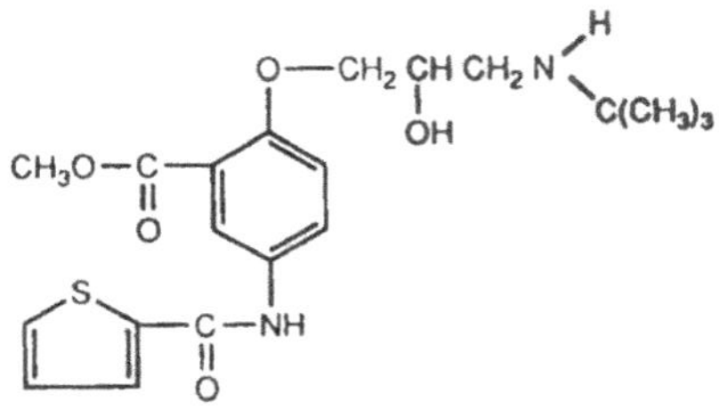

Additional diuretic activity

BW B385C

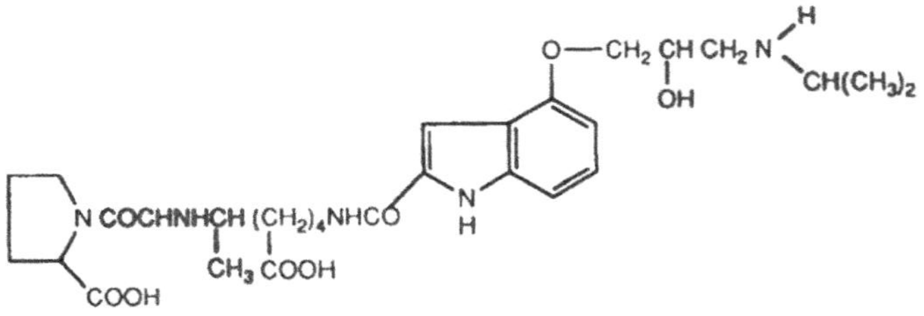

Additional ACE inhibition

Notes: [a] Intrinsic sympathomimetic activity.
[b] Wood & Robinson (1981).
[c] Inactive prodrug, hydrolysed at position 2 to produce active β-blocker.
[d] Site of radiolabel indicated.
[e] Correct structure according to original publication (Engel *et al.* 1981) but differs from pindolol itself.
[f] Asterisks indicate two assymetric centres of labetalol.
G, Topical formulation for glaucoma.

grouping into non-polar sites. They do appear to be involved in specific polar interactions with the receptor site. In the phenoxypropanolamines, substitution in the 2 and 3 positions with a second aryl ring structure (propranolol, nadolol, pindolol) or a simple group, such as NO_2, methoxy or methyl, produces favourable blocking activity. Compounds with polar groups in positions 2 or 4 are also potent antagonists in which these groups are probably involved in specific interactions with receptor components (Triggle 1976). Substitution in the *para*-position (4) of phenoxypropanolamines with an amide (practolol, atenolol) or ether-containing group (metoprolol, betaxolol, bisoprolol, cicloprolol) imparts β_1-adrenoceptor cardioselectivity. *Para*-substituted amides in compounds with another group in the 2 position (acebutolol, celiprolol) also retain cardioselectivity. All compounds where the amide is linked directly to the aromatic ring are weak partial agonists (acebutolol), while those having the amide separated from the ring have no intrinsic sympathomimetic activity (ISA) (atenolol)(Main 1990).

Substitution on the α-carbon of the side-chain reduces activity. However, as with agonists (see Chapter 4), it also confers β_2-adrenoceptor selectivity. The α-methyl derivative of DCI is relatively selective for β_2-adrenoceptors. Butoxamine, the N-*tert*-butyl analogue of methoxamine, displays selectivity for certain β_2-adrenoceptor-mediated responses, blocking vascular and uterine smooth muscle more than cardiac responses. Note that methoxamine is a selective α-adrenoceptor agonist which also has some β-blocking properties. N-Isopropylmethoxamine is not an α-agonist but produces the same activity *in vivo* as does methoxamine, to which it is metabolized in the body. Butoxamine, however, is not dealkylated to methoxamine. H35/25 has been used as a selective β_2-adrenoceptor antagonist, blocking the depressor responses to isoprenaline but not the cardiac effects. These agents have now been replaced by ICI 118,551 which also bears the α-methyl group and is the most selective β_2-antagonist currently available. This compound has widespread use as a pharmacological tool in the classification and characterization of β_2-adrenoceptor populations in different organs, but at present has no therapeutic application. The close similarity between structural requirements for agonist and antagonist activity at the β-adrenoceptor suggests that they interact with common regions of the receptor. They probably bridge two or more of the membrane-spanning α-helices within the cylinder of the receptor itself (see Chapter 13).

5.3.2 *Properties of β-Adrenoceptor Antagonists*

There are several important aspects of the pharmacology of β-adrenoceptor antagonists that are of relevance to their pharmacological actions and uses. First, they are competitive reversible antagonists of β-adrenoceptor-mediated responses. The shift of the dose–response curves for directly acting sympathomimetic amines is characteristically parallel (Figure 3.10). With indirectly acting sympathomimetic amines, however, there is non-competitive blockade, the maximum responses being depressed in the presence of the β-blocker (Kenakin 1984). The main consideration with regard to their competitive blocking activity is whether they are non-selective for β_1- or β_2-adrenoceptors or whether they have selectivity for β_1-adrenoceptors. The selective antagonism of the tachycardia induced by isoprenaline to leave the fall in blood pressure (β_2-mediated vasodilatation) unaffected is shown in Figure 5.2 for the prototypical β_1-adrenoceptor antagonist, practolol. In addition to the selectivity and

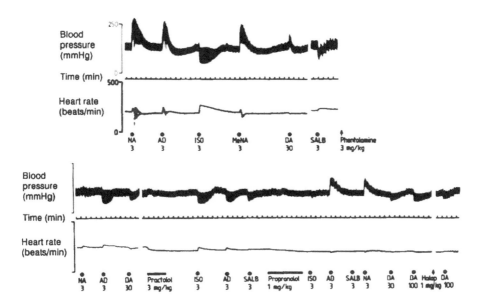

Figure 5.2 Effects of the non-selective β-adrenoceptor antagonist, propranolol, and the β₁-adrenoceptor antagonist, practolol, on responses of the blood pressure and heart rate of a cat (2.33 kg), anaesthetized with chloralose (80 mg/kg iv) after induction with halothane and nitrous oxide. Responses to intravenous doses (in micrograms) of noradrenaline (NA), adrenaline (AD), isoprenaline (ISO), α-methylnoradrenaline (MeNA), dopamine (DA) and salbutamol (SALB) were obtained before the α-adrenoceptor antagonist phentolamine and repeated after practolol, propranolol and the dopamine receptor antagonist, haloperidol (Halop). Note that practolol lowers resting heart rate and blocks the heart rate response (β₁) but not the vasodilatation (β₂) to ISO, AD or SALB. Propranolol blocks the vasodilatation to ISO and SALB but not to DA. The vasodilatation to a higher dose of DA was reduced by haloperidol.

potency of β-blockade, there are several other properties including intrinsic sympathomimetic activity (ISA), membrane stabilizing properties and their lipid solubility.

5.3.2.1 *Intrinsic sympathomimetic activity (ISA)*

ISA is the ability of the blocker to produce an initial stimulation of the receptor in the absence of β-agonists. This is classical partial agonist activity which may be profound in the case of DCI and prenalterol, compounds which should be regarded as agonists with blocking activity. The true β-blockers which possess some ISA include pindolol and oxprenolol (Table 5.3). Of interest is the observation that blockade often occurs at lower concentrations (high potency) than for stimulation (low potency). This dissociation occurs mainly with the racemates and may be explained by different populations of β₁/β₂-adrenoceptors mediating the blockade and ISA of the (−)-enantiomer (Walter *et al.* 1984).

Whether ISA is beneficial or detrimental depends upon the use being made of the β-blocker. Generally, it is undesirable to induce a response that the drug is intended to antagonize. This is particularly true in the management of myocardial infarction where initial cardiac stimulation may increase the myocardial oxygen demands and thus cause further deterioration. Some degree of cardiac stimulation may be beneficial to prevent precipitous falls in heart rate and reduction in cardiac output. This may be advantageous in the treatment of hypertension, in patients who have a tendency towards bradycardia or who have a reduced cardiac reserve. It is not known whether drugs with ISA are less prone to exacerbating cardiac failure in patients with heart failure compensated by increased sympathetic drive.

The ISA may also be relevant to the up-regulation of β-adrenoceptors that occurs with the chronic treatment with β-blockers. The up-regulation of β-adrenoceptors and accompanying increase in sensitivity to catecholamines is thought to explain the 'propranolol-withdrawal phenomenon' whereby patients experience an increased incidence of angina or even sudden death after abrupt discontinuation of β-blockade (Shand & Wood 1978). Chronic treatment with β-antagonists that have ISA, such as pindolol, do not appear to cause up-regulation and supersensitivity to β-agonists (Hedberg *et al.* 1986). In heart failure-free patients, treatment with a selective β_1-antagonist causes an increase in β_1-adrenoceptor binding sites but no change in β_2-adrenoceptors, although functional sensitivity to β_2-adrenoceptor stimulation is enhanced (Brodde 1991, Kaumann 1991). In contrast, patients treated with a β-blocker having ISA (pindolol), showed a reduced number of β_2-adrenoceptors and elevated number of β_1-adrenoceptors. This is probably because pindolol is a non-selective β-blocker but has greater ISA at β_2-adrenoceptors, leading to selective down-regulation of β_2-adrenoceptors. The selectivity of ISA for the β_2-adrenoceptor may reside in the $(+)$-enantiomer of pindolol (Walter *et al.* 1984). This introduces the potential for manipulation of agonist and antagonist activity to control up- and down-regulation of receptors with the chronic use of β-blockers. For example, in the management of heart failure a β_1-adrenoceptor antagonist could protect the heart from sympathetic overdrive and prevent the consequent noradrenaline-induced desensitization of β-adrenoceptors while allowing an up-regulation of β_1-binding and β_2-sensitivity. The combination of this approach with intermittent administration of β_2-agonists is a future possibility.

5.3.2.2 *Membrane stabilization*

Several of the β-adrenoceptor antagonists also possess membrane stabilizing or local anaesthetic activity. This is most marked in propranolol and, where it occurs, it resides equally in the $(+)$ and $(-)$ isomers, whereas β-blockade displays marked stereoselectivity towards the $(-)$ enantiomer. The presence of polar substituents on the phenyl ring increases β-adrenoceptor activity but these compounds usually lack membrane stabilizing activity (Table 5.3). The presence of these groups reduces the interaction with sites that accommodate only non-polar residues which are involved in membrane stabilizing activity. The membrane stabilizing properties may contribute to the antiarrhythmic activity of β-blockers. This activity is due to inhibition of the inward Na^+ current (i_{Na}) and is therefore akin to that of quinidine, a class IA antiarrhythmic drug. The electrophysiological action is characterized by a depression of the Phase 0 depolarization (Figure 3.12), a slowing of conduction velocity in myocardial Purkinje fibres and prolongation of repolarization. Pro-pranolol and acebutolol in particular have this property and are indicated for the treatment of cardiac arrhythmias. β-Blockade, however, remains of primary importance in their antiarrhythmic activity.

The membrane stabilizing properties of β-blockers may be undesirable in preparations used topically on the eye for the treatment of glaucoma, because of the local anaesthetic action; timolol is used for this purpose and has no membrane stabilizing activity.

5.3.2.3   Lipid solubility

Lipid solubility affects penetration of the blood–brain barrier and therefore the more lipid soluble β-blockers, such as propranolol, pindolol and timolol, can exert central effects. This may be a component of the antihypertensive activity of these agents but may also be correlated with the central side-effects including depression, fatigue, sleep disturbances and vivid dreams (Drayer 1987). Low lipid-soluble blockers, such as atenolol, bisoprolol, celiprolol, nadolol and sotalol, may be less likely to cause CNS side-effects. High lipid solubility also favours placental transport.

The other factor that lipid solubility controls is the metabolism of β-blockers. The lipid soluble agents are extensively metabolized to water-soluble products, which may retain β-blocking activity, but which are eliminated by the kidneys. They are subject to rapid first-pass metabolism in the liver after oral administration. Propranolol has a plasma half-life ($t_{1/2}$) of ~4 hr, whereas nadolol has a half-life of ⩾20 hr. The rate of elimination depends upon hepatic blood flow and considerable variations occur between subjects. Also, the β-blocker itself may reduce hepatic blood flow in the short term by blocking vasodilator $β_2$-adrenoceptors, which will impair its metabolism. This effect is less marked with partial agonists, and with chronic use peripheral blood flow generally returns to normal.

Esmolol is a very short-acting $β_1$-selective antagonist, having a half-life of ~8 min after intravenous administration. This is because it is rapidly metabolized via esterases in erythrocytes which break the ester linkage of the *para*-substitution on the ring structure (Table 5.3). The carboxylic acid metabolite has only weak β-blocking activity, although its half-life is longer. Esmolol is used only for the treatment of supraventricular tachycardias. Stepped increases in infusion concentrations for 5 min periods are made, followed by maintenance infusions of ⩾48 hr (Benfield & Sorkin 1987).

Water-soluble β-blockers, such as atenolol, are not metabolized in the liver and plasma concentrations are more predictable. They are excreted largely unchanged via the kidneys in the urine and the elimination half-life is ~5-8 hr. As a consequence, such drugs are best not administered to patients with renal failure in whom they accumulate. Elderly patients are therefore generally more susceptible to such compromised elimination.

The pharmacological effects of β-adrenoceptor antagonists last longer than the pharmacokinetics would suggest so that the antihypertensive activity of propranolol lasts considerably longer than 4 hr. The reasons for this are complex. For example, the (−)-enantiomer is metabolized more slowly than is the inactive enantiomer. The (−)-form of several β-adrenoceptor blockers is taken up into sympathetic nerve endings and then released by sympathetic nerve activity (Walle *et al.* 1988). Propranolol is metabolized to the 4-hydroxy derivative, which retains substantial β-blocking activity. Thus, pharmacological activity outlasts propranolol plasma levels which therefore serve as a poor indication of clinical efficacy.

Metoprolol is extensively metabolized by hepatic oxidative pathways. These are α-hydroxylation to α-hydroxymetoprolol, O-demethylation with subsequent rapid oxidation to yield a carboxylic acid group in the *para*-ring position, and finally N-dealkylation. The elimination half-life of metoprolol is between 3 and 4 hr.

Sustained release formulations are therefore preferred for once daily dosing. Plasma levels of α-hydroxymetoprolol may be about half those of the parent compound, but it has only one-tenth of the β_1-blocking activity (Plosker & Clissold 1992).

Hydrophilic β-adrenoceptor antagonists, such as CGP 12177 (Table 5.3), have been used as radioligands in binding studies to label extracellular receptors only. This compound fails to penetrate the cell membrane and therefore in whole cells does not label intracellular receptors, such as those that have become sequestered during desensitization (Chapter 14) (Staehelin *et al.* 1983).

5.3.3 Pharmacological Actions of β-Blockers

The pharmacology of β-adrenoceptor antagonists is generally that expected from blockade of the responses to sympathetic nerve stimulation that are mediated via β-adrenoceptors (Table 3.2). Thus, if sympathetic tone in an organ such as the heart is high, β-blockade will exert an inhibitory effect and heart rate will slow. If sympathetic tone is minimal then β-blockade will have little effect.

5.3.3.1 Cardiovascular effects

The cardiovascular effects are of importance since these account for the major therapeutic applications of β-blockers. The action of sympathetic stimulation to the heart, either through neuronally released noradrenaline or circulating adrenaline, is a β-adrenoceptor-mediated positive inotropy and chronotropy. These responses are antagonized competitively by the β-blockers (Figure 5.2). If there is high sympathetic drive, for example during exercise, then heart rate and myocardial contractility fall. The immediate effect of β-blockade is a fall in cardiac output accompanied by some reflex sympathetic vasoconstriction via α-adrenoceptors. Blockade of vascular β_2-adrenoceptors, which normally produce vasodilatation particularly in skeletal muscle, also results in vasoconstriction. The β_2-adrenoceptor-mediated vasodilatation and fall in blood pressure induced by isoprenaline or selective β_2-agonists such as salbutamol is blocked (Figure 5.2). Blood flow to all tissues with the exception of the brain is reduced. The peripheral resistance therefore increases, an effect tending to oppose the antihypertensive action. This response partly explains the impaired blood flow to the skin which gives rise to the symptoms of cold extremities experienced by many patients taking β-blockers. Since cold extremities is also a side-effect of β_1-antagonists, although a less commonly observed one, other factors such as the reduced cardiac output are also probably involved. β-Blockers should therefore be avoided in patients with Raynaud's syndrome. Reduced blood flow to skeletal muscle will also exacerbate intermittent claudication – impaired function and pain in skeletal muscle due to ischaemia that occurs with exercise. β_1-Selective blockers may impair exercise performance less than do non-selective agents (Tesch 1985). Antagonists with ISA are of theoretical advantage since they cause vasodilatation. However, it is possible that vasodilatation to non-ischaemic areas of mucle may cause a 'steal' phenomenon which diverts blood flow away from the ischaemic region. Thus, β_2-adrenoceptor blockade may in fact be of value in the treatment of intermittent claudication, but this question has yet to be resolved. With long-term administration of β-blockers the total peripheral resistance returns to normal values (Mimran & Ducailar 1988), and indeed ultimately there may be a *fall* in peripheral resistance (Man in't Veld *et al.* 1988).

β-Adrenoceptor antagonists block exercise-induced tachycardia which is a measure of sympathetic nerve activity via noradrenaline release onto β_1-adrenoceptors. They also block isoprenaline-induced tachycardia (Figure 5.2), a response mediated via both β_1- and β_2-adrenoceptors in man (see Chapter 3). β_1-Adrenoceptor-selective antagonists such as atenolol are more effective against exercise-induced tachycardia than are non-selective antagonists. β-Adrenoceptor antagonists with no ISA increase atrioventricular conduction time, sinus node recovery, sinus cycle length and atrial refractoriness. These effects are revealed on the ECG as increases in PR interval and QRS duration, and additionally there is an increase in RR interval indicative of increased vagal tone (Wadworth *et al.* 1991).

Renal blood flow and glomerular filtration rate are generally reduced by non-selective β-antagonists, but β_1-selective antagonists may have no effect. The catecholamine-induced release of renin from the juxtaglomerular apparatus via β_1-adrenoceptors is also blocked, thus plasma renin levels are reduced (Wadworth *et al.* 1991).

5.3.3.2 *Pulmonary responses*

β_2-Adrenoceptor-mediated bronchodilatation is blocked by non-selective antagonists such as propranolol, oxprenolol and timolol. In normal individuals this may not cause bronchoconstriction, however in those with impaired bronchial function such as asthmatics they may cause life-threatening bronchospasm. Selective β_1-antagonists and those with ISA should have less effect, but selectivity is only relative and such agents should still be avoided in these patients.

5.3.3.3 *Metabolic effects*

β-Adrenoceptor antagonists affect plasma lipid and carbohydrate levels. Since catecholamines mobilize free fatty acid from adipose tissue through stimulation of β_3-adrenoceptors (see Chapter 3), blockade of this effect will reduce the release of FFA. However, there is an opposing action of β-blockers on total triglycerides, very low density lipoprotein (VLDL) triglyceride and total cholesterol, the levels of which may increase after chronic administration of β-blockers. These effects are less marked with the β_1-selective antagonists such as atenolol, as compared with propranolol. There is some concern that such changes in plasma lipid profile may have a detrimental effect in patients with cardiovascular disease, since they may exacerbate any atherosclerotic lesions.

On carbohydrate metabolism, β_2-adrenoceptor stimulation causes glycogenolysis and an increase in blood glucose levels, but an opposing increase in insulin secretion via β_3-adrenoceptors (see Chapter 4). Thus, non-selective β-blockers may reduce blood glucose levels and delay recovery from hypoglycaemia in insulin-dependent diabetics, an effect that is less pronounced with β_1-selective antagonists. The effects upon insulin levels are inconclusive; but rather than falling, they may even increase under resting conditions and during glucose tolerance testing. There is also evidence for a reduced sensitivity to insulin. β-blockers should therefore be used with great caution in diabetic patients. They also prevent the reflex tachycardia that accompanies hypoglycaemia in the diabetic and which serves as a warning sign. The sympathetic sweating persists and there may be an unopposed increase in blood pressure mediated via α-vasoconstriction and due to release of adrenaline.

5.3.4 *Therapeutic Uses of β-Adrenoceptor Antagonists*

5.3.4.1 *Hypertension*

β-Blockers lower blood pressure in hypertensive subjects. Since the (+)-isomer of propranolol is ineffective, their antihypertensive activity can be attributed to β-blockade. All β-blockers irrespective of whether they have cardioselectivity, ISA or membrane stabilizing properties are effective antihypertensives. They have assumed an important place in the management of hypertension because of their relative lack of serious side effects compared with the drugs available prior to their introduction (Nadelmann & Frishman 1990). The chief advantage is that they lower blood pressure without blocking compensatory reflexes, thus there is no block of exercise- or postural-induced changes in blood pressure. Postural hypotension does not therefore occur. They also have the advantage of being compatible with other antihypertensives, such as hydrallazine, with which their actions are usually additive. Hydrallazine lowers blood pressure but induces a β-adrenoceptor-mediated reflex tachycardia which, with concomitant administration of a β-blocker, is prevented. They may usually be administered once daily for satisfactory control of blood pressure and thus improve patient compliance.

With acute administration, there is an immediate fall in cardiac output and a rise in peripheral resistance, the latter being due to blockade of vascular β_2-adrenoceptors and to reflex vasoconstriction. Blood pressure falls relatively quickly (within hours or days) but there may be a further more gradual fall over several weeks. With chronic use, the peripheral resistance returns to normal or may be reduced with β-blockers having ISA, due to stimulation of vascular β_2-adrenoceptors.

In spite of the widespread use of β-blockers for many years, their mechanism of action in hypertension remains uncertain. There are several likely modes of action and the resultant effect is probably a combination of more than one of these.

Reduced cardiac output   Reduced cardiac output as a result of β_1-adrenoceptor blockade is of major importance with all of the β-blockers, since this action persists throughout the treatment.

Reduced renin release   Renin release from the juxtaglomerular cells of the kidney is mediated via β_1-adrenoceptors. Their blockade should lower the levels of renin and consequently of angiotensin and aldosterone. This should lower blood pressure by virtue of a reduced pressor activity from angiotensin and a lower plasma volume because of impaired aldosterone-mediated Na^+ and water reabsorption in the kidney tubules. There is some evidence that propranolol is more effective in patients with renovascular hypertension in which plasma renin levels are raised (Buhler *et al.* 1972). However, several β-blockers, including pindolol (Frishman 1983) and atenolol (Valvo *et al.* 1982), may reduce blood pressure without affecting plasma renin levels. Thus, this mechanism is considered of little importance in the antihypertensive action of β-blockers.

Central mechanisms Administration of a β-adrenoceptor agonist directly into the medulla oblongata of the brain of anaesthetized animals causes an increase in blood pressure indicating the presence of central β-adrenoceptors that modulate vasomotor output and blood pressure. Thus, the possibility arises that blockade of these central receptors will produce hypotensive effects (Day & Roach 1974). There is no doubt that some β-blockers readily cross the blood–brain barrier and cause CNS side-effects such as fatigue, sleep disturbances and depression. For example, the lipid-soluble agents such as propranolol readily enter the CNS, however, no clear relationship has been established between lipid solubility and incidence of CNS side-effects. A central mechanism of action is therefore only possible with those β-blockers that can cross the blood–brain barrier. Highly polar agents such as atenolol penetrate the brain only slowly, yet they are effective antihypertensives and must therefore have alternative mechanisms.

Blockade of prejunctional β-adrenoceptors The role of prejunctional β-adrenoceptors will be discussed in greater detail in Chapter 6. Their stimulation results in facilitation of transmitter release from sympathetic nerve endings. There is some evidence for raised levels of circulating adrenaline in spontaneously hypertensive rats (SHR) at 5–7 weeks of age and these are raised by stress. The development of hypertension is prevented by β-blockers, adrenalectomy and destruction of sympathetic nerves by 6-OHDA. Infusion of adrenaline also increases blood pressure in normotensive rats, an effect also prevented by β-adrenoceptor blockade with metoprolol rather than α-blockade. Similarly, increased plasma levels of adrenaline have been observed in patients with labile and sustained benign essential hypertension. Since patients with labile or borderline hypertension usually progress to established hypertension, it seems reasonable to conclude that raised adrenaline levels and its interaction with prejunctional β-adrenoceptors plays a major role in the aetiology of essential hypertension. Indeed, plasma levels of adrenaline are raised in some patients with established essential hypertension (Misu & Kubo 1986). There is also evidence for a raised sensitivity of prejunctional β-adrenoceptors in hypertension, so that facilitatory effects of circulating adrenaline on transmitter release may help to maintain the hypertension (Borkowski 1988, Marin 1993).

A possible mechanism for β-blockers in lowering blood pressure is therefore inhibition of this facilitatory process. The question is whether the prejunctional receptors are of the β_1- or β_2-subtype. Most evidence points to them being of the β_2-type. Indeed, the β_2-selective antagonist, ICI 118,551, has been shown to reduce blood pressure in hypertensive patients (Vincent *et al.* 1985). However, this is at variance with a report by Dahlof *et al.* (1983). In contrast, the β_1-selective antagonist, metoprolol, has been shown to have effects at prejunctional β-adrenoceptors and to prevent the increase in blood pressure in rats due to infusions of adrenaline. It is possible that either β_1-selective antagonists behave as non-selective antagonists and block prejunctional β-adrenoceptors or that there is a population of prejunctional β_1-adrenoceptors (Misu & Kubo 1986). Notwithstanding these inconsistencies, a prejunctional site of action of β-antagonists remains a viable mechanism in their antihypertensive action.

Miscellaneous mechanisms Of minor importance, there may be additional sites of action of certain β-adrenoceptor antagonists. For example, propranolol may block peripheral sympathetic nerves either by an adrenergic neurone blocking activity

237

similar to that of guanethidine, or by a local anaesthetic action (Day *et al.* 1968). Attempts to distinguish this property from prejunctional β-adrenoceptor blockade have been made on the basis of frequency-dependence, stereoselectivity and reversal of blockade by ISAs, such as amphetamine (see Chapter 6). The guanethidine-like activity is selective for lower frequencies, similar to activity at prejunctional β-adrenoceptors. Some reports show reversal of β-blockers by amphetamine while others do not, but there are reports of the (+)-isomers of β-blockers having this property (Misu & Kubo 1986). This mechanism remains a possible minor component of the antihypertensive action, although there is little evidence of side effects resulting from sympathetic blockade such as nasal stuffiness or postural hypotension.

Interference with prostaglandin synthesis is another possible target, since non-steroidal anti-inflammatory agents such as indomethacin reduce the antihypertensive action of propranolol. This may be due to inhibition of synthesis of prostacyclin in the vascular wall which propranolol increases (Beckmann *et al.* 1988).

β-Adrenoceptor antagonists may inhibit sympathetic function at the level of a local renin–angiotensin system in the vascular wall. The local release of angiotensin II (AII) may be induced by β_2-adrenoceptor stimulation, for example by circulating adrenaline. The AII has a facilitatory effect upon noradrenaline release from sympathetic neurones (see Chapter 6). This local facilitatory system is more sensitive in the mesenteric vasculature of SHR rats (Draper *et al.* 1989) and suggests that it may be a further target for β-blockers in lowering sympathetic tone to the peripheral vasculature.

Systemic arterial hypertension is associated with left ventricular hypertrophy (LVH) as an adaptive change of the heart to the sustained overload (after-load). Patients with LVH have a poor prognosis and are at greater risk of developing CHF, ischaemic heart disease and sudden death. In LVH, the myocardial cells increase in mass (hypertrophy) rather than in number (hyperplasia), with an increase in contractile protein. LVH may also be idiopathic and its onset quickened with age and obesity. Angiotensin II and circulating catecholamines (acting on α_1-adrenoceptors) may be involved in the cell growth. The coronary circulation does not develop to match the increase in myocardial mass and therefore coronary reserve is reduced with a consequent risk of ischaemia. It is therefore a desirable objective in the treatment of hypertension to prevent or reverse LVH and several classes of antihypertensive drugs (eg ACE inhibitors, β-blockers and centrally acting sympatholytics, such as clonidine) have been shown to reduce left ventricular mass by 10% or more. Most β-adrenoceptor antagonists have been shown to produce some reduction in LV mass. Although β-blockade reduces myocardial contractility, systolic function is usually improved after LVH regression with this class of agent. Whether the reduced myocardial mass by β-blockers and other antihypertensive agents leads to improved mortality rates and morbidity remains to be established (Lavie *et al.* 1991, Baxter & Yellon 1993).

5.3.4.2 Angina

Angina is a cardiac disease in which there is an inbalance between the demand of the myocardium for oxygen and its supply by the coronary circulation. In stable angina there is usually coronary vessel narrowing due to atherosclerosis, and the attack of chest pain may be induced by exercise or emotional stress resulting in myocardial ischaemia. Unstable angina is a sudden increase in the incidence of ischaemic symptoms in patients with pre-existing stable angina. Silent myocardial ischaemia occurs without angina or stress, patients being asymptomatic for long periods but with occasional transient ischaemic attacks. In Prinzmetal's or variant angina there may be no evidence of atheroma and attacks may occur at any time, probably due to coronary artery vasospasm. The coronary vasoconstriction may be induced by noradrenaline (via α-adrenoceptors), 5-HT or histamine. A possible mechanism is through local damage to the coronary vascular endothelium which therefore has diminished capacity for release of EDRF (nitric oxide)(see Chapter 3, vascular α-adrenoceptors). Reduced levels of the endogenous vasodilator therefore allows for unopposed vasoconstriction by noradrenaline.

β-Blockers are well established as effective drugs for the treatment of stable angina. They are used prophylactically to prevent the onset of an angina attack with exertion or stress. Their effectiveness is assessed by improved exercise duration, reduction in frequency of anginal attacks and of nitroglycerin consumption. On the ECG, lowering of the ST segment (≥ 1 mm) is also a characteristic of both stable and unstable angina and 'silent' myocardial ischaemia. The frequency of these episodes during 24-hr ECG monitoring is reduced and the time to 1 mm ST segment depression on exercise is prolonged by β-blockade. In Prinzmetal's variant angina the ST segment may be elevated. β-Blockers are generally ineffective in this type of angina, and may even worsen the severity of the attack because of unopposed α-vasoconstriction of coronary vessels.

The mechanism of action is derived from the reduction in heart rate, which reduces the oxygen consumption by the myocardium at rest and protects the heart from excessive sympathetic stimulation during exercise and stress. The lowered heart rate allows more relative effective time in diastole, during which time coronary flow occurs. However, there is an opposing increase in systolic ejection time, which may increase oxygen demands. With chronic use, β-blockers may also lower peripheral resistance and blood pressure, which reduces the after-load on the heart and consequently the oxygen demands.

The blockade of coronary vascular β_2-adrenoceptors may also diminish the vasodilator actions of circulating catecholamines. This vasoconstriction in the non-ischaemic myocardium may redistribute blood flow to the ischaemic areas ('reverse steal'). This effect may be greater with those β-adrenoceptor antagonists having ISA at vascular β_2-adrenoceptors.

It might be expected that ISA at the cardiac β_1-adrenoceptor is undesirable and that cardioselective (β_1) blockers without ISA should be preferred. While there is little difference in efficacy of β-blockers in angina, atenolol (cardioselective without ISA) has been shown to have a more favourable effect when compared with celiprolol or pindolol, both with ISA (Wadworth *et al.* 1991).

5.3.4.3 *Myocardial infarction (MI)*

A myocardial infarction (MI) is the consequence of obstruction of a coronary artery by thrombus formation, usually at the site of an atherosclerotic lesion. The resulting myocardial ischaemia distal to the obstruction induces biochemical, electrophysiological and contractility changes that result in cardiac output being very low or absent. In the USA alone, the annual admissions of patients with acute MI is >750 000. After such an event, the risk of dying is greatest during the first 24 hr when ~85% of deaths occur. Thereafter, the chances of survival improve with time.

The treatment of MI falls into two categories. Firstly, the acute treatment is aimed at improving the chances of survival and to relieve the patient's symptoms and distress. Treatments initiated immediately after the infarct are designed to limit the infarct size, prevent further occlusion of the coronary artery, and prevent cardiac failure and ventricular arrhythmias. Second, post-infarct management is aimed at prevention of a re-infarct. The major life-threatening event both during an infarct and subsequently is the occurrence of ventricular fibrillation. The immediate treatment of patients presenting with chest pain and regional ST segment elevation associated with acute MI should be with thrombolytic agents to disperse the presumed clot. Agents include streptokinase, recombinant tissue-type plasminogen activator (rt-TPA, Alteplase) and the anisolyated complex of plasminogen and streptokinase activator (anistreplase, Eminase[R]). Their use as early as possible after the infarct reduces mortality by as much as 50% when treatment is commenced in the first hour. There is little difference between the three agents apart from cost, streptokinase being the least expensive.

β-Adrenoceptor antagonists also have an important place in the management of MI, both in the immediate post-infarct period and long-term use after recovery from MI. MI is usually associated with severe autonomic disturbances, which may be manifest as bradycardia, hypotension and AV block through increased parasympathetic activity, and is susceptible to atropine. More likely, there is increased sympathetic activity which appears as sinus tachycardia and is therefore susceptible to β-blockade. This group of drugs also owe their beneficial effects to reduced oxygen demand, reduced infarct size and reduced myocardial wall stress.

A further potential advantage of β-blockers in MI is derived from their action against the metabolic effects of catecholamines. Myocardial ischaemia releases catecholamines from the intracardiac stores, and the fall in cardiac output leads to increased noradrenaline release via reflex sympathetic activity (Schömig 1988). This stimulates the release of FFAs by local lipolysis through β_3-adrenoceptors (see Chapter 3, lipolysis), an effect likely to be susceptible to β-blockade. In the ischaemic myocardium, FFAs exert a detrimental effect, the favoured energy substrate being carbohydrate. However, as described above (metabolic effects), β-blockers have an opposing undesirable profile of action on other plasma lipids by raising total triglyceride, VLDL and cholesterol. There is concern that these changes in the long term may be potentially atherogenic. There is, however, no direct evidence of risk from these changes in serum lipid or lipoprotein levels. Indeed, there is mounting evidence that some β-blockers, including metoprolol, may induce haemodynamic and biochemical changes that could provide an *anti*-atherogenic effect (Cruickshank 1990). The biochemical changes include increased synthesis of prostacyclin, which is thought to prevent cholesterol accumulation in the arterial intima. Also, β-blockers may prevent binding of LDL to the arterial proteoglycans, which would otherwise lead to extracellular deposition of LDL in atherosclerotic plaques (Camejo *et al.* 1991). The haemodynamic effects of reduced heart rate and pulse pressure may also contribute to an impaired disposition of lipid into the atheroma (Plosker & Clissold 1992).

Starting treatment early within the first few hours with intravenous administration of β-blockers reduces ischaemic pain, prevents dangerous tachyarrhythmias and limits infarct size, thereby reducing subsequent morbidity and mortality. Long-term treatment orally for 2–3 years with propranolol, acebutolol, metoprolol or timolol has shown significant reductions (~20–50%) in mortality compared with placebo

(Hinstridge & Speight 1991). Contraindications for the early use of β-blockers in MI are second- or third-degree AV block, sinus bradycardia, sick sinus syndrome, uncompensated left ventricular failure and hypotension.

5.3.4.4 *Cardiac arrhythmias*

β-Adrenoceptor antagonists belong to the class II antiarrhythmic agents. They inhibit the effects of catecholamines upon the electrophysiological properties of the heart. They reduce catecholamine-induced increase in slope of Phase 4 depolarization of the action potential and rate of discharge of the SA node, and therefore slow heart rate (Figure 3.12). They also reduce catecholamine-maintained automaticity of the Purkinje fibres. β-Blockers increase the effective refractory period of the AV node, which prevents the propagation of re-entrant arrhythmias in the AV node. This explains the effectiveness of β-blockers in the treatment of supraventricular (atrial) arrhythmias. Thus the atrial origin of the tachyarrhythmia is prevented from reaching the ventricles, which therefore slow and cardiac output is maintained. Normal sinus rhythm is rarely restored. In the ischaemic myocardium catecholamines are released (Schömig 1988) and these may facilitate ventricular arrhythmias. β-Blockers are effective against these ventricular arrhythmias because they prevent or reduce the ischaemia rather than through an antiarrhythmic action. Similarly, when ventricular arrhythmias are induced by catecholamines released by exercise and emotional stress, they may respond to β-blockers. However, large doses of propranolol are likely to be required to alleviate ventricular arrhythmias induced by other mechanisms. An antiarrhythmic action may also be derived from the effects upon K^+ levels. Catecholamines induce hypokalaemia through β_2-adrenoceptor stimulation (see Chapter 4, metabolic effects), which is proarrhythmic and this is prevented by β-blockers.

The beneficial effects of β-blockers in the management of MI, described above, is largely due to their prevention of cardiac arrhythmias.

Propranolol and other β-blockers with membrane stabilizing activity, in addition to their β-blocking class II antiarrhythmic properties, have a class I action on the phase-0 of the action potential. They slow the influx through Na^+ channels. In addition to use in supraventricular arrhythmias and to prevent ventricular arrhythmias in MI, β-blockers are also used to treat digoxin-induced arrhythmias and Wolff-Parkinson-White syndrome. In the latter condition, there is an abnormal atrioventricular conduction pathway. Atrial fibrillation arising from AV re-entrant conduction is prevented by the increased AV refractoriness imparted by propranolol and other β-blockers.

5.3.4.5 *Idiopathic dilated cardiomyopathy*

This is a specific form of heart failure not associated with ischaemia. It is of unknown origin and is the underlying disease in ~40-50% of heart transplant procedures. Traditionally, β-blockers are contraindicated in heart failure because of their negative inotropic effects. Several small-scale studies, however, have suggested a surprising benefit in certain patients, particularly those with idiopathic dilated cardiomyopathy. In a recent trial, metoprolol over 6 weeks was shown to reduce mortality or the need for a heart transplant by 34% (not quite significant) compared with placebo. The mechanism is probably due to blockade of the adverse effects of compensatory sympathetic stimulation in heart failure, including tachycardia, serum sodium levels and arrhythmias. Other β-blockers including carvedilol and bucindolol are also under trial.

5.3.4.6 *Hyperthyroidism*

Many of the undesirable side-effects of hyperthyroidism are due to the enhanced sympathetic activity. It was shown in Chapter 3 that thyroid hormone affects β-adrenoceptor sensitivity. Thus, as an interim measure while the effects of antithyroid treatments are instigated, β-blockers may be useful to improve the patient's comfort. They are, however, contraindicated where thyrotoxicosis has induced heart failure, since in these patients the heart is supported by sympathetic drive.

5.3.4.7 *Glaucoma*

Several β-adrenoceptor antagonists including timolol, carteolol, levobunolol and betaxolol (Buckley *et al.* 1990) are used topically for the treatment of open-angle glaucoma (Table 5.3). In open-angle glaucoma, drainage of aqueous humour is not impaired by narrowing of the drainage angle but is related to the diffusion through the trabecular network to the canal of Schlemm (Figure 1.7). In Chapter 4, it was noted that adrenaline is also effective in the treatment of open-angle glaucoma, part of this activity being due to β-adrenoceptor-mediated vasodilatation of the veins draining the canal of Schlemm. It is therefore anomalous that β-blockers should also be useful. Their mechanism of action is probably due to impaired production of aqueous humour by the ciliary body. The precise site of action is still uncertain, but β-adrenoceptor blockade rather than ISA or membrane stabilizing properties is involved; these properties being absent in the most active agent, timolol. There may be blockade of β_2-adrenoceptors in the epithelial cells of the ciliary body that control active transport and secretion of aqueous humour (Lotti *et al.* 1984). β_1-Selective antagonists in general are not effective, although betaxolol is an exception. Additionally, there may be a reduction in blood flow to the ciliary body and therefore a decreased formation of ultrafiltrate (Lesar 1987). The advantages of β-blockers in the treatment of glaucoma is that they produce minimal local adverse effects. Care must nevertheless be exercised in their use in patients with bronchial asthma or heart disease.

5.3.4.8 *Migraine headache*

The pathophysiology of this headache has been briefly considered when the use of ergot alkaloids was described. The β-adrenoceptor antagonists are used only in the prophylaxis of migraine. Their mechanism of action is still unclear, but is thought to involve the balance of vasodilatation and vasoconstriction of the cranial vasculature. Vasodilatation and increased blood flow of the cranial vessels appear to be the cause of the throbbing headache. Propranolol, metoprolol, nadolol and timolol are used in migraine treatment. Thus β_1-selectivity does not appear to be a requirement, but agents without ISA are necessary. Whether it is only β-blocking activity that is responsible for the beneficial effects is unsure, since the ($+$)-isomer of propranolol is also effective. The results of several clinical trials of β-blockers in migraine have been critically reviewed (Tfelt-Hansen 1986).

5.3.4.9 Anxiety

Propranolol and other β-blockers are used to relieve anxiety. They are particularly effective where there are physical manifestations of sympathetic overactivity such as tremor, palpitations, diarrhoea and localized sweating (palms of the hands). The benzodiazepines are the first-choice drug in the treatment of chronic anxiety without these peripheral autonomic symptoms. The awareness of these physical signs of stress by the individual may further add to the anxiety, creating a positive feedback loop. The β-blockers have a dual action in relieving anxiety. First, they block the β-adrenoceptor-mediated peripheral effects of tachycardia (β_1) and tremor (β_2). Second, the lipophilic β-blockers cross the blood–brain barrier and exert a central sedative action. In general, the lipophilic agents, propranolol and oxprenolol, have a greater effect.

5.3.4.10 Sporting and social uses of the anxiolytic activity

The anxiolytic action of β-blockers described above may be used to advantage in the performance of stressful activities associated with these symptoms of sympathetic overactivity. The calming effect may improve performance in such varied activities as public speaking, musical performances and car driving, particularly during the driving test for which β-blockers may be prescribed. The heart rates of surgeons and of car, rail and aircraft travellers have been shown to be substantially reduced after taking oxprenolol (Figure 5.3). The performances of musicians and public speakers and of students taking examinations were judged to be enhanced after β-blockade (Krishnan 1976, James *et al.* 1977).

The improvements to muscle control by reducing tremor and from slowing of the heart rate in aiming sports has led to the use of β-blockers in sporting events such as shooting, archery, bowling and snooker. They appear to be of most value to the more anxious competitors and to the inexperienced and less competent shooters. A significant improvement in scores has been observed in slow-fire shooting events but not in rapid-fire events (Antal & Good 1980). The slowing of heart rate enables the marksman to time the firing to the heart rate.

β-Blockers are of no value in endurance events which require high levels of oxygen utilization. They reduce exercise-induced increases in heart rate and contractility and in oxygen consumption. The ability to sustain maximal and near-maximal work in athletes and healthy volunteers is therefore reduced. This of course contrasts with patients suffering from ischaemic heart disease, where exercise tolerance is improved. β-Blockers can result in early termination of exercise testing due to fatigue or leg pain, presumably due to peripheral vasoconstriction (Van Baak 1988). A further detrimental effect in athletes is a reduced energy supply through inhibition of β-adrenoceptor-mediated glyco-genolysis and, with longer endurance events, the provision of FFAs through lipolysis. There is also greater sweating and hence stricter fluid replacement is necessary.

Thus β-blockers have potential for abuse in sports requiring calm and sensitive muscle control, and where physical activity is of little or no importance. The IOC Medical Commission has therefore included all β-blockers in its list of banned substances and reserves the right to test for their presence in the urine in those sports which it deems appropriate (Mottram 1988, Badewitz-Dodd 1991).

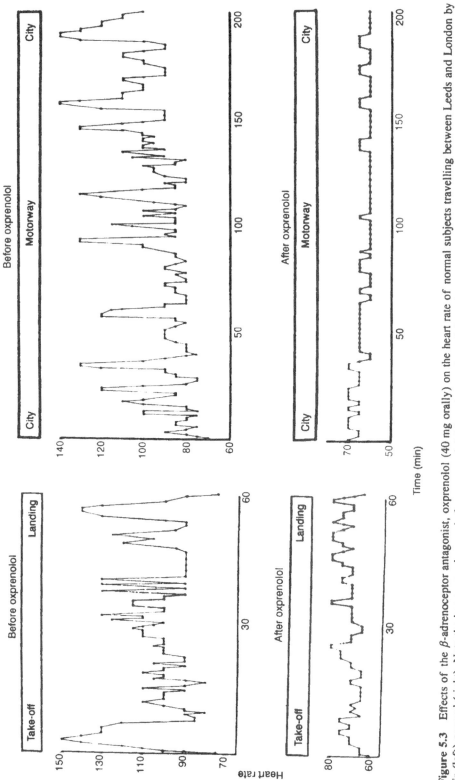

Figure 5.3 Effects of the β-adrenoceptor antagonist, oxprenolol (40 mg orally) on the heart rate of normal subjects travelling between Leeds and London by air (left) or road (right). Note the heart rate changes before oxprenolol (above) are substantially reduced after oxprenolol (below). Reproduced with permission from Taylor & Meeran (1972).

5.3.5 *Adverse Effects of β-Blockers*

The popularity of β-blockers and the rapid growth in their numbers is largely due to their relative freedom from major side-effects as compared with the drugs available for hypertension prior to their introduction in the early 1960s. Side-effects that do arise are mostly the result of β-adrenoceptor blockade. The major side-effects and their origins are shown in Table 5.4. The main consequence of a side-effect is to render them contraindicated in certain groups of patients. For example, bronchoconstriction is a major adverse effect of β-blockers which precludes their use in patients with obstructive airway disease such as asthma. The occurrence of impaired peripheral circulation leading to a sensation of coldness of the extremeties prevents their use in patients predisposed to Raynaud's syndrome. They may also induce intermittent claudication. The metabolic effect of β-blockade on glucose metabolism antagonizes the glycogenolytic effect of catecholamines and lowers blood sugar levels, thus enhancing the hypoglycaemic action of insulin. The β-blockers are therefore contrainduced in diabetics. They also block the sympathetic β-adrenoceptor-mediated reflexes such as tachycardia, arising from hypoglycaemia which is normally a useful warning sign to the diabetic. They are contraindicated in patients with heart block, sinus bradycardia, sick sinus syndrome, cardiogenic shock and heart failure because of their cardiac depressant actions. Heart failure is a potential adverse reaction. Side-effects arising from the CNS-depressant action, thought to largely occur with the more lipophilic β-blockers, include fatigue, depression and sleep disturbances. They may therefore interact with CNS depressant drugs. Special precautions are advised in pregnancy and lactation, since the β-blockers can cross into the breast milk and induce neonatal hypoglycaemia and bradycardia.

Oculomucocutaneous syndrome is the condition whereby the eye dries and damage occurs to the conjunctiva, often under the eyelids. This was identified in a small but significant number of patients who were taking practolol (Eraldin[R]), the first cardioselective β-blocker. Some of these patients became blind and some required surgery; as a consequence, practolol was withdrawn from general use. This condition was not predicted by the preclinical animal tests. It has not been identified with other β-blockers, although it remains a possibility. Thus the drug should be withdrawn gradually if there are any signs of eye dryness and, in the UK, the Committee for the Safety of Medicines (CSM) is notified by the Yellow Card system. The precise cause of the condition remains uncertain, but it appears to be of immunological origin.

Finally, an adverse reaction that occurs on ceasing treatment is the 'withdrawal phenomenon', which is manifest as an increased risk of angina attacks or of sudden death from cardiac failure. This is due to the up-regulation of β-adrenoceptors that occurs with the chronic use of β-blockers and which has been described earlier in this chapter. There is a raised sensitivity to β-adrenoceptor agonists in patients and animals after chronic administration of β-blockers. This effect does not occur with blockers having ISA, such as pindolol, presumably because the desensitizing effect of agonist activity offsets the up-regulation by the antagonism. β-Blockers should therefore be withdrawn gradually.

Drug interactions that occur are both pharmacodynamic and pharmacokinetic. The latter includes reduced plasma levels of those β-blockers that undergo hepatic metabolism (eg propranolol but not nadolol) because of enzyme induction by other

Table 5.4 Adverse effects and interactions of β-blockers

Adverse reaction	Mechanism	Comments/interactions	Contraindications
Bronchoconstriction	Block of airways β_2-adrenoceptors	Less if β_1-selective	Asthma
Cold extremities	Peripheral vasoconstriction, block of vascular β_2-adrenoceptors, reflex α-vasoconstriction, reduced CO	ISA of theoretical advantage	Raynaud's syndrome Intermittent claudication
Cardiac failure, heart block, bradycardia	Block of cardiac β_1-adrenoceptors, slow AV conduction	Failure compensated by $\uparrow$ sympathetic activity Interact with drugs slowing AV conduction (verapamil), antiarrhythmics, cardiodepressant anaesthetics	Heart failure, AV block
Hypoglycaemia	Block β-mediated glycogenolysis	Block warning tachycardia of hypoglycaemia Interact with hypoglycaemics	Diabetics

(continued)

CNS depression	Possibly more frequent with lipophilic blockers (eg propranolol)	Interact with CNS depressants	
Withdrawal phenomenon—angina, sudden death	Up-regulation of β-adrenoceptors	Withdraw gradually by reducing dosage over several weeks	
Fatigue (exertional)	Reduced muscle blood flow		
Gastrointestinal upset	Non-specific effects		
Dry eyes (oculomucocutaneous syndrome)	Unknown toxicity of practolol	Withdraw if observed	
Foetal and neonatal responses	Possibly more frequent with lipophilic blockers	Cross placenta and into breast milk	Pregnancy and lactation
Drug interactions Sympathomimetics Liver enzyme inducers Cytochrome P450 inhibitors	Enhanced α-vasoconstriction Enhanced hepatic metabolism of susceptible β-blockers Reduced hepatic metabolism	Hypertension	Cold remedies Phenytoin, alcohol, smoking Cimetidine
Cyclooxygenase inhibitors	Block antihypertensive action		Indomethacin

drugs such as phenytoin or phenobarbitone and by smoking or alcohol. Cimetidine is also recognized as interacting with β-blockers that are metabolized via hepatic oxidases, because it inhibits cytochrome P450 activity. This leads to elevated and prolonged plasma levels of the β-blocker. Non-steroidal anti-inflammatory drugs such as indomethacin oppose the antihypertensive activity of β-blockers. As explained under miscellaneous mechanisms of antihypertensive activity, this has been taken to indicate that β-blockers may increase the synthesis and release of prostacyclin from the vascular wall. This effect would be prevented by blockade of cyclo-oxygenase by indomethacin. Sympathomimetic amines in cold remedies may interact in hypertensive patients because their α-vasoconstrictor effect will be unblocked and therefore lead to undesirable pressor effects.

5.4   Irreversible β-Adrenoceptor Antagonists

Among the α-adrenoceptor blockers, the β-haloalkylamines (such as dibenamine) are irreversible antagonists which form covalent bonds with the receptor by alkylation. The value of these compounds in measuring receptor numbers and for calculating the affinity of agonists was described in Chapter 3. The need for comparable irreversible antagonists at β-adrenoceptors was therefore apparent. Such compounds could be used to inactivate β-adrenoceptors and thus study the turnover of receptors after their inactivation. Calculation of agonist affinities from functional responses and of receptor reserves would also be possible. The criteria for such compounds are as follows.

1   High affinity for the β-adrenoceptor so that non-specific interactions with other membrane components are reduced and low concentrations can be used for blockade.

2   Low lipid solubility which reduces its uptake into cells and tissues but facilitates washout from isolated tissues and membrane preparations.

A number of compounds have been developed with this aim and some of these are shown in Table 5.5. One group of compounds are the photoaffinity labels which includes p-azidocarazolol. These antagonists reversibly interact with the receptor in the dark but form covalent bonds when photoactivated by exposure to ultraviolet light (Lavin *et al.* 1981). However, their use has not been entirely successful since only a small fraction of receptors appear to be labelled. Furthermore, the photoactivation procedure cannot be applied *in vivo* and in isolated intact tissues; uniform exposure to light for a constant period of time is not possible.

By analogy with dibenamine, substitution of a halogen into a competitive β-blocker was seen as a way towards achieving reactive compounds capable of alkylating the receptor. Chloropractolol was the first such compound which was reputed to be an irreversible antagonist based on the long-lasting antagonism and depression of the maximum response to isoprenaline of rat isolated atria. Such evidence alone, however, showed a major pitfall in interpreting concentration–response curves. It transpired that chloropractolol was a partial agonist which raised the baseline rate of atrial contraction. After exposure to chloropractolol, isoprenaline could still produce the same *total* maximum rate of contraction, but when expressed as the *increase* above resting, there was an apparent reduction of the maximum (Kenakin & Black 1977). Thus, chloropractolol is not an irreversible antagonist.

Table 5.5 Irreversible β-adrenoceptor antagonists

Compound	Structure	Notes
Bromoacetyl alprenolol menthane (BAAM) (mixture of two compounds)	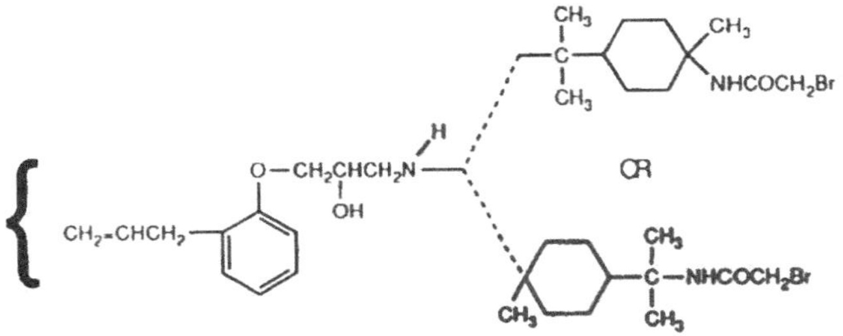	Established irreversible antagonist (β_1 and β_2)
Bromoacetyl pindolol	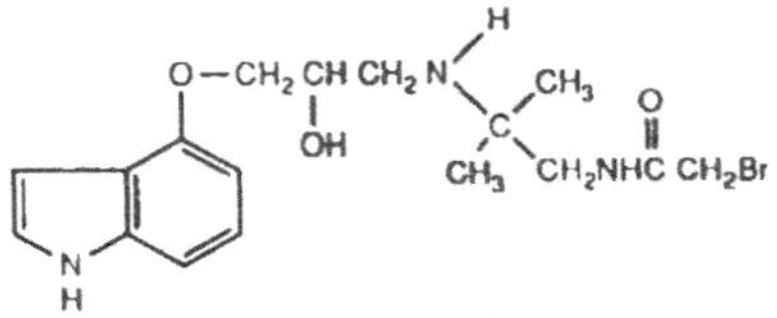	Irreversible β_1 and β_2 antagonist
Bromoacetyl propranolols (NHNP-NBE)	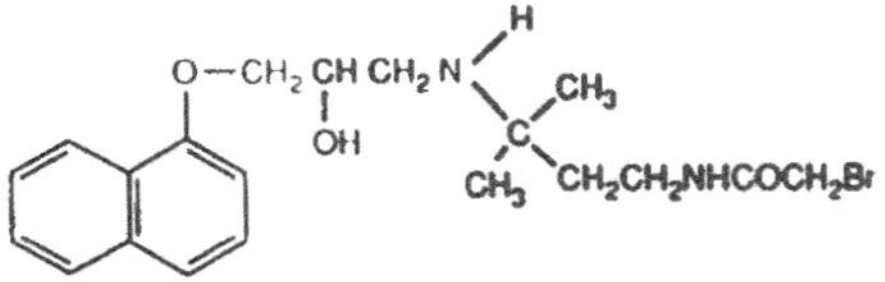	Mainly competitive, but irreversible inhibition of adenylyl cyclase and binding at high concentrations
Ro 03-7894	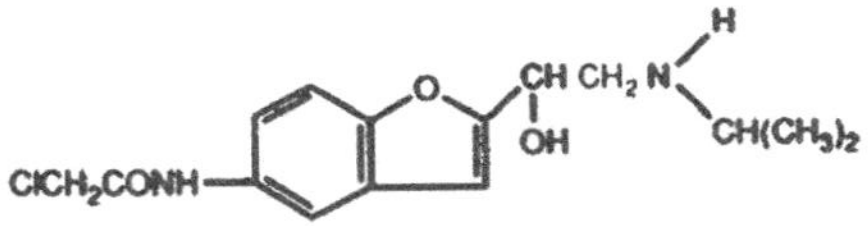	Irreversible, but this action at β-adrenoceptors disputed

(*Continued*)

Table 5.5 *Continued*

Chloropractolol	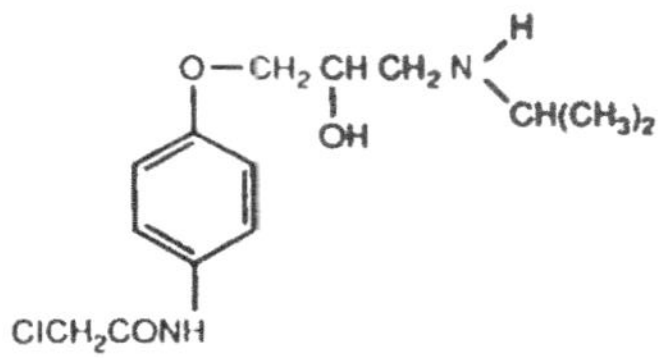	Not irreversible

O—CH₂ CH CH₂ N⟨H⟩ CH(CH₃)₂ / OH

ClCH₂CONH

FM 24	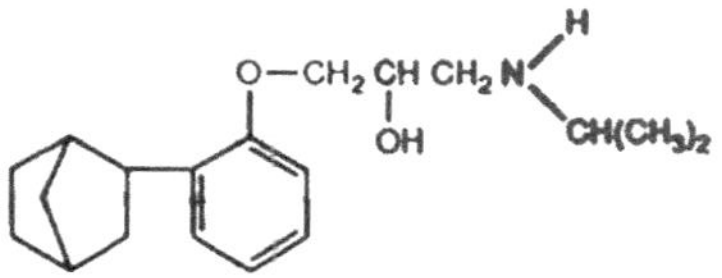	Prolonged irreversible block without alkylation — hydrophobic property

O—CH₂ CH CH₂ N⟨H⟩ CH(CH₃)₂ / OH

Bromoacetyl — carbostyril analogue	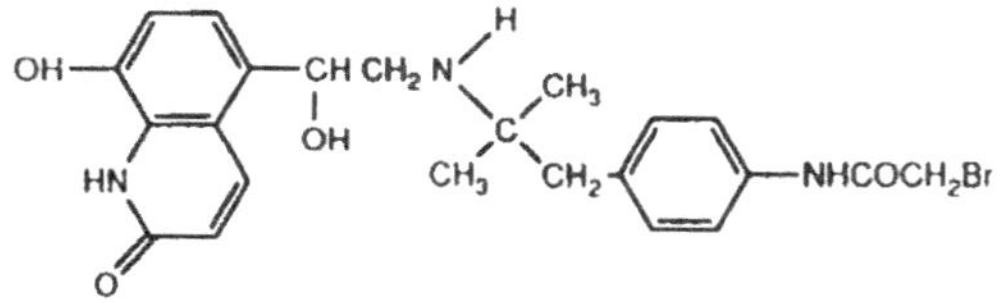	Initial full *agonist* on adenylyl cyclase with subsequent irreversible binding

p-Azidobenzylcarazolol	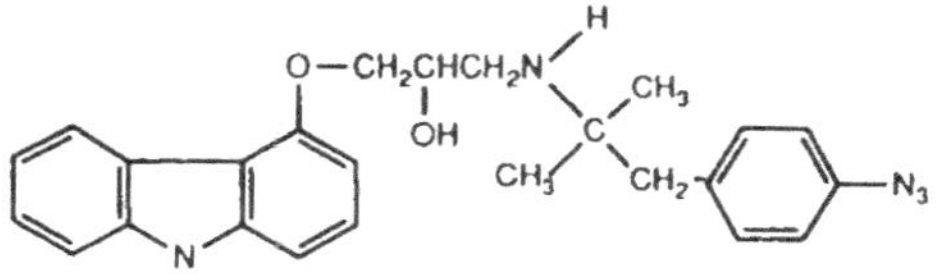	Photoaffinity label forms covalent link with β-adrenoceptor on exposure to UV light

Subsequently other halogenated derivatives of competitive antagonists have been introduced (Table 5.5). The degree of irreversible binding to the β-adrenoceptor of some of these (NHNP-NBE and Ro 03-7894) has been disputed. Currently, BAAM appears to be the most useful agent for irreversible inhibition of both β_1- and β_2-adrenoceptors in radioligand binding and functional studies with isolated tissues (Minneman & Mowry 1986).

5.5 β-Adrenoceptor Antagonists with Additional Properties

The regulatory authorities are demanding increasingly rigorous testing of new drugs and a preference for single isomers to avoid side-effects and toxicity from the inactive isomer. Preparations containing several active constituents with different mechanisms are also less popular since they give rise to a corresponding increase in the breadth of potential adverse effects. These limitations can be overcome by combining the desired pharmacological activities into the same drug molecule. β-Blockers are an example of this approach whereby β-blockade is combined with other properties that are of value in the treatment of cardiovascular disease. Labetalol, the combined α/β-blocker, was one of the first examples, and the structures of several compounds are shown in Table 5.3(C).

5.5.1 *Antagonists with α- and β-Adrenoceptor Blocking Activity*

Labetalol The rationale for the introduction of labetalol was the fact that α-blockade was known to induce a reflex tachycardia via β-adrenoceptors, which limited the effectiveness of α-adrenoceptor antagonists in hypertension. Second, the peripheral vasoconstriction and cold extremities was an undesirable side-effect of the non-selective β-blockers in particular. Thus, it was reasoned that blockade of both receptor types should lower blood pressure with minimal physiological disturbance. Labetalol blocks α_1- and both β-adrenoceptor subtypes. The ratio of β- to α-adrenoceptor blocking activity of labetalol is $\sim 5-10$-fold. It additionally has some inhibitory action upon neuronal uptake of noradrenaline (Drew *et al.* 1978). There is extensive first-pass metabolism of labetalol in the liver by oxidation and glucuronidation, its half-life being 4–6 hr.

 Dilevalol (Unicard[R]). Labetalol contains two asymmetric centres and therefore four stereoisomers exist. The clinically used labetalol consists of a mixture in equal proportions of the four isomers, RR, SR, SS and RS. In an attempt to improve the clinical profile, the activities at α- and β-adrenoceptors of these isomers have been examined and dilevalol is the RR isomer. The SS and RS isomers are devoid of α_1- or β-antagonistic activity, while the α_1-adrenoceptor blocking activity resides primarily in the SR isomer. β-Adrenoceptor blocking activity lies in the RR isomer (dilevalol), which is non-selective for β_1- and β_2-adrenoceptors. Dilevalol also has ISA at β_2-adrenoceptors which is responsible for peripheral vasodilatation. This property allows peripheral blood flow to be sustained and to increase with exercise. It makes the profile of dilevalol particularly attractive for use in physically active hypertensive subjects and in those with peripheral vascular disease or who complain of cold extremities while using conventional β-blockers. Furthermore, there is no complaint of bronchoconstriction. Clinical experience with dilevalol has shown it to effectively lower blood pressure in hypertensive patients. It has a half-life of 15–18 hr and has a favourable side-effects profile as compared with other antihypertensives, which includes a neutral or beneficial effect on plasma lipid (Wallin & Frishman 1989). Unfortunately, the compound has recently been withdrawn from the market due to reports of liver toxicity. The lack of this effect with labetalol suggests that a mixture of the RR and SR isomers of labetalol might be a useful antihypertensive combination, since this would additionally have the α_1-adrenoceptor blocking activity (Riva *et al.* 1991).

Carvedilol is a selective β_1-adrenoceptor antagonist with additional α_1-adrenoceptor blocking properties. The latter accounts for most of its vasodilator activity, although at higher concentrations Ca^{2+} channel blockade may also accur. Carvedilol is an effective antihypertensive agent. It increases exercise capacity of patients with chronic stable angina and reduces their myocardial oxygen consumption. In addition, it has beneficial effects in patients with chronic congestive heart failure (NYHA class II or III) secondary to ischaemic heart disease. It is highly lipophilic and undergoes rapid first-pass hepatic metabolism. It appears to have few adverse effects on lipid metabolism (McTavish *et al.* 1993). Carvedilol provides marked cardioprotection in models of MI and affords protection against the growth of new vascular intima after balloon angioplasty. This novel action appears to be independent of β-adrenoceptor or L-type Ca^{2+} channel blockade and indicates a potential use in the treatment of vascular restenosis following balloon angioplasty (Douglas *et al.* 1994).

Oxprenolol also has some α_1-adrenoceptor blocking activity which may contribute to its antihypertensive activity (Vila *et al.* 1982).

5.5.2   β-Adrenoceptor Blockade with Vasodilator Activity

Celiprolol (Celectol[R]) has been refered to as a 'third generation' β_1-selective antagonist, in that it has additional vasodilator properties. The vasodilator activity appears to be due to β_2-adrenoceptor partial agonist activity, making it a novel compound if it selectively blocks at β_1-adrenoceptors yet selectively stimulates at β_2-adrenoceptors (Dhein *et al.* 1992). It effectively lowers blood pressure but there is some doubt as to whether the ISA is truly selective for β_2-adrenoceptors in man (Wheeldon *et al.* 1992).

Bucindolol is also a β-adrenoceptor antagonist with no selectivity for β_1- or β_2-adrenoceptors, but has vasodilator activity.

5.5.3   β-Adrenoceptor Blockade with Diuretic Activity

Tienoxolol has β-blocking, diuretic and natriuretic properties combined in the same molecule (Table 5.3). It induces the urinary excretion of 6-keto-$PGF_{1\alpha}$, the stable hydrolysis product of prostacyclin and, at high doses, prostaglandin PGE_2. Propranolol and atenolol do not stimulate urinary prostaglandin secretion. Although these prostaglandins exert diuretic and natriuretic effects by regulating water, sodium and chloride excretion, the effect of tienoxolol is not thought to be due to this action on prostaglandin excretion. Instead, it has been shown to bind to adenosine A_1 and A_2 receptors and thus exert diuresis through a theophylline-like antagonism of these receptors (Caussade & Cloarec 1993).

5.5.4 β-Adrenoceptor Blockade with Class III Antiarrhythmic Activity

Sotalol (Sotacor[R]) was originally described in 1964 as a β-blocker (Larsen & Lish 1964) but took many years before introduction to therapeutic use. It is non-selective between β_1- and β_2-adrenoceptors and has no ISA or membrane stabilizing activity; the latter property eliminates any class I antiarrhythmic activity. Its effectiveness, however, is equal to or greater than that of other β-blockers against supraventricular tachyarrhythmias and it shows promise in the treatment of ventricular arrhythmias. This action is attributed to a class III antiarrhythmic property characterized by prolongation of the cardiac action potential and of the QTc interval. There is no change in the depolarization rate which eliminates a class I action. Since this property is shared by both the (+) and (−) enantiomers, it is unrelated to the β-blocking properties. It is consistent with an inhibition of the outward K^+ current (i_K) that is largely responsible for repolarization in sinus node fibres (Campbell 1987). The lengthening of the action potential duration tends to augment contractility and therefore offsets the cardiac depressant effects of β-blockade. Sotalol is an effective antihypertensive and antianginal agent. There is no first-pass hepatic metabolism and the low lipid solubility prevents CNS penetration and central side-effects (Singh *et al.* 1987).

5.5.5 β-Adrenoceptor Blockade with ACE Inhibition

Inhibition of antiotensin converting enzyme (ACE) prevents the conversion of angiotensin I to angiotensin II which is a powerful vasoconstrictor and releases aldosterone to enhance Na^+ and water retention by the kidneys. Thus ACE inhibitors, such as enalapril, are now established in the treatment of congestive heart failure and hypertension. Inhibition of ACE consistently elevates plasma levels of both angiotensin I and plasma renin activity. This response is sensitive to β-adrenoceptor blockade, suggesting the involvement of the renal sympathetic innervation in the raising of renin activity after ACE inhibition. It has therefore been argued that a combination of ACE inhibition with β-blockade might produce a more complete control of the renin-angiotensin system. To this end, the combination of ACE inhibition and β-blockade in a single entity (BW B385C) has been shown to produce blood pressure lowering with attenuated plasma renin elevation (Cambridge *et al.* 1992).

Drugs Affecting Storage and Release from Sympathetic Neurones

The synthesis, storage and release of noradrenaline from noradrenergic nerve endings was described in Chapter 2. In this chapter, the effects of drugs upon the storage and release of noradrenaline are described. Three groups of agents are considered: drugs that modulate noradrenaline release by an action on prejunctional receptors of the postganglionic neurone terminal; drugs that impair sympathetic neurotransmission by depletion of transmitter; and, finally, the noradrenergic neurone blockers which interfere with noradrenaline release independently of prejunctional receptors.

6.1 Prejunctional Receptors on Sympathetic Nerve Endings

6.1.1 *General*

Prejunctional receptors are located at the axon terminal and modulate transmitter release. Activation of these receptors may either inhibit or facilitate neurotransmitter release. They may be stimulated by the transmitter itself, in which case they are termed autoreceptors. If the transmitter is inhibitory upon its own release (eg α_2-adrenoceptors at sympathetic nerve terminals) then there is said to be a negative feedback. If the transmitter acts on facilitatory receptors (eg β-adrenoceptors) then there may be positive feedback. Prejunctional receptors on sympathetic neurones may also be stimulated by transmitters from other adjacent neurones (eg Ach from parasympathetic nerves) or by circulating or locally released autacoids (eg histamine, angiotensin II). In this case they may be termed heteroreceptors.

The first indication of the existence of prejunctional receptors at sympathetic nerve endings was the observation that the α-blocker, phenoxybenzamine (PBZ), could enhance the overflow of radiolabelled noradrenaline from the cat perfused spleen during sympathetic nerve stimulation. At first, this was thought to be due to the known inhibitory effect of PBZ upon noradrenaline uptake (see Chapter 2). However, when a similar degree of neuronal uptake block was produced by cocaine or DMI, agents which are not recognized as α-adrenoceptor antagonists, little or no increase in transmitter overflow was found during nerve stimulation. Furthermore, the increase in noradrenaline overflow by PBZ and by other α-blockers, including phentolamine, during nerve stimulation, occurred at concentrations which do not inhibit either neuronal or extraneuronal uptake. These results led to the now widely accepted concept of prejunctional α-adrenoceptors located on the sympathetic nerve ending.

6.1.2   *Prejunctional α-Adrenoceptors*

The prejunctional α-adrenoceptors were proposed to be stimulated by noradrenaline released during nerve activity, resulting in negative feedback which inhibits further transmitter release. Thus, α-blockade by PBZ will prevent this inhibitory feedback and enhance transmitter release. In support of these proposals is the fact that α-adrenoceptor agonists also inhibit transmitter release (Langer 1977) (Figure 6.1). The effect of prejunctional α-adrenoceptor stimulation may be demonstrated by the tissue response to sympathetic nerve stimulation, in addition to measurement of radiolabelled noradrenaline overflow (Doxey & Roach 1980). However, demonstration from the pharmacological response is only possible when the tissue response is mediated via a different receptor subtype. Thus, in the cat spleen, the contractile response to nerve stimulation is also α-adrenoceptor mediated ($α_1$; Table 3.4) and therefore blocked by PBZ. In isolated cardiac preparations, however, the increases in rate of contraction induced by field stimulation of sympathetic nerves is β-adrenoceptor mediated. This response is enhanced by PBZ or phentolamine because of the increased transmitter release (Langer *et al.* 1977) (Figure 6.2). Similarly, under *in vivo* conditions, the response to cardioaccelerans nerve stimulation in the anaesthetized dog is potentiated by α-blockade. This indicates that negative feedback mechanisms operate under physiological conditions. Further evidence that there is background autoinhibition of noradrenaline release via prejunctional α-adrenoceptors *in vivo* comes from the fact that α-blockers increase the levels of noradrenaline

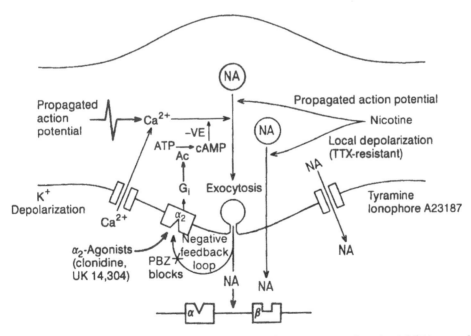

Figure 6.1 Mechanism of prejunctional $α_2$-adrenoceptor-mediated inhibition of noradrenaline (NA) release. Only exocytotic release by propagated action potentials and K^+ depolarization are inhibited (left side). Indirectly acting sympathomimetic amines (eg tyramine), Ca^{2+} ionophore A23187 and nicotine (right side) are unaffected. Nicotine releases NA by a propagated action potential and by a tetrodotoxin (TT_X)-resistant local depolarization.

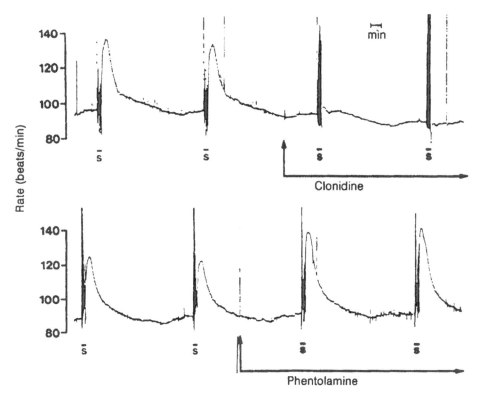

Figure 6.2 Isolated perfused guinea-pig atrial pairs. Effects of α_2-adrenoceptor stimulation by clonidine (0.11 μM) and blockade by phentolamine (0.26 μM) on the positive chronotropic effects following field stimulation (S; 0.8 Hz, 3 msec, 130 V for 30 sec every 10 min). Note that clonidine blocks the response to released noradrenaline and phentolamine potentiates, while neither affect basal atrial rate. Reproduced with permission from Doxey & Roach (1980).

in venous blood and of noradrenaline metabolites (eg DOPEG) in cerebrospinal fluid (indicating central modulation). Presynaptic α-adrenoceptors become supersensitive after chronic blockade, which again indicates their constant activation under physiological conditions (Starke *et al.* 1989).

α-Adrenoceptor antagonists are only effective in enhancing release of transmitter if the prejunctional receptors are being activated by the neurotransmitter. Thus, a threshold concentration of noradrenaline in the synaptic cleft is required for the activation of presynaptic negative feedback. PBZ only enhances transmitter release from guinea-pig atria when a train of pulses is applied but not with a single pulse. When two consecutive single stimuli are applied to a tissue, such as the vas deferens, the effect of transmitter released by the first stimulus can be examined upon the responses of the second. The contractile response is biphasic (see Chapter 3, vas deferens), and when the gap between the two pulses is only 3 sec, the second noradrenergic phase of the second response is missing, indicating negative feedback on noradrenaline release. Early studies indicated that this was restored by an α_2-adrenoceptor antagonist (Brown *et al.* 1983) and therefore due to prejunctional α_2-adrenoceptor autoregulation. This has not, however, always been supported by other researchers (Mallard *et al.* 1992).

The effect of antagonists is therefore frequency-dependent, with no increase in transmitter overflow by yohimbine being observed with trains of pulses at 1 Hz frequency. Yet the blockade by yohimbine was sufficient to shift dose-response curves for the inhibitory action of an α-agonist (Fuder *et al.* 1983). As the frequency increases, antagonists become more effective, until finally at high frequency (30 Hz) their effectiveness again declines. This latter effect may be unrelated to the concentration of noradrenaline in the synaptic cleft, but due to greatly elevated levels of Ca^{2+} intraneuronally occurring at high frequency.

Frequency-dependence of agonists is also a well-known phenomenon. When there is very little endogenous noradrenaline released to cause autoinhibition at low frequency, exogenously added agonists are particularly effective. As frequency increases, more noradrenaline is released and there is a greater degree of autoinhibition. Exogenously added α_2-agonists then become less effective as the frequency of nerve stimulation increases. This frequency-dependence becomes particularly important for partial agonists such as clonidine. They behave as agonists at low frequency (0.3 Hz) causing inhibition of transmitter release, but become antagonists at higher frequency (3 Hz).

Another way in which the concentration of noradrenaline in the synaptic cleft may be altered and thereby affect prejunctional receptor modulation of transmitter release is by inhibition of neuronal uptake, for example, with cocaine. In the presence of cocaine, the raised intrasynaptic noradrenaline causes greater autoinhibition and thus α-antagonists become more effective. Conversely, exogenously added agonists are less effective inhibitors of neurotransmitter release in the presence of uptake blockade. This, however, appears to apply only to imidazoline α_2-agonists (eg, clonidine) and not phenylethylamines (eg, α-methylnoradrenaline). The latter group are in fact apparently potentiated. One explanation for this is that they are substrates for uptake into the neurone and therefore displace the preloaded radiolabelled noradrenaline. When uptake is blocked this effect is prevented and only the inhibition of transmitter release via prejunctional α_2-adrenoceptors is observed (Langer 1981). A more recent suggestion is that there is a link between the neuronal uptake site and the imidazole site of the prejunctional α_2-adrenoceptor. The phenylethylamines and imidazolines may consequently interact differently with the prejunctional α_2-adrenoceptor (see also Chapter 3) (Starke *et al.* 1989).

For the reasons discussed above, the importance of negative feedback appears to be greater in tissues with a narrow synaptic cleft. In these tissues there is a higher concentration of noradrenaline in the vicinity of the prejunctional α_2-adrenoceptors.

6.1.2.1 *Receptor type*

The subclassification of prejunctional α-adrenoceptors as being of the α_2-subtype was first proposed by Berthelsen & Pettinger (1977). This was based on the agonist potency orders for inhibition of noradrenaline release differing from that for postjunctional smooth muscle contractile activity (see Chapter 3, Figure 3.18). Methoxamine, phenylephrine, cirazoline and amidephrine are selective for the postjunctional α_1-adrenoceptor-mediated effect whereas UK-14,304, clonidine and α-methylnoradrenaline occupy the other extreme of the spectrum (Table 4.3). Subsequently, the selective inhibition of this response by antagonists such as yohimbine and idazoxan has confirmed that these prejunctional α-adrenoceptors are predominantly of the α_2-subtype (Table 5.2). The distribution of α_2-adrenoceptors at other sites, including their postjunctional location in blood vessels as extrasynaptic receptors stimulated by circulating adrenaline rather than neurotransmitter, has been discussed in Chapter 3. The identification of α_2-adrenoceptors by radioligand binding and their further subclassification into α_{2A}- to α_{2C}-subtypes have also been described earlier. As discussed in Chapter 3, at present there is no consistent pattern as to whether the prejunctional α_2-adrenoceptors modulating transmitter release, as determined by functional studies, are of the the α_{2A}- or α_{2B}-subtype. It appears to vary from tissue to tissue (eg atria as compared with salivary gland).

It is now clear that a minor population of α_1-adrenoceptors also exist prejunction-ally and these may also inhibit transmitter release. Thus, selective α_1-adrenoceptor agonists, such as methoxamine, have also been shown to inhibit the release of noradrenaline by sympathetic nerve stimulation or to reduce the postjunctional response to sympathetic nerve stimulation in a wide range of isolated tissues. This inhibition is abolished by α_1-adrenoceptor antagonists (prazosin) but not by α_2-adrenoceptor antagonists (rauwolcine).

6.1.2.2   *Site of α_2-adrenoceptor-mediated transmitter inhibition*

The site of inhibition of transmitter release by α_2-adrenoceptor stimulation has been located to the axon terminal. An effect on the conduction of the propagated action potential down the axon or at the cell body–dendritic region has been discounted. The influence of the cell body and propagated action potential can be eliminated by treating with tetrodotoxin which blocks Na^+-dependent nerve action potentials; transmitter may then be released simply by raising extracellular K^+ to depolarize the nerve terminal. Under these conditions, α_2-adrenoceptor agonists still have an inhibitory effect upon transmitter release. Further evidence that α_2-adrenoceptors are located on the varicosity itself rather than the intervaricosity membrane is their functional inhibition of [^{3}H]noradrenaline release from synaptosomes; the pinched-off nerve terminals from the brain. Further indirect evidence is from radioligand binding experiments, after destruction of the sympathetic neurone by, for example, 6-hydroxydopamine (see later in this chapter). In theory, this should lead to a loss of the prejunctional α_2-adrenoceptor binding sites, and in many studies this is the case. However, the data is inconsistent, possibly because of a compensatory up-regulation of postjunctional α-adrenoceptors. Finally, it is worth noting that the α_2-adrenoceptors probably reach the nerve terminal varicosities by transport down the axon after synthesis in the cell body (Starke *et al.* 1989).

6.1.2.3   *Mechanisms of α_2-adrenoceptor-mediated inhibition of transmitter release*

To understand the mechanism of inhibition of transmitter release, an appreciation of the release process is necessary and this has been described in Chapter 2. Noradrenaline release occurs by exocytosis associated with raised intraneuronal Ca^{2+} levels, following influx through voltage-sensitive N-type Ca^{2+} channels. In Ca^{2+}-free medium there is no exocytosis but noradrenaline can be released by raising intracellular Na^+ with ouabain (non-exocytotic). This form of release is not, however, inhibited by the α_2-agonist, clonidine, in the vas deferens, indicating that prejunctional α_2-adrenoceptors are only linked to exocytotic release (Bönisch & Trendelenburg 1988). A further indication that it is exocytotic release that is inhibited is obtained when peripheral sympathetic nerves are depleted of noradrenaline but exocytosis is measured as the release of the vesicular enzyme, DBH (see Chapter 2). Under these conditions, a_2-adrenoceptor agonists inhibit release of DBH and also other co-stored transmitters such as neuropeptide Y (see Chapter 12). Furthermore, the depleted levels of noradrenaline reduce synaptic transmitter levels and thereby enhance sensitivity to exogenously added a_2-agonists.

Because it is only exocytotic release that is inhibited by a_2-agonists, they do not prevent the release of noradrenaline nor the pharmacological response induced by indirectly acting sympathomimetic amines. As discussed in Chapter 4, indirect activity does not involve membrane depolarization, does not require extracellular Ca^{2+} and cannot be prevented by tetrodotoxin (TT_X) (ie is independent of any propagated action potential). Noradrenaline is released from the vesicle into the cytoplasmic pool and thence via the neuronal carrier process by facilitated exchange diffusion. Whether a_2-adrenoceptor agonists inhibit the release of noradrenaline induced by nicotine is a matter of dispute. This may arise because of the uncertainty regarding the mechanism whereby nicotine releases noradrenaline. The effects of nicotine are inhibited by tetrodotoxin, indicating that it sets up a propagated action potential. At higher concentrations of nicotine, however, the release of noradrenaline becomes tetrodotoxin-resistant, indicating a possible local membrane depolarization and influx of Ca^{2+} into the sympathetic varicosity. Clonidine has been shown to be ineffective in reducing positive inotropic responses of isolated atria and output of radiolabelled noradrenaline induced by nicotine. This suggests that stimulation of prejunctional a_2-adrenoceptors does not interfere with the spread of the action potential to the terminal varicosities (Brasch 1991) (Figure 6.1). It is of interest that selective a_2-adrenoceptor antagonists enhance the vasoconstrictor responses of dog mesenteric arteries to electrical stimulation of the sympathetic nerves, a response mediated by both noradrenaline release onto postjunctional a_1-adrenoceptors and ATP release onto P_2-purinoceptors (see Chapter 12). This indicates that both the release of noradrenaline and its co-transmitter, ATP, are subject to prejunctional a_2-adrenoceptor autoinhibition (Muramatsu *et al.* 1989).

The dependence of a_2-adrenoceptor-mediated inhibition of transmitter release upon adequate extracellular Ca^{2+} levels suggests that the inhibition is due to reduced availability of Ca^{2+} for excitation–secretion coupling. Synaptosomes have been used to study the role of Ca^{2+} in the inhibition of [^{3}H]noradrenaline release by a_2-adrenoceptor stimulation. a_2-Adrenoceptor agonists inhibit the release of [^{3}H]-noradrenaline due to Ca^{2+} influx arising from depolarization induced by raising extracellular K^+. However, the release induced by Ca^{2+} ionophore, A23187, which raises intracellular Ca^{2+} but bypasses the voltage-gated Ca^{2+} channel, is not inhibited by a_2-agonists (Fillenz 1992). Finally, a_2-adrenoceptor agonists have been shown to inhibit veratrine-induced release of noradrenaline from brain synaptosomes in a Ca^{2+}-free medium; this is presumably due to the mobilization of Ca^{2+} from intracellular stores following influx of Na^+. This observation is, however, at odds with previous findings and suggests that the effect of a_2-adrenoceptor activation lies somewhere beyond Ca^{2+} influx. Apart from this last observation, the majority of results point to the site of inhibition by a_2-adrenoceptor activation being with the voltage-dependent Ca^{2+} channels (Figure 6.1). An enhanced efflux of K^+ is a possible alternative explanation for the a_2-adrenoceptor-mediated inhibition of transmitter release, but Ca^{2+} channels remain the favoured site of interaction (Starke *et al.* 1989).

The phosphorylation of these Ca^{2+} channels by cAMP-dependent protein kinase increases their probability of opening upon depolarization of the neuronal membrane (Reuter 1983). Thus, increases in neuronal cAMP levels, for example by forskolin and phosphodiesterase inhibitors or the introduction of the stable analogue, 8-bromo-cAMP, promote transmitter release. α_2-Adrenoceptors are negatively coupled to adenylyl cyclase through a G_i regulatory protein. Stimulation of prejunctional α_2-adrenoceptors therefore exerts an *inhibitory* effect of adenylyl cyclase and cAMP production. This reduction in cAMP levels would therefore explain the inhibition of Ca^{2+} influx through voltage-dependent Ca^{2+} channels and reduction in exocytotic noradrenaline release. Evidence for the role of G_i regulatory protein is derived from experiments with pertussis toxin (islet activating protein) and N-ethylmaleimide. Pertussis toxin catalyses the adenosine 5'-diphosphate (ADP)-ribosylation of several guanine nucleotide binding regulatory proteins (G proteins) including G_i and G_o, which are linked to *inhibition* of adenylyl cyclase (G_i) and directly to the opening of K^+ channels (G_o) (Dolphin 1987). There is, however, considerable dispute over whether pertussis toxin inhibits agonist action at prejunctional α_2-adrenoceptors of peripheral sympathetic nerves. Although pertussis toxin inhibits α_2-adrenoceptor function in slices or synaptosomes of CNS origin, there are now many observations that fail to demonstrate inhibition in peripheral tissues such as isolated atria (eg Musgrave *et al.* 1987, and reviewed by Starke *et al.* 1989). N-Ethylmaleimide (NEM) is an alkylating agent that is thought to inactivate the same G_i proteins that are affected by pertussis toxin, but its actions are probably less selective. NEM has in fact been shown to inhibit prejunctional α_2-adrenoceptor-mediated responses, while in the same study it was the α_1- and not the α_2-adrenoceptor-mediated inhibition of transmitter release that was attenuated by pertussis toxin (Rump *et al.* 1992). Another phosphorylating enzyme that has been suggested to be involved and which may cause Ca^{2+} channel opening is protein kinase C (PKC). This is activated by phorbol esters which have been shown to increase the release of noradrenaline from both peripheral and central neurones. Further support for a general role of PKC in release of noradrenaline is that inhibitors of PKC, such as polymyxin B, inhibit noradrenaline release induced by high frequency sympathetic stimulation and by blockade of K^+ channels (Musgrave & Majewski 1991).

The currently favoured view is that prejunctional α_2-adrenoceptors mediate inhibition of transmitter release by interacting with a G protein (G_i or G_o) to inhibit voltage-sensitive Ca^{2+} channels. The G protein may be coupled to the Ca^{2+} channel directly or via adenylyl cyclase. The precise details of this chain of events, however, remain to be fully established (Figure 6.1) (Starke *et al.* 1989).

6.1.2.4 *Prejunctional α_2-adrenoceptors and noradrenaline synthesis*

Tyrosine hydroxylase is the rate-limiting step of noradrenaline synthesis from tyrosine to dopa (Figure 2.4; Chapter 2). Synthesis of noradrenaline in brain synaptosomes is inhibited by activation of prejunctional α_2-adrenoceptors through inhibition of this enzyme. The mechanism is thought to be unrelated to the negative coupling to adenylyl cyclase, but at a step beyond Ca^{2+} influx, as discussed in Chapter 2 (Figure 2.6). There is, however, no evidence to date that prejunctional autoreceptors modulate noradrenaline synthesis in peripheral noradrenergic neurones (Fillenz 1990, 1992)

6.1.2.5 *Prejunctional α-adrenoceptors on non-sympathetic nerve endings*

In addition to sympathetic nerves, parasympathetic nerves also have prejunctional α-adrenoceptors, stimulation of which also inhibits release of neurotransmitter, in this case Ach. The classic preparation in which this effect has been demonstrated is the guinea-pig transmurally stimulated ileum (Paton preparation). Twitch contractions to low frequency stimulation are inhibited by noradrenaline and selective α_2-adrenoceptor agonists. This is attributed to the prevention of Ach release onto postjunctional muscarinic receptors (see Chapter 3, ileal adrenoceptors Figure 3.17). The potency orders of agonists at inhibiting the response is consistent with the receptor being of the α_2-subtype, with clonidine showing high potency and methoxamine being at the opposite extreme. Clonidine, however, does display partial agonist activity on this preparation in that it does not consistently produce complete inhibition of the twitch response. Selective α_2-adrenoceptor antagonists display high pA_2 values against this inhibitory response, whereas the α_1-selective antagonist, prazosin, has low potency. The evidence points to the receptor being of the same α_2-subtype as found on the prejunctional site of sympathetic neurones (Doxey & Roach 1980). The pA_2 value for the selective α_2-antagonist, idazoxan, against inhibition by UK 14,304 of contractions of the rabbit distal colon induced by pelvic parasympathetic nerve stimulation was lower (7.1 ± 0.4) than obtained in guinea-pig ileum and vas deferens. This is evidence for the heterogeneity of prejunctional α_2-adrenoceptors which may differ between species rather than between nerve types (Thaina *et al.* 1993). Like many tissues, the contractile response to pelvic parasympathetic nerve stimulation in the rabbit colon is due to release of both Ach and non-cholinergic transmitter(s) since there is an atropine-resistant component (Figure 6.3). α_2-Adrenoceptor agonists inhibit both the cholinergic and non-cholinergic (atropine-resistant) components. The nature of the

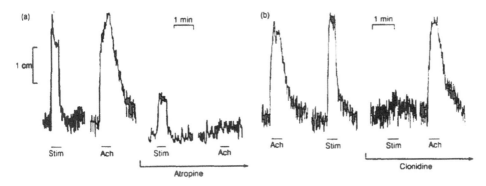

Figure 6.3 Contractions of the rabbit isolated distal colon to pelvic nerve (parasympathetic) stimulation (stim; 1 msec pulses, 20 V/cm at 2 Hz for 30 sec) and to acetylcholine [Ach; 3 μM (a) or 1 μM (b) for 30 sec]. (a) Partial blockade of the response to nerve stimulation and complete block of Ach by atropine (100 nM) indicating the presence of a non-adrenergic non-cholinergic (NANC) component of the nerve-mediated response. (b) Complete blockade of the nerve-mediated contraction and no effect on Ach by clonidine (3 μM). Vertical axis shows isotonic shortening of the colon. Reproduced with permission from Thaina *et al.* (1993).

transmitter of the latter component is still uncertain, but by analogy with closely related visceral structures, such as the urinary bladder or the cat colon, it would appear that ATP acting on P_2-receptors is the most likely non-adrenergic, non-cholinergic motor transmitter (Hoyle & Burnstock 1985) (Chapter 12). The release of the co-transmitter, ATP, is therefore also inhibited by prejunctional α_2-adrenoceptors.

In the gastrointestinal tract, α-adrenoceptor agonists can of course induce a direct relaxation of the longitudinal smooth muscle which could account for the inhibition of the contractile response to nerve stimulation. This may occur with non-selective agonists such as noradrenaline and is probably due to α_1-adrenoceptor stimulation (see Figure 3.17, Chapter 3). However, to discount this possibility with the α_2-adrenoceptor-selective agonist and to confirm a prejunctional site of action, it is necessary to show preferential blockade of the response to nerve stimulation but not of exogenously added Ach (Figure 6.3). This is the classic demonstration of a prejunctional effect of an agonist.

The first phase of the biphasic contraction of the rat vas deferens to sympathetic nerve stimulation is also attributed to the release of ATP and this response is reduced by exogenously added α_2-adrenoceptor agonists, an effect that is antagonized by idazoxan (Mallard *et al.* 1992). The exocytotic release of the co-transmitter, neuropeptide Y, by stimulation of sympathetic nerves to the guinea-pig heart is also reduced by agonists at prejunctional α_2-adrenoceptors and enhanced by the antagonist, yohimbine (Haas *et al.* 1989).

6.1.2.6   *Pharmacology of α_2-adrenoceptor agonists (α-methyldopa, clonidine, guanabenz and guanfacine)*

The antihypertensive action of α-methyldopa (see Table 2.5) was described in the early 1960s (Oates *et al.* 1960), and the hypothesis to explain the mechanism of this action provides a good example of pharmacological detective work. α-Methyldopa is the methylated analogue of the precursor L-dopa in noradrenaline biosynthesis (see Figure 2.4, Chapter 2). An early property to be demonstrated was that it inhibited the enzyme, dopa decarboxylase, which is responsible for decarboxylation of L-dopa and tryptophan (Sourkes 1954). Thus developed the hypothesis that by inhibition of the synthesis of neuronal noradrenaline, there might be a reduction of stored noradrenaline and impaired sympathetic function which would result in a fall in blood pressure. Indeed, it is only the L ($-$)-isomer that is capable of being decarboxylated and this is the form used therapeutically to lower blood pressure. However, treatment of experimental animals with α-methyldopa fails to attenuate responses to either sympathetic nerve stimulation or the indirectly acting sympathomimetic amines. Furthermore, the more potent dopa decarboxylase inhibitor, benserazide (Figure 2.5), has little antihypertensive activity. Benserazide, however, did prevent the antihypertensive action of α-methyldopa. This indicated that the effect of the latter is due to its conversion by dopa decarboxylase to α-methyldopamine or α-methylnoradrenaline (Figure 6.4), increased levels of which were found in the hearts and brains of animals receiving α-methyldopa. Thus developed the 'false transmitter' hypothesis (Day & Rand 1961). This hypothesis proposed that after administration of α-methyldopa, it would be incorporated into the sympathetic pathway for noradrenaline and α-methylnoradrenaline would be formed in the sympathetic storage vesicles to be released as a false transmitter. The proposal was that this was a less potent pressor agent and the effect of it replacing

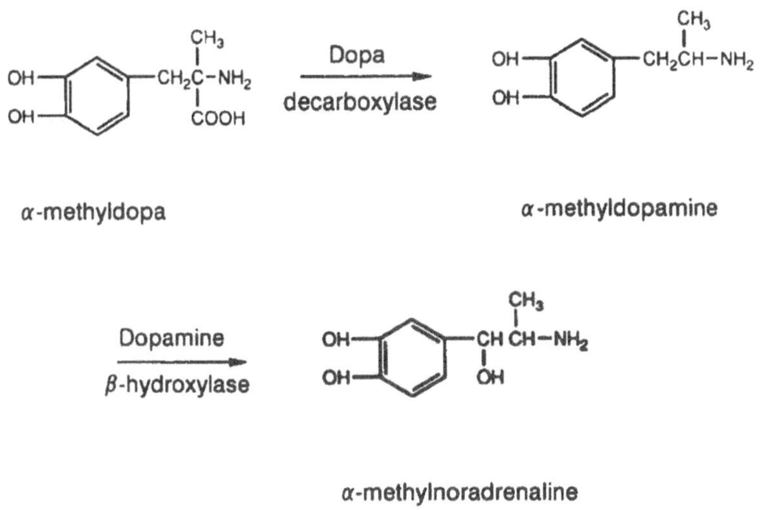

Figure 6.4 Conversion of α-methyldopa to α-methylnoradrenaline.

noradrenaline would be to reduce the effects of sympathetic nerve activity to the vasculature. Certainly, α-methylnoradrenaline was shown to be released along with noradrenaline by sympathetic nerve stimulation after pretreatment with α-methyldopa. Second, α-methyldopa restored responses to indirectly acting sympathomimetic amines after reserpine-induced depletion of noradrenaline.

The main objection to the 'false transmitter' hypothesis is the fact that there is little difference in potency between α-methylnoradrenaline and noradrenaline (Figure 5.2) (see also Chapter 3, structure–activity relationships). This is consistent with the minor reduction of response to sympathetic stimulation after α-methyldopa treatment of animals. Finally, the related α-methylated precursor of noradrenaline, α-methyl-*m*-tyrosine, is also capable of conversion to a false transmitter, metaraminol (Table 4.1). This is much less potent than is noradrenaline at α-adrenoceptors, but α-methyl-*m*-tyrosine is a weaker antihypertensive agent than α-methyldopa. Thus an alternative explanation for the antihypertensive action of α-methyldopa had to be sought. It was clear that the activity still depended upon its conversion to the false transmitter, α-methylnoradrenaline, but that this interacted at a site additional to vascular postjunctional α-adrenoceptors.

The first evidence that it was a central site of action was provided by Henning & van Zwieten (1968) who showed that α-methyldopa lowered blood pressure more when administered to anaesthetized cats via the vertebral arteries supplying the brain than when administered intravenously. When the decarboxylation of α-methyldopa was prevented with dopa decarboxylase inhibitors that have access to the CNS, the hypotensive effect was blocked but not by inhibitors that poorly cross the blood–brain barrier (eg benserazide) (Henning 1969). The hypotensive action is also abolished by inhibitors of dopamine β-hydroxylase, which prevents the conversion of α-methyldopamine to α-methylnoradrenaline. Thus, the hypotensive effect of α-methyldopa can now be attributed to its conversion to α-methylnoradrenaline primarily in central noradrenergic neurones. Here it stimulates α_2-adrenoceptors to inhibit sympathetic outflow from the brain stem to the heart and vasculature (see Chapter 1).

Stimulation of central, possibly prejunctional, α_2-adrenoceptors in the brain stem is therefore regarded as the primary site of the antihypertensive action of α-methyldopa and other α_2-adrenoceptor agonists including the imidazolines clonidine, guanabenz and guanfacine (Table 4.3). Microinjections of these α_2-adrenoceptor agonists into the region of central blood pressure regulation in the solitary tract nuclei of the medulla produces decreases in arterial blood pressure and heart rate. The fall in blood pressure is accompanied by blockade of the carotid occlusion pressor response (see Chapter 1), which is an index of the baroreceptor reflex vasoconstriction via the medullary blood pressure control centres (Kobinger & Walland 1967). When administered into the arterial inflow of neurally intact, but vascularly isolated hearts of recipient dogs, clonidine reduced blood pressure of both the recipient and donor dogs. Since the drug could not enter the peripheral circulation of the recipient, the action must have been due to a central effect mediated through reduced sympathetic output (Sherman *et al.* 1968). The decrease in sympathetic outflow is associated with a reduced level of noradrenaline in the plasma of experimental animals. In hypertensive patients, the hypotensive effects of α-methyldopa and guanfacin are well correlated with the fall in plasma noradrenaline levels (Sorkin & Heel 1986). The hypotensive effects of clonidine and guanfacin are abolished by administration of α-adrenoceptor antagonists. These selective α_2-adrenoceptor agonists also decrease the synthesis of cerebral noradrenaline, producing a relative depletion which may also be involved in the central mechanism of blood pressure reduction (Holmes *et al.* 1983, Sorkin & Heel 1986). The effect on tyrosine hydroxylase activity through prejunctional α_2-adrenoceptors has been described earlier. A minor component is probably due to inhibition of noradrenaline output from peripheral sympathetic nerves through prejunctional α_2-adrenoceptors. It is also possible that a contribution arises from the conversion of α-methyldopa to α-methyladrenaline in the adrenal medulla and brain. This is more potent than is α-methylnoradrenaline at lowering blood pressure when injected into the cerebral ventricles, although the levels in the brain are probably too low to exert a major influence on blood pressure. The corresponding depletion of brain adrenaline levels, however, may contribute to the antihypertensive effect of α-methyldopa (Tung *et al.* 1988). Clonidine also stimulates imidazoline receptors (I_1), although guanabenz only activates α_2-adrenoceptors. Since the fall in blood pressure of anaesthetized spontaneously hypertensive rats was blocked by both idazoxan (α_2- and I_2-antagonist) and by SK&F 86466 (6-chloro-3-methyl-2,3,4,5-tetrahydro-3-benzazepine)(α_2-antagonist only), it is concluded that the effect is due to α_2-agonist activity with no significant effect from I_1-receptor activation (Hieble & Kolpak 1993).

The hypotensive action of these agents therefore arises from their selectivity for α_2-adrenoceptors. Brief pressor effects may precede the fall in blood pressure after intravenous administration or may occur with higher doses of clonidine. These responses are due to agonist activity at the minor population of postjunctional α_2-adrenoceptors in the vasculature (see Chapter 3). Such activity may account for the loss of therapeutic efficacy that may occur with these agents (eg guanfacin; Sorkin & Heel 1986).

6.1.2.7 *Pharmacological actions and uses*

α-Methyldopa (Aldomet[R]) The maximum fall in arterial blood pressure with α-methyldopa occurs 6–8 hr after oral or intravenous administration, presumably due

to the time required for transport into the brain and its conversion to α-methyl-noradrenaline. It is rapidly absorbed after oral administration by an active amino acid transporter, peak plasma concentrations occurring after several hours. However, because of its prodrug nature, blood levels are a poor index of its hypotensive activity. The fall in blood pressure is associated with a decrease in peripheral resistance with little or no change in cardiac output. The blood pressure is lowered almost as much in the supine position as in the upright, when sympathetic reflex vascular control is greater. Interference with reflex control of blood pressure due to changes in posture is minimal and orthostatic (postural) hypotension is only an occassional problem. Renal and coronary blood flow are well maintained. Plasma renin levels fall, an effect probably due to an action of α-methylnoradrenaline on α_2-adrenoceptors of the juxtaglomerular apparatus inhibiting sympathetic nerve-mediated renin release (see Chapter 1, kidney; and Chapter 3, Table 3.4). There is also probably a reduced sympathetic output from the brain to the kidney. The net result will be reduced activation of the angiotensin–aldosterone mechanisms and therefore salt and water loss. How much this contributes to the antihypertensive action is uncertain. With prolonged use there is in fact the opposite effect of sodium and water retention which probably explains the refractoriness to α-methyldopa that often occurs. This can be overcome by the concurrent use of a diuretic such as bendrofluazide. Indeed, it is a far more effective antihypertensive agent when used in combination with a diuretic.

Side-effects of α-methyldopa arise most commonly from its central site of action and include sedation and drowsiness. This is particularly troublesome in individuals requiring a high level of mental concentration in their work and may lead to absentmindedness and mental confusion. Other rare signs of disturbed CNS metabolism of noradrenaline include Parkinsonism, nightmares, headache and depression. Mild block of peripheral sympathetic output may also lead to uncommon side-effects such as dry mouth, nasal stuffiness, ejaculation failure, bradycardia and postural hypotension. A rare but serious side-effect is haemolytic anaemia due to the development of an autoimmune response. The antibodies are directed against the patient's erythrocytes which are destroyed. Twenty percent of patients receiving α-methyldopa for a year produced a positive Coomb's test, which detects the antibodies, but only 1–5% of these patients develop haemolytic anaemia. Leucopaenia, thrombocytopaenia, hepatitis and fever are also rare side-effects. Gynaecomastia (breast enlargement) and lactation occur due to interference with dopaminergic suppression of prolactin secretion. α-Methyldopa is contraindicated in liver disease, depression and phaeochromocytoma, the latter being exacerbated because of further increases in plasma catecholamines. Because of these adverse effects, α-methyldopa is no longer a drug of first choice in the routine management of hypertension.

Clonidine (Catapres^R) Clonidine was first synthesized as a local vasoconstrictor for use as a nasal decongestant, but was found to produce a fall in blood pressure and bradycardia. It is an imidazoline sympathomimetic (Table 4.3) which initially stimulates postjunctional α_2-adrenoceptors to induce peripheral vasoconstriction when given intravenously, but then causes hypotension. The fall in arterial blood pressure is associated with both reduced peripheral resistance and cardiac output. It is equally effective in the standing and supine positions but the major effect when upright is on vascular resistance. There is a modest degree of orthostatic hypotension because of reduced venous return but impairement of cardiovascular reflexes to exercise is uncommon. As with α-methyldopa, there is a reduction in renin secretion, but salt and water retention may require concurrent use of a diuretic. Clonidine has no effect, or a favourable profile of activity, on plasma lipids.

Clonidine is a potent antihypertensive agent and is well absorbed orally. Its duration of action is relatively short, requiring three times daily (0.1 mg) dosage. The dose is usually increased every second or third day until the desired fall in blood pressure is attained. Alternatively, it may be administered by means of a transdermal patch which requires 3–4 days to reach equilibrium plasma concentrations. On removal of the patch, plasma levels remain stable for 8 hr and then slowly decline over the next few days.

Other uses of clonidine include the prophylactic treatment of migraine and of menopausal flushing (Nagamani *et al.* 1987) as Dixarit[R], 25 μg tablets. The mechanism of action appears to be related to the α_2-adrenoceptor-mediated vasoconstrictor action. Clonidine has also found use in suppressing some of the adverse effects of opioid withdrawal from addicts. Its main effect is to reduce the autonomic signs of withdrawal such as the nausea, vomiting and diarrhoea, with little action against the anxiety, insomnia and restlessness. These actions are mainly upon peripheral sympathetic neurones (Ruffolo *et al.* 1993). Another α_2-agonist recently introduced for opiate detoxification is lofexidine (Britlofex[R]) which has less hypotensive and sedative effects than does clonidine (Table 4.3).

The antidiarrhoeal activity of clonidine and other α_2-adrenoceptor agonists can be attributed to an inhibitory action on motility of the colon through stimulation of prejunctional α_2-adrenoceptors on parasympathetic nerve endings, to inhibit motor transmitter release (see Figure 6.3). It also increases fluid and electrolyte absorption by an action on α_2-adrenoceptors on the intestinal epithelial cells (see gastrointestinal tract, Chapter 3; Table 3.4). The side-effects, however, need to be eliminated by developing more selective agents before this activity can be utilized routinely. An inhibitory effect on gastric acid secretion may be due to stimulation of imidazoline I_1 receptors either centrally or peripherally (see Chapter 3) (Glavin & Smyth 1995).

The *para*-amino derivative of clonidine, apraclonidine, has greater hydrophilicity and therefore reduced access to the CNS. In common with other α_2-agonists it lowers intraocular pressure by decreasing the production of aqueous humour. It may also influence secondary routes of aqueous outflow, such as uveoscleral outflow, and may also affect conjunctival and episcleral vascular flow. It is approved for use in the US in preventing elevated intraocular pressure after argon laser iridotomy and trabeculoplasty (Hurvitz *et al.* 1991). Finally, it is worth pointing out that clonidine is ineffective in lowering blood pressure of phaeochromocytoma and indeed may be used diagnostically since, unlike primary hypertension, it fails to lower plasma noradrenaline levels in this form of secondary hypertension.

Side effects of clonidine are due mainly to the central site of action and include sedation and bradycardia. Dry mouth (xerostomia) associated with dry nasal mucosa and dry eyes occur infrequently on initiation of treatment, but may diminish after several weeks. Sudden discontinuation of treatment with clonidine may cause a withdrawal syndrome consisting of tachycardia, overshoot hypertension above pretreatment levels, headache, sweating and tremors. This appears to be due to an increased sympathetic discharge since plasma catecholamine levels are elevated. It is rare, but when it occurs treatment should involve the combined use of α- and β-antagonists. β-Blockers should not be used alone since they will accentuate the hypertension by allowing unopposed α-vasoconstriction.

Drug interactions of clonidine warn against concurrent use of tricyclic antidepressants, which will oppose the antihypertensive effect of clonidine. This is probably not because of the α-adrenoceptor blocking activity of these drugs (see Chapter 2, uptake blockers), since they are selective against α_1-adrenoceptors. However, it is possible that neuronal uptake block will enhance levels of noradrenaline in the vicinity of the α_2-adrenoceptor and thus compete with clonidine. Care should be exercised in using clonidine in patients with peripheral vascular disease and during lactation. Thus, although clonidine is an effective antihypertensive agent, the incidence of side-effects has reduced it to a second-line drug only for resistant cases.

Guanabenz (Wytensin[US]*) and guanfacine (Tenex*[US]*)* These α_2-adrenoceptor agonists are imidazolines structurally related to clonidine (Table 4.3) and have essentially the same profile of activity as clonidine. They are available in the US but not the UK for the treatment of hypertension. Guanfacine is more selective for α_2-adrenoceptors than is clonidine and has a longer half-life of 12–24 hr such that it may be given once daily. The side-effects are similar but perhaps less severe than with clonidine and a withdrawal phenomenon occurs less frequently (Sorkin & Heel 1986).

Guanabenz undergoes extensive first-pass metabolism in the liver and has a half-life of <12 hr. The side-effects are similar to those of clonidine, with dry mouth and sedation being sufficiently troublesome to necessitate discontinuation of therapy. It is therefore of little advantage over other newer antihypertensive agents (Holmes *et al.* 1983).

6.1.3. *Prejunctional β-Adrenoceptors*

The first demonstration that prejunctional β-adrenoceptors could modulate transmitter release from sympathetic neurones was the observation that the β-adrenoceptor agonist, isoprenaline, could enhance the overflow of [^{3}H]noradrenaline induced by nerve stimulation in isolated atria (Adler-Graschinsky & Langer 1975). This response has been demonstrated extensively in other tissues and the role of β-adrenoceptors is confirmed by its susceptibility to β-adrenoceptor blockade (Figure 6.5). The overflow of dopamine-β-hydroxylase from the perfused spleen is also enhanced by isoprenaline in tissues depleted of noradrenaline with reserpine. This proves that the facilitatory effect of β-adrenoceptors is linked to exocytotic release. The facilitation of transmitter release is stereospecific, activity residing in the (−)-enantiomer of isoprenaline. Similarly, the response is preferentially blocked by the (−)-enantiomer of propranolol in most studies. Blockade of the facilitatory effect of isoprenaline on transmitter release by 0.1 μM of both enantiomers in guinea-pig atria led Kalsner (1980) to suggest a non-specific action. The bulk of evidence, however, convincingly points to a specific action at cell surface β-adrenoceptors located on the prejunctional neuronal membrane (Figure 6.6).

The facilitation of release of [^{3}H]noradrenaline occurs whether the postjunctional response to nerve stimulation is mediated via α- or β-adrenoceptors. However, potentiation of the tissue response to sympathetic nerve stimulation by a β-agonist will depend upon the postjunctional receptor subtypes present in the tissue. If postjunctional β-adrenoceptors are present, then stimulation of these may mask the facilitatory effect on transmitter release. For example, in vascular preparations, the

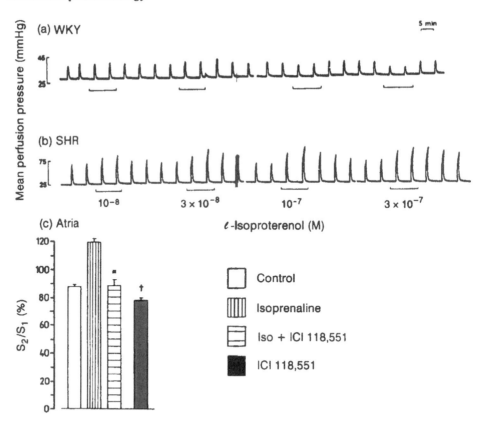

Figure 6.5 Prejunctional facilitatory effects of the β-adrenoceptor agonist, isoprenaline (isoproterenol), on responses to sympathetic nerve stimulation (a and b) and release of [^{3}H]noradrenaline (c). Isoprenaline in increasing molar concentrations potentiates the pressor responses to periarterial nerve stimulation of isolated mesenteric vascular beds of (a) WKY normotensive and (b) spontaneously hypertensive rats (SHR). In guinea-pig isolated spontaneously beating paired atria (c), isoprenaline (10 nM) enhances the efflux of [^{3}H]noradrenaline from sympathetic nerves when added between two periods of field stimulation, S_1 and S_2. This effect is significantly ($P < 0.05$,*) blocked by the β_2-selective antagonist, ICI 118,551 (0.1 μM), which alone causes a significant reduction of noradrenaline release (■). Parts (a) and (b) reproduced with permission from Kawasaki *et al.* (1982); part (c) unpublished results.

vasodilator effects of isoprenaline at postjunctional β_2-adrenoceptors may *oppose* the vasoconstriction induced by stimulation of sympathetic nerves. Low concentrations of isoprenaline, however, usually potentiate the vasoconstriction in tissues such as the perfused mesenteric bed when exposed to periarterial nerve stimulation (Kawasaki *et al.* 1982) (Figure 6.5). The prejunctional mechanism of action is supported by the fact that vasoconstriction to exogenously added noradrenaline is not potentiated, but may be reduced. At higher doses, isoprenaline does inhibit the response to periarterial nerve stimulation (Misu & Kubo 1986).

An early suggestion for the physiological relevance of the facilitation of transmitter release via prejunctional β-adrenoceptors was that it served as a positive feedback loop. This may operate at low frequencies of nerve stimulation, the released noradrenaline enhancing the further release of neurotransmitter. There is

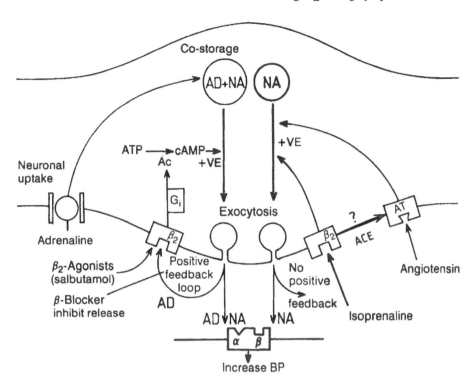

Figure 6.6 Diagrammatic representation of prejunctional β_2-adrenoceptors facilitating noradrenaline (NA) release from a sympathetic nerve varicosity. There is no positive feedback loop for noradrenaline (right side). When adrenaline (AD) is taken up, co-stored (*in vivo* or after perfusion of isolated tissues with AD) and released, it can exert a positive feedback loop which is blocked by β-adrenoceptor antagonists (left side). Angiotensin II also facilitates release. A possible link is suggested between the prejunctional β-adrenoceptor and local production of angiotensin since facilitation of NA release is blocked by angiotensin converting enzyme (ACE) inhibitors.

some evidence that facilitation of noradrenaline efflux by isoprenaline is inversely related to stimulation frequency, a greater effect being observed at low frequencies. This inverse frequency-dependence is in fact similar to that for the prejunctional α_2-adrenoceptors described earlier, suggesting that it is unlikely that prejunctional α_2- and β-adrenoceptors have opposing roles at high and low frequencies, respectively. There are two possible explanations for the inverse relationship between frequency and facilitation of transmitter release. First, as already noted with the prejunctional α_2-adrenoceptors, if there is already a high level of endogenous activation of the prejunctional receptor at high frequencies, then exogenously added isoprenaline will have reduced effectiveness. The second explanation is the more likely and is due to the opposing α_2-adrenoceptor inhibition of transmitter release which will increase as stimulation frequency is raised. In the presence of an α-adrenoceptor antagonist, phentolamine, the facilitatory action of isoprenaline is revealed (Majewski 1983).

The positive feedback loop hypothesis for prejunctional β-adrenoceptors depends upon two conditions. First, exogenously added noradrenaline should effectively enhance transmitter release. Even in the presence of phentolamine or phenoxybenzamine to block prejunctional inhibitory α-adrenoceptors, however, there is no evidence of a substantial potentiation of the response or efflux of noradrenaline with nerve stimulation. The potency of noradrenaline in producing modest facilitation appears to be 500 times less than that of adrenaline. The second prerequisite for a positive feedback loop by released noradrenaline is that β-blockade alone should inhibit transmitter release and responses induced by sympathetic nerve stimulation. Such evidence, however, is controversial and inconsistent. Generally, *in vitro* studies have shown a lack of inhibitory effect of β-blockers. The presence of neuronal uptake inhibitors (eg cocaine or desmethylimipramine), which might increase the levels of noradrenaline in the synaptic cleft and therefore exert a greater positive feedback, has also failed to reveal inhibitory actions of β-blockade on tissue responses and noradrenaline efflux.

In vivo experiments, however, have revealed inhibitory effects of β-adrenoceptor antagonists upon the effects of sympathetic nerve stimulation. For example, in the pithed rat, the pressor responses to stimulation of sympathetic outflow have been shown to be attenuated by β-blockade. Furthermore, the overflow of noradrenaline into the circulation by sympathetic nerve stimulation is reduced by β-blockade. An explanation for the lack of effectiveness of noradrenaline in potentiating, and for β-blockade in inhibiting, transmitter release *in vitro* is that adrenaline rather than noradrenaline is the physiological activator of prejunctional β-adrenoceptors. Circulating adrenomedullary adrenaline can stimulate the prejunctional β-adrenoceptors directly, but additionally it may be taken up into neurones and co-released along with noradrenaline. This co-released adrenaline would then exert a positive feedback loop (Figure 6.6). This is confirmed by *in vitro* studies in which isolated atria are first incubated with adrenaline and then the release of radiolabelled noradrenaline by sympathetic nerve stimulation monitored. In this situation, β-adrenoceptor antagonists are effective inhibitors of noradrenaline release. Also, *in vivo* administration of adrenaline causes an increase in the rate of noradrenaline release into the plasma. This occurred after adrenaline had disappeared from the plasma, but at a time when the adrenaline levels in the sympathetically innervated tissues were raised. The facilitation of noradrenergic transmission was maintained for 48 hr after an infusion of adrenaline (Coppes *et al.* 1994). The effect was prevented by inhibition of uptake of the adrenaline into the neurone with desmethylimipramine. Chronic administration of adrenaline to rats also causes an elevation of blood pressure, not by a direct α-adrenoceptor-mediated vasoconstriction, but through β-adrenoceptors since the effect is blocked by concomitant treatment with a β-blocker (Majewski 1983, Borkowski 1988). The presynaptic β-adrenoceptors therefore appear to be activated by adrenaline rather than noradrenaline *in vivo*. There is evidence in humans for facilitated sympathetic activity and of elevated noradrenaline levels after infusions of adrenaline, isoprenaline or selective β_2-adrenoceptor agonists such as salbutamol (Majewski 1983, Misu & Kubo 1986).

The classification of the prejunctional β-adrenoceptor in early experiments yielded inconsistent results. In particular, metoprolol which is assumed to be a β_1-selective antagonist (Table 5.3), has been shown to inhibit the vasoconstrictor responses and release of [³H]noradrenaline induced by nerve stimulation, suggesting a β_1-subtype. The bulk of evidence now points, however, to it being of the β_2-subtype. β_2-Adrenoceptor agonists, salbutamol and terbutaline, increased impulse-evoked [³H]noradrenaline release, whereas tazolol and dobutamine (β_1-selective) did not. The β_2-selective antagonist, ICI 118,551, potently inhibits the facilitatory effects of isoprenaline on transmitter release (Figure 6.5). Metoprolol was probably behaving as a non-selective antagonist in the earlier studies. Whether there is a small population of prejunctional β_1-adrenoceptors remains to be established.

The location of the prejunctional β_2-adrenoceptor is probably on the noradrenergic nerve terminal; synaptosomes and tissue slices from the hippocampus show enhanced release of [^{3}H]-noradrenaline after β-adrenoceptor stimulation (Ueda *et al.* 1983, Fillenz 1990).

It is worth noting that isoprenaline also increases the synthesis of noradrenaline in synaptosomes, presumably through a mechanism common to effects on noradrenaline release (Fillenz 1990, see Figure 2.6). The coupling of the β_2-adrenoceptor to the increased noradrenaline release and synthesis appears to be via increased levels of cAMP. Phosphodiesterase inhibitors which elevate intracellular cAMP levels and membrane-permeating derivatives of cAMP, such as 8-Br-cAMP, facilitate the nerve stimulation-elicited release of noradrenaline from sympathetic nerve endings in a range of isolated tissues. Phosphodiesterase inhibitors also potentiate the facilitatory actions of isoprenaline (Misu & Kubo 1986).

6.1.3.1 *Relationship between prejunctional β_2-adrenoceptors and angiotensin (AII) receptors*

Angiotensin II, the active octapeptide arising from the endopeptidase activity of angiotensin converting enzyme (ACE) on angiotensin I, also exerts a facilitatory action on responses to sympathetic nerve stimulation and efflux of noradrenaline (Ziogas *et al.* 1984). This effect is mediated via prejunctional angiotensin II receptors. There is now evidence that the facilitatory actions of β_2-agonists on noradrenaline release from sympathetic nerves may in fact be due to a local release of angiotensin II. This is based on the observations that the ACE inhibitor, captopril, and the angiotensin II receptor antagonist, Sar^1-Ile^8-angiotensin II (Sar), both inhibited the potentiation of vasoconstriction induced by periarterial sympathetic nerve stimulation to the perfused mesenteric vascular bed (Figure 6.6). The reverse situation, of angiotensin II potentiation being linked to β_2-adrenoceptors, however, is discounted by the fact that the β_2-adrenoceptor antagonist, ICI 118,551, does not inhibit the response to angiotensin II (Draper *et al.* 1989). The antagonism by captopril and Sar was at the prejunctional level since they failed to influence the vasoconstrictor response to exogenous noradrenaline and to nerve stimulation. The precise location of the local renin–angiotensin system and of the associated β_2-adrenoceptors in these vascular systems is still uncertain.

6.1.3.2 *Role of prejunctional β_2-adrenoceptors in hypertension and its treatment*

There is evidence that prejunctional β_2-adrenoceptors may display a raised sensitivity in hypertension (Marin 1993). If circulating adrenaline levels are correspondingly elevated in hypertensive states, there will be an increased degree of prejunctional facilitation of noradrenaline release from sympathetic nerves, which will contribute to the raised peripheral resistance and blood pressure. The potentiation by β_2-agonists of vasoconstrictor responses to sympathetic nerve stimulation to the perfused mesenteric bed of the rat is greater in tissues from spontaneously hypertensive rats (SHR) (Figure 6.5) (Kawasaki *et al.* 1982). The facilitation by angiotensin II is also elevated in hypertension (Draper *et al.* 1989). Prejunctional β_2-adrenoceptors are therefore a target site of action of the β-blockers in the treatment of hypertension, as was discussed in greater detail in Chapter 5. The blockade of prejunctional β-adrenoceptor-mediated facilitation of transmitter release may contribute to their antihypertensive activity (Borkowski 1988). Support for this role was provided by the fact that the selective β_2-adrenoceptor antagonist, ICI 118,551, lowered blood pressure in hypertensive patients (Vincent *et al.* 1985). A central site of action of β-adrenoceptor antagonists, whose lipid solubility permits penetration of the blood–brain barrier, would depend upon the presence of prejunctional β-adrenoceptors in the cardiovascular regulatory centres of the hypothalamus and brain stem. There is evidence for isoprenaline facilitating the release of noradrenaline from brain stem slices, which is blocked by propranolol. Also, when the cat posterior hypothalamus was superfused and electrically stimulated, isoprenaline caused potentiation of the rise in blood pressure during electrical stimulation (Misu & Kubo 1986).

6.1.4 Other Prejunctional Receptors at Sympathetic Nerve Terminals

So far, the modulation of noradrenaline release by inhibitory α_2-adrenoceptors and facilitatory β_2-adrenoceptors and angiotensin II receptors has been considered. Additionally, a large number of other prejunctional receptor types have been identified which may modulate transmitter release. Inhibition of K^+-induced release of noradrenaline from hippocampal synaptosomes has been demonstrated by agonists at adenosine A_1 receptors, muscarinic receptors, μ opiate receptors and $GABA_B$ receptors; the latter is not blocked by the $GABA_A$ receptor antagonist, bicuculline. They probably have a common action of inhibition of Ca^{2+} channels through inhibition of adenylyl cyclase. Adenosine A_1 receptors are located prejunctionally on peripheral sympathetic nerve endings and adenosine and its analogues produce concentration-dependent inhibition of responses to sympathetic nerve stimulation, for example, in the vas deferens (Broadley *et al.* 1986), an effect blocked by the P_1 purinoceptor antagonist, 8-phenyltheophylline. The release of the co-transmitter, ATP, from noradrenergic neurones may exert prejunctional negative feedback through its rapid breakdown to adenosine, which may then exert inhibition of noradrenaline release via the A_1 receptors. There is recent evidence for A_2 receptors mediating facilitation of noradrenaline release and P_2 purinoceptors, stimulated by ATP, which inhibits its release (von Kügelgen 1994). The sympathetic co-transmitter, neuropeptide Y (NPY), also exerts an inhibitory effect upon efflux of noradrenaline, ATP and NPY from many tissues. However, opposing this action is an enhancement of the postjunctional effects of noradrenaline on smooth muscle contractions, so that NPY often potentiates the vasoconstrictor effects of noradrenaline and sympathetic nerve stimulation (Morris & Gibbins 1992) (see Chapter 12).

The prejunctional muscarinic receptor at peripheral sympathetic nerve endings provides a means whereby one division of the autonomic nervous system (parasympathetic) can modulate the activity of the other (sympathetic). Released acetylcholine from adjacent parasympathetic nerve endings exerts an inhibitory effect on sympathetic outflow. In the heart, this action occurs within 3–10 msec after vagal stimulation and the inhibitory muscarinic receptor appears to be of the M_2 subtype (see Chapter 7). A facilitatory M_1 receptor may also be present, being revealed when the M_2 subtype is blocked with AF-DX 116 and operating at higher concentrations of acetylcholine (Muscholl *et al.* 1991).

Another prejunctional receptor mediating inhibition of noradrenaline release from peripheral neurones is the histamine H_3 receptor (Ishikawa & Sperelakis 1987). This response is susceptible to blockade by the selective H_3 receptor antagonists, impromidine and thioperamide. Also inhibitory towards noradrenaline release from peripheral sympathetic nerves is dopamine, which stimulates dopamine receptors of the D_2 subtype. Although dopamine is the immediate precursor of noradrenaline in sympathetic nerves, it is also released from specific dopaminergic neurones which are found in certain organs such as the kidney and mesenteric vasculature (see Chapter 12). Thus, prejunctional dopamine receptors do not serve as autoreceptors but as heteroreceptors, activated by release of dopamine from neighbouring dopaminergic neurones. The vasoconstrictor responses to periarterial nerve stimulation are inhibited by dopamine, in the presence of the D_1 receptor antagonist, SCH 23390, to block the postjunctional effects of dopamine. It is unlikely that prejunctional D_2 receptors, however, serve an important modulatory role under physiological conditions since their blockade does not significantly enhance the adrenergic response to nerve stimulation (Willems *et al.* 1985).

5-HT receptors (5-HT_3) are located on sympathetic postganglionic neurones where they cause depolarization and release of noradrenaline. It appears that 5-HT receptors of the 5-HT_1-like type may also be present at the nerve terminal where their activation causes inhibition of noradrenaline release, although stimulation of α_2-adrenoceptors by 5-HT is still a possibility (Wallis 1989, Barrús *et al.* 1993). Finally, prostaglandins, in particular PGE_2 (Hedqvist 1977), and endothelin (Reid *et al.* 1989) exert neuroinhibitory effects on transmitter release from sympathetic nerves. PGE_2 inhibits noradrenaline release through EP_3 receptors whereas PGD_2 facilitates its release via DP receptors (Molderings *et al.* 1994).

6.2 Noradrenergic Neurone Blocking Drugs

Drugs whose primary action is to prevent the release of noradrenaline, not by an action on prejunctional receptors but directly on the neurone release mechanisms, are known as adrenergic (or, more correctly, noradrenergic) neurone blockers. The forerunner of this class of drug was xylocholine (TM10), which is the 2,6-xylyl ether of choline (Table 6.1), and is thus structurally related to acetylcholine. It displayed the characteristics of noradrenergic neurone blockade by preventing responses to sympathetic nerve stimulation without affecting exogenously added adrenaline. However, its wide spectrum of other pharmacological actions as a local anaesthetic and upon cholinergic receptors limited its usefulness and further development was curtailed. Nevertheless, TM10 did lead the way to the synthesis of several compounds that blocked sympathetic nerves, the first of which was bretylium (Table 6.1). This compound was briefly used clinically for the treatment of hypertension but it was soon abandoned because of side-effects, development of tolerance that necessitated increasing dosage to impractical levels, and unreliable absorption from the intestine which yielded unpredictable blood pressure reduction. Almost simultaneously, but independently, guanethidine was first described (Maxwell *et al.* 1959) and this has become the prototype noradrenergic neurone blocking drug.

Table 6.1 Adrenergic neurone blockers

Generic	Trade name	Structure
TMIO (xylocholine)		
Bretylium tosylate	(No longer available)	
Guanethidine	Ismelin[R] Ganda[R] (G)	
Bethanidine	Esbatal[R]	
Guanoxan	Envacar[R] (withdrawn)	
Debrisoquine	Declinax[R]	
Guanoclor	Vatensol[R] (withdrawn)	
Guanadrel	Hylorel[R]	

6.2.1 Pharmacology of Guanethidine and Related Drugs

6.2.1.1 Characteristics of neurone blockade

Guanethidine selectively prevents the release of noradrenaline from sympathetic nerves, reducing the responses to sympathetic nerve stimulation *in vivo*. For example, the contractions of the nictitating membrane to both pre- and postganglionic cervical sympathetic nerve stimulation are blocked, indicating no action at the sympathetic ganglion. The contractile responses of the spleen to splanchnic nerve stimulation and of the vas deferens to hypogastric nerve stimulation are also blocked (Figure 6.7). The responses to indirectly acting sympathomimetic amines such as tyramine are also blocked but the responses to noradrenaline are potentiated (Figure 6.7). This evidence

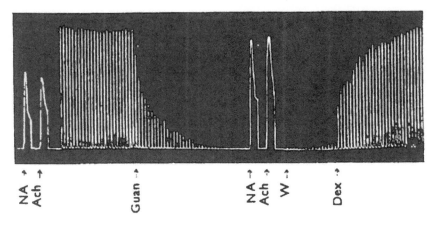

Figure 6.7 Effect of the adrenergic neurone blocking agent, guanethidine (Guan, 4×10^{-6} M), upon the contractions of the guinea-pig vas deferens induced by transmural electrical stimulation (sympathetic hypogastric nerve) and, in the absence of stimulation, to noradrenaline (NA, 5×10^{-4} M) and acetylcholine (Ach, 2×10^{-4} M) for 1 min. Guanethidine gradually abolishes the contractions to transmural stimulation but potentiates the responses to NA and Ach. Washing thirty times at W produced poor reversal, but dexamphetamine (5×10^{-8} M) produced gradual recovery. Reproduced with permission from Birmingham & Wilson (1963).

alone distinguishes this class of drug from the prejunctional α_2-adrenoceptor agonists, which admittedly do not block adrenaline and noradrenaline, but they do not potentiate their action. The α_2-agonists, however, only inhibit exocytotic release of noradrenaline and therefore do not block tyramine. Responses to parasympathetic nerve stimulation are not blocked and neither is the release of adrenaline from the adrenal medulla by splanchnic nerve stimulation or by the nicotinic agonist DMPP (see Chapter 11). Only the initial component of the biphasic pressor response to DMPP, due to stimulation of sympathetic ganglia and release of noradrenaline from sympathetic neurones, is inhibited by neurone blocking agents. The secondary delayed response due to release of adrenaline is unaffected or even potentiated. The sympathetic cholinergic nerves supplying the sweat glands and certain blood vessels (see Chapter 1) are not blocked by the neurone blockers, indicating their selectivity for noradrenaline release mechanisms.

Other characteristic features of adrenergic neurone blockade are the effects of frequency of nerve stimulation and interactions with other drugs acting on the neurone. The blockade of responses to sympathetic nerve stimulation by guanethidine is frequency–dependent, with preferential inhibition of lower frequencies of stimulation. The frequency-response curves are therefore shifted to the right in an approximately parallel manner. Bretylium, however, depresses the maximum of the frequency–response curve. Interactions with other drugs include the reversal of blockade by indirectly acting sympathomimetic amines, such as dexamphetamine (Figure 6.7), phenylethylamine and mephentermine, although tyramine may be active only *in vitro*. The blockade is not consistently reversed or prevented by directly acting sympathomimetic amines, such as the catecholamines noradrenaline and adrenaline, nor by MAO inhibitors, providing they do not have indirect activity. These features contrast with blockade by reserpine (see later). The blockade is reversed by neuronal uptake inhibitors, such as cocaine and imipramine (Boura & Green 1965). In view of these interactions, the concomitant use of sympathomimetics, for example as nasal decongestants, and of the tricyclic antidepressants is contraindicated.

6.2.1.2 *Pharmacological properties*

The pharmacological effects of noradrenergic neurone blockade by guanethidine may be seen on the cardiovascular system. Immediately after administration there is a pressor effect due to the indirect sympathomimetic properties. This is followed by a fall in arterial blood pressure and bradycardia due to blockade of sympathetic vasomotor tone. The bradycardia is atropine-sensitive and therefore due to unmasking of the parasympathetic innervation of the heart. Cardiovascular reflexes are impaired and there is exertional and orthostatic hypotension in humans (Chapter 1). The carotid occlusion pressor response (Chapter 1) is reduced in anaesthetized cats and dogs and the overshoot pressor response to tilting from the supine to vertical position in humans is reduced. Because adrenal medullary function is not inhibited, neurone blocking agents do not completely abolish cardiovascular pressor reflexes which are in part due to release of adrenaline from the adrenal medulla. Recent studies confirm that the postural hypotension is due to frequency-related blockade of peripheral sympathetic neurones by guanethidine and not to a central mechanism (Park *et al.* 1991).

The pharmacological actions at other organs can generally be attributed to the release of parasympathetic influences after sympathetic blockade. These give rise to unwanted side-effects including increased bowel activity and loss of sphincter control, leading to diarrhoea. This appears to be more severe with guanethidine and rare with bethanidine. A possible explanation is that guanethidine preferentially inhibits the low frequencies of stimulation which occur physiologically. Other side-effects include failure of ejaculation (block of vas deferens contractile response to sympathetic nerve stimulation, Figure 6.7), miosis and the syncope arising from orthostatic or exertional hypotension.

6.2.1.3 *Mechanisms of neurone blockade*

Evidence has been provided to show that in addition to the tissue responses to sympathetic nerve stimulation, the release of radiolabelled noradrenaline is also inhibited. The release of co-transmitters from sympathetic nerves also appears to be inhibited by guanethidine. Guanethidine abolishes the entire contraction of the vas deferens to sympathetic nerve stimulation (Figure 6.7), whereas reserpine only reduces the second phase of the response which is attributed to noradrenaline release. It must therefore be assumed that guanethidine also blocks the co-release of ATP from sympathetic nerves. In the guinea-pig vas deferens, excitatory junction potentials (EJPs) set up by nerve stimulation are not blocked by α-adrenoceptor antagonists, suggesting that they are not due to noradrenaline release. They are, however, blocked by guanethidine and bretylium, which indicates that they are due to transmitter release from sympathetic nerves. Periarterial sympathetic nerve stimulation-induced release of ATP and relaxation responses of isolated blood vessels are also blocked by guanethidine. Release of neurotransmitters from non-adrenergic non-cholinergic, (NANC) nerves, for example in the intestine and urinary bladder, is not inhibited by guanethidine; the excitatory transmitter is probably ATP. Non-adrenergic, non-cholinergic relaxation responses and IJPs to electrical stimulation of the intestine have also been observed, which are resistant to guanethidine and these may be due to VIP release (Brock & Cunnane 1992). Thus, the release of neurotransmitters from NANC nerves and from cholinergic nerves is not prevented by guanethidine. However, guanethidine does inhibit the release of co-transmitters of sympathetic nerves, noradrenaline, ATP and NPY, which are all subject to exocytosis (Chapter 12). The neurone blockers therefore inhibit both exocytotic release of noradrenaline and its non-exocytotic release by ISAs from sympathetic nerve endings. They gain access to the neurone by means of the neuronal uptake process for which they are substrates (Chapter 2). Like ISAs, they bind to the noradrenaline transporter in a Na^+- and Cl^--dependent manner. After transport into the neurone, because they have affinity for the storage vesicle, noradrenaline is released. It is then transported out of the neurone by facilitated exchange diffusion via a reversal of the uptake process (Figure 2.22). This explains the indirect sympathomimetic activity of the neurone blocking agents. As a result of this affinity for the noradrenaline carrier, neurone blockers also behave as uptake inhibitors and prevent subsequent uptake of sympathomimetic amines. As a consequence, they block ISAs but potentiate directly acting amines such as noradrenaline and adrenaline (Figures 2.24, 2.25 and 6.7). Access of neurone blockers to the neurone is therefore inhibited by the true uptake[1] inhibitors, such as DMI, which therefore prevent both their neurone blocking and ISA activity. Further evidence for the binding of neurone blockers to the noradrenaline carrier is the fact that ISAs, such as dexamphetamine, characteristically prevent or reverse the neurone blockade (Figure 6.7).

One proposal for the mechanism of inhibition of transmitter release by guanethidine is via an action on Ca^{2+}-dependent K^+ channels of the noradrenergic neuronal membrane. This is based on the observation that the K^+ channel blockers, 4-aminopyridine (4-AP), apamin and tetraethylammonium (TEA), antagonize the inhibitory effect of guanethidine on responses of the vas deferens to noradrenergic nerve stimulation. Guanethidine was proposed to open the outward K^+ channels, the resulting neurone hyperpolarization causing inhibition of neurotransmitter release (Berry *et al.* 1992).

The release of noradrenaline by guanethidine and bretylium ultimately leads to depletion of noradrenaline stores with regular administration. This is a reserpine-like property which is additional to the indirect sympathomimetic action. Whereas the indirect activity is due to facilitated exchange diffusion, via the noradrenaline carrier mechanism, depletion involves destabilization of the storage vesicle and exposure of the noradrenaline to MAO, which is released as inactive deaminated metabolites. Of interest is the finding that the neurone blockers prevent depletion by reserpine. The degree of depletion in peripheral tissues such as the heart, spleen and sympathetic ganglia varies between different neurone blockers, with guanethidine causing consistent depletion of noradrenaline. The variability may be due to MAO inhibitory properties of some of the blockers (eg bretylium and bethanidine); this would offset the reserpine-like activity. Indeed, MAO inhibitors reduce guanethidine-induced depletion. The noradrenergic neurone blocking action that occurs with acute administration of this class of drug cannot be attributed to their depleting effect upon noradrenaline stores. The onset of sympathetic blockade precedes any loss of transmitter from the nerves (Boura & Green 1965). Further evidence that this is the case is the fact that after chronic treatment of cats with guanethidine to produce depletion, responses of the nictitating membrane could still be restored by dexamphetamine. Thus, even in the presence of depletion, neurone blockade could be reversed indicating an independent mechanism (Day 1979).

Much of the recent work on neurone blocking agents has involved the chronic administration of high doses of guanethidine (50 mg/kg) to induce sympathectomy. This is a neurotoxic action which results in a permanent loss of sympathetic neurones and their cell bodies. There is widespread depletion of noradrenaline with loss of neurones and ganglia, in particular the vertebral ganglia and notably the superior cervical ganglion (Chapter 1). The depletion of dopamine from the superior cervical ganglion is much less than that of noradrenaline. The dopamine-containing small intensely fluorescent (SIF) cells of the ganglia are tyrosine hydroxylase (TH)-positive but dopamine β-hydroxylase (DBH)-negative, indicating the minimal conversion of dopamine to noradrenaline. These are also not reduced by guanethidine. Thus, guanethidine only destroys the noradrenaline-containing sympathetic neurones where dopamine serves as a precursor but spares the dopamine-containing neurones and SIF cells of the ganglia (Favre-Maurice *et al.* 1992). The destruction of sympathetic neurones is now thought to involve an immune-mediated mechanism since it is associated with infiltration of lymphocytes into the ganglionic parenchyma. These are mainly natural killer T-cells (CD8-NIC), identified by means of monoclonal antibodies (Hougen *et al.* 1992).

6.2.1.4 *Therapeutic uses and side-effects of neurone blockers*

The effective blood pressure lowering action of the neurone blockers and protection from pressor reflexes made them the drugs of choice for the treatment of all grades of hypertension in the early 1970s. However, because of the incidence of undesirable side-effects, their use is now restricted to severe hypertension which fails to respond to the newer agents and to patients who have tolerated the drugs for many years. Guanethidine does not cross the blood–brain barrier and does not cause CNS side-effects. The neurone blockers, however, do have a range of unwanted effects arising from sympathetic blockade, including diarrhoea, ejaculation failure, postural hypotension and lasitude due to poor muscle blood flow. They produce sodium and water retention leading to oedema which may be corrected by concurrent use of a diuretic. The thiazide diuretics usefully potentiate the antihypertensive activity and thus allow the dosage of the neurone blocker to be reduced with a consequent reduction of the side-effects. Drug interactions with other agents acting on the noradrenergic neurone are abundant, as discussed above.

The principle contraindication of neurone blockers is in phaeochromocytoma; the pressor effect of circulating catecholamines will be enhanced. Because renal blood flow is modestly reduced, renal failure or insufficiency is also a contraindication.

The antihypertensive actions, side-effects and contraindications of the currently available noradrenergic neurone blockers shown in Table 6.1 are essentially the same. The main differences between compounds appear to be in their pharmacokinetics. Bethanidine has a short duration of action and is well absorbed orally. It has less tendency to produce diarrhoea and early morning postural hypotension. Guanoxan appears to have some additional postjunctional action as an α-adrenoceptor antagonist, its structure resembling the benzodioxane antagonists such as idazoxan and WB 4101 (Table 5.2). It also has central noradrenaline depleting actions but, more importantly, it causes liver damage with long-term use and is no longer available. Guanadrel has a high bioavailability and relatively short half-life (Finnerty & Brogden 1985).

Finally, debrisoquine is well absorbed orally and mostly eliminated within 24 hr, which necessitates twice-daily dosing. Debrisoquine is an interesting drug because of its metabolism to 4-hydroxydebrisoquine by the liver cytochrome P450 mixed-function oxidase system (Table 6.1). There is a marked variation in the ability of individuals to metabolize debrisoquine which has been found to be of genetic origin. The affected cytochrome P450 is CYP2D6. A small (10%) population of poor metabolizers exists who display high sensitivity to the antihypertensive effect of debrisoquine and a low level of the 4-hydroxy metabolite. The genetic control is by a single gene pair and poor metabolism is a recessive trait arising in homozygous recessive children of heterozygous extensive-metabolizing parents. Debrisoquine is therefore a useful model for evaluating whether an individual has a genetic difficiency in the metabolism of related drugs that undergo 4-hydroxylation. Of relevance to drugs in this book, these include guanoxan, metoprolol, timolol, nontriptyline and bufanolol. In poor metabolizers, there is potential for increased potency and side-effects of several β-blockers; however in practice this may be only reflected in more patient discomfort leading to poor compliance (Meyer 1994, Tucker 1994).

Glaucoma is an indication for the use of noradrenergic neurone blockers. Guanethidine is used for open-angle glaucoma; narrow-angle glaucoma is a contraindication (see sympathomimetic amines, Chapter 4; and the eye, Chapter 1). It is also used to relax the retracted eyelids in exophthalmos (or staring appearance) associated with thyrotoxicosis, presumably by inhibition of the sympathetic innervation to smooth muscle of the eyelids.

Guanethidine applied locally to the eye induces a transient mydriasis and rise in intraocular pressure by α-adrenoceptor-mediated contraction of the radial muscles of the iris through its indirect sympathomimetic action. This is followed by a more prolonged miosis and fall in intraocular pressure. With repeated use, only the latter effects are produced. It appears that sympathetic blockade of the α-adrenoceptor-mediated effects (mydriasis and block of vasoconstriction) combine with potentiation of the β-effects of adrenaline through uptake inhibition to exert the beneficial action. The β-adrenoceptor-mediated actions diminish production of aqueous humour (see adrenaline, Chapter 4).

6.3 Noradrenergic Depleting Agents

6.3.1 *Reserpine and Other Rauwolfia Alkaloids*

Reserpine is an alkaloid from *Rauwolfia serpentina* (Benth), a flowering shrub that is indigenous to India and surrounding countries. Many alkaloids have been isolated from the various species of rauwolfia, but reserpine is the most important and has been the most widely studied. The most active alkaloids are the esterified alkaloids, reserpine and rescinnamine, although non-esterified reserpic acid, and rauwolfine (ajmaline) are also present. The highest concentrations of reserpine are found in the roots (0.15% in dried root) and extracts have been used in Indian folk medicine for hundreds of years to treat a wide range of conditions. Rauwolfia alkaloids were only introduced into Western medicine in the 1950s, following the isolation of reserpine which became one of the early drugs for the effective treatment of hypertension. The chemical structures of reserpine and related alkaloids are shown in Table 6.2.

Table 6.2 Noradrenergic depletors

Rauwolfia alkaloids

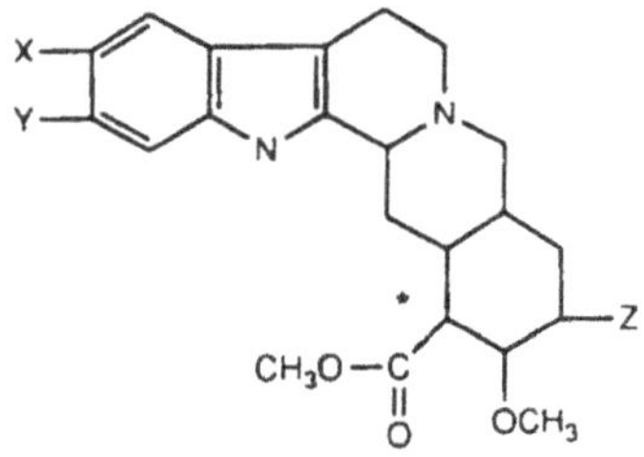

	X	Y	Z
Reserpine (Serpasil[R])	-H	-OCH$_3$	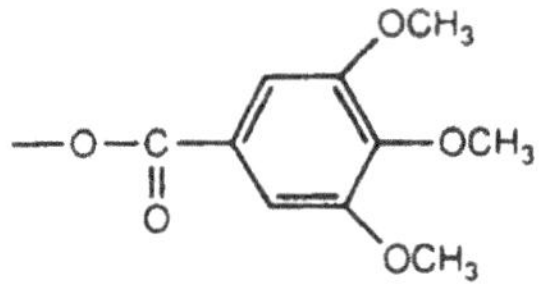
Methoserpidine (Enduronyl[US] Harmonyl, no longer available)	-H	-H	ditto
Deserpidine (Decaserpyl[R])	-OCH$_3$	-H	ditto
Syrosingopine	-OCH$_3$	-H	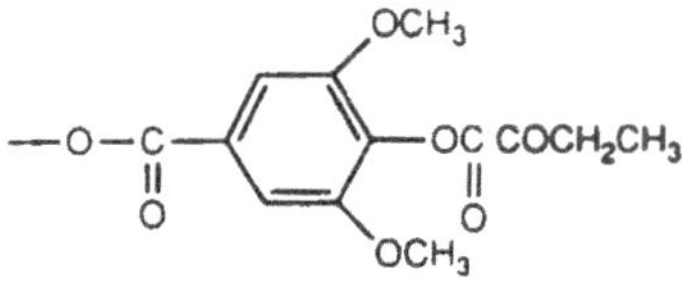
Rescinnamine	-H	-OCH$_3$	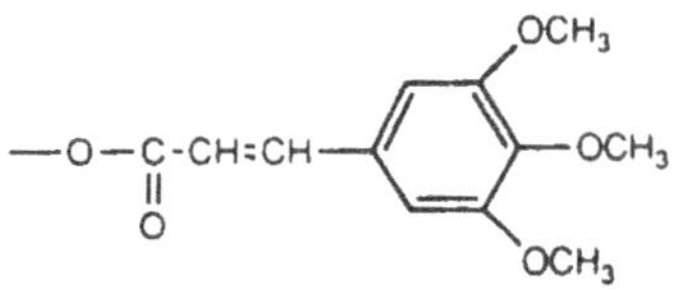
Rescimetol	-H	-OCH$_3$	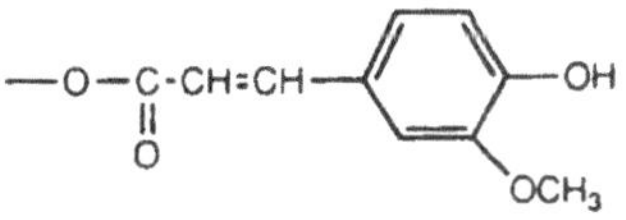
Reserpic acid	-H	-OCH$_3$	-OH (-COOH at [*])

Ajmaline (rauwolfine)

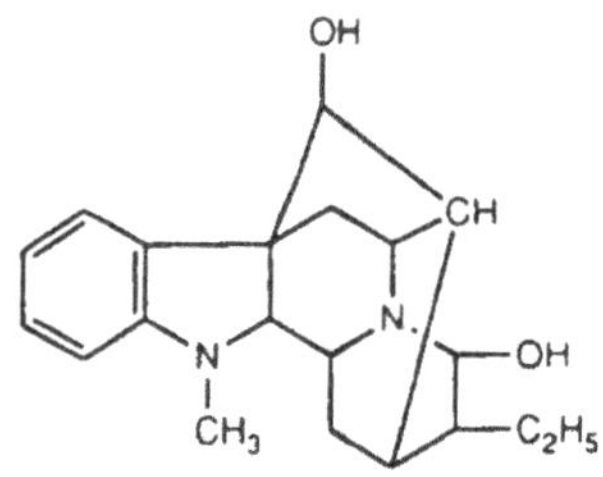

(continued)

Table 6.2 *Continued*

Benzoquinolizines

Tetrabenazine

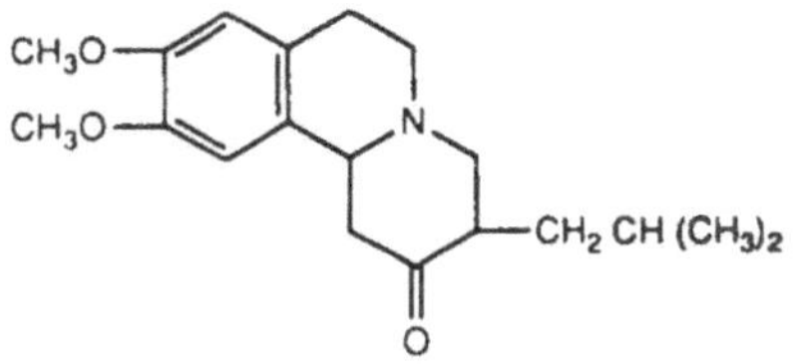

CH_3O—
CH_3O—

—$CH_2 CH (CH_3)_2$

O

Short-acting central
noradrenaline depletor

Ro-4-1284

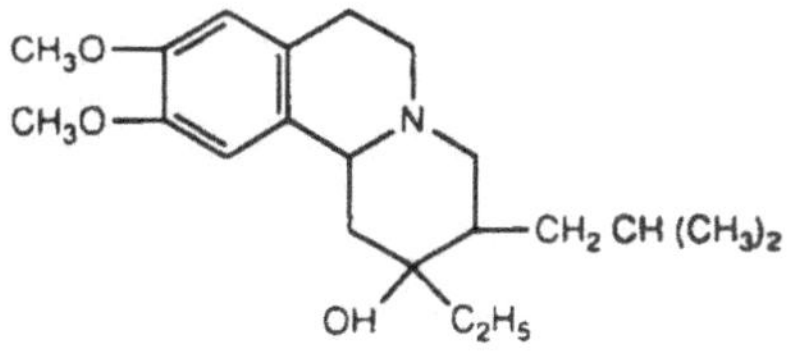

CH_3O—
CH_3O—

—$CH_2 CH (CH_3)_2$

OH C_2H_5

5-Hydroxydopamine

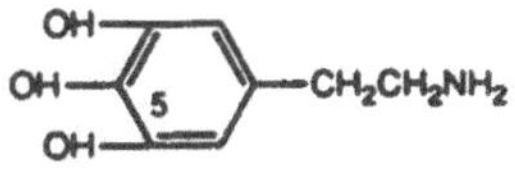

OH—
OH—
OH—

5

—$CH_2CH_2NH_2$

Depletor

6-Hydroxydopamine

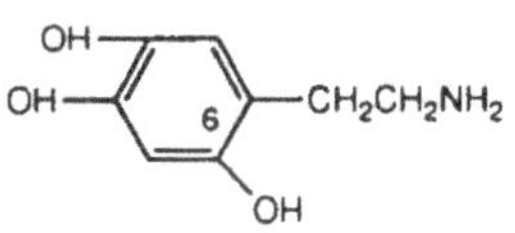

OH—
OH—

6

—$CH_2CH_2NH_2$

OH

Chemical
neurotoxin

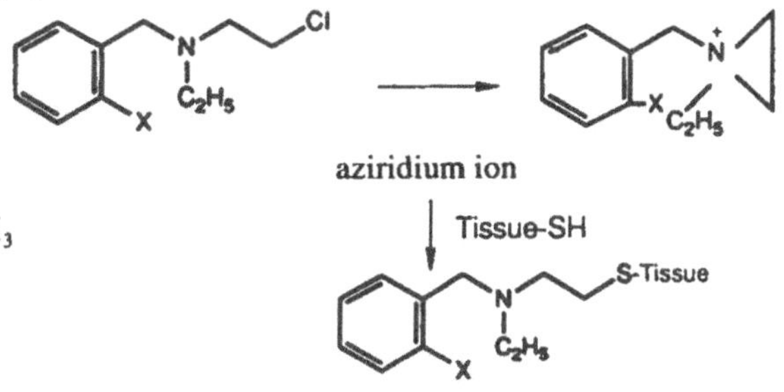

Xylamine and DSP-4

Cl

N

X C_2H_5

→

N$^+$

X C_2H_5

Irreversible

aziridium ion

Xylamine X = CH_3
DSP-4   X = Br

Tissue-SH

S-Tissue

N

X C_2H_5

6.3.1.1 Pharmacological actions

Reserpine causes depletion of noradrenaline from noradrenergic nerve endings both peripherally and centrally, leading to loss of sympathetic function. This results in a fall in blood pressure and heart rate and impairment of the cardiovascular pressor reflexes, resulting in some orthostatic hypotension. It also causes depletion of 5-HT from central and peripheral serotonergic neurones and release from the enterochromaffin cells of the gastrointestinal tract. The latter effect may contribute to the diarrhoea caused by reserpine through direct stimulation of the gut by the released 5-HT.

Reserpine treatment causes reduction of the responses to sympathetic nerve stimulation and indirectly acting sympathomimetic amines, both in intact animals and in isolated tissues (Figure 6.8). For example, the contractions of the nictitating membrane to cervical sympathetic nerve stimulation and to tyramine are greatly

283

(a) (b) (c)

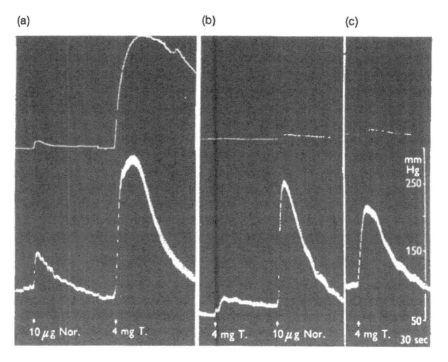

Figure 6.8 Effect of reserpine pretreatment (2.5–5.0 mg/kg on two successive days) upon the responses of the nictitating membrane (upper traces) and blood pressure (lower traces) to intravenous injections of noradrenaline (Nor, 10 μg) and tyramine (T, 4 mg) in the spinal cat. (a) Normal untreated cat. (b) Reserpine pretreated cat showing reduced responses to tyramine and potentiated pressor response to noradrenaline. (c) Same experiment as (b) but after the intravenous infusion of noradrenaline (0.12 mg) which partially restored the response to tyramine. Reproduced with permission from Burn & Rand (1958).

reduced (Burn & Rand 1958). The response to directly acting sympathomimetic amines, however, is usually unaffected. After short-term treatment there is no effect upon directly acting sympathomimetic amines; reserpine has no affinity for the neuronal uptake$_1$ system and does not inhibit uptake or potentiate exogenously added noradrenaline. After chronic treatment with reserpine for $\geqslant 3$ days, however, α- and β_1-adrenoceptor-mediated responses to noradrenaline are potentiated. This effect is the result of receptor upregulation arising from the loss of sympathetic traffic to the effector organ (see Chapter 14). Only those tissues with innervated adrenoceptors (α_1 and β_1) display reserpine-induced supersensitivity. β_2-adrenoceptor-mediated bronchorelaxation, for example, is not affected (Grassby & Broadley 1986).

The subsensitivity to ISAs may be partially reversed by infusion with noradrenaline or one of its precursors, including dopamine or L-dopa. This illustrates that some storage capacity for noradrenaline must be retained after reserpine depletion and that noradrenaline is still capable of synthesis. The level of depletion by reserpine has been shown to vary greatly between tissues. The heart shows the greatest reduction of noradrenaline content because of its higher turnover rate. In the adrenal medulla high doses are required to achieve any depletion. The loss of noradrenaline is both dose- and time-dependent. A single dose of reserpine (1 mg/kg) takes over >12 hr to achieve maximal depletion of cardiac noradrenaline stores.

The recovery is slow and may take 7–14 days. Low doses (0.01 mg/kg) given over the course of 1 week produce equivalent depletion. To achieve any reduction of sympathetic function, however, the noradrenaline levels have to be depleted by >80%.

Reserpine does have actions additional to its neuronal depletion properties, particularly at higher dose levels ($\geqslant$1 mg/kg). These include effects on tissue calcium content, mitochondrial damage resulting in uncoupling of oxidative phosphorylation, and interference with noradrenaline uptake and efflux mechanisms at the neuronal membrane (Maxwell *et al.* 1976). More recently, reserpine has been shown to have Ca^{2+} channel blocking activity in isolated smooth muscle preparations (Satoh *et al.* 1992). A further property of reserpine is its ability to supress delayed hypersensitivity reactions. This may be due to depletion of mast cells of 5-HT; however, more recent studies indicate that reserpine inhibits T-cell proliferation and interleukin IL-2 production after mitogen stimulation (Mekori *et al.* 1989).

6.3.1.2 *Mechanism of the depleting action of reserpine*

The depletion of noradrenaline by reserpine is largely prevented by inhibition of MAO with drugs such as nialamide and pargyline (Chapter 2). The accompanying pharmacological effects are also prevented by MAO inhibitors, illustrating that these responses to reserpine are due to noradrenaline depletion. The antagonism by MAO inhibitors indicates that the intraneuronal deamination of noradrenaline by MAO is an essential requirement for the depletion of noradrenaline by reserpine. Reserpine inhibits the uptake and binding of noradrenaline into the storage vesicles. The binding of reserpine to the vesicular membrane is rapid, specific and irreversible. Indeed, a reserpine-sensitive vesicular noradrenaline transporter has now been cloned (Erickson *et al.* 1992). As a result of the inhibition of vesicular storage of noradrenaline, it is exposed to cytoplasmic deamination and loss. The major deaminated metabolite of noradrenaline, DOPEG (Chapter 2, Figure 2.8), is produced intraneuronally by reserpine and related agents such as Ro 4-1284 (Langer 1974). DOPEG is not transported into the storage vesicle but is rapidly lost into the synaptic cleft where it is inactive at α- or β-adrenoceptors. Therefore, reserpine does not induce a pharmacological response; it releases noradrenaline in inactive form and is without indirect activity. Unlike ISAs or guanethidine, it does not enter the neurone via the uptake$_1$ transporter mechanism (Figure 2.21), but by passive diffusion. It does not therefore make the noradrenaline carrier available intra-neuronally for outwards transport of noradrenaline in active form (Bönisch & Trendelenburg 1988). After inhibition of MAO, however, reserpine may be observed to induce a brief sympathomimetic response.

The intracellularly accumulated DOPEG initially impairs synthesis of noradrena-line by inhibition of tyrosine hydroxylase (TH), the rate-limiting step (Figure 2.4, Chapter 2). This is compensated for *in vivo* by an induction of the enzyme because of the increased nerve traffic resulting from reflex activation of sympathetic nerves. Also, in the longer term (10–18 hr), reserpine pretreatment increases the amount of TH by induction through protein synthesis (see Chapter 2). In the locus coeruleus region of the brain, the induction of TH by acute reserpine treatment is associated with increased levels of mRNA for TH and for the neuropeptides, NPY and galanin (Gundlach *et al.* 1990). These peptides are possibly co-stored with noradrenaline in sympathetic neurones, galanin being a 29 amino acid peptide located in the intrinsic neurones of the intestine and adrenal medulla. In spite of the inhibition of noradrenaline synthesis by reserpine, it does not prevent the synthesis of dopamine but it actually increases dopamine levels in the peripheral neurones. The increased levels of dopamine in dopaminergic nerves are then available for release by ISAs. Reserpine blocks the active uptake of both noradrenaline and dopamine into the storage vesicles, thus dopamine fails to be converted to noradrenaline (Fillenz 1992).

The release of noradrenaline from the vesicles by reserpine is not accompanied by the release of DBH (the vesicular synthetic enzyme) and therefore does not involve exocytosis, a point further illustrated by the fact that the release of noradrenaline by reserpine does not require extracellular Ca^{2+}. The levels of DBH in the neurone do, however, initially fall due to loss at the nerve terminal but then rise due to replacement down the axon in newly synthesized vesicles (Fillenz 1990). Reserpine affects the two types of storage vesicle identified in the varicosities of sympathetic nerves – large and small electron-dense vesicles – differently. Reserpine causes loss of the electron-dense core of small vesicles but only a reduction of electron density of the large vesicle core. This is consistent with the core of the small vesicles being composed mainly of noradrenaline, while additional vesicular components are present in the core of the large vesicles. Since the ATP content of noradrenergic neurones is not depleted by reserpine (Kirkpatrick & Burnstock 1987), this would suggest that ATP is a major component in the large vesicle core. The fast components of the contractile response and EJPs to sympathetic nerve stimulation of various isolated blood vessels and of the vas deferens are not prevented by reserpine, indicating that they are mediated via ATP. However, the noradrenaline-mediated slow contraction and depolarization are reduced by reserpine. Although ATP is not depleted by reserpine, the other major co-transmitter of sympathetic nerves, neuropeptide Y (NPY), is lost. This occurs more slowly than does depletion of noradrenaline and may be secondary to the reflex increase in nerve traffic which occurs in response to the primary depletion of noradrenaline, since it can be prevented by section of the preganglionic nerve (Morris & Gibbins 1992). NPY in rat interscapular brown fat was not lost when measured immunohistochemically, suggesting the absence of reflex nerve activity to this tissue (Mukherjee *et al.* 1989).

Recovery from reserpine depletion involves the transport of new vesicles from the cell body down the axon to the terminal varicosities.

6.3.1.3 Uses of reserpine

Hypertension Although reserpine causes depletion of noradrenaline from both peripheral and central noradrenergic neurones, the central effects are thought to play a minor part in the reduction of blood pressure. While the central depressant actions of reserpine may be thought to reduce the output of sympathetic traffic from the vasomotor centre to peripheral cardiovascular structures, in fact there is a compensatory reflex *increase* in sympathetic traffic in peripheral sympathetic nerves, presumably arising from the fall in blood pressure. Thus, it appears to be the peripheral depletion of noradrenaline and the consequent reduced release onto α- and β-adrenoceptors of the heart and vasculature that mainly accounts for the fall in blood pressure. This is supported by the fact that alkaloid depletors such as syrosingopine and methoserpidine that do not deplete central noradrenaline are still effective at lowering blood pressure. Reserpine may also have a direct vasodilator action which contributes to the antihypertensive action (Day 1979).

Reserpine also causes a reduction of renin secretion, but salt and water are retained, which may result in a pseudo-tolerance to the blood pressure lowering activity. Combination with a diuretic overcomes this tolerance and enhances the antihypertensive activity so that the dose may be reduced. This has the added advantage that the incidence of unwanted side-effects is reduced. The occurrence of unpleasant side-effects and the low potency in patients with severe hypertension of long-standing has resulted in a considerable decline in the use of reserpine. It has largely been replaced by more effective and better tolerated drugs that have fewer side-effects. Some use of low doses remains where its low cost is a consideration.

The major unwanted actions of reserpine arise from depletion of central and peripheral noradrenaline, the latter allowing unopposed parasympathetic activity to predominate. The peripheral manifestations of sympathetic neuronal depletion include flushing of the skin, nasal congestion, failure of ejaculation and overactivity of the gastrointestinal tract. The latter causes diarrhoea and excessive secretion of gastric acid which precludes the use of reserpine-like drugs in patients with gastric ulcers or ulcerative colitis.

If these side-effects are not enough to depress the patient, then the major effect on central noradrenaline levels will. The depletion of central monoamine levels results in lethargy and drowsiness. This may progress to mental depression which may result in suicide. Reserpine and related drugs are therefore contraindicated in patients with a history of mental depression. Indeed, reserpine formerly found use in the treatment of psychiatric disorders including schizophrenia and manic states. Since reserpine also causes depletion of central stores of dopamine (note that peripheral levels do not appear to fall), in particular from the corpus striatum, it can cause extrapyramidal side-effects of Parkinson's-like hypokinesia. A further contraindication for reserpine-like drugs is therefore in Parkinson's disease. Dopamine is also thought to be the inhibitory factor normally supressing prolactin secretion from the pituitary gland. Depletion of dopamine by reserpine may result in elevated prolactin secretion and plasma levels and the occasional occurrence of lactation (galactorrhoea) in women and swelling of the breasts in men (gynaecomastia). Whether this is the cause of a suspected higher incidence of breast cancer among postmenopausal women who have received reserpine is now uncertain. It now appears that this arose because of poor control of patient selection and that reserpine may not be responsible for these cases of breast cancer (Feinstein 1988).

Other potential uses of reserpine include the treatment of Cushing's disease and to reduce the manifestations of morphine withdrawal syndrome and to suppress tolerance (Stroescu *et al.* 1991). Finally, it is of interest that reserpine treatment of animals has been used to mimic cystic fibrosis by increasing the production of intestinal mucin rich in glycoprotein. Pancreatic weight, glycoprotein GP-2 content and amylase activity are also reduced in this model (Leblond *et al.* 1989).

6.3.2   *Other Reserpine-like Agents*

Methoserpidine and *syrosingopine* are synthetic derivatives of reserpine (Table 6.2) which have a greater depleting action on catecholamines from peripheral noradrenergic neurones than from the brain. They therefore have a theoretical advantage over reserpine in not having the central depressant side-effects. However, being less potent, larger doses have to be used, which may then produce central effects. They are nevertheless useful pharmacological tools for selective peripheral catecholamine depletion.

Tetrabenazine, in contrast, causes short-lasting depletion of noradrenaline and dopamine selectively from central sites (Table 6.2). It interacts with the same binding sites on catecholamine storage vesicles as does reserpine, which it may displace in competition binding experiments. Reserpine does not cause further depletion of brain catecholamines when administered after tetrabenazine, indicating the same sites of action. Tetrabenazine produces similar effects of sedation and depression, and extrapyramidal signs of Parkinson's-like hypokinesia. It may therefore be used in the treatment of Huntington's chorea, a genetic disorder of hyperkinesia or jerking movement associated with elevated dopamine levels of the corpus striatum.

Ajmaline is a rauwolfia alkaloid that also has class IA antiarrhythmic activity (Bembilla-Perrot & Terrier de la Chaise 1992).

6.3.3 6-Hydroxydopamine (6-OHDA)

The 6-hydroxy derivative of dopamine (Table 6.2) causes a selective destruction of sympathetic nerves known as chemical sympathectomy. Thus it is more than a depletor since it inhibits both neuronal uptake and vesicular storage capacity of noradrenaline. The selectivity of 6-OHDA for noradrenergic neurones is due to its accumulation by the uptake process which can be inhibited by uptake blockers such as DMI. Although 6-OHDA causes depletion of both noradrenaline and dopamine from sympathetic neurones, in the presence of DMI which selectively inhibits uptake of noradrenaline, dopaminergic neurones are depleted of dopamine leaving noradrenaline levels unchanged. Conversely, selective inhibitors of dopamine uptake will protect dopaminergic neurones. After uptake of 6-OHDA, it accumulates in the storage vesicle where it may serve a false transmitter role. It is unlikely that any conversion to 6-hydroxynoradrenaline accounts for the chemical sympathectomy since it is not prevented by inhibitors of DBH. When critical intraneuronal levels of 6-OHDA or its metabolites are reached, destruction of the nerve terminal begins. There is loss of storage capacity for noradrenaline which is released into the synaptic cleft to produce indirect sympathomimetic responses such as a rise in blood pressure. The nerve terminal loses its ability to conduct action potentials and energy-producing cytochromes are destroyed.

6-OHDA is therefore regarded as a neurotoxin causing a slow loss of neuronal function, a condition known as plasticity of the noradrenergic neurone (Fillenz 1990). Ultimately, there is marked depletion of noradrenaline so that responses to ISAs are abolished. Directly acting sympathomimetic amines that are substrates for uptake$_1$ (eg noradrenaline) are potentiated because of inhibition of neuronal uptake. In the long term, there is potentiation of those amines that are not taken up (eg isoprenaline at β-adrenoceptors) because of receptor upregulation due to the loss of sympathetic nerve traffic (Chapter 14). The depletion of noradrenaline varies considerably from tissue to tissue, with the heart being the most readily depleted and sympathetic ganglia and the adrenal medulla showing the least. Unlike reserpine, 6-OHDA causes reduction of all components of the contractions to electrical stimulation of the vas deferens − both the purinergic and noradrenergic. It also prevents the release of both noradrenaline and ATP from blood vessels and the vas deferens (Kirkpatrick & Burnstock 1987). 6-OHDA may be used either by pretreating animals before tissue removal or by addition directly to the organ bath.

The cellular mechanism for the destructive effect of 6-OHDA upon sympathetic neurones is due to its intraneuronal conversion to reactive oxygen species. It is unstable in neutral or alkaline solution and is rapidly oxidized to yield a red colouration due to the presence of quinones. Stability for >2 hr can be achieved with solutions of pH 3 and ascorbic acid is often used to prevent its oxidation. In addition to quinones, other reactive species arising from 6-OHDA include peroxides and superoxide radical. Evidence for their involvement includes the observation that catalase, the enzyme responsible for destruction of peroxide, partially reverses the inhibition of noradrenaline uptake by 6-OHDA. The oxygen radicals react with the structural lipids, membrane and enzyme suphydryl groups (-SH) and protein amino groups ($-NH_2$) to induce generalized destruction of the sympathetic nerve endings (Kostrzewa & Jacobowitz 1974).

The 6-hydroxy derivative of L-dopa is converted by dopa decarboxylase to 6-OHDA, and its peripheral administration leads to depletion of noradrenaline from both central and peripheral noradrenergic neurones. It has the advantage over 6-OHDA that it readily crosses the blood–brain barrier. Its action is prevented by inhibitors of dopa decarboxylase and of uptake$_1$. The depletions of noradrenaline by 6-OHdopa and 6-OHDA are enhanced by MAO inhibitors, indicating that they are removed by MAO which exerts a protective effect (Kostrzewa & Jacobowitz 1974). Another mechanism for protection against the depletion by 6-OHDA is the glutathione antioxidant system. Reduction of glutathione synthesis by treatment of rats with L-buthionine sulfoximine (L-BSO), an inhibitor of τ-glutamylcysteine synthetase, has been shown to potentiate the catecholamine depleting action of 6-OHDA in the striatum of rat brains. Thus, glutathione appears to protect noradrenergic neurones from the oxidant effect of endogenous catecholamines and 6-OHDA by scavenging reactive oxygen species (Pileblad *et al.* 1989).

5-OHDA is a potent noradrenaline depleting agent but does not destroy the noradrenergic nerve endings in the same way as does 6-OHDA. This is probably because it forms the peroxide at one-twelfth the rate of 6-OHDA (Kostrzewa & Jacobowitz 1974). 5-OHDA is readily taken up by uptake$_1$. 5-OHdopa is rapidly converted to 5-OHDA and 5-hydroxynoradrenaline, which replaces noradrenaline and serves as a false transmitter.

6.4 Miscellaneous Agents Affecting Noradrenaline Release

6.4.1 Nerve Growth Factor Antisera

Nerve growth factor (NGF) is a polypeptide that is also essential for the normal development of the sympathetic nervous system and for maintenance of function. Immunization of rats with mouse NGF results in the production of antibodies against both mouse NGF and the endogenous NGF. The consequent loss of NGF results in suppression of peripheral sympathetic function as demonstrated by reduced ganglionic tyrosine hydroxylase activity and depletion of noradrenaline (Chess-Williams *et al.* 1994).

6.4.2 *Gangliosides*

These are naturally occurring sialoglycosphingolipids that are considered to be involved in neuronal processes including synaptic transmission. They have an important role in neural growth and regeneration and may therefore serve a protective role against damage to the cell membrane lipids induced by ischaemia and neurotoxic agents. Gangliosides have been shown to protect against the release of noradrenaline from isolated atria caused by DMSO through disruption of the neuronal membrane organization (Lorenzo & Adler-Graschinsky 1992). Exogenously administered gangliosides have been used as therapeutic agents in Italy (Cronassial[R]) and Spain (Nevrotal[R]) for several disorders of the nervous system including peripheral neuropathy. However, recent concern has been expressed over an increase in the incidence of Guillain-Barré syndrome, an accute inflammatory demyelinating polyradiculoneuropathy, associated with dysesthesia followed by muscular weakness. This has resulted in the withdrawal of these drugs.

6.4.3 *Xylamine and DSP-4*

These are haloalkylamines related to dibenamine.They form an aziridium ion which is highly reactive towards nucleophilic groups such as thiols and amines. They have high affinity for the neuronal uptake transporter and are accumulated in noradrenergic neurones where they interfere with storage and transport of noradrenaline, probably by reducing mitochondrial function (Dudley *et al.* 1990) (Table 6.2).

6.4.4 *ω-Conotoxin*

This toxin selectively blocks N-type Ca^{2+} channels and therefore prevents Ca^{2+} influx into sympathetic neurones and the consequent exocytotic release of noradrenaline. It has been discussed earlier (Chapter 2) but is included here as a further example of an agent that interferes with noradrenaline release by a prejunctional mechanism.

Neurotransmission at Parasympathetic Nerve Endings

7.1 Introduction

The neurotransmitter at the parasympathetic nerve terminal with smooth muscle, cardiac muscle or glandular tissue is acetylcholine (Ach). Ach is also the transmitter at both sympathetic and parasympathetic ganglia, at the neuromuscular junction between somatic nerves and skeletal muscle, and is located in the brain. These are all termed cholinergic nerves and the receptors with which Ach interacts are termed *cholinoceptors*. Schmiedeberg & Koppe (1869) first reported the similarity between the effects of the alkaloid muscarine, from the fungus *Amanita muscaria*, and the response to vagal parasympathetic nerve stimulation. Dixon (1907) suggested that the vagus nerve might release a muscarine-like substance that served as a chemical transmitter and he attempted to extract such a substance from the heart following vagal stimulation. Dale (1914) was struck by the coincidence between the activity of Ach and responses to parasympathetic nerve stimulation and coined the term *parasympathomimetic* to describe its effects. Loewi (1921) was the first to convincingly demonstrate the release of a neurotransmitter upon nerve stimulation and this was from the vagus nerve of the frog heart. On stimulation of the vago-sympathetic trunk of a frog isolated perfused heart, the rate and force of contraction were reduced, indicating predominance of the parasympathetic effect. When the perfusate was taken to a recipient heart after stimulation, this heart was also inhibited. Thus Loewi proposed that the inhibition was due to the liberation of a chemical substance which he called *vagusstoff* and later identified as Ach (Loewi & Navratil 1926). The concepts of neurohumoral transmission and the receptor sites for the neurotransmitter are illustrated more thoroughly with particular reference to sympathetic neurones in Chapter 2.

Arising from Dale's (1914) study of the action of Ach came the first suggestion that Ach produced two separate types of action in the body. The primary response to the transmitter in the anaesthetized cat is a fall in blood pressure which, like that to muscarine, was antagonized by atropine (see Figure 11.1). This effect was referred to as the muscarine-like (or *muscarinic*) action and is now attributed to the interaction of Ach with muscarinic (M) receptors. After blockade by atropine, larger doses of Ach cause pressor responses, stimulate the heart (see Figure 11.1) and cause twitching of skeletal muscle. Since these responses were mimicked by the alkaloid (nicotine) from *Nicotiana tabacum* and were unaffected by atropine, they were termed the nicotine-like (or *nicotinic*) actions of Ach. The cardiovascular effects are due to stimulation of autonomic ganglia. They are ascribed to an interaction with nicotinic cholinoceptors (N). The nicotinic actions of Ach at skeletal muscle and autonomic ganglia are, however, due to interactions with two distinct subtypes of nicotinic receptor. The neuronal nicotinic receptor in ganglia, the adrenal medulla and the CNS is designated as the N_N receptor in contrast to the skeletal muscle nicotinic receptor (N_M).

The nicotinic receptor has been extensively studied in skeletal muscle and the electric organ of the electric ray (*Torpedo*) and identified as a pentamer, the five units arranged symmetrically around an ion channel (see Figure 11.2). Purification of the receptor has led to the isolation of cDNAs that encode for these subunits and this has facilitated the cloning of genes of the receptor subunits from mammalian neurones and skeletal muscle. Differences between the N_M and N_N nicotinic receptors have therefore been confirmed by molecular biology, each being composed of different combinations of subunits (α, β, τ and δ) in their pentomer structure. These subunits consist of amino acid sequences probably arranged into α-helices as four transmembrane domains. The α-subunits carry the recognition site for Ach and are pseudo-irreversibly bound by the α-toxin, α-bungarotoxin, from the venom of a snake, the Taiwan banded krait (*Bungarus multicinctus*). This toxin selectively inhibits neuromuscular (N_M) nicotinic receptors. The α-subunits in skeletal muscle nicotinic receptors are α_1, and two distinct types, α_{1a} and α_{1b}, have been cloned. Neuronal nicotinic receptor α-subunits are α_2 but multiple subtypes (α_2 to α_8) have been cloned from chick and rat autonomic and central nervous systems. The neuronal nicotinic receptor is blocked by a minor component of the *Bungarus* venom, known as *n*-bungarotoxin and equivalent to *K*-bungarotoxin (Watson & Girdlestone 1994). The structure and function of the nicotinic neuronal receptor are described in more detail in Chapter 11.

The synthesis, storage and release of Ach has been examined in tissues that are readily accessible and contain a high density of cholinergic neurones. These include synaptosomes (essentially pinched-off nerve endings) from brain homogenates and the electric organ of electric rays (*Torpedo*). The latter receives only a cholinergic innervation. Sympathetic ganglia and adrenal medullary chromaffin tissue are also used, but the sympathetic nerve endings form a diffuse innervation of the effector organ and are therefore less appropriate for study. It is clear, however, that the mechanisms of synthesis and release of Ach are essentially the same in all cholinergic neurones.

The terminal portion of the autonomic parasympathetic neurone, in common with the sympathetic nerve (see Chapter 2, Figure 2.1), has varicosities distributed along its length which are filled with agranular electron-luscent vesicles. These are the sites of Ach release rather than at a true terminal synapse, as in skeletal muscle. The junction between a varicosity and the smooth or cardiac muscle cell is wide and variable (up to micrometers in width) and there is a lack of discrete postjunctional specialization. The cholinoceptors are distributed over the entire effector cell, unlike in skeletal muscle where they are concentrated at the synaptic cleft. The release of Ach from parasympathetic nerve endings induces a response of the effector by interacting with cholinoceptors of the muscarinic subtype. The effect is rapidly terminated by the dissociation of the Ach from the receptor and hydrolysis to choline and acetate by means of acetylcholinesterase, which is located both pre- and postjunctionally (see Chapter 10).

7.2 Synthesis, Storage and Release of Ach at Parasympathetic Neurones

The basic mechanisms of synthesis and release of Ach have been examined in the vagally stimulated heart and myenteric plexus of the ileum, and appear to be the same as those mechanisms that have been extensively examined in synaptosomes and torpedo axons. The sequence of events is shown diagramatically in Figure 7.1. Ach is synthesized from choline and acetylcoenzyme A (AcCoA) by the action of the enzyme, choline acetyltransferase (ChAT):

$$CH_3\text{-}N^+CH_2CH_2OH + AcCoA \xrightleftharpoons{ChAT} CH_3\text{-}N^+\text{-}CH_2CH_2\text{-}O\text{-}C\text{-}CH_3 + CoA.$$

choline   acetylcholine

Choline acetyltransferase (EC 2.3.1.6) is a globular protein of relative molecular mass in the region of 73 kDa. It is located in the axonal cytoplasm in predominantly soluble

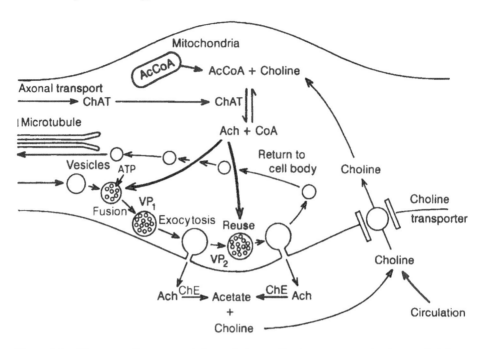

Figure 7.1 Diagrammatic representation of the synthesis, storage and release of Ach at a terminal varicosity of a parasympathetic nerve. ChAT, choline acetyltransferase; AcCoA, acetylcoenzyme A; ChE, cholinesterase.

form but also ionically bound to membrane lipids. It is transported down the axon after synthesis in the ribosomes of the cell body (perikaryon) and is concentrated at the nerve terminal. There are probably multiple forms of this enzyme and the primary structure has now been identified from molecular cloning (Berrard *et al.* 1989). The reaction sequence probably occurs with AcCoA binding initially to the enzyme, before choline, and then Ach dissociating first. The binding of AcCoA to the enzyme is probably by the 3'-phosphate to an arginine residue at the active site. The choline appears to bind with its cationic head to an anionic site. The hydroxyl group of choline attacks the thiol-ester groups of AcCoA to release the acetyl group; this is facilitated by the imidazole group of a histidine residue within the active centre (Figure 7.2).

The AcCoA substrate is formed in the mitochondria, the membrane of which is relatively impermeable. It therefore has to pass to the cytoplasm in the form of pyruvate, citrate or acetate depending upon the location of the nerve. AcCoA is then reformed. Pyruvate is generated within the mitochondria by the action of pyruvate dehydrogenase. Citrate is derived from oxaloacetic acetate and AcCoA by citrate synthase and transported from the mitochondria. AcCoA is produced in the cytoplasm from citrate and ATP by the action of ATP citrate lysase. Acetate reacts with ATP under the influence of acetate thiokinase to form an enzyme-bound acetyl AMP, which in the presence of CoA undergoes transacetylation to form AcCoA.

The second substrate, choline, is derived from phosphatidylcholine, a phospholipid formed mainly in the liver and supplied to the tissue via the circulation. There are four mechanisms of breakdown, the principal one being by deacylation to glycerylphosphorylcholine and subsequent cleavage to choline and glycerophosphoric acid. Phospholipase D may also directly cleave it to choline

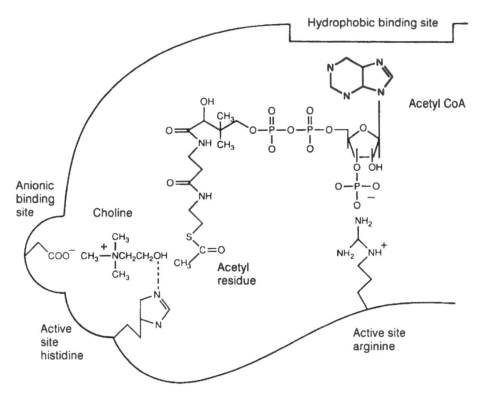

Figure 7.2 Proposed interaction of choline and acetylcoenzyme A (acetyl CoA) with the active site of choline acetyltransferase in the synthesis of acetylcholine.

and diacylphosphatidate (phosphatidic acid) (see Figure 13.5). The second major source of choline (50–60%) for synthesis of Ach is that derived from the breakdown of released Ach by acetylcholinesterase. Finally, activation of muscarinic receptors causes release of choline into the extracellular spaces, probably by activation of phospholipase D (see Chapter 13).

The rate of synthesis of Ach adapts to the rate of release so that relatively constant levels are maintained within the neurone. The enzyme activity of ChAT appears to operate well within its capabilities since the substrate concentrations are substantially below the K_m values. It is not therefore rate-limiting and does not control the rate of synthesis. The main mechanism of control is by regulation of choline transport into the neurone. Increased synaptic activity increases uptake of choline initially by neuronal depolarization and Ca^{2+} influx. This process also increases the supply of AcCoA, probably by increasing permeability of the mitochondria. A poststimulation increase in choline uptake is attributed to the raised activity of Na^+,K^+-ATPase and resting hyperpolarization (Tuček 1988).

7.2.1 Choline Transport

The choline for Ach synthesis is provided by a high affinity ($K_m = 1–5 \ \mu M$) uptake mechanism that carries the charged molecule across the neuronal membrane (Ducis

1988). This process is temperature-, energy- and sodium-dependent and is blocked by hemicholinium-3 (HC-3) (Figure 7.3). A low affinity uptake ($K_m > 20$ μM) is also present. The high affinity uptake is the rate-limiting step for Ach synthesis and is associated with increased ChAT activity and synthesis. It requires an intact metabolically active nerve since it is lost by denervation and is blocked by the metabolic uncoupler of oxidative phosphorylation, 2,4-dinitrophenol.

The sodium dependence suggests that choline may be co-transported with Na$^+$ in addition to Cl$^-$, two Na$^+$ molecules being carried with each choline molecule. The inwardly directed Na$^+$ gradient is maintained by the ouabain-sensitive Na$^+$,K$^+$-ATPase and, in synapsosomes, the uptake of choline is inhibited by ouabain. Uptake of choline is also enhanced by a high internal K$^+$. The transport of choline is

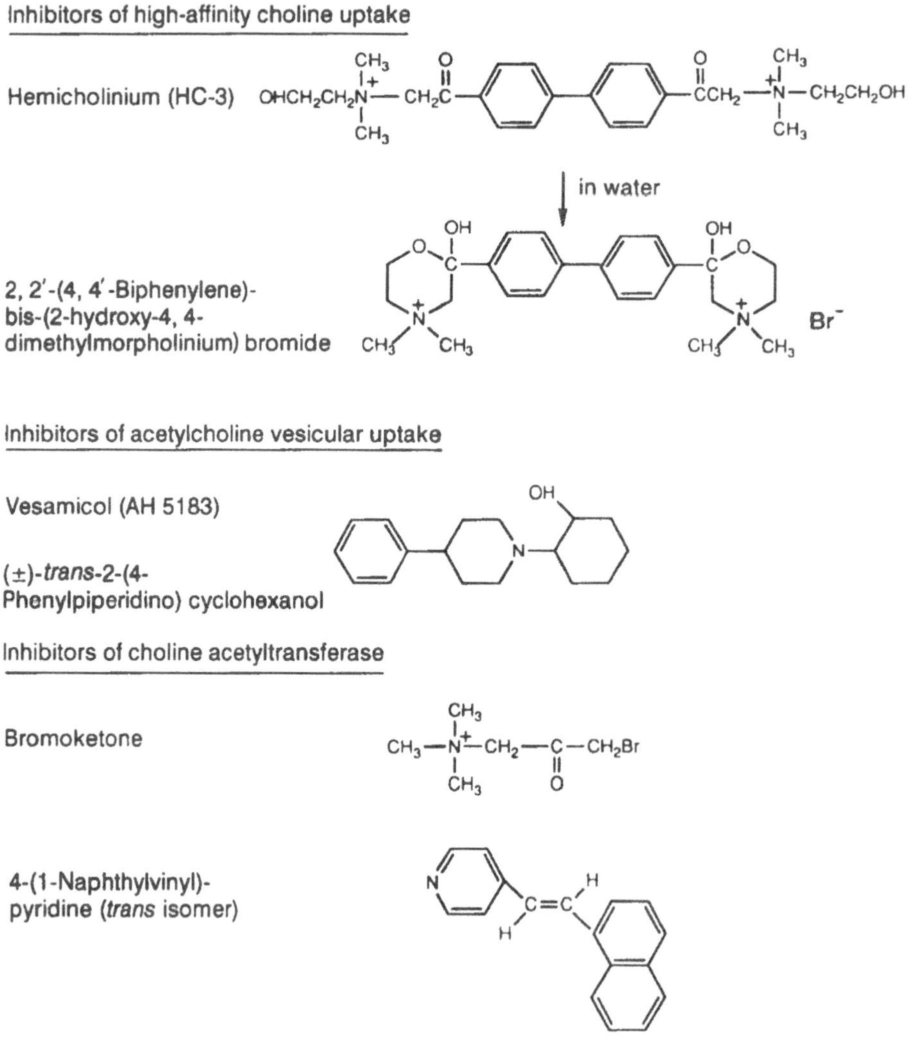

Inhibitors of high-affinity choline uptake

Hemicholinium (HC-3)

2, 2'-(4, 4'-Biphenylene)-bis-(2-hydroxy-4, 4-dimethylmorpholinium) bromide

in water

Inhibitors of acetylcholine vesicular uptake

Vesamicol (AH 5183)

(±)-*trans*-2-(4-Phenylpiperidino) cyclohexanol

Inhibitors of choline acetyltransferase

Bromoketone

4-(1-Naphthylvinyl)-pyridine (*trans* isomer)

Figure 7.3 Structures of agents affecting the uptake, storage and synthesis of acetylcholine.

therefore an electrogenic process similar to that of noradrenaline (see Chapter 2, Figure 2.21). Knowledge does not appear to be as well advanced as for noradrenaline uptake and there is little information on efflux of choline. A specific choline binding site associated with the carrier is characterized by $[^3H]$hemicholinium-3, which exhibits saturable, reversible binding with high specificity. Binding requires both Na^+ and Cl^- and is competitively displaced by choline.

The structural requirements for choline uptake are relatively restrictive since the quaternary head and free hydroxyl group are essential. Ach itself is therefore not taken up into the neurone and this process does not serve as a means of eliminating Ach from the synapse. The N-substituents can only be extended by one carbon atom and the chain length by only one methylene group to give homocholine (Figure 7.4).

Choline analogues

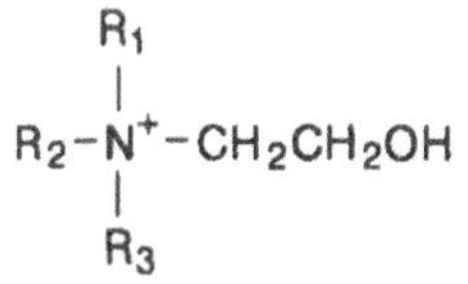

$$R_2-\overset{\overset{\displaystyle R_1}{|}}{\underset{\underset{\displaystyle R_3}{|}}{N^+}}-CH_2CH_2OH$$

	R_1	R_2	R_3
Choline	CH_3	CH_3	CH_3
Monoethylcholine	C_2H_5	CH_3	CH_3
Diethylcholine	C_2H_5	C_2H_5	CH_3
Triethylcholine (TEC)	C_2H_5	C_2H_5	C_2H_5
Pyrrolidinecholine	$-CH_2-CH_2-$		CH_3
N-Aminodeanol	NH_2	CH_3	CH_3

Homocholine and analogues

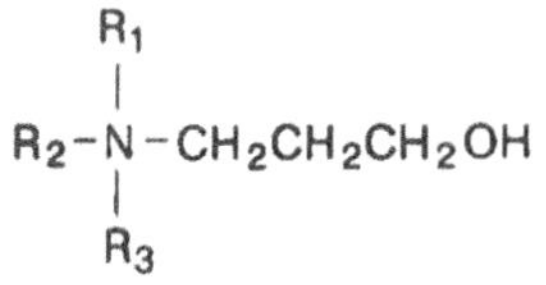

$$R_2-\overset{\overset{\displaystyle R_1}{|}}{\underset{\underset{\displaystyle R_3}{|}}{N}}-CH_2CH_2CH_2OH$$

	R_1	R_2	R_3
Homocholine	CH_3	CH_3	CH_3
Monoethylhomocholine	C_2H_5	CH_3	CH_3
Diethylhomocholine	C_2H_5	C_2H_5	CH_3
Triethylhomocholine	C_2H_5	C_2H_5	C_2H_5

Figure 7.4 Structures of cholinergic false transmitters.

Homocholine, triethylcholine (TEC) and related analogues are readily transported into the parasympathetic neurone where they are acetylated and taken into the storage vesicles to replace Ach and act as false transmitters (see later).

7.2.2 Storage of Ach

After its synthesis from choline, Ach is taken up by the storage vesicles principally at the nerve terminal (Figure 7.1). The vesicles are transported anterogradely from the cell body via the microtubules (see Chapter 1), any Ach incorporation during this process being minor (Tytell & Stadler 1988). The vesicles appear to be of two types: electron-lucent vesicles (40–50 nm diameter) and a small number of dense-cored vesicles (80–150 nm). The vesicular core contains Ach and ATP (in a ratio of 10:1) mainly dissolved in the fluid phase, metal ions (Ca^{2+}, Mg^{2+}), and proteoglycan (vesiculin) (Figure 7.5). The latter is negatively charged and may sequester the Ca^{2+} or Ach. It is bound within the vesicle, the protein moiety anchoring it to the vesicular membrane. The proteoglycan is highly antigenic and antibodies may be used for immunohistochemistry of the cholinergic nerves to trace vesicle movements. Additionally, peptides including vasoactive intestinal peptide (VIP) (70:1 with Ach) are found, usually in the dense-cored vesicles. The vesicle membrane is rich in lipids, principally cholesterol and phospholipids, and proteins. The proteins include ATPase (ouabain-sensitive) which may be involved in proton pumping and

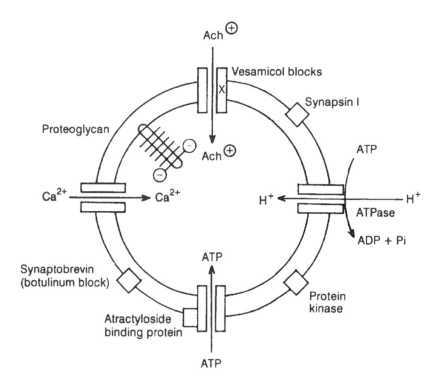

Figure 7.5 Structure and function of a cholinergic storage vesicle. Sites of acetylcholine (Ach) uptake and influx of Ca^{2+}, H^+ and ATP are shown. Influx of H^+ is driven by a ouabain-sensitive ATPase.

in vesicular inward transport of Ca^{2+}, protein kinase (involved in phosphorylation mechanisms of Ca^{2+} uptake), calmodulin (drives vesicular Ca^{2+} transport), atractyloside-binding protein (ATP carrier) and synapsin (may control exocytosis, see also Chapter 2).

The uptake of Ach into the vesicle has a four-fold concentrating power, is saturable and is ATPase-dependent. Uptake of choline and its analogues (except homocholine and diethylhomocholine) into vesicles does not take place. The process is inhibited by vesamicol (AH 5183, (±)-*trans*-2-(4-phenylpiperidino)cyclohexanol) (Figure 7.3). The inhibition is non-competitive and reversible with no effect on vesicular ATPase. The (−)-isomer is the active form which probably binds to the Ach transporter complex (Prior *et al*. 1992).

There is also a store of free cytoplasmic Ach but this is susceptible to hydrolysis by acetylcholinesterases which are present in the cytoplasm and associated with the outer surface of vesicular membranes (Zimmermann 1988).

7.2.3 *Release of Ach*

In common with the release of noradrenaline from sympathetic nerves, Ach release primarily involves exocytosis (Zimmermann 1988). After high-frequency stimulation of the nerve supplying the *Torpedo* electric organ, the electronmicrograph reveals loss of vesicles and an increase in the area of the nerve terminal plasma membrane, which becomes convoluted. The vesicles fuse with the membrane at the active zone. They become incorporated into the membrane, as demonstrated by immunofluorescence for the proteoglycan at the nerve terminal. After stimulation, there is retrieval of the vesicle by endocytosis (Figure 7.1). Exocytosis of dense-cored vesicles, additionally releasing peptides, may occur outside the active zone.

Lower frequency of stimulation does not result in loss of vesicles but the appearance of a population of vesicles of reduced size and increased density. These are regarded as being of endocytotic origin on the basis that they take up extracellular dextran. These recycled vesicles (VP_2) have a high capacity for uptake of newly synthesized Ach and ATP. From experiments with radiolabelled choline, it is concluded that newly synthesized Ach is released in preference to Ach stored previously in mature vesicles (VP_1). Thus, recycled vesicles appear to be the major source of released Ach. Vesamicol, which inhibits vesicular uptake of Ach, inhibits the release of newly synthesized Ach but not of preexisting stores. This inhibition occurs without affecting influx of Ca^{2+} or of choline transport. When added *after* the accumulation of radiolabelled Ach, it does not affect the release of labelled Ach. Therefore Ach has to be taken up into vesicles (principally of the recycled type) before it can be released by nerve stimulation. After reuse of vesicles, it is probable that they are ultimately returned to the cell body by retrograde axonal transport.

These processes of synthesis, storage and release of Ach appear to apply to the mammalian parasympathetic nerve ending. In contrast to other cholinergic nerves, however, high rates of stimulation can be tolerated better and do not result in significant loss of vesicles unless the synthesis of Ach is impaired by HC-3. Whether the contents of a single vesicle represents the quanta of transmitter release responsible for the spontaneous excitatory junction potentials (SEJPs) recorded at parasympathetic nerve endings is uncertain. The detailed electrophysiological recording that has been performed with sympathetic nerves has not been possible so far with parasympathetic nerves. There is no reason to believe, however, that the situation is any different. Therefore the entire content of a vesicle would be expected to be released and an intermittency of release, in that exocytosis does not occur from all varicosities even though the action potential actively invades the entire nerve terminal. This is attributed to the low probability of release from the invaded varicosity (Brock & Cunnane 1992) (see Chapter 2 for a more detailed discussion as it applies to the sympathetic neurone). That release of Ach is due to a propagated action potential is indicated by its inhibition with tetrodotoxin.

The membrane depolarization associated with invasion of the terminal varicosities by the action potential results in rapid influx of Ca^{2+}. This facilitates the fusion of the vesicle with the active zone of the neuronal plasma membrane to trigger exocytosis. The soluble contents of the vesicles, Ach and ATP, are extruded into the synaptic cleft. This process is also induced by depolarization with high extracellular K^+, where Ca^{2+} influx is likewise responsible. Ca^{2+} ionophore, A23187, which raises intracellular Ca^{2+} levels by bypassing the voltage-gated Ca^{2+} channel, also stimulates release of Ach. Reduction of extracellular Ca^{2+} inhibits the release of Ach and the response to parasympathetic nerve stimulation, for example in the rat urinary bladder (Bhat *et al.* 1989). The type of Ca^{2+} channel involved appears to be of the N-type, as in sympathetic neurones (see Chapter 2 for a discussion of Ca^{2+} channel types). The evidence for this is that ω-conotoxin inhibits the cholinergic contractile response of the rat urinary bladder to field stimulation of the parasympathetic nerves (Figure 7.6) (Maggi 1991b). This study also shows the dual response to electrical stimulation; the early contraction is atropine-insensitive and mediated primarily by the release of ATP, while the late contraction is blocked by atropine

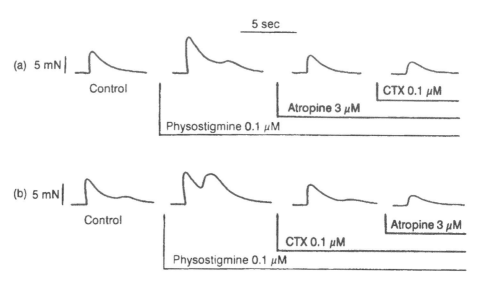

Figure 7.6 Contractions of the rat isolated bladder to electrical field stimulation. The anticholinesterase, physostigmine, potentiates the secondary component preferentially. Administration of atropine then abolished the secondary phase (a) indicating that this component was due to acetylcholine (Ach) release. An early atropine-resistant component was therefore of non-adrenergic non-cholinergic (NANC) origin. Both components were reduced by the N-type Ca^{2+} channel blocker, ω-conotoxin (CTX), the primary component more effectively when given before atropine (b). Reproduced with permission from Maggi (1991b).

and therefore due to Ach release. ω-Conotoxin preferentially blocked the Ach-mediated response.

Also co-released from certain cholinergic nerves is VIP, for example, after stimulation of the parasympathetic nerve to the cat submandibular salivary gland. The Ach mediates the secretory response while VIP causes vasodilatation (see Chapter 12).

7.3 Drugs Affecting Synthesis, Storage and Release of Ach

7.3.1 *False Transmitters*

There has been some debate as to whether Ach may also be released from the cytoplasmic stores. This has been studied by use of cholinergic false transmitters. By definition, these substances are analogues of the endogenous transmitter which are incorporated into the releasable pool of transmitters and are released on nerve stimulation in the same Ca^{2+}-dependent manner (Kopin 1968). This usually results in reduced activity because of the weaker activity of the false transmitter. In the case of cholinergic nerves, this is limited to precursors since acetylated analogues of choline are not taken up by the choline transporter. The precursor has then to be acetylated before take up by the vesicle. Consequently, there are three structural determinants for a false transmitter to reach the storage vesicle: uptake by the choline carrier, acetylation by ChAT, and vesicular uptake. The false transmitters include the acetyl derivatives of triethylcholine (TEC) and homocholine. Acetylpyrrolecholine is preferentially taken up by recycling vesicles (Figure 7.4). To determine the role, if any, of cytoplasmic release of Ach, a false transmitter precursor is selected that is taken into the nerve but its acetylated product is poorly incorporated into the vesicle. TEC fulfils these criteria. After loading with [^{3}H]acetate, the radiolabelled esters were introduced into recycling vesicles by low-frequency stimulation. On applying a high-frequency release stimulus, a higher proportion of [^{3}H]Ach was released than [^{3}H]acetylTEC and this was in the same ratio as found in the vesicles. This result indicates that release from the cytoplasmic pool is only minor.

False transmitters for the cholinergic system are therefore useful tools for examining the dynamics of transmitter release and storage. As pharmacological agents they have not proved clinically useful. TEC gradually reduces the response to repeated nerve stimulation but this is more readily observed in skeletal muscle. N-Aminodeanol (Figure 7.4) appears to be incorporated into the brain where it is acetylated and has some behavioural effects (Newton & Jenden 1986).

7.3.2 *Inhibitors of Choline Acetyltransferase (ChAT)*

Bromoketone and 4-(1-naphthylvinyl)-pyridine (4-NVP) (Figure 7.3) are inhibitors of ChAT, the former blocking the enzyme irreversibly. Unlike most of the agents examined, these two do not inhibit cholinesterase but they do inhibit several other enzymes. Since this enzyme is not rate-limiting, inhibitors have a weak action on Ach levels and they have not found therapeutic application.

7.3.3 **Inhibitors of Choline Uptake**

Hemicholinium-3 (HC-3) (Figure 7.3) is the prototype inhibitor of choline uptake and since uptake is the rate-limiting step for synthesis of Ach, it effectively blocks the synthesis and release of Ach. Other inhibitors include the alkaloid neurotoxins, aconitine and veratridine, which maintain Na^+ channels in the open state and thus dissipate the Na^+ gradient across the neuronal membrane. The choline mustard aziridinium ions (CMA) are also potent irreversible inhibitors of choline transport but do not have absolute selectivity.

7.3.4 **Inhibitors of Vesicular Transport**

Vesamicol (Figure 7.3) blocks transport of Ach into the vesicle and will therefore affect its availability for release by the nerve action potential. The pharmacology of vesamicol has already been described. It has inhibitory effects upon all cholinergic nerves and has no therapeutic uses.

7.3.5 **Inhibitors of Ach Release**

7.3.5.1 *Botulinum toxin*

The toxin from the anaerobic, spore-forming micro-organism, *Clostridium botulinum*, is one of the most toxic substances known. Ingestion of the preformed toxin on contaminated food results in botulism. Uncooked meats, fish products or improperly prepared canned fruits and vegetables are the major sources of food-borne botulism. Several types of organism (A–G) have been identified, each yielding an antigenically distinguishable form of neurotoxin. The spores of *C. botulinum* are heat resistant, but the toxins are heat labile, being inactivated by boiling food (100 °C) for 10 min. Botulinum toxins are not inactivated by gastric acid or proteolytic enzymes of the gut, and after absorption they are bound to the external surface of cholinergic nerve terminals. The toxin does not affect nerve conduction, cholinesterase or the synthesis of Ach but produces prolonged (several months) block of Ach release by exocytosis. The binding site may be to gangliosides (see also Chapter 6). These contain sialic acid and trisialoganglioside G_1 has been found to bind actively with the polar toxin. Mixing of the ganglioside and toxin before injection to mice renders it inactive. Sialic acids are also present in erythrocyte membranes and it is possible that the agglutination of blood by the toxins is due to this binding. The toxins consist of a disulphide-linked heavy and light chain. After binding on the cholinergic nerve terminal through the heavy chain, the toxin is internalized where the light chain exerts the inhibitory effect on transmitter release. The target proteins that are cleaved by the individual toxins are synaptobrevin (Figure 7.5), which lies in the vesicle membrane (toxins B, D and F), and SNAP-25 (25 kDa protein) (toxins A and E) and syntaxin (toxin C), both of which are associated with the inner face of the neuronal membrane. These three target proteins occur as a complex and may be involved in the docking and fusion of the vesicle with the neuronal membrane (Wonnacott & Dajas 1994).

The clinical symptoms of botulism usually appear 18–36 hr after ingestion and are due to cholinergic blockade. Muscle weakness due to the block of somatic nerves occurs and the acute lethal effect is due to paralysis of the respiratory muscles (intercostal and diaphragm). Autonomic parasympathetic dysfunction is manifest by an intensely dry mouth and inability to swallow, blurred vision (paralysis of the control of ciliary muscle for near vision, Figure 1.7, Chapter 1), dilated pupils (blockade of parasympathetic innervation of the circular muscle of the iris) and urinary retention. Treatment involves administration of the antitoxin to neutralize any unbound toxin. Toxin already bound to nerve endings cannot be neutralized. Pharmacological treatment with anticholinesterases or drugs that release Ach cannot restore transmission, although the use of the K^+ channel antagonist, 4-aminopyridine (4-AP), has been advocated since it can increase the release of any residual Ach.

Botulinum toxin itself has been used by local injection to interfere with neurotransmission to skeletal muscles in the management of involuntary muscle spasm including blepharospasm, strabismus and facial tics. No applications at autonomic sites are evident.

7.3.5.2 *Black widow spider venom*

The venom of the black widow spider (*Latrodectus mactans tredecimguttatus*) (BWSV) and the β-bungarotoxin component of the venom from the Taiwan banded krait (see nicotinic binding of the α-bungarotoxin) also exert powerful effects on Ach release from cholinergic nerves. They initially cause clumping of vesicles at the neuronal membrane, many showing fusion with the membrane. Miniature postjunctional excitatory potentials occur. Subsequently, these potentials cease, and vesicles become permanently fused with the axonal membrane which becomes expanded and convoluted. At this point synaptic transmission is blocked. BWSV has a similar action on noradrenaline release from sympathetic nerve terminals.

7.4 Prejunctional Receptors

Modulation of Ach release from parasympathetic neurones can be elicited by the actions of several neurotransmitters and endogenous substances. The receptors for these agents are located on the prejunctional membrane of the terminal varicosity. Modulation of Ach release has been monitored from the release of radiolabelled Ach during electrical stimulation of intact tissues, from isolated synaptosomes or by changes in the functional response of the tissue. Released Ach may exert regulatory actions on its own further release by interacting with prejunctional cholinoceptors, which are defined as autoreceptors. Receptors for other autacoids are heteroreceptors.

7.4.1 *Prejunctional Muscarinic Receptors (Autoreceptors)*

Muscarinic receptor agonists inhibit the release of Ach from parasympathetic nerves evoked by electrical stimulation and by depolarization induced by high K^+. This effect has been observed in a wide range of tissues including the heart, gastrointestinal tract, bronchi, bladder and iris. In many of these tissues it is difficult to eliminate the ganglionic sites since the parasympathetic cell bodies are often present (see Chapter 1, Figures 1.5 and 1.6). The iris, however, is ganglion-free. Muscarinic receptors may also be located on the cell bodies of the postganglionic neurone (soma-dendritic receptors) and thereby modulate release of Ach if there is any *preganglionic* nerve stimulation. Such nerve stimulation would of course be susceptible to ganglionic blockade by hexamethonium. Muscarinic receptors do appear on the cholinergic cell bodies of the myenteric plexus of the gut (see Figure 1.5) but here their stimulation *facilitates* transmission and Ach release at the nerve terminal. Furthermore, the receptor is of a different subtype (M_1) to the prejunctional receptor at the nerve terminal (M_2) (see later and Chapter 11) (Figure 7.7). Presynaptic nicotinic receptors are almost certainly absent.

Demonstration of measurable levels of released Ach from parasympathetic neurones usually requires inhibition of acetylcholinesterase to prevent its breakdown. However, the raised level of Ach itself inhibits release by stimulation of the prejunctional muscarinic receptor. Therefore exogenously added muscarinic agonists exert only small inhibition in the presence of anticholinesterases. It is probably the prejunctionally located cholinesterase that protects the prejunctional muscarinic receptors from activation by Ach. In contrast, antagonists of prejunctional muscarinic receptors (atropine) are more effective at enhancing transmitter release than if used in the absence of cholinesterase inhibition. This facilitatory effect of muscarinic antagonists indicates that prejunctional autoreceptors are functional under normal conditions. The release of Ach is probably controlled by the Ach release from a previous nerve impulse. Pulse-to-pulse modulation is indicated by the gradual decline in Ach release per pulse with high-frequency stimulation in the myenteric plexus. This decline is retarded by a muscarinic antagonist.

In parasympathetic neurones, the release of both endogenous and newly synthesized labelled Ach is inhibited equally. Release of the co-transmitters, ATP

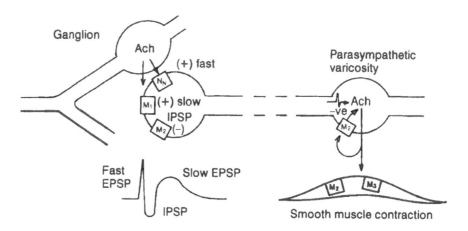

Figure 7.7 Muscarinic receptors at different locations along the autonomic parasympathetic efferent pathway. Acetylcholine (Ach) released from the preganglionic varicosity can interact with nicotinic N_N, muscarinic M_1 and muscarinic M_2 receptors to produce a fast depolarization (fast EPSP), a slow depolarization (slow EPSP) and hyperpolarization (IPSP), respectively. Nicotinic receptor stimulation sets up the action potential in the postganglionic neurone to release Ach from parasympathetic nerve terminal varicosities. This can interact with M_2 or M_3 receptors of the effector organ. Released Ach may also stimulate prejunctional M_2 receptors which inhibits Ach release (negative feedback).

and VIP, is also inhibited. The synthesis of Ach is also inhibited in parallel with effects on its release, probably as a result of a common inhibition of Ca^{2+} influx which is required for both synthesis and exocytosis.

The inhibition of Ach release due to a propagated action potential by electrical stimulation is greater than that due to K^+ which causes depolarization of the entire nerve ending without impulse propagation. This led to the suggestion that autoinhibition by muscarinic prejunctional receptors is due to restriction of impulse propagation to the terminal varicosities (Alberts *et al.* 1982). It is possible that muscarinic agonists do not reduce the number of varicosities invaded by the action potential but decrease the fraction of Ach released by each action potential reaching a varicosity. The possibility that K^+ and electrical stimulation release Ach from separate stores which are differently modulated cannot be discounted.

7.4.1.1 Mechanism of autoinhibition by Ach

The mechanism of inhibition of transmitter release by Ach appears to involve a reduced availability of intraneuronal Ca^{2+} for exocytosis. The evidence for a decrease in voltage-dependent Ca^{2+} influx is as follows:

1   Reduced extracellular Ca^{2+} enhances the inhibitory effects of muscarinic agonists.
2   Muscarinic agonists become less effective as frequency of stimulation increases, possibly because intracellular levels of Ca^{2+} accumulate at higher frequencies and thus overcome muscarinic inhibition of transmitter release.
3   Only Ca^{2+}-dependent release mechanisms are inhibited by muscarinic agonists, that is, electrical stimulation, K^+ and nicotinic agents, but not ouabain-induced release. The situation with Ca^{2+} ionophore-induced Ach release is variable but in brain slices it is not enhanced by muscarinic antagonists.

The signal transduction pathway between activation of the prejunctional muscarinic receptor (M_2, see later) and inhibition of exocytosis in peripheral parasympathetic neurones has not been thoroughly elucidated. It does not appear to involve the accumulation of cGMP or activation of protein kinase C.

The prejunctional muscarinic receptor undergoes desensitization when chronically exposed to Ach (see Chapter 14). This can be achieved by chronic treatment of animals with anticholinesterases. Tissues removed from these animals display reduced sensitivity to prejunctional inhibition of Ach release by electrical stimulation. As a consequence, electrical stimulation of the myenteric plexus elicits a greater release of Ach after chronic treatment with a cholinesterase inhibitor than after a single acute dose (Kilbinger 1984, Starke *et al.* 1989).

There does not appear to be any therapeutic application of agonists and antagonists acting at the prejunctional muscarinic autoreceptors. The receptor is of the M_2 subtype and currently cannot be distinguished by antagonists from the M_2 receptors found postjunctionally in many tissues. It has been differentiated in tissues where the postjunctional functional responses are mediated via M_3 receptors. For example, in the guinea-pig trachea, the release of [^{3}H]Ach by parasympathetic vagal stimulation is enhanced by M_2-selective antagonists (methoctramine and AF-DX 116, see Tables 9.7 and 9.6), while the postsynaptic contractile responses to nerve stimulation were preferentially blocked by an M_3-selective antagonist (4-DAMP, see Table 9.5)

(Kilbinger *et al.* 1991). Generally, any facilitation of transmitter release by a muscarinic antagonist will be offset by the blockade of the postjunctional action and rarely will the muscarinic antagonist yield *enhanced* parasympathetic responses. Low doses of atropine can enhance the vagal influences to the heart and cause bradycardia. Although this has been attributed to a central action, it also occurs with methyl-atropine (see Table 9.1), which has a quaternary nitrogen and therefore poorly penetrates the blood–brain barrier. The facilitated vagal slowing could therefore be due to presynaptic muscarinic blockade occurring before blockade of the postsynaptic effects (Starke *et al.* 1989). There is also evidence of facilitated vagal bronchoconstriction *in vivo* by muscarinic antagonists (eg methoctramine) in spite of some inhibition of Ach-induced bronchoconstriction. Similarly, in isolated tracheal preparations, methoctramine enhances responses to nerve stimulation, leaving the contractile response to exogenously applied Ach unaffected (Watson *et al.* 1992).

7.4.2 *Heteroreceptors on Parasympathetic Neurones*

7.4.2.1 *Adenosine A_1-receptors*

Adenosine released from hypoxic tissues or arising from the breakdown of ATP co-released from autonomic nerves (see Chapter 12) exerts an inhibitory effect on Ach release by stimulation of the A_1 subtype of P_1-purinoceptor. Thus, adenosine and its analogues inhibit responses to parasympathetic nerve stimulation. For example, in the guinea-pig transmurally stimulated ileum (Paton preparation 1957), adenosine inhibits the twitch responses but does not inhibit contractions due to exogenously added Ach. It also inhibits release of Ach (Ribeiro 1991).

The inhibition of transmitter release is probably due to interference with Ca^{2+}-influx through the voltage-dependent channels. Inhibition of Ach release from hippocampal neurones is associated with decreasing levels of cAMP. Although the prejunctional A_1-receptor probably belongs to the family of G protein-linked receptors, there has been no evidence of it being sensitive to pertussis toxin which usually blocks responses mediated via G_i regulatory proteins (Fredholm *et al.* 1991) (see Chapter 13).

7.4.2.2 *Histamine receptors*

Prejunctional histamine receptors of the H_3 subtype have been identified in the ileum (Menkveld & Timmerman 1991) and human airways (Ichinose & Barnes 1989). Histamine and the H_3-selective agonist, (R)-α-methylhistamine, cause dose-related inhibition of the twitch responses of these isolated preparations to stimulation of the cholinergic parasympathetic neurones. This inhibitory response is not blocked by mepyramine or cimetidine at H_1 and H_2 receptor blocking concentrations (10^{-6} and 10^{-4} M, respectively) but is prevented by thioperamide, which is regarded as an H_3 receptor antagonist. The function of these receptors in the airways is uncertain; histamine would appear to inhibit vagally-mediated bronchoconstriction perhaps at low doses, yet it exerts powerful bronchoconstriction as the concentration increases.

7.4.2.3 *α_2-Adrenoceptors*

These receptors are located on parasympathetic neurones, stimulation also causing inhibition of Ach release. Their properties have already been described extensively in Chapter 6. Sympathetic nerve stimulation inhibits excitatory gastric smooth muscle responses produced by vagal stimulation but not those produced by exogenously added Ach. Noradrenaline inhibits field-stimulated ileal preparations by a combination of postjunctional α_1-adrenoceptor-mediated smooth muscle relaxation and prejunctional α_2-adrenoceptor-mediated inhibition of Ach release (Broadley & Grassby 1985, see Figure 3.17). Selective α_2-agonists such as clonidine exert inhibitory effects on the contractions of the gastrointestinal tract to parasympathetic nerve stimulation without affecting the responses to Ach (Thaina *et al.* 1993, Figure 6.3). α_2-Adrenoceptor agonists also inhibit vagally-mediated gastric acid secretion from parietal cells in pylorus-ligated rats. All of these effects are attributed to inhibition of Ach release from parasympathetic nerve endings. This effect partly explains the antidiarrhoeal action of α_2-adrenoceptor agonists and partial agonists, like clonidine (Ruffolo *et al.* 1993).

7.4.2.4 β-Adrenoceptors

β-Adrenoceptors with a prejunctional location have been extensively described for sympathetic neurones where they facilitate noradrenaline release (see Chapter 6). However, there is considerably less information regarding their presence on parasympathetic nerve terminals. In Figure 3.17 isoprenaline can be seen to cause an initial increase in twitch height of the guinea-pig transmurally stimulated ileum preparation, before the more usual inhibition occurs due to the smooth muscle relaxant activity at postjunctional β-adrenoceptors. This is indicative of prejunctional facilitatory β-adrenoceptors being present in parasympathetic neurones (Broadley & Grassby 1985). β-Adrenoceptors facilitating Ach release have been detected on the prejunctional nerves of autonomic ganglia. In rat isolated superior cervical (sympathetic) ganglia, isoprenaline has been shown to increase the amplitude of the compound action potential elicited by preganglionic nerve stimulation and to increase the release of [³H]Ach from the preganglionic neurone. The effects are blocked by propranolol (Misu & Kubo 1986).

7.4.2.5 Opioid receptors

The inhibitory effect of morphine upon the contractile response of the guinea-pig ileum and release of Ach due to transmural stimulation of the parasympathetic nerve endings was demonstrated many years ago (Paton 1957). This response has subsequently been characterized in several gastrointestinal tract preparations from various species. The inhibition is attributed to the activation of prejunctional opioid receptors preventing release of Ach; Ach itself is not inhibited. This may occur under physiological conditions due to the co-release of endogenous opiate peptides from sympathetic nerves (Morris & Gibbins 1992). Enkephalins such as D-Ala²-D-Leu⁵-enkephalin (DADLE) have inhibitory actions on cardiac vagus nerve stimulation without affecting exogenous Ach (Wong-Dusting & Rand 1985). The sympathetic nerve-mediated inhibition of the vagal response of the heart can be blocked by the opioid receptor antagonist, naloxone, suggesting the release of opiate peptides (Koyanagawa *et al.* 1989). The opioid receptors are subdivided into the μ (mu), δ (delta) and κ (Kappa) types, all of which are blocked by naloxone. In the guinea-pig ileum, the inhibitory action of morphine is mediated via μ opioid receptors. In the rat ileum the atropine-sensitive (parasympathetic) response to electrical stimulation is only inhibited by morphine at high concentrations. It is, however, inhibited by the δ opioid receptor agonist, DADLE, which failed to inhibit responses to exogenous Ach, suggesting the presence of prejunctional δ receptors in this species. This effect was also blocked by naloxone (Coupar & De Luca 1994).

This inhibitory action on Ach release by opioids is the basis of the long established use of morphine-containing preparations in the symptomatic treatment of diarrhoea. Perhaps one of the most widely dispensed antidiarrhoeal preparations is Kaolin and Morphine Mixture (Mist. Kaolin Sed.). Opium is more constipating than morphine because of the smooth muscle relaxing properties of the additional papaverine and noscapine. Opium Tincture (Laudanum) or Opium Liquid Extract have been widely used in the past, for example, in Mixture of Aromatic Chalk with Opium and Collis Browne's Compound[R], respectively. Morphine reduces gastrointestinal propulsive motility and slows transit time through the gut to permit better fluid reabsorption. Intestinal secretion is also impaired by a naloxone-sensitive action on the intestinal mucosa. As with the analgesic action of morphine, there may be tolerance to the inhibition of Ach release via prejunctional opioid receptors (Kromer 1988). Thus, morphine induces constipation, a common and troublesome side-effect of its use as an analgesic in severe pain. A related effect is on the urinary bladder, where morphine inhibits the parasympathetic voiding reflex (Chapter 1) leading to urinary retention. One might expect an inhibitory effect on the parasympathetic innervation of the structures of the eye, resulting in mydriasis and block of accommodation of the lense for near vision. However, morphine causes a pinpoint pupil due to central stimulation of the autonomic segment of the nucleus of the oculomotor nerve.

7.5 Muscarinic Receptors

Ach released from parasympathetic nerve endings induces a functional response of the tissue by interaction with muscarinic receptors. The muscarinic receptors are readily distinguished from their nicotinic counterparts even though they share the same neurotransmitter, Ach. Muscarinic, but not nicotinic, receptors are selectively stimulated by muscarine and blocked by atropine. Molecular biological studies have shown them to have separate genetic origins and structures and, whereas muscarinic receptors are linked through guanosine triphosphate regulatory proteins (guanine nucleotide or G proteins) to second messenger transduction pathways, the nicotinic receptor has no such linking. It is now clear that muscarinic receptors can be subdivided into several types (Table 7.1).

A major influence in the recognition that multiple muscarinic receptors might exist was the antagonist, pirenzepine (see Table 9.6). This drug was shown to inhibit gastric acid secretion in animals and humans at lower doses than the inhibition of salivation and the cholinergically stimulated smooth muscle of the ileum and urinary bladder and the induction of tachycardia (blockade of vagal tone). Binding experiments certainly proved that pirenzepine could discriminate differences in subtypes of muscarinic binding sites (Hammer *et al.* 1980). However, the preferential binding was not for the gastric mucosa of the stomach wall but for certain areas of the brain. Thus, the *in vivo* selectivity of pirenzepine could not be attributed to selective binding to muscarinic receptors on the secretory parietal (oxyntic) cells. Indeed, the pA_2 values for antagonism of agonist-induced

Table 7.1 Muscarinic receptor classification

	Subtypes				
Functional response / binding	M_1	M_2	M_3	M_4	—
Cloned receptors	m_1	m_2	m_3	m_4	m_5
Previous nomenclature	M_1	M_2 cardiac $M_{2\alpha}$	M_2 glandular M_2 muscle $M_{2\beta}$		
Selective antagonists [a]	Pirenzepine Telenzepine Dicyclomine	AF-DX116 Methoctramine Himbacine Gallamine [b]	4-DAMP HHSiD p-fHHSiD	Labelled by pirenzepine R-(+)-hyoscyamine	—

(*Continued*)

Table 7.1 *Continued*

	Subtypes			
Location-functional				—
Tissue	Autonomic ganglia	Heart	Smooth muscle	Guinea-pig uterus
Response	Slow excitatory Postsynaptic potential (S-EPSP)	Negative inotropy and chronotropy	Contraction	Contraction
Tissue	Rabbit vas deferens	Postganglionic neurones	Endothelium	
Response	Inhibition of twitch to nerve stimulation	Prejunctional inhibition of transmitter release	Release of EDRF	
Location-binding	Brain regions: cortex, hippocampus NB-OK 1 cells	Heart Smooth muscle of trachea, ileum, colon	Glands: lacrymal, salivary, pancreas Smooth muscle	Striatum of brain Rabbit lung NG 108–15 cells
Coupling to intracellular second messenger[c]	Phosphoinositide (PI) hydrolysis via PLC to IP_3	Inhibition of adenylyl cyclase ↓cAMP	PI	Inhibition of adenylyl cyclase ↓cAMP
Minor pathway	↑cAMP			PI

Notes: [a] Structures of antagonists are shown in Tables 9.4 and 9.5 (Chapter 9).
[b] Gallamine is normally regarded as a nicotinic receptor antagonist, N_M.
[c] Second messenger pathways discussed in Chapter 13

bradycardia of isolated atria (6.45), contraction of the ileum (6.66) and acid secretion from the mouse perfused stomach (6.32) were not different (Szelenyi 1982). Similar pA_2 values for pirenzepine against tracheal contraction and acid secretion led Black & Shankley (1985) to conclude that the muscarinic receptors mediating the two responses were the same. They concluded that the selective inhibition of acid secretion *in vivo* compared to atropine could be due to its lack of loss through the oxyntic cells and hence an effective blocking concentration would occur at much lower plasma levels. The pharmacology and therapeutic uses of pirenzepine are further considered in Chapter 9.

The preferential binding characteristics of pirenzepine to neuronal tissue rather than smooth and cardiac muscle led to the proposal for the existence of more than one muscarinic receptor subtype. By pharmacological means, using functional tissue responses or radioligand binding, there are now three firmly identified subtypes referred to as M_1, M_2, and M_3, with evidence accumulating for a fourth M_4 receptor (Table 7.1). Molecular biological approaches have resulted in the cloning of five muscarinic receptor cDNAs, designated m_1 to m_5, which encode the corresponding muscarinic receptors. When expressed in mammalian cell lines, such as Chinese hamster oocytes (CHO), these receptors have the characteristics of the M_1 to M_4 receptors defined pharmacologically. An M_5 receptor has yet to be identified as a discrete pharmacological entity. The amino acid sequences of the m_1 to m_5 receptors have been determined. There are a number of amino acid residues that are conserved in the m_1, m_3 and m_5 receptors but are substituted by a different residue in the m_2 and m_4 receptors. The m_1 and m_3 receptor sequences are more closely related to each other than to the m_5 receptor. The similarities between the m_1, m_3 and m_5 receptors are reflected in their coupling to the second messenger since they are, in common, preferentially linked to phosphoinositide hydrolysis. In contrast, the m_2 and m_4 receptors are preferentially coupled to adenylyl cyclase, their activation resulting in inhibition of this enzyme and a fall in cAMP levels (Table 7.1) (see also Chapter 13). In common with other related G protein-linked receptors, muscarinic receptors are thought to span the cell plasma membrane in seven hydrophobic transmembrane domains arranged in α-helixes. This creates four extracellular loops and four intracellular loops (Goyal 1989, Hulme *et al.* 1990, Caulfield 1993) (see Figures 13.1 and 13.2, Chapter 13).

The classification of muscarinic receptors into the M_1, M_2 and M_3 subtypes is based upon the selectivity of antagonists in blocking the functional responses or displacement of a radiolabelled antagonist ligand from binding sites.

7.5.1 *Muscarinic Receptors Mediating Functional Responses*

The principles of receptor classification have been discussed thoroughly in Chapter 3 with special emphasis on adrenoceptors. In the case of muscarinic receptors, however, agonists have only played a minor role in receptor subclassification since, with the exception of McN-A-343 (see Table 8.3), selective agonists are only recently becoming available. Even so, these still do not display the marked selectivity for muscarinic receptor subtypes that occurs with adrenoceptors. McN-A-343 was shown to selectively stimulate autonomic ganglia and induce a depolarization of the postganglionic neurone. This response, however, was not mediated as expected via nicotinic receptors but via a muscarinic receptor since it was blocked by atropine. This receptor mediates the *slow* excitatory postsynaptic potential (s-EPSP) involved in ganglionic transmission (see Chapter 11) and is defined as the M_1 receptor (Figure 7.7). The M_1 receptor is selectively stimulated by McN-A-343 and preferentially blocked by pirenzepine in ganglia (eg the rat superior cervical ganglion). As a result of these observations, two subtypes of muscarinic receptor were initially proposed: M_1 and M_2 receptors. The M_1 receptor was located to certain brain regions (cerebral cortex and hippocampus) and autonomic ganglia. More recently, the M_1 receptor has also been identified as the muscarinic receptor subtype mediating inhibition of noradrenaline release and of the twitch response to electrical stimulation of the rabbit vas deferens (Grimm *et al.* 1994). Two toxins from the venom of the Eastern green mamba (*Dendroaspis angusticeps*), MTx1 and MTx2, display selective agonist activity at M_1 receptors and inhibit twitch responses of the rabbit vas deferens. A further toxin, m_1-toxin, binds irreversibly and selectively to M_1 receptors and behaves as an allosteric antagonist (Jerusalinsky & Harvey 1994).

According to the original subdivision, the M_2 receptor was located principally in autonomic effector tissues including smooth and cardiac muscle and exocrine glandular tissue (Hammer & Giachetti 1982). Subsequently, the muscarinic antagonist, AF-DX 116, (see Table 9.6) was shown to distinguish between cardiac and glandular muscarinic receptors, its cardioselectivity leading to subdivision of the M_2 receptor into M_2 cardiac (M_{2a}) and M_2 glandular or M_2 muscle ($M_{2\beta}$). The latter subtype is now termed M_3 and this remains the basis of muscarinic receptor classification, with an additional functional M_4 subtype now becoming recognized (Table 7.1).

As a broad generalization, responses mediated via M_1 receptors are selectively antagonized by pirenzepine, with McN-A-343 or AHR-602 being somewhat selective agonists. Responses mediated via M_2 receptors are preferentially antagonized by AF-DX 116, methoctramine, the alkaloid himbacine and the nicotinic neuromuscular blocking agent, gallamine (see Tables 9.6 and 9.7). Responses mediated via M_3 receptors are preferentially antagonized by 4-DAMP, hexahy-drosiladifenidol (HHSiD) and p-fHHSiD (see Table 9.5). The receptor subtype selectivity of muscarinic antagonists is, unlike that with adrenoceptor antagonists, relatively poor and generally in the order of 10-fold between any two subtypes. Thus, the receptor subtype mediating a response is determined from the pA_2 values of these selective antagonists. The pA_2 values for a range of tissues and their responses are shown in Table 7.2. Although the antagonists have been ranked in potency order, this is meaningless for receptor classification since, for example, methoctramine will always appear near the end of the ranking because of its poor affinity for *all* receptor subtypes. However, inspection of actual pA_2 (or pK_B) values will show a higher affinity of methoctramine for tissue responses mediated via M_2 receptors (7.8) than for M_3 receptors (6.2). Similarly, 4-DAMP usually appears high in the ranking order, being a potent antagonist, but it has approximately 10-fold lower pA_2 values for M_2-mediated responses (7.9) than for M_3-mediated responses (9.0). It does not, however, discriminate between M_3 and M_1 receptors (Hulme *et al.* 1990, Caulfield 1993).

The more recent identification of the M_4 muscarinic receptor subtype mediating contraction of the guinea-pig uterus was based upon a different profile of antagonist activity compared with the other three types (Figure 7.8) (Dörje *et al.* 1990). This figure illustrates how muscarinic receptor subtypes in different tissues may be distinguished. The close correlation between pK_i values for displacement of

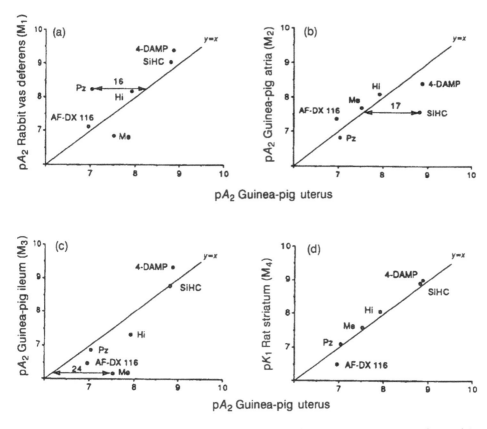

Figure 7.8   Comparisons of the pA₂ values for a range of muscarinic antagonists obtained for the contraction of the guinea-pig uterus (M₄ muscarinic receptors) with the pA₂ values for (a) inhibition of twitch responses of rabbit vas deferens (M₁), (b) inhibition of guinea-pig left atrial tension (M₂), (c) contraction of guinea-pig ileum (M₃) and (d) pK₁ values for binding displacement in rat striatum (M₄). pA₂ values for antagonism of carbachol in guinea-pig uterus are plotted on the abscissa in each case. Note that the points are distributed close to the theoretical equality regression line ($y = x$) in the case of the rat striatum, indicating the good correlation and similarity of pA₂ and pK₁ values in these two tissues. The muscarinic receptor of the uterus may therefore be classified as the M₄ subtype. Marked deviations are shown for pirenzepine (Pz) in (a), for sila-hexocyclium (SiHC) in (b), and for methoctramine (Me) in (c). Reproduced with permission from Dörje et al. (1990).

radioligand binding in rat striatum and pA₂ values for guinea-pig uterus suggest that the receptors are of the same subclass. The weaker correlation with values obtained for responses of the rabbit vas deferens (M₁), guinea-pig ileum (M₃) and atrium (M₂) indicates that these may not be of the M₁, M₂ or M₃ subtypes. Selectivity of antagonists for the M₄ site is now becoming evident. R-(+)-Hyoscyamine, one of the isomers making up the racemate, atropine, may be a selective M₄ antagonist (Ghelardini *et al.* 1993). Secoverine and tropicamide also display modest M₄ selectivity compared with all other subtypes (Table 9.7).

Discrimination between M₂ and M₄ receptor-mediated responses may be achieved with AF-DX 116 which shows M₂ selectivity compared with all other subtypes. Himbacine may separate M₄ from M₁ receptors, having higher affinity for M₄

313

Table 7.2 Muscarinic receptor types mediating functional tissue responses based on antagonist pA_2 values (arranged in rank order)

Tissue	Response	Antagonist pA_2 values (rank order)	Receptor
Blood vessels			
Renal vasculature (rat)	Vasodilatation (+endothelium)	Atr ≥ DAMP > HHSiD > pirenz > p-fHHSiD ≥ himb > AF-DX384 (8·4) (8·3) (7·3) (6·2) (6·0) (5·9) (5·7)	M_3
Mesenteric vessels (rat)	Vasodilatation (+endothelium)	Atr > DAMP > p-fHHSiD > pirenz > AF-DX116 (9·9) (9·4) (8·0) (7·1) (6·1)	M_3
Cerebral artery	Vasodilatation (+endothelium)	Atr ≥ DAMP > HHSiD ≥ dicyc > pirenz > methoc > AF-DX116 (9·4) (9·3) (8·7) (8·5) (7·5) (6·4) (6·0)	M_3
Pulmonary artery	Vasodilatation (+endothelium)	DAMP > pirenz > methoc (9·2) (6·9) (5·5)	M_3
Aorta (rabbit)	Vasodilatation (+endothelium)	Atr > DAMP > pirenz > methoc (9·1) (8·7) (6·4) (5·8)	M_3
Aorta (rabbit)	Contraction (−endothelium)	Atr > DAMP > p-fHHSiD > himb > pirenz > methoc (9·6) (9·4) (7·4) (7·2) (6·6) (5·4)	M_3
Coronary artery (pig)	Contraction (−endothelium)	Atr > DAMP > pirenz > AF-DX116 > methoc (9·5) (9·1) (7·3) (6·2) (5·6)	M_3
Portal vein (rat)	Contraction and ↑ size of rhythmic action	Atr > DAMP > p-fHHSiD > pirenz > AF-DX116 (9·7, 10·0) (9·5, 9·8) (7·5, 7·9) (6·9, 7·2) (6·5, 6·8)	M_3
Saphenous vein (dog)	Contraction and ↑ size of rhythmic action	Atr > pirenz > p-fHHSiD > methoc (9·5) (8·1) (7·0) (6·2)	M_1

Heart			
Left atria (guinea-pig)	Negative inotropy	DAMP > himb > methoc ⩾ SiHC > AF-DX116 > pirenz (8·4) (8·1) (7·7) (7·6) (7·4) (6·8)	M$_2$
Right Atrium Ventricle	Negative chronotropy Anti-adrenergic (no direct action)	Atr > DAMP > AF-DX116 > pirenz	M$_2$
Gastrointestinal tract			
Ileum (guinea-pig)	Contraction (longitudinal muscle)	Atr > DAMP > HHSiD > p-fHHSiD > himb ⩾ AQ-RA 741 (9·0) (8·8) (7·8) (7·5) (7·1) (7·0)	M$_3$
Jejunum (rabbit)	Contraction	Atr > DAMP > pirenz ⩾ methoc (9·1) (8·0) (6·5) (6·4)	M$_3$?
Colon (human)	Contraction (circular)	DAMP > Atr > himb ⩾ AF-DX116 > pirenz = HHSiD > p-fHHSiD (9·4) (8·7) (7·5) (7·4) (7·2) (6·9)	M$_3$
Gall bladder (guinea-pig)	Contraction	DAMP > pirenz ⩾ methoc ⩾ p-fHHSiD > AF-DX116 (8·3) (7·9) (7·7) (7·6) (6·7)	Not M$_1$, M$_2$ or M$_3$
Glandular tissue			
Gastric acid (oxyntic cell)	Secretion of acid	NMA > Atr > pirenz (9·7) (7·8) (6·7)	M$_3$
Pancreas (β cells)	Insulin release	Atr > SiHC > pirenz ⩾ methoc	M$_3$
parotid salivary	Amylase release	HHSiD > pirenz > AF-DX116 (8·1) (7·1) (6·2)	M$_3$

(continued)

Table 7.2 *Continued*

Tissue	Response	Antagonist pA_2 values (rank order)	Receptor
Genito-urinary tract			
Urinary bladder (rat)	Contraction	Atr > DAMP > pirenz > AF-DX116 > methoc (9·4) (9·2) (6·8) (6·2) (5·8)	M_3
Seminal vesicle (guinea-pig)	Contraction	Atr > DAMP > p-fHHSiD > himb = pirenz > methoc (10·3)(9·6) (8·1) (7·2) (7·2) (6·0)	M_3
Uterus (guinea-pig)	Contraction	DAMP ≥ SiHC > himb > methoc > pirenz = AF-DX116 (8·9) (8·8) (7·9) (7·5) (7·0) (7·0)	M_4
Tracheal muscle (mouse)	Contraction	DAMP ≥ Atr > pirenz > AF-DX116 (8·7) (8·6) (6·5) (6·3)	M_3
Guinea-pig	Contraction	DAMP > pirenz > methoc (epithelium-independent) (9·0) (7·6) (5·6)	M_3
Bovine	Contraction	Atr ≥ DAMP > pirenz > AF-DX116 (9·0) (8·9) (6·4)	M_3
Eye			
Iris circular muscle	Contraction (rabbit)	Atr > HHSiD > pirenz > AF-DX116 ≥ himb > methoc (9·1) (7·7) (7·2) (6·5) (6·4) (5·9)	M_3
radial muscle	No response		M_2
Ciliary muscle	Contraction		M_3

Autonomic ganglia			
Rat superior cervical	Depolarization (by muscarine)	NMA > pirenz > himb > methoc (9·4) (8·4) (7·1) (6·5)	M_1
	Hyperpolarization	methoc (7·4)	M_2
Myenteric plexus	Ach release	pirenz (8·6)	M_1
	Depolarization	pirenz (8·4)	M_1
Prejunctional at autonomic nerve ending			
Ear artery (rabbit)	Reduced noradrenaline output on nerve stimulation	Atr > Secov > DAMP > pirenz (9·0) (8·1) (7·7) (6·2)	M_2
Iris (rabbit /rat)	Reduced noradrenaline output on nerve stimulation	Atr > himb > DAMP > methoc > AF-DX116 > pirenz (9·0) (8·2) (8·0) (7·6) (7·4) (7·2)	M_2
Trachea	↓Ach output ↓Twitch height	methoc > AF-DX116 > gallamine	M_2
Vas deferens (rabbit)	↓Twitch height	Atr ⩾ DAMP > SiHC > pirenz = himb > AF-DX116 > methoc (9·5) (9·4) (9·0) (8·2) (8·2) (7·1) (6·8)	M_1
	↓NA output	Atr > himb > pirenz > methoc (9·5) (8·3) (7·7) (6·9)	M_1

Notes: Antagonists and abbreviations: AF-DX116; AF-DX384; AQ-RA741; Atr, atropine; DAMP, 4-DAMP (4-diphenyl-acetoxy-N-methyl-piperidine); dicyc, dicyclomine; HHSiD, hexahydro-sila-difenidol; himb, himbacine; methoc, methoctramine; NMA, N-methylatropine; *p*-fHHSiD, *p*-fluoro-hexahydro-sila-difenidol; pirenz, pirenzepine; secov, secoverine; SiHC, sila-hexocyclium. Structures are shown in Tables 9.1 and 9.4 –9.7.

(Lazareno *et al.* 1990). Recent studies suggest that himbacine can also distinguish different M_1 subtypes in the rabbit vas deferens ($pA_2 = 8.17$) and rat superior cervical ganglion ($pA_2 = 7.14$) (Sagrada *et al.* 1993). The K_i value for displacement of binding in the hippocampus or cerebral cortex was 7.0. Muscarinic receptors in the hippocampus have been designated $M_{1\alpha}$ while those of the ganglia may be $M_{1\beta}$.

7.5.2 Mixed Receptor Populations

Unlike the situation with adrenoceptors, there is no evidence from functional responses that the pA_2 values of selective antagonists vary when different agonists are used (Table 7.3). Schild plots for different agonists are usually superimposable (for comparison, see Figure 3.13 and Table 3.3). Separated Schild plots for β-adrenoceptor antagonists indicated heterogeneous populations of functional β_1- and β_2-adrenoceptors. Whether mixed populations of muscarinic receptors mediate functional responses has so far proved difficult to establish. This is because of the limited selectivity of the agonists and antagonists available at present. Orders of selectivity in excess of 10-fold for both agonist and antagonist would be necessary. The similar pA_2 values with a range of agonists may therefore arise because the responses are mediated via homogeneous receptor populations or because of the poor selectivity of the agents employed.

7.5.3 Radioligand Binding Experiments

Identification of the muscarinic receptor subtypes in different tissues by radioligand binding has confused rather than clarified the reasonably straightforward generalizations outlined above. The dominant receptor subtype present in a tissue has been revealed by the affinities (K_i values) of the selective antagonists in displacing a nonselective radiolabelled antagonist, such as [^{3}H]-quinuclidinyl benzilate ([^{3}H]QNB) or [^{3}H]N-methylscopolamine ([^{3}H]NMS) (Table 7.4). For example, AF-DX 116 shows approximately ten-fold higher affinity (7.3) for displacing [^{3}H]NMS in

Table 7.3 pA_2 or pK_B values for antagonism of contractile responses of guinea-pig trachea to a range of muscarinic agonists by pirenzepine (M_1 selective), methoctramine (M_2 selective) and 4-DAMP (M_3 selective)

	Antagonists		
Agonist	Pirenzepine pA_2	Methoctramine pK_B	4-DAMP pA_2
Acetylcholine	7.5 ± 0.08	5.2 ± 0.12	9.0 ± 0.05
Methacholine	7.5 ± 0.06	5.0 ± 0.14	9.0 ± 0.08
Carbachol	7.6 ± 0.10	5.6 ± 0.04	9.1 ± 0.06
Bethanechol	7.6 ± 0.09	5.6 ± 0.04	8.8 ± 0.07
Oxotremorine	7.8 ± 0.04	5.5 ± 0.06	9.0 ± 0.03

Notes: Epithelium was absent, but its presence did not affect pA_2 values. From Morrison & Vanhoutte (1992).

Table 7.4 Muscarinic receptor subtypes present in various tissues determined by radioligand binding

Tissue		pK_i Values (rank order)	Receptor subtype
Brain			
Cerebral cortex (rat)		Atr ≥ DAMP > pirenz > methoc > AF-DX116 (9·2) (9·1) (8·0) (7·6) (6·1) DAMP = dicyc ≥ secov > pirenz > methoc > AF-DX116 (8·5) (8·5) (8·4) (7·7) (6·9) (6·6)	M_1
Striatum		Atr > DAMP > himb > methoc > pirenz > AF-DX116 (9·6) (9·1) (8·0) (7·6) (7·1) (6·3)	M_4
Smooth muscle			
Trachea (dog)	High affinity	Atr ≥ DAMP > methoc = HHSiD > pirenz (8·8) (8·7) (7·6) (7·6) (6·2)	(55%) M_2
	Low affinity	DAMP > HHSiD > methoc (7·3) (6·7) (5·3)	(45%) M_3
(bovine)		Atr > DAMP > pirenz (8·7) (7·8) (6·2)	M_3
Ileum (rat)	Major site	methoc > AF-DX116 > HHSiD ≥ p-fHHSiD > pirenz (7·2) (7·0) (6·1) (6·0) (5·8)	(84%) M_2
	Minor site	HHSiD > p-fHHSiD > pirenz > AF-DX116 > methoc (8·3) (7·5) (7·1) (5·6) (5·2)	(16%) M_3
Lung (rabbit)		Atr > secov > dicyc = DAMP > himb > HHSiD > methoc > pirenz (9·6) (9·3) (8·8) (8·8) (8·2) (8·0) (7·8) (7·5)	M_4

Cardiac muscle			
Heart (rat)	Atr > DAMP = himb ⩾ methoc ⩾ AF-DX116 > pirenz		M_2
	(9·0) (8·1) (8·1) (7·9) (7·3) (6·6)		
	secov = DAMP > methoc > himb > dicyc > AF-DX116 > pirenz		
	(8·0) (8·0) (7·8) (7·5) (7·2) (7·0) (6·4)		
Glandular tissue			
Submaxillary salivary	DAMP > dicyc > secov > pirenz ⩾ himb > methoc > AF-DX116		M_3
gland (rat)	(8·7) (8·0) (7·7) (6·8) (6·7) (6·3) (6·0)		
Pancreas (rat)	Atr > DAMP > pirenz = himb > methoc ⩾ AF-DX116		M_3
	(9·4) (9·0) (6·6) (6·5) (6·0) (5·8)		
Lacrymal			
seminal vesicles	mRNA expressing m_3 by Northern blot analysis		M_3
Cell Lines			
Human neuroblastoma	Atr > DAMP > pirenz > methoc ⩾ himb > AF-DX116		M_1
(NB-OK 1 cells)	(9·6) (9·1) (8·2) (7·3) (7·1) (6·3)		
Mouse-rat neuroblastoma-glioma	secov > himb > methoc > pirenz		M_4
(NG 108-15 cells)	(9·3) (8·5) (8·1) (7·2)		

Notes: Values in parenthesis are K_i values determined from displacement of the non-selective [^{3}H]antagonists, [^{3}H]NMS and [^{3}H]QNB, or the M_1 receptor-selective [^{3}H]pirenzepine. Abbreviations as for Table 7.2.

cardiac tissue homogenates (M$_2$) than in glandular tissue such as the pancreas (M$_3$, 5.8), the striatum (M$_4$, 6.3) or NB-OK1 cells (M$_1$, 6.3) (Waelbroeck *et al.* 1992). Generally, only monophasic displacement curves have been observed, suggesting homogeneous receptor populations (see Chapter 3 for details of methodology of radioligand binding). For example, AF-DX 116 and 4-DAMP show monophasic displacement in smooth muscle homogenates and isolated smooth muscle cells from canine colon, consistent with an M$_2$ population (Table 7.4). This contrasts with the functional data which reveals M$_3$-mediated responses. Biphasic displacement of [^{3}H]QNB binding by carbachol and pirenzepine in these tissues was, however, observed with 80% of binding sites being M$_2$ and 20% being M$_3$ receptors. The monophasic displacement with AF-DX 116 and 4-DAMP was attributed to their selectivities between M$_2$ and M$_3$ receptors being insufficient to show up both receptor populations (Zhang *et al.* 1991). Similar two-site displacement has been obtained in homogenized longitudinal muscle of rat ileum by AF-DX 116 and methoctramine (72 and 91%, respectively, M$_2$) and by HHSiD and *p*-fHHSiD (19 and 14%, respectively, M$_3$). The M$_2$ receptor population was associated with inhibition of cAMP and the minor M$_3$ population associated with phosphoinositide hydrolysis and the contractile response (Candell *et al.* 1990). Both M$_2$ and M$_3$ binding sites have also been identified in dog trachea (Yang 1991). Because the assays in trachea and colon were performed on isolated smooth muscle cells, it is concluded that the M$_2$ and M$_3$ receptors are both located on these cells and do not arise from different components of the tissue.

The displacement of radioligand binding by agonists generally displays even poorer selectivity between receptor subtypes than with antagonists. McN-A-343 is selective for M$_1$ receptors in functional tests, but in binding studies only minor differences in its affinity for the receptor types have been demonstrated. At best, six-fold selectivity of agonist binding has been observed. As with β-adrenoceptors (see Chapter 3), displacement of binding by agonists is shifted to a low affinity state by the presence of GTP or its stable analogue, GppNHp. This indicates that the high affinity state of the receptor is the agonist–receptor–G protein complex.

Radiolabelled agonists, such as [^{3}H]methyl-oxotremorine (Oxo-M) or [^{3}H]*cis*-methyl-dioxolane (CMD) (see Table 8.3), have been used to predict the efficacy of muscarinic agonists. These label the high affinity state of the muscarinic receptor. Radiolabelled antagonists such as [^{3}H]NMS or [^{3}H]QNB preferentially label the low affinity state of the receptor. Agonists such as carbachol recognize preferentially the high affinity state of the receptor and therefore display high affinity in displacing Oxo-M, whereas antagonist drugs show similar displacing affinity for Oxo-M and NMS. The NMS/Oxo-M or QNB/CMD ratio of K_i values of a compound has been shown to correlate closely with its ability to induce a functional response. This ratio has therefore been used to predict the efficacy of a compound at muscarinic receptors. The higher the NMS/Oxo-M ratio, the more effective the compound will be as an agonist at the muscarinic receptor under study. At the M$_1$ receptor of the rat cerebral cortex, carbachol has an NMS/Oxo-M ratio of 4600, whereas the value for atropine is 2.1 (Hargreaves *et al.* 1992). High efficacy agonists display poor selectivity between receptor subtypes. Therefore newer approaches to achieving selectivity of agonists are aimed at producing partial agonists with intermediate NMS/Oxo-M ratios. The selectivity of newer compounds such as L-689,660 (see Chapter 8, Table 8.1) for M$_1$ receptor-mediated responses is then achieved by utilizing different receptor reserves for the different receptor subtypes. Because of a higher receptor reserve for M_1-mediated responses, compounds such as L-689,660 (NMS/Oxo-M ratio = 28) will exhibit full agonism, whereas at M_2 (atria) and M_3 (ileum) receptors, where there are low reserves, it is respectively a partial agonist and antagonist (Freedman *et al.* 1993).

The rate of dissociation of a radioligand from the muscarinic receptor binding sites may be used to differentiate between receptor subtypes. The dissociation of [^{3}H]NMS from the M_4 binding sites in the rat striatum appears to be slower than for the other subtypes (Waelbroeck *et al.* 1992). Another feature that has emerged from examination of dissociation rates of the radioligand from muscarinic receptors is the presence of allosteric binding sites. These sites are regarded as lying outside the principal site of binding of agonists with the muscarinic receptor. It was first identified by the use of the neuromuscular nicotinic antagonist, gallamine (see Table 9.1), which also displays weak muscarinic blocking activity, with some M_2 selectivity. Gallamine and certain other antagonists at high concentrations, including atropine, slow the rate of dissociation of [^{3}H]NMS from muscarinic binding sites, an effect attributed to interaction with an allosteric site. The allosteric binding site probably has an extracellular location. Binding to it by gallamine and other neuromuscular antagonists including pancuronium and alcuronium creates a steric obstacle for the passage of ligands to the normal ligand binding site (orthosteric site) in the cleft formed by the seven transmembrane domains (see Chapter 13). This effect is seen primarily at M_2 receptors where gallamine causes blockade of Ach-induced inhibitory responses of isolated atria characterized by progressively less blockade as the concentration of gallamine increases. Schild plots for the antagonism (see Chapter 3) have low slopes and may be curved. Amino acid residues forming the allosteric site have not yet been identified, but the sequence Glu-172–Asp-173–Gly-174–Glu-175 in the second extracellular loop of the receptor may be important and is unique to the M_2 receptor. Methoctramine may bind to the M_2 receptor at both the allosteric and orthosteric sites. The possibility of an endogenous ligand that binds to the allosteric site must be considered, an example could be eosinophil major basic protein (Tuček & Proška 1995).

Further support for the existence of multiple muscarinic receptor subtypes comes from the availability of radiolabelled cDNA probes for each of the receptor genes. The tissue mRNA is identified by Northern blot analysis. This usually involves extraction of the mRNA from the tissue and separation on an agarose/formaldehyde gel by electrophoresis. This is electroblotted onto filters to which the mRNA is covalently bound and immobilized by baking. *In situ* hybridization between mRNA and cDNA is then effected by overnight exposure to ^{32}P-labelled cDNA oligonucleotide probes for the m_1 to m_5 receptors. After washing, retention of the radiolabel is visualized by apposition to X-ray film (Figure 7.9).

Identification of receptor protein has also been made by producing subtype-specific antibodies to the third intracellular loop (i_3) amino acid sequences, which deviate considerably between the five cloned receptor types. The m_1 to m_5 receptors are then identified by immunoprecipitation. In common with the binding data, the mRNAs responsible for the synthesis of the m_2 and m_3 receptors have been identified in the colon (Figure 7.9) (Zhang *et al.* 1991) and the ileum. In the heart, as with functional studies and radioligand binding, there is only m_2 mRNA, and in the rabbit lung m_4 mRNA and M_4 binding sites have been identified (Lazereno *et al.* 1990). The occurrence of receptors and mRNA may not always coincide in a tissue. mRNA is located in the cell body where receptor synthesis takes place in the ribosomes; in

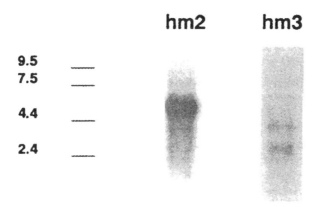

Figure 7.9 Northern blot analysis of colonic smooth muscle RNA showing the presence of both m_2 and m_3 receptor RNA, the former in greater density. The Northern blot containing poly (A)$^+$ RNA from canine colonic circular smooth muscle was hydridized to the muscarinic cDNA probes hm2 and hm3, human m_2 and m_3 receptor cDNA fragments. Numbers on the left represent molecular weight markers (in kilobases) obtained by running an RNA ladder alongside the tissue-derived RNAs. Reproduced with permission from Zhang et al. (1991).

nerves, the receptors are transported down the axon to the nerve terminal (Levey 1993).

Although both M_2 and M_3 receptors have been identified in a range of smooth muscle tissues in the proportion of ~4 : 1, from pA_2 values of selective antagonists the receptor subtype involved in the contractile response appears to be of the M_3 type (Table 7.2). The functional roles of each receptor in the ileum have been further assessed by receptor protection experiments using the irreversible antagonist, phenoxybenzamine. In addition to alkylating α-adrenoceptors (Chapter 5), phenoxybenzamine also alkylates muscarinic receptors. Protection from alkylation by incubation with the selective competitive M_2 antagonist, methoctramine, failed to prevent the block of the contractile response by phenoxybenzamine. Incubation with the M_3 selective antagonist, *p*-fHHSiD, (see Table 9.5), however, protected against alkylation and block of the contractile response to a muscarinic agonist (see Figure 9.4). This further shows that only M_2 receptors are involved in the contractile response (Eglen & Harris 1993). If the contractile response of the ileum and other smooth muscle tissues is mediated solely by M_3 receptors, what is the function of the M_2 receptor population which outnumbers the M_3 receptors by 4 : 1? The M_2 receptors are linked via guanine nucleotide binding proteins (G_i or G_k) to inhibition of adenylyl cyclase (Chapter 13). Their stimulation may therefore oppose smooth muscle relaxation induced by β-adrenoceptor stimulation and elevation of intracellular cAMP levels (Fernandes *et al.* 1992). Whether functional antagonism between β-adrenoceptor agonists and muscarinic agonists involves M_2 receptors, however, remains a matter of debate (Roffel *et al.* 1995).

7.6 Pathological and Other Changes in Muscarinic Receptors

The amount of information on changes in muscarinic receptors in different pathological states is considerably less than has been described for adrenoceptors. A few changes in muscarinic receptor and cholinergic nerve function have been observed which may have important implications in the therapeutic uses of agonists and antagonists. In the lungs, defective parasympathetic nerve pathways may contribute to the airway narrowing in asthma and chronic obstructive airways disease (COPD). Reflex cholinergic bronchoconstriction may be triggered in asthma through the activation of sensitized sensory nerves by inflammatory mediators such as bradykinin (see Chapter 12, Figure 12.4). Asthma is associated with a non-specific elevation of sensitivity of the airways to the bronchoconstrictor action of Ach released from the parasympathetic vagal nerve endings. Additionally, there is evidence for an increased release of Ach from the parasympathetic nerves. Recent evidence suggests that this may be due to defective prejunctional M_2 receptors. These normally exert negative feedback control on Ach release (see earlier, Figure 7.7). Any dysfunction induced by viral infection, ozone or antigen exposure will result in enhanced Ach release and exacerbation of airway bronchoconstriction. The loss of prejunctional M_2 receptor function appears to be due to a decrease in affinity for the agonists. Positively charged proteins (eg major basic protein) released by inflammatory challenge may act as selective allosteric antagonists at the M_2 receptors (Barnes 1993a, Fryer & Jacoby 1993).

In dogs with heart failure induced by pressure overload from aortic constriction, dysfunction of the cardiac muscarinic receptors has been observed (Vatner *et al.* 1988). In the human heart, however, the muscarinic M_2 receptor-mediated responses and their coupling to adenylyl cyclase do not appear to be modified in patients with heart failure of the idiopathic dilated cardiomyopathy type. This contrasts with the population of β-adrenoceptors which are reduced in the hearts of these patients, together with an increase in guanine nucleotide (G_i) binding protein. G protein links the muscarinic M_2 receptor to the functional response and one might therefore expect the muscarinic responses to be enhanced; however, responses to the G_i protein-linked adenosine A_1 or muscarinic M_2 receptors are not affected (Brodde 1991).

The effect of hypertension upon cardiac muscarinic receptors has not been extensively studied. Inconsistent changes in the density of muscarinic receptor binding sites have been observed in the hearts of spontaneously hypertensive rats with either no change (Robberecht *et al.* 1981) or a decrease (Grammas *et al.* 1989) having been observed. The vascular response to Ach is altered in spontaneously hypertensive rats; the aorta reveals an M_3 receptor-mediated contractile component that is dependent upon the endothelium and attributed to the release of cyclo-oxygenase products (Boulanger *et al.* 1994).

In Alzheimer's disease there is a correlation between the loss of cholinergic neurones projecting into the cerebral cortex and hippocampus and the degree of cognitive memory impairment. This is associated with a loss of M_2 receptors in cortical areas and the hippocampus of brains from Alzheimer's patients. The loss of M_2 receptors is correlated with the reduction of cholinergic neurones measured from ChAT activity. These M_2 receptors are therefore probably located presynaptically on the cholinergic nerve terminals, their stimulation normally causing inhibition of transmitter (Ach) release. The cortical M_1 receptors are usually preserved and may show minor increases; these receptors are thought to lie postsynaptically (Quirion *et al.* 1989).

The muscarinic response of several tissues is reduced by aging. In the lung the contractile response to muscarinic receptor stimulation is decreased during senescence, but this effect is not associated with a decrease in binding affinity or density of receptors. The effect appears to be due to an altered coupling of the muscarinic receptor to the guanine nucleotide (G) binding protein (Wills & Douglas 1988, Wills-Karp & Gavett 1993).

The endocrine status also appears to influence muscarinic receptors. For example, in diabetic rats an initial reduction in sensitivity of the heart to muscarinic receptor stimulation (>100 days) followed by a supersensitivity has been observed. These changes in sensitivity were not correlated with receptor density, for example, during the supersensitive phase muscarinic receptor density *decreased* (McNeill 1985). It is possible that muscarinic receptors are also influenced by the oestrous cycle, since oestrogen treatment reduces the density of muscarinic binding sites in the uterus (Matucci *et al.* 1993).

Parasympathomimetic Amines

Parasympathomimetic amines may be defined as agonists that mimic the effects of stimulating the parasympathetic nerve innervating an organ or tissue. This was the original term used by Dale (1914) to describe the effects of Ach. Since the tissue responses to parasympathetic nerve activity are mediated via the release of Ach onto muscarinic receptors, the actions of parasympathomimetic amines are essentially their muscarinic properties. Ach and its analogues do, however, also stimulate nicotinic receptors at the somatic nerve–skeletal muscle junction, at autonomic ganglia and the adrenal medulla and in the brain. These effects are usually overwhelmed and masked by the muscarinic actions of parasympathomimetic amines.

The pharmacological actions of parasympathomimetic amines may therefore, in general, be predicted from the knowledge of the body's responses to parasympathetic nerve stimulation (Table 1.1, Chapter 1). Some tissues, however, such as the nictitating membrane of the cat and blood vessels of the skin, do not receive a parasympathetic innervation (Table 1.1), yet muscarinic receptors may be present. In these instances the tissues may produce responses – muscarinic responses. Another exception is the autonomic response of the sweat glands, which is mediated via Ach but is anatomically a sympathetic nerve (see Chapter 1).

The classification of muscarinic receptors mediating functional tissue responses into the M_1 to M_3 subtypes, with a possible M_4 subtype now recognized, was described in Chapter 7. The tissue responses to muscarinic receptor stimulation and the muscarinic subtypes involved are shown in Table 7.2. While the previous chapter was receptor-orientated, the emphasis of this chapter is on the parasympathomimetic drugs and their actions and uses.

8.1 Pharmacological Properties of Muscarinic Agonists

The muscarinic agonists currently available display relatively poor selectivity towards the different receptor subtypes. With the exception of the M_1-selective agonist, McN-A-343 (see Table 8.3), only recently have compounds with selectivity for the different muscarinic receptor subtypes been developed. Thus, in general the different agonists exhibit common pharmacological profiles. Selectivity between organs appears to depend more upon distribution to the tissue (ie their pharmacokinetics) than upon any receptor selectivity. For example, choline ester analogues of Ach that are resistant to hydrolysis by either acetylcholinesterase or butyrylcholinesterase (carbachol, bethanechol) are stable enough to be distributed to areas of low blood flow where they can exert a pharmacological response.

8.1.1 *Cardiovascular System*

8.1.1.1 *The heart*

The parasympathetic innervation of the heart is concentrated primarily in the atria, the nodal pacemaker tissue and atrioventricular conducting tissue (Chapter 1). There is little innervation of the ventricles, although Ach, ChAT and acetylcholinesterase are present. The principle response to muscarinic stimulation is therefore slowing of heart rate (negative chronotropy) (Figure 8.1 and 11.1). The force of ventricular contraction is only reduced when it is elevated by sympathetic tone from nerve activity or exogenously added β-adrenoceptor agonists. The force of atrial contractions is reduced without any prestimulation.

The direct negative chronotropic and inotropic (in atria) actions of muscarinic stimulation by Ach involves at least two mechanisms. Ach causes an increase in potassium conductance in pacemaker and atrial muscle cells, resulting in hyperpolarization and a reduction in action potential duration. This decreases the time available for Ca^2 influx and thus reduces force of contraction. The increase in

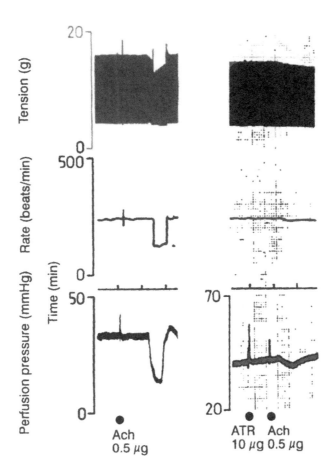

Figure 8.1 Effects of Ach (0.5 μg) on the force of contraction (tension, upper record), rate of contraction (middle record) and coronary perfusion pressure (lower record) of a spontaneously beating guinea-pig isolated perfused heart (Langendorff heart). The cardiac slowing (negative chronotropy) and coronary vasodilatation are abolished by atropine (ATR, 10 μg), which exerts a small vasodilator effect just after the injection artefact of the second dose of Ach. Note that Ach does not reduce force of ventricular contraction directly, but the small depression of force observed here is probably an indirect result of the fall in perfusion pressure.

outward movement of K$^+$ has been measured from increases in ^{42}K or ^{86}Rb efflux from atrial tissue and by means of patch-clamping studies. The latter have indicated that the K$^+$ channels activated by Ach (i$_{KAch}$) are not identical to the inward-rectifying K$^+$ channels of nodal tissue (i$_K$) that operate to shorten the action potential of the spontaneous firing of the pacemaker (see Figure 3.11, Chapter 3). The activation of K$^+$ channels by Ach is attenuated by pertussis toxin, an inactivator of guanine nucleotide binding proteins. Furthermore, GTP must be present on the inside of the cell membrane for Ach to be effective and it is independent of the levels of cAMP. The activation of K$^+$ channels is therefore mediated via a pertussis toxin-sensitive G protein without involvement of cyclic nucleotide second messengers. The muscarinic receptor (M$_2$) is linked through the G protein directly to the K^+ channel without intermediate second messengers. The G protein (G_K) controlling K^+ channel activation (i_{KAch}) cleaves off an α-subunit of the G protein, which is involved in channel opening (see Chapter 13, Table 13.1). Low (picomolar) concentrations of purified α-subunits applied to atrial patches cause K^+ channel opening. High concentrations (nanomolar) of the β_y-subunits also cause opening, although inhibition of K^+ channels in certain tissues has also now been observed (Buckley & Caulfield 1992, DiFrancesco 1993). The role of K^+ channel opening in the functional negative inotropic and chronotropic responses, however, appears to be minimal. K^+ channel blockade by 4-aminopyridine, while abolishing ^{86}Rb efflux, has only minor inhibitory effects upon the negative inotropic action of muscarinic agonists attributable to K^+ channel blockade. Most of its blocking activity is in fact due to muscarinic receptor blockade (Urquhart *et al.* 1993).

The second mechanism for the direct negative chronotropy and atrial inotropy by muscarinic agonists is through inhibition of Ca^{2+} influx via L-type Ca^{2+} channels. This is usually believed to occur only where Ca^{2+} influx is elevated, for example, by catecholamines. However, there is evidence that Ach can also reduce basal activity. The intracellular second messenger of this action has not been identified, but an elevation of the levels of cGMP has been proposed. This may activate phosphodiesterase which then reduces cAMP levels. Interference with elevated cAMP levels certainly appears to be the main mechanism for the *indirect* action of muscarinic receptor stimulation in all areas of the heart. Raised rate and force of contraction by β-adrenoceptor stimulation is effectively reduced by Ach. It is only cAMP-dependent positive inotropes that are inhibited by Ach, indicating the involvement of cAMP. The increase in force of ventricular contraction by the Ca^{2+} channel opener, BayK 8644, is not affected by muscarinic agonists (Urquhart & Broadley 1992). Although the increased force of cardiac contraction by forskolin (a direct activator of adenylyl cyclase) and phosphodiesterase inhibitors are inhibited by Ach, they are not always associated with changes in cAMP levels (Buckley & Caulfield 1992).

Changes in heart rate induced by Ach are also brought about by interaction with the pacemaker current, i_f. This is the inward current carried by Na^+ and K^+ ions, which is activated by diastolic hyperpolarization and explains the spontaneous firing of pacemaker tissue (see Chapter 3). Ach inhibits the i_f pacemaker current by a mechanism opposite to that of β-adrenoceptor agonists, probably involving reduced cAMP production. There is a shift to more negative voltages. This process operates at lower concentrations of Ach than does activation of the outward K^+ channel. The inhibition of i_f probably involves linkage of the M_2 muscarinic receptor through a G protein to adenylyl cyclase. Direct linkage, however, to the ion channels via G_o regulatory protein is also a possibility; Ach is ineffective after pertussis toxin treatment (Buckley & Cauldfield 1992, DiFrancesco 1993).

Finally, muscarinic receptors have also been shown to mediate *positive* inotropic responses. In particular, when the negative effects of muscarinic agonists have been blocked by pertussis toxin, carbachol induces increases in atrial force of contraction. This response is associated with depolarization of cell membranes, a rise in intracellular Ca^{2+}, and may be due to increases in phosphoinositide turnover. The response is mediated via M_2 receptors identical to those for the negative inotropy, displaying the same pA_2 values for the selective antagonists, 4-DAMP and AF-DX 116. Thus, the two responses appear to be due to differences in G protein coupling (Kenakin & Boselli 1990). The physiological role of this response is not clear, however it is interesting that it has yet to be demonstrated with vagal stimulation.

Thus, muscarinic receptor stimulation of the heart has been shown to exert a wide range of effects upon several ion channels, intracellular second messengers and different actions in the various regions of the heart. Which of these are involved in the functional negative inotropic and chronotropic responses remains to be fully elucidated. It is also still unclear whether the physiological responses to vagal nerve stimulation involve the same processes as do those with exogenously added Ach.

8.1.1.2   The vasculature

Many blood vessels do not receive a parasympathetic innervation, yet they respond to Ach with both vasoconstriction and vasodilatation, indicating the presence of muscarinic receptors. Other vessels, such as the arteries of the facial skin, hind limb and coronary circulation, do respond to stimulation of the nerve trunks. The vasodilatation to nerve stimulation and to Ach is blocked by the muscarinic antagonist, atropine (Figure 8.1) (Caulfield 1993).

Administration of Ach and other muscarinic agonists intravenously produces generalized vasodilatation and a fall in arterial blood pressure. This is associated with the fall in heart rate (bradycardia) and cardiac output at higher doses, but at low doses may be accompanied by a reflex tachycardia. After atropine, larger doses of Ach or carbachol cause a rise in blood pressure and heart rate due to stimulation of nicotinic receptors in autonomic ganglia, inducing sympathetic vasoconstriction and release of adrenaline from the adrenal medulla (see Figure 11.1).

In isolated vascular preparations, vasoconstrictor responses were usually obtained in experiments performed up to the late 1970s. This was always difficult to explain in view of the *in vivo* pharmacology. However, Furchgott & Zawadzki (1980) showed that if ring preparations were prepared in which care was taken to preserve the vascular endothelium, then Ach would produce the expected relaxation of precontracted preparations (Figure 8.2). Removal of the endothelium mechanically, by passing bubbles of air through a vascular bed or by an amphoteric detergent such as CHAPS (3-[(3-cholamidopropropyl)-dimethylammonio]-1-propane-sulphonate), eliminates the vasodilatation and then Ach causes constriction. It is now known that Ach stimulates muscarinic receptors of the M_3 subtype located on the endothelium (Table 7.2), which releases an endothelium-derived relaxing factor (EDRF), nitric oxide (NO), which in turn relaxes the adjacent vascular smooth muscle. Nitric oxide is probably released from the terminal guinidinium nitrogen of L-arginine by the action of nitric oxide synthase (NOS) (see also Chapter 3, vascular a-adrenocep-tors, and Chapter 12). The inhibitor of NO production, N^G-monomethyl-L-arginine (L-NMMA), abolished the release of NO by Ach from perfused rabbit coronary vasculature but only inhibited the vasorelaxation by 55–70%. This suggests the additional existence of an NO-independent mechanism in the vasodilatation by Ach (Rees *et al.* 1989). The mechanism for the smooth muscle relaxation by NO is probably through increased levels of cyclic GMP through activation of guanylyl cyclase. Like other NO-releasing vasodilators, such as sodium nitroprusside, the response is prevented by oxyhaemoglobin which binds to NO, and by methylene blue which inhibits guanylyl cyclase (Moncada *et al.* 1991). Superoxide anions (O_2^-) accelerate the inactivation of NO and therefore limit the effect. In contrast, the enzymatic removal of superoxide by superoxide dismutase (SOD) therefore prolongs the activity of NO and its vasodilator responses. Whether the response to Ach is prolonged by SOD is a matter of debate (Miller & Vanhoutte 1989).

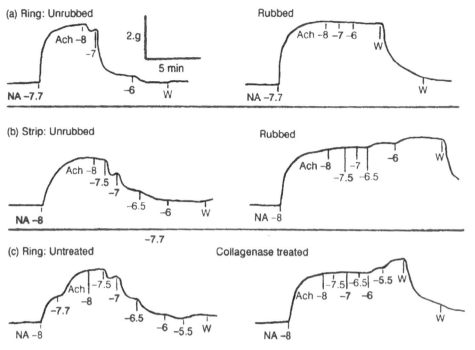

Figure 8.2 Loss of relaxation response to Ach in rabbit aortic ring or strip preparations after rubbing the intimal surface (a, b) or exposure to collagenase (c) to remove endothelium. In intact preparations (left panels) Ach causes concentration-related relaxations of noradrenaline-contracted tissues (NA) attributed to release of endothelium-derived relaxant factor (nitric oxide). After endothelium removal, Ach causes no effect or a small additional contraction. Reproduced with permission from Furchgott & Zawadzki (1980).

The endothelium-dependent vasodilatation is accompanied by hyperpolarization of the smooth muscle, probably because of an increase in K^+ conductance. The Ach-induced hyperpolarization is lost when the endothelium is removed. Relaxation, however, can occur without hyperpolarization. Furthermore, in some arteries the hyperpolarization is not inhibited by oxyhaemoglobin, methylene blue or NOS inhibitors. It has therefore been suggested that another substance, termed endothelium-derived hyperpolarizing factor (EDHF), is released from endothelium in addition to NO. While most blood vessels do not have a parasympathetic innervation, vasodilatation mediated via NO has been demonstrated after stimulation of non-adrenergic non-cholinergic (NANC) nerves (Sneddon & Graham 1992) (Chapter 12).

The muscarinic receptor subtype mediating endothelium-dependent vasodilatation has been characterized as the M_3 subtype (Table 7.2). The response is not inhibited by pertussis toxin which ADP-ribosylates the G protein coupling receptors to adenylyl cyclase. Thus, responses mediated via M_3 receptors are not prevented by pertussis toxin, while M_2 receptor function is (Eglen & Whiting 1990).

A contractile response occurs at high concentrations of Ach and when endothelium is removed. In contrast to the relaxation response, this is inhibited by pertussis toxin. The receptor type has been poorly characterized, but appears to vary between blood vessels of different types, regions and species. In canine saphenous and femoral veins they are of the M_1 subtype, whereas the contractile responses of porcine coronary artery, portal vein and rabbit aorta appear to be mediated via M_3 receptors (Table 7.2) (Eglen & Whiting 1990).

Finally, vasodilatation by Ach may be mediated indirectly through prejunctional muscarinic receptors on sympathetic nerve endings. Activation of these causes inhibition of noradrenaline release from sympathetic neurones which normally causes vasoconstriction via α-adrenoceptors. Thus, Ach has an inhibitory action on sympathetic α-vasoconstrictor tone to blood vessels. The involvement of M_2 muscarinic receptors has been identified both from inhibition of prelabelled [^{3}H]noradrenaline release and from inhibition of the vasoconstrictor responses to field stimulation of both venous and arterial preparations (Table 7.2) (Eglen & Whiting 1990).

8.1.2   Gastrointestinal Tract

The role of the parasympathetic vagal innervation in the control of gastrointestinal motility has been described under the enteric nervous system in Chapter 1. The complex interactions between ascending and descending nerve pathways innervating the circular and longitudinal muscles of the intestine are illustrated in Figure 1.5. Although excitatory 5-HT and non-adrenergic non-cholinergic (NANC) inhibitory transmitters (ATP and VIP) probably have a role in the peristaltic reflex, Ach has the most clearly defined excitatory role in the gut. Parasympathetic nerve stimulation and exogenously added Ach increase peristaltic activity, raise the tone and amplitude of contractions, and enhance the secretory activity of the gastrointestinal tract. The enhanced activity may be accompanied by vomiting, intestinal cramps, defaecation and flatulence.

The Ach-induced contraction of the gut is accompanied by depolarization and Na^+ influx, arising from the opening of a non-selective cation channel. The initial phase of contraction of longitudinal muscle is due to an increase in intracellular Ca^{2+} due to release of Ca^{2+} from intracellular stores. The tonic phase of maintained contraction is due to Ca^{2+} influx through receptor-operated L-type Ca^{2+} channels. The former mechanism is illustrated by the ability of Ach to induce contraction of isolated ileum depolarized by high K^+ medium, which opens voltage-operated channels. Muscarinic agonists also cause contractions in Ca^{2+}-free medium.

The muscarinic receptors initiating smooth muscle contraction of the gut are solely of the M_3 type, which appears to be linked to the various transduction mechanisms but with different efficacies (Parekh & Brading 1992). However, as discussed in Chapter 7, the most abundant receptor identified by radioligand binding is the M_2 receptor (Tables 7.2 and 7.4). The M_3 receptor is coupled to the functional response most likely through a G protein, since pertussis toxin blocks the Ach-induced inward current. The second messenger linking receptor activation to elevation of intracellular Ca^{2+} levels is probably increased phosphoinositide turnover and raised inositol triphosphate (IP_3) levels (Buckley & Cauldfield 1992) (Chapter 13). The M_2 receptors are linked through G_i regulatory proteins to adenylyl cyclase, which is inhibited leading to reduced cAMP levels. Thus, activation of these receptors may induce gastrointestinal contraction indirectly by opposing sympathetic relaxation mediated via β-adrenoceptor activation of adenylyl cyclase (Eglen & Harris 1993).

Ach increases the secretory activity of the exocrine glands of the gastrointestinal tract. The increase in output of acid from the parietal (oxyntic) cells of the stomach has received most attention. Vagal parasympathetic nerve stimulation to the stomach causes an increase in both the acid and pepsin output from the stomach. These responses are blocked by atropine and are therefore mediated via muscarinic receptors, which are of the M_3 subtype (Table 7.2). Pirenzepine antagonizes the response to vagal stimulation according to its M_1 selectivity but blocks exogenous muscarinic agonists weakly consistent with the involvement of M_3 receptors. This reflects blockade of M_1 receptors at the parasympathetic ganglia in the gastric intramural plexus when stimulation is via the vagus nerve, but weak blockade of M_3 receptors located on the oxyntic cells when Ach is the agonist (Figure 8.3) (Kromer & Eltze 1991).

Ach also modulates gastric acid secretion indirectly through other endogenous secretagogues. It increases the release of a neuropeptide of the bombesin family (gastrin releasing peptide, GRP). Bombesin-like immunoreactivity has been identified in the myenteric neurones of the mammalian gastrointestinal tract and bombesin causes marked increases in acid secretion and motility. Bombesin stimulates gastrin secretion from the mucosa of the stomach through activation of a BB_2 receptor. In contrast, activation of muscarinic receptors causes inhibition of another local peptide, somatostatin. Somatostatin normally exerts inhibitory effects upon the release of several transmitters through stimulation of SS receptors and this includes the release of gastrin. Thus, the Ach-induced inhibition of somatostatin release removes its inhibitory effect on gastrin release, which is in turn raised. As a result, acid secretion is also raised (Figure 8.3) (Makhlouf 1984). Vagal stimulation

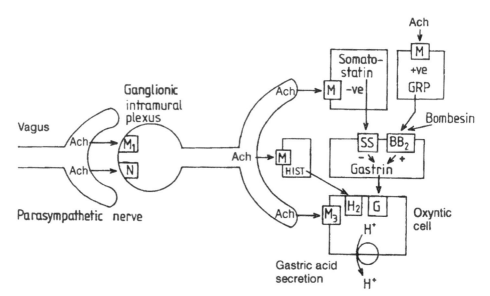

Figure 8.3 Diagrammatic representation of the control of gastric acid secretion by Ach and vagal parasympathetic nerve stimulation. Somatostatin exerts an inhibitory effect on gastrin release onto gastrin receptors of the parietal (oxyntic) cell, thereby inhibiting acid secretion. Gastrin releasing peptide (GRP) stimulates gastrin release via bombesin receptors (BB_2). Ach stimulates gastric acid secretion by acting directly on the oxyntic cell of the fundus or indirectly via somatostatin and GRP and by histamine (HIST) release onto H_2 receptors (see text for details).

in the isolated stomach of rodents also appears to induce gastric acid secretion indirectly via histamine release since the response is also sensitive to H_2 receptor blockade with tiotidine (Welsh *et al.* 1994).

8.1.3 *Airways*

Vagal parasympathetic nerve stimulation and both inhaled and intravenously administered muscarinic agonists induce bronchoconstriction and narrowing of the airways. There is a reduction of specific airway conductance (sG_{aw}) and, in the artificially respired anaesthetized guinea-pig, an increase in pulmonary inflation pressure (PIP) (Figure 8.4). This is a direct result of contraction of the airway smooth muscle through activation of M_3 receptors. The parasympathetic nerves also supply the submucosal glands where Ach released onto M_3 receptors promotes an increased volume of mucus secretion but not of its composition. M_3 receptors are also localized to epithelial cells lining the airways, the function of which is

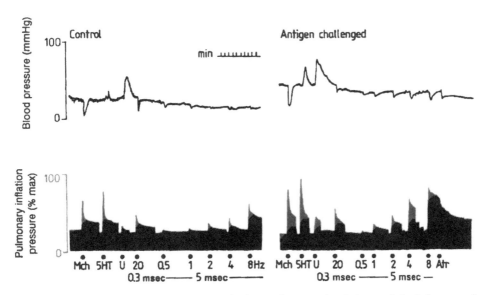

Figure 8.4 Airway responses to parasympathetic vagal nerve stimulation and the influence of airway hyperreactivity. The muscarinic agonist, methacholine (Mch, 6 µg/kg), 5-hydroxytryptamine (5HT, 3 µg/kg) and the thromboxane A_2 agonist, U46619 (U, 0.3 µg/kg), were administered intravenously to guinea-pigs anaesthetized with pentobarbitone (60 mg/kg) and maintained on artificial respiration. The cut descending (efferent) vagal nerves were electrically stimulated at 20 Hz and 0.3 msec pulse width, which releases acetylcholine (Ach) only, and at 0.5–8 Hz and 5 msec which releases both Ach and neuropeptides (substance P). Bronchoconstrictor responses are observed for each challenge, measured as increases in pulmonary inflation pressure; blood pressure is also shown. Animals were sensitized to ovalbumen (10 µg with Al(OH)₃ ip) 14–21 days beforehand. Twenty-four hours before the experiment shown, animals were challenged with aerosols (2 min) of saline (control) or ovalbumen (10 mg/ml, antigen challenged). Protection from fatal anaphylaxis was prevented with mepyramine (10 mg/kg ip 30 min before challenge). Antigen challenge causes increased bronchoconstriction to the spasmogens and vagal stimulation. Results provided by Alex Johnson working in the author's laboratory.

uncertain. Transepithelial ion and water transport and goblet cell secretion are not under direct cholinergic control. The cholinergic nerves may control the rate of ciliary beating to promote mucociliary clearance of particulates. Methacholine increases the rate of beating and clearance in the upper airways, an effect that is blocked by atropine (Gross 1988, Barnes 1993b).

In asthmatics, there is raised sensitivity to the bronchoconstrictor effect of many inhaled spasmogens (hyperreactivity) including muscarinic agonists. Similarly, in several animal models of airway hyperreactivity induced by antigen challenge of sensitized animals and by exposure to ozone, influenza virus or platelet activating factor (PAF), the bronchoconstriction by muscarinic agonists is enhanced (Figure 8.4) (Pretolani & Vargaftig 1993). Reflex vagal bronchoconstriction may be triggered in asthma by the presence of inflammatory mediators, such as bradykinin, that stimulate sensory nerves in the airways (Barnes *et al.* 1988). Although responsiveness to cholinergic agents is raised in asthma, there is no evidence of altered cholinergic innervation of the airways. In asthmatics, only a small proportion of the resting airway obstruction is mediated through cholinergic efferents. When given an atropinic agent, airflow increases only slightly more than in normal subjects (Gross 1988).

M_2 receptors located prejunctionally on the vagal nerve endings are stimulated by released Ach and exert a damping effect on transmitter release. Administration of muscarinic agonists such as pilocarpine can have an inhibitory effect upon bronchoconstriction induced by vagal stimulation. This effect is mediated via M_2 receptors since it is attenuated by M_2-selective antagonists (eg gallamine) which also enhance the airway bronchoconstriction to vagal stimulation. As discussed earlier, these prejunctional M_2 receptors appear to be defective in asthmatics since inhaled pilocarpine fails to suppress the reflex bronchoconstriction. This may be an important mechanism for accentuated vagally-mediated reflex bronchoconstriction observed in asthma (Fryer & Jacoby 1993).

8.1.4   Other Smooth Muscle

Smooth muscle in other organs is generally contracted by parasympathetic stimulation and by muscarinic agonists. The mechanisms are similar to those of the airway and gastrointestinal smooth muscle. M_3 receptors, linked through regulatory G proteins to phospholipase C, phosphoinositide breakdown and increases in intracellular Ca^{2+} levels, are common pathways for these contractile responses. The tissues include the uterus, detrusor muscle of the urinary bladder, urethra, gall bladder and ducts, seminal vesicles, vas deferens (NB. This is the same response as produced by noradrenaline.) and nictitating membrane (non-innervated). The trigone and sphincter of the urinary bladder and sphincters of the gastrointestinal tract do not contract but usually relax. The trigone is probably non-innervated. The response of the urinary bladder to muscarinic agonists results in voiding of the bladder (micturition).

8.1.5   Glandular Secretions

Exocrine glands are uniformly stimulated by muscarinic receptor agonists, which is associated with their cholinergic innervation. This applies to lacrymal glands, submandibular and parotid salivary glands, pancreatic acinar cells secreting digestive enzymes, sweat glands, the gastric mucosal parietal cells (see above) and mucosal cells of the airways and mucus membranes. Although not all exocrine cells have so far been examined in detail, the muscarinic receptor subtype mediating increased secretory activity is of the M_3 subtype (Table 7.2) (Goyal 1989). The secretion of sweat is anatomically and physiologically a sympathetic response, but is mediated via the release of Ach from sympathetic nerve endings onto muscarinic receptors (see Chapter 1). Hence, parasympathomimetic amines cause profuse sweating (diaphoresis) in humans.

The mechanisms involved in elevated secretory functions are probably similar in each of the glands. The stimulation of muscarinic M_3 receptors on the basolateral membrane of the gland cell results in raised intracellular levels of inositol triphosphate (IP_3) due to activation of phospholipase C, which increases intracellular Ca^{2+} levels from the SR. The raised Ca^{2+} levels open Ca^{2+}-dependent K^+ channels promoting K^+ efflux. At higher concentrations of Ach, there is opening of Ca^{2+}-dependent Cl^- channels allowing Cl^- to also leave the exocrine cells. Secondary ion fluxes then occur as a result of the amiloride-sensitive $K^+/Na^+/Cl^-$ co-transporter, whereby fluid and electrolyte (Na^+, K^+ and Cl^-) are secreted. These changes are accompanied by secretion of digestive amylase from salivary glands and of lipase, amylase and protease precursors from the pancreas (Petersen & Findlay 1987).

8.1.6 *Adrenal Medulla*

The chromaffin cells of the adrenal medulla have the same embryonic origin do as the autonomic ganglion cell bodies. They therefore secrete catecholamines in response to splanchnic nerve activity, primarily by the release of Ach onto nicotinic receptors. This causes depolarization and opening of voltage-operated Ca^{2+} channels, the increased intracellular levels of Ca^{2+} inducing not only release of catecholamines but also of neuropeptide Y, atrial natriuretic peptide and enkephalin. As with autonomic ganglia, however, the mammalian chromaffin cells also possess muscarinic receptors (see Chapter 11). Stimulation of these muscarinic receptors induces catecholamine release probably via increase in IP_3 formation and release of Ca^{2+} from intracellular stores. The receptors are, in common with those of autonomic ganglia, probably of the M_1 subtype since pirenzepine is a potent antagonist in guinea-pig perfused adrenal glands (Nakazato *et al.* 1988).

8.1.7 *The Eye*

Parasympathomimetics interact with M_3 muscarinic receptors on the circular (sphincter pupillae) muscle around the pupil margin of the iris and cause contraction. The pupil may be constricted (miosis) to 'pinpoint' dimension. The radial muscle is not innervated by parasympathetic nerves and does not respond to muscarinic agonists. The ciliary muscle contracts in response to muscarinic agonists acting on M_3 receptors and the lense is accommodated for near vision (see Chapter 1, Figure 1.7). Spasm of accommodation in this mode results in blurred vision. The contraction of the pupil away from the canal of Schlemm opens the drainage angle, facilitating the outflow of aqueous humour. This reduces intraocular pressure when raised in glaucoma (see later).

8.2 Parasympathomimetic Amines: Structures and Sources

The pharmacological properties described in the preceding paragraphs are common to the muscarinic receptor agonists because there is little selectivity of agonists between receptor subtypes. Thus, in considering the different agents available, it is convenient to divide them into three groups: choline esters, naturally occurring muscarinic agonists, and newer synthetics with some receptor subtype selectivity.

8.2.1 Choline Esters

The choline esters are well established derivatives of Ach. They were synthesized to achieve greater stability to hydrolysis by acetylcholinesterase (AchE) and plasma butyrylcholinesterase (Table 8.1). Acetyl-β-methylcholine (methacholine) was synthesized as early as 1911 but not extensively examined until the 1930s. Its hydrolysis by AchE is considerably slower than Ach and is almost completely resistant to hydrolysis by butyrylcholinesterase. It has predominantly muscarinic activity which occurs mainly on the cardiovascular system. The muscarinic activity of the (+)-S-isomer is about equal to that of Ach, but the (−)-R-isomer is 100–200 times less active. The (+)-S-isomer is hydrolysed by AchE at ~50% the rate of Ach hydrolysis but the (−)-R-isomer is only very weakly hydrolysed. Acetyl-α-methylcholine has less muscarinic activity than Ach but retains nicotinic activity.

Carbachol (carbamoylcholine) and bethanechol are carbamoyl esters of choline which are totally resistant to hydrolysis by either AchE or non-specific (pseudo) cholinesterases. Their greater stability therefore permits distribution to areas of low blood flow and they exert selective effects upon the gastrointestinal and urinary tracts. They produce only slight falls in blood pressure with reflex tachycardia after oral dosing. This selectivity is lost with intravenous administration. Bethanechol, by virtue of its β-methyl substitution, has little or no nicotinic activity, whereas carbachol retains some nicotinic activity but is a potent muscarinic stimulant. In binding studies, carbachol is selective for M_2 receptors relative to M_1 by a factor of 0.03 (Tecle *et al.* 1993).

8.2.2 Naturally Occurring Muscarinic Alkaloids

The parent muscarinic agonist, after which the receptor is named, is muscarine. It is isolated from the poisonous mushroom, *Amanita muscaria*, but the levels are insufficient to account for the toxicity of this fungus. The toxicity is attributed to anticholinergic and hallucinogenic isoxazole derivatives. Other richer sources of muscarine include species of *Inocybe* and *Clitocybe*, poisoning by which causes symptoms that may be predicted from the muscarinic agonist activity. These include salivation and lacrymation, nausea, vomiting and abdominal pain, diarrhoea, bronchospasm, hypotension and bradycardia (slow weak pulse) and visual disturbances. Muscarine is a quaternary ammonium compound that occurs naturally as the L-(+)-isomer and is highly selective for muscarinic receptors (Table 8.2).

338

Table 8.1 Choline esters

		Hydrolysis by cholinesterase		Musc	Nic	Cardiovasc	GI/urinary
		AchE	Non-specific				
Acetylcholine	$(CH_3)_3\overset{+}{N}CH_2CH_2OCCH_3$ (O)	++++	+++	+++	++	++	++
Methacholine	$(CH_3)_3\overset{+}{N}CH_2CHOCCH_3$ (O), CH_3 (+)-S-	++	–	+++	–	+++	+
	(–)-R-	–	–	+	–		
Carbachol (Isopto carbachol^R)	$(CH_3)_3\overset{+}{N}CH_2CH_2OCNH_2$ (O)	–	–	+++	+	+	++
Bethanechol (Myotonine^R) (Urecholine^{US})	$(CH_3)_3\overset{+}{N}CH_2CHOCNH_2$ (O), CH_3	–	–	++	–	+	++

Notes: **Approximate muscarinic (musc) and nicotinic (nic) activity relative to acetylcholine is shown. Cardiovascular (cardiovasc) and gastrointestinal/ urinary tract (GI/urinary) selectivity is only observed after oral or subcutaneous administration**

Autonomic pharmacology

Table 8.2 Natural muscarinic alkaloid agonists

	Source	
Muscarine	*Amanita muscaria*	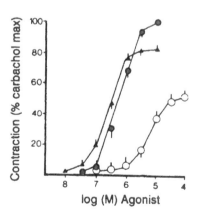
Pilocarpine	*Pilocarpus jaborandi* *P. microphyllus*	
Arecoline	*Areca catechu*	

Note:- $CH_2CH_2CH_3$ group at [+] or methyl group at [*] converts arecoline to antagonist.

Pilocarpine is a tertiary amine and is therefore readily absorbed (Table 8.2). It is the main constituent of the leaflets of the South American shrub *Pilocarpus jaborandi*, which are chewed by local natives to promote salivation. Pilocarpine has predominantly muscarinic activity but for many M_3 receptor-mediated responses (trachea, $\alpha = 0.6$; urinary bladder, $\alpha = 0.3$; ileum, $\alpha = 0.8$) it is a partial agonist, although not in the atria (M_2, $\alpha = 1$) (Eglen *et al.* 1988a) (Figure 8.5). Pilocarpine produces profound sweating and salivation, and jaborandi has been added to hair tonics for its supposed effect on stimulating hair growth. The main use of pilocarpine, however, is as an aqueous solution of the hydrochloride in eye drops for the treatment of glaucoma (see later).

Arecoline is a tertiary amine obtained from the betel or areca nut (*Areca catechu*) (Table 8.2). It has muscarinic activity that is dependent upon the pH of the medium;

Figure 8.5 Concentration–response curves for the contractile responses of rat urinary bladder to the muscarinic agonists carbachol (●), oxotremorine (▲) and pilocarpine (○). Mean responses (±SEM) are plotted as a percent of the maximum to carbachol to illustrate full and partial agonist activity. Reproduced with permission from Lucchelli *et al.* (1992).

340

at the pH of plasma and most physiological salt solutions (7.4) it is 60% protonated and has ~80% of the muscarinic activity of Ach. The non-protonated form is virtually inactive. Betel nuts have been chewed in many parts of Asia to produce salivation and for its mild central stimulant action. They are not used therapeutically in human medicine but arecoline has been employed in veterinary medicine as a purgative and vermifuge for *Taenia*, although it is not taenicidal. It probably expels the worm by intense intestinal contractions.

8.2.3 Newer Synthetic Agonists

8.2.3.1 Agonists not selective for muscarinic receptor subtypes

The starting points for synthesis of more potent and stable muscarinic agonists with potential selectivity between the receptor subtypes have been either the choline esters (Table 8.3) or arecoline (Table 8.4).

Muscarinic agonists resembling the choline esters include furtrethonium and (+)-*cis*-dioxolane and their methyl derivatives. They are all stable and potent non-

Table 8.3 Muscarinic agonist resembling choline esters

Furtrethonium iodide (Furmethide^R)		Full agonist, non-selective
5-Methyl furtrethonium	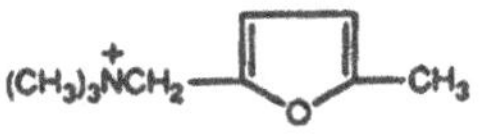	Potent full agonist, non-selective
(+)-*cis*-Dioxolane	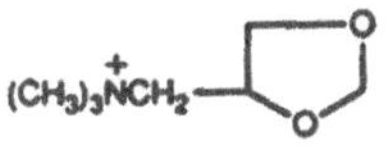	Potent full agonist, non-selective
cis-Methyldioxolane (CMD)	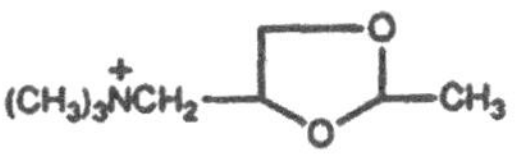	Highly potent, non-selective (used as radioligand)
McN-A-343	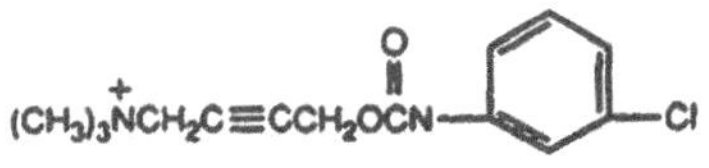	M$_1$ selective
Oxotremorine (oxotremorine sesquifumarate)	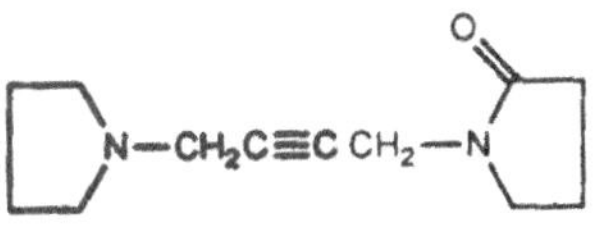	Non-selective, full agonist
Oxotremorine-M (oxotremorine methiodide)	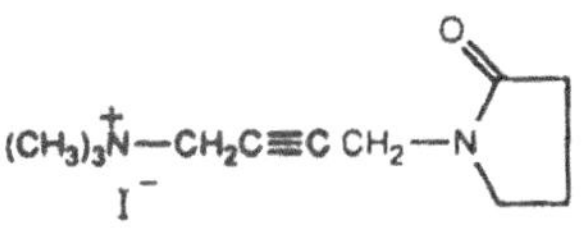	Potent agonist with peripheral actions only (used as radioligand)
AHR-602	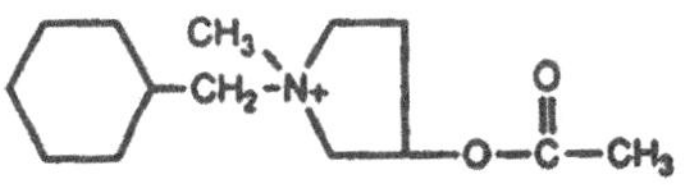	M$_1$ selective

341

Table 8.4 Muscarinic agonists resembling arecoline

Arecaidine propargyl ester (APE)		Highly potent, slight M_2 preference
Aceclidine (Glaucostat[FR])		Non-selective
RS-86		Non-selective between m_1 to m_5
Oxidiazoles		
L-658,825		
L-658,903	—C_2H_5 antagonist	Highly potent, non-selective full agonists, penetrate blood–brain barrier
L-670,548	$X = CH_3$	
L-687,306	$X = \triangleleft$	$\begin{cases} M_1 \text{ partial agonist} \\ M_2/M_3 \text{ antagonist} \end{cases}$
Chloropyrazines		
L-689,660		M_1 agonist, M_3 partial agonist, M_2 antagonist, equal binding affinity
Azabicyclo-oximes		M_1 versus M_2 selectivity
WAL 2014		M_1 selective (pD_2 5.3 vas deferens, 4.6 atria)
AF102B (cis)		M_1 selective

selective full agonists. Oxotremorine and its N-methyl quaternary derivative (oxotremorine-M) are all non-selective full agonists. Oxotremorine is of interest as a research tool since it induces Parkinson's-like central effects including tremor and ataxia. It is a tertiary amine and therefore crosses the blood–brain barrier; oxotremorine-M does not cross the blood–brain barrier and therefore has peripheral muscarinic properties only and is a more potent agonist. [^{3}H]Oxotremorine-M is used to bind to high affinity state muscarinic receptors, thus displacement by non-labelled agents with high affinity is indicative of potent full agonist activity (eg carbachol, muscarine and methylfurtrethonium). Pilocarpine and McN-A-343 are weaker, indicating their partial agonist activity. The ratio of affinities (K_{app}) for displacement of [^{3}H]oxotremorine binding and [^{3}H]NMS binding (low affinity antagonist binding) is high (>1,000) for full agonists such as muscarine, carbachol, oxotremorine-M, oxotremorine and methylfurtrethonium. Intermediate values are obtained for arecoline and pilocarpine, whereas values close to unity are obtained for antagonists (Freedman *et al.* 1988).

8.2.3.2 *Muscarinic receptor subtype-selective agonists*

McN-A-343 was the earliest compound to be identified which suggested the existence of muscarinic receptor subtypes (Roszkowski 1961). This agonist is a quaternary compound containing a propargyl group. It has a selective stimulant action at muscarinic receptors of autonomic ganglia. Thus, after systemic administration, it increases heart rate, blood pressure and peripheral resistance due to activation of sympathetic pathways, rather than reducing these parameters like other parasympathomimetic amines. These ganglionic muscarinic receptors have subsequently been classified as M_1 receptors. Being a quaternary nitrogen compound, it does not penetrate the blood–brain barrier. It produces negligible responses mediated via M_2 (atria) or M_3 receptors (contraction of the trachea, ileum or urinary bladder), possibly because of high receptor reserve in these tissues (Eglen *et al.* 1987). Low Ca^{2+} medium and previous exposure to muscarinic full agonists may also account for impaired responses to McN-A-343 in these tissues (Rubinstein & Cohen 1992). The selectivity displayed by McN-A-343 in functional tests does not, however, extend to radioligand binding where only two-fold selectivity for M_1 receptors is observed. Many analogues of McN-A-343 have been prepared but none show improved selectivity or potency nor the ability to cross the blood–brain barrier.

Arecoline has been the basis for many newer synthetic muscarinic agonists since it has a rigid structure with little opportunity for conformational flexibility at the muscarinic receptor binding sites. In general, muscarinic agonists do not tolerate addition of steric bulk. With arecoline, addition of a methyl group at the 4-position or lengthening of the side-chain to an O-*n*-propyl group converts it to antagonist molecules (Table 8.2). This lack of tolerance may be due to the fact that agonist binding to aspartate residues (eg Asp-147) within the seven membrane-spanning domains of the muscarinic receptor is fairly precise. Differences in amino acid sequence and topography between receptor types is also very limited so that agonist selectivity is difficult to obtain. An exception to this rule with increasing size of substitution is the introduction of bulky side-chains with a propargyl moiety. Such a group was present in the first M_1 selective agonist, McN-A-343. Addition of propargyl-containing side-chains to arecoline has generated several analogues of which arecaidine propargyl ester (APE) is a potent agonist with three-fold functional selectivity for M_2 receptor-mediated atrial negative inotropic responses compared with the ileum (M_3). It also has potent M_1 agonist activity, like McN-A-343, and *increases* heart rate in the pithed rat (Moser *et al.* 1989). Aceclidine is a synthetic muscarinic agonist resembling arecoline which was first used over 20 years ago to reduce intraocular pressure in glaucoma with equal effectiveness to pilocarpine.

Recent attempts have been made to develop high potency derivatives related to arecoline that can penetrate the blood–brain barrier for potential use in Alzheimer's disease (see later). The oxidiazole substituted compounds, such as L-670,548, were found to be highly potent full agonists which, however, lacked any selectivity for the M_1, M_2 or M_3 muscarinic subtypes. Their peripheral effects, such as bradycardia (M_2) and salivation (M_3), however, have made them too hazardous for clinical use. Because of the difficulty in achieving the desired M_1 selectivity, from the affinity of binding, an alternative approach has been to develop agonists with low efficacy. These could display selectivity based upon the differences in receptor reserve for the various muscarinic receptor subtypes and their functional responses. Examples of arecoline derivatives include the oxidiazole L-687,306 and the related chloropyrazine, L-689,660. The former has M_1 partial agonist and M_2/M_3 antagonist activity, while L-689,660 is an M_1 agonist, M_3 partial agonist and M_2 antagonist based upon functional tests in isolated tissues. By radioligand binding analysis, however, they show no selectivity of binding affinity (Hargreaves *et al.* 1992, Freedman *et al.* 1993). Attempts to produce receptor selectivity have been made by introducing extended propargyl side-chains to similar azabicyclo-oximes. Several of these show modest selectivity of binding to M_1 compared with M_2 receptors (Tecle *et al.* 1993) (Table 8.4).

8.3 Therapeutic Uses of Muscarinic Agonists

8.3.1 Gastrointestinal Tract

The M_3 receptor-mediated stimulation of gastrointestinal motility is utilized in cases of postoperative gastric distension and atony and non-obstructive adynamic (paralytic) ileus. Bethanechol is used orally, except where complete gastric retention occurs and absorption from the stomach is impaired, when the subcutaneous route is used. Bethanechol may also be used to expel gas from the intestine prior to X-ray examinations. In reflux oesophagitis, bethanechol improves the tone of the lower oesophageal sphincter and oesophageal peristalsis. Agents that promote gastric emptying in cases of gastric hypomotility are known as prokinetic agents, but bethanechol is not particularly effective in accelerating gastric emptying. Metoclopramide (Reglan[US], Maxalon[UK]) and cisapride (Alimix[UK]) are prokinetic agents used for gastric hypomotility and reflux oesophagitis. Their action is prevented by muscarinic antagonists and is therefore attributed to the release of Ach from the myenteric neurones of the gastric wall (see Figure 1.5). This improves co-ordinated contractile activity rather than merely increased tone, which is the case with bethanechol (Goyal 1989).

8.3.2 Urinary Bladder

Postoperative and postpartum non-obstructive urinary retention may be relieved by the M_3 receptor agonist activity of bethanechol on the detrusor muscle. It may also be useful in certain cases of neurogenic bladder and in paralysis of the bladder after spinal injury.

8.3.3 The Eye

Muscarinic agonists applied topically to the eye cause M_3-mediated constriction of the pupil (miosis). This opens the drainage angle and the conventional flow of aqueous humour from the anterior chamber of the eye is facilitated by altering the configuration of the trabecular network and allowing better drainage into the canal of Schlemm. The effect is almost entirely a mechanical consequence of the contractions of the sphincter muscle of the iris and of the ciliary muscle. Direct effects on the trabecular meshwork and canal of Schlemm are unknown and of minor importance. This effect reduces the raised intraocular pressure that occurs in glaucoma. Pilocarpine is the chief muscarinic agonist used, although Ach (Miochol[R]) and carbachol have occasionally been employed in the past, and in France, aceclidine is available. There may be an initial transient increase in intraocular pressure due to the vasodilatation and congestion of the vascular beds, together with an increase in aqueous humour production. The unconventional or uveoscleral outflow of aqueous humour via the spaces between the muscle bundles into the sclera also appears to be diminished by muscarinic agonists (Kaufman *et al.* 1984).

Pilocarpine is used in primary glaucoma of both the narrow-angle (acute) and open- or wide-angle (chronic simple) types. In open-angle glaucoma, there is no physical obstruction to aqueous humour drainage as there is in the acute narrow-angle form. However, miotics are equally effective since the contraction of the circular muscle of the iris and of the ciliary body enhances tone of the trebecular network and lines up this meshwork of pores to permit improved passage of aqueous humour (Kaufman *et al.* 1984). A disadvantage of pilocarpine is that it causes blurred vision and it may cause initial irritation. These may be reduced when pilocarpine is administered in a sustained release reservoir composed of a co-polymer bilayer (Ocusert pilo[R]). This is placed under the eyelids for constant release over a period of 7 days. Care should be exercised with pilocarpine in patients prone to retinal detachment. Pilocarpine, alternating with mydriatics, may be used to break adhesions between the iris and lens.

8.3.4 Side-effects and Contraindications of Parasympathomimetic Amines

On no account should muscarinic agonists be administered intravenously because of the risk of severe cardiovascular collapse, which may be overcome by adrenaline (1 mg im or sc) or atropine (1 mg sc or iv). Methacholine is no longer used for its cardiovascular vasodilator actions in the treatment of peripheral vascular insufficiency, such as Raynaud's syndrome, because of the unreliable activity due to its partial hydrolysis by AchE.

Other side-effects with normal usage of muscarinic agonists include bronchoconstriction, gastric disturbances including abdominal cholic and enhanced risk of peptic ulcer, visual discomfort, salivation, sweating and flushing. The latter action may be associated with headache. They are therefore contraindicated in asthma and patients with peptic ulcers or a history of hyperthyroidism, who may develop atrial fibrillation. In patients with myocardial ischaemia, the fall in arterial blood pressure produced by parasympathomimetics may further reduce coronary blood flow and compromise the heart. Because of the central arousal induced by muscarinic agonists, especially those that penetrate the blood–brain barrier, they should also be used with caution in patients with epilepsy and Parkinson's disease.

8.4 Potential Applications of Muscarinic Agonists

Use is not currently made of the cardiovascular properties of muscarinic agonists. However, stable selective M_2 agonists with negative chronotropic and antiadrenergic properties may have potential in treating cardiac hyperactivity states and in atrial arrhythmias. Indeed, Ach itself and methacholine have been used in the past to overcome supraventricular paroxysmal tachycardias.

The CNS offers great potential for the use of muscarinic agonists. Intravenous administration of muscarinic agonists causes cortical arousal, an effect that is inhibited by atropine and therefore due to muscarinic receptor stimulation. Oxotremorine, for example, causes Parkinson's disease-like responses and tremor associated with activation of muscarinic receptors in brain areas including the basal ganglia. These properties are of potential for the management of Alzheimer's disease. In senile dementia of the Alzheimer's type there is neurodegeneration of the cerebral cortex associated with deposition of numerous amyloid plaques, together with neurofibrillary tangles in dead and dying neurones. Additionally, there is damage to cholinergic neurones in the basal forebrain which project into the cerebral cortex and hippocampus. These cholinergic projections form vital components of the ascending reticular activating system which is involved in memory function and attention. There is a correlation between loss of these neurones at *post mortem* and the degree of memory impairment measured *pre mortem*. These observations led to the cholinergic hypothesis which could explain the intellectual decline and memory loss of Alzheimer's disease and provide a basis for its treatment by cholinergic replacement therapy. The postsynaptic muscarinic receptors involved in cognitive function are of the M_1 type and, although the neurones are damaged, the M_1 receptors are not significantly reduced. Thus, M_1 agonists are of potential application in the symptomatic improvement of cognitive function in Alzheimer's disease.

Non-selective muscarinic agonists have not been particularly effective and suffer the disadvantage of unacceptable peripheral effects through M_2 and M_3 receptors. These latter effects, however, would be largely eliminated if substantial M_1 selectivity could be achieved. The compound must readily penetrate the blood–brain barrier, which eliminates many potent agonists which tend to be charged quaternary compounds (Table 8.3). Newer arecoline derivatives were potent and could penetrate to the brain but selectivity has been elusive. Azabicyclo compounds (Table 8.4), however, have been developed with some M_1 selectivity. These have low efficacy in an attempt to achieve selectivity through differences in receptor reserve (Tecle *et al.* 1993). These compounds have shown improvement in animal tests of memory function and reverse defects in performance induced by the muscarinic antagonist, scopolamine (Hargreaves *et al.* 1992). The prospects for this class of drug in the long-term treatment of Alzheimer's disease have yet to be fully evaluated. For example, it is not known how chronic use of muscarinic agonists will affect muscarinic receptor function or the progress of the disease itself.

Antagonists at Muscarinic Cholinergic Receptors

9.1 Introduction

The muscarinic receptor antagonists interact with the muscarinic receptors, produce no stimulant or agonist effects by themselves, but inhibit the effects of endogenously released Ach or exogenously introduced muscarinic agonists. They therefore inhibit the known physiological responses mediated via muscarinic receptors that have been described in the previous chapter. The principal targets for antagonism are the postjunctional M_2 and M_3 muscarinic receptors on cardiac and smooth muscle and glandular tissues, which are normally activated by the release of Ach from postganglionic parasympathetic nerve terminals. Additionally, however, the muscarinic receptor located on the cell bodies of postganglionic nerves in autonomic ganglia (M_1), the prejunctional receptor (M_2) controlling transmitter release from sympathetic and parasympathetic neurones, and the cholinergic sympathetically innervated sweat glands will also be affected. Muscarinic receptors located on non-innervated structures such as certain blood vessels are also antagonized. Finally, muscarinic receptors are distributed throughout the brain and, provided that the antagonist is capable of crossing the blood–brain barrier, the central effects upon mood, behaviour and cognitive powers with therapeutic doses are primarily due to antagonism of muscarinic receptors. In general, the available antagonists have high selectivity for muscarinic compared with nicotinic receptors for Ach; they do not therefore have blocking activity at the somatic nerve–skeletal muscle junction nor at the nicotinic receptors involved in ganglionic transmission, unless doses considerably in excess of muscarinic blocking doses are administered.

The prototype muscarinic blocking drug is the naturally occurring alkaloid, atropine, which displays classic competitive antagonism towards Ach (Chapter 3). It exerts no stimulant effect through its interaction with the muscarinic receptor but prevents the access of Ach to the receptor, thereby inhibiting agonist activity. The blockade may, however, be completely overcome by addition of higher concentrations of the muscarinic agonist. Atropine is non-selective towards the different subtypes of muscarinic receptors identified in the previous chapters. Indeed, apart from pirenzepine and its derivatives, receptor subtype-selective antagonists are still not in regular clinical use.

9.2 Sources and Chemistry of Muscarinic Antagonists

The chief source of the naturally occurring alkaloids with muscarinic blocking activity is among the Solanaceous plants. Atropine is the racemate, (±)-hyoscyamine (Table 9.1), (−)-hyoscyamine being found in the leaves, root and berries of the Deadly Nightshade (*Atropa belladonna*). The properties of belladonna preparations as poisons and medicines were known in the Middle Ages and in the times of the Roman Empire. The name *Atropa belladonna* was given by Linnaeus after Atropos of Greek mythology, the oldest of the Three Fates, whose task was to cut the thread of life. The name belladonna is reputedly derived from the Italian for 'beautiful woman' and attributed to the practice of instilling belladonna extracts into the eyes; the dilated pupils making the eyes appear larger and more attractive. As medicinal agents, plants containing atropine were first introduced into Western medicine in the early 1800s. The root, leaves and seeds of *Datura stramonium* (known as Jimson or Jamestown weed, thornapple, stinkweed and devil's apple), which are also rich

Table 9.1 Naturally occurring muscarinic antagonists and their derivatives

Atropine	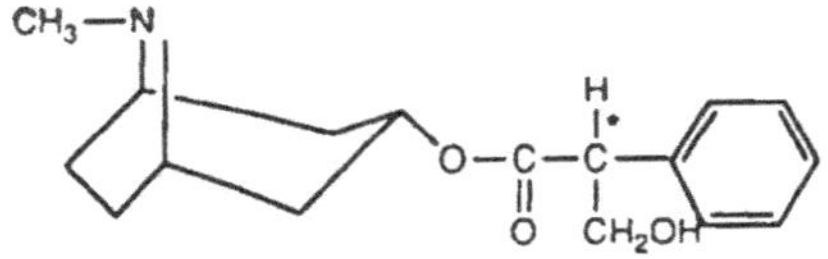	HBr or sulphate
(S-(−)-Hyoscyamine and R-(+)-hyoscyamine)		
Atropine methonitrate (or methiodide)	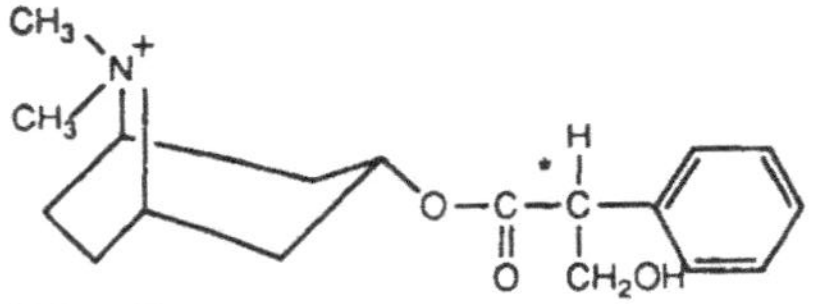	Quaternary derivative, no CNS penetration
Hyoscine (Scopolamine)	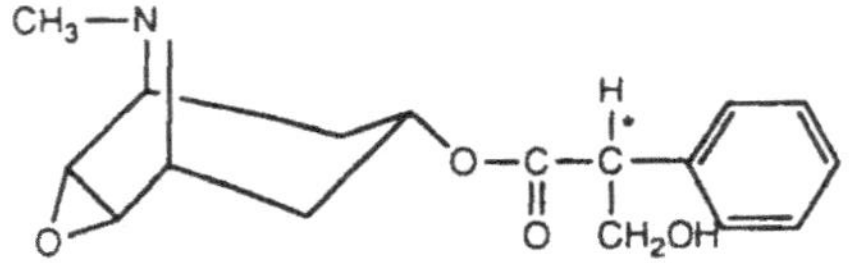	CNS penetration good, sedative
Homatropine	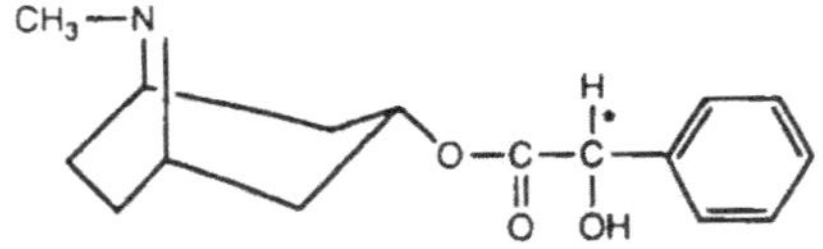	Short-acting mydriatic
Ipratropium (Atrovent^R)	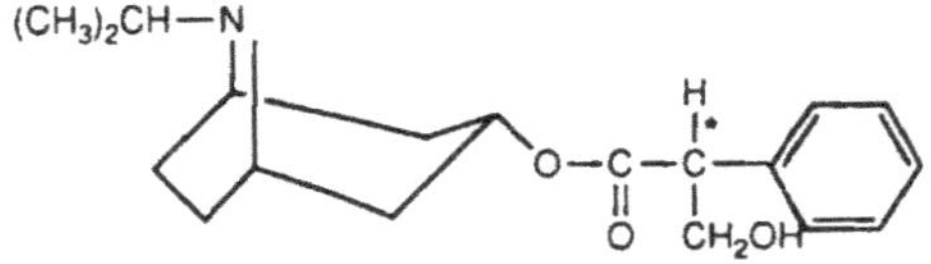	Inhaled bronchodilator
Oxitropium (Oxivent^R)	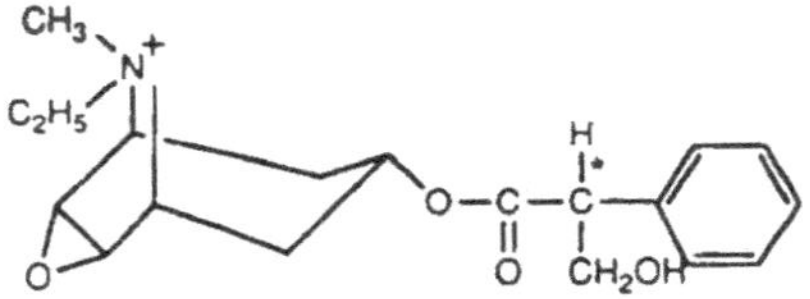	Inhaled bronchodilator Br^-

sources of (−)-hyoscyamine, had been burned and the smoke inhaled for relief of asthma. However, it was not until the 1920s that atropine was used for premedication prior to surgery. The closely related alkaloid, hyoscine (scopolamine), is found in the leaves and flowering top of the shrub *Hyoscyamus niger* (black henbane), which also contains substantial levels of (−)-hyoscyamine (Martindale 1989).

(−)-Hyoscyamine and (−)-hyoscine are aromatic esters of tropic acid and the organic bases, tropine (tropan-3-α-ol) or scopine, respectively. These are optically active and occur naturally as the (−)-isomers. The racemization of (−)-hyoscyamine to atropine occurs by heating or by the action of alkali during extraction. They exhibit a high degree of stereoselectivity, S-(−)-hyoscyamine having an affinity over two orders of magnitude greater than the (+)-isomer (pK_A values 9.38 and 6.86, respectively, isomeric ratio 330). Quaternization of the amine group to the N-methyl derivative increases muscarinic receptor blocking activity (eg S-(−)-atropine methiodide, p$K_A = 9.67$) but also increases nicotinic receptor blocking properties at sympathetic ganglia (10-fold) and at the neuromuscular junction (Table 9.1). The stereoselectivity, however, declines as the size of the N-alkyl quaternizing substituent increases, and at atropine butyliodide the isomeric ratio is 17 (Triggle & Triggle 1976).

In spite of the synthesis of a vast range of synthetic muscarinic antagonists, atropine remains one of the most potent agents. S-(−)-Hyoscyamine does not display selectivity between muscarinic receptor subtypes. In functional studies the pA_2 values in rat ileum (M_3), rabbit vas deferens (M_1, inhibition of twitch) and guinea-pig uterus (M_4) were 9.01, 9.28 and 8.71, respectively. The less active isomer, R-(+)-hyoscyamine, may have selectivity towards M_4 receptors (pA_2 in guinea-pig uterus of 9.56 compared with 7.0−7.1 at M_1, M_2 and M_3 receptors) (Ghelardini *et al.* 1993).

(−)-Hyoscine (scopolamine) is the ester of tropic acid and scopine, the latter differing from tropine by the addition of an oxygen bridge (Table 9.1). The potency of hyoscine is equivalent to that of atropine, having pA_2 values of 9.5 and 9.0, respectively, for antagonism of the contractile responses of guinea-pig ileum to Ach (Bowman & Rand 1980). It is almost completely metabolized in the liver. The quaternary derivative, N-methylscopolamine, is a potent antagonist with no selectivity between muscarinic receptor subtypes. It is widely used as [³H]N-methylscopolamine ([³H]NMS) for radioligand binding to the low affinity state of the muscarinic receptor. Homatropine is the semisynthetic ester of tropine and mandelic acid (Table 9.1) which has approximately one tenth of the muscarinic blocking potency of (−)-hyoscyamine. The addition of an N-methyl quaternizing group to homatropine increases muscarinic activity three-fold and nicotinic ganglionic blocking activity 10-fold.

9.2.1 Synthetic Muscarinic Antagonists

In view of the wide therapeutic applications of muscarinic antagonists (see later), it is not surprising that a diverse range of synthetic muscarinic antagonists have been produced. In general, it has been the search for agents more selective for the gastrointestinal tract, to inhibit secretion and therefore for application in treating peptic ulcer, that has been the driving force behind the synthetic programmes. Muscarinic antagonist activity has been shown to occur when the size of the non-polar substitution of muscarinic agonist molecules increases. As illustrated in the previous chapter, agonists do not tolerate bulky substitution without becoming antagonists. For example, replacement of the ring 2-methyl substitution of the 1,3-dioxolane, (+)-*cis*-dioxolane (Table 8.3) with two phenyl groups yields a muscarinic antagonist with a pA_2 value of 7.6. The dioxolane with a cyclohexyl and phenyl substitution in this position is more potent, with a pA_2 value of 8.0 or more depending on the enantiomer. Similar substitution of oxotremorine produces antagonist molecules (Table 9.2) (Triggle & Triggle 1976). Introduction of phenyl, cyclohexyl or similar polycyclic groups into the choline esters also produces antagonist activity (Table 9.2).

The earlier synthetic antagonists generally have an asymmetric benzylic carbon atom (indicated in the tables by ★) and may be conveniently divided into tertiary amines and quaternary ammonium compounds. Although large numbers of these compounds have been produced, only representative examples are shown in Tables 9.3 and 9.4 and these are mainly compounds that are in therapeutic use.

Table 9.2 Conversion of agonists to antagonists at muscarinic receptors by addition of bulky aryl groups

Agonists	Antagonists
$(CH_3)_3\overset{+}{N}-CH_2CH_2-O-\overset{\overset{O}{\|\|}}{C}-CH_3$ Acetylcholine	$CH_3-\overset{\overset{C_2H_5}{\|}}{\underset{\underset{C_2H_5}{\|}}{\overset{+}{N}}}-CH_2CH_2-O-\overset{\overset{O}{\|\|}}{C}-\overset{\overset{OH}{\|}}{\overset{★}{C}}-\text{(cyclohexyl)(phenyl)}$ Oxyphenonium
	$CH_3-\overset{\overset{CH_3}{\|}}{\underset{\underset{C_2H_5}{\|}}{\overset{+}{N}}}-CH_2CH_2-O-\overset{\overset{O}{\|\|}}{C}-\overset{\overset{OH}{\|}}{C}-\text{(phenyl)(phenyl)}$ Lachesine
$(CH_3)_3\overset{+}{N}-CH_2-\text{(dioxolane ring)}-CH_3$ *cis*-Methyldioxolane pD_2 7.6 ± 0.02 (guinea-pig ileum)	$(CH_3)_3\overset{+}{N}-CH_2-\text{(dioxolane ring with cyclohexyl and phenyl)}$ 2S4R enantiomer pA_2 8.26 ± 0.03
$\text{(pyrrolidine)}N-CH_2C\equiv CCH_2-N\text{(pyrrolidinone)}$ Oxotremorine	$\text{(pyrrolidine)}N-CH_2C\equiv CCH_2-\overset{\overset{OH}{\|}}{C}-\text{(phenyl)(phenyl)}$

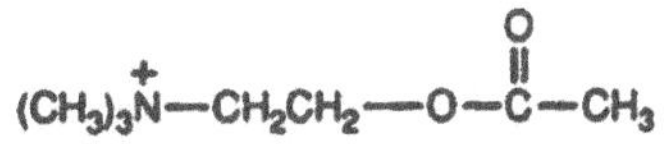

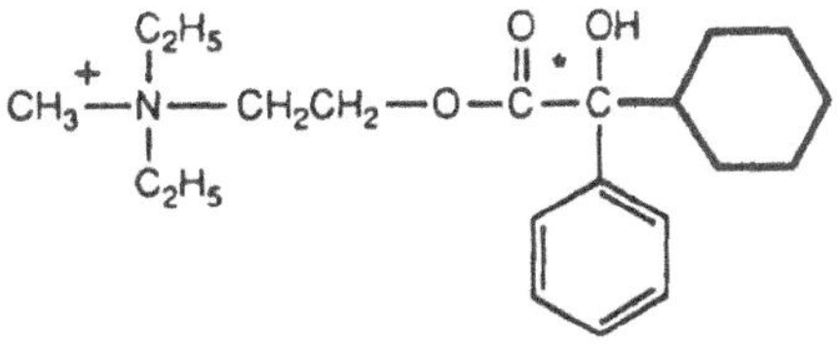

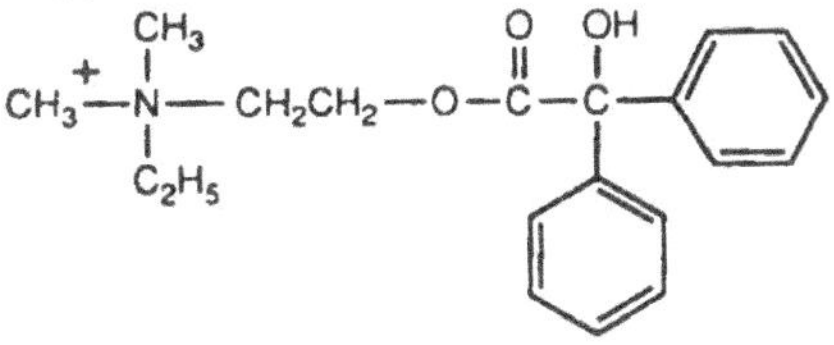

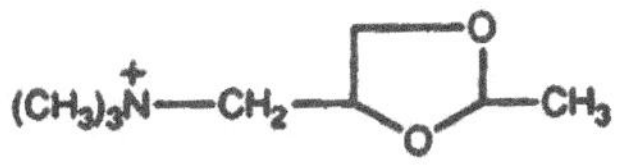

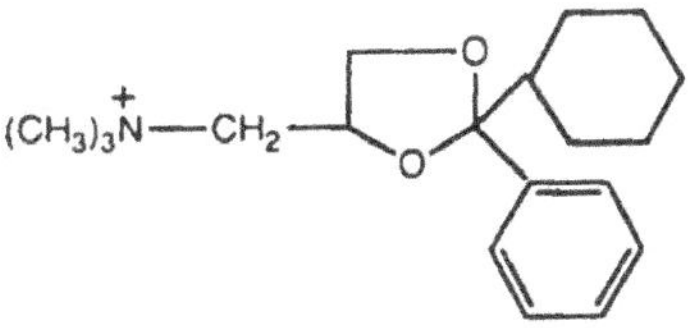

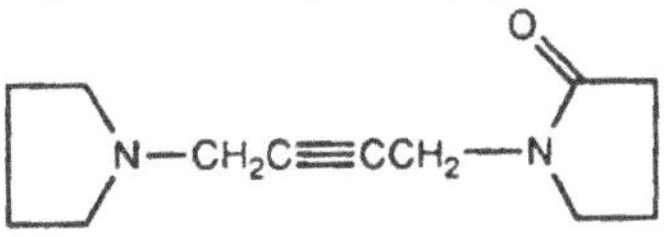

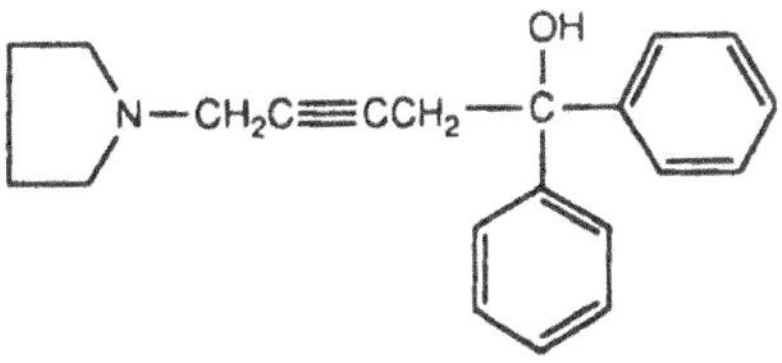

Table 9.3 Some synthetic muscarinic antagonists in current use (quaternary ammonium compounds)

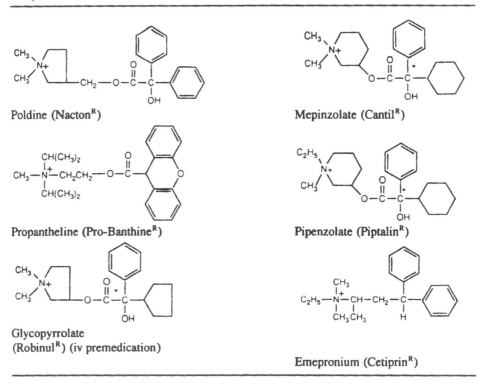

Poldine (Nacton^R)

Mepinzolate (Cantil^R)

Propantheline (Pro-Banthine^R)

Pipenzolate (Piptalin^R)

Glycopyrrolate
(Robinul^R) (iv premedication)

Emepronium (Cetiprin^R)

Note: An asterisk indicates an asymmetric chiral centre.

The quaternary compounds are equivalent to the quaternized derivatives of the naturally occurring alkaloids, the salts atropine methonitrate, hyoscine methonitrate and hyoscine N-butylbromide. The quaternary ammonium synthetic agents are generally more potent, but their absorption from the gut is poorer than for the tertiary amines. They are consequently more active on the intestine and have fewer side-effects elsewhere. The ganglionic blocking activity introduced by quaternization also contributes to the inhibition of gastrointestinal motility and their antispasmodic activity. The quaternary compounds do not readily cross the blood–brain barrier and therefore have little or no central actions. Examples of these compounds include poldine (Nacton^R), mepinzolate (Cantil^R), pipenzolate (Piptalin^R), propantheline (Pro-Banthine^R) and glycopyrrolate (Robinul^R), the latter being a potent antagonist ($pA_2 = 9.4$) in the isolated trachea. The tertiary amines include tropicamide (Mydriacyl^R) and cyclopentolate (Alnide^R, Mydrilate^R) which are used in eye drops for dilating the pupil and have short durations of action. Others include the antispasmodic, dicyclomine (Merbentyl^R), and a range of anticholinergic agents used for their central actions in the treatment of Parkinson's disease (procyclidine, trihexyphenidyl, biperiden, orphenadrine and benztropine) (Table 9.4). More recently, several of these centrally active agents have been shown to possess selectivity for the M_1 receptor as measured by their ability to displace [^{3}H]NMS from cortical binding sites.

Table 9.4 Synthetic muscarinic antagonists with current uses (tertiary amines)

<u>Gastrointestinal</u>

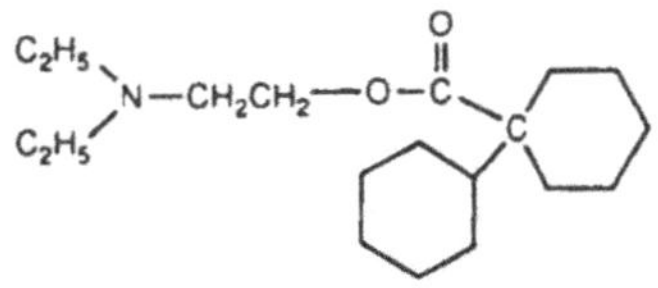

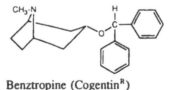

Dicyclomine (Merbentyl^R)
M_1 Selective

<u>Urinary tract – urge incontinence</u>

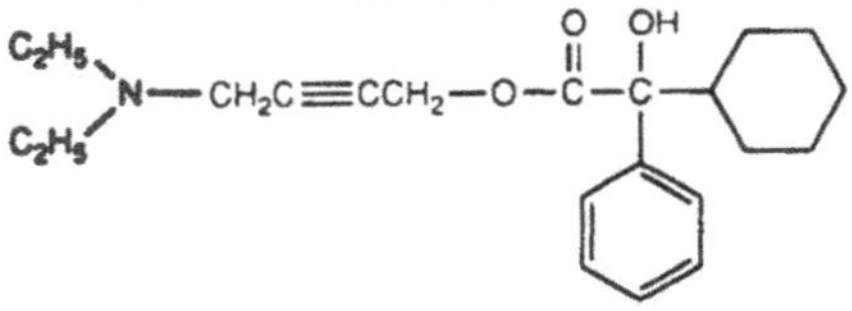

Oxybutynin (Cystrin^R, Ditropan^R)
Additional smooth muscle relaxant

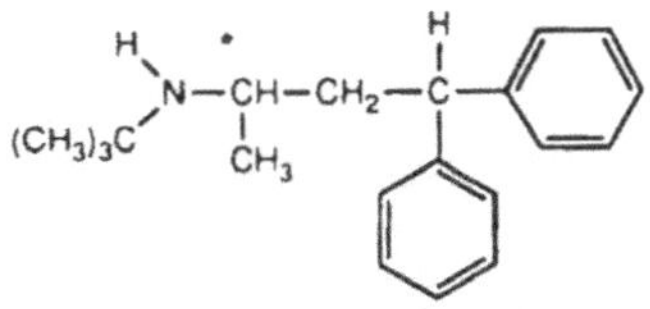

Terodiline (Micturin^R, withdrawn)
Additional Ca^{2+} antagonist

<u>Mydriatics</u>

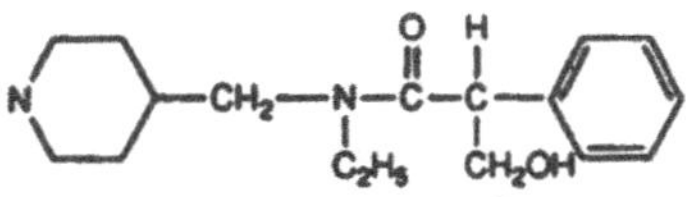

Tropicamide (Mydriacyl^R)
Modest M_4 selectivity

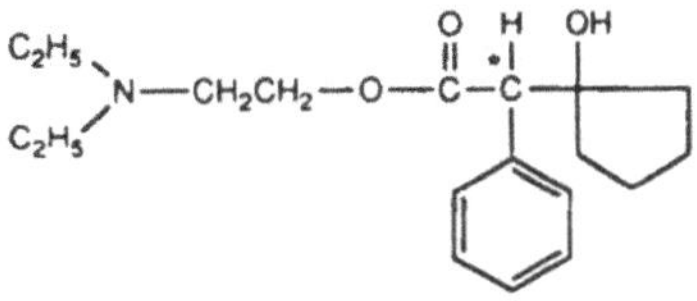

Cyclopentolate (Mydrilate^R)

<u>Anti-Parkinsonian drugs</u>

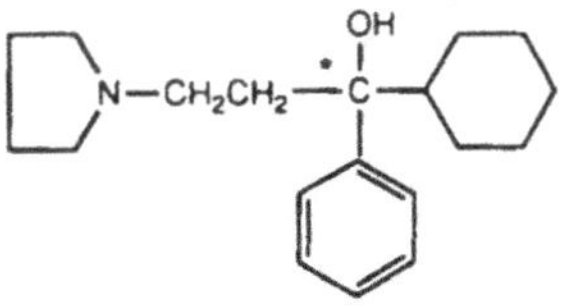

Procyclidine (Kemadrin^R)
$M_1 > M_3 > M_2$
pK_i 8.1 : 7.6 : 7.1

Benztropine (Cogentin^R)

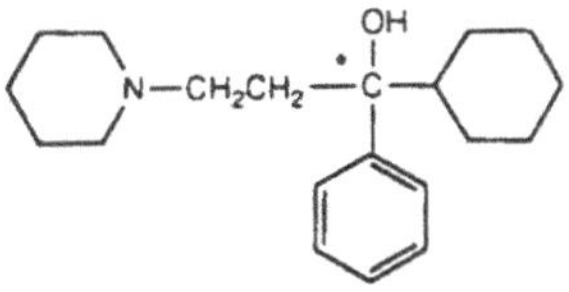

Benzhexol (trihexylphenidyl)
(Artane^R) $M_1 > M_3 > M_2$

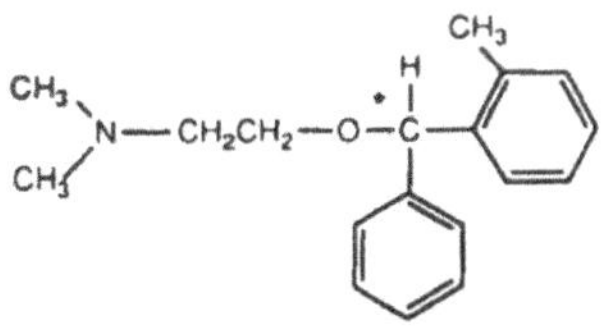

Orphenadrine (Disipal^R, Biorphen^R)

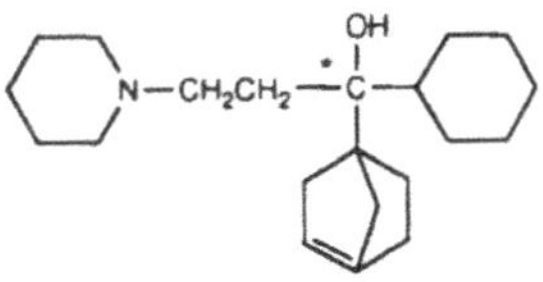

Biperiden (Akineton^R)
$M_1 > M_3 > M_2$

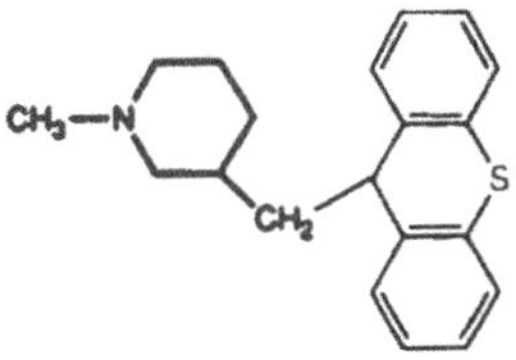

Methixene (Tremonil^R)

Note: An asterisk indicates an asymmetric chiral centre.

355

Procyclidine has become a reference antagonist for many derivatives. It displays marked stereoselectivity, the (R)-isomer having greater affinity, with a pA_2 value of 8.04 compared with 5.4 for the (S)-isomer for antagonism of contractile responses of the guinea-pig ileum (M_3). The derivatives of procyclidine having either two phenyl substituents (pyrrinol) or two cyclohexyl substituents (hexahydro-procyclidine) on the central carbon atom are non-chimeric. Their affinities (6.91 and 6.37, respectively) are still greater at ileal M_3 receptors than for the (S)-isomer of procyclidine. This indicates the importance of the configuration of the phenyl and cyclohexyl substitutions for their optimum interaction with the receptor site. The lack of potency of compounds with such substituents (see later, eg HHSiD and *p*-f-HHSiD) at M_2 receptors suggests that these groups are poorly bound to the M_2 receptor (Table 9.5) (Waelbroeck *et al.* 1989).

9.2.1.1 *Selective muscarinic antagonists*

The drugs used for Parkinson's disease, which have a central asymmetric carbon atom, have served as the starting point in synthetic programmes to develop antagonists selective for the major muscarinic subtypes (M_1, M_2 and M_3). Procyclidine and trihexyphenidyl display selectivity for displacing [^{3}H]NMS in the order $M_1 > M_3 > M_2$. Replacement of the central carbon atom with silicon formed a series of sila-derivatives with the same potency order and only modest changes in affinity. Increasing the chain length of trihexyphenidyl by one methylene group (hexahydrodifenidol, HHD) produced derivatives with reduced activity at atrial M_2 receptors (Table 9.5). This selectivity was retained with the sila-analogue, hexahydro-siladifenidol (HHSiD). Of the various substitutions on the phenyl ring of these compounds, a *para*-fluoro was found to produce the most selectivity in favour of M_3 and away from M_2 receptors (*p*-f-HHSiD). These compounds display marked stereoselectivity, the (R)-isomer being 10–500-fold more potent. Quaternization increases affinity more at M_1 receptors and little at M_3 receptors, so that the potency order for *p*-f-HHSiD methiodide becomes $M_1 > M_3 = M_2$. Stabilization of the flexible side-chain of this series of compounds has been achieved by introducing double or triple bonds. The side-chain of the acetylenic ($\equiv$) analogues of HHD is constrained to be linear. A triple bond in (R)-HHD to produce (R)-hexbutinol (Table 9.5) increased affinity for M_2 and M_3 receptors but not M_1 receptors; the selectivity profile was therefore diminished (pA_2 values of 8.8, 7.8 and 8.8 for M_1, M_2 and M_3, respectively). Stereoisomerism of hexbutinol is lower than for (R)-HHD, but is still greatest at M_1 receptors. It is also greater at M_1 receptors in the derivative of HHD with a double bond at the asymmetric carbon end of the chain, *trans*- being more potent than *cis*-hexbutenol (Lambrecht *et al.* 1989).

4-DAMP was one of the earliest muscarinic antagonists to reveal selectivity between tissues (Table 9.5); it preferentially blocked the ileal muscarinic response (M_3) by ~15-fold compared with the cardiac responses (M_2) (Barlow *et al.* 1976). It is therefore a potent antagonist selective for M_3 receptors ($pA_2 = 9.0$), although it does not discriminate between M_1 and M_3 receptors (Hulme *et al.* 1990). For example, 4-DAMP had the same pA_2 values for antagonism of relaxation of rabbit aorta (M_3, 8.70) and inhibition of electrical stimulation-evoked contraction of the rabbit vas deferens (M_1, 8.90) (Angeli *et al.* 1991). It blocks a range of agonists with equal affinity at the M_3 receptors of the trachea (Table 7.3). A comparison of the blocking potency of 4-DAMP for a wide range of tissue responses can be seen in Table 7.2.

Table 9.5 Experimental muscarinic antagonists aimed at subtype selectivity

Derivatives of established anti-Parkinsonian agents

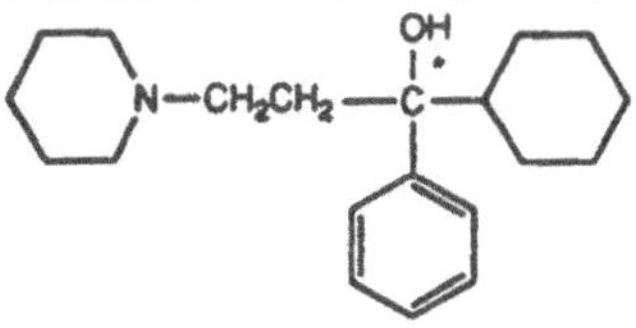

Trihexyphenidyl
$M_1 > M_3 > M_2$

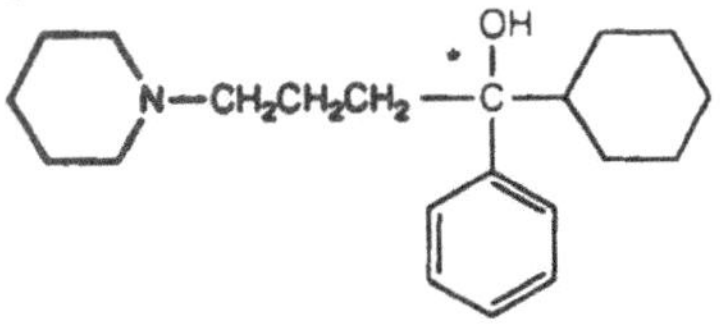

Hexahydro-difenidol (HHD)
$M_3 = M_1 > M_2$
pA$_2$ 8.0 7.8 6.7

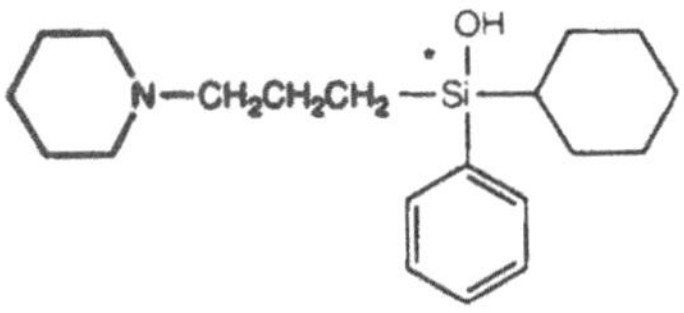

(R)-Hexahydro-sila-difenidol (HHSiD)
$M_1 = M_3 > M_2$
pA$_2$ 7.9 8.1 6.5
R/S ratio   501 : 16 : 200

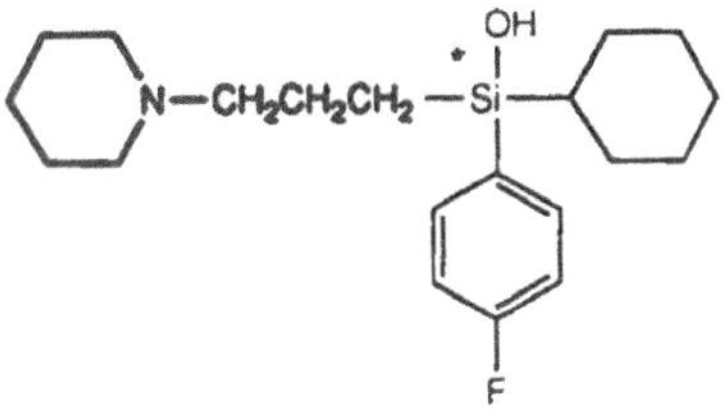

p-Fluoro-hexahydro-sila-difenidol (p-f-HHSiD)
$M_3 > M_1 > M_2$
pA$_2$ 7.84 6.68 6.01

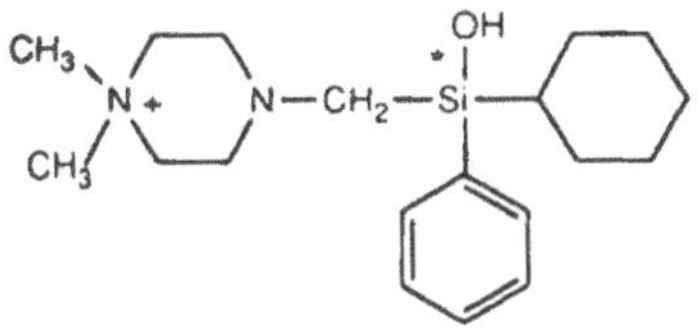

Sila-hexocyclium

M_1 selective $M_1 > M_4 > M_2$
pA$_2$ 9.0 8.8 7.6

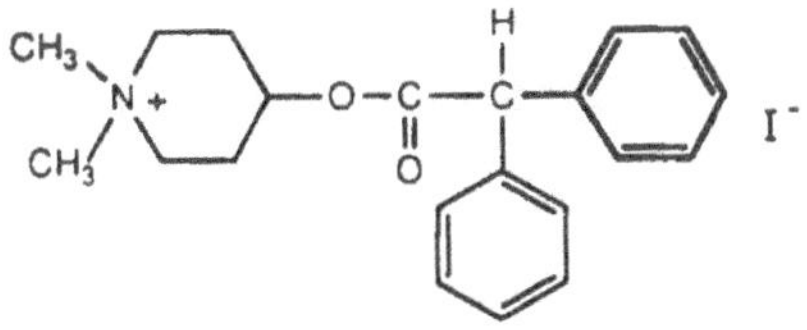

4-DAMP (4-diphenyl-acetoxy-N-methyl-piperidine) (methiodide)
M_3 selective $M_3 = M_1 > M_2$
pA$_2$ 8.8–9.6 9.4 8.0

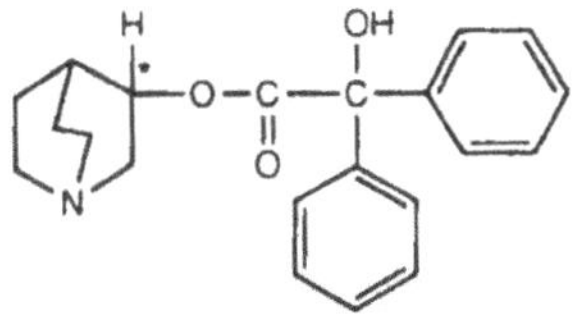

Quinuclidinyl benzilate (QNB)
Non-selective, highly potent

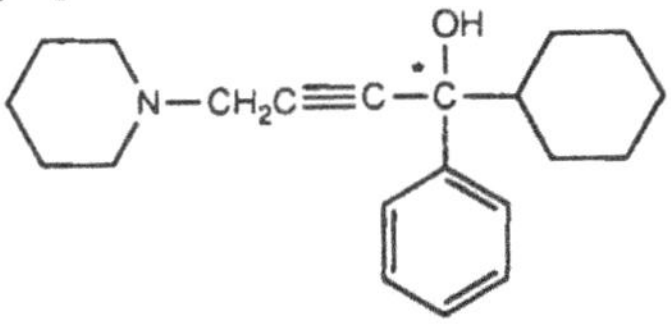

(R)-Hexbutinol
$M_3 = M_1 > M_2$
pA$_2$ 8.8 8.8 7.8

Note: An asterisk indicates an asymmetric chiral centre.

Polymethylene tetra-amines formed the basis of another approach in the search for selectivity among the muscarinic antagonists (Melchiorre *et al.* 1989). These consist of two discrete N-substituted phenyl groups separated by a spacer containing two amino groups. The optimum methylene chain length between the two amino groups is eight for antagonism at guinea-pig atrial M_2 receptors. The optimum methylene chain length between the amino substitution on the phenyl ring and each linking amino group is six. The resulting compound is methoctramine (see Table 9.7). This antagonist has selectivity for M_2 receptors (pA$_2$ in guinea-pig left atria, 7.97) compared with the M_3 receptor mediating contractile responses of the guinea-pig ileum ($pA_2 = 5.92$). Inspection of Table 7.2 shows that pA_2 values for M_1 and M_4 receptor-mediated responses are approximately 6.2 and 7.5; it does not therefore discriminate between M_2 and M_4 receptors particularly well or between M_1 and M_3 receptors in functional studies. The four basic nitrogen atoms in the spacer chain are required for optimum activity, which suggests that these interact with four corresponding nucleophilic sites on the muscarinic receptor. The primary candidates as anionic receptor sites for binding to the protonated amine groups are the four conserved aspartate residues (67, 97 103 and 120) on transmembrane domains 2 and 3 of the receptor. Alternatively, the extracellular loop between transmembrane α-helixes 4 and 5 has Asp-173 and Glu-172 and Glu-175 which are found only on the m_2 receptor and may therefore be possible binding sites. The antagonism by methoctramine is always competitive at the M_2 receptors in isolated atria, although at high concentrations the Schild plots deviate from unity slope (Eglen *et al.* 1988b). In displacing radioligands, low concentrations ($\leqslant 1\ \mu M$) are competitive, whereas high concentrations ($>1\ \mu M$) show an allosteric interaction. In the presence of high concentrations, [^{3}H]NMS or [^{3}H]dexetimide show slow dissociation from cardiac membranes, indicating a non-competitive interaction. This is similar to the allosteric interactions of AF-DX 116, gallamine and other quaternary compounds. The allosteric binding site of the M_2 receptor may account for the selectivity of these compounds for the M_2 receptor. The allosteric binding of methoctramine may involve one end of the molecule, possibly to the anionic groups located in the extracellular loop between transmembrane helixes 4 and 5, indicated above.

9.2.2 *Pirenzepine and its Analogues*

Pirenzepine is characterized by the tricyclic pyrido-benzo-diazepinone system with a semi-rigid side-chain (Table 9.6). Pirenzepine was the first antagonist to show that muscarinic receptor binding sites in the cerebral cortex, hippocampus and peripheral sympathetic ganglia (M_1) could be distinguished from those in the salivary and lacrymal glands, the heart, stomach and ileum (Hammer *et al.* 1980). Radiolabelled pirenzepine displayed a 20- to 50-fold variation in affinity between tissues and between the M_1, M_2 and M_3 receptor sites. In numerous functional studies, the selectivity of pirenzepine for M_1 receptor-mediated responses has been clearly demonstrated. Antagonism of the slow depolarization of the rat superior cervical ganglion and inhibition of the stimulation-induced contraction of the rabbit vas deferens by pirenzepine both show higher pA_2 values (8.4 and 8.2) than for responses mediated via any of the other three functional muscarinic receptor subtypes (Table 7.2). pA_2 values for M_2, M_3 and M_4 receptor-mediated responses lie in the range 6.2–7.2 (Dörje *et al.* 1990, Angeli *et al.* 1991).

The tricyclic ring nucleus of pirenzepine has been used as the basis for structural modifications to achieve different selectivities. The most significant changes have occurred by alteration of the side-chain; for example, removal of the most basic of the nitrogens from the side-chain ring.This has yielded antagonists with selectivity for cardiac M_2 receptors compared with ileal M_3 receptors. The first of these compounds was AF-DX 116, which has low affinity but modest selectivity for M_2 receptors in binding and functional studies (Hammer *et al.* 1986). Improved affinity and selectivity is claimed for AF-DX 384 and AQ-RA 741. The latter compound, however, is protonated and hydrophilic which limits its penetration of the

Table 9.6 Muscarinic antagonists based on pyrido-benzo-diazepinones

					Selectivity
Pirenzepine (Gastrozepine[R])					M_1
					$M_1 > M_2 = M_3 = M_4$
					pA_2 8.4 6.8 6.9 7.0
Telenzepine					M_1
					$M_1 > M_2 = M_3$
					pA_2 7.9 7.7

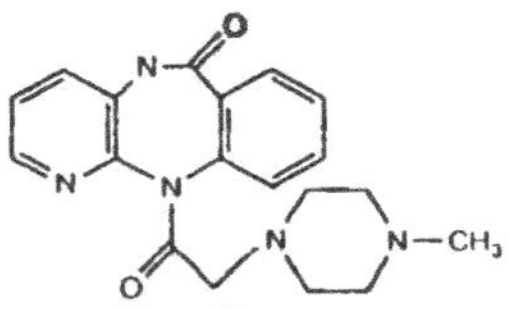

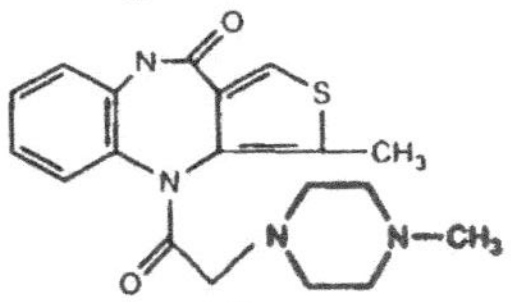

	X	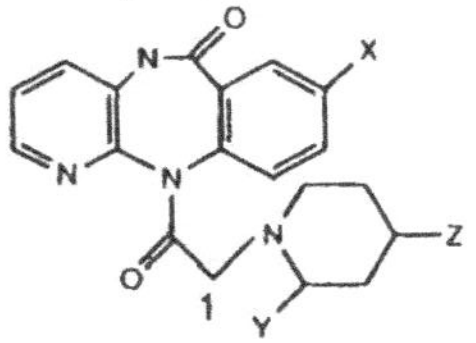 Y	Z		
AF-DX 116	-H	-CH$_2$N(C$_2$H$_5$)$_2$	– H		M_2
					$M_2 > M_1 = M_4 = M_3$
					pA_2 7.4 7.1 7.0 6.5
AF-DX 384	-H	-CH$_2$N(C$_2$H$_5$)$_2$ (-NHCH$_2$CH$_2$ – at 1)	– H		M_2
					$M_2 > M_1 \gg M_3$
					pK_i 8.2 7.5 7.0
AQ-RA 741	-H	-H	-(CH$_2$)$_4$ N(C$_2$H$_5$)$_2$		M_2
					(protonated, hydrophilic)
					$M_2 > M_1 > M_3$
					pK_i 8.3 7.7 6.9
BIBN 99	-Cl	-H			M_2
					(lipophilic)
					$M_2 > M_1 = M_3$
					pK_i 7.6 6.2 6.1

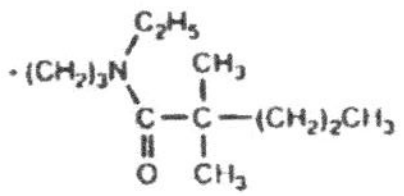

blood–brain barrier. The search for a lipophilic agent that would cross into the brain for potential use in Alzheimer's disease (see later) has yielded BIBN 140 and BIBN 99. These have improved selectivity for cardiac M_2 receptors (pA_2 of 7.2 for BIBN 99) compared with M_1 (6.2) and M_3 (6.1) receptor binding sites (Doods *et al.* 1993).

9.2.3 *Miscellaneous Antagonists*

Finally, there are several antagonists of miscellaneous structures which have not been dealt with so far but which are of interest or have been found to display selectivity. These include the naturally occurring alkaloid, himbacine, which is obtained from the stem bark of *Galbulimima belgraveana* (Table 9.7). It has approximately a 10-fold selectivity in blocking atrial M_2 receptors ($pA_2 = 8.52$) compared with guinea-pig ileal ($pA_2 = 7.2$), tracheal (7.61) and oesophageal muscularis mucosal (7.57) M_3 receptors. The antagonism is competitive with no evidence of allosteric interactions. Himbacine is also able to discriminate between M_1 and M_4 and between M_1 and M_3 receptors in functional tests (Eglen *et al.* 1988b; Dörje *et al.* 1990; Caulfield 1993). Secoverine is an M_4-selective antagonist. Zamifenacin is a novel muscarinic antagonist (Table 9.7) with M_3 selectivity but also appears to preferentially block M_3 receptors of the gut compared with the trachea. This may indicate a further subdivision of the M_3 receptor in these tissues or that the tracheal contraction is mediated via more than one receptor for which zamifenacin has different affinities (Caulfield 1993, Eglen *et al.* 1994). Gallamine has already been mentioned as a nicotinic antagonist of the neuromuscular junction N_M nicotinic receptor. It has three quaternary heads and displays weak blocking activity at muscarinic receptors which is selective for the M_2 receptor. Like methoctramine, it binds to an allosteric site of the M_2 receptor. The competition curves of gallamine with [^{3}H]NMS binding are shifted to higher concentrations by D-tubocurarine which therefore shares the same allosteric site. The atropine, AF-DX 116 and dexetimide competition curves, however, are not affected by D-tubocurarine and these antagonists do not therefore recognize this allosteric site (Waelbroeck *et al.* 1993).

Quinuclidinyl benzilate (QNB) (Table 9.5) and dexetimide (Table 9.7) are non-selective antagonists with high affinity for all receptor subtypes and are therefore widely used as radioligands in binding studies. They both display high stereoselectivity which has been applied to investigations of the binding geometry to the muscarinic receptor. High stereoselectivity of an antagonist indicates that in the weaker enantiomer there is reduced interaction with the receptor arising through three possible causes.

1   Loss of a single interaction between drug and receptor. Dexetimide serves as an example, its very high stereoselectivity indicating that all four moieties at the chiral centre probably interact strongly with the receptor. Inappropriate alignment of one of the binding groups in the weaker enantiomer leads to only three points of interaction.

2   Steric hindrance. An example of this type of stereoselectivity is probably QNB, the (R)-form having greater affinity ($pK_i = 10$) with an isomeric ratio of ~100 at M_2 receptors. It is suggested that the protonated amino group in the quinuclidinyl ring in the (R)-isomer forms a strong ionic bond with the receptor. The (S)-configuration of this ring, however, causes steric interference and reduced activity.

Table 9.7 Other muscarinic antagonists with diverse structures

Polymethylene tetra-amines

(eg methoctramine)

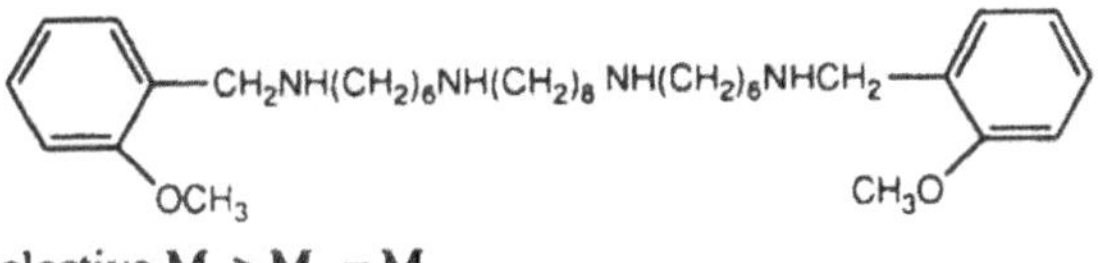

$-CH_2NH(CH_2)_6NH(CH_2)_8 \; NH(CH_2)_6NHCH_2-$

OCH$_3$ CH$_3$O

M_2 selective $M_2 > M_3 = M_1$
pA$_2$ 8.0   5.9   6.2

Secoverine

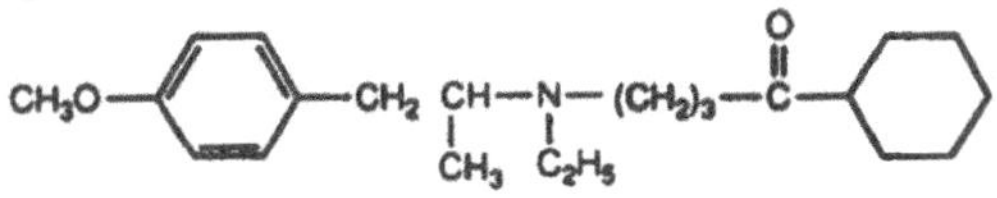

CH$_3$O— —CH$_2$ CH—N—(CH$_2$)$_3$—C

CH$_3$ C$_2$H$_5$

M_4 selective $M_4 > M_1 > M_2 > M_3$
pK$_i$ 9.3   8.4   8.0   7.7

Gallamine
(Flaxedil^R)

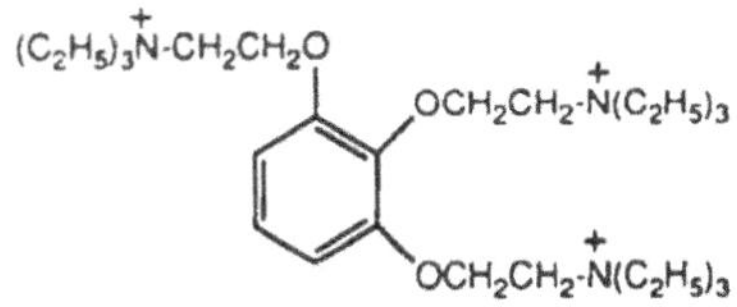

$(C_2H_5)_3\overset{+}{N}$-CH$_2$CH$_2$O

OCH$_2$CH$_2$-$\overset{+}{N}$(C$_2$H$_5$)$_3$ M_2 selective

OCH$_2$CH$_2$-$\overset{+}{N}$(C$_2$H$_5$)$_3$

(Principally nicotinic, N$_M$, non-depolaring skeletal muscle relaxant)

Himbacine

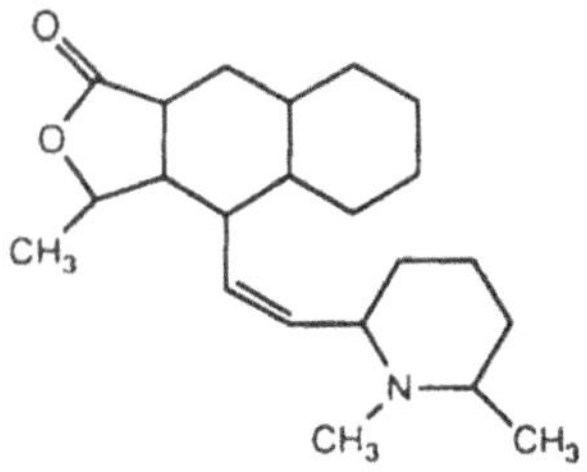

CH$_3$

CH$_3$ CH$_3$

$M_2 > M_3$, also discriminates between M_1 and M_4

$M_2 = M_4 > M_1 > M_3$
pK$_i$ 8.1   8.0   7.1   6.5

Dexetimide
((S)-benzetimide)

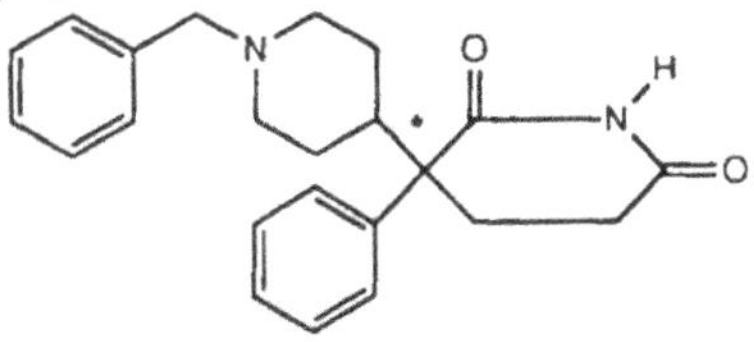

Non-selective, high stereoselectivity ratio (1000-fold)

Zamifenacin

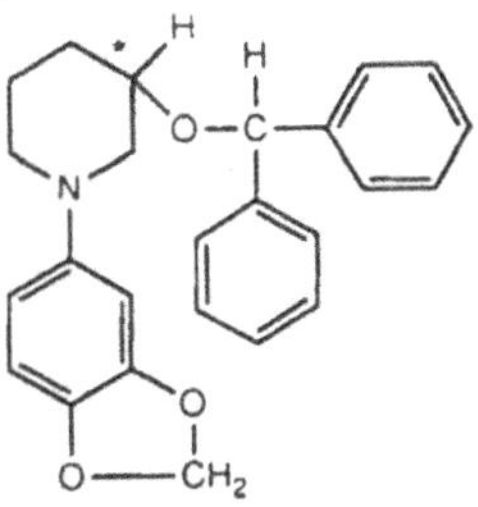

Ileum M_3 selective

M_3 (gut) $> M_3$ (trachea) $> M_1 > M_2$
pA$_2$ 9.09   8.06   7.38   7.14

3   Exchange of position of two groups with similar physicochemical properties. This form of stereoisomerism is illustrated by procyclidine (Table 9.4), the (R)-enantiomer of which is more active than is the (S)-enantiomer (pA_2 values at ileal M_3 receptors and 8.04 and 5.4, respectively). The two analogues having either two phenyl groups (pyrrinol) or two cyclohexyl groups (hexahydro-procyclidine) on the central carbon atom are non-chimeric. Their affinities ($pA_2 = 6.91$ and 6.37) at ileal M_3 receptors are still more than the equivalent figure for (S)-procyclidine. Thus, the cyclohexyl and phenyl groups are interchangeable in their receptor interactions. In (S)-procyclidine, a four-group interaction is preferred to one in which three of the groups are 'correctly' bound, leaving the fourth group pointing away from the receptor. Similar arguments would apply to (R)-HHD and its derivatives. The weak M_2 activity of these compounds bearing a cyclohexyl and phenyl group at the chiral centre suggests that these groups are poorly bound to the M_2 receptor.

9.2.4   *Irreversible Muscarinic Receptor Antagonists*

The β-haloalkylamines, dibenamine and phenoxybenzamine (PBZ), have already been described as irreversible α-adrenoceptor antagonists (Chapter 5, Table 5.2). Their selectivity is, however, low and they also have the ability to irreversibly alkylate other receptors including muscarinic receptors. PBZ is less selective for α-adrenoceptors than is dibenamine, and exhibits irreversible muscarinic blockade at all subtypes of the receptor. This activity is believed to involve conversion to the intermediate ethyleniminium (aziridium) ion which then alkylates the muscarinic receptor in a similar fashion to α-adrenoceptors (see Figure 5.1). Support for a specific interaction with the muscarinic receptor comes from receptor protection and cross-protection experiments. The blockade by the β-haloalkylamines of muscarinic receptor-mediated responses is not prevented by co-incubation with noradrenaline. The blockade is, however, prevented by prior equilibration with atropine. The possibility exists that this class of irreversible antagonist interacts with at least two sites, one of which is the noradrenaline-binding site of the α-adrenoceptor, the other being less specific and common to several receptors (Triggle & Triggle 1976). In the guinea-pig ileum, the irreversible blockade of the M_3 receptor-mediated contractile responses to (+)-*cis*-dioxolane are prevented by prior equilibration with the competitive M_3 selective antagonist, p-f-HHSiD, but not with methoctramine (M_2 selective). In contrast, the blockade by PBZ of the negative inotropic response of guinea-pig atria is prevented by incubation with methoctramine but not with p-f-HHSiD (see Figure 9.4). Thus, PBZ blocks both M_2 and M_3 receptors (Eglen & Harris 1993).

The competitive tertiary amine class of muscarinic antagonists (Table 9.4) has been combined with the β-haloalkylamine structure to produce irreversible muscarinic antagonists. These include benzilylcholine mustard (BCM) (Table 9.8), a chloroethyl derivative of lachesine (Table 9.2), which produces a large rightward shift of the carbachol dose–response curve of guinea-pig ileum, followed by depression of the maximum response (Triggle & Triggle 1976). Such agents are proving useful in the elucidation of the precise receptor sites with which muscarinic agonists and antagonists interact. Propylbenzilylcholine mustard (PrBCM) is non-selective between muscarinic receptors and is presumed to bind irreversibly to the muscarinic site in an

Table 9.8   Irreversible antagonists at muscarinic receptor

	Aziridinium ion	
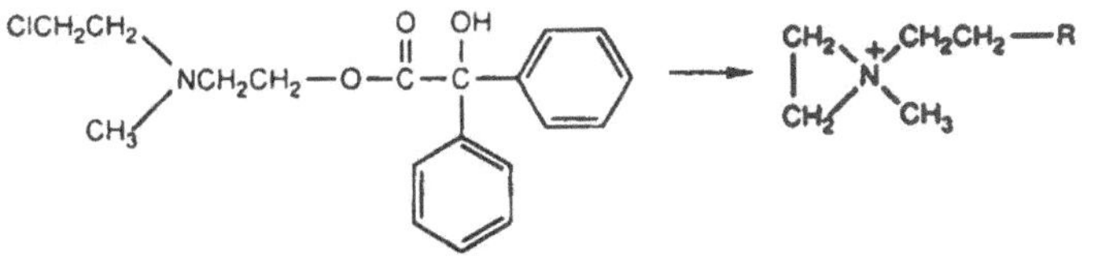 Benzilycholine mustard (BCM)		Non-selective
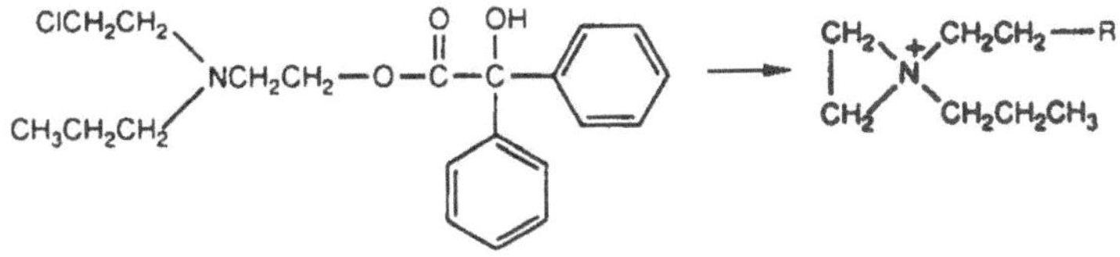 Propylbenzilylcholine mustard (PrBCM)		Non-selective
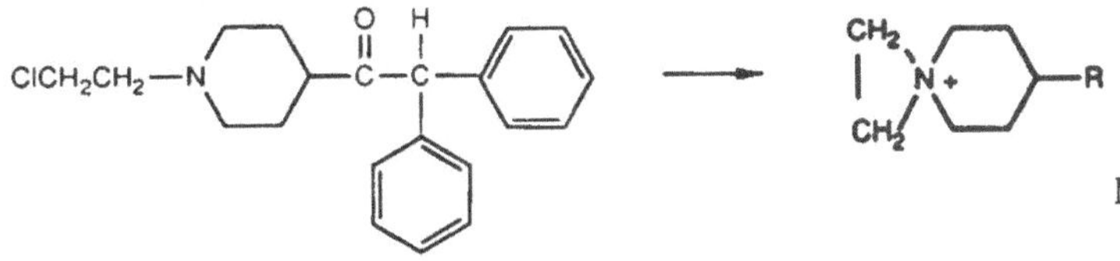 4-Diphenyl-acetoxy-N-(2-chloroethyl)-piperidine (4-DAMP mustard)		Modest M$_3$ selectivity

identical manner to the competitive form of the ligand (Table 9.8). The reaction involves a nucleophilic attack by an amino acid side-chain of the receptor sequence on the reactive aziridinium ion of the alkylating agent. The link formed between [^{3}H]PrBCM and the muscarinic receptor can be cleaved by hydroxylamine under denaturing conditions, which is indicative of an ester link with a carboxylic acid group rather than thioether bonds with cysteine residues. The likely candidates for alkylation of muscarinic receptors by [^{3}H]PrBCM are Asp-105 (M$_1$) and Asp-111 (M$_4$), according to sequencing of the cleavage products of the receptor amino acids. The groups associated with these residues are located in transmembrane helix 3 of the receptor (see Chapter 13) (Hulme *et al.* 1989). Ach mustard also appears to bind irreversibly to Asp-105, suggesting it is the binding site of Ach (Birdsall *et al.* 1993).

More recently, there has been a search for irreversible antagonists that would selectively alkylate the muscarinic receptor subtypes. These agents would additionally assist in the characterization of the physiological role of the subtypes. 4-DAMP mustard, like its competitive precursor, is claimed to have selectivity for M$_3$ over the other subtypes (Birdsall *et al.* 1993) (Table 9.8). The selectivity, however, has been disputed and found to be only moderate at seven-fold (Eglen & Harris 1993). 4-DAMP mustard penetrates the blood–brain barrier and alkylates both peripheral and central muscarinic receptors. 4-DAMP bromomustard, however, rapidly cyclizes to the aziridinium ion and does not penetrate to the brain, alkylating only peripheral muscarinic receptors (Griffin *et al.* 1993). An alternative approach has been to introduce an isothiocyanate group into the M$_1$-selective competitive antagonists, pirenzepine and telenzepine. The resulting derivative of pirenzepine was non-selective as an irreversible muscarinic inhibitor, whereas the corresponding telenzepine product was five-fold selective for M_1 compared with M_2 receptors (Baumgold *et al.* 1992).

9.3 Pharmacology of Atropine and Muscarinic Antagonists

The prototype antimuscarinic agent is atropine and a description of its pharmacology serves to illustrate the pharmacological effects of muscarinic blockade in general and remains relevant to the majority of therapeutically useful agents. Like any generalization, however, there are exceptions to the rule and atropine is no exception. Atropine does have distinctive properties that may not be shared by other muscarinic antagonists; these will be indicated where appropriate.

Atropine is notably less effective in antagonizing the responses to parasympathetic nerve stimulation than to exogenously administered Ach. In the past, explanations for this phenomenon have included the suggestion that Ach is released from parasympathetic nerves into very close proximity to the receptors and diffusion may limit the concentration of atropine at these receptor sites. Differences also occur in the susceptibilities of various organs and tissues to the blocking action of atropine. For example, salivary and bronchial secretions are depressed at low doses of atropine. As the dose of atropine is increased, so progressively more organs are affected, the hierachial order being the pupil, heart, bladder and gastrointestinal tract. Finally, much larger doses are required to inhibit gastric acid secretion and motility. Both of the phenomena – different susceptibilities of Ach and nerve stimulation to atropine and the variable organ sensitivity to blockade – may have a common and more satisfactory explanation. The bladder exhibits one of the greatest differences in sensitivity to exogenous Ach and nerve stimulation and micturition is one of the least sensitive to blockade by atropine. Two explanations can be identified. Firstly, because of the presence of prejunctional M_2 muscarinic receptors that inhibit Ach release, atropine, being non-selective, will facilitate Ach release. This will offset the blockade of the postjunctional muscarinic receptors (M_2 or M_3) and could explain the reduced effectiveness against responses mediated via nerve stimulation. A second reason is the likelihood that electrical stimulation to some organs involves non-adrenergic non-cholinergic (NANC) pathways (Chapter 12). Since these release alternative transmitters including ATP, which are not blocked by muscarinic antagonists, it is not surprising that there should be atropine-resistant components of the response (eg Figure 7.6). The involvement of NANC pathways is particularly evident in organs such as the bladder and gastrointestinal tract.

In the description of the pharmacological profile of atropinic drugs that follows, the activity on each body system will be related to (1) the muscarinic receptor subtype involved (see Table 7.2), (2) the symptoms of belladonna poisoning (Table 9.9) and (3) the therapeutic uses. The therapeutic applications of muscarinic antagonists and of individual agents with specific uses will be considered separately.

9.3.1 Cardiovascular System

The heart rate is under parasympathetic vagal tone via M_2 receptors on the SA node, which are blocked by atropinic drugs to cause tachycardia (see Figure 11.1). There may be a paradoxical transient slowing of heart rate in humans after slow intravenous administration. The most likely explanation for this is blockade of prejunctional M_2 receptors on parasympathetic nerve endings which normally inhibit transmitter release (Chapter 7). Their blockade will facilitate Ach release and hence induce a brief bradycardia. Alternatively, M_1 receptors involved in sympathetic ganglionic transmission to the heart may be blocked since pirenzepine, the M_1 selective antagonist, is also quite effective at reducing heart rate and then prevents any further bradycardiac action of atropine. However, since the effect persists in the presence of propranolol, which would prevent the sympathetic effect to the heart, a prejunctional M_1 receptor inhibiting transmitter release has also been implicated. This, however, awaits confirmation (Wellstein & Pitschner 1988).

The predominant tachycardia after atropine is most pronounced in healthy young adults who have a greater degree of vagal tone. Hyoscine surprisingly causes only the bradycardiac response. Atropine has important effects upon atrioventricular conduction. Since the normal vagal influence on the heart is removed, conduction velocity is increased by atropine, as indicated by shortening of the P–R interval on the ECG. This facilitation of transmission of excitation through the AV node will increase ventricular rate in patients with supraventricular tachyarrhythmias. In patients with heart block, for example induced by cardiac glycoside toxicity, the improved AV conduction by atropine may reduce the degree of block. The inherent ventricular rate (idioventricular rate) of patients with complete heart block may be accelerated by atropine.

The changes in heart rate induced by atropine are not usually accompanied by alterations in blood pressure or cardiac output because there is adequate compensatory reflex adjustment. Atropine and hyoscine have little effect upon blood pressure or the calibre of blood vessels since they are under no significant parasympathetic tone. Those blood vessels that receive a sympathetic cholinergic innervation in skeletal muscle (see Chapter 1) do not appear to have a major role in the control of vascular tone and therefore clinical doses of atropine are similarly ineffective. Larger doses approaching toxic levels, however, cause an anomalous peripheral vasodilatation and fall in blood pressure. Cutaneous blood vessels, particularly in blush areas, are dilated to produce a profound flushing of the skin. The mechanism for this remains obscure, but it is unrelated to muscarinic receptor blockade and is not produced by other muscarinic antagonists. It may be due to a direct vasodilator action or related to the inhibition of sweating that accompanies atropine administration and compensation for the consequent increase in skin temperature. The latter, however, would be expected to occur with other M_3 antagonists. The dry flushed skin is a characteristic sign of belladonna poisoning, together with the rapid but weak pulse.

In isolated perfused hearts, atropine alone has little effect other than a small coronary vasodilatation, but it abolishes the negative chronotropy and coronary vasodilatation induced by Ach (Figure 8.1).

9.3.2 Gastrointestinal Tract

9.3.2.1 Contractility

Atropine abolishes the contractile responses of the gastrointestinal tract to Ach and other muscarinic agonists. This is classically demonstrated in the guinea-pig ileum where it has a pA_2 value of 9.0 and the receptor is of the M_3 subtype (Table 7.2). Muscarinic antagonists are not always as effective in inhibiting the contractile responses to nerve stimulation as they are the effects of Ach. For example, in the rabbit colon there is only partial inhibition of the response to pelvic nerve stimulation by atropine (100 nM), although increasing the concentration to 100 μM does abolish this response (Figure 6.3). This resistance of nerve stimulation to blockade by atropine may be due to the simultaneous blockade of prejunctional M_2 receptors. The response to single pulses of transmural stimulation of the guinea-pig ileum preparation (Paton preparation) is also abolished by atropine, indicating that this response is due entirely to the release of Ach from postganglionic parasympathetic nerves (Paton 1957) (Figure 9.1). At high frequency stimulation, however, there is another contractile response of the circular muscle of the ileum which is not reduced by atropine. Since blockade of substance P receptors causes almost complete inhibition of this atropine-resistant response, it is probably mediated through the release of substance P (Chapter 12). Increasing the intraluminal pressure of the isolated segment of guinea-pig ileum initiates the local myenteric reflex and causes a contractile response. This is susceptible to muscarinic antagonists and therefore largely due to Ach release. In the maintained presence of the muscarinic antagonist, further increasing the distention pressure within the ileum restores the reflex contraction. Since this is unaffected by muscarinic antagonists, but is inhibited by a substance P antagonist, it is attributed to the neuronal release of the neuropeptide, substance P (Morris & Gibbins 1992) (Chapter 12). This co-transmission in the gastrointestinal tract probably explains the relative ineffectiveness of atropine in reducing gut motility in humans. Larger doses than required for the ocular responses have to be employed to produce sustained reductions in motor activity of the stomach, duodenum, ileum and colon. The frequency and amplitude of peristaltic contractions are reduced and there is inhibition of smooth muscle tone. The reason for the insensitivity appears to be because the parasympathetic nerves synapse in the intramural plexus with both cholinergic postganglionic fibres and with non-cholinergic fibres (Figure 1.5). The partial inhibition of gastrointestinal motility by atropine and synthetic substitutes has led to their widespread use in the treatment of various conditions associated with intestinal hypermotility.

9.3.2.2 Gastric acid secretions

Gastric acid secretion induced by parasympathetic nerve activity is due to a direct effect upon parietal cells of the fundus of the stomach and an indirect effect through

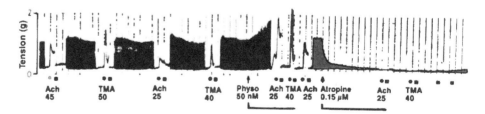

Figure 9.1 Contractions of guinea-pig isolated ileum transmurally stimulated at one pulse every 10 sec with square-wave pulses of maximal voltage (20 V) and 5 msec pulse width. Acetylcholine (Ach, 25 or 45 nM, O) or tetramethylammonium (TMA, 40 or 50 μM, ●) were added with the stimulator turned off, followed by washout (■). Physostigmine (50 nM) potentiates the contractions induced by transmural (parasympathetic) stimulation, Ach and TMA. Atropine blocks the contractions to all three stimuli washout (■), causing slow recovery.

release of gastrin from the antrum of the stomach (Figure 8.3). Isolated fundic parietal cells are stimulated by muscarinic agonists, and atropine is 100-fold more potent than is the M_1 selective agonist, pirenzepine. This relative weakness in the action of pirenzepine is proof that the direct action of Ach on gastric acid secretion is not mediated via M_1 receptors. Similarly, acid release from the mouse isolated stomach induced by muscarinic agonists is blocked by pirenzepine with an identical pA_2 value to that in the trachea, atrium or ileum (Szelenyi 1982, Black & Shankley 1985). M_1 receptors are therefore not involved in the direct effect at the level of the parietal cell.

Gastrin release induced by nerve activity or by placing food in the stomach is only partly blocked by atropine, whereas the dose-dependent increase in gastrin release by muscarinic agonists is blocked, consistent with M_2 receptors being involved. As shown in Figure 8.3, part of this effect may be due to removal of the inhibitory effect of somatostatin. The inhibition of somatostatin release by cholinergic agents is also mediated via muscarinic receptors not of the M_1 subtype (Soll 1984). *In vivo*, vagally-mediated gastric acid and pepsin secretion from the fundic region of the stomach (after selective vagotomy of the antrum) is equally antagonized by atropine and pirenzepine, indicating that M_1 receptors *are* involved. Bethanechol also stimulated gastric acid and pepsin secretion and was antagonized by pirenzepine but less effectively than was vagal stimulation (Hirschowitz & Molina 1984). These results suggest that pirenzepine probably inhibits the M_1 receptors located on the soma-dendritic portion of the postganglionic nerve of the myenteric plexus. The release of Ach from postganglionic nerves of the guinea-pig myenteric neurones has been shown to be susceptible to pirenzepine, with a pA_2 value consistent with it being via an M_1 receptor (Kilbinger 1984).

The control of gastric acid secretion occurs in three overlapping phases. The cephalic phase occurs when food is smelled, seen or swallowed and vagal parasympathetic activity causes a small degree of gastric acid secretion *and* gastrin secretion from the antrum. The gastric phase involves gastrin and stimulation of the intrinsic nerves by the presence of food in the stomach. The intestinal phase consists of both an excitatory component due to gastrin release and inhibitory mechanisms mediated via cholecystokinin (CCK-8), gastric inhibitory polypeptide (GIP) and secretin initiated by the presence of acid in the duodenum. Secretion during the cephalic phase is therefore more markedly reduced by atropinic drugs, as is basal secretion. The intestinal phase, however, is less affected. The volume of secretion is usually reduced by atropine, but pH is not necessarily altered, possibly because secretion of HCO_3^- in addition to H^+ is reduced. Vagally-mediated secretion of mucin and proteolytic enzymes are also reduced by atropine.

As a consequence of these properties of atropine and related antagonists, they have applications in the management of peptic ulcers.

9.3.3 *The Eye*

Muscarinic antagonists prevent the contraction of the circular muscle of the iris in response to light, mediated via the parasympathetic innervation. The iris is therefore dilated (mydriasis) and fails to respond to bright light. The contraction of the ciliary muscle which normally increases the curvature of the lens in accommodating for near vision is blocked (Figure 1.7). Thus, accommodation is blocked (cycloplegia) and vision is blurred. These effects occur with both systemic and local administration and tend to be an early sign of belladonna poisoning (after drying of secretions) (see Table 9.9). The mydriasis differs from that of sympathomimetic amines (Chapter 4) which do not block the light reflex or cause cycloplegia. These effects are due to blockade of M_3 receptors on the smooth muscle (Table 7.2).

The dilation of the pupil and relaxation of the iris back into the drainage angle of the anterior chamber of the eye interferes with the drainage of aqueous humour via the trabecular network into the canal of Schlemm (Figure 1.7). Thus, there is a risk of raising intraocular pressure and precipitating glaucoma in susceptible individuals. Patients with narrow-angle glaucoma are in greater danger than those with the more common open-angle condition. Caution should therefore be exercised in the topical use of atropinic drugs as mydriatics, for eye examinations and for inflammatory disorders of the eye.

9.3.4 Secretions

Exocrine secretion from lacrymal, salivary and mucous glands is induced by parasympathetic activation via release of Ach onto M_3 receptors. Atropinic drugs therefore dry up mucous membranes of the respiratory tract. This is the basis of their use in preanaesthetic medication since excessive bronchial secretion can cause reflex laryngospasm. Irritation from the older gaseous anaesthetics induced hypersecretion and a risk of the patient choking.

Salivary secretions are particularly sensitive to atropinic agents and a dry mouth, with difficulty in talking and swallowing, is the first sign of belladonna poisoning. It is also a common side-effect of the muscarinic antagonists.

Sweat glands are under sympathetic cholinergic innervation, with Ach serving as the transmitter. Thus, after small doses of atropine and hyoscine, the skin becomes dry (and hot, because of cutaneous vasodilatation). The reduced sweating causes body temperature to rise giving symptoms of fever; atropine poisoning differing from other forms of fever by the dry skin. Secretions of milk, bile and from the pancreas are unaffected by atropine.

9.3.5 Respiratory Tract

The tone of smooth muscle in the respiratory tract is under vagal parasympathetic control. These nerves are activated by a wide range of stimuli, including stimulation of afferent sensory pathways triggered by noxious agents such as sulphur dioxide or interactions with specific receptors, such as bradykinin (Chapter 12). The Ach released at the efferent nerve terminals stimulates M_3 receptors on airway smooth muscle and submucosal glands, resulting in bronchoconstriction and mucous secretion, respectively. Airway submucosal glands are of two distinct secretory types: mucous and serous cells. Both types of cell appear to be stimulated to secrete by Ach, the mucous cells secreting a protein-rich mucus with increased viscosity, whereas secretion from serous cells has reduced viscosity. Ach and parasympathetic nerve activity result in increased volume of output from submucosal glands without affecting viscosity. Atropine-like drugs inhibit these responses and cause bronchodilatation and dried mucous secretions. The bronchoconstrictor responses to a wide range of stimuli are attenuated by atropine, indicating the role of parasympathetic reflexes in their responses. For example, for bronchoconstriction induced by inhalations of 5-HT, bradykinin and histamine are reduced by atropine, albeit to a lesser extent than the direct effect of muscarinic agonists (Figure 9.2). This action is utilized in the treatment of asthma and chronic obstructive pulmonary disease (COPD) (Sheppard 1987, Barnes 1993b).

9.3.6 Urinary Bladder

The detrusor muscle of the bladder receives an abundant parasympathetic innervation, contraction being induced by the release of Ach onto M_3 receptors (Table 7.2). The trigone has a limited muscarinic receptor population. The contraction of the detrusor muscle occurs when intraluminal pressure reaches a critical level, with consequent micturition. This response is inhibited by atropine (Figure 9.3). Continuous infusion of saline directly into the bladder causes cyclical waves of voiding (cystometry) which are blocked by atropinic drugs (Maggi *et al.*

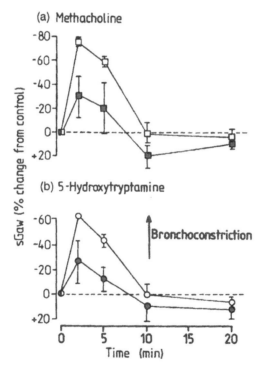

Figure 9.2 Bronchoconstriction of conscious guinea-pigs measured as the fall in specific airway conductance (sGaw, minus values) by whole body plethysmography. (a) Methacholine (□, 300 μg/ml) or (b) 5-hydroxytryptamine (5-HT, O, 500 μg/ml) were administered by aerosol inhalation (1 min) and repeated 24 hr later (solid symbols) at ½ to 1½ hr after atropine (100 μg/kg, ip). Note that muscarinic blockade by atropine inhibits the bronchoconstriction by both the muscarinic agonist methacholine and 5-HT, indicating that the latter is also mediated via parasympathetic nerve activation. Results of experiments performed by Dr Christine Lewis in the author's laboratory.

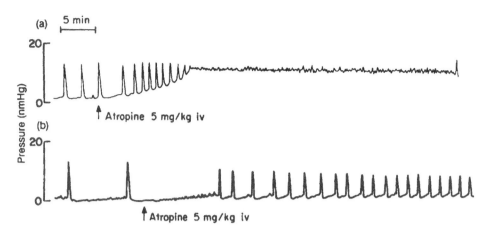

Figure 9.3 Inhibitory effect of atropine on the voiding cycle of the urinary bladder of anaesthetized guinea-pigs. Pressure waves are induced by the saline filling of the bladder, followed by voiding and return of pressure to basal level. Atropine (5 mg/kg, iv) produced overflow incontinence (a) or impairment of voiding with reduced amplitude of micturition contraction (b). Reproduced with permission from Maggi *et al.* (1987).

1987, Petersen *et al.* 1990). The early component of the increased bladder pressure occurring prior to voiding is resistant to atropine. This atropine-resistance was also observed in the contractile response of isolated bladder preparations to electrical field stimulation (Figure 7.6). The resistant contraction to the neurogenic response (blocked by tetrodotoxin) is believed to be mediated via the release of ATP (Maggi 1991b) (Chapter 12). The atropine resistance of the bladder does not seem to occur in the normal human bladder muscle. The reduced tone and amplitude of bladder contractions result in urinary retention and is therefore of value in the management of urge incontinence (Andersson 1988).

Bile duct, gall bladder and urethra spasm may be reduced to some extent by muscarinic antagonists.

9.3.7 Central Nervous System (CNS)

It is the central effects of muscarinic antagonists that vary most between the different agents. This is the major difference between atropine and hyoscine (scopolamine). In therapeutic doses, atropine has little effect on the CNS, whereas hyoscine has a substantial depressant action, producing drowsiness, amnesia, fatigue and dreamless sleep with less rapid-eye-movement (REM) sleep. This difference is probably due to limited penetration of atropine through the blood–brain barrier. The reduced cognitive function of Alzheimer's disease can be mimicked in experimental animals by administration of hyoscine. Since in Alzheimer's disease it is the input to unchanged postjunctional M_1 receptors from defective cortical neurones that is impaired, performance deficits in animal models of learning and memory induced by hyoscine are probably due to blockade of these postjunctional M_1 receptors. It is worth noting that depletion of Ach from central cholinergic neurones by intracerebroventricular (ICV) injections of hemicholinium-3 (HC-3, see Chapter 7) produces a similar impairment of learning. HC-3 mimicks Alzheimer's pathology better than

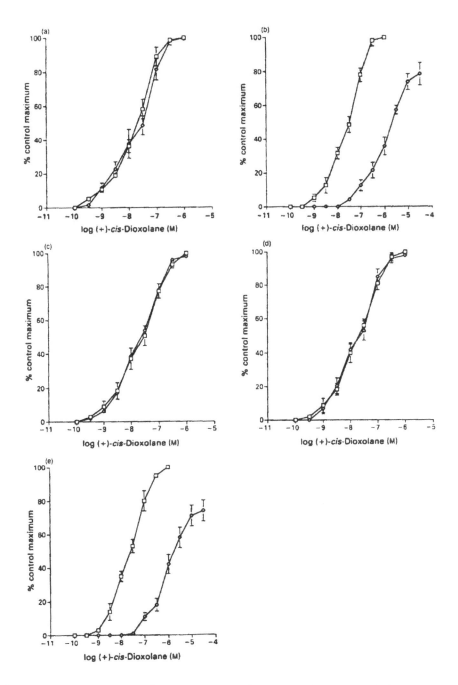

Figure 9.4 Receptor protection experiments by competitive muscarinic antagonists. Concentration–response curves for contractile responses of guinea-pig ileum to the agonist (+)-*cis*-dioxolane are shown before (□) and after (O) exposure to the irreversible antagonist, phenoxybenzamine (10 μM, for 20 min followed by washout at 10 min intervals for 180 min). Time-matched controls without phenoxybenzamine are shown in (a). The irreversible blockade with depression of maximum is shown in (b). Pre-equilibrium with competitive antagonists shows protection from irreversible blockade by atropine (c, 30 nM, non-selective) and p-f-HHSiD (d, 3 μM, M_3-selective) but not by methoctramine (e, 0.1 μM, M_2-selective). M_3 receptors therefore mediate the functional contraction of guinea-pig ileum. Reproduced with permission from Eglen & Harris (1993).

371

hyoscine; in Alzheimer's there is no block of the postjunctional M_1 receptor (Andrews *et al.* 1993, Freedman *et al.* 1993).

The depressant effect of hyoscine may also be associated with euphoria. These properties are utilized in preanaesthetic medication. The use of hyoscine to suppress motion sickness is also due to its central depressant action combined with an effect on the vestibular balance mechanisms. Finally, the non-quaternary muscarinic antagonists with ability to penetrate the CNS have long been used for the treatment of Parkinson's disease.

Higher toxic doses of atropine do cross to the CNS and produce excitation. The characteristics of belladonna poisoning are therefore restlessness, irritability, hallucination and delirium. At near-lethal dose levels, this is followed by depression, coma, circulatory collapse and respiratory failure. Hyoscine at high doses produces the excitatory effects seen with atropine poisoning.

9.4 Therapeutic Uses of Muscarinic Antagonists – Established and Potential

The pharmacological actions of atropine and related agents have been described and where these may lead to a therapeutic application has been indicated. The structures of a wide range of atropinic drugs have been considered, including newer products which may still be experimental or yet to reach the commercial market. We will now deal with the established therapeutic uses of muscarinic antagonists, together with any potential applications or improvements in strategy of their uses.

9.4.1 Peptic Ulcer and Gastrointestinal Spasm

Antimuscarinic agents are effective in reducing gastric acid secretion and motility, and as a result they aid the healing of peptic ulcers. They do, however, cause considerable side-effects arising from muscarinic receptor blockade elsewhere and have been largely superceded by the histamine H_2 receptor antagonists such as cimetidine and ranitidine. The latter agents are relatively free from side-effects and show substantial improvements in treatment of the ulcer.

The M_1-selective antagonist, pirenzepine (Gastrozepin[R]), shows selectivity for inhibition of gastric acid secretion with a lower incidence of muscarinic side-effects than the non-selective antimuscarinics. It has been shown to be as effective as the H_2 receptor antagonist, cimetidine, at promoting wound healing in patients with duodenal ulcers. The dose required is approximately one-tenth that of cimetidine. The patients experience reduced day and night pain and antacid consumption. Pirenzepine also performs equally with cimetidine in benign gastric ulceration. The addition of pirenzepine to the H_2 receptor antagonists inhibits gastric acid secretion to a greater extent than do the individual drugs alone, and their combined use appears to be more effective in treatment of gastric and duodenal ulceration. Muscarinic side-effects are minimal and are only the cause of withdrawing the treatment in ~2% of patients. Pirenzepine is not a quaternary compound but is hydrophilic with low lipid solubility and therefore does not penetrate into the CNS. It therefore has no central side-effects at the normal twice daily therapeutic dosing. The site of action of pirenzepine has already been discussed. The M_1 receptor selectivity of pirenzepine is clearly not targeted at parietal cell secretion, which is M_2 receptor-mediated. It appears to be due to inhibition of ganglionic M_1 receptors in the myenteric plexus of the gut. Telenzepine is structurally related to pirenzepine (Table 9.6) and has a similar profile of action, being 4–10 times more potent as an antisecretory agent (Eltze *et al.* 1985).

The other indication for muscarinic antagonists on the gastrointestinal tract is disorders associated with hypermotility. They reduce motility and tone of the gut and may be of benefit in conditions such as irritable bowel syndrome, although their effectiveness is questionable. A wide range of agents of the quaternary ammonium and tertiary amine type are available. Dicyclomine is a tertiary amine with M_1 selectivity that is widely used for relief of intestinal colic due to muscle spasm, particularly in children (Table 9.4). Newer selective M_3 receptor antagonists, such as zamifenacin (Table 9.7), have recently been developed for irritable bowel syndrome and are currently under clinical evaluation (Eglen *et al.* 1994). Established antimuscarinics, including dicyclomine, are also useful in symptomatic treatment of diarrhoea associated with irritation of the lower bowel, such as mild dysenteries and diverticulitis. The belladonna alkaloids have long been used for the treatment of gastrointestinal upset, Tincture of Belladonna having formerly been a common constituent of several widely used mixtures along with the adsorbent, kaolin (eg Mist. Bellad. Alk.) or the antacid, magnesium trisilicate (eg Mist. Mag. Trisil. Bellad.).

While considering old-fashioned remedies, it is worth noting here that extracts of belladonna root or herb have been applied to the skin as a liniment or plaster. These have been used to relieve the pain of neuralgia, rheumatism or lumbago. The rationale for this application is obscure; they may have acted as counterirritants, sometimes aided by the inclusion of capsicum (Belladonna & Capsicum Self-adhesive Plaster BPC 1963) (see Chapter 12).

9.4.2 *Ophthalmology*

Atropinic drugs are instilled into the eyes as drops for their mydriatic and cycloplegic effects. The dilation of the pupil and abolition of the response to light permits examination of the retina and optic disc. Cycloplegia occurs with higher doses, the paralysis of accommodation for near vision leading to blurred sight. The poor response to light (photophobia) necessitates the patient having to wear dark glasses and this, combined with the blurred vision, are particularly hazardous for patients leaving the examination room when long-acting agents like hyoscine and atropine are used. It is therefore preferred to employ shorter-acting atropinics including homatropine (Table 9.1), cyclopentolate and tropicamide (Table 9.4).

This class of drug is also used to prevent or break adhesions of tissue between iris and lens when they are alternated with miotic drugs. The inflammatory disorders, iridocyclitis and keratitis, may also be treated, complete cycloplegia being necessary.

The major adverse effect of atropinic drugs is the precipitation of glaucoma of the narrow-angle type in susceptible individuals. The rise in intraocular pressure can induce an attack of acute glaucoma with a threat of possible blindness. Thus, the possible presence of a narrowed filtration angle should be carefully evaluated before commencing use of antimuscarinic agents. The mydriasis can be reversed by topical application of a muscarinic agonist such as pilocarpine, although this may be incomplete with hyoscine or atropine. A further complication occurring only with atropine is that it may cause local irritation of the eye and conjunctivitis which may be relieved with an antihistamine. Systemic side-effects are rare since absorption from the conjunctival sac is poor, although absorption after entry into the nasolacrymal duct (Chapter 1) can occur. This may be prevented by application of pressure to the inner canthus of the eye after each instillation.

9.4.3 Asthma and Chronic Obstructive Pulmonary Disease (COPD)

The bronchodilator and antisecretory effects of atropine and other belladonna alkaloids have long been used in the treatment of asthma, bronchitis and coughs and colds. Mixtures containing Tincture of Belladonna, such as Mist. Bellad. c Ipecac. Paed., have been old remedies for asthma and whooping cough in children, the ipecacuanha serving as an expectorant. Atropine and its derivatives reduce secretions in the upper and lower airways and produce bronchodilatation. They may therefore be useful for the symptomatic relief of acute rhinitis, although not affecting the natural course of the condition. The effectiveness of cough and cold remedies that include an antihistamine may be due in part to the weak anticholinergic activity of such compounds in addition to the anti-allergic component.

Atropinic drugs administered by aerosol are effective in treating acute exacerbations of asthma rather than acting as bronchodilators in the chronic condition. This suggests that cholinergic mechanisms become more important during the asthma attack. They are particularly useful in COPD associated with emphysema and chronic bronchitis. These conditions are often regarded as having fixed or non-reversible obstruction of the airways, however vagal tone is probably the only reversible element (Barnes 1993b). One possible limitation of the atropinic drugs is the inhibition of beating of the ciliated epithelium. Coupled with the reduced volume of mucous secretions, which does not become viscid, this impairs mucociliary clearance.

The newer synthetic atropinic agent, ipratropium, is a quaternary ammonium compound (Table 9.1) which is more potent as a bronchodilator and antisecretory agent than is atropine. Administered by inhalation, it is virtually devoid of systemic effects on the heart, eye or bladder. The charged nitrogen atom prevents absorption from the gastrointestinal tract and airways and, if administered systemically, it fails to cross the blood–brain barrier and exert any CNS effects. An important and beneficial difference from atropine is that ipratropium has little effect upon mucociliary clearance. The reason for this property is still unexplained, but it clearly avoids the build up of secretions in the lower airways (Gross 1992). A related quaternary ammonium derivative of hyoscine is oxitropium (Table 9.1), which is less potent but appears to have similar properties. Ipratropium and oxitropium are recommended for use in chronic asthma where short-acting β_2-adrenoceptor agonists (see Chapter 4) and maximum anti-inflammatory medication is inadequate. They have a slow onset of action but a prolonged duration. Ipratropium lasts up to 8 hr and oxitropium lasts for 8–12 hr. They should be taken regularly by inhalation, initially on a trial basis. Side-effects are uncommon but care should be exercised in patients with glaucoma or prostatic hypertrophy, and in children the saliva viscosity may increase.

These agents are non-selective for muscarinic receptor subtypes. It would be of theoretical advantage to block the M_3 receptors mediating the bronchial secretion and smooth muscle contraction of the airways, together with the M_1 receptors involved in ganglionic transmission of the vagal reflex bronchoconstrictor pathways. The latter appears to be triggered in asthma by the presence of inflammatory mediators. Indeed, the selective M_1 receptor antagonist, pirenzepine, has bronchodilator properties when given intravenously to humans, although at the dose used it may have been blocking non-selectively at smooth muscle M_3 receptors. Lower doses had no effect of forced expiratory volume (FEV_1) but increased expired air flow at low lung volumes (Barnes 1993a). Blockade of M_2 receptors, however, could have an opposing detrimental effect on the airways since these are the prejunctional receptors, stimulation of which causes inhibition of Ach release from vagal nerve terminals. Blockade of these would therefore facilitate Ach release and thus offset any beneficial effects of blockade at M_3 receptors.

A paradoxical increase in vagal activity has been demonstrated after low doses of the selective M_2 antagonist, methoctramine, which inhibits the bronchoconstriction induced by exogenous Ach but facilitates the vagally-mediated response (Watson *et al.* 1992). Conversely, stimulation of these inhibitory M_2 receptors with pilocarpine decreases the bronchoconstriction induced by vagal stimulation in anaesthetized guinea-pigs (Fryer & Jacoby 1993). It is therefore suggested that muscarinic antagonists selective for M_1 and M_3 receptors and without M_2 blocking activity might prove to be more effective in the treatment of airway disease (Barnes 1993a).

9.4.4 *Urinary Incontinence*

Urinary bladder dysfunction associated with a failure to store urine may be divided into two forms of urinary incontinence: urge incontinence and stress incontinence. Urge incontinence is an involuntary loss of urine associated with a strong desire to void. It may be subdivided into motor and sensory forms. Motor urge incontinence occurs where there is increased contractile activity of the bladder brought on by outflow obstruction, inflammation or irritation, as might occur with the infection associated with cystitis. Stress incontinence is not due to bladder hypertonicity, but more likely to sphincter dysfunction. This may arise from weakness of the skeletal muscle of the pelvic floor which is associated with the external sphincter at the neck of the bladder. It may also be caused by damage to the nerves innervating this musculature. The anticholinergic drugs are therefore not effective in this condition. They do have widespread application in urge incontinence in the elderly, in children suffering from nocturnal enuresis, to reduce urinary frequency in spastic paraplegia, and where bladder hyperactivity is associated with irritation (cystitis). The anticholinergic drugs in current use include propantheline and emepronium. Since they are quaternary ammonium compounds, bioavailability is poor, they are poorly transported to the CNS, and have only peripheral side-effects including dry mouth, blurred vision and tachycardia. Dicyclomine has also produced favourable results in urinary incontinence.

Other anticholinergic drugs used for urinary incontinence have additional properties; these include oxybutynin and terodiline (Table 9.4).

Oxybutynin combines antimuscarinic activity with a direct smooth muscle relaxant property. It is now widely used for unstable bladder conditions, in spite of characteristic peripheral side-effects arising from muscarinic receptor blockade (Andersson 1988).

Terodiline exhibits muscarinic blocking activity in animal and human isolated detrusor muscle, the (−)-isomer being more potent. Additionally, it has Ca^{2+} blocking properties, since it inhibits K^+-induced contractions which are due to Ca^{2+} influx through voltage-operated L-type Ca^{2+} channels. This effect appears to be specific for the bladder. Terodiline is well tolerated, the only side-effects arising from muscarinic receptor blockade. It is rapidly absorbed after oral administration and has an elimination half-life of ~60 hr. It has been found to be effective in reducing diurnal and nocturnal micturition frequency and episodes of incontinence, and has gained a valuable place in therapy (Langtry & McTavish 1990). However, it has now been withdrawn worldwide after reports of serious heart rhythm disturbances in people taking the drug. It appears to depress the SA node and is therefore probably unrelated to muscarinic receptor blockade and not likely to occur with other agents in this class.

9.4.5 Parkinson's Disease

The excitatory effects of the cholinergic system in the striatum of the brain are exaggerated by the defect of the dopaminergic pathways that occurs in patients with Parkinson's disease. Suppression of these cholinergic pathways by muscarinic antagonists has long been a mode of treatment of this disease. The muscarinic antagonists producing greater benefit are the tertiary amines having M_1 receptor selectivity, as exemplified by benzhexol (trihexyphenidyl) (Table 9.4). They have been largely superseded by L-dopa (Chapter 2), but they still serve a useful role as adjuncts to L-dopa or alone in patients with minimal symptoms who are unable to tolerate L-dopa. These agents reduce the tremor, rigidity and bradykinesia of Parkinson's disease, although they are less effective than L-dopa. One additional property, however, is that the antimuscarinic activity reduces the excessive salivation associated with Parkinson's disease.

This class of drug does produce the unwanted atropine-related side-effects of cycloplegia, urinary retention and constipation which may be troublesome in the elderly. Furthermore, the blockade of central M_1 receptors may induce Alzheimer's-like effects of mental confusion, delirium and hallucinations.

Alzheimer's disease is now in fact a potential target for treatment with muscarinic antagonists. As described in the previous chapter, centrally acting M_1-selective agonists have been developed to improve cognitive function, but so far the selectivity is disappointing. An alternative approach to stimulating the defective postjunctional M_1 receptors is to enhance the release of endogenous Ach. This may be achieved by blockade of the prejunctional receptors located in the hippocampus and cortex, stimulation of which normally inhibits transmitter release. Thus, selective M_2 blockade in the absence of M_1 blocking properties could enhance Ach output without detrimental effects upon the already impaired postjunctional M_1 receptors. Indeed, selective M_2 antagonists such as the pirenzepine derivative, AF-DX 116 (Table 9.6), have been shown to enhance the release of Ach from certain brain areas both *in vivo* and *in vitro*. For such an agent to be therapeutically effective it must readily penetrate the blood–brain barrier and have high M_2 potency and selectivity. Structural modification of the AF-DX 116 side-chain has yielded first AQ-RA 741 and later BIBN 99 (Table 9.6) which show the desired selectivity for M_2 receptors. The former compound has the disadvantage of being protonated and hydrophobic and poorly penetrates into the brain. BIBN 99, however, is lipophilic and in animal trials has shown improved performance in learning capacity of age-impaired rats. Such compounds may therefore provide a basis for development of M_2-selective muscarinic antagonists for the management of Alzheimer's disease (Doods *et al*. 1993).

9.4.6 Motion Sickness

The central depressant effect of hyoscine was one of the earliest applications in the prevention of motion sickness. Hyoscine is very effective and probably the drug of choice for the prophylaxis of motion sickness arising from short-term exposures to severe motion. It does not stop nausea and vomiting that has already developed. Tablet forms are readily available (eg Kwells[R]), but for long-term relief transdermal administration has been shown to be particularly effective. The patches (Transderm Scop[US], Scopaderm TTS[UK]) are applied behind the ear in the postauricular mastoid area, where efficient absorption takes place. Patches are changed every 72 hr, alternating the sides of application. The side-effects are predictably dry mouth, blurred vision and drowsiness. Nausea and dizziness may occur 3 days after withdrawal, perhaps due to a delay in the adaptation to motion because of the use of hyoscine (Price *et al*. 1981). Other motion sickness preventatives also owe their action largely to muscarinic receptor blockade; these include cyclizine, meclizine, dimenhydrinate and cinnarizine.

9.4.7 Surgical Premedication

Hyoscine has been used as a premedicant 30–60 min prior to induction of general anaesthesia to reduce bronchial and salivary secretions which may be increased by inhalation anaesthetics and by intubation. It also provides a degree of amnesia and sedation to relieve the anxiety of impending surgery. Hyoscine was formerly used in obstectrics with morphine or pethidine to induce a state of partial analgesia and amnesia known as 'twilight sleep'. Glycopyrrolate (Table 9.3) is an alternative premedicant given by intramuscular or intravenous injection, which induces less tachycardia than does atropine. The use of antimuscarinics as premedicants is less common nowadays, particularly as the modern anaesthetics are less likely to induce secretion. Antimuscarinics are used to counteract the muscarinic side-effects (eg excessive salivation, bradycardia) of anticholinesterases such as neostigmine, when used in surgery to reverse neuromuscular blockade.

9.4.8 Cardiovascular Uses

Atropine is useful in the emergency treatment of acute myocardial infarction where the dominant autonomic influence on the heart is via the vagus nerve, causing sinus or nodal bradycardia. Severe bradycardia may lead to hypotension and atrioventricular block. Atropine should be given carefully to avoid the low-dose bradycardiac effect and to avoid excessive dosing which would induce harmful tachycardia. The latter effect will increase oxygen demands and exacerbate the ischaemia. Since the cardiac muscarinic receptor is of the M_2 subtype, it would seem reasonable to develop M_2-selective antagonists for the treatment of sinus or nodal bradycardia or AV block (Goyal 1989).

9.5 Adverse Drug Reactions, Contraindications and Poisoning by Muscarinic Antagonists

The adverse reactions to atropinic drugs are related to their muscarinic receptor blocking actions and have been dealt with during the discussions of their actions and uses. They are listed in Table 9.9. together with the types of patient and their conditions where such drugs are contraindicated. The symptoms of poisoning from ingestion of belladonna alkaloids are essentially an extension of these adverse reactions to toxic dose levels. The berries of the Deadly Nightshade are attractive to children and are a potential source of poisoning. Children are especially susceptible to the toxic effects of atropinic drugs, and doses as low as 10 mg have been fatal. This is an unusual reaction, however, and doses of this level usually produce a severe response which may be reversed by the anticholinesterase, physostigmine (Chapter 10). Such reversal is in fact a good diagnostic measure for atropine poisoning.

Treatment of poisoning, where poison has been taken orally, should also include limitation of absorption from the gut, by emptying the stomach. Ice packs will help to reduce the fever. If atropine poisoning has progressed to the central excitation stage with hallucination and delerium, diazepam is suitable to control convulsions and for sedation. However, this should be avoided if the later stages of poisoning, associated with depression and coma, have been reached. At this stage, artificial respiration may be necessary.

Poisoning may also occur through mishandling or accidental systemic absorption of other atropinic products such as eye drops or patches. Toxicity due to the antimuscarinic properties of other drugs, such as tricyclic antidepressants, may also occur with suicide attempts. The symptoms of poisoning by atropine are listed sequentially in Table 9.9.

Table 9.9 Adverse drug reactions and contraindications of antimuscarinics and symptoms of poisoning by atropine

Adverse drug reactions	Contraindications	Poisoning by atropine
Tachycardia	Tachycardia	Increased pulse, palpitations (1 mg)
Dry mouth, thirst		Dry mouth, thirst
Visual disturbances	Glaucoma	Dilated pupil (2 mg)
Dermal flushing		Hot, dry skin (5 mg)
Dizziness (hypotension)		Headache, fatigue, difficulty in
Constipation	Intestinal obstruction	swallowing, talking and
Urinary retention	Urinary obstruction	micturition, blurred vision
	Ulcerative colitis	Restlessness, excitement (10 mg)
	Myasthenia gravis	Delirium, hallucination, coma
		Respiratory failure

Atropine is rapidly absorbed from the gastrointestinal tract, lacrymal ducts and mucous membranes, but less readily from the eye and skin. It has a half-life of ~4 hr, about half of a dose being eliminated in the urine unchanged, the remainder being subjected to hepatic metabolism. Some species have an esterase in the liver and serum, known as atropinase, which rapidly hydrolyses atropine. In rabbits, the amount of this enzyme is genetically determined, and some animals can eat belladonna leaves without showing any adverse effects. This is relevant to experimental pharmacology with rabbits where muscarinic blockade by atropine may be short-lived.

10

Anticholinesterases

10.1   Introduction

Following its release from cholinergic neurones, Ach must be rapidly inactivated to terminate the physiological effect of nerve activity. The transmitter must be inactivated or dissipated to prevent continued stimulation of cholinoceptors. As discussed in Chapter 7, there is no uptake of Ach back into the neurone, the uptake pathways being relatively specific for choline. Removal of Ach must therefore be via its breakdown to inactive products. As with so many aspects of autonomic pharmacology, it was Sir Henry Dale who first proposed the existence of an enzyme for this breakdown of Ach in 1914. It was not until 1932, however, that an enzyme with such activity was found in horse serum and named cholinesterase (Stedman *et al.* 1932). This enzyme, however, was shown to hydrolyse butyrylcholine and propionylcholine at faster rates than Ach. Subsequently, horse serum was shown to contain two enzymes, most of the activity being attributed to a non-specific enzyme that could breakdown not only cholinesters but also non-cholinesters. The remaining activity was due to a specific enzyme for cholinesters (Mendel & Rudney 1943). We now know that there are two types of cholinesterase: acetylcholinesterase (AchE), also called 'true' or specific cholinesterase, which is found in the vicinity of cholinergic synapses, and butyrylcholinesterase (BuchE), which has been called 'pseudo' or non-specific cholinesterase. It was this latter enzyme that was found in the serum (serum esterase) as well as in the liver and intestine, although it occurs only to a limited extent in the neuronal components of the peripheral and central nervous systems.

Drugs which inhibit the action of cholinesterase, the anticholinesterases, have widespread pharmacological actions on autonomic effector cells. Because hydrolysis by cholinesterase is the major route of inactivation of Ach, inhibition of this enzyme will result in accumulation of endogenous Ach in the synaptic cleft. The physiological effects at all sites of Ach release are therefore enhanced. This explains most of the pharmacological effects of the anticholinesterases, which are quite widespread and reflect the extensive locations of Ach as a transmitter in skeletal muscle, autonomic ganglia, parasympathetic nerve endings in the effector organ and in the CNS. Anticholinesterases therefore have major actions and uses at somatic nerve−skeletal muscle junctions which are not relevant to the autonomic nervous system and will not therefore be dealt with in this book. Similarly, the CNS applications will only be considered briefly as they relate to the toxicity and developments of newer drugs. The autonomic sites of action that will be considered in most detail are the autonomic ganglia and at the parasympathetic effector organ where nicotinic and muscarinic effects, respectively, are enhanced.

Most of the drugs that are now known to inhibit cholinesterases were first described many years before the enzyme was identified and isolated. Physostigmine (eserine) (Table 10.1) was isolated from the Ordeal or Calabar bean in 1864, the

Table 10.1 Slowly reversible anticholinesterases

Carbamate compounds

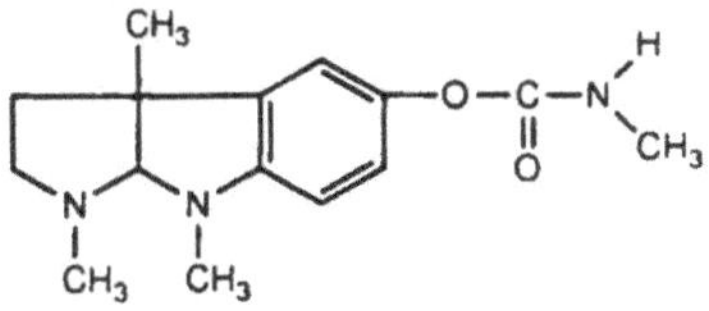

Physostigmine or eserine

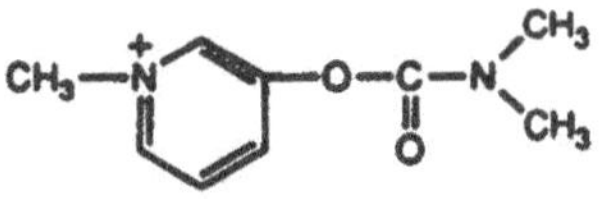

Pyridostigmine (Mestinon^R)

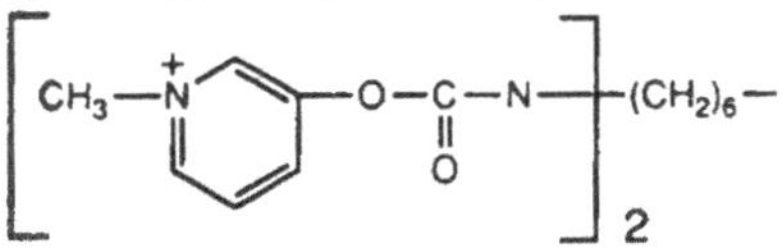

Distigmine (Ubretid^R)

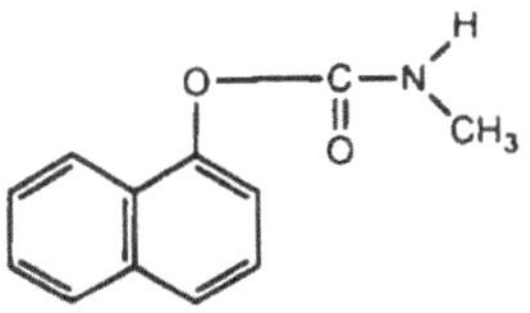

Carbaryl (sevin)
(Carylderm^R, Derbac-C^R,
Suleo-C^R)

Non-carbamates

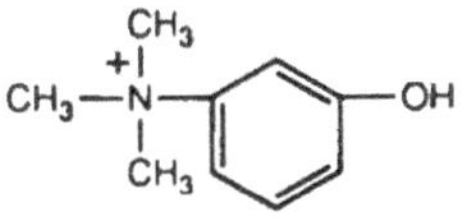

Edrophonium (Tensilon^R)
Selective binding to anionic site

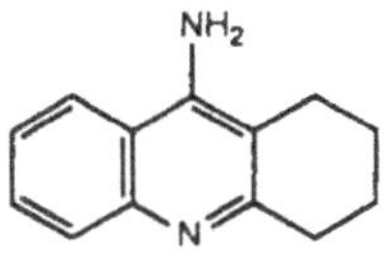

Tacrine (Cognex^R)

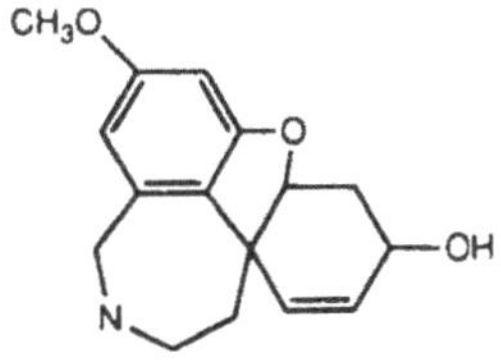

Galantamine

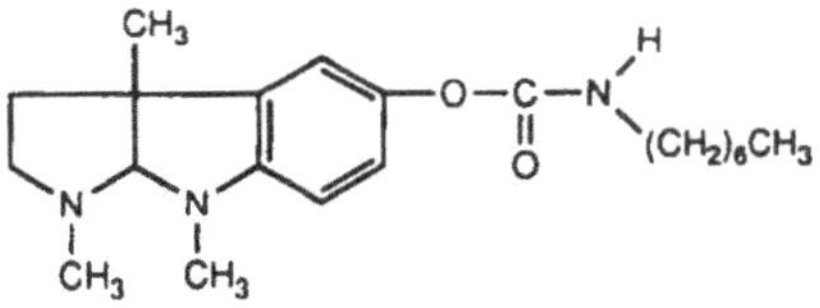

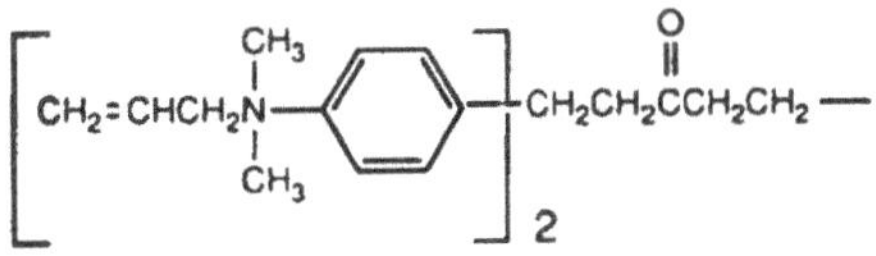

Eptastigmine (heptylphysostigmine)

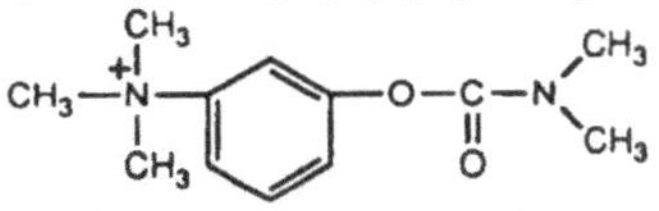

Neostigmine (Prostigmine^R)

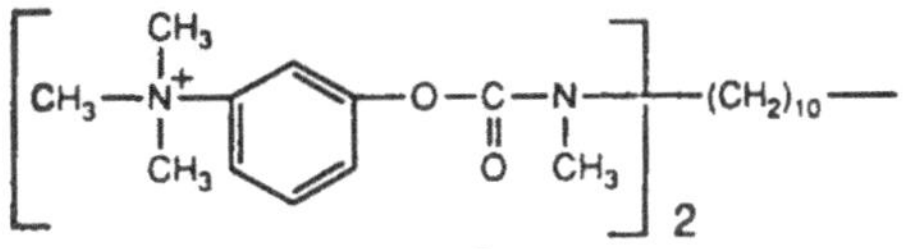

Demecarium (Humorsol^R)

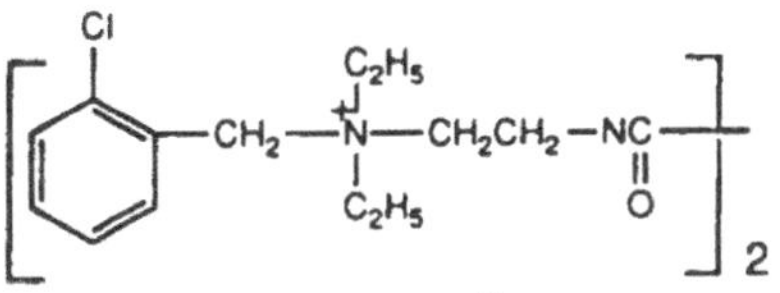

Ambenonium (Mytelase^R)

BW284c51
Selective for AChE

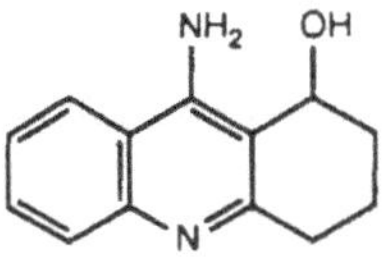

Velnacrine (Mentane^R)

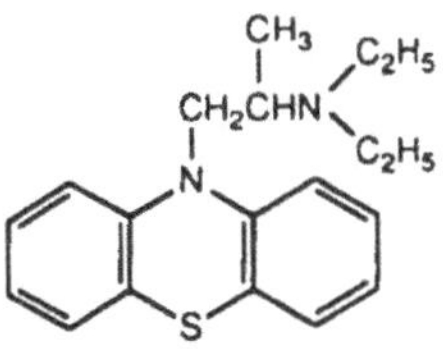

Ethopropadine
Selective for BuchE

Note: Carbamate compounds bind to both esteratic and anionic sites of the enzyme.

383

Table 10.2 'Irreversible' anticholinesterases – organophosphorus compounds

Basic structure:

Common name	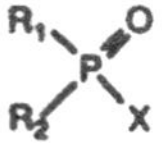 Chemical name	Structure	Potency/use	Reversal	Comments/form
DFP	Diisopropylphosphorofluoridate (Dyflos)	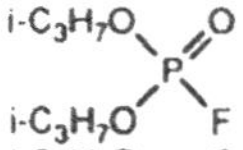	Potent early inhibitor		Selective for BuChE
Sarin	Isopropylmethylphosphonofluoridate (GB)	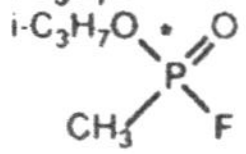	Extremely toxic nerve gas	Oxime-reversible	Highly volatile liquid (non-persistent)
Tabun	Ethyl-N,N-dimethyl-phosphoramidocyanidate (GA)	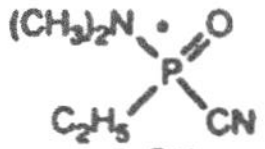	Extremely toxic nerve gas	Oxime-reversible	Volatile liquid
Soman	Pinacolyl methylphosphonofluoridate (GD)	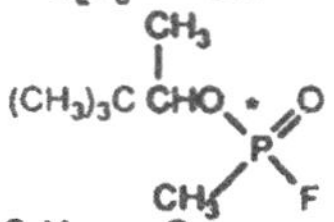	Extremely toxic nerve gas	Rapidly ages, not oxime-reversible	Volatile liquid
VX	Ethyl-S-2-N,N-diisopropylaminoethyl methylphosphonofluoridate	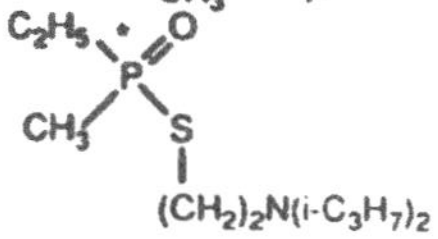	Nerve agent	Little ageing	Heavy viscous liquid, persistent cutaneous and inhalation absorption

	Compound	Structure	Notes	Reversibility	Solubility/use
TEPP	Tetraethylpyrophosphate	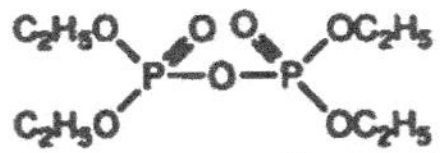 C_2H_5O, OC_2H_5 pyrophosphate; $(C_2H_5O)_2P(O)-O-P(O)(OC_2H_5)_2$	First synthesized early insecticide	Oxime-reversible	Highly lipid soluble
Iso-OMPA	Tetramonoisopropyl pyrophosphortetramide	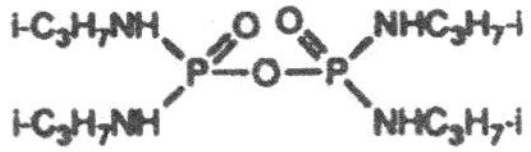 $i\text{-}C_3H_7NH$, $NHC_3H_7\text{-}i$; pyrophosphortetramide		Oxime-reversible	Selective for BuchE
Parathion	O,O-Diethyl-O-(4-nitrophenyl)-phosphorothiolate	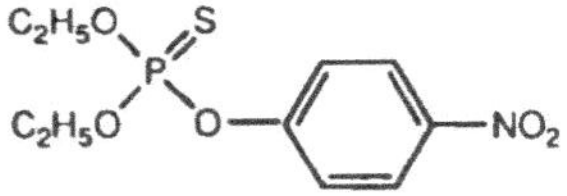 C_2H_5O, C_2H_5O, $P=S$, $O-C_6H_4-NO_2$	Potent insecticide		Dispersed in aqueous solution for spraying as aerosols
Paraoxon	O,O-Diethyl-O-(4-nitrophenyl)-phosphate	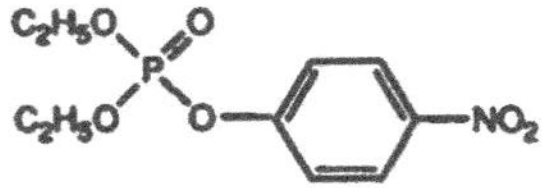 C_2H_5O, C_2H_5O, $P=O$, $O-C_6H_4-NO_2$	Active metabolite of parathion	Oxime-reversible	Dispersed in aqueous solution for spraying as aerosols
Malathion	O,O-Dimethyl-S-(1,2-dicarbethoxyethyl) phosphorothiolate	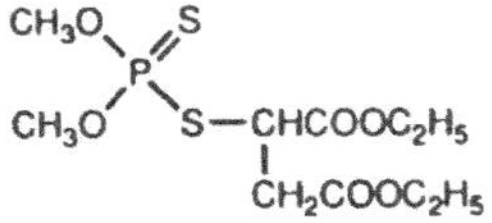 CH_3O, CH_3O, $P=S$, $S-CHCOOC_2H_5$, $CH_2COOC_2H_5$	Inactive precursor insecticide	Oxime-reversible, rapid metabolism in animals	Dispersed in aqueous solution for spraying as aerosols
Dimefox	bis-N,N-Dimethylamido-phosphorofluoridate	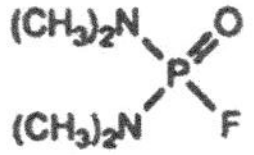 $(CH_3)_2N$, $(CH_3)_2N$, $P=O$, F	Converted by plants to highly active inhibitor		Dispersed in aqueous solution for spraying as aerosols
Ecothiopate (Phospholine^{US})	Diethoxyphosphinylthiocholine iodide	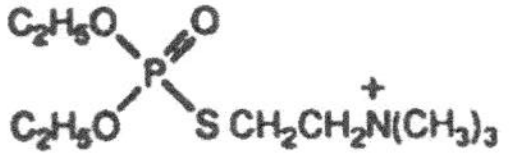 C_2H_5O, C_2H_5O, $P=O$, $S\,CH_2CH_2\overset{+}{N}(CH_3)_3$	Clinical use in glaucoma	Combines with anionic and esteratic site	Water soluble

pharmacological properties of the seed having already been described some 10 years earlier by Christioson (1855) and by Fraser (1863) and Argyll-Robertson (1863). The first therapeutic use of the pure alkaloid was for the treatment of glaucoma in 1877 (reviewed by Karczmar 1970). Even earlier the synthesis of a highly potent member of the organophosphorus anticholinesterases, tetraethylpyrophosphate (TEPP), was described by deClermont in 1854 (Table 10.2). In 1932, Lange & von Krueger synthesized dimethyl and diethyl phosphorofluoridates and described the persistent choking sensation on their inhalation. German scientists became interested in the potential of this class of agent as insecticides. Commencing in the mid-1930s at Farbenfabriken Bayer, Schrader (1952) performed pioneering work culminating in the synthesis of tabun, one of the most toxic organophosphorus compounds known, and sarin and soman. Schrader synthesized a wide range of compounds in search of improved insecticides. One of the most effective of these early organophosphorus compounds that became a widely used insecticide was parathion. Schrader's group recognized the potential application of the toxicity of these compounds in chemical warfare, and immediately before and during World War II their efforts were directed towards the synthesis of nerve gases. Tabun and sarin were kept secret for several years. During the war, Germany held large quantities of tabun and sarin, although they were never used during the conflict (Holmstedt 1959).

Although the toxic effects of the organophosphorus compounds are made use of in insecticides and have warfare implications, anticholinesterases do have therapeutic applications. Before considering the pharmacological actions and uses of anticholinesterases as they apply to the autonomic nervous system, the locations, structures and properties of the cholinesterase enzyme will be described.

10.2 Cholinesterases

Most of the work on the structure and function of cholinesterase has been performed on the enzyme extracted from the electric organs of the electric eel (*Electrophorus electricus*) and electric ray (*Torpedo marmorata, T. californica, T. torpedo*). These organs are phylogenetically derived from muscle which has lost its contractile apparatus and are rich sources of cholinesterase, in addition to Ach and cholinergic receptors. The distribution, structure and properties of cholinesterase in autonomic nerves has received far less study.The following account applies generally to cholinesterases found throughout mammalian species.

10.2.1 *The Two Cholinesterases*

Cholinesterases hydrolyse cholinesters specifically and rapidly by an acylation–deacylation reaction (Figure 10.1). There are two distinct cholinesterases, acetylcholinesterase (AchE, EC 3.1.1.7) and butyrylcholinesterase (BuchE, EC 3.1.1.8), the fundamental difference being their substrate specificities.

10.2.1.1 *AchE*

AchE hydrolyses Ach at a faster rate than it does other cholinesters which have acyl groups larger than acetate or propionate. It does hydrolyse methacholine; the (+)-S-

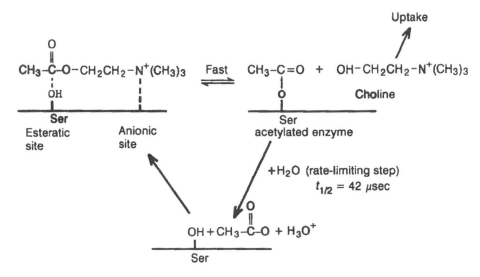

Figure 10.1 Basic reaction for hydrolysis of acetylcholine (Ach) by cholinesterase.

enantiomer but not the (−)-R-enantiomer (Table 8.1), although it does not hydrolyse butyrylcholine at a significant rate. AchE is found primarily at the synapses of cholinergic nerves in the periphery and brain. It is highly concentrated at the junction between somatic nerves and skeletal muscle, where it is localized at the surface and infolds of the postjunctional membrane, but activity also appears presynaptically. It is synthesized in the nerve cell body and occurs in the dendrites, perikarya and axons of cholinergic nerves and is also synthesized in muscle cells. AchE is also found in non-cholinergic tissues such as erythrocytes. A single gene encodes for AchE which is located on the long arm of chromosome 7 at 7q22 and in the mouse on chromosome 5.

10.2.1.2 BuchE

This enzyme hydrolyses Ach and other cholinesters, but the preferred substrate is butyrylcholine. It does not hydrolyse methacholine and has been alternatively called propionylcholinesterase, non-specific cholinesterase or pseudocholinesterase. BuchE is found to a more limited extent in nervous tissue in glial or satellite cells, its main location being the serum and the liver where it is synthesized. Whereas AchE is inhibited by high concentrations of Ach (>3 mM), BuchE is not. The replacement time for serum BuchE is ~50 days. The levels show considerable variation over the year by as much as 50%, and activity is lower in women than in men (Sidell 1992). It is regarded as a scavenging enzyme, in contrast to AchE which is a synaptic enzyme. The scavenging role may be for the hydrolysis of certain dietry esters, which is consistent with its prevalence in the liver and plasma (Vellom *et al.* 1993).

Examples of drugs of clinical relevance that are hydrolysed by BuchE include heroin, the β-adrenoceptor agonist prodrug, bambuterol (Chapter 4), and suxamethonium. The depolarizing neuromuscular blocking agent, suxamethomium (succinyldicholine), is hydrolysed first to succinylmonocholine and then to succinic acid and choline by BuchE. The blocking activity of suxamethonium at skeletal muscles is relatively short at ~5 min.

It was through the use of suxamethonium that the observation of genetic polymorphism of plasma BuchE was made. The catalytic subunit of human BuchE is encoded by a single gene on chromosome 3, with an additional locus on chromosome 16 being identified by *in situ* hydridization. A large number of allelic variants of the human BuchE gene have been identified at the single locus for enzyme encoding. These allelomorphic genes are autosomal and co-dominant, that is, both alleles of the chromosome pair contribute to the formation of the enzyme. The homozygote (uu), where both alleles are identical and encode for the typical form of enzyme, has normal enzyme activity. The *atypical* enzyme (homozygote aa) has a lower rate of activity and a decreased affinity for substrates and inhibitors and is characterized by its resistance to inhibition by the local anaesthetic, dibucaine. This has amino acid Val-70 replaced by Gly (genotype $BCHE^*70G$). A *fluoride-resistant* enzyme is produced by homozygote individuals (ff) and heterozygote individuals (uf and af), the former genotype having substantially lower activity. A fourth type of deficiency is the *silent enzyme* (homozygote, ss) which is almost completely devoid of BuchE activity; several mutations have been identified which involve substitution of a single amino acid residue (eg Pro-37 replaced by Ser, genotype $BCHE^*375$), creation of a stop codon or frameshift mutations. In these silent enzyme types the enzyme may be absent. Homozygote individuals belonging to the silent, fluoride-resistant and atypical phenotypes exhibit prolongation of neuromuscular blockade by suxamethonium. Despite this lack of BuchE activity they do not, however, show any untoward symptoms, which suggests that the enzyme is not vital for normal physiological function (McGuire *et al.* 1989, Massoulié *et al.* 1993). Another genetic variant of BuchE, the C^+5 variant, has been suggested to arise from the second genetic locus (CHE2) on chromosome 16, in which the enzyme displays greater activity. However, it has more recently been shown that this variant consists of a combination of BuchE subunits with another protein and is associated with the long arm of chromosome 2 (Massoulié *et al.* 1993).

Separate genes are therefore responsible for encoding AchE and BuchE. The two enzymes differ in their immunoreactivity. Chemically they are glycoproteins, the AchE and BuchE of human erythrocytes and serum containing approximately 15 and 25% of carbohydrate, respectively. It is the polypeptide component of the catalytic subunit of the enzyme that confers the major antigenic properties, the tail of the enzyme (see later) is less antigenic. Some immunological reaction may be directed against the carbohydrate component, which is lost after deglycosylation of the enzyme. Antibodies raised against the enzymes show no immunological cross-reactivity between AchE and BuchE. Antibodies prepared against the AchE of one species generally do not show equal inhibitory activity against the AchE of other species. This suggests that the structures in the close vicinity of the active site of the enzyme are not conserved throughout different species, although the active sites themselves appear to be very similar (see structures later in chapter).

10.2.2   Molecular Form of Cholinesterase

Both enzymes, AchE and BuchE, exist in several molecular forms arranged into two classes of either *asymmetric* or *globular types*. The asymmetric molecules consist of one (A_4), two (A_8) or three (A_{12}) tetramers (groups of four) of catalytic subunits attached to a collagen-containing tail arranged as a triple helix. The globular forms consist of monomers (G_1), dimers (G_2) and tetramers (G_4) of the catalytic subunit with no tail (Figure 10.2). The tail of the asymmetric forms may be cleaved by collagenase or trypsin to yield tetramers of the globular forms. The catalytic subunits are identical within the asymmetric forms but differ from those of the globular forms. The catalytic subunits are linked by disulphide bonds to form the external dimers and also to the tail chains by disulphide bonds (Figure 10.2). The tail chains are also linked along their length by several disulphide bonds. BuchE exists in asymmetric form less abundantly than does AchE.

The asymmetric form of the enzyme is associated only with innervated tissues and is found in muscles, the CNS and peripheral nerves. It is localized to the synaptic cleft, where it is anchored to the basal lamina of the synapse by means of the collagen tail (Figure 10.2). Extraction of the asymmetric form is only achieved after collagenase digestion. The insertion into the extracellular structure is probably through ionic interactions involving polyanions (heparin sulphate proteoglycan) and divalent ions.

The globular forms of AchE and BuchE are ubiquitous and occur in cholinergic and non-cholinergic structures. Human erythrocytes contain membrane-bound dimers of globular AchE while the serum contains globular BuchE (G_4). The globular forms occur in several fractions with different solubilities. They may be insoluble hydrophilic forms such as the BuchE of human serum. Most native globular cholinesterase exists as amphiphilic molecules, that is, they interact with non-denaturing detergent and have dual solubility. These include the G_2 forms of human erythrocyte. Dimeric (G_2) forms also exist as hydrophobic molecules, known as type I amphiphilic G_2 molecules. They have a small hydrophobic domain that consists of a phosphatidylinositol glycolipid covalently attached to the carboxyl terminal. The hydrophobic portion may be removed by proteolytic cleavage, converting the enzyme to a hydrophilic form. It probably serves to bind these globular forms of the enzyme to the outer surface of the cell membrane. For example, hydrophobic dimers are located at the prejunctional membrane in the neuromuscular junction and in erythrocytes. *In vitro*, the hydrophobic part of the anchor may be removed by phosphatidylinositol-specific phospholipase C (PI-PLC) or phospholipase D (PLD), although there are resistant forms of the enzyme (Massoulie *et al.* 1993).

The distribution of these molecular forms of cholinesterase in autonomic structures is less well understood compared with the brain and the neuromuscular junction. Autonomic ganglia, such as the superior cervical ganglion of the rat, have been shown to contain AchE mainly as G_1 and G_4, but with some G_2 and a trace of A_{12}. The axon contains a higher proportion of A_2, with G_4 attached to the outer surface of the axonal membrane. Denervation results in a decrease of ganglionic AchE, whether they are sympathetic or parasympathetic ganglia. Little information is available on the cholinesterases of the parasympathetic nerves innervating effector organs.

10.2.3 Formation of Cholinesterases

Synthesis occurs in both muscle and nerve, initially as monomer G_1 molecules of the catalytic subunit. A single gene encodes the catalytic subunit of AchE which is identical for all molecular forms (A_4 to A_{12} and G_1 to G_4). A separate gene encodes

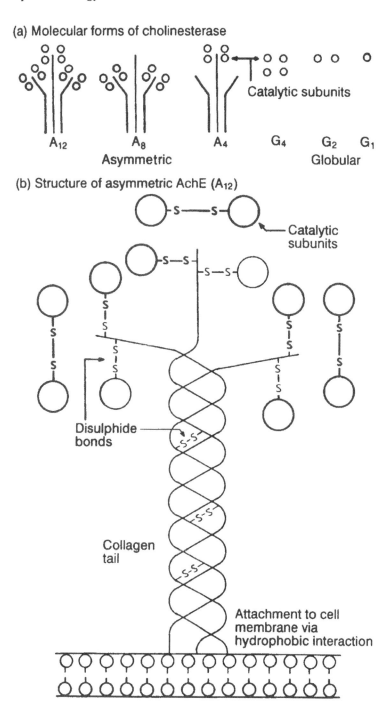

(a) Molecular forms of cholinesterase

A_{12} A_8 A_4 Asymmetric

Catalytic subunits

G_4 G_2 G_1 Globular

(b) Structure of asymmetric AchE (A_{12})

Catalytic subunits

Disulphide bonds

Collagen tail

Attachment to cell membrane via hydrophobic interaction

Figure 10.2 Structural forms of cholinesterases. (a) Molecular forms of cholinesterase showing the asymmetric forms with catalytic subunits in groups of four attached to collagen tails and the globular forms as monomers (G_1), dimers (G_2) and tetramers (G_4) of catalytic subunits with no collagen tail. (b) Structure of an asymmetric acetylcholinesterase (AchE) molecule with three tetramers of catalytic subunits (A_{12}) attached by the collagen tail to the cell membrane.

the catalytic subunit for BuchE. The G_1 monomer precursors are formed in the nucleus and used as the basis for all of the molecular forms. Multiple mRNAs have been identified, which suggests that the common cDNA undergoes transcription into variant forms possibly by alternative splicing. Further differentiation occurs at the posttranscriptional stage of polypeptide synthesis from mRNA, whereby the catalytic subunits are formed into dimers, hydrophobic glycolipid anchors are added, and asymmetric forms are assembled. These processes take place in the endoplasmic reticulum and Golgi apparatus. The different forms are then transported to the cell surface of muscle and nerve cells. Transport of AchE down the axon (anterograde transport) has been studied by ligaturing and noting the accumulation behind the ligature. A small portion is also transported retrogradely. The majority of AchE in transit seems to be the membrane-bound forms, G_4 and A_{12}. In common with the transport of Ach, AchE is transported via the microtubules (Tytell & Stadler 1988). At the nerve ending, the AchE is externalized and secreted, probably by a process of exocytosis of intracellular vesicles. This process is blocked by colchicine and is controlled by intracellular Ca^{2+} levels. Extracellular AchE exists primarily as the G_4 and A_{12} forms (Toutant & Massoulie 1988; Massoulie *et al.* 1993).

10.2.4   *Catalytic Subunit of Cholinesterase*

The catalytic centre of the enzyme has long been regarded as consisting of two sites for interaction with the cholinester: an *esteratic* and an *anionic* site. The esteratic site is involved in interaction with the acyl group of Ach and during hydrolysis this site becomes acylated. The anionic site has been thought for many years to assist in the binding of the quaternary ammonium group of Ach to the enzyme (Figure 10.1).

The primary amino acid sequence of the catalytic subunit of the cholinesterase enzyme was first deduced for AchE from the sequence of DNA obtained from *T. californica* electric organ. It has also been directly sequenced. The catalytic unit consists of 575 amino acid residues with a $M_r = 65\ 612$. The sequence is similar to that now identified for human BuchE and other sources of the enzyme, but is different from that of the muscarinic and nicotinic receptor. The homology among different cholinesterases is quite good, with certain regions highly conserved, including those associated with the esteratic site.

The *esteratic site* has long been known to contain a serine hydroxyl group which becomes acetylated during the hydrolysis of Ach. The enzyme is therefore a member of a larger group of serine hydrolases; these are all blocked by organophosphorus compounds such as DFP, which covalently binds to the serine residue. The serine amino acid has been identified at position 200 of the primary structure of AchE from *T. californica* and several other sources. A six amino acid sequence (197–202) containing the active site serine residue is highly conserved among the cholinesterases: Phe-Gly-Glu-SER-Ala-Gly. Also conserved is a histidine at position 440, which has been shown by site-directed mutagenesis to be essential for hydrolysis of Ach.

The *anionic site*, which binds the quaternary ammonium head of Ach, has not been identified with such certainty. A tetrapeptide residue, Gly-Ser-Phe-Phe (328–331), has been identified in electric eel by labelling experiments (Massoulie & Toutant 1988). An alternative tetrapeptide, Asp-Leu-Phe-Arg (217–220), has been proposed on the basis of prevention of binding of N,N-dimethyl-2-phenyl-aziridinium (DPA) to the active centre of the enzyme. The aziridinium ion was shown to irreversibly inhibit at α-adrenoceptors, being the product of the β-haloalkylamine, DMPE (see Chapter 5, Table 5.2). The labelling of the enzyme with the alkylating agent may be prevented by protecting with specific ligands. Edrophonium, for example, is an inhibitor of cholinesterase that binds selectively to the anionic subsite of the active centre (Table 10.1). In the presence of edrophonium, no labelling of the peptide in positions 217–220 of *Torpedo* AchE could be identified. Asp-217 carries a negative charge and could constitute a component of the anionic site. However, this sequence is not found in close proximity to the active site according to X-ray analysis of the AchE from *T. californica*. A better bet for the identity of the anionic binding residue is Trp-84 which lies in the sequence Ser-Gly-Ser-Glu-Met-TRP-Asn-Pro-Asn (79–87) of *T. californica* AchE (Figure 10.3) (Hucho *et al.* 1991).

10.2.4.1 Peripheral binding sites

The existence of additional anionic binding sites is indicated by non-competitive inhibition of AchE by the neuromuscular blocking agents, D-tubocurarine and gallamine. Propidium is a fluorescent ligand that binds to this peripheral site rather than to the anionic site at the active centre of the enzyme. Binding to this peripheral site is regarded as allosteric, in that it allows conformational changes of the enzyme to occur, which assists binding at the active centre. Labelling of this site by DPA can be prevented by propidium and a peptide sequence, Lys-Pro-Gln-Glu-Leu-Ile-Asp-Val-Glu (270–278), has been identified as forming part of this site (Figure 10.3). Only marginal binding to this site can be demonstrated under physiological conditions and its significance is therefore uncertain. It may be involved in inhibition of catalytic activity when high concentrations of the Ach substrate are present, which can then bind to this peripheral site. This process may be associated with substrate inhibition which occurs with high concentrations of substrate at AchE. No substrate inhibition or peripheral binding sites have been reported for BuchE (Hucho *et al.* 1991, Massoulie *et al.* 1993).

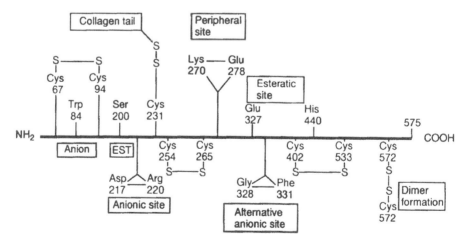

Figure 10.3 Amino acid sequence of acetylcholinesterases and possible binding sites of Ach. Note the disulphide (-S-S-) bonds involved in cross-bridging and linking to the collagen tail and the paired unit in dimers.

The amino acid residues involved in dimer formation and in attachment of catalytic subunits to the collagen tail have also been identified. These linkages are through disulphide bonds between cysteine residues. The disulphide bond linking two catalytic subunits into dimers is probably Cys-572 near the carboxyl terminal of the polypeptide. The linkage to the peptide tail is probably through Cys-231. Cysteine residues are also involved in disulphide bridging within the catalytic subunit and these, together with the proposed esteratic and anionic sites, are shown in Figure 10.3 (Massoulie & Toutant 1988, Hucho *et al.* 1991).

10.3 Mechanism of Hydrolysis of Ach

Ach is attracted to the catalytic subunit of the cholinesterase enzyme by the esteratic and anionic sites. The quaternary ammonium group is attracted to the anionic site through an ionic interaction. However, as the number of methyl groups are increased in ammonium compounds, their potency as inhibitors of the enzyme increases. This suggests that hydrophobic binding through Van der Waals forces also occurs. It appears that a charge cloud spread over several amino acids at the anionic site is involved rather than a single negatively charged amino acid side chain such as glutamate ion (Hucho *et al.* 1991, Somani *et al.* 1992). The esteratic site serine hydroxyl group provides the point of interaction with the acyl group of the Ach. Nucleophilic attack on the acyl carbon results in the formation of acetylated enzyme. In this group of serine esterases, the serine OH group is part of a 'catalytic triad' in which serine, histidine and aspartate, at different positions of the enzyme molecule, are cross-linked by hydrogen bonds. Ser-200 and His-440 have already been described as being important for catalytic activity, but an aspartate component has not been identified. Glu-327 provides an alternative site and according to the X-ray structure completes the triad. These three sites are located in an active site gorge with Ser-200 situated at the bottom of this gorge. The peripheral site is located at the surface of the protein near the entrance to the active site gorge. Peripheral site ligands may inhibit activity by obstructing the entrance to the gorge. Ach may bind initially to the peripheral site which contains negative charges ('landing site') and then slide down the aromatic lined gorge into the active site (Massoulie *et al.* 1993).

The three groups of the catalytic triad are clearly not adjacent in the amino acid sequence, however the conformation of the molecule may be such that they come into close contact linked by hydrogen bonding (Figure 10.3). The hydrogen bonding between the glutamate, serine and histidine imidazolium may increase the electron density at the serine oxygen, facilitating nucleophilic attack on the substrate and enhancing proton transfer. Following the initial interaction, an intermediate tetrahedral structure is formed between the enzyme and the choline ester. This breaks down to leave the esteratic serine hydroxyl group acetylated and the liberation of free choline. The acetylated enzyme rapidly hydrolyses to regenerate the enzyme and release acetate (Figure 10.4). This latter step is the rate-limiting stage of Ach breakdown, having a half-life of 42 μsec turnover rate for each molecule. At a turnover rate of 25 000 molecules per second, this is one of the body's most efficient catalytic processes. A high proportion of the choline is transported back into the nerve ending for resynthesis into Ach which is stored in vesicles (Somani *et al.* 1992) (see Chapter 7).

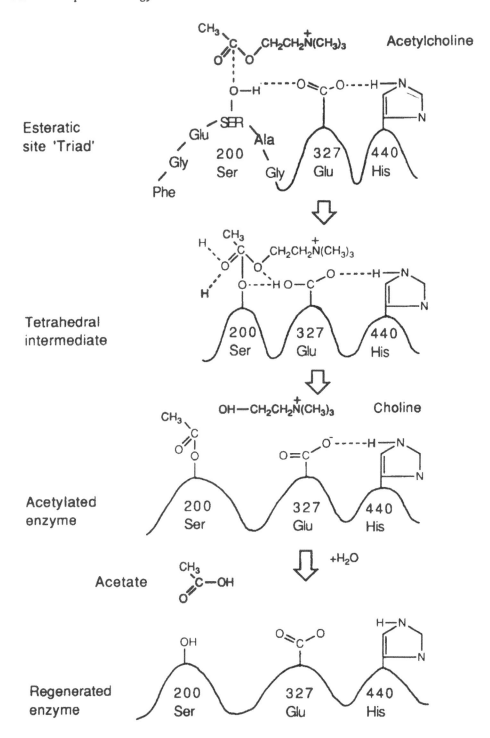

Figure 10.4 Breakdown of acetylcholine at the active site of acetylcholinesterase. The essential amino acid residues serine (Ser-200), glutamate (Glu-327) and histidine (His-440) form the esteratic site 'Triad' joined by hydrogen bonding. A tetrahedral intermediate is formed with acetylcholine before release of choline and acetylation of the enzyme at Ser-200.

Other cholinesters that are susceptible to hydrolysis by cholinesterase undergo a similar reaction with the enzyme. The esteratic site of BuchE becomes butyrylated when butyrylcholine is hydrolysed.

10.4 Other Functions of Cholinesterases

In addition to the catalytic activity in hydrolysing cholinesters, cholinesterases may be involved in the growth and development of nervous tissue. BuchE also appears to have a function in the regulation of cellular expression of AchE. The selective inhibition of BuchE by iso-OMPA causes inhibition of AchE activity which is not due to cross-inhibition directly of the enzyme. There is also a close homology of the enzyme structure with that of the cell adhesion molecule, neurotactin, suggesting an additional role of cholinesterases in confering adhesive properties on cells (Massoulie *et al.* 1993).

Cholinesterases are found intracellularly in noradrenergic neurones of the substantia nigra, where they are secreted along with dopamine. They may hydrolyse other non-cholinergic substrates such as substance P and enkephalins. The esteratic site of AchE and BuchE has been shown to differ from the peptidasic site since there is a difference in susceptibility to the organophosphorus agent, DFP (Toutant & Massoulie 1988).

10.5 Anticholinesterases

Drugs which inhibit cholinesterases may be divided into two types based upon the nature of their interaction with the enzyme. The *reversible* inhibitors interact with both the anionic and esteratic site of the enzyme and *in vivo* may be reversed slowly over a period of 3–4 hr. This group comprises agents structurally related to Ach, having an ester linkage and a quaternary or tertiary amine that is charged at physiological pH. Most of the agents are carbamyl esters such as physostigmine and neostigmine. The *irreversible* anticholinesterases are the organophosphorus compounds which usually interact only with the esteratic site. The resulting phosphorylated or phosphonylated enzyme is extremely stable and recovery is dependent mainly upon regeneration of new enzyme (Elbein 1990).

10.5.1 *'Reversible' Anticholinesterases – Carbamyl Esters*

10.5.1.1 *Physostigmine*

The prototype reversible anticholinesterase is physostigmine (or eserine), a naturally occurring alkaloid obtained from the Calabar or Ordeal bean, the dried seed of *Physostigma venenosum* (Table 10.1). These beans were once chewed by natives of tropical West Africa as an 'ordeal' poison in trials for witchcraft. Physostigmine is a carbamyl ester and tertiary amine which is charged a physiological pH. It serves as an alternative substrate to Ach, the protonated amine being held to the anionic site and the esteratic site being carbamoylated (see Figure 10.5, as for neostigmine). The hydrolysis of physostigmine is much slower than for Ach. The carbamoylated enzyme is also much

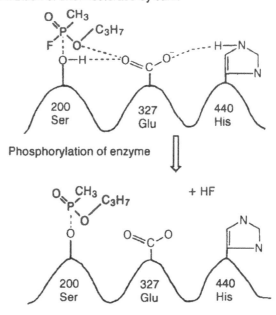

Figure 10.5 Interactions of (a) neostigmine and (b) the organophosphorus compound, sarin, with the cholinesterase esteratic site.

more stable than the acetylated enzyme. Its slow hydrolysis effectively prevents catalytic activity upon Ach. Physostigmine, being a tertiary amine, crosses the blood–brain barrier and has central effects. It is also absorbed orally from the gastrointestinal tract.

10.5.1.2 *Neostigmine and related agents*

Neostigmine is a synthetic derivative having a quaternary ammonium group which increases potency and prevents penetration of the blood–brain barrier. In pyridostigmine, the quaternary nitrogen is included in a pyridyl ring, while in distigmine, two pyridostigmine molecules are connected by a six-carbon methylene chain through the carbamate nitrogens. Demecarium is formed by a similar linkage of a ten-carbon methylene chain between two neostigmine molecules (Table 10.1). The bisquaternary compounds are generally more potent than their monoquaternary counterparts, possibly because they interact with two catalytic subunits of the cholinesterase (Holmstedt 1959). These compounds are often selective for AchE; ambenonium, for example, has little or no activity against BuchE.

The carbamyl esters are slowly hydrolysed by cholinesterases and therefore remain in contact with the enzyme for longer than Ach, so that they effectively protect the enzyme from other substrates such as Ach. Second, once the esteratic site is carbamoylated by these carbamyl esters, it undergoes very slow hydrolysis. The methylcarbamoyl (from physostigmine) or dimethylcarbamoyl enzyme (from neostigmine or pyridostigmine) is more stable than the acetylated enzyme; the half-life for hydrolysis of the latter is 15–30 min (Wilson & Harrison 1961).

Edrophonium is a derivative of neostigmine that lacks the carbamoyl group; it therefore combines only with the anionic site. Edrophonium has a rapid onset of action and its duration of action is very short. The enzyme is neither acetylated nor carbamoylated and edrophonium is not hydrolysed during its inhibitory action, which may therefore be regarded as strictly reversible. The naturally occurring alkaloid, galantamine, which is obtained from the bulbs of the Caucasian snowdrop (*Galanthus woronowii*) is also a reversible anticholinesterase. This agent has occasionally been used in Russia for the treatment of myasthenia gravis.

Finally, it is worth noting that simple aliphatic carbamates may also exhibit weak anticholinesterase activity. An example is urethane (ethyl carbamate, $C_2H_5OOCNH_2$), an anaesthetic still used in experimental animals. It is often difficult to demonstrate the effects of anticholinesterases in animals anaesthetized with urethane.

Neostigmine and physostigmine are structurally related to Ach and have actions at the nicotinic receptor. Large doses block the nicotinic receptor-mediated effects at autonomic ganglia, the quaternary compound, neostigmine, causing an initial depolarizing effect. A blocking action on nicotinic Ach-induced inward currents of ganglion cells has been demonstrated, neostigmine being suggested to interact with the receptor but physostigmine interacting directly with the Ach-gated cation channel (Sadoshima *et al.* 1988). Neostigmine may also induce the release of Ach from parasympathetic neurones (Norel *et al.* 1993).

10.5.2 *'Irreversible' Anticholinesterases – Organophosphorus Compounds*

The organophosphorus compounds interact with the esteratic site of the catalytic subunit only. They phosphorylate or phosphonylate the serine hydroxyl group of the esteratic site. This covalent binding is extremely stable and results in long-term irreversible blockade of the enzyme. These agents have the same general structure (Table 10.2), and in forming the phosphorylated esteratic site the group X is lost. Examples of group X include a halide (fluoride in diisopropylphosphofluoridate, DFP), cyanide (tabun), phenoxy (parathion) or a second phosphate (tetraethylpyrophosphate, TEPP). The phosphorylating group forms a tetrahedral arrangement with the esteratic site, which is similar to the intermediate of the acylated enzyme and increases the reactivity and stability of the complex (Figure 10.5). The groups R_1 and R_2 in the general formula have been varied extensively; most frequently they are alkyl (sarin, soman), alkoxy (TEPP, parathion, DFP) and amido (tabun), but many other groups have been substituted (Table 10.2). If the alkyl groups are methyl or ethyl, regeneration of the enzyme occurs after several hours. Rates of recovery of the enzyme are the same for different compounds having R_1 and R_2 in common, but differing in their X residue. This was early proof of phosphorylation of the enzyme by a common dialkylphosphorus residue. Secondary (DFP) or tertiary alkyl groups increase the stability of the phosphorylated complex so that reversal is extremely slow. Recovery from these irreversible anticholinesterases is then due to *de novo* synthesis of the enzyme. In some organophosphorus compounds, such as parathion and malathion, a sulphur atom is substituted for the oxygen atom. *In vitro*, these are inactive as inhibitors of cholinesterases. An oxygen group must be restored to the compound *in vivo* for it to be an active inhibitor. This occurs in the liver by mixed-function oxygenases. The active metabolite of parathion is paraoxon, which is an inhibitor *in vitro*.

Many of the organophosphorus compounds contain asymmetric centres, which in the case of the nerve agents is the phosphorus atom. In soman there is an additional chiral carbon atom which results in there being four stereoisomers in the synthetic product. The stereoisomers have different toxicities, with C(−),P(−)-soman and C(+),P(−)-soman being the more toxic by >20-fold. Reactivation by oximes is also stereoselective, with inhibition by C(+),P(−)-soman being more readily reversed compared with C(−),P(−)-soman. Where there is a single chiral centre due to the phosphorus atom, it is the (−)-enantiomer that is more toxic (Somani *et al.* 1992).

The above mentioned organophosphorus compounds interact with only the esteratic site of the enzyme. There is also a group of compounds that contain a quaternary nitrogen atom, and these additionally interact with the anionic site. These compounds, the phosphostigmines, are highly potent inhibitors and ecothiopate is used clinically.

10.5.2.1 *Reversal of organophosphorus inhibitors*

The spontaneous hydrolysis of the phosphorylated enzyme occurs extremely slowly. Reversal of the inhibition may, however, be effected by nucleophilic agents including hydroxylamine (NH_2OH), hydroxamic acids (RCONH-OH) and oximes (RCH = NOH). By adding a quaternary nitrogen that could interact with the anionic site of the enzyme, 2-pyridine aldoxime methiodide (2-PAM or pralidoxime) was developed as an effective oxime for enzyme reactivation (Holmstedt 1959) (Table 10.3). The oxime group combines with the phosphate or phosphonate group that is covalently bound to the serine hydroxyl group at the esteratic site. The link is broken by nucleophilic attack, which splits off the oxime–phosphorus moiety,

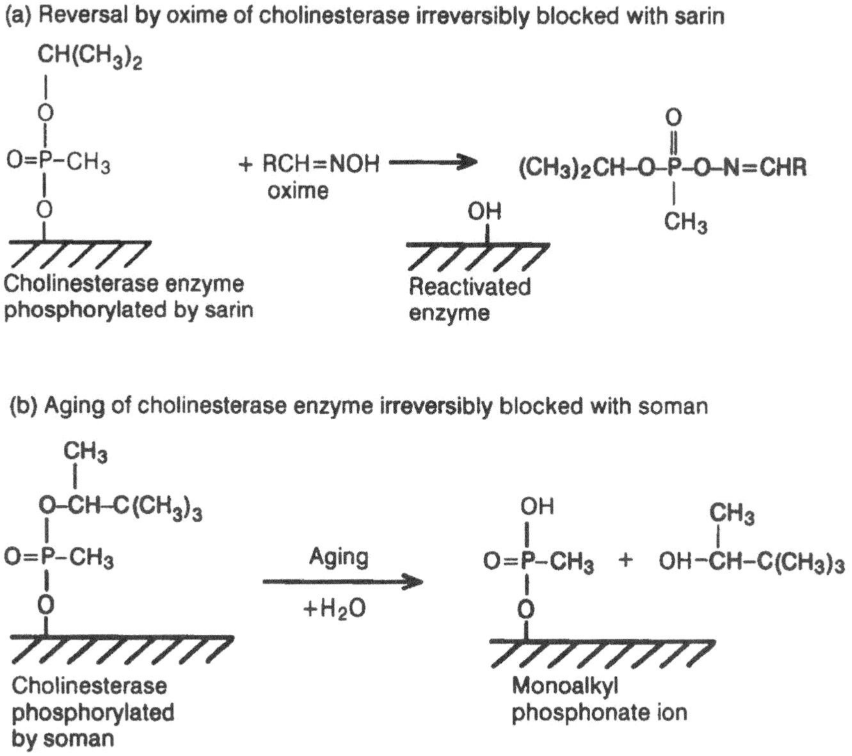

(a) Reversal by oxime of cholinesterase irreversibly blocked with sarin

(b) Aging of cholinesterase enzyme irreversibly blocked with soman

Figure 10.6 Reactivation by oximes (a) and aging (b) of cholinesterase irreversibly blocked by the organophosphorus compounds, sarin and soman.

leaving the regenerated enzyme (Figure 10.6). The rate of reactivation of the phosphorylated enzyme by the oxime depends upon the type of group that phosphorylates the enzyme. The order of decreasing reversal rates is essentially the same as for the spontaneous recovery, which is dimethylphosphoryl (eg malathion) > diethylphosphoryl (eg paraoxon) > diisopropylphosphoryl (DFP), and so forth. Several bisquaternary oximes, such as obidoxime, have been shown to have greater potency as enzyme reactivators and antidotes to nerve gas poisoning (Table 10.3). Pralidoxime is effective against tabun and sarin poisoning, but not against soman. The reason for this is that the inhibited enzyme is said to 'age'. Aging occurs quite rapidly, within minutes or hours, depending upon the phosphorylating moiety. Phosphonates containing tertiary alkoxy groups (eg soman) age more readily than do those with secondary (eg sarin) or primary alkoxy groups (eg tabun).

The process of aging involves the loss of an alkyl group from the oxygen that is attached to the phosphorus atom, to leave a phosphonate or phosphate anion bound to the enzyme. The rate at which the inhibited enzyme ages is dependent upon the structure of the inhibitor. It is also temperature- and pH-dependent. Soman ages particularly rapidly, being essentially complete within 10 min of the initial inhibition. Aging of soman involves loss of the pinacolyl group, probably by an acid-catalysed process. The imidazolium ion of the esteratic site may play a critical role in initiating the aging process, so that the enzyme itself is involved in the aging mechanism. The resulting monoalkyl- or monoalkoxy-phosphonate ion is even more resistant to hydrolysis than is the neutral ester and cannot be removed by the nucleophilic oximes (Figure 10.6). Protection of the anionic site of AchE by quaternary ammonium compounds (decamethonium, tubocurarine) may prevent aging of soman-inhibited enzyme, since they have been reported to be effective with non-human AchE inhibited by sarin. The commercially available oximes, pralidoxime and obidoxime, are ineffective against the aged enzyme inhibited by soman. However, a series of oximes known as the H-series, after Hagedorn, have been synthesized and show reactivating potency *in vitro* and promising activity *in vivo*, in various animal species (Table 10.3). Hl-6 is more active against enzyme blocked by sarin and may produce a small amount of reactivation of soman-inhibited human AchE (Somani *et al.* 1992).

The oximes may also protect against organophosphorus poisoning by reacting directly with the inhibitor before it can phosphorylate the enzyme and additionally with the enzyme to protect it from phosphorylation (Holmstedt 1959). The latter property is revealed in pralidoxime as weak anticholinesterase activity. Because of this, the oximes are not suitable for treatment of overdose with the 'reversible' inhibitors, physostigmine and neostigmine, that carbamoylate the enzyme, against which they are ineffective. In the case of carbaryl, the inhibition of the enzyme is even potentiated (Sidell 1992).

Reversal of skeletal muscle paralysis by oximes, even after aging of the enzyme, may also occur by a direct pharmacological effect on the nicotinic receptors (N_M). Several compounds, including some without the oxime group, have been shown to block nicotinic receptor ion channels in isolated diaphragm preparations. This action would counteract the overstimulation of the nicotinic receptors at the endplate resulting from Ach accumulation (Tattersall 1993).

10.5.2.2 *Interactions of organophosphorus compounds with other enzymes and their metabolism*

Other enzymes belonging to the group of carboxyesterases are also susceptible to inhibition by the organophosphorus compounds. These include aliphatic esterases, chymotrypsin, trypsin and thrombin (B-type esterases). All have a serine hydroxyl group that is phosphorylated by the organophosphorus compound, but probably only have the esteratic site. When BuchE is phosphorylated by $DF^{32}P$, the serine is labelled with a monoisopropyl phosphoryl group and not the diisopropyl-containing group, indicating aging of the enzyme. In the other enzymes, the serine residues carry the diisopropyl group, suggesting a difference in the type of interaction and lack of aging (Holmstedt 1959). Phosphorylphosphatases are a broad group of enzymes, known as A-type esterases, which hydrolyse phosphorus esters. They are found in a number of tissues including the liver, kidney and plasma. Unlike the B-esterases, these enzymes are in fact not irreversibly inhibited by the organophosphorus compounds, probably because the phosphorylated esteratic site is rapidly reversed by reaction with water. However, these enzymes may break down both the phosphorylated enzyme and the phosphorus ester link of the inhibitor. They are therefore the major route of metabolism and detoxification of many of the organophosphorus inhibitors, including DFP, TEPP, tabun, sarin and paraoxon, in which the A-esterase also splits the P-F or P-CN bonds. Malathion is rapidly metabolized by hydrolysis of the diethylsuccinate moiety by these carboxyesterases and of the P-S-C link. This metabolism is irrespective of whether it is malathion or the active oxygen-carrying derivative. Malathion is widely used as an insecticide for crop spraying; its metabolism in mammals is more rapid than in insects and it is therefore much less toxic to mammals than to insects. The acquired resistance of insects to these insecticides is due to adaptive development of the carboxyesterase metabolic enzymes. C-Type esterases do not react with organophosphorus compounds.

Other enzymes inhibited *in vivo* by the organophosphorus compounds include Na^+,K^+-ATPase, aldolase, succinate dehydrogenase, tyrosine hydroxylase and triglyceride lipase. *In vitro*, the organophosphorus compounds also inhibit some of these enzymes and a wide range of others including cyclic AMP phosphodiesterase (see Chapter 13). They also inhibit the binding of [^{3}H]L-phenylisopropyladenosine ([^{3}H]L-PIA), indicating that they could bind to the A_1 adenosine receptors and thereby produce pharmacological effects including changes in K^+ permeability in central synaptic membranes (Somani *et al.* 1992).

10.5.3   *Selectivity of Anticholinesterases for AchE or BuchE*

DFP and iso-OMPA selectively inhibit the plasma enzyme (BuchE), the former by 90% compared with only 20% of the erythrocyte AchE. DFP is highly permeant and can inactivate the intracellular compartment of cholinesterase. Parathion and malathion selectively inhibit the plasma enzyme, whereas others such as dimefox, mevinphos and the potent nerve agent, VX, initially inhibit the erythrocyte enzyme selectively (Sidell 1992). Selectivity among the 'reversible' inhibitors has also been reported, with bis(3-trimethylammonium-5-hydroxyphenoxyl) 1,3-propane (3116CT) being 250 000 times more potent against AchE than BuchE and 10-(1-diethylaminopropionyl) phenothiazine (Astra 1397) being 10 000 times more potent against the plasma enzyme (Bowman & Rand 1980). BW284c51 is a non-permeant selective inhibitor of AchE that inactivates only extracellular AchE (Atack *et al.* 1989); it also has some muscarinic blocking activity (Norel *et al.* 1993). Ethopropazine selectively inhibits BuchE (Massoulie *et al.* 1993, Vellom *et al.* 1993). These selective inhibitors have been used to demonstrate that BuchE has a role in the control of parasympathetic activity in tissues such as the airways. BW284c51 causes contraction of human isolated bronchial preparations whereas iso-OMPA does not, indicating that AchE is mainly responsible for degradation of endogenous Ach. However, iso-OMPA does potentiate the contractile effects of exogenous Ach. This indicates that BuchE, probably located in the smooth muscle, may have a secondary role to protect against an increased release of Ach or against the extraneuronal appearance of Ach (Norel *et al.* 1993).

10.6   **Pharmacological Properties of Anticholinesterases**

The pharmacological properties of the anticholinesterase agents are generally the result of accumulation of Ach at the various sites of its release. These agents therefore have actions at the neuromuscular junction (N_M nicotinic effects), autonomic ganglia (N_N nicotinic effects), in the CNS (mainly muscarinic) and at the parasympathetic effector organs (muscarinic).

10.6.1 *The Neuromuscular Junction of Skeletal Muscle*

Skeletal muscle targets are peripheral to the subject of this book and will be dealt with only briefly. Accumulation of Ach at the endplate allows multiple activation of nicotinic receptors (N_M) so that the decay of the endplate potential is delayed beyond the refractory period of the muscle fibre membrane. As a result, at the end of the muscle fibre action potential, the endplate potential triggers off a series of action potentials. This leads to repetitive firing and fibrillation of the muscle fibres. Fasciculations also arise from stimulation of prejunctional sites on the somatic nerve. These effects are seen as rippling or twitching of superficial muscles (Miyamoto 1978).

The 'reversible' anticholinesterases, neostigmine, pyridostigmine and edrophonium, are used to terminate the neuromuscular paralysis by competitive neuromuscular blocking agents, such as tubocurarine. The raised levels of Ach induced by agents such as neostigmine, by competing with the antagonist for the N_M receptor, displace it and allow normal neuromuscular transmission to resume. Neostigmine also has direct effects at nicotinic receptors. The depolarizing blockers such as suxamethonium are not, however, reversed; instead the blockade may be deepened by the additional depolarizing action of the accumulating Ach. Continued activation of nicotinic receptors and depolarization of the endplate of the neuromuscular junction eventually leads to flaccid muscle paralysis. The respiratory muscles (intercostal and diaphragm) are the most vulnerable and thus respiratory failure, also attributed to pulmonary oedema, is the primary cause of death from anticholinesterase overdose and poisoning.

10.6.2 *Autonomic Ganglia*

The elevated levels of endogenous Ach at the synapse of autonomic ganglia causes spontaneous discharge of electrical activity in the postganglionic fibre. This is due to the stimulation of both the dominant nicotinic (N_N) receptor and the minor muscarinic receptor (M_1) of the autonomic ganglion cell (Figure 7.7). The stimulatory effect on opposing sympathetic and parasympathetic ganglia is likely to yield a complex resultant response depending upon which division of the autonomic nervous system is dominant. A further complication is the fact that at higher doses of an anticholinesterase, the persistent activation of cholinoceptors and depolarization of the ganglionic cell membrane induces ganglionic blockade (see Chapter 11). These ganglionic effects are also influenced by the actions at muscarinic receptors of the parasympathetic nerve endings, which in fact give rise to the major pharmacological responses of most organs.

10.6.3 *Central Nervous System (CNS)*

Responses of the CNS depend upon access of the anticholinesterase to the brain and its ability to cross the blood–brain barrier. The organophosphorus compounds are generally highly lipid soluble and therefore penetrate the blood–brain barrier readily, whereas the charged quaternary compounds (neostigmine, edrophonium, pyridostigmine and ecothiopate) do not. At low, pharmacologically active dose levels, acute exposure to anticholinesterases produces central stimulation. Subjects exposed to organophosphorus compounds percutaneously display anxiety, sleep disturbances including insomnia and unusual dreams, mood changes and nervousness. These effects may be antagonized by atropine, indicating that they are due to stimulation of central muscarinic receptors. Surprisingly, in view of the possible applications in the management of Alzheimer's disease (see later), there are reports of forgetfulness, impaired judgement and poor comprehension. With larger toxic levels of exposure, there is CNS depression. Drowsiness and weakness are soon followed at higher doses by a rapid loss of consciousness. Within seconds of losing consciousness, there is onset of seizure activity followed, after several minutes, by cessation of breathing. There is a close correlation between the convulsions and brain damage. A major factor contributing to these central effects after high level exposure is probably hypoxia. Convulsions and brain damage induced experimentally in animals and convulsions arising from accidental exposure of humans to organophosphorus insecticides are reduced by diazepam. Depression of the medullary respiratory centre contributes to the respiratory paralysis, and depression of the vasomotor centre leads to hypotension (Sidell 1992).

10.6.4 *Muscarinic Effects of Anticholinesterases at Autonomic Effectors*

These responses represent the primary pharmacological actions of the anticholinesterase agents. The responses due to accumulation of Ach at the parasympathetic nerve endings are observed with low levels of exposure and in the usual therapeutic dose ranges. They will also be the first effects seen with high levels of exposure, soon followed by additional toxic symptoms due to actions at other sites.

The pharmacology of anticholinesterases has been well established for some time and relatively few reports occur nowadays of the basic pharmacology of these agents at peripheral autonomic sites. The scientific literature on drugs such as physostigmine mainly involves their use as tool drugs in evaluating the role of Ach in tissue responses, for example, to autonomic nerve stimulation. In any such studies it is essential to show the potentiating effect of physostigmine upon the response to exogenously added Ach in the same tissue as its effect upon the response to nerve stimulation. A classic example is shown in Figure 10.7, where Remak's nerve to the chicken rectum is stimulated and the effluent from the tissue is passed over a preparation of longitudinal muscle of the guinea-pig ileum. A small contraction of the ileal muscle on nerve stimulation was potentiated by physostigmine, indicating the release of Ach from Remak's nerve. The response was also blocked by the muscarinic antagonist, atropine. Confirmation of the anticholinesterase activity of physostigmine is shown by the potentiation of exogenous Ach, but no effect upon a non-choline ester, the peptide substance P (Komori & Ohashi 1984).

The responses of autonomically innervated tissues to anticholinesterases can be predicted from a knowledge of the effects of parasympathetic nerve stimulation (Table 1.1) and are essentially the same as those of the muscarinic receptor agonists described in Chapter 8. Responses do depend upon there being an adequate parasympathetic discharge to the organ. Only effects on organs having therapeutic implications or providing diagnostic features of intoxication will therefore be described in detail below.

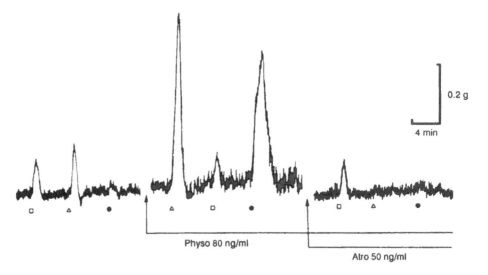

Figure 10.7 Effects of physostigmine (Physo) alone and with atropine (Atro) on the responses of the longitudinal muscle of the guinea-pig ileum to the venous effluent from the isolated, perfused chicken rectum during stimulation of Remak's nerve (0.2 msec, 30 Hz for 30 sec, ●) and to added substance P (3 ng/ml, □) or acetylcholine (5 ng/ml, △). Potentiation of responses to Ach and Remak's nerve stimulation by physostigmine and blockade by atropine indicates the cholinergic nature of the nerve transmission. The response to substance P is unaffected indicating that neuropeptides are not involved. Reproduced with permission from Komori & Ohashi (1984).

10.6.4.1 The eye

Responses of the eye are often the first to appear after systemic absorption of anticholinesterases since it is quite sensitive to their action. Topical application also causes contraction of the circular muscle of the iris leading to a constricted pupil (miosis) which has been described as 'pinpoint' in size. Further constriction may occur with exposure to bright light. The ciliary muscles are contracted to accommodate the lens for near vision. Complaints of dim vision are usually attributed to the decrease in pupil size. An additional impairment of vision, independent of the miosis, may be due to a central action or to accumulation of Ach in the retina. Contraction of the iris opens up the drainage angle for the passage of aqueous humour into the canal of Schlemm by alignment of the trabecular network (Figure 1.7). As a result of improved drainage, intraocular pressure may fall, especially if it is already elevated (Kaufman *et al.* 1984).

10.6.4.2 Gastrointestinal tract

Anticholinesterases cause enhanced motility and tone of the stomach and large and small intestines, which is largely dependent upon the presence of an intact vagal innervation. Gastric acid secretion is promoted. These effects are due to the accumulation of Ach at both the muscarinic receptors located in smooth muscle at the parasympathetic nerve ending and at nicotinic receptors of ganglion cells of the parasympathetic pathways in the intramural plexus (Figure 1.5). These actions produce diarrhoea, intestinal colic, nausea and stomach cramps.

In the transmurally stimulated guinea-pig ileum (Paton preparation), the twitch response is enhanced by physostigmine, indicating that the electrical stimulation is due to the release of Ach from parasympathetic nerve endings (Figure 9.1). The response to the ganglion stimulant, TMA, is also enhanced. This is not because TMA is a substrate for cholinesterase, but because its activity is due to the release of Ach at the parasympathetic nerve endings in the longitudinal smooth muscle. Stripping the innervation from the longitudinal muscle removes the activity of TMA and the direct contractile activity of physostigmine, which depends upon the presence of endogenous Ach (Paton & Zar 1968).

Related to actions on the gastrointestinal tract is the effect on the bladder, where the parasympathetic tone is enhanced and the detrusor muscle contracts. This results in voiding of the bladder. In the rat isolated bladder, field stimulation induces a monophasic contraction which is potentiated by physostigmine and converted to a biphasic response. The late phase was revealed by physostigmine and was more sensitive to blockade by atropine. This indicates that the late phase of contraction is due to Ach release from parasympathetic nerves, whereas the early contraction is non-adrenergic non-cholinergic, possibly purinergic and due to ATP release (Figure 7.6) (Maggi 1991b) (Chapter 12).

10.6.4.3 Cardiovascular system

The accumulation of Ach at parasympathetic vagal nerve endings in the presence of anticholinesterases results in bradycardia. At higher doses arising from insecticide poisoning, however, there may be tachycardia probably due to ganglionic stimulation of the sympathetic pathways. Thus, heart rate is an unreliable diagnostic sign of anticholinesterase poisoning. In atrial muscle the effective refractory period is shortened, but in the SA and AV nodal tissue the refractory period and conduction time is increased. Bradyarrhythmias may occur, while heart block and idioventricular rhythm (inherent ventricular rate) have been reported after exposure to nerve agents.

There is usually a fall in blood pressure, with vasodilatation of most vascular beds, except the coronary and pulmonary circulations, which may show the opposite effect. Increases in blood pressure, however, may also occur. Indeed, intravenous injections of physostigmine into anaesthetized rats causes a hypertensive effect which is due to a centrally mediated activation of peripheral sympathetic mechanisms. This effect is shared by the organophosphorus compounds. The hypertensive response also occurs in the conscious rat, is unaffected by adrenalectomy, but is blocked by atropine, suggesting an effect on central muscarinic receptors. The site of activation appears to be on mechanisms in the hypothalamus that integrate sympathetic outflow (Varagić & Krstić 1966). Intracarotid injections of the opioids, [Leu]-enkephalin, [Met]-enkephalin and β-endorphin, depress or abolish this hypertensive response to physostigmine. This opioid response was reversed by the opioid antagonist, naloxone. The opioids appear to inhibit Ach and/or noradrenaline release in the central cholinergic link (Varagic & Stojanovic 1987). Prostigmine enhances the flushing reaction (cutaneous vasodilatation) in the 'Chinese restaurant syndrome' (Ghadimi *et al.* 1971), which is attributed to local release of Ach after ingestion of large amounts of monosodium glutamate. The possible involvement of cholinergic pathways in local vasodilatation via antidromic stimulation of *sensory* nerves (see Chapter 1) has been proposed on the basis of this potentiation by anticholinesterases (Couture *et al.* 1985).

10.6.4.4 Actions at other sites

Secretory glands innervated by postganglionic cholinergic nerves (Figure 1.3) produce increased secretion in response to nerve activity, and at higher doses of anticholinesterases there is an increase in basal secretion. The glands affected include those in the airways, the increased bronchial secretion together with the bronchoconstriction, contributes to respiratory difficulty (dyspnoea) and ultimately to respiratory failure. Lacrymal secretion produces conjunctival congestion and, combined with the increased nasal mucous gland secretion, causes severe rhinorrhoea. Salivation, sweating and increased secretion in the gastrointestinal tract also occur. Smooth muscle at sites not already mentioned, such as the ureters, usually shows increased tone and spontaneous activity.

10.7 Uses of Anticholinesterase Agents

Anticholinesterases have uses both in therapeutics and as insecticides and nerve agents. The therapeutic applications are due to the accumulation of Ach both at muscarinic sites and at nicotinic receptors of skeletal muscle; the latter uses are not directly concerned with the autonomic nervous system and will only be dealt with briefly.

10.7.1 Reversal of Neuromuscular Blockade

Anaesthetists tend to use neostigmine or edrophonium intravenously to reverse or decrease the duration of competitive skeletal muscle paralysis by tubocurarine. This activity is due to the accumulation of Ach competing with tubocurarine for the nicotinic receptors. Neostigmine is superior to physostigmine since its onset of action is more rapid, probably because it is charged and has poor lipid solubility. Edrophonium has an extremely short duration of action. The quaternary ammonium compounds are generally more selective for the neuromuscular junction. It is necessary to use a muscarinic blocker simultaneously with the anticholinesterase in order to prevent the muscarinic effects of Ach, such as bronchoconstriction, bronchosecretion and salivation. Neostigmine is often combined with glycopyrronium for this purpose (Table 9.3).

10.7.2 Myasthenia Gravis

Myasthenia gravis is a disease of the neuromuscular junction characterized by weakness and fatigue of the skeletal muscles. There is a defect in neuromuscular transmission in which the number of postjunctional nicotinic receptors (N_M) is reduced. It is now regarded as an autoimmune disease in which antireceptor antibodies present in the plasma of myasthenia gravis patients induce receptor degradation (Havard & Fonseca 1990). Diagnosis is assisted by the edrophonium test, whereby intravenous injections of the short-acting edrophonium (2 mg) leads to a brief improvement of muscle strength. The anticholinesterases, neostigmine, pyridostigmine, distigmine and ambenonium (Table 10.1), are the standard drugs for symptomatic treatment of myasthenia gravis. Their action results from the increased availability of Ach at the diminished receptors. They are given orally, despite the poor adsorption via the gastrointestinal tract. The muscarinic side-effects can be controlled with atropine, although these signs may be of value in recognizing overdose, which will have detrimental effects upon neuromuscular transmission arising from depolarizing block.

10.7.3 Glaucoma

Glaucoma is associated with a rise in intraocular pressure which leads to optic nerve damage and blindness if untreated. The anticholinesterases, by causing accummulation of Ach, reduce intraocular pressure by mechanisms similar to those already described for the parasympathomimetic amines (Chapter 8). These agents lower intraocular pressure chiefly by their actions on the circular muscles of the iris, contraction of which, away from the canal of Schlemm, improves the drainage of aqueous humour. Secondary effects on production of aqueous humour in the ciliary body also arise from vascular actions of anticholinesterases, with an initial rise in pressure sometimes preceding the fall. Both narrow- (closed) angle and wide- (open) angle forms of primary glaucoma are treated with anticholinesterase drugs. The narrow-angle type is relieved by opening of the restricted drainage angle at the entrance to the canal of Schlemm. In open-angle glaucoma there is no physical obstruction but the anticholinesterases, by contraction of the circular muscle of the iris and the ciliary body, increase the tone and alignment of the trabecular network. This improves the outflow and drainage to the canal of Schlemm (Kaufman *et al.* 1984).

Physostigmine is readily absorbed after topical application to the eye. Indeed, conjunctival instillation may result in systemic absorption via the nasal mucosa, which may be prevented by application of pressure on the inner canthus of the eye during and immediately following instillation. It may cause considerable irritation and pain due to the muscle spasm. A disadvantage of anticholinesterases is that accommodation of the lense is fixed for near vision, resulting in blurred far vision. This block of accommodation occurs with relatively large doses and may diminish with long term administration.

The organophosphorus compounds are highly potent and long-acting, but agents like DFP have to be dissolved in an oily vehicle which patients find unpleasant. The only organophosphorus compound now in use is ecothiopate which is water soluble and used in the USA for open-angle glaucoma. There is a high risk of developing cataracts with ecothiopate, which has been withdrawn from the UK market. Use of the other long-acting phosphorus compounds is also associated with an increased lenticular opacity, although the formation of spontaneous cataracts are quite common in the groups of patients usually involved. The parasympathomimetic amine, pilocarpine, is also to some extent cataractogenic. This effect appears to be specific for muscarinic activity derived either from the biochemical or mechanical effects, since it can be prevented by atropine; the mechanism, however, remains uncertain (Kaufman *et al.* 1984). For this reason, long-term control of glaucoma is better achieved with agents having other mechanisms of action, such as the β-adrenoceptor antagonists (see Chapter 5).

10.7.4 Gastrointestinal Tract and Urinary Bladder

As described in Chapter 8, parasympathomimetic amines with muscarinic activity are employed in the treatment of postoperative paralytic ileus or atony of the urinary bladder (urinary retention). By the same mechanism, anticholinesterases such as neostigmine, pyridostigmine and distigmine may be used. Neostigmine is used orally or, for more rapid onset of peristaltic activity, subcutaneously. They are contraindicated where there is intestinal or urinary bladder obstruction or where dysfunction is a result of inflammatory disease.

10.7.5   Alzheimer's disease

The deficiency of the cholinergic subcortical neurones projecting into the cortex and hippocampus of the brain in patients suffering from senile dementia of the Alzheimer's type (SDAT) and the correlation with the degree of cognitive impairment has been discussed in Chapter 8. The development of muscarinic agonists with selectivity for the central M_1 muscarinic receptors for the treatment of Alzheimer's disease was described. An alternative approach is to raise cholinergic activity by enhancing the limited levels of Ach by means of an anticholinesterase. The short-acting cholinesterase inhibitor, physostigmine, can penetrate the blood–brain barrier and has been reported to induce small transient improvements in cognitive function in Alzheimer's patients. Tacrine (tetrahydroaminoacridine, THA) (Table 10.1) is a reversible non-competitive inhibitor of cholinesterase with three-fold selectivity for BuchE (Atack *et al.* 1989) which shows greater promise than does physostigmine (Hunter *et al.* 1989). It may owe its greater activity to additional properties such as potassium channel blockade and release of central monoamines, although this is now questioned (Baldwin *et al.* 1991). The greater activity in Alzheimer's patients may be a result of its longer half-life and muscarinic receptor blocking activity (Hunter *et al.* 1989) which may be at the prejunctional receptors, thus facilitating Ach release (Zhang *et al.* 1993). A close derivative is velnacrine (Mentane[R]). Whether selectivity for AchE or BuchE is desirable is unclear; it is necessary to achieve high levels of inhibition of centrally located AchE, while BuchE inhibition leads to peripheral effects (Giacobini 1992).

A potential disadvantage of cholinesterase inhibitors is that they may negate their own activity. The build up of synaptic Ach may stimulate prejunctional receptors, thereby inhibiting release of Ach (Doods *et al.* 1993). This possibility is therefore minimized with tacrine. One disadvantage of tacrine, however, seems to be its hepatotoxicity, and there are doubts about the experimental design of some initial clinical studies (FDA Report 1991). A derivative of physostigmine, eptastigmine (heptylphysostigmine), has been shown to have a greater duration of action than physostigmine and an improved safety margin. It has been shown to reverse the cognitive defects induced by hyoscine in rodent and primate models (Dawson *et al.* 1991). The clinical prospects of this agent and its toxicity remain to be fully evaluated.

Whether cholinesterase levels are altered in the cortex of patients with Alzheimer's disease is not clear. A decrease in membrane-bound G_4 AchE has been reported which may reflect the selective destruction of cholinergic axonal processes (Massoulie *et al.* 1993). The modest but statistically significant clinical improvements seen with cholinesterase inhibitors, however, offers hope that this approach will be the major symptomatic therapy for Alzheimer's disease into the next century when disease modifiers may reach the clinic. Many new compounds are under development by the pharmaceutical industry.

Adverse drug reactions for cholinesterase inhibitors include nausea, salivation, diarrhoea and intestinal cholic. They should be used with caution in patients with asthma, due to their bronchoconstrictor activity, and those with cardiac disease, epilepsy and Parkinson's disease.

10.7.6   Anticholinesterases as Insecticides

The organophosphorus compounds are highly lipid-soluble liquids, the less volatile agents such as parathion and malathion having been developed as insecticides for crop protection. They are dispersed by spraying as aerosols or dusts, their high lipid solubility allowing ready penetration of the insect cuticle. Death of the insect arises from respiratory paralysis. Malathion selectively inhibits insect cholinesterase because it is metabolized in mammals by hydrolysis of the carboxyl ester linkage. In insects, oxidation to the active metabolite is the preferred route. In addition to agricultural uses, it is widely used to treat infestations of body lice (*Pediculus humanus*); the adult forms, larvae and eggs are all susceptible. Carbaryl has also become a widely used topical insecticide for the treatment of head and body lice. Being a carbamate, it has a reversible action and has low toxicity to humans via the dermal route.

Parathion is more toxic than malathion and, although used extensively in agriculture, it has been replaced for domestic use as an insecticide by less hazardous alternatives. It has been a widespread source of accidental death through mishandling. Absorption of all the organophosphorus compounds readily occurs by all routes – inhalation of aerosols, cutaneous penetration and via the gastrointestinal tract.

Certain insects have developed resistance to anticholinesterase pesticides at three levels: increased impermeability of the cuticle, presence of increased levels of enzymes metabolizing the insecticide or, most efficiently, resistance by mutation of the cholinesterase itself. Malathion resistance appears to be due to a single amino acid mutation at residue 331.

10.7.7   Nerve Agents in Chemical Warfare

The volatile organophosphorus compounds, tabun, sarin and soman, were developed in secrecy before and during World War II. Such compounds are now known to be in the military weaponry of several nations. In particular, Iraq had a manufacturing capability and used these agents against Iranian troops and against a civilian population of Iraqi Kurds. Nerve agents were therefore considered to be a major threat during the war with Iraq and remain a major cause for concern in devising measures for protecting military personnel and providing suitable antidotes against the toxic effects.

The G-agents, tabun (GA), sarin (GB) and soman (GD), are volatile liquids at moderate temperature and humidity, but their vapour pressures permit evaporation and therefore a major inhalation hazard. They are, however, readily dispersed in the air and are said to be non-persistent. The V-agents such as VX are persistent non-volatile liquids which cause casualties through absorption via the skin and by inhalation. Germany held large stockpiles of nerve agents during World War II but for some reason Hitler failed to use them; had he done so, antidotes would not have been available.

The signs and symptoms of nerve agent exposure are the predicted pharmacological effects of cholinesterase inhibition already described. Miosis and blurred vision occur first, with rhinorrhoea, salivation and bronchoconstriction soon following. Sweating, particularly at the site of absorption after dermal contact, and gastrointestinal and cardiovascular effects occur. Systemic effects occur within minutes of inhalation after vapour exposure, but take longer after percutaneous absorption.

Battlefield exposure will have a two-fold effect of debilitating and thus minimizing the effectiveness of military personnel before eliminating them by death. Death results from anoxia due to central respiratory paralysis, bronchoconstriction and paralysis of accessory muscles for breathing.

10.7.7.1 Neurotoxicity of organophosphorus compounds

After chronic low dose or acute high dose exposure, there is a significant neuronal degeneration and necrosis of different brain regions. This may be due to the anoxia of the brain or a direct neurotoxic effect arising from the enhanced stimulation of cholinergic neurones. This latter effect will open Ca^{2+} channels, increase intracellular free Ca^{2+}, which in turn releases excitatory amino acid neurotransmitters such as glutamate and aspartate. These induce convulsions and cell death. A delayed neurotoxicity has been demonstrated with fluorine-containing organophosphorus compounds. This is associated with muscle weakness and ataxia, accompanied by muscle twitching and fasciculations which progress to paralysis and muscle wasting. This condition may not be associated with the anticholinesterase activity of these agents. The neuropathy involves covalent inhibition of the enzyme, neurotoxic esterase (NTE), with the requirement of subsequent aging of the enzyme through dealkylation.

10.7.7.2 Treatment of intoxication

Care of the intoxicated person involves termination of the exposure by decontamination, maintaining ventilation and circulation, and administration of antidotes. These procedures apply equally to individuals poisoned by insecticide exposure and nerve agents. Atropine is the drug of choice for reversal of the muscarinic effects of anticholinesterases. An initial dose of 2 mg for self- or colleague-administration is used after exposure or for protection. This dose has minimal effects on the performance of military personnel, although the drying of secretions, including sweating, may cause raised body temperature when soldiers are performing hard work such as a long hike. Larger doses should then be administered as necessary to control the muscarinic symptoms. Intravenous administration should be avoided until the subject is ventilated. Diazepam should be used in patients with seizures and in those in a preconvulsive state. To reverse the organophosphorus compound, an oxime such as pralidoxime may be used; however this will be relatively ineffective if aging of the phosphorylated enzyme has occurred, as is the case with soman poisoning. A combination of pralidoxime and atropine is more effective than either drug administered alone. Being a quaternary ammonium compound, pralidoxime does not penetrate the blood–brain barrier and is therefore ineffective against irreversible blockade of central cholinesterases. Penetration to the CNS is improved in 2-PAD (Table 10.3). Pralidoxime causes hypertension, which can be reversed with the α-adrenoceptor antagonist, phentolamine.

Table 10.3 Reactivators of organophosphorus-poisoned cholinesterase

Compound	Structure	Notes
Pralidoxime (2-PAM, 2-pyridine aldoxime methiodide)	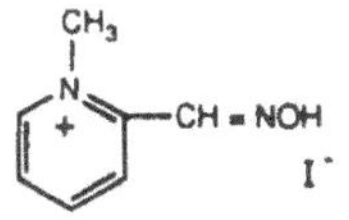 CH₃ / N⁺ —CH = NOH I⁻	Reverses blocked enyzme in periphery but not CNS
2-PAMCl	As above but Cl⁻	
2-PAD (2-pyridine-aldoxime dodecyl iodide)	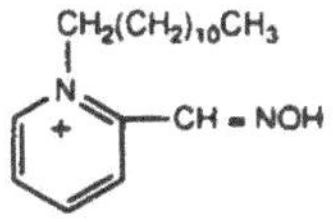 CH₂(CH₂)₁₀CH₃ / N⁺ —CH = NOH	Improved CNS penetration
Trimedoxime (TMB-4)	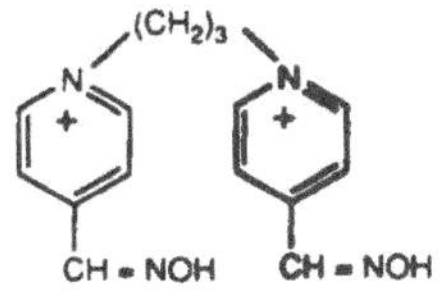 (CH₂)₃ bridging two N⁺ pyridine rings; CH = NOH CH = NOH	More potent than 2-PAM
Obidoxime (toxogonin)	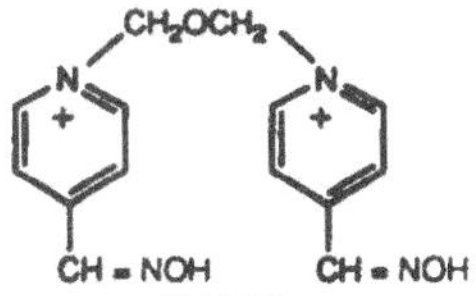 CH₂OCH₂ bridging two N⁺ pyridine rings; CH = NOH CH = NOH	More potent than 2-PAM
Hagedorn compound, HI-6	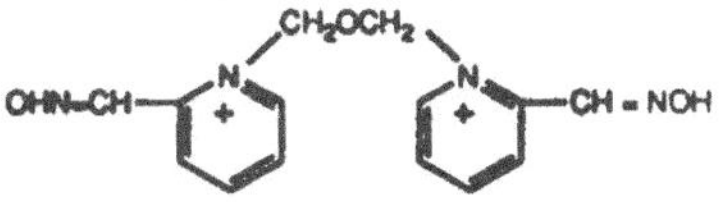 CH₂OCH₂ bridging; OHN=CH— N⁺ N⁺ —CH = NOH	Some reversal of aged enzyme after soman

Protection in the anticipation of exposure to nerve agents can be afforded by pretreatment with a reversible cholinesterase inhibitor of the carbamate type. The carbamoylation of the esteratic site of the enzyme protects it from irreversible phosphorylation and subsequent aging. The enzyme activity is then slowly restored by hydrolysis, hopefully when the risk of exposure has passed. Pyridostigmine has been approved by the US Army as a pretreatment drug against nerve agents. It is a quaternary ammonium compound and does not readily cross the blood–brain barrier and protect the central cholinesterases. Pyridostigmine is orally effective when given before challenge; it is not an antidote. There is no evidence that pyridostigmine is effective against nerve agents with slower aging times, such as sarin or VX. It should only be used when the threat is soman. Physostigmine protects against sarin and soman but its disadvantage is that it has a short duration of action necessitating frequent dosing. As it is a tertiary amine, it crosses the blood–brain barrier and protects central cholinesterases. It does carry the risk of causing some impairment of critical performance.

To summarize, treatment after nerve agent exposure requires a combination therapy comprising a muscarinic antagonist such as atropine, pralidoxime and diazepam as an anticonvulsant. In the US military, injectors containing pralidoxime are packaged with atropine for self-administration. With the threat of exposure, pyridostigmine or physostigmine should be added (Sidell 1992, Somani *et al.* 1992). There is recent concern that the excessive use of these preventative measures may be responsible for the impaired mental and physical states of military personnel who

411

have served in the conflict against Iraq, known as 'Gulf War Syndrome'. One possibility is that those individuals experiencing these symptoms may have a genetic variant of BuchE (Lotti & Moretto 1995).

Drugs Affecting Autonomic Ganglia (Including the Adrenal Medulla)

11.1 Introduction

All autonomic nerves synapse outside the CNS once, at an autonomic ganglion, before reaching the target organ. Each preganglionic nerve fibre loses its myelin sheath and divides to form numerous fine branches ($0.1-0.3$ μm diameter) with swellings called terminal buttons located along their length. These buttons contain the vesicles for transmitter synthesis and storage and form synaptic contact mainly with the dendrites of the postganglionic neurone cell body. Synaptic contact may occur at more than one point and a ganglion cell body may receive terminals from many preganglionic fibres. The ratio of preganglionic to postganglionic fibre varies considerably; in most sympathetic ganglia the ratio is $\sim 1:30$. In most parasympathetic ganglia the ratio is much smaller, but in the parasympathetic ganglia of the intestine the ratio may be over >100. There are also interconnecting fibres between ganglion cells known as interneurones. The synapses are surrounded by satellite cells which form a continuous sheath with the Schwann cells of the presynaptic nerve axon. Also present in ganglia are small intensely fluorescent (SIF) cells which are catecholamine-containing chromaffin cells. Some sympathetic ganglia contain peripheral reflex pathways which begin with afferent fibres whose cell bodies are located peripherally to the ganglion. These afferent fibres enter the ganglion and terminate on its neurones (Skok 1980).

Autonomic ganglia consist of clusters of postganglionic cell bodies which, in the case of the sympathetic division of the autonomic nervous system, are located alongside the vertebral column as the sympathetic chain (vertebral ganglia) or more distally in the body cavities as discrete peripheral ganglia. The parasympathetic ganglia are located usually within the organ that is innervated. They form more diffuse networks or plexuses of cells, such as the myenteric plexus of the gastrointestinal tract (Figure 1.5). The anatomical differences between parasympathetic and sympathetic ganglia are described in Chapter 1 and are illustrated in Figure 1.3 and 1.4. The sympathetic ganglia form more discrete structures and are therefore more accessible for study, since electrodes may be placed pre- and postganglionically to examine the pharmacological effects of drugs upon transmission through the ganglion and upon the end organ response to nerve stimulation.

The superior cervical ganglia of the cat and rat have been widely studied and provide much of the data on ganglionic transmission. The superior cervical ganglia of the cat were first studied by Langley in the last century. They are bilaterally situated at the base of the jaw and may be located by deep careful dissection. Pre- and postganglionic electrical stimulation produces an α-adrenoceptor-mediated contraction of the nictitating membrane (see Figure 11.1 and 11.4). Drugs may be

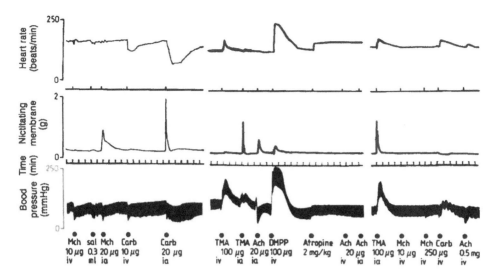

Figure 11.1 Effects of ganglion stimulants on the heart rate (upper record), nictitating membrane (middle record) and blood pressure (lower record) of a cat (3.8 kg) anaesthetized with chloralose (80 mg/kg iv) after induction with halothane: nitrous oxide. The cholinergic agonists, methacholine (Mch), carbachol (Carb), tetramethylammonium (TMA), acetylcholine (Ach) and dimethylphenylpiperazinium (DMPP) were administered intravenously (iv) or close arterially to the superior cervical ganglion via a cannula in the lingual artery (ia). The contractions of the nictitating membrane induced by the close arterial injections of Mch, Carb, Ach and TMA are abolished by atropine and therefore mediated via muscarinic receptors. The contraction to TMA remains and is therefore due to stimulation of nicotinic (N_N) ganglion receptors. Increasing intravenous doses of Mch, Carb and Ach in the presence of atropine produce tachycardia or pressor responses indicating stimulation of nicotinic receptors in sympathetic ganglia.

administered intravenously or close-arterially directly to the ganglion via a cannula placed into the lingual artery. This artery is a branch of the common carotid artery located close to the blood supply of the ganglion. By pulling ligatures around the carotid artery at points below the ganglion and above the lingual artery, to occlude blood flow headwards to the nictitating membrane, the injection is directed only to the ganglion. Thus, drug effects upon ganglionic transmission may be measured from the contractile response of the nictitating membrane (Figure 11.1) or from the electrical activity of the postganglionic nerve by suitably placed recording electrodes. Furthermore, the ganglion may be perfused and release of the transmitter, acetylcholine (Ach), may be assayed in the effluent after preganglionic nerve stimulation. The ganglion may also be isolated and perfused for recording the end-organ response when left *in situ*. For recording postganglionic nerve activity with surface or intracellular electrodes, the ganglion may be removed to a perfusion chamber (Bowman & Rand 1980, Gyermek 1980).

The elements of Ach synthesis, storage and release, already discussed in Chapter 7, are found at the preganglionic nerve terminals (Figure 7.1). Choline is taken up into the nerve terminal by an active uptake process and synthesis to Ach by the action of choline acetyltransferase (ChAT) occurs in the neuronal cytoplasm. The Ach is then taken up by the storage vesicles which are concentrated at the synaptic varicosities that make contact with the dendrites of the postganglionic nerve cell

415

body. The release of Ach is induced by depolarization of the preganglionic nerve either by a nerve action potential or by high concentrations of K^+. The release is exocytotic. Acetylcholinesterases of both the globular (G^4) and asymmetric collagen-tailed form (A_{12}) are found in ganglia and are located both pre- and postjunctionally. Butyrylcholinesterase is found postsynaptically (Buckley & Caulfield 1992). High concentrations of nicotinic Ach receptors are located at the synapse, their aggregation being directed by a protein, agrin. This is present in the cell body and undergoes orthograde transport along the axon to the nerve terminal (Fallon 1991).

11.2 Ganglionic Transmission

The arrival of a propagated action potential at the terminal varicosity causes the exocytotic release of Ach. The Ach interacts with cholinoceptors on the postsynaptic nerve cell body to produce multiphasic potential changes in the postganglionic neurone. There is an initial fast excitatory postsynaptic potential (EPSP) followed by an after hyperpolarization (IPSP), attributed to an increase in K^+ conductance of the postsynaptic membrane, and finally a delayed EPSP. In the presence of partial nicotinic receptor blockade (eg with D-tubocurarine), the three components become clearly visible by intracellular recording in the postganglionic neurone (Figure 7.7). Each component is blocked by botulinum toxin, which prevents the release of Ach, and therefore all are due to the released Ach. The fast depolariz-ation is the major component associated with ganglionic transmission arising from the Ach released by activation of the preganglionic nerve. The secondary components are seen to have a modulatory function on this primary synaptic transmission (Volle 1966, Skok 1980).

11.2.1 *Primary Fast Ganglionic Transmission*

The primary depolarization of the postsynaptic membrane results from the interaction of released Ach with nicotinic receptors (N_N). This produces a localized non-propagated depolarization of the postsynaptic membrane, EPSP. When this local EPSP reaches a critical value (area or amplitude), it sets up a propagated action potential which passes down the postganglionic nerve fibre. Intracellular recordings have shown that the onset is rapid, beginning to rise after $1-2$ msec and the duration of this phase is $50-100$ msec. The depolarizing EPSP is due to the influx of Na^+ through channels linked to nicotinic receptors of the neuronal (N_N) type. The rapid primary action potential of the postganglionic neurone is generated when the EPSP activates sufficient voltage-sensitive Na^+ channels. This action potential is dependent upon Ach release and an action potential in the preganglionic neurone, since it is abolished by prevention of Ach synthesis with the choline uptake blocker, HC-3 (Chapter 7), and with tetrodotoxin. There is some entry of Ca^{2+} during the action potential. Whether this is via Ca^{2+} channels secondary to the influx of Na^+, or via the nicotinic receptor Na^+ channel, awaits confirmation. The mechanisms involved in ion channel opening have been extensively studied by voltage clamping of postganglionic neurones, and single channel conductances have been measured in membrane patches of the ganglionic neurone. In rat sympathetic ganglia, the channel opening occurs in short bursts ($0.32-0.42$ msec) which are associated with the opening of a single channel, and longer burst durations (8.5–11.9 msec) which may be due to the repeated opening of a channel by the activation of a single nicotinic receptor (Buckley & Caulfield 1992).

11.2.1.1 The nicotinic receptor of autonomic ganglia

Nicotinic receptors for Ach are characterized by their insensitivity to blockade by atropine. They have been subdivided into those that mediate the contraction of skeletal muscle which are blocked by D-tubocurarine and decamethonium (N_M) and those present on ganglion cells, the adrenal medulla and in the CNS (N_N). These are also blocked by D-tubocurarine, albeit less potently, but, as will be seen later, they are preferentially antagonized by ganglion blocking agents such as hexamethonium, mecamylamine and trimethaphan (Brown 1980). The differences between the two nicotinic receptor subtypes are now evident from their molecular structures, which have been determined following their isolation and subsequent molecular cloning. As discussed in Chapter 7, the electric organ of the electric ray (*Torpedo* species) has the same embryonic origin as does skeletal muscle and is a rich source of nicotinic receptors of the N_M type. The nicotinic receptor of the *Torpedo* was the first neuroreceptor to be isolated and sequenced. The electrophysiological properties arising from its stimulation by Ach and its physical characteristics were also the first to be identified. Most of our knowledge has been obtained with the N_M receptor of the *Torpedo*, and differences between this and the neuronal (N_N) receptor of autonomic ganglia have subsequently become apparent.

The nicotinic receptor consists of a pentamer of five subunits which have been shown by electron-density mapping of the crystallized *Torpedo* postsynaptic membranes to be arranged symmetrically around the ion channel. There are four subunits (α, β, γ and δ) which have distinct molecular structures that have been cloned. Their primary amino acid sequences have been determined from peptide mapping and the cDNAs encoding the different subunits from *T. californica* have been isolated. The primary sequences show close homology between the four subunit proteins and it is suggested that they have evolved by gene duplication from a primordial subunit, the α-subunit having evolved more slowly. The N-terminal and C-terminal are located extracellularly, with the major part of the receptor being located on the extracellular side of the cell membrane. The protein spans the cell membrane four times, and from X-ray diffraction studies these membrane-spanning sections have an α-helical arrangement (Maelicke 1988). The five subunits of the N_M receptor of *T. californica* are in the stoichiometric ratio, $\alpha_2\beta\gamma\delta$ (Figure 11.2). The N_M receptor pentamer therefore has two α-subunits which contain all, or the major part, of the Ach binding sites.

The Ach binding sites have been located to Cys-192 and Cys-193, which are linked by a disulphide bridge. The Ach quaternary head has generally been assumed to be attached to the side-chain carboxylate of a glutamate or aspartate residue. The α-subunit also carries the recognition sites for the reversible antagonists (eg mecamylamine) and α-bungarotoxin, a snake venom from the Taiwan banded krait (*Bungarus multicinctus*). This toxin is a high molecular weight peptide that binds non-covalently to the nicotinic receptor with high affinity and has a slow dissociation rate. It blocks neuromuscular transmission of skeletal muscle and, as a radiolabel, has been used to identify N_M nicotinic receptors (Changeux *et al.* 1984). α-Bungarotoxin does not, however, block the N_N receptors of autonomic ganglia, the

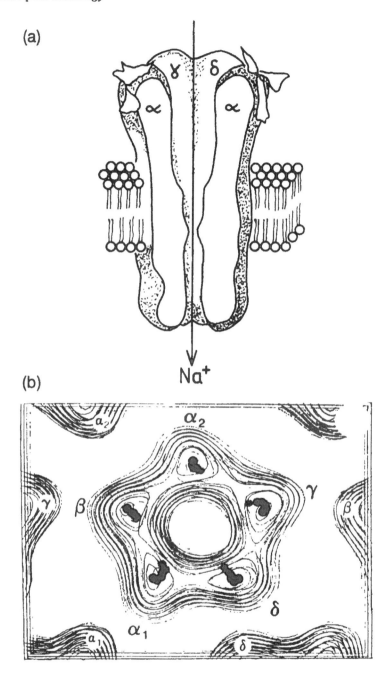

Figure 11.2 Structure of the nicotinic cholinergic receptor. (a) Side view model of the receptor cut through the two α-subunits and with the β-subunit removed. The Na$^+$ channel is shown running through the centre of the pentamer. The leaf-like structures binding to the α-subunits are molecules of bungarotoxin, which in the case of the ganglionic (N_N) receptor would be K-bungarotoxin. (b) Cross-section of an electron density map of the nicotinic receptor from *Torpedo marmorata* (N_M) taken 50–60 Å above the extracellular membrane showing the pentamer arrangement around the ion channel. Lower map reproduced with permission from Unwin *et al.* (1988).

CNS or the prototype ganglion cell in culture, PC12 cells. Furthermore, it does not block synaptic transmission in mammalian ganglia. The fact that a degree of blockade of ganglionic transmission could be obtained with some batches of commercially available α-bungarotoxin was attributed to a contaminant, K-bungarotoxin. This toxin, which is now known as neuronal bungarotoxin, is a potent blocker of ganglionic transmission in the chick ciliary ganglion (parasympathetic), and at higher concentrations in the rat sympathetic ganglia. It appears to interact with the Ach binding sites of the nicotinic receptor since the blockade can be prevented by preincubation with a competitive nicotinic receptor antagonist (Loring & Zigmond 1988). Radiolabelled neuronal bungarotoxin has been used to identify the ganglionic nicotinic receptor.

Other toxins which selectively block the nicotinic effects in autonomic ganglia include lophotoxin and K-flavitoxin. Surugatoxin, a non-polypeptide toxin from shellfish, also selectively blocks carbachol-induced depolarization of rat sympathetic ganglia and Ach responses of parasympathetic ganglia (Brown *et al.* 1976).

A population of α-bungarotoxin binding sites has been identified in sympathetic autonomic ganglia and PC12 cells. These sites are able to modulate Ca^{2+} movement, since α-bungarotoxin attenuates nicotine-evoked increases in intracellular Ca^{2+} in chick ciliary ganglion cells. This is probably not via a Ca^{2+} channel, but possibly via Ca^{2+}-sensitive-Cl^- channels, since there is no increased Ca^{2+} current. The function of these neuronal α-bungarotoxin binding sites remains to be determined (Clarke 1992).

An α-subunit is defined as having conserved cysteine residues in the presumed extracellular domain of the subunit that are homologous with the Cys-192 and Cys-193 of the *Torpedo* α-subunit. The α-subunit of skeletal muscle has the α_1 annotation, but subsequently a series of α-subunits have been encoded by cloned cDNAs and expressed in oocytes following insertion of mRNA. Various combinations of cloned subunits have been examined to determine which produce a functional receptor capable of opening the oocyte ion channel. At least two kinds, α and β, arranged as pentamers are necessary. The α-subunit of neuronal nicotinic receptors is not the same as the α_1-subunit of muscle receptors. Screening of the PC12 cell cDNA library has resulted in the cloning of several α-subunits (α_2 to α_8) and β-subunits (β_2 to β_4). Of the cloned nicotinic subunit combinations expressed in oocytes, neuronal bungarotoxin blocks receptors formed from the α_3/β_2 pairing most effectively. This suggests that this is the combination found in peripheral autonomic ganglia; presumably the required pair of α-subunits bears the Ach binding sites. Comparative studies of these cloned receptors, expressed in *Xenopus* oocytes, with the properties of the receptor on neuronal cells show single channel properties that do not correlate fully. This suggests that the assembled subunits may still not be the identical combinations to the natural neuronal receptor, N_N (Deneris *et al.* 1991). The identity of the subunits making up the nicotinic receptor (N_N) responsible for the primary fast EPSP of autonomic ganglia is therefore still uncertain.

The nicotinic receptor is activated by Ach released across the ganglionic synapse. The Ach binding sites are closely coupled to the ion channel, the binding of two Ach molecules resulting in conformational changes that open the channel. A single quantum of released Ach from the exocytotic discharge of one vesicle opens ~100 channels in ganglionic neurones. The chances of the channel opening without occupation by Ach is 10^7 lower than dual occupancy, while the probability of the channel opening by single occupancy is 100 times less. The open-time of the channel depends upon the structure of the agonist in question (Maelicke 1988).

11.2.2 Secondary Cholinergic Modulation of Ganglionic Transmission

Stimulation of the preganglionic neurone to autonomic ganglia results not only in the fast synaptic transmission associated with nicotinic (N_N) receptors, but also in slow potential changes in the postganglionic neurone which modulate neuronal excitability. These secondary components of the potential changes become more marked in the presence of nicotinic receptor blocking agents such as D-tubocurarine, to depress the EPSP and also with repetitive stimulation of the preganglionic nerve. The complex potential changes occurring after preganglionic nerve stimulation show the initial EPSP, a secondary hyperpolarization or IPSP, and finally a slow delayed EPSP. Application of Ach to the isolated superior cervical ganglion of the rat produces a three-phased potential change in the postganglionic neurone. This differs from that which follows stimulation of the preganglionic neurone, and constitutes an initial hyperpolarization followed by the fast predominant depolarization and then a secondary slower depolarization (Figure 11.3) (Newberry & Priestley 1987). The initial hyperpolarization and the delayed EPSP are susceptible to blockade by atropine, whereas the predominant fast EPSP is attenuated by the nicotine receptor antagonist, hexamethonium. The former components are therefore mediated via muscarinic receptors.

11.2.2.1 Slow muscarinic depolarization of autonomic ganglia (slow EPSP)

The delayed depolarization of sympathetic ganglia has been recorded from the cell bodies of isolated ganglia by means of intracellular electrodes or by extracellular

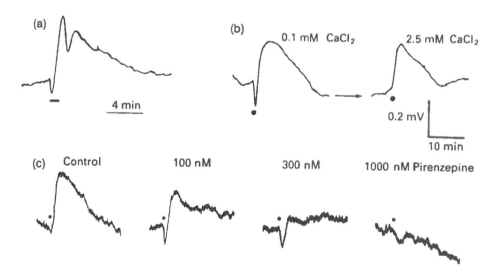

Figure 11.3 Records of potential changes of rat isolated superior cervical ganglia induced by application of muscarinic agonists. (a) Acetylcholine (10 μM) evoked initial hyperpolarization followed by a fast and then slower depolarization (medium-contained 0.1 mM $CaCl_2$, 10 μM physostigmine and 0.1 μM tetrodotoxin). (b) Low calcium medium (0.1 mM) facilitates the hyperpolarizing muscarinic response to carbachol (1 μM). (c) The slow depolarizing response to carbachol (1 μM) in low $CaCl_2$ medium (0.1 mM) is abolished by the M_1 muscarinic receptor antagonist, pirenzepine, to reveal the M_2 receptor-mediated hyperpolarization. Adapted with permission from Newberry & Priestley (1987).

surface air–gap electrodes. Muscarinic agonists, including Ach, cause a depolarization which is attributed to inhibition of a voltage-dependent K^+ current. This has been described as the M current (I_M) (Brown 1988). This current is activated slowly at potentials more positive than -70 mV and normally acts as a brake on neuronal excitation as the membrane depolarizes through nicotinic receptor activation. In the presence of a muscarinic agonist to inhibit this M-current, the brake is released and a slow secondary depolarization ensues.

The depolarization is produced by a range of muscarinic agonists including muscarine, methylfurmethide, AHR-602 and pilocarpine. The cholinesters, Ach, methacholine and bethanechol, produce a rapid depolarization, presumably due to activation of nicotinic receptors since it is converted to a slower depolarization in the presence of blockade of nicotinic receptors with hexamethonium (Brown *et al.* 1980). Neither the slow EPSP evoked by a single preganglionic stimulus nor the slow Ach potential can normally evoke a discharge from the ganglionic neurone, probably because of their small amplitude and slow rising phase. Neurones can only be fired through muscarinic receptors if high frequency (>60/sec) is used. This is higher than the natural firing frequency of preganglionic fibres which is maximal in sympathetic ganglia at ~30–35/sec. It is therefore unlikely that muscarinic receptors normally transmit natural impulses through the ganglia (Skok 1980).

The slow depolarization of the rat superior cervical ganglion induced by muscarinic agonists is antagonized by the M_1-selective antagonists, pirenzepine (Figure 11.3) and hexahydrodifenidol (HHD), but not by AF-DX 116. Gallamine and methoctramine also showed poor antagonism of this response (Table 9.6). It is therefore mediated via muscarinic receptors of the M_1 type. The presence of these antagonists revealed only the hyperpolarizing response to muscarine (Newberry & Priestley 1987, Field & Newberry 1989). Similar responses have been recorded by intracellular recording techniques in intracardiac neurones in culture (Allen & Burnstock 1990) and the parasympathetic ganglia of the myenteric and submucous plexus of the guinea-pig ileum (North *et al.* 1985).

The muscarinic inhibition of the M-current involved in the slow EPSP is mediated by G-protein activation. This is not susceptible to ADP-ribosylation by pertussis toxin, suggesting that G_i or G_o are not involved. Recent evidence suggests that G_q and/or G_{11} may be responsible (Caulfield *et al.* 1994). Changes in cAMP or cGMP are probably not involved but increased production of inositol phosphates, including IP_3, has been observed. Thus, G protein-coupled stimulation of phospholipase C activity is likely to be the mediator of M-current supression by muscarinic agonists (Buckley & Caulfield 1992) (see Chapter 13).

11.2.2.2 *Slow muscarinic hyperpolarization of autonomic ganglia (slow IPSP)*

In the presence of nicotinic receptor blockade, application of exogenous muscarinic agonists to isolated ganglia causes an initial hyperpolarization. This precedes the slow muscarinic receptor-mediated depolarization (slow EPSP). The slow hyperpolarization may be enhanced by using a low Ca^{2+} medium (0.1 nM $CaCl_2$) (Figure 11.3). This hyperpolarizing activity of muscarinic agonists is mediated via receptors of the M_2 subtype. The response in rat isolated superior cervical ganglia is preferentially blocked by AF-DX 116, gallamine and methoctramine (Newberry & Priestley 1987, Field & Newberry 1989).

421

Early studies indicated that the hyperpolarizing effect of muscarinic agonists was susceptible to blockade by α-adrenoceptor antagonists (Skok 1980). This led to the suggestion that the muscarinic (M_2) receptors were probably located in the catecholamine-containing small intensely fluorescent (SIF) cells. These in turn released dopamine or adrenaline onto the neuronal cell which was inhibited via an α-adrenoceptor-mediated action leading to the slow IPSP, possible by an action on α_2-adrenoceptors (Brown & Caulfield 1979). Conflicting results have been obtained with the selective α_2-adrenoceptor antagonist, idazoxan (Chapter 5). In the rat perfused superior cervical ganglion, idazoxan failed to antagonize the muscarine-induced hyperpolarization, although the expected antagonism of the noradrenaline-induced hyperpolarization occurred (Newberry & Priestley 1987). In the submucous plexus, neurones of the guinea-pig ileum (Mihara & Nishi 1989) and rabbit sympathetic ganglia (Mochida *et al.* 1988), however, the slow IPSP was sensitive to α_2-adrenoceptor blockade.

The underlying ionic mechanisms for the slow-IPSP remain confusing and appear to depend upon the conditions under which it is measured. The possibility exists for two mechanisms. The fast IPSP which follows the rapid spike potential induced by a single preganglionic stimulus is known as the after-hyperpolarization (AHP) and is due to increased K^+ and Cl^- conductances. The slow IPSP has a slower latency but is also probably due to increased K^+ permeability together with activation of the electrogenic sodium pump (Skok 1980). More recently, inhibition of Ca^{2+} influx has also been implicated in the hyperpolarizing effect of muscarinic receptors in autonomic ganglia. This may be linked to the prejunctional inhibitory effects on transmitter release (Buckley & Caulfield 1992). Whether this inhibitory effect of the slow IPSP mediated via muscarinic receptors operates during normal ganglionic transmission is doubtful. It may only take on relevance when the normal fast nicotinic transmission is impaired. The slow IPSP has not been observed during natural spontaneous activity of mammalian sympathetic ganglia (Skok 1980).

The catecholamines have long been known to exert depressant effects upon ganglionic transmission. The depressant effect of exogenously applied noradrenaline is antagonized by α- but not by β-adrenoceptor antagonists. In contrast, isoprenaline facilitates ganglionic transmission through β-adrenoceptor activation (Volle 1966, 1980). Isoprenaline has no effect on ganglionic discharges caused by nicotinic receptor stimulation but causes a marked enhancement of ganglionic firing produced by muscarinic drugs or by anticholinesterases. Adrenaline causes variable effects attributed to dominance of either α- or β-adrenoceptor activation. The catecholamines can have direct postsynaptic effects upon the ganglion cell as discussed above for the slow IPSP. Additionally, at concentrations that have no direct effects, they depress the EPSP, reduce the frequency of miniature EPSPs and block ganglionic transmission. Ganglionic responses to applied Ach, however, are unaffected which indicates that this is due to prejunctional depression of Ach release. Thus, sympathetic nerves appear to terminate at autonomic ganglia and transmission, particularly in the parasympathetic ganglia, is inhibited by sympathetic nerve discharge (Volle 1980).

11.2.3 *Non-cholinergic Ganglionic Transmission*

5-Hydroxytryptamine (5-HT) has a stimulant effect upon postganglionic neurones causing excitation after close-arterial injection and enhancing responses of the nictitating membrane to Ach, nicotine or submaximal preganglionic electrical stimulation. The depolarizing effect of 5-HT displays rapid tachyphylaxis. The effect is not blocked by hexamethonium or atropine and shows poor blockade with the 5-HT antagonist, methysergide (Haefely 1980). The depolarization of postganglionic sympathetic neurones causing release of noradrenaline is mediated via $5-HT_3$ receptors. Both fast and slow depolarizations have been observed, the fast depolarization being the result of $5-HT_3$ receptor-gated conductance increases to Na^+ and K^+ ions. The slow depolarization shows different receptor characteristics (Wallis 1989). A presynaptic facilitation of cholinergic transmission has also been reported through a $5-HT_1$-like receptor (Hills & Jessen 1992).

Polypeptides, including angiotensin II, bradykinin, substance P and vasoactive intestinal polypeptide (VIP), have been shown to have effects on autonomic transmission. In common with 5-HT, they have been identified in autonomic ganglia by means of immunofluorescence and appear to be released by preganglionic nerve stimulation from certain ganglia. Substance P was the first biologically active peptide to be localized to enteric neurones of the gastrointestinal tract. It is a member of the tachykinin group of peptides, together with neurokinin A and neurokinin B, and shows preferential activity at NK_1 receptors. Substance P probably has a role in the non-cholinergic component of peristalsis through an excitatory action on ganglia of the myenteric plexus. In the guinea-pig inferior mesenteric ganglion, substance P has been shown to mimic the slow EPSP seen after preganglionic nerve stimulation in the presence of nicotinic receptor blockade. A late slow EPSP lasting several minutes may be mediated via substance P because of its greater stability as compared with Ach. VIP may also be involved as an inhibitory mediator of the intestinal peristaltic reflex in the enteric nervous system both at the ganglionic level and on the smooth muscle directly (see Chapter 1, Figure 1.5). Endogenous opiates inhibit ganglionic function by a prejunctional effect on Ach release, opioid peptide immunoreactivity having been identified in the preganglionic innervation of prevertebral ganglia (Dockray 1992). Angiotensin II generally has stimulatory or facilitatory effects upon ganglionic transmission, although sensitizing procedures are usually required. Bradykinin resembles angiotensin very closely; both agents have no effects on cholinergic postganglionic neurones (Haefely 1980).

These secondary non-cholinergic agents and the muscarinic receptor-mediated effects of Ach only have a modulating role on normal ganglionic transmission. The primary transmission is via Ach release onto nicotinic receptors (N_N) which induces the fast EPSP and spike potential of postganglionic neurones. Blockade of these nicotinic receptor-mediated responses by conventional ganglion blockers effectively abolishes ganglionic transmission and the responses of the effector to preganglionic nerve stimulation (Figure 11.4). Receptors having modulating functions can be shown to influence synaptic transmission when activated or blocked by selective agonists or antagonists. For example, the M_1 receptor-mediated slow EPSP can be initiated by means of the putative selective M_1 agonist, McN-A-343 (Chapter 8, Table 8.3). Administration of this agonist produces pressor responses which are blocked, not by ganglionic blockers, but by atropine and thus provided the earliest evidence for the presence of muscarinic receptors in autonomic ganglia (Roszkowski 1961). The prolonged pressor effect was due to stimulation of sympathetic ganglia, although McN-A-343 does not appear to stimulate parasympathetic ganglia. AHR-602 is also a selective M_1 agonist which resembles McN-A-343 in its ganglionic stimulant activity (Haefely 1980). Close-arterial administration of Ach to the

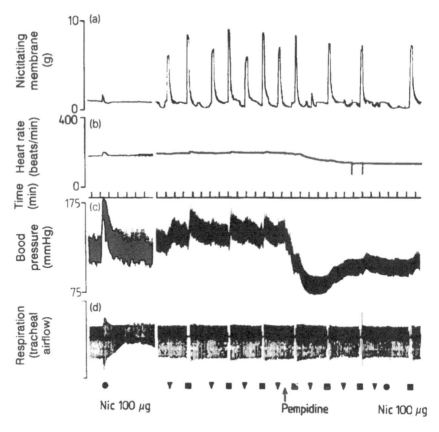

Figure 11.4 Effect of the ganglionic blocking drug, pempidine (1 mg/kg iv), on the responses of (a) the nictitating membrane, (b) heart rate, (c) blood pressure and (d) respiration (tracheal flow, measured with a Fleisch tube) of a cat anaesthetized with chloralose (80 mg/kg iv) after halothane: nitrous oxide induction. Contractions of the nictitating membrane were induced by alternating pre- (▼, 5 V, 5 Hz for 15 sec) and postganglionic (■, 2 V, 5 Hz for 15 sec) stimulation of the superior cervical sympathetic trunk. Only preganglionic nerve stimulation is blocked by pempidine which also abolishes the contraction of the nictitating membrane, tachycardia and increases in blood pressure and respiration (nicotinic 'gasp') in response to nicotine (iv).

superior cervical ganglion can cause activation of postganglionic nerves and contraction of the nictitating membrane via muscarinic receptors since it is blocked by atropine (Figure 11.1).

11.3 Ganglion Stimulants

Although a wide range of agents can serve as ganglion stimulants, including the M_1 muscarinic receptor agonists McN-A-343 and AHR-602, only those agents which activate nicotinic neuronal receptors (N_N) are, by the strictest definition, regarded here as ganglionic stimulants. This group of drugs are exemplified by nicotine and mimic the initial rapid EPSP induced by preganglionic nerve stimulation and thereby set up an action potential in the postganglionic neurone. Stimulation of postsynaptic nicotinic receptors of autonomic ganglia results in widespread pharmacological effects at the effector organs innervated by both the sympathetic and parasympathetic divisions. These responses are complex and unpredictable because of the generally opposing actions upon any one organ of the two divisions, both sets of ganglia being stimulated.

Two further complications add to the complexity of the pharmacological actions of nicotine and similar agents. These are the fact that nicotinic receptors occur at many sites other than autonomic ganglia and, second, that nicotine produces an initial stimulant action at these receptors which on repeated or prolonged exposure is followed by blockade. These aspects will be dealt with next, first the other sites of nicotinic receptors which appear to be common with the nicotinic receptors in ganglia, since they are in general blocked equally well by the ganglion blocking agent, hexamethonium. They differ, however, from the nicotinic receptor (N_M) of the skeletal muscle neuromuscular junction, although nicotine itself has activity at all receptor types.

11.3.1 Nicotinic Receptors at Non-ganglionic Sites

11.3.1.1 Extrajunctional receptors on sympathetic neurones

The extrajunctional sites are located outside the ganglia on the postganglionic sympathetic neurones; their stimulation causes local depolarizations sufficient to release noradrenaline. This effect can be induced by low concentrations of nicotine and is associated with exocytosis. Whether this action of nicotine is sensitive to blockade by tetrodotoxin (TTX), and is therefore due to a propagated action potential, is a matter of dispute. In isolated atria, release of noradrenaline by low concentrations of nicotine is TTX-sensitive (Sarantos-Laska *et al.* 1981, Brasch 1991). It has been suggested that nicotine initiates a propagated action potential by a local action on extrajunctional receptors (sympathetic ganglia are presumed to be absent in the isolated atrial preparation). At higher concentrations, the release of noradrenaline becomes TTX resistant and is possibly caused by a local membrane depolarization and massive influx of Ca^{2+} into the sympathetic varicosities (see also Chapter 4, indirectly acting sympathomimetics).

11.3.1.2 The adrenal medullae

Adrenal medullary chromaffin cells have the same embryonic origins as do ganglion cells, being formed from cells of the thoracic portion of the neural crest. These cells migrate to prevertebral regions to form the adrenal medullae together with the cells that become sympathetic ganglion cells. The mature adrenal medulla contains two types of chromaffin cells, namely adrenaline-containing (A-cells) and noradrenaline-containing (N-cells). At birth the cells are primarily of the N-type. They become A-type cells after birth due to the influence of the surrounding adrenal cortex mediated via the vasculature. The medulla is supplied directly by arteries from the capsule of the gland and by cortical veins which drain into the peripheral radicles of the central vein. The A-cells are arranged alongside these routes, which have high concentrations of adrenocortical hormones. These cortical steroids favour the development of adrenaline-containing cells because they induce the enzyme phenylethanolamine-N-methyltransferase (PNMT) that converts noradrenaline to adrenaline (see Chapter 2).

425

Thus, adrenaline is only formed in species (eg humans and rats) where there is a close vascular link between the adrenal cortex and medulla. The N-cells are usually located near the medullary arteries. In humans, the ratio of A: N cells is about ~7 : 1, with the proportion of adrenaline being about ~80%. Chromaffin cells outside the adrenal medulla predominantly contain noradrenaline but also with dopamine in autonomic ganglia (eg SIF cells). Bovine chromaffin cells of the A- and N-type may be separated by centrifugation on a Percoll gradient and their secretory activity measured in cells grown in primary culture. Other study systems include cultured PC12 cells, which are phaeochromocytoma cells from the rat adrenal medulla.

The adrenal medullae are innervated by the preganglionic sympathetic splanchnic nerves (Figure 1.3). These lose their myelin sheath as they enter the medulla and the finely branching nerve terminals form close contact with the chromaffin cells. The release of Ach from preganglionic nerves is the same as described for ganglionic transmission. It interacts with nicotinic (N_N) receptors on chromaffin cells and adrenaline (or noradrenaline) is released from storage vesicles by Ca^{2+}-dependent exocytosis. Indeed, the chromaffin cell has been widely used as a model for the study of exocytosis.

Noradrenaline and adrenaline are accumulated in chromaffin cells by a saturable transport system that is dependent upon extracellular Na^+, energy and temperature. This process is equivalent to the uptake$_1$ carrier of noradrenergic neurones and is inhibited by cocaine and desmethylimipramine (Bönisch & Michael-Hepp 1990). Ascorbic acid is transported into the cells by a similar process and both are released concomitantly when the cells are stimulated. The uptake of catecholamines into the storage vesicles involves a proton gradient and a translocator protein. This protein appears to provide binding sites for reserpine which causes depletion of adrenomedullary catecholamines, but only at higher doses than are required for depletion of neuronal noradrenaline (see Chapter 6). Denervation of the adrenal medulla does not affect the ability of chromaffin cells to synthesize and store catecholamines.

Phaeochromocytoma is a tumour of the adrenal medullary chromaffin cells and affects mostly the adrenaline-secreting cells. The spontaneous release of adrenaline from the increased bulk of chromaffin tissue can be sufficient to raise blood pressure and lead to elevated levels of adrenaline and noradrenaline in the urine of these patients.

The adrenal medullary chromaffin cells also contain peptides within the storage vesicles and cytosol, principally opioid peptides. Neuropeptide Y, vasoactive intestinal peptide (VIP) and substance P have also been identified in the adrenal medulla, but the latter two are probably mainly located within the splanchnic nerve terminals. Substance P inhibits release of catecholamines induced by Ach and nicotine, while VIP may be a co-transmitter with Ach which induces catecholamine release. Antagonists of VIP have been shown to block catecholamine release evoked by splanchnic nerve stimulation and by VIP (Wakade *et al.* 1990). The opioid peptides may be co-released with the catecholamines during stress and may act locally on adrenal cortical cells to stimulate corticosteroid secretion. The main opioids of the human adrenal medulla are the enkephalins, leu-enkephalin (0.2 ng/ mg) and met-enkephalin (0.5 ng/mg), the former being located mainly in the vesicular fraction. These are derived from a large precursor molecule known as proenkephalin. The opioid peptides interact with μ-, k- or δ-binding sites and inhibit Ach-induced release of catecholamines.

The release of adrenaline from the adrenal medulla is induced by Ach released from the preganglionic nerve ending or by nicotinic receptor stimulants. Because anticholinesterases have little effect on release of catecholamines, it is concluded that cholinesterases are not closely associated with the synapse between nerve terminal and chromaffin cell. Stimulation of the nicotinic receptor results in secretion not only of catecholamines but also of the other contents of the chromaffin cell vesicle, including ATP and enkephalins. Release of vesicular contents can also be achieved *in vitro* by means of high extracellular K^+. Ach-induced release of catecholamines is antagonized by D-tubocurarine and by K-bungarotoxin (Carmichael 1986).

Muscarinic receptor stimulation also causes catecholamine release, the action being additive with that of nicotinic receptor stimulation, the latter representing the physiological process. Distinct mechanisms are involved and they are presumably mediated via receptors of the M_1 subtype, as in autonomic ganglia. Agents which cause low-level cyclic AMP accumulation, such as adenosine, forskolin, dopamine D_1 agonists and VIP, stimulate basal catecholamine release from chromaffin cells by an increase in intracellular Ca^{2+} entry through L-type voltage-dependent Ca^{2+} channels (Parramón *et al.* 1995).

11.3.1.3 *Non-myelenated vagal C fibres and sensory nerve endings*

Conduction of impulses in non-myelenated vagal C fibres of the nodose ganglion are transiently blocked by nicotine and related drugs. Since this blockade is antagonized by hexamethonium, it is assumed that nicotinic receptors exist on the vagal fibres. Their stimulation probably results in depolarization of the nerve fibre which then prevents nerve conduction (Volle 1980).

Sensory nerve endings also have nicotinic receptors which appear to have no physiological function. Sensory receptors stimulated by nicotine include the chemoreceptors of the carotid body, stretch, pressure, temperature and pain receptors. Chemoreceptors do in fact have the same embryonic origins as ganglionic cells. Nicotine causes discharge in the afferent pathways from these sensory nerve endings. For example, nicotine stimulates respiration and blood pressure changes via chemoreceptor reflex pathways. Cutaneous sensory pathways are also stimulated by nicotine (as well as by 5-HT and histamine). These receptors are apparently not stimulated by muscarinic agonists such as McN-A-343 and are therefore primarily nicotinic in nature (Gyermek 1980).

11.3.2 *Receptor Blockade by Nicotinic Receptor Stimulants*

In the continued presence of nicotinic ganglion stimulants, there is blockade of ganglionic transmission. This occurs in two phases, the first coinciding with depolarization of the ganglion and the second when depolarization has subsided but the nicotinic agent remains present. During depolarization induced by nicotine, the action potentials activated by stimulation are reduced, as are the ganglionic cell discharges produced by non-nicotinic agents such as K^+ and 5-HT. This non-specific blockade is presumed to be due to inactivation of the sodium conductance in the ganglion cell membrane adjacent to the synaptic and extrajunctional nicotinic receptors. As cell depolarization declines, so the responsiveness to these non-nicotinic agents returns but block of orthodromic nerve stimulation persists. This

427

second phase of prolonged ganglionic blockade is therefore due to a specific action at nicotinic receptors. A similar situation occurs at the neuromuscular junction of skeletal muscle, in which desensitization of the nicotinic receptor is thought to be associated with receptor deformation. The deformed receptor is still able to combine with Ach or nicotine but is unable to activate the electrogenic process. It is unlikely that this persistent blockade is due to a competitive type of receptor blockade. The desensitization of the receptor appears to require receptor stimulation but does not necessarily require activation of the electrogenic mechanisms. Doses of nicotinic agents too small to produce depolarization have been shown to block ganglionic transmission. The desensitized state of the nicotinic receptor–ion channel complex results in a non-competitive blockade. It may be partially surmountable if there are sufficient undesensitized receptors available and if the amount of Ach normally released does not occupy all of the receptors (a receptor reserve).

Two other possible mechanisms may also contribute to the secondary suppression of ganglionic transmission induced by nicotinic stimulants. First, there may be a prejunctional inhibition of stimulation-evoked transmitter release due to depolarization of the preganglionic nerve terminal by an action on nicotinic receptors. Additionally, hyperpolarization of the ganglionic cell body immediately follows the excitation potential and firing of the postganglionic nerve. This effect is due to the operation of an electrogenic Na^+ pump which extrudes the Na^+ ions that have entered the cell during the agonist-induced depolarization. During this period of hyperpolarization, transmission block probably occurs because, at the greater membrane potential, the EPSP fails to reach the threshold for postganglionic nerve firing.

Receptor desensitization, however, remains the most likely major component of the loss of excitability in ganglion cells. A similar desensitization appears to occur for the non-ganglionic nicotinic receptors in sensory nerve endings and in the adrenal medulla. Thus, repeated chronic administration of nicotine or TMA, or their prolonged application, results in a blockade of nicotinic receptor function (Brown 1980, Volle 1980).

Studies on nicotinic receptor binding sites after chronic administration of nicotine have shown their up-regulation in the brain, especially in the hippocampus, striatum and frontal cortex. This is contrary to the view of a desensitization of the receptors, but could be due to nicotine or a metabolite (eg cotinine) behaving as an antagonist. Alternatively, nicotine may be reducing the availability of Ach for release in the brain (Schwartz & Kellar 1983). Another possibility is that nicotine may be acting as an agonist which on chronic treatment causes an increase in the number of receptors in the desensitized state rather than the non-desensitized receptors (Marks *et al.* 1983). Abstinence from nicotine intake by smokers will eventually result in nicotinic receptor numbers returning to control levels and will restore most of the pharmacological effects of nicotine (Schwartz & Kellar 1985).

11.4 Pharmacology of Nicotinic Agonists

The prototype of the nicotinic receptor stimulants is nicotine, whose pharmacological properties will be described. The best known other members of this group are tetramethylammonium (TMA), lobeline and DMPP, which have similar properties. Their structures are shown, together with that of nicotine, in Table 11.1. A brief

Table 11.1 Ganglion stimulants of nicotinic receptors

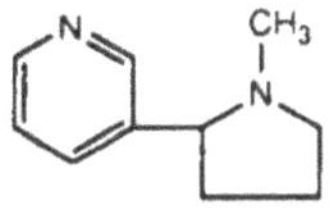 Nicotine	(−)-S-Isomer is naturally occurring and more potent
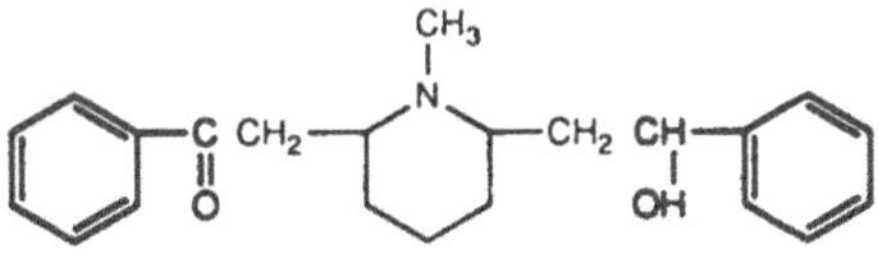	Lobeline from *Lobelia inflata* (Indian tobacco)
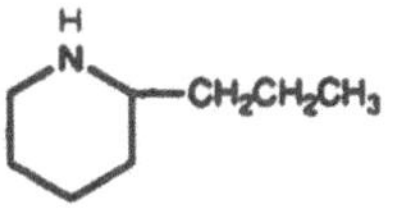	Coniine from hemlock (*Conium maculatum*)
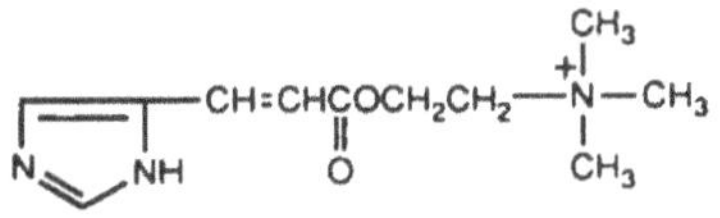	Murexine from molluscs
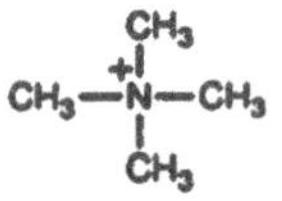 Tetramethylammonium (TMA)	Synthetic onium compound
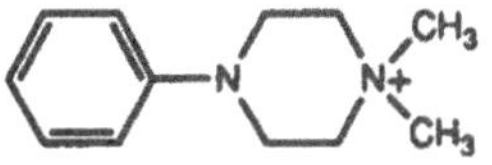	1,1-Dimethyl-4-phenylpiperazinium (DMPP)

indication of any differences in their activity compared with nicotine will be given later.

Nicotine is a naturally occurring pyridine-pyrrolidine alkaloid obtained from the leaves of tobacco, *Nicotiana tabacum*. The natural alkaloid occurs as the base and is a colourless caustic volatile liquid ($pK_A = 8.5$) that turns brown on exposure to air. The neutral salts, such as the sulphate, are crystalline solids. At physiological pH, it exists as the protonated nicotinium ion. The (−)-S-isomer is the natural form and is more active than the (+)-R-isomer. Protonation to form N-methylnicotine causes some loss of activity, probably because of reduced ability to penetrate membranes. Nicotine is currently of no therapeutic value but is of considerable pharmacological interest because of its presence in tobacco and its toxicity from accidental poisoning from ingestion of nicotine-containing insecticides and in children who ingest tobacco products. The actions of nicotine at autonomic ganglia have been recognized since the early demonstration by Langley & Dickinson (1889) that nicotine (1%) painted onto the superior cervical ganglion of the cat produced an initial stimulation of the postganglionic nerve. They excluded an action on the pre- or postganglionic fibres and later suggested the existence of a receptive substance for nicotine.

11.4.1   *Pharmacology of Nicotine*

The effects of nicotine are initially stimulatory at the nicotinic receptors in autonomic ganglia. The stimulatory effect in different organs depends upon which division of the autonomic nervous system is dominant. As a general rule the effect of parasympathetic stimulation predominates, except for the cardiovascular blood pressure effects. It is also influenced by the release of adrenaline from the adrenal medulla and by actions at extrajunctional sites. Larger doses and repeated exposure to nicotine ultimately leads to opposing effects due to ganglionic blockade.

11.4.1.1 Cardiovascular system

Nicotine produces an increase in blood pressure due to stimulation of sympathetic ganglia causing peripheral vasoconstriction mediated via the release of noradrenaline onto α-adrenoceptors. It is therefore susceptible to α-adrenoceptor blockade by phentolamine. Also contributing to the pressor effects are: (1) tachycardia through β-adrenoceptors due to sympathetic ganglion stimulation, although an initial fall in heart rate due to vagal ganglion stimulation often occurs; (2) release of adrenaline from the adrenal medulla; (3) stimulation of noradrenaline release from vascular sympathetic nerve endings via extrajunctional nicotinic receptors; and (4) activation of chemoreceptors of the aortic and carotid bodies, resulting in reflex vasoconstriction and tachycardia.

Nicotine causes coronary vasoconstriction and reduced coronary flow, particularly in patients with atherosclerosis. Small quantities of nicotine in healthy subjects may increase coronary blood flow, an effect that is converted to coronary vasoconstriction by β-adrenoceptor blockade. Vasoconstriction may be induced by release of arginine vasopressin (anti-diuretic hormone, ADH) from the neurohypophysis of the pituitary gland, by release of thromboxane A_2 and, most likely, through α_1-adrenoceptor stimulation by the released catecholamines. Vasopressin is released into local capillaries from nerve endings of cells which have their cell bodies in the supraoptic nucleus and paraventricular nucleus of the anterior hypothalamus. Thus, nicotine and Ach stimulate these cell bodies and induce vasopressin (ADH) release into the circulation. The release of vasopressin may also occur secondary to the stimulation of the medullary chemosensitive trigger zone and the associated nausea caused by nicotine (Bassenge 1988).

11.4.1.2 Gastrointestinal tract

Parasympathetic stimulation of enteric ganglia results in increased tone and motility of the intestine. Isolated ileal preparations provide a simple model for examining the effects of ganglion stimulants since they contain the parasympathetic ganglia of the enteric neurones. Nicotine and other ganglion stimulants produce bell-shaped dose–response curves for the contractile responses of the guinea-pig ileum (Figure 11.5) (Brown 1980). Tobacco smoking and the *in vivo* administration of nicotine, however, usually causes inhibition of gut motility. Since this effect can be inhibited by α- and β-adrenoceptor blockade, it is due to the catecholamine releasing actions of nicotine. Tobacco smoking relaxes the lower oesophageal sphincter, the main barrier to gastro-oesophageal reflux, which may explain the aggravation of heartburn often experienced by smokers. Variable effects are reported for nicotine on gastric acid secretion, but the main effect appears to be inhibitory (Yeomans 1988). Appetite is suppressed by nicotine, particularly for sweet foods, with a consequent loss of weight with chronic intake in smokers (Grunberg 1988). CNS stimulation by nicotine causes nausea and vomiting.

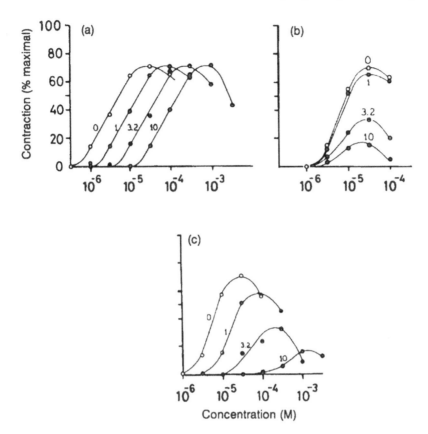

Figure 11.5 Non-cumulative concentration–response curves for the contraction of guinea-pig isolated ileum (jejunum) induced by nicotine in the absence (O) and presence of increasing concentrations of (a) hexamethonium ($\times 10^{-6}$ M), (b) chlorisondamine ($\times 10^{-7}$ M) and (c) mecamylamine ($\times 10^{-6}$ M). Contraction height is expressed as the percentage of that produced by pyridine-3-methyltrimethylammonium. Note the parallel displacement by hexamethonium and depression of maxima by chlorisondamine and mecamylamine. Reproduced with permission from van Rossum (1962).

11.4.1.3 Secretions

Nicotine initially promotes secretion of saliva, sweat and bronchial mucus. The hypersecretion of bronchial mucus is linked to bronchoconstriction and is due to activation of parasympathetic pathways since they are blocked by atropine (Lang *et al.* 1988). The salivation occurring with smoking may be caused reflexly from inhalation of the irritant smoke rather than from the systemic effects of nicotine. Sweating and cutaneous vasoconstriction combine to produce the 'cold sweat' effect of nicotine in non-smokers.

11.4.1.4 Central effects

Nicotine has a wide range of central stimulant actions. The inhalation of nicotine in tobacco smoke produces subjective states of euphoria, cognitive enhancement and changed adaptation to stress. These contribute to making nicotine an addictive substance. The primary expression of this addiction is the appearance of persistent drug-seeking behaviour. This behaviour is manifest in self-administration studies where animals learn that lever pressing will deliver a dose of nicotine. Nicotine is then said to be a positive reinforcer (Stolerman & Shoaib 1991).

Nicotine also induces peripheral autonomic changes through stimulation of brain centres. It stimulates respiration by a direct action on the medullary respiratory centre, in addition to reflexly by activation of chemoreceptors of the carotid and aortic bodies. Injection of nicotine into anaesthetized animals causes a deep inspiration known as a 'nicotinic gasp' (Figure 11.4). Blood pressure is increased by stimulation of the medullary vasomotor centre but the bradycardia results from stimulation of the vagal cardioinhibitory centre. Vomiting and nausea result from stimulation of the chemosensitive trigger zone (CTZ) of the medulla oblongata, but also from stimulation of vagal afferents that form the sensory pathways for the vomiting reflex. Nicotine also causes release of antidiuretic hormone through the stimulation of the supraoptic nucleus of the hypothalamus, which reduces urine outflow. Finally, nicotine in low doses causes tremor and convulsions, which at higher toxic doses is followed by depression. Death in nicotine poisoning results from failure of respiration due to both central depression of the respiratory centre and to blockade of the skeletal muscles associated with breathing.

Preliminary indications suggest that stimulation of nicotinic receptors in the brain may be beneficial in Alzheimer's disease, whereas their blockade may impair cognitive function (Newhouse *et al.* 1992). It has also been found that the density of nicotinic binding sites is reduced in the brains of Alzheimer's disease sufferers (Kellar *et al.* 1988). It is therefore important to ensure that agents of potential application in Alzheimer's disease, such as anticholinesterases (see Chapter 10), do not have nicotinic receptor blocking activity (Clarke *et al.* 1994).

11.4.1.5 *Metabolic effects*

Plasma noradrenaline and adrenaline levels are increased by nicotine, as are plasma free fatty acid (FFA) levels. The elevated FFA levels are probably secondary to the increased secretion of noradrenaline and adrenaline from the adrenal medulla and postganglionic sympathetic nerve endings (Bassenge 1988). These activate lipolysis of triglycerides to FFA in adipocytes via β_3-adrenoceptors (see Chapter 4). Nicotine also promotes the aggregation of platelets. One mechanism for this property appears to be due to a reduced production of the antiaggregatory prostacyclin (PGI_2) from arachidonic acid. In aortic rings and rabbit heart, nicotine inhibits the formation of 6-keto-prostaglandin F_{1a} (6-keto-PGF_{1a}), which is the stable metabolite of PGI_2. The fact that it did not inhibit the conversion of prostaglandin endoperoxide to PGD_2, PGE_2, PGF_2 or 6-keto-PGF_{1a} indicates that the inhibitory action of nicotine is on cyclo-oxygenase (Alster & Wennmalm 1983). A second mechanism whereby nicotine may increase reactivity of platelets to aggregation is via α_2-adrenoceptors, through the elevated catecholamine levels (see Chapter 3). Another effect of raised catecholamines is the release of thromboxane A_2, possibly from activated platelets, which in turn has further proaggregatory properties (Dusting & Li 1986).

11.4.2 Absorption, Excretion and Blood Levels of Nicotine – Relationships to Smoking

Nicotine is readily absorbed across the respiratory and buccal mucous membranes and transdermally, hence its recent introduction for delivery in transdermal patches. The nicotine in cigarette smoke is suspended in an aerosol of small particles as the protonated form because of the low pH. This low pH (5.5) favours inhalation and absorption from the lungs into the circulation, reaching the brain within 8 sec after inhalation. Pipe and cigar smoke is more alkaline (pH 8.5) than cigarette smoke and the nicotine is in the unprotonated form, making it more readily absorbed in the buccal cavity. The nicotine in chewing tobacco and nicotine gums (Nicotine^R, 2 and 4 mg), is also readily absorbed from the mouth. The lower pH of the stomach results in slower gastric absorption after ingestion but there is more rapid intestinal absorption.

The nicotine content of smoking tobacco is between 1 and 2% which, in an average cigarette, amounts to ~20 mg of nicotine. The amount of nicotine delivered by smoking the cigarette depends upon smoking technique, but in the mainstream smoke received by the smoker it is between 0.05 and 2.0 mg. Plasma levels after smoking a single cigarette reach 25–50 ng/ml, but rapidly decline over a period of 5–10 min. The half-life of nicotine is ~2 hr. These levels are sufficient to produce pharmacological effects in the periphery and centrally, although the peripheral effects are primarily due to induction of sensory reflexes (Benowitz 1988a).

The lethal dose of nicotine is about 60 mg. Although it is theoretically possible to reach such levels by ingesting tobacco (for example, a child would only need to consume three cigarettes), the amount reaching the circulation would be substantially less. The nicotine initially absorbed would delay gastric emptying and then induce vomiting, which would remove most of the tobacco remaining in the stomach.

About 90% of nicotine is metabolized to cotinine by oxidation, mainly in the liver, but also in the lungs and kidney.

Nicotine   Cotinine

(* indicates position of OH group on trans-3'-hydroxycotinine)

Cotinine is virtually devoid of nicotinic receptor activity and therefore has no subjective or cardiovascular effects. It is cleared by the kidneys more slowly than nicotine (half-life ~18 hr) and is a better measure of overall nicotine intake than nicotine itself. Cotinine is further metabolized to trans-3'-hydroxycotinine, which is the most abundant metabolite (40%) in the urine. Only ~10% of the absorbed nicotine is excreted unchanged in the urine. Small amounts of nicotine-N-oxide are also produced by N-oxidation of the pyrrolidine ring. It is unlikely that metabolites of nicotine contribute to the central effects derived from tobacco consumption. Nicotine is excreted in the breast milk of women smokers. There is some induction of microsomal enzymes leading to greater metabolism of nicotine and various drugs by these enzymes in smokers. However, this effect is small and does not account for the tolerance that develops to the peripheral pharmacological effects of nicotine. The metabolism of nicotine is too slow to appreciably affect plasma levels immediately after inhalation of tobacco smoke (Benowitz 1988b).

11.4.3 Nicotine Poisoning and Toxicity

11.4.3.1 Acute poisoning

The main sources of poisoning arise from ingestion of tobacco or nicotine insecticide sprays. The early symptoms include nausea, salivation, cold sweat, vomiting and abdominal pain. These are followed by faintness and a rapid, weak and irregular pulse due to the fall in blood pressure. Difficulty in breathing, collapse and finally convulsions precede death from respiratory failure. Treatment of poisoning should be instigated rapidly by removal of any residual gastric nicotine by inducing vomiting with an emetic or by gastric lavage. A slurry of activated charcoal should be introduced into the stomach and left there. Respiratory assistance and treatment of shock may be necessary (Holmstedt 1988).

11.4.3.2 Chronic toxicity

Chronic toxicity arising from tobacco smoking is due to both the nicotine content, the carcinogens in the smoke and the raised carbon monoxide levels (and related hypoxia) of the blood. Cancers of the lung, larynx, buccal cavity and bladder are probably due to the carcinogens. The increased incidence of cardiovascular disease in smokers is, however, related to the elevated levels of nicotine and carbon monoxide.

The increased morbidity and mortality in smokers from cardiovascular diseases such as ischaemic heart diseases (angina and myocardial infarction) and peripheral vascular disorders, such as intermittent claudication, is associated with the development of atherosclerosis. The main pathogenesis of atherosclerosis is increased levels of lipids in the plasma. Low density lipoproteins (LDL) are cholesterol-rich and appear to be the main sources of atheromatous plaques. LDLs contain ~45% of cholesterol together with FFAs. Chronic nicotine administration increases the levels of low density (LDL) and very low density lipoprotein (VLDL) in the plasma, together with a shift to a low HDL/LDL ratio (Bassenge 1988). An increase in FFAs has been demonstrated after cigarette smoking. This is due to the elevated levels of catecholamines since there is a close relationship between plasma catecholamines and FFA after cigarette smoking. In adrenalectomized subjects, or after blockade of nicotinic receptors with a ganglion blocker, there is no increase in FFA levels (Kershbaum & Bellet 1964). The levels of LDL are raised by stimulation of their synthesis and delay in their clearance, apparently due to impaired liver function. The latter effect may be due to conitine and other by-products of nicotine affecting hepatic lipid and lipoprotein metabolism. Nicotine may accelerate lipid transfer protein (LTP) which is secreted by the liver and which results in the transfer of high density lipoprotein (HDL) cholesteryl ester from HDL to VLDL and LDL. Thus, a diminished removal of LDL and increased lipid transfer to LDL will elevate the concentration of LDL in the circulation and raise the likelihood of LDL being deposited in the arterial wall to result in atherosclerosis (Hojnacki *et al.* 1986).

Nicotine also increases the ability of platelets to aggregate, thus raising the chance of a thrombus formation at the site of atherosclerotic plaques and the development of myocardial infarction. In addition to the mechanisms described earlier (metabolic effects), a further one may arise from the elevated levels of LDL which inhibit endothelium-derived relaxant factor (EDRF, nitric oxide) (see Chapters 3 and 12), probably through an LDL receptor (Andrews *et al.* 1987). EDRF normally maintains vasodilator tone and inhibits platelet aggregation; loss of these may further contribute to the pathophysiology of coronary atherosclorosis and occlusion (Bassenge 1988).

A further risk factor in the development of cardiovascular disease with chronic nicotine intake is an elevated fibrinogen level. This glycoprotein induces platelet aggregation and is the precursor of fibrin monomers that interconnect to form the initial stages of a haemostatic plug in the blood coagulation cascade. A relationship has been found between raised plasma fibrinogen levels and increased severity of atherosclerosis. Fibrinogen and its degradation products are found in atherosclerotic lesions. Smoking tends to raise plasma fibrinogen levels and it is thought that nicotine is the causative factor. Nicotine may increase fibrinogen levels through the increased FFA levels, which may stimulate its synthesis in the liver or through the elevated prostaglandin levels (Cook & Ubben 1990). Cigarette smoking also causes coronary vasoconstriction, the nicotine probably acting through local release of catecholamines to stimulate vascular α_1-adrenoceptors since the effect is blocked by phentolamine and exacerbated by β-adrenoceptor blockade with propranolol. Thus, in long-term smokers with coronary artery disease, the nicotine content of cigarette smoke increases myocardial oxygen demands through β-adrenoceptor stimulation but reduces the coronary blood flow and myocardial oxygen supply through vasoconstriction. These combine to increase the risk of myocardial infarction (Winniford *et al.* 1986).

11.4.4 *Tolerance to, and Dependence on, Nicotine*

After chronic administration of nicotine, tolerance can build up to its effects leading to an increased dose being required to produce the same pharmacological effects as on the first exposure. On inhaling a first cigarette, dizziness, nausea and vomiting, sweating and piloerection are experienced and these disappear in the chronic smoker. Tolerance does not appear to build up to the cardiovascular actions of nicotine as the chronic smoker still experiences increases in blood pressure, pulse rate and plasma concentrations of catecholamines. The tolerance or tachyphylaxis is attributed to pharmacodynamic changes at the nicotinic receptor rather than to induction of microsomal metabolic enzymes (see earlier). The tolerance is related to the blockade of nicotinic receptor activity that occurs with prolonged or repeated exposure to nicotinic stimulants. The underlying mechanisms have already been discussed, but desensitization of the receptor–ion channel complex appears to be the primary causitive factor. Tolerance seems to decline over ~24–48 hr after abstinence from smoking. The first cigarette of the day usually produces more profound effects on the cardiovascular system and CNS.

Physical dependence occurs with regular intake of nicotine. Cessation of tobacco use may be followed by withdrawal symptoms including irritability, anxiety, restlessness and depression. Headache, increased appetite, insomnia and difficulty in concentrating are also common. Heart rate, blood pressure and plasma adrenaline levels all decrease. These changes occur within 24 hr of the last cigarette and can persist for weeks or months. The central effects appear to be associated with an up-regulation of nicotinic receptors in the brain (see earlier) (Marks *et al.* 1983, Stolerman & Shoaib 1991).

11.4.5 Uses of Nicotine

Nicotine is used as an insecticide spray in horticulture and is one of the most toxic agents in use. It readily penetrates the chitinous exterior of the insect and causes death by prolonged blockade of nicotinic receptors of the respiratory skeletal muscle.

Nicotine has human use as an aid to giving up smoking. Nicotine is supplied as chewing gum or more recently in patches for transdermal delivery. The symptoms of withdrawal from tobacco can be reduced to some extent by these products. The irritability, anxiety and difficulty in concentrating are suppressed. Each piece of chewing gum contains 2 or 4 mg of nicotine (Nicorette[R]) bound to an ion-exchange resin. Chewing the gum releases the nicotine which is absorbed via the buccal mucosa. It also provides oral activity to overcome the psychological habit. The patient must chew slowly for 30 min when there is an urge to smoke and should avoid swallowing. The lower dose does not cause noticeable cardiovascular effects, but the 4 mg preparation may cause the expected cardiovascular responses, hiccoughs, nausea and vomiting. Clearly the adverse effects of chronic nicotine intake upon the cardiovascular system will persist, but since the carbon monoxide and carcinogenic products of smoking are absent, on balance the risks of chewing gum are less. The aim is for gradual withdrawal of the gum after 3–6 months anyway.

Patches have provided a more recent alternative, with much advertising hype indicating favourable results in achieving smoking abstinence. Patches are placed on the trunk or upper arm in the morning and left in place for 24 (Nicoderm[US], Habitrol[US], Nicotinell[R], Nacabate[R]) or 16 hr (Nicotrol[US], Nicorette[UK]). The latter is claimed to avoid risk of sleep disturbances and reduce tolerance. The initial strength is determined by the former cigarette usage; at >20 cigarettes per day the commencing dose is 21 mg per 24 hr, reducing to 14 mg after 4–6 weeks, and to 7 mg after a further 2–4 weeks. The maximum treatment period is 3 months.

11.4.6 Drug Interactions and Contraindications

The major drug interactions with the above nicotine products arise from the induction of mitochondrial enzymes that are also involved in the metabolism of other drugs. These include theophylline, caffeine, pentazocine, propranolol, imipramine and the benzodiazepines, oxazepam and nordazapam (Benowitz 1988a). Thus, smokers may require larger doses of benzodiazepines to achieve satisfactory relief of anxiety. Nicotine products also interact with warfarin and insulin. The use of nicotine products is contraindicated during pregnancy and lactation. Special precautions must be observed in patients with a history of peptic ulceration, of angina, recent myocardial infarction, cerebrovascular accident, serious cardiac arrhythmias, hypertension and peripheral vascular disease. They should also be avoided in diabetes and hyperthyroidism. The chewing gum should not be used concomitantly with liquids like coffee or carbonated drinks which lower the pH of the buccal cavity since this will ionize the nicotine and reduce its rate of absorption.

11.4.7 Other Nicotinic Receptor Agonists

TMA, lobeline and DMPP (Table 11.1) have similar properties to nicotine. Unlike nicotine, they are exclusively active at nicotinic N_N receptors with little or no neuromuscular activity. DMPP may have some preference for the adrenal medulla, and pressor responses are substantially reduced by demedullation. However, it clearly stimulates autonomic ganglia and causes contraction of the guinea-pig ileum and conduction block of vagal non-myelinated C fibres. It produces a rise in blood pressure, tachycardia and contraction of the nictitating membrane after intravenous administration to the anaesthetized cat (Figure 11.1). DMPP and TMA are charged quaternary compounds and are therefore poorly absorbed orally and do not cross the blood–brain barrier. They therefore do not exert the central effects of nicotine. These compounds are also less prone to the development of blockade of ganglionic transmission with repeated administration, larger doses and more frequent administration being required. The delayed blockade associated with desensitization and hyperpolarization is most marked with nicotine and lobeline. This may be due to their non-quaternary nature which allows them to accumulate intracellularly.

Lobeline stimulates respiration by an action on carotid body chemoreceptors. The reflex increases in respiration and bronchodilatation have been used in the past for treating asthma and chronic bronchitis. The effects are, however, transient and unreliable and, when smoked, lobeline fumes aggravate the condition. Lobeline has also been used to stimulate respiration of the newborn when other measures have failed. Finally, lobeline by mouth, as the hydrochloride or sulphate, has been claimed to be of value as a smoking deterrent and several proprietry preparations of lobelia or lobeline salts have recently been available worldwide (eg Lobidan, Nikoban) (Martindale 1989). The results of controlled trials, however, are disappointing and it produces all the unpleasant side-effects of nicotine, including nausea, vomiting, coughing, headache and dizziness.

11.5 Ganglion Blockers

The ganglion blockers are here defined as those agents which produce blockade of the actions of Ach at the nicotinic N_N receptor without causing initial depolarization. The structures of the more important compounds are shown in Table 11.2. Tetraethylammonium (TEA) was the first ganglion blocker to be described by Marshall (1913) and by Burn & Dale (1915). It was virtually ignored until interest was revived by Acheson and colleagues (Acheson & Pereira 1946), and subsequent tests in man showed that it lowered blood pressure in hypertensive patients. Unfortunately, its poor absorption by mouth and very short duration of action made it unsuitable for routine clinical use. This simple onium compound did, however, lead the way to the study of a range of bis-quaternary ammonium compounds independently by Paton & Zaimis (1949) and by Barlow & Ing (1948). These compounds are symmetrical molecules with two

Table 11.2 Ganglion blockers

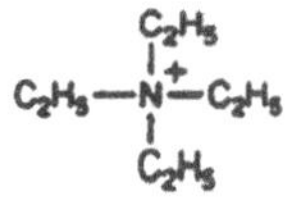

Tetraethylammonium (TEA)

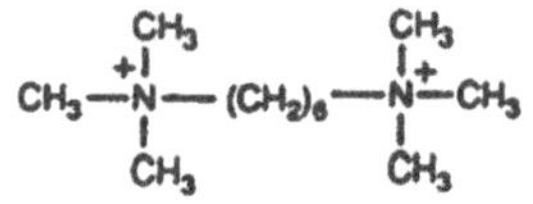

Hexamethonium (C$_6$)

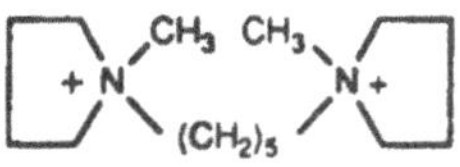

Pentolinium (Ansolysen^R) (obsolete)

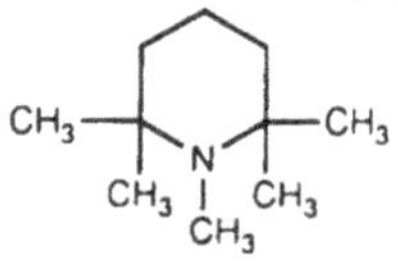

Pempidine (Perolysen^R) (obsolete)

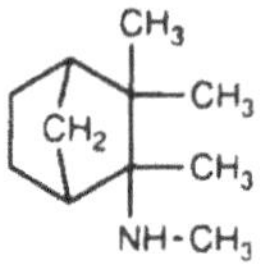

Mecamylamine (Inversine^{US})

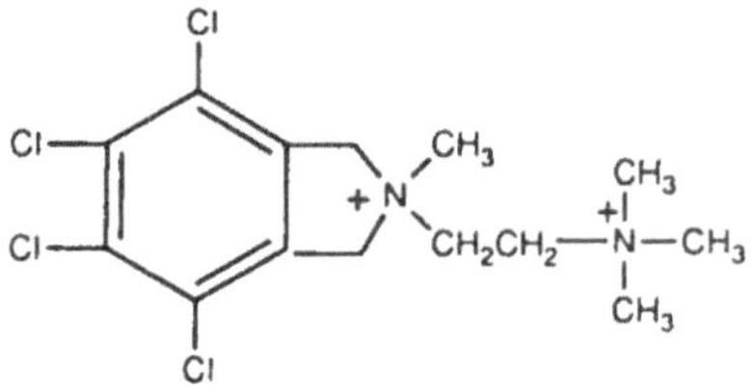

Chlorisondamine (obsolete)

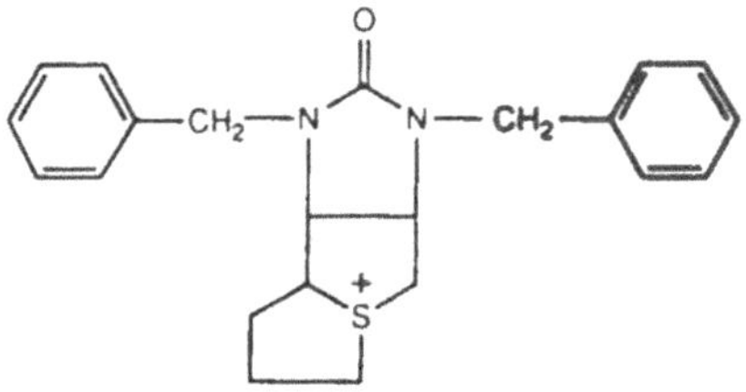

Trimethaphan or trimetaphan (Arfonad^R)
as the camsylate

quaternary nitrogen groups separated by a polymethylene chain of variable length. Optimum ganglion blocking activity appears with five or six methylene groups (Figure 11.6) and hexamethonium (C$_6$) became the first successful drug treatment of essential hypertension. An increase in the chain length results in loss of ganglion blocking activity, but blockade at the neuromuscular junction increases to become greatest at decamethonium (C$_{10}$) (Figure 11.6). This agent causes initial stimulation of the neuromuscular junction nicotinic receptor (N$_M$) which subsequently progresses to a depolarizing-type blockade. In contrast, penta- and hexamethonium do not stimulate the N$_N$ nicotinic receptor or depolarize ganglionic cells, and they do not cause blockade at the neuromuscular junction. The main disadvantage of hexamethonium for clinical use, however, was that its charged nature prevented effective and uniform absorption when given by mouth; only 10–15% of an oral dose reaches the circulation. Predictable blood levels are possible only after subcutaneous or intramuscular injection.

Subsequently, several series of bisquaternary ammonium compounds have been developed as ganglion blockers. Optimum activity when the charged nitrogen atoms are separated by five or six methylene groups is a consistent finding, and in the bispyrollidinium series pentolinium shows greatest activity. This was used clinically in the 1950s, having a longer duration of action and being five-fold more potent than C$_6$. In common with all quaternary ammonium ganglion blockers, including

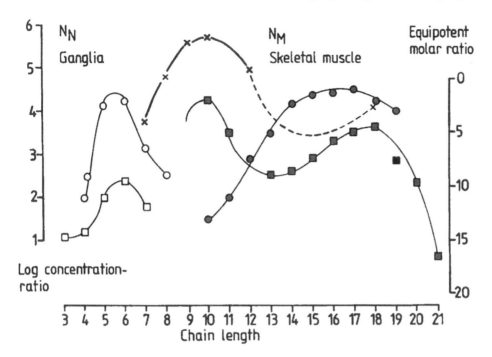

Figure 11.6 Effect of chain lengths of polymethylene *bis*-trimethyl (O, □, X, ■) or *bis*-triethylammonium (●) compounds on relative blocking activities at nicotinic receptors of autonomic ganglia (N_N) and the neuromuscular junction of skeletal muscle (N_M). Ganglionic responses were determined from rat superior cervical ganglion (O) and the contractile responses of rabbit ileum (□) (Paton & Zaimis 1949). Neuromuscular responses were determined from the cat tibialis muscle for the *bis*-trimethylammonium (Paton & Zaimis 1949, X; Triggle & Triggle 1976, ■) and the *bis*-triethylammonium compounds (Triggle & Triggle 1976, ●). Note that optimum N_N blocking activity occurs at C_6 while two peaks for N_M activity occur at C_{10} and C_{15-17}.

chlorisondamine, it has no central effects, but poor absorption and all the other disadvantages of ganglion blockers has led to their obsolescence.

Secondary (eg mecamylamine) or tertiary (eg pempidine) amines which are orally effective ganglion blockers of longer duration were introduced in the late 1950s. They are protonated at physiological pH following their absorption. Mecamylamine remains available for limited clinical use. Trimetaphan is not a quaternary ammonium compound but a sulphonium compound. It is a potent ganglion blocking agent with a rapid onset of action and brief duration that is administered by intravenous drip. Its main use is to lower blood pressure during surgery to provide a bloodless operative field. Trimetaphan may also have some direct vasodilator activity and it releases histamine.

11.5.1 Pharmacological Properties of Ganglion Blockers

Ganglion blocking agents of the quaternary and non-quaternary types have common pharmacological properties attributable to blockade of N_N type nicotinic receptor mechanisms. They therefore block the effects of nerve stimulation applied pre- but not postganglionically (Figure 11.4). Transmission in both sympathetic and parasympathetic ganglia is blocked, and the resultant effect is dependent upon which division is in predominant control of the various organs. In general, parasympathetic tone dominates and therefore the effect of parasympathetic blockade is seen in most organs. The exception is the predominant vasoconstrictor sympathetic tone to the arterioles and veins.

11.5.1.1   Cardiovascular system

Blockade of sympathetic tone to arterioles by the ganglion blockers results in vasodilatation, increased peripheral blood flow, reduced total peripheral resistance and a fall in arterial blood pressure (Figure 11.4). Vascular resistance and blood flow changes in individual vascular beds, however, does vary depending upon the degree of sympathetic vasoconstrictor tone. Skeletal muscle blood flow is unaltered (β-adrenoceptor vasodilator tone) while splanchnic and renal blood flow decrease. Cardiovascular sympathetic tone is high in subjects in the erect, standing and sitting positions due to activation of baroreceptor mechanisms (see Chapter 1). Ganglion blockers therefore have a greater effect on standing blood pressure than in the supine position. Interruption of the sympathetic vasoconstrictor pathways at the autonomic ganglia results in a blunting of the cardiovascular reflex adjustments in response to changes in posture and exercise. Thus, the fall in blood pressure in the carotid sinus consequent upon abrupt standing does not receive adequate compensatory peripheral vasoconstriction and the subject experiences dizziness or fainting. This is postural hypotension, which represents a major problem in the use of ganglion blocking drugs. Other sympathetically-mediated reflexes are also inhibited such as the cold pressor response and the carotid occlusion response (see Chapter 1).

Venous sympathetic tone is reduced by ganglion blockade, leading to venodilatation, venous pooling and decreased venous return to the heart. This results in a reduced cardiac output in subjects with normal cardiac output. In heart failure patients, however, cardiac output may be improved due to the reduced venous return and reduced peripheral resistance. In hypertensive patients, cardiac output and stroke volume are diminished. Heart rate usually increases because of blockade of dominant vagal tone.

Pressor responses to ganglion stimulants such as nicotine, TMA and DMPP are blocked by ganglion blockers due to inhibition of nicotinic receptor effects at sympathetic ganglia, the adrenal medulla and at sympathetic nerve endings (Figure 11.4).

11.5.1.2   Other sites

At sites other than the cardiovascular system, it is parasympathetic blockade that predominates and the response to ganglionic blockade may be predicted from knowledge of organ responses to stimulation of the parasympathetic cholinergic innervation (Table 1.1). The response of the gastrointestinal tract to ganglion blockade is a slowing of motility and reduced tone resulting in constipation. Atony of the bladder occurs with impaired voiding reflexes and urinary retention. The autonomic responses of the eye mediated via the ciliary ganglion in the parasympathic pathways are blocked so that contraction of the circular muscle of the iris in response to light does not occur, resulting in mydriasis. Contraction of the ciliary body is also impaired and accommodation for near vision is partially lost (cyclo-plegia). Secretions of saliva and sweat are reduced. Sexual function is impaired by a combination of sympathetic blockade of ejaculation and parasympathetic blockade of vasodilatation in the penis which normally causes erection.

It is the effects at other sites that give rise to the untoward side-effects of ganglion blockers which has resulted in them becoming obsolete in the treatment of essential hypertension. An inability to block the sympathetic ganglia preferentially – which could only be possible if the nicotinic receptor sites were different – and the introduction of superior agents with other mechanisms of action and a superior side-effect profile has led to the virtual cessation of development work on ganglion blockers over the last 20 years. The major troublesome side-effects are visual disturbances, dry mouth, urinary retention, constipation and dizziness and syncope (fainting) due to postural hypotension. Central side-effects including tremors, mental confusion, seizures and depression occur with the non-quaternary compounds such as mecamylamine. The side-effects and the blood pressure lowering activity tend to decline, that is tolerance develops, as administration of ganglion blockers continues.

11.5.2 Therapeutic Uses of Ganglion Blockers

The introduction of the ganglion blockers represented a major advance in the treatment of essential hypertension in the 1950s and showed that lowering blood pressure would prevent the development of secondary complications from the elevated blood pressure and thus improve the prognosis. The unpleasant side-effects and improvements in treatment with other drugs has led to their virtual replacement. They are nevertheless a very effective means of producing blood pressure lowering and remain available for specialist applications. For example, trimetaphan may be used for the initial control of blood pressure in patients with acute dissecting aortic aneurysm. Intravenous infusions are administered to reduce the blood pressure and sympathetic discharge to the site of the aortic tear. Second, ganglion blockade may be used to lower blood pressure during surgery; to reduce haemorrhage in the operating field, to reduce blood loss in orthopaedic procedures and to facilitate surgery on blood vessels.

11.5.3 Mechanisms of Blockade of Nicotinic Receptor-mediated Responses

The bisquaternary ammonium compounds showed optimum ganglion blocking activity when the chain separating the charged nitrogen heads was five or six methylene groups (Figure 11.6). This suggested to Barlow & Ing (1948) and Paton & Zaimis (1949) that their effectiveness might be related to their ability to simultaneously occupy two nicotinic receptor sites. Since in the trimethylammonium series (eg hexamethonium) the corresponding monoquaternary compound is a nicotinic receptor *agonist* (TMA), it is clear that the presence of two quaternary groups is essential for antagonistic activity. At the N_M nicotinic receptor of skeletal muscle, decamethonium (C_{10}) has optimum blocking activity but, unlike hex-amethonium, produces an initial depolarization. The nicotinic receptors of ganglia

441

(N_N) and skeletal muscle (N_M) have clear similarities. Nicotine stimulates at both sites. Furthermore, D-tubocurarine is a competitive antagonist at the neuromuscular junction but it is also an effective antagonist at autonomic ganglia, being five times more potent than hexamethonium. That the receptors are different, however, is indicated by the ability of α-bungarotoxin to block the neuromuscular junction but not transmission through the sympathetic ganglia (Brown 1980) (see earlier). As the methylene chain-length of the bisquaternary ammonium compounds increases beyond ten, a second optimum for ganglion blockade and neuromuscular blockade occurs at between C_{15} and C_{17} (Figure 11.6). This suggests interaction with a third binding site. Early studies concluded that the two sites of interaction of hexamethonium were not both with nicotinic receptor sites since asymmetrical compounds have been shown to produce blockade. In these agents, one quaternary group may be trimethyl substituted while the other may be triethylammonium. Thus, different binding characteristics may prevail for the two binding sites (Triggle & Triggle 1976). The binding of various nicotinic blockers to the nicotinic (N_N) receptor has been examined from their ability to displace [^{3}H]Ach (in the presence of atropine to block muscarinic receptors and an anticholinesterase to prevent its hydrolysis) or [^{3}H]-(−)-nicotine from brain homogenates. The antagonists, including C_6, trimethaphan, chlorisondamine and mecamylamine, have low affinity. This indicates that they may bind primarily at a position on the receptor–ion channel complex that does not overlap with the domain of the agonist recognition site (Kellar *et al.* 1988).

The bisquaternary ammonium compounds do not produce initial stimulation of the receptor but cause competitive antagonism. The dose–response curves for contractile responses of guinea-pig ileum to nicotine or DMPP are displaced in a parallel manner with no depression of maximum by hexamethonium, pentamethonium, trimethaphan and TEA. In contrast, chlorisondamine produces non-competitive blockade, depressing the maximum with little or no rightwards displacement. Mecamylamine and pempidine produce both shift and depression of the maximum (Figure 11.5). The competitive antagonists shifted both the ascending and descending limbs of the dose–response curve in parallel manner, implying that the bell-shaped curve arises from autoinhibition of the response to receptor activation rather than autoinhibition at the receptors.

Other differences between the mode of blockade of various ganglion antagonists have been described. For example, the blockade by TEA and C_6 of transmission through the cat superior cervical ganglion were shown to be additive. C_6 and mecamylamine, however, were less than additive and C_6 reduced the blockade by mecamylamine. These may be due to differences in dissociation rate constants. Alternatively, these antagonists may have different affinities for the resting state of the receptor and for the receptor in its agonist transformed state. Membrane hyperpolarization has also been shown to affect C_6 and tubocurarine differently in guinea-pig sympathetic ganglion cells. Hyperpolarization increased the blocking activity of C_6 but not that of tubocurarine. This suggested that the binding of C_6 was voltage-dependent or, more likely, that C_6 might influence the ionic conductance changes induced by the transmitter. The ionic channel is intimately linked to the nicotinic receptor and the rate constants for its opening and closing during transmitter action are voltage-dependent. They are slowed by membrane hyperpolarization. The possibility therefore exists that C_6 may insert itself into the channel or combine with the receptor–channel complex in its activated or open state. This would explain why ganglion blockers display a marked frequency-dependence or sensitization by a prior period of preganglionic nerve stimulation. Both of these procedures would favour the opened/activated state of the ion channel (Brown 1980).

These observations have led to the conclusion that ganglion blockers have two modes of blockade: receptor blockade or interaction with the ion channel. The ion channel blocking activity has been examined in rat submandibular ganglion cells by using the model:

$$R \underset{\alpha}{\overset{\beta'}{\rightleftharpoons}} R^* \underset{k_-}{\overset{k_+}{\rightleftharpoons}} BR^*,$$

where R and R^* represent the closed and open states of the ion channel, respectively. According to this model, the antagonist B combines with the open state of the channel which cannot close with the antagonist bound to it and is non-conducting. The opening of the channel is controlled by the rate constants β', which is a function of the agonist or transmitter concentration, and the rate constant, α. The blocking action is controlled by the forward (k_+) and backward (k_-) rate constants, which will be constants for an individual antagonist. The validity of this model has been evaluated by examining the rate of decay of the ganglionic cell excitatory synaptic currents (equivalent to the EPSP) in voltage clamped ganglion cells. Normally the decay from the peak is exponential and represents closure of the channel, with a rate constant equal to α. The model predicts that in the presence of an agent that blocks the channel in the open state, the initial rate of decay will be increased because recovery can occur by both closure of the channel (rate constant α) and by it becoming unblocked. A further prediction is that repeated activation of the receptor with agonist in the presence of a slowly dissociating antagonist should result in cumulative blockade. This use- or frequency-dependence was noted above and was observed in the early experiments of Paton & Zaimis (1949).

The effects of different blockers on the rate of decay are not as clear cut as the simple model predicts since there are multiple phases of decay and evidence for both fast and slow channels. The results with the methonium compounds show an initial shortening of decay of the synaptic current and frequency-dependence. This is consistent with them interacting not with the nicotinic receptor but with the ion channel itself in the open position. The reduced activity of compounds smaller than C_5 may be due to them passing straight through the channel without blocking, while compounds larger than C_8 may prevent channel closure and thus escape quickly. Tubocurarine behaved in a similar fashion, which contrasts with its competitive receptor blocking activity at the neuronmuscular junction. Mecamylamine and trimetaphan, however, blocked in a manner consistent with a competitive interaction with the nicotinic receptor. They did not affect the decay, but reduced the amplitude of the excitatory synaptic current (Gurney & Rang 1984).

These differences did not, however, explain the variable characteristics seen in the guinea-pig ileum where C_6 behaves competitively. Other differences between antagonists include an ability to exert presynaptic effects upon transmitter release. The quaternary compounds act entirely postsynaptically and do not depress Ach release (Brown 1980, Gurney & Rang 1984). Mecamylamine, however, may exert a prejunctional inhibitory effect possibly only at high stimulation frequencies. TEA exerts some excitatory effects not seen with other ganglion blockers, which are due to enhanced release of Ach from the preganglionic nerve and an increase in ganglion cell excitability by a lowering of the excitability threshold (Brown 1980). These effects may be related to the well-known potassium channel blocking activity of TEA (Stanfield 1983). It is now quite widely used as a tool drug for its K^+ channel inhibitory actions, although it is relatively non-specific and blocks a wide range of K^+ channel types. Blockade of K^+ efflux results in depolarization of nervous tissue causing an increase in electrically evoked [^{3}H]Ach release (Drukarch *et al.* 1989).

The mechanisms for the tolerance to the pharmacological effects of ganglion blockers are probably two-fold. First, the block of nicotinic receptor-mediated ganglionic transmission may allow the muscarinic receptor-mediated (M_1) component to become physiologically relevant. Thus, Ach released from the prejunctional neurone may still cause depolarization of the ganglionic cell body, but via the slow excitatory postsynaptic potential (slow EPSP). As pointed out earlier, muscarinic ganglionic transmission probably only occurs at high frequency, which may occur with reflexes induced by the low blood pressure caused by ganglion blockade. A second mechanism for any tolerance may be due to sensitization of the vasculature to vasoconstriction induced by noradrenaline released from sympathetic nerve endings. This may arise because of the reduced output of noradrenaline due to sympathetic ganglion blockade, which allows α-adrenoceptors to up-regulate (see Chapter 14).

Non-adrenergic Non-cholinergic Autonomic Transmission

12.1 Introduction

The sympathetic and parasympathetic divisions of the autonomic nervous system have been shown to exert control over the function of the target organs and tissues by the release of noradrenaline and Ach, respectively. These are the primary neurotransmitters of the efferent pathways of these divisions of the autonomic nervous system, however, it has been shown in previous chapters that other transmitters co-exist in the nerve terminals and their release may accompany that of the principal transmitters. This chapter considers the role of the non-adrenergic non-cholinergic (NANC) transmitters of autonomic nerves. They are generally co-transmitters, which by definition are two or more transmitters released from the same neurone by the invading action potential and which influence the activity of an effector cell, being the target cell or the neurone itself. In general, the NANC transmitters co-exist with either noradrenaline or Ach. Noradrenaline and Ach do not usually serve as co-transmitters from the same neurone. In a very few instances, namely certain enteric neurones of the gastrointestinal tract, release of only NANC peptide transmitters may occur. The main types of co-transmitters of the autonomic nerves are peptides, purines and nitric oxide (NO), and the locations and pharmacology of these will be considered in this chapter.

Other materials may be found stored in autonomic nerve endings which have not been synthesized by the neurone, but which have been taken up from the circulation. Such agents may then be co-released along with the principal neurotransmitters and the co-transmitters. Examples of this type of co-released material include 5-hydroxytryptamine (5-HT) and adrenaline. Adrenaline is carried in the circulation from the adrenal medullae and co-released with noradrenaline; its role has been discussed in Chapter 6. 5-HT may also be taken up by the neuronal uptake pathway of sympathetic neurones (see Chapter 2) after being released from circulating platelets. It may then be released from the neurone as a transmitter, to produce effects on postjunctional 5-HT$_2$ receptors and cause the primary vascular response of vasoconstriction (Kupfermann 1991). Neurones that synthesize and release 5-HT are located in the myenteric plexus of the guinea-pig and human intestine. Most of the 5-HT of the gastrointestinal tract is found in the enterochromaffin cells, and it is difficult to assess whether 5-HT that is released and assigned to a neuronal location originates from these cells. The 5-HT-containing neurones have their cell bodies in the myenteric plexus (Figure 1.5). The axons travel in an anal direction and project through the circular muscle to innervate submucosal ganglia. This is consistent with an interneuronal function within the myenteric plexus and a function on modulating intestinal secretion in the submucous plexus (Hills & Jessen 1992).

A second type of NANC activity displayed by autonomic nerves is the release of mediators from the *sensory* nerve terminals, the so-called efferent function of sensory nerves. These mediators are the peptides, tachykinins and CGRP, which are also co-released from certain of the efferent nerves. The activity of these sensory nerve endings will also be considered.

Finally, the neurotransmitter role of dopamine will be described. In addition to being an intermediate in the synthetic pathways of noradrenaline, dopamine has a transmitter role in its own right, particularly in the brain but also in selected peripheral organs. Peripheral neurones releasing dopamine rather than noradrenaline – dopaminergic neurones – have been identified in certain organs such as the kidney and mesenteric vascular bed where dopamine serves a regulatory function.

For all of these agents to serve a true neurotransmitter role, certain criteria have to be met and these were described in Chapter 2.

12.2   Co-transmission in Autonomic Efferent Pathways

The co-existence of NANC transmitters with Ach in parasympathetic, or with noradrenaline in sympathetic, nerves has been identified by means of immunoh-istochemical techniques, in particular with the neuropeptides. Measurement of the co-transmitters in the effluent from an organ during stimulation of the autonomic nerve or by field stimulation, together with the production of a pharmacological response of the organ, confirms the release of the neurotransmitter and its association with a tissue response. Tetrodotoxin (TT_X) sensitivity of the release and of the functional response substantiates that the transmitter is released as a result of a propagated action potential in the efferent nerve pathway. Confirmation that a functional response is mediated by the co-transmitter release is provided by the mimickry of the response by exogenous addition of the transmitter under study, or a close analogue. Finally, where antagonists are available, blockade of the responses to both exogenously added transmitter and to nerve stimulation provides further evidence for a neurotransmitter role for the substance identified in the neurones. Studies with agents that destroy autonomic neurones or deplete neurotransmitters are less well established. As will be seen, neurodegeneration of sympathetic nerves with 6-OHDA (see Chapter 6) prevents storage of all transmitters and thus eliminates the release of both noradrenaline and the major co-transmitter of sympathetic nerves of the vas deferens, ATP. This type of evidence confirms the co-existence and co-transmitter function of noradrenaline and ATP in these sympathetic neurones. ATP also co-exists in parasympathetic neurones together with Ach. This is less readily confirmed, although use of botulinum toxin which inhibits the exocytotic release of Ach (Chapter 7) suggests that ATP co-transmits with Ach in the urinary bladder (MacKenzie *et al.* 1982). The atropine-resistant contraction of the bladder is blocked by botulinum toxin. In addition to ATP, the co-transmitter roles of the neuropeptides, including substance P (SP), neuropeptide Y (NPY) and vasoactive intestinal peptide (VIP), and of nitric oxide (NO) will be described. The locations, main transmitter, tissue responses to each co-transmitter and the receptors involved are shown in Tables 12.1 and 12.2.

12.2.1 *Purinergic Transmission*

Purinergic nerves are characterized by the fact that the nucleotide, adenosine 5'-triphosphate (ATP), is the co-transmitter, which may occur in either parasympathetic neurones together with Ach or in sympathetic neurones co-stored with noradrenaline. Neuropeptides may additionally be present.

ATP is co-stored with noradrenaline in sympathetic nerve terminals in both the large and small dense-cored vesicles. The ratio of noradrenaline to ATP ranges from 4 : 1 to 60 : 1 and it is probably this store of ATP that is released by the nerve action potential together with noradrenaline. ATP has been identified histochemically by the uranaffin reaction in which uranium dioxide complexes with the phosphate groups of the nucleotide to form an electron-dense precipitate. It has been observed in several tissues including the vas deferens and in adrenal medullary chromaffin cells. The evidence for the co-existence of ATP with Ach in parasympathetic neurones is less obvious. The main support comes from its detection in tissue effluents after stimulation of the parasympathetic nerves to tissues such as the guinea-pig urinary bladder, taenia coli and gall bladder, and the rabbit portal vein.

The ATP is released from the vesicles of sympathetic neurones by Ca^{2+}-dependent exocytosis when invaded by the action potential. The overflow of [^{3}H]ATP after sympathetic nerve stimulation of the vas deferens and of many blood vessels is blocked by TT_X, indicating the requirement of a propagated action potential. The release and contractile response of the vas deferens is also prevented by 6-OHDA and guanethidine (Chapter 6) which cause degeneration of sympathetic nerves (Hoyle 1992).

Table 12.1 Tissues in which ATP serves as a co-transmitter with noradrenaline (NA) or Ach

Tissue	Response	Receptor type	Primary transmitter	
			Receptor	Response
Vas deferens	Initial contraction and initial EJP	P_{2x}	NA (α_1)	Secondary contraction
Blood vessels	Initial constriction and EJP	P_{2x}	NA (α_1)	Secondary contraction
Arteries Rabbit ear Mesenteric of Most species				
Urinary bladder	Primary contraction and EJP	P_{2x}	Ach (M_3)	Secondary contraction
Distal colon (rat)	Contraction	P_{2x}	Ach (M_3)	Contraction
Taenia coli	Relaxation	P_{2y}	Ach	Contraction
Rat duodenum	Relaxation	P_{2v}	Ach	Contraction

Table 12.2 Tissues in which neuropeptides serve as co-transmitters with noradrenaline (NA) or Ach

Peptide	Tissue	Receptor type	Response	Primary transmitter (receptor)
NPY	Blood vessels			
	Large	Y_1	Direct contraction	NA
	Small	Y_2	Enhances contraction by NA	(α_1)
NPY	Vas deferens	Y_1	Enhances contraction by NA	NA (α_1)
NPY	Heart/blood vessels/ intestine/vas deferens	Y_2	Inhibits release of NA and ATP	NA
SP/NKA	Gastrointestinal tract:			
	Enteric neurones	NK_3	Release of Ach-contraction	Ach
	Circular muscle (human and guinea-pig)	NK_2 (NK_{2A})	Contraction ⎫ peristaltic	Ach
	Longitudinal muscle	NK_1	Contraction ⎭ reflex	Ach
SP/NKA	rat parotid salivary gland	NK_1	Salivation/amylase	Ach
VIP/PHI	Salivary gland			
	Vasculature	VIP	Vasodilatation	Ach
	Acinar cells	VIP	Enhances Ach-induced salivation	Ach
VIP	Gastrointestinal tract			
	Enteric neurones		Relaxation and inhibitory	
	Circular muscle	VIP	component of peristalsis	?
VIP	Blood vessels			
	Penis/uterus	VIP	Vasodilatation; erection	Ach
VIP	Ovarian follicle	VIP	Inhibits NA-induced contraction	NA
CCK-8	Guinea-pig ileum Enteric neurones	CCK_A	Releases Ach and SP; contraction	Ach

Notes: Neuropeptides: NPY, neuropeptide Y; SP, substance P; NKA, neurokinin A; VIP, vasoactive intestinal peptide; PHI, peptide histidine isoleucineamide; CCK-8, cholecystokinin.

448

After its release, ATP is broken down by the ectoenzyme, 5'-nucleotidase, which is located on the smooth muscle cell membrane. The product of this degradation is adenosine, which is then rapidly taken up by the neurone or by the effector cells through facilitated diffusion or the 'adenosine transporter' system. This is carrier-mediated and capable of competitive inhibition by drugs such as dipyridamole and dilazep, which generally affect both the influx (uptake) and efflux of adenosine. Thus, adenosine is conserved and resynthesized to ATP within the neurone, firstly to 5'-nucleotide via adenosine kinase (Broadley 1995). The adenosine produced by breakdown of the released ATP may therefore exert some of the pharmacological effects of NANC nerve stimulation.

ATP produces its physiological effects by interacting with purine receptors. These have been subdivided into P_1 and P_2 subtypes, according to the rank orders of potency of ATP, ADP, AMP and adenosine. P_1 receptors are more responsive to adenosine than to the nucleotides, whereas ATP has the greater affinity for P_2 purinoceptors. Furthermore, P_1 but not P_2 receptors are blocked by methylxanthines such as theophylline. P_2 receptors are, however, selectively blocked by the trypanocidal drug, suramin. Unlike P_1 purinoceptors, the P_2 subtype do not exert pharmacological effects via inhibition or stimulation of adenylyl cyclase, but their occupation may lead to prostaglandin synthesis (Stone 1991, Collis & Hourani 1993). Thus, ATP co-released from autonomic nerves interacts with postjunctional P_2 purinoceptors which have been further subdivided into the P_{2x} and P_{2y} major subtypes. P_{2x} receptors are not coupled to a G protein but are linked directly to an ion channel. P_{2T} and P_{2Z} receptors have also been described as members of this class of ionotropic purinoceptor. P_{2y} receptors are coupled to a G protein ($G_{q/11}$) as is the related P_{2u} receptor which is activated by both nucleotides (ATP) and pyrimidines (eg, uridine triphosphate, UTP) and also referred to as a pyrimidinoceptor or nucleotide receptor. Responses mediated via P_{2x} purinoceptors are characterized by the greater potency of analogues of ATP in which the phosphate chain is modified, such as α,β-methylene ATP or β,γ-methylene ATP, compared with analogues in which the purine nucleus is altered (2-methylthio ATP) or ATP itself. P_{2x} purinoceptors are also readily desensitized by α,β-methylene ATP, which acts as a selective antagonist after prolonged exposure. At P_{2y} purinoceptors, 2-methylthio ATP is more potent than is α,β-methylene ATP or ATP. Suramin is non-selective between P_{2x} and P_{2y} receptors. The photoaffinity analogue of ATP, arylazidoaminopropionyl ATP (ANAPP$_3$), covalently binds to the purinoceptor when activated by UV light. It blocks ATP-induced responses mediated via P_{2x} receptors and has only weak activity at P_{2y} receptors.

P_{2x} purinoceptors mediate contraction of the vas deferens, urinary bladder and blood vessels. P_{2y} purinoceptors mediate relaxation of the guinea-pig taenia coli and of blood vessels and also the release of endothelium-derived relaxant factor (EDRF), now identified as NO, to cause vasodilatation (Hoyle 1992). P_{2u} receptors are also found on endothelial cells and mediate vasodilatation by release of NO.

12.2.1.1 Tissues in which ATP has a co-transmitter role

Sympathetic nerves. Stimulation of sympathetic nerves to the isolated vas deferens of rats and guinea-pigs causes contraction. The initial fast contraction is associated with an excitatory junction potential (EJP) and depolarization, which are abolished by sympathetic degeneration induced by 6-OHDA and guanethidine. This phase of the response is not affected by catecholamine depletion with reserpine or by α-

adrenoceptor blockade (Figure 3.21). It is mimicked by ATP or α,β-methylene ATP but not by neuropeptide Y. It is blocked by desensitization of the P_{2x} receptors with α,β-methylene ATP and by the P_2 purinoceptor antagonist, suramin, and by photoactivated ANAPP$_3$. ATP is released by sympathetic nerve stimulation together with noradrenaline and neuropeptide Y and has been shown to be co-stored with noradrenaline. The slower phase of the contraction of the vas deferens is sensitive to α-adrenoceptor blockade by prazosin and therefore due to noradrenaline release (Figure 3.21), while the initial phase appears to be mediated via ATP.

Stimulation of the sympathetic nerves supplying a wide range of isolated blood vessels (periarterial nerve stimulation) may cause three phases of contraction. These include the rat tail artery, the rabbit ear, mesenteric, pulmonary and saphenous arteries, and the dog mesenteric artery. An initial rapid contraction is followed by a secondary slower vasoconstriction with repetitive stimulation, and a third very slow contraction, peaking at <2 min may occur with prolonged stimulation, mainly in guinea-pig blood vessels. The fast phase of the contraction is affected by purinergic antagonists in the same way as in the vas deferens and is mimicked by ATP and its analogues. It is therefore attributed to the release of ATP (Figure 12.1). The second slower phase of vasoconstriction occurs with repetitive stimulation, is often associated with slow depolarization lasting 20-40 sec, and is blocked by reserpine and α-adrenoceptor antagonists. As in the vas deferens, this component is also due to noradrenaline release. The very slow contraction is blocked by guanethidine and therefore due to sympathetic nerve activation. It is not blocked by prazosin or other non-selective α-adrenoceptor antagonists. Since neuropeptide Y causes a slow

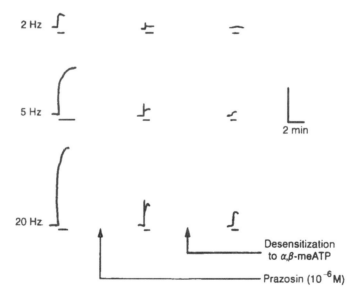

Figure 12.1 Illustration of co-transmission by noradrenaline and ATP in the rabbit isolated central ear artery. Contractions are induced by periarterial sympathetic nerve stimulation (0.1 msec pulse width, supramaximal voltage) at 2, 5 and 20 Hz until the maximum response was reached. The α_1-adrenoceptor antagonist, prazosin (10^{-6} M), reduces the plateau response which is mediated via noradrenaline. An initial spike contraction remains which is then abolished by desensitization with α,β-methylene ATP and therefore attributed to release of ATP. Reproduced with permission from Kennedy et al. (1986).

contraction in those vessels that display this third phase of contraction, and it has been shown to be co-stored with noradrenaline and ATP, probably together in large vesicles, this phase is possibly due to the release of neuropeptide Y. The three components of contraction do not occur in all blood vessels when electrically stimulated *in vitro*. It is likely that the three transmitters exert different effects depending on the location of the vessel and upon the frequency of nerve stimulation. ATP probably only mediates the initial vasoconstriction to single pulses or low levels of stimulation. In contrast, noradrenaline is probably the major transmitter of vasoconstriction by trains of pulses. Neuropeptide Y appears to be located to larger vessels and veins rather than resistance vessels and is probably only released at high levels of sustained sympathetic nerve discharge.

Sympathetic nerve stimulation of other tissues does not appear to produce responses mediated via the release of ATP. For example, field stimulation of the rat isolated anococcygeus muscle causes a contraction and EJPs that are sensitive to α-adrenoceptor blockade and unaffected by α,β-methylene ATP (Cunnane *et al.* 1987). The contraction is therefore not due to co-released noradrenaline and ATP, although ATP may produce a small contraction. Differences in the sensitivity to ATP along the length of the vas deferens explains the relative roles of co-released ATP and noradrenaline in the contractile response. The prostatic end shows high sensitivity to ATP and low sensitivity to noradrenaline. Contractions to sympathetic nerve stimulation of the prostatic portion are therefore predominantly due to ATP, whereas in the epididymal end the contractions are mainly noradrenaline-mediated. The EJPs, however, appear to be exclusively ATP-mediated in both ends of the vas deferens (Sneddon & Machaly 1992). Thus, although ATP may be co-released with noradrenaline from most sympathetic nerve endings, whether it mediates a functional response depends upon the presence of purinoceptors and the sensitivity to ATP.

Parasympathetic nerves. Pelvic nerve stimulation to the urinary bladder causes a contraction. This response is largely atropine-resistant, indicating that Ach is not the principal mediator. The biphasic nature of the twitch response to field stimulation of the rat urinary bladder may be enhanced by physostigmine (Figure 7.6). The secondary component is atropine-sensitive and therefore due to Ach release onto M_3 muscarinic receptors. The primary component and the EJPs are blocked by α,β-methylene ATP desensitization of the P_{2x} receptors (Maggi 1991b). The contraction of the bladder is also blocked by suramin (Hoyle *et al.* 1990) and mimicked by ATP; it is therefore attributed to the co-release of ATP onto P_{2x} receptors. Immunochemistry suggests that parasympathetic neurones co-storing ATP and noradrenaline form a homogeneous population in the urinary bladder. Neuropeptides (SP and NPY) are also located in the neurones of the bladder, the released neuropeptide probably serving a modulating role on the cholinergic and purinergic transmission. The atropine-resistant component of the response is less evident in the human bladder than in other species (Cowan & Daniel 1983). This component appears more readily at low stimulation frequencies but by itself is probably insufficient to generate bladder emptying (Morris & Gibbins 1992).

Pelvic nerve stimulation to the rabbit distal colon (Figure 6.3) or rat distal colon *in vivo* also causes contraction which persists in the presence of atropine. This is mimicked by α,β-methylene ATP and blocked by desensitization of P_{2x} receptors, indicating an involvement of ATP as a co-transmitter in this response to parasympathetic nerve stimulation.

The usual response to field stimulation of parasympathetic nerve endings in the gastrointestinal tract is a contraction which can be abolished by atropine (see Figure 9.1). However, if the tone is raised in guinea-pig taenia coli by carbachol, field stimulation induces a relaxation response. ATP also causes a relaxation, the response being blocked by suramin while the relaxation to parasympathetic nerve stimulation is reduced (Hoyle *et al.* 1990) (Figure 12.2). Similarly, the rat duodenum exhibits inhibition of muscle tone and spontaneous activity in response to nerve stimulation and to ATP (Manzini *et al.* 1985). The relaxation responses of the rat duodenum and guinea-pig taenia coli are therefore attributed to NANC transmission through the release of ATP. This relaxation is mediated via P_{2y} purinoceptors. The alternative mediator of these inhibitory NANC pathways is vasoactive intestinal peptide (VIP, see later). The relaxation responses of intestinal tissues to ATP, but not VIP, are inhibited by an antagonist of Ca^{2+}-dependent K^+ channels, apamin. Nerve stimulation-mediated relaxations in several gastrointestinal preparations of the guinea-pig are inhibited by apamin, although in other areas (fundus circular muscle) the relaxation response was not influenced by apamin (Costa *et al.* 1986). In the rat gastric fundus, apamin also blocked the relaxation response to ATP but not that to transmural electrical stimulation or to VIP (Lefebvre *et al.* 1991).

The conclusion is that the inhibitory pathways of the gastrointestinal tract involve both VIP and ATP, the latter probably being responsible for the maintained phase of mechanical relaxation. ATP may therefore be involved in the inhibitory reflex upon

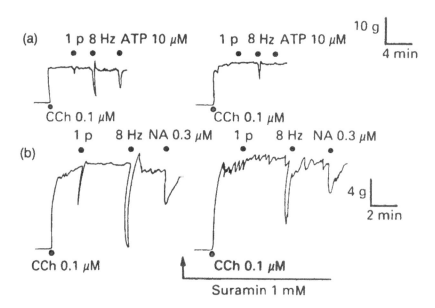

Figure 12.2 Relaxation responses of the guinea-pig isolated taenia coli induced by electrical field stimulation of parasympathetic nerves (single pulse 1 p or 8 Hz for 10 sec) and attributed to the release of ATP. Contractile tone was induced by carbachol (CCh, 0.1 μM) before field stimulation or application of ATP (10 μM) in (a) or of noradrenaline (NA) in (b) which relaxes presumably via α- and β-adrenoceptor stimulation. The response to ATP is abolished by the P_2 receptor antagonist, suramin, and the relaxation to field stimulation is reduced indicating the transmitter role of ATP. Noradrenaline is not blocked by suramin. Reproduced with permission from Hoyle *et al.* (1990).

circular muscle of the intestine induced by gut distension and by electrical stimulation. The initial phase of relaxation and the inhibitory junctional potential may be due to ATP (see Figure 1.5). Due to the limited availability of antagonists and the diversity of action of ATP, the possible therapeutic applications of drugs acting via these purinergic autonomic pathways remains uncertain.

While it was once believed that the lungs might also receive a purinergic innervation (Coleman & Levy 1974) and thus have functional receptors for ATP (P_2-purinoceptors), there is no evidence for such an innervation. Early studies had shown that adenosine and ATP could exert equi-effective relaxation of isolated trachea. The α, β-methylene analogue of ATP, which is resistant to metabolic degradation to adenosine, however is inactive. It is therefore concluded that the effects of ATP could be attributed to its conversion to adenosine by ectonucleotidases and thus only P_1-purinoceptors exist in the airways (Christie & Satchell 1980).

12.2.2 *Peptidergic Transmission*

Neuropeptides are synthesized only in the cell bodies of autonomic neurones, unlike the conventional transmitters or ATP, which are synthesized at the nerve terminal in the cytoplasm and within the storage vesicle. The neuropeptides are initially synthesized in the Golgi apparatus, as large precursor molecules which may consist of >100 amino acid residues. A single gene may encode the precursors for several neuropeptides. The coding DNA sequences produce several mRNA species by alternative splicing. These mRNAs produce the precursors which undergo extensive cleavage, the first step involving removal of an N-terminal signal sequence which ensures its sequestration into the endoplasmic reticulum. The subsequent posttranslational steps of cleavage occur within the Golgi system and the immature secretory vesicles. The preprotachykinin A gene produces three mRNAs, translation of which yields the three precursor peptides, the α-, β- and γ-forms of preprotachykinin-A (PPT-A). β-PPT-A is cleaved to form three major neuropeptides, namely substance P (SP), neurokinin A and neuropeptide K (Otsuka & Yoshioka 1993).

The storage vesicles are transported down the axon to the nerve terminal via the microtubules. The peptides are stored at the terminal varicosities in large vesicles which are identified by their electron-dense core and are associated with immunoreactivity for specific neuropeptides. The release of the neuropeptides by nerve terminal depolarization following the arrival of a propagated action potential is by Ca^{2+}-dependent exocytosis. This may be achieved *in vitro* by nerve or field stimulation or by K^+-induced depolarization. The release of neuropeptide following nerve stimulation may be measured immunochemically. It has become generally accepted dogma that low frequencies of nerve stimulation probably release the classical neurotransmitters, while high stimulation frequency releases the peptide-containing vesicles, presumably from the large vesicles (Lundberg & Hokfelt 1983). However, there are examples of low frequency stimulation also being able to release peptides (eg Santicioli & Maggi 1994). Exocytosis of the neuropeptide-containing vesicles occurs away from the active zone of the synaptic membrane. The Ca^{2+} influx is probably via L-type channels since SP release is inhibited by dihydropyridines rather than by the N-type channels associated with exocytosis of noradrenaline which are blocked by ω-conotoxin (Figure 2.2, see Chapter 2) (Agoston 1988, De Camilli & Jahn 1990, Trimble *et al.* 1991). In some systems however, such as the excitatory innervation to the circular muscle of the guinea-pig colon, ω-conotoxin does have an inhibitory effect upon both the cholinergic (ACH) and peptidergic contractions (Maggi *et al.* 1994a).

After release of neuropeptides and interaction with their specific receptors, they diffuse from the receptor but are not taken back into the neurone as are the classical transmitters and ATP. Instead, they are metabolized by proteolytic enzymes, including endopeptidase 24.11 (enkephalinase, EC 3.4.24.11), which cleaves the N-terminus of hydrophobic residues in many neuropeptides. Inhibitors of this enzyme, including phosphoramidon and thiorphan, shift dose–response curves for the effects of neuropeptides to the left, indicating the importance of this enzyme in terminating the action of these biologically active peptides. Angiotensin converting enzyme (ACE) is a peptidyl dipeptidase (EC 3.4.15.1) that cleaves C-terminal dipeptides from not only angiotensin I in its conversion to angiotensin II, but also from enkephalin, SP and bradykinin. Thus, the inhibitor captopril also prevents their breakdown. The main tissues where neuropeptides are involved as co-transmitters in efferent autonomic nerves are shown in Table 12.2 (Dockray 1992).

12.2.2.1 *Tachykinins*

Tachykinins are a group of neuropeptides which share common C-terminal sequences of Phe-X-Gly-Leu-Met-NH$_2$. Substance P (SP) was the first to be isolated in 1931 (Table 12.3) and subsequently others have been identified, the two major mammalian tachykinins being neurokinins A and B. Three tachykinin precursor molecules are generated by a single gene, α-, β- and γ-preprotachykinin-A. SP is generated from all three precursors and neurokinin A (NKA) from only β- and γ-, the latter two additionally yield neuropeptide K and neuropeptide γ. A separate gene produces the mRNA encoding preprotachykinin-B, the precursor for neurokinin B (NKB). In most neurones where NKA is located, it co-exists with SP, although SP may occur alone. NKB is barely detected in peripheral neurones (Otsuka & Yoshioka 1993). All natural tachykinins are amidated at their C-terminus (hence the -NH$_2$ shown at the end of the terminal sequence for SP above and in Table 12.3). This is important for their biological activity. The mature amidated form of SP is produced by the enzyme, peptidylglycine α-amidating mono-oxygenase, which, being copper-dependent, may be inhibited by the copper chelator, disulfuram (see also Chapter 2), leading to depletion of SP in the brain and an increase in the C-terminally extended form of SP.

The principal location of tachykinins is neurones, but they are also found in chromaffin cells of the adrenal medulla, endothelial cells and the carotid body. The two main neuronal sites of distribution are primary afferent neurones and the intrinsic neurones of the gastrointestinal tract. Tachykinins present in the peripheral nerve endings of sensory afferent neurones are involved in neurogenic inflammatory responses rather than efferent autonomic function and will be considered later under the efferent function of sensory nerves. The intrinsic neurones of the gastrointestinal tract contain tachykinins in the cell bodies of the myenteric and submucous (Meisner's) plexuses, although some sensory afferent neurones may also be present at this site. Consequently, capsaicin, which selectively depletes neuropeptides from sensory nerves, has no significant effect upon the amount of tachykinin-like immunoreactivity in the gut. SP was the first biologically active peptide to be localized to the enteric neurones of the gut. It has been identified in several different populations

Table 12.3 Tachykinins and some antagonists

Substance P	H-Arg-Pro-Lys-Pro-Gln-Gln-<u>Phe</u>-Phe-<u>Gly-Leu-Met</u>-NH$_2$
Neurokinin A	H-His-Lys-Thr-Asp-Ser-<u>Phe</u>-Val-<u>Gly-Leu-Met</u>-NH$_2$
Neurokinin B	H-Asp-Met-His-Asp-Phe-<u>Phe</u>-Val-<u>Gly-Leu-Met-NH</u>$_2$

<u>First generation antagonists</u> – peptide derivatives of substance P
[D-Pro^2, D-Trp^{7,9}] Substance P
[D-Arg$^+$, D-Trp^{7,9}, Leu^{11}] Substance P   Spantide I

<u>Second generation antagonists</u> – conformationally constrained peptide derivatives

Cyclo(Gln, D-Trp, (NMe)Phe(R) Gly[ANC-2] Leu, Met)$_2$	L 668,169	NK$_1$ Selective
[D-Pro^9 [spiro-γ-lactam] Leu^{10}, Trp^{11}] Substance P	GR 71,251	NK$_1$ Selective
Cyclo(Gln-Trp-Phe-Gly-Leu-Met)	L 659,877	NK$_2$ Selective
Polycyclo(Asp-Trp-Phe-Dap-Leu-Met)	MEN 10,627	NK$_2$ Selective
[Tyr^5, D-Trp^{6,8,9}, Lys^{10}] Neurokinin A (4–10)	MEN 10,376	NK$_2$ Selective
Ac-Leu-Asp-Gln-Trp-Phe-Gly-NH$_2$	R 396	NK$_2$ Selective

<u>Third generation antagonists</u> – non-peptides

CP 96,345   NK$_1$ Selective

of enteric neurones in combination with other transmitters. NKA is also found, but NKB appears to be present in only very small amounts. Additional peripheral neuronal sources of tachykinins are the parasympathetic nerves innervating the salivary glands and a small proportion of intramural neurones of the urinary bladder. In the lung and urinary tract, however, it is the capsaicin-sensitive primary afferents that account for almost all of the tachykinin immunoreactivity (Maggi *et al.* 1993).

Tachykinins are released from the enteric neurones of the gut, for example, by electrical stimulation of the myenteric plexus-longitudinal muscle preparation of the guinea-pig ileum. SP is also released under more physiological conditions of myenteric reflex peristaltic contraction induced by high levels of distention of the gut. Stimulation of the ileum with single pulses causes contraction which is blocked by atropine (see Figure 9.1). At high frequency stimulation, however, an atropine resistant contractile response of the circular muscle appears, which is probably due to SP release or a related tachykinin (eg NKA). It is therefore suggested that the ascending excitation is largely due to the parasympathetic pathways and when this is blocked by atropine, a remaining excitatory pathway is probably mediated via SP. The contractile response is blocked by tachykinin antagonists and therefore due to the release of SP onto specific tachykinin receptors.

Tachykinin receptors have been subdivided into three types: NK$_1$, NK$_2$ and NK$_3$. NK$_1$ receptors mediate responses for which SP shows greater potency than NKA or NKB (SP>NKA = NKB). NK$_2$ receptors show higher affinity for NKA (NKA>NKB⪢SP) and NK$_3$ receptors show higher affinity for NKB (NKB>NKA>SP). All three of these naturally occurring tachykinins display full agonist activity at the three receptor subtypes, but with different affinities. The suggestion is that SP, NKA and NKB are the endogenous ligands for the three receptor subtypes, although there is clearly some overlap. More selective agonists have been developed which are often stable to degradation by endopeptidases. Three distinct genes have now been isolated which encode the NK_1, NK_2 and NK_3 receptors. The receptors have been cloned and the amino acid sequences identified. The receptors have the same structural features as do other G protein-coupled receptors, probably having the seven membrane spanning hydrophobic domains. The second messengers for the pharmacological effects are increases in inositol 1,4,5-triphosphate (IP_3) and diacylglycerol (DAG) arising from the breakdown of phosphoinositide, due to stimulation of phospholipase C. The increase in IP_3 triggers an increase in intracellular Ca^{2+} (see Chapter 13) (Maggi *et al.* 1993, Otsuka & Yoshioka 1993, Regoli *et al.* 1994).

Further subdivision of the three receptor subtypes has been proposed according to the activities of selective antagonists. The early peptide antagonists were produced by inserting D-amino acid residues into the backbone of the natural peptides (Table 12.3). Their low potencies and problems such as partial agonist activity, and actions at other sites including mast cell degranulation, local anaesthetic properties and other peptide receptors has resulted in the development of improved antagonists. The second generation of antagonists were also peptides but with improved potency and selectivity. These include conformationally constrained analogues achieved by cyclization or lactam bridges between Gly-9 and Leu-10, analogues containing residues in the D-configuration or with the Gly in the 9 position replaced with L- or D-Pro, and pseudopeptides containing the reduced amide bond. NK_1-selective antagonists include L 668,169 and GR 71,251, whereas NK_2-selective antagonists include L 659,877 and MEN 10,376 (Table 12.3). Being peptides, these agents suffer the disadvantage of poor oral bioavailability since they will be subject to proteolytic digestion in the gut. A further intramolecular bridge in L 659,877 has yielded MEN 10,627 which shows greater potency and stability with NK_2-receptor selectivity (Maggi *et al.* 1994b). A third generation of non-peptide antagonists has also been developed, commencing with CP 96,345, a selective NK_1 receptor antagonist. This compound has been instrumental in detecting subtypes of peripheral NK_1 receptors since it displays 100 times greater affinity in guinea-pig ileum, vas deferens, rabbit jugular vein and aorta (endothelium-dependent vasodilatation) than in the rat urinary bladder. This probably reflects a species difference in the primary sequence of the NK_1 receptor rather than a true differentiation of two receptor subtypes. The NK_2 receptor has also been subclassified into NK_{2A} and NK_{2B} subtypes, based upon antagonist potency orders MEN 10,376 > L 659,877 > R 396 for NK_{2A} and L 659,877 > R 396 > MEN 10,376 for NK_{2B}. Furthermore, MDL 28,564 behaves as an agonist at NK_{2A} receptors but as an antagonist at NK_{2B} receptors (Maggi *et al.* 1993). As with the NK_1 receptor subtypes, however, these probably represent a species heterogeneity of the NK_2 receptor. The NK_2 receptors of the rat and hamster (NK_{2r}) can be distinguished from those of the rabbit, guinea-pig and human (NK_{2rb}) by several NK_2 receptor antagonists (Regoli *et al.* 1994).

In the guinea-pig ileum, where SP has been proposed as an autonomic co-transmitter, the atropine-resistant contractile response to field stimulation is mediated via both NK_1 and NK_2 receptors. Only NK_2 antagonists, however, depress the atropine-resistant ascending excitation reflex activated by radial stretch of the gut through balloon distension. In the longitudinal muscle of the guinea-pig ileum, only the NK_1 receptor antagonists, GR 82,334 or GR 71,251, were effective, while the NK_2 antagonist, MEN 10,207, was ineffective. In the human ileal circular muscle, the atropine-resistant contraction to electrical stimulation is mediated via NK_2 receptors. These results therefore suggest differences in the distribution of tachykinin receptors between the smooth muscle layers of the ileum. The NK_2 receptor of the circular muscle of the human colon and small intestine appears to fit into the same subtype as in the guinea-pig bronchi and rabbit vas deferens (NK_{2b}), while the NK_{2r} receptor is probably found in several rat tissues such as the urinary bladder, vas deferens, ileum and colon (Maggi *et al.* 1993, Regoli *et al.* 1994). In most smooth muscle preparations, tachykinins produce contractile responses, usually by a direct action but also by indirect release of neurotransmitter(s). For example, in the guinea-pig ileum the direct contractile response of the circular muscle is mediated via NK_2 receptors, whereas an w-conotoxin- and TTX-sensitive indirect contractile response, attributed to release of transmitters from the enteric neurones, is mediated via NK_3 receptors (Maggi *et al.* 1994c). SP and related tachykinins also cause bronchoconstriction and contractions of the urinary bladder, urethra, ureter and uterus. In addition, they lower blood pressure by endothelium-dependent arterial dilatation (NK_1), cause plasma extravasation (NK_1) and modulate transmitter release from autonomic nerve terminals but not the release of neuropeptides from sensory nerve endings (Regoli *et al.* 1994). In view of the role of tachykinins as excitatory transmitters in the human ileum, antagonists may have potential in the control of exaggerated gut motility in certain diseases such as irritable bowel syndrome.

Substance P is also a secretagogue in salivary glands, pancreas and seromucous glands of the respiratory tract. In the rat parotid salivary gland, SP may be the mediator of non-cholinergic secretion in response to stimulation of parasympathetic nerves having their cell bodies in the otic ganglion. These neurones display SP immunoreactivity and are localized around the acini of the gland. The otic ganglion of the rat also shows immunoreactivity for VIP and ChAT. There is also some NKA in the rat parotid salivary gland. The atropine-resistant salivation of the rat and ferret is therefore attributed to the release primarily of SP, since it is abolished by a tachykinin antagonist. Furthermore, in the absence of atropine, the response to parasympathetic nerve stimulation is reduced by >20%. SP stimulates both fluid and amylase secretion in the rat parotid gland. The non-cholinergic response is probably due to the release of both SP and NKA; the latter, although not as potent as a secretagogue, can produce a saliva rich in amylase. NK_1 receptors are involved in salivation. They are coupled to phospholipase C activation via a G protein, leading to increased phosphatidylinositol turnover and mobilization of intracellular Ca^{2+}. This opens Ca^{2+}-dependent K^+ channels leading to secretion. Simultaneous increases in diacylglycerol cause activation of PKC which appears to be involved in amylase secretion (Morris & Gibbins 1992, Otsuka & Yoshioka 1993). NK_1 receptors are also involved in mucus secretion from the respiratory tract induced by SP and stimulation of sensory efferent nerves (Ramnarine *et al.* 1994).

12.2.2.2   *Vasoactive intestinal peptide (VIP)*

VIP is found in many parasympathetic postganglionic neurones co-existing with Ach. It is also found in certain intrinsic enteric neurones of the gastrointestinal tract, apparently independent of Ach. The two major sites of efferent pathway co-transmission by VIP are in the salivary glands, where it is involved in both control of secretion and blood flow, and in the gut.

Salivary glands. Stimulation of the preganglionic parasympathetic nerves to the cat submandibular gland was shown to induce increased salivary secretion and vasodilatation. The former response was abolished by atropine and therefore due to Ach release, however, the vasodilatation was largely resistant to atropine. Atropine almost abolishes the fast dilatation and at low frequency stimulation it is predominantly Ach-mediated. The sustained vasodilatation induced by moderate frequency nerve stimulation, however, is only slightly reduced by atropine. The localization of the neuropeptide, VIP, to the parasympathetic neurones supplying the salivary glands, the biochemical demonstration of its release into the venous circulation on stimulating the chorda tympani nerve, and the induction of a slow vasodilatation by infused VIP indicated that it might be the mediator of the non-cholinergic component of the vasodilatation (Lundberg *et al.* 1982). At high frequency stimulation, atropine in fact may *increase* vasodilatation and release of VIP, indicating a normal presynaptic inhibitory effect of Ach upon VIP release from these neurones.

The VIP in the submandibular gland occurs exclusively in the large dense-cored vesicles, probably co-stored with Ach which occurs in both large and small vesicles. VIP immunoreactivity is found in the nerves surrounding secretory acinar cells and blood vessels of salivary glands and in the cell bodies of the sphenopalatine ganglion innervating the nasal mucosa. Salivary secretion is not induced by VIP alone and therefore no receptors appear to be involved, although VIP is clearly released into close proximity of the acinar cells. The secretory effects of Ach are, however, enhanced by VIP, probably by increasing the sensitivity of the muscarinic receptors. VIP produces a 10 000-fold increase in the displacement of muscarinic radioligands by Ach in membrane preparations from cat salivary gland.

Also present in the salivary gland and co-released by parasympathetic nerve stimulation is peptide histidine isoleucineamide (PHI), a structurally related peptide derived from the same precursor. PHI has similar actions to VIP but is a weaker vasodilator (Dockray 1992, Kupfermann 1992, Morris & Gibbins 1992).

Gastrointestinal tract. Field stimulation of the myenteric plexus-longitudinal muscle preparation of the guinea-pig ileum releases VIP together with Ach. While Ach is released maximally at low frequency stimulation (<1 Hz), VIP is preferentially released at high frequency stimulation (as for salivary gland vasodilatation). Atropine increases the release of VIP, indicating the existence of prejunctional muscarinic receptors which are inhibitory on VIP release. Vagus nerve stimulation to the stomach also releases VIP into the venous outflow, an effect that is TT_x-sensitive and Ca^{2+}-dependent. The response of the gastrointestinal tract to VIP is inhibitory. It may therefore mediate the descending inhibitory component of the intestinal peristaltic reflex to distension by an action on the circular smooth muscle (Figure 1.5). The response can be distinguished from that of ATP, which is also inhibitory, by the fact that the K^+ channel antagonist, apamin, blocks the relaxation response to ATP but not VIP (Lefebvre *et al.* 1991). Although ATP may mediate part of the reflex inhibitory response of the gut, probably the initial phase of relaxation, VIP causes a slow relaxation and may contribute to the maintained phase of relaxation (Agoston 1988, Dockray 1992, Morris & Gibbins 1992). Recent evidence suggests that NO may be more important than VIP as the neurotransmitter of vagally induced relaxation of the guinea-pig stomach (Desai *et al.* 1994).

Other organs. VIP may also mediate the entirely atropine-resistant vasodilatation of the vasculature of the penis and uterus upon pelvic nerve (parasympathetic) stimulation. VIP causes vasodilatation, but both Ach and VIP appear to be necessary for the full erection of the penis when injected into the penile circulation (Carati *et al.* 1988). The blockade of the slow response to nerve stimulation by trypsin also suggests involvement of a neuropeptide such as VIP. The fast response is probably due to NO, since NO synthase inhibitors have been shown to reduce NANC relaxations of vascular smooth muscle of the human corpus cavernosum of the penis (Rajfer *et al.* 1992) (see later).

VIP also relaxes airway smooth muscle (Palmer *et al.* 1986). It may be released from vagal nerve endings where it is co-stored with Ach, or from sensory nerve endings. It may then exert an inhibitory effect at the postsynaptic level upon neuronally-mediated bronchoconstriction. No evidence of a direct bronchodilator effect has been observed (Yu-Hong *et al.* 1993). Release of VIP and PHI from vagal nerves into the airways may mediate vasodilatation (Widdicombe 1990).

VIP has also been located in sympathetic nerves co-stored with noradrenaline, an example being the innervation of the ovary, where it probably exerts a postjunctional inhibitory effect on noradrenaline-induced contraction of the ovarian follicle (Liedeberg *et al.* 1993).

VIP produces its inhibitory effects upon smooth muscle in the vasculature and the gastrointestinal tract by interaction with VIP receptors which have separate binding characteristics from those of other peptide receptors. Activation of these receptors leads to stimulation of adenylyl cyclase, accumulation of intracellular cAMP resulting in activation of protein kinases. No established selective agonists or antagonists are yet available.

12.2.2.3 Neuropeptide Y (NPY)

Neuropeptide Y is a 36-residue peptide with a tyrosine (designated Y in the single letter notation and hence the naming of the peptide) at the N-terminus and a tyrosine amide at the C-terminus. It is released from large dense-cored vesicles by electrical stimulation of sympathetic nerves supplying blood vessels, vas deferens, the spleen and from the adrenal medullary chromaffin cells. It has been located in sympathetic nerve endings of large blood vessels. Reserpine causes depletion of both noradrenaline and NPY. Exogenous NPY causes a slow contraction of these blood vessels which display a very slow contraction with prolonged transmural stimulation, mainly in the guinea-pig. NPY probably has a three-fold action in blood vessels: in some vessels it causes a direct weak contractile effect, the constrictor effects of noradrenaline and sympathetic nerve stimulation are potentiated through a postjunctional effect which is thought to involve an enhanced influx of Ca^{2+} (Oshita *et al.* 1989) or release from intracellular stores (Vu *et al.* 1989), and noradrenaline release from sympathetic neurones is inhibited by stimulation of prejunctional receptors. The latter two responses also appear to be the effects of NPY released from sympathetic nerves to the vas deferens. NPY selectively reduces the ATP-mediated twitch response of the prostatic portion of the vas deferens, probably because there are more NPY receptors in this section of the rat vas deferens (Kupfermann 1991).

In the mammalian heart, NPY has been identified in sympathetic fibres and both NPY and noradrenaline may be co-released by stellate ganglion stimulation. NPY is a coronary vasoconstrictor, can exert negative inotropy and chronotropy, and inhibits transmitter release from both parasympathetic and sympathetic nerves (Franco-Cereceda *et al.* 1985). There does not appear to be a NANC component to the effects of autonomic nerve stimulation in the heart and consequently NPY may serve not as a neurotransmitter, but as a neuromodulator of released noradrenaline. In the frog heart, however, co-transmission by ATP and *adrenaline* may explain a positive chronotropic response to sympathetic stimulation in this species. Similarly in the gut, NPY-containing neurones have been found in non-adrenergic intrinsic neurones where it may have a neuromodulatory inhibitory effect, since NPY has little direct effect on the smooth muscle of the gut. *In vivo*, reflex sympathetic discharge by haemorrhagic shock, cold pressor tests and intense exercise can cause α-adrenoceptor-independent vasoconstriction, and there is evidence for the release of NPY in addition to noradrenaline into the circulation (Morris & Gibbins 1992).

NPY interacts with specific receptors negatively linked to adenylyl cyclase resulting in a fall in cAMP. Two subtypes have been proposed, Y_1 and Y_2, based on limited availability of selective peptide analogues with agonist activity. The Y_1 receptor may correspond to the postjunctional, while the Y_2 may be the prejunctional site.

12.2.2.4 *Other neuropeptides primarily with neuromodulatory functions*

The opioid, enkephalin, has been identified in perfusates from the field-stimulated guinea-pig ileum longitudinal muscle-myenteric plexus preparation. The contraction of this preparation displays a gradual fade with prolonged stimulation, which may be reversed by the opioid antagonist naloxone. Thus, it is possible that the released opioids exert an inhibitory effect on Ach release through stimulation of μ opioid receptors, the guinea-pig myenteric plexus being rich in μ receptors and the neurones therein displaying opioid immunoreactivity. Opioids may also have a direct effect on the submucous neurones controlling secretory function, having an inhibitory effect upon secretion. They are also released from the adrenal medulla (see Chapter 11).

Bombesin and gastrin-releasing peptide (GRP) are related 27-residue peptides with a common C-terminal heptapeptide. GRP is released into the venous effluent by vagal stimulation to the stomach where it is involved in the release of gastrin. Stimulation of the vagus nerve or the presence of protein in the stomach causes release of gastrin via an atropine-resistant nervous pathway (see Figure 8.3, Chapter 8, gastrointestinal tract). Bombesin does not occur naturally in mammals, being first isolated from the frog skin. Bombesin-like immunoreactivity, however, has been identified in mammalian peripheral tissues. Furthermore, bombesin itself does activate bombesin (BB_1) receptors in various tissues including the guinea-pig and rat urinary bladder to cause contraction. Bombesin-like peptides, however, currently do not appear to serve as excitatory transmitters in this tissue (Maggi *et al.* 1992). The predominant second messengers for BB receptors are inositol triphosphate (IP_3) and diacylglycerol, and the receptor for GRP is the BB_2 subtype.

Cholecystokinin (CCK-8) is a sulphated octapeptide found in enteric neurones of the ileum and released by ileal distension. Its primary role is in the control of secretion from the pancreas and gall bladder but additionally it contracts the gut through release of Ach and SP. These three effects are mediated via CCK_A receptors. CCK shares a common C-terminal pentapeptide with gastrin and also stimulates the gastrin receptors on parietal cells to promote gastric acid secretion (see Figure 8.3). Gastrin, however, does not have high affinity for CCK_A receptors.

Somatostatin (SS) is mainly released from endocrine cells where it has a paracrine function to inhibit adjacent endo- or exocrine cells (see Figure 8.3 for gastrin release). Additionally, SS immunoreactivity has been identified in the gut, its release from which is induced by distension. Since this is only partly inhibited by TT_x, it is uncertain how much of this is from endocrine cells rather than from neurones. Finally, calcitonin gene-related peptide (CGRP) and galanin are two peptides located in the intrinsic neurones of the gastrointestinal tract. CGRP is also located in primary afferent neurones, is a potent vasodilator, and contributes to the neurogenic vasodilatation in inflammatory responses (see later). CGRP and galanin are probably not involved in efferent autonomic functions.

12.2.2.5 *Role of peptidergic co-transmission and differences from conventional transmitters*

The neuropeptides identified above as having possible co-transmitter functions are the tachykinins (SP), VIP and NPY. They may be regarded as being involved in slow postsynaptic events of the target cell because of the slow synthesis and provision of peptides to the nerve terminal. There are also no mechanisms for their re-use after release leading to them being readily exhausted. This has been called metabotropic transmission as opposed to the ionotropic transmission of the classical transmitters which are linked to the rapid opening of ion channels. Co-transmitters probably exert their effect over different time courses. NANC fast EJPs and contractions can usually be attributed to ATP. Intermediate responses are due to noradrenaline or Ach, are of greatest magnitude, and occur at low to moderate frequencies. The very slow responses occurring at higher frequencies and with prolonged stimulation appear to be mediated via the neuropeptides. The contribution of this latter response at physiologically relevant frequencies of nerve activity is questionable, but it may occur under intense reflex activity such as during sustained exercise in humans. The role of the co-transmitter may therefore be in the fine tuning of autonomic activity or at the extremes of activity. For example, there may be amplification of the effect of the primary transmitter by synergism such as that occurring between noradrenaline and NPY for vasoconstriction. Modulation of release may occur and there is evidence that the co-transmitter may reduce the degree of desensitization to the primary transmitter. First, the levels of primary transmitter required for release onto its receptor to achieve the desired physiological effect are lower than in the absence of the amplifying co-transmitter. Second, the co-transmitter may exert a direct action to minimize receptor desensitization (Kupfermann 1991) (see Chapter 14).

12.2.3 *Nitric Oxide*

Nitric oxide (NO) was first recognized as the mediator of endothelium-dependent vasodilatation induced by a range of agonists through their respective receptors (Moncada *et al.* 1991). This aspect of the autonomic pharmacology of NO was described in Chapters 3 and 8 for α_2-adrenoceptor and muscarinic receptor-mediated vasodilatation. The endothelium-dependent vasorelaxation in response to noradrenaline and ACH are shown in Figure 3.20 and 8.2, respectively. More recently, NO has also been identified as a co-transmitter released from peripheral autonomic nerves (Sneddon & Graham 1992).

NO is formed by the action of the enzyme, nitric oxide synthase (NOS, EC 1.14.23), on the amino acid, L-arginine, which is converted to L-citrulline. Three isoforms of NOS have been identified. Two constitutive isoforms, NOS I and NOS II, are normally present in neuronal tissue and endothelial cells, respectively.

461

Isoform II, however, is not usually constitutively expressed but can be induced by bacterial lipopolysaccharide and cytokines in macrophages, endothelial cells and smooth muscle. This form appears to exert the cytotoxic and antimicrobial effects of NO, for example, in the inflamed asthmatic airways. NOS II is Ca^{2+}-independent whereas the constituitive forms are Ca^{2+}-calmodulin-dependent (Marin & Rodriguez-Martinez 1995). NADPH is an essential co-factor for both inducible and constitutive forms, the reaction also requiring molecular oxygen. Both forms are competitively inhibited by various guanidino substituted analogues of L-arginine, including N^G-monomethyl-L-arginine (L-NMMA), N^G-nitro-L-arginine methylester (L-NAME) and nitro-L-arginine (NOARG). The inhibition can be overcome by the exogenous addition of the substrate, L-arginine. The NOS in nerves is a constitutive Ca^{2+}-calmodulin-dependent form which has been identified in a number of autonomically innervated tissues including the stomach and small intestine and in the neurones of the myenteric plexus. Whether free NO exists in neurones or whether it occurs as a stable intermediate, such as nitrosothiol, is not yet certain. NO exerts its relaxant effects upon smooth muscle by activating guanylyl cyclase by binding to the haem group of this enzyme (Ignarro 1990). This leads to increased intracellular levels of cGMP which are presumed to activate cGMP-dependent protein kinases. The role of NO as a co-transmitter may then be evaluated in isolated tissues from antagonism of the responses to nerve stimulation by NOS inhibitors such as L-NAME. Additionally, NO-mediated NANC responses are inhibited by oxyhaemoglobin, which binds to the NO, and they are enhanced by superoxide dismutase (SOD), which reduces superoxide anions (O_2^-). These anions accelerate the inactivation of NO and may be generated by pyrogallol. Indirect evidence that the relaxation response is guanylyl cyclase-mediated is its susceptibility to methylene blue which inhibits this enzyme. The response is also potentiated by the phosphodiesterase inhibitor, zaprinast, which selectively inhibits the breakdown of cGMP (Chapter 13).

The relaxation of smooth muscle by NO appears to be associated with hyperpolarization which could directly contribute to the relaxation by inhibition of spontaneous action potentials and by inactivation of voltage-dependent Ca^{2+} channels. The NO-induced hyperpolarization is achieved by increasing K^+ conductance by the opening of Ca^{2+}-activated K^+ channels. NANC inhibitory junction potentials (IJPs) are also associated with hyperpolarization and a role of NO in these has been indicated by the use of L-NAME or L-NMMA. Apamin, an inhibitor of K^+ channels, also inhibits NANC IJPs which are associated with NANC relaxation in some tissues.

NOS inhibitors have been shown to inhibit the NANC relaxation responses of a wide range of isolated tissues to electrical stimulation. These include all regions of the gastrointestinal tract studied, vascular smooth muscle preparations, airway smooth muscles such as the guinea-pig trachea, the urethra, the anococcygeus muscle and the analogous muscle in the ox, the bovine retractor penis. These NANC relaxation responses therefore appear to be mediated in part by the release of NO. That this release is due to a propagated action potential in the stimulated autonomic nerves is indicated by the blockade of the response to nerve stimulation by TT_X. These effects are illustrated in Figure 12.3 for the anococcygeus muscle of the rabbit, where the primary effects of field stimulation are contractile due to noradrenaline release from the lumbar sympathetic nerves. In the presence of guanethidine to block the release of noradrenaline and of atropine to prevent cholinergic responses, and with tone raised by histamine, field stimulation then causes frequency-dependent relaxation. In the rat anococcygeus muscle with elevated tone and atropine present, Ach also causes a TT_X-

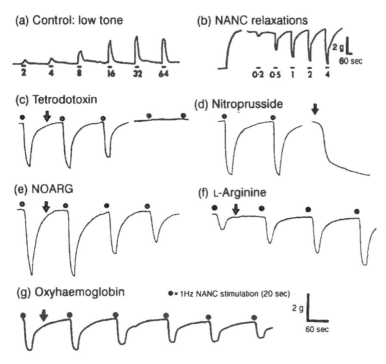

(a) Control: low tone

(b) NANC relaxations

2 4 8 16 32 64

0·2 0·5 1 2 4

2 g
60 sec

(c) Tetrodotoxin

(d) Nitroprusside

(e) NOARG

(f) L-Arginine

(g) Oxyhaemoglobin

●▪ 1Hz NANC stimulation (20 sec)

2 g
60 sec

Figure 12.3 Non-adrenergic non-cholinergic (NANC) relaxation of the rabbit anococcygeus muscle mediated via nitric oxide (NO). Field stimulation at low tone (a) produced the expected sympathetic nerve-mediated contraction which was blocked by the adrenergic neurone blocker, guanethidine, or α-adrenoceptor antagonists. When tone was raised (b) by histamine (10^{-6} M), NANC relaxations were obtained in the presence of guanethidine (10^{-5} M) and the muscarinic receptor antagonist, atropine (10^{-6} M). These NANC relaxations were abolished by tetrodotoxin (5×10^{-7} M) (c) and are therefore due to a propagated action potential in NANC nerves. The NANC responses were mimicked by the nitric oxide generating compound, sodium nitroprusside (10^{-5} M) (d). NANC relaxation was inhibited by the nitric oxide synthase inhibitor, NOARG (10^{-4} M) (e), which was reversed by the substrate L-arginine (10^{-4} M) (f). The response was also inhibited by oxyhaemoglobin which binds NO (g). Reproduced with permission from Sneddon & Graham (1992).

sensitive relaxation through NO release by stimulation of nicotinic receptors presumably on the nerve ending (Rand & Li 1993).

In addition to having a direct relaxant action on various smooth muscles, NO may mediate prejunctional inhibition of conventional transmitter release. In the rat trachea, contractions and release of Ach induced by field stimulation of cholinergic nerves were potentiated by L-NMMA and methylene blue without affecting responses to exogenous Ach. In contrast, NO added as an acidified solution of $NaNO_2$ decreased the response to field stimulation without affecting exogenous Ach. That NO was released from efferent nerves (probably cholinergic) and not sensory nerves was indicated by the lack of effect of capsaicin pretreatment on the potentiating action of methylene blue (Sekizawa *et al.* 1993). In the rat anococcygeus muscle (Brave *et al.* 1993, Kasakov *et al.* 1994) and guinea-pig trachea (Brave *et al.* 1991), however, L-NOARG had no effect on release of [^{3}H]noradrenaline or Ach, respectively, from the tissues. This indicates that the NO is released from the autonomic efferent nerves but exerts a postjunctional effect to depress the contractile effects of Ach or noradrenaline.

The precise location of the NO released by field stimulation of isolated tissues is uncertain. The assumption is that it is released from the parasympathetic or sympathetic nerve terminals together with the classical transmitters since NANC responses are TTX-sensitive. The release is non-vesicular, although it is Ca^{2+}-dependent because of the Ca^{2+}-calmodulin-dependence of the NOS, and is inhibited by prejunctional α_2-adrenoceptors (De Man *et al.* 1994). It is probably the mediator of fast relaxant NANC responses in several tissues, including atropine-resistant vasodilatation to parasympathetic nerve stimulation.

12.3   Neuropeptides Released from Sensory Autonomic Nerves (Efferent Function of Sensory Nerves)

The sensory nerve endings of primary afferent nerves have two functions. In addition to transmitting afferent inputs to the brain arising from painful nociceptive stimuli and from mechanical and chemical stimuli, they release bioactive substances locally from their terminals. Stimulation of the sensory nerve endings by irritating chemicals such as lactic acid, tissue damage, infectious agents and mechanical stimulation such as tissue distension, causes the release of stored mediators. The principal mediators are the neuropeptides, substance P (SP), neurokinin A and CGRP. This forms the basis of the neurogenic inflammatory response to noxious stimuli. An example of such a response is the *triple response* that occurs when histamine is injected locally into the skin and is associated with nettle stings and insect bites. At the site of injection is an immediate *flush* or reddening due to dilation of blood vessels by injected histamine or following its release from mast cells. There is a raised area of skin, known as a *wheal*, which is due to oedema arising from leakage of fluid and plasma proteins from capillaries. This arises from an increase in vascular permeability due to histamine and to SP released from the sensory nerve endings. The stimulation of the afferent nerve by histamine also induces an autodromically transmitted nerve impulse towards the spinal cord and the dorsal root ganglia, which then passes antidromically down other branches of the sensory nerve – an axon reflex. These antidromic impulses release neuropeptides which produce vasodilatation distant from the site of irritation, which explains the third phase of the triple response, the *flare* or surrounding reddening (Figure 12.4). The released transmitters produce a range of biological actions including vasodilatation, an increase in vascular (venular) permeability leading to plasma protein extravasation, changes in smooth muscle contractility, degranulation of mast cells to release further histamine and activation of lymphocytes, macrophages and granulocytes. These effects all contribute to the neurogenic inflammation (Kupfermann 1991, Dockray 1992).

Antidromic impulses in sensory nerves may be produced by electrical stimulation and responses of a wide range of isolated tissues to field stimulation have been shown to be due to the release of these mediators of neurogenic inflammation. Responses include contractions of smooth muscle, such as the iris sphincter muscle, ureter, bronchi, rabbit pulmonary artery and increased contractility of isolated atria. An important agent for determining whether responses are due to release from *sensory* nerves or from *efferent* autonomic nerves, is capsaicin.

Capsaicin is the pungent constituent of hot peppers (chillies, paprika, *Capsicum*

464

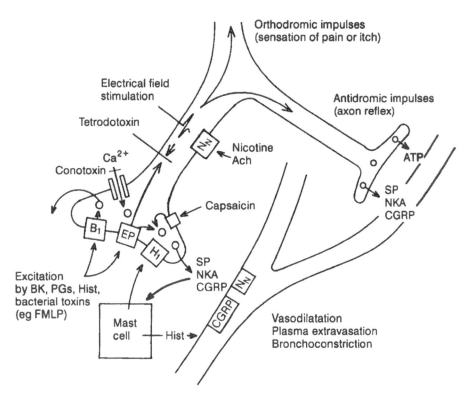

Orthodromic impulses
(sensation of pain or itch)

Electrical field
stimulation

Tetrodotoxin

Antidromic impulses
(axon reflex)

Nicotine
Ach

Ca^{2+}

Conotoxin

ATP

Capsaicin

SP
NKA
CGRP

B_1 EP

H_1

SP
NKA
CGRP

Excitation
by BK, PGs, Hist,
bacterial toxins
(eg FMLP)

Mast
cell — Hist →

Vasodilatation
Plasma extravasation
Bronchoconstriction

Figure 12.4 Diagrammatic representation of the release of the tachykinins, substance P (SP), neurokinin A (NKA) and calcitonin gene-related peptide (CGRP), from sensory nerve endings induced by inflammatory mediators bradykinin (BK), prostaglandins (PGs) and histamine (Hist). This stimulation and that induced by nicotine, electrical field stimulation (EFS) and capsaicin sets up antidromic impulses which also releases mediators distal to the site of stimulation. Responses to EFS and BK are generally blocked by tetrodotoxin and ω-conotoxin, indicating involvement of Ca^{2+} influx through N-type Ca^{2+} channels. Inflammatory responses include vasodilatation, plasma extravasation and bronchoconstriction.

annum) which, as preparations of capsicum extract, is still employed as a counter-irritant and rubefacient for painful musculoskeletal disorders (Algipan[R], Cremalgin[R], Cremalgex[R]). Capsaicin is a potent stimulant of sensory nerve endings and induces a wide range of responses in isolated tissues that are attributed to the release of neuropeptides. Capsaicin exposure also leads to depletion of tissues of CGRP and other neuropeptides in their sensory nerve endings and therefore to blockade of neurogenic responses. These responses are therefore described as being mediated via capsaicin-sensitive primary afferents. Other pungent components of natural occurrence that are also capable of stimulating sensory pathways include piperine from black peppers (*Piper nigrum*), curcumin from turmeric (*Curcuma longa*) and the principal constituent of curry powder, eugenol from oil of cloves (*Eugenia caryophyllus*), mustard oil from *Brassica nigra* (which yields allyl isothiocyanate), and resiniferatoxin (RTx), an irritant diterpene from the latex of the plants of the genus, *Euphorbia*. RTx has greater potency than does capsaicin and radiolabelled RTx binds selectively to sites on membranes from dorsal root ganglion cells. This binding is displaced competitively by capsaicin, indicating the possibility of a receptor site on sensory nerves for all of these agents (Maggi 1991a).

The sensory nerve endings are activated by the mediators of inflammation. Bradykinin is considered to be a major mediator of inflammatory pain and specific receptors (B_2) located on nociceptive primary afferent nerves are involved in mediating pain and reflex stimulation. Bradykinin induces sensory impulses and local release of neuropeptides from sensory nerve endings (Figure 12.4). Bradykinin-induced responses, such as protein extravasation, tachycardia and bronchoconstriction, are blocked by capsaicin. This release does not entirely depend upon a propagated action potential or antidromic nerve stimulation since it is not fully blocked by TT_X. In certain tissues, however, bradykinin may induce distant release of mediators by antidromic activity. The response is blocked by ω-conotoxin, indicating that it depends upon influx of Ca^{2+} through N-type calcium channels. Capsaicin-induced stimulation and mediator release, however, is not usually blocked by either TT_X or ω-conotoxin, indicating a direct effect on local cation channels which are blocked by ruthenium red (Maggi 1991a). Low concentrations of capsaicin may, however, induce release by a TT_X-sensitive mechanism suggesting that it can induce a propagated action potential in a similar manner to bradykinin (Lou *et al.* 1991).

Prostaglandins potentiate both the sensory and efferent functions of primary sensory nerves, PGE_1 being more potent than PGE_2. Leukotriene B_4 also excites or sensitizes primary afferents. The actions of several sensory stimulants are probably mediated in part or fully through the release of prostaglandins since their responses are blocked or reduced by cyclo-oxygenase inhibition with indomethacin. Examples include bradykinin, interleukin 1, FMLP (N-formyl-methionyl-leucyl-phenylalanine), a synthetic analogue of the chemotactic peptide derived from bacteria, and toluene diisocyanate, a polyurethane precursor that causes airway hyperreactivity and occupational asthma. Other stimulants of sensory nerves include 5-HT via $5-HT_3$ receptors, histamine via H_1 receptors (although release is *inhibited* via H_3 receptors) and nicotine via the nicotinic N_N receptor (see Chapter 11) (Maggi 1991a).

The primary transmitters located in sensory nerve endings are the tachykinins, substance P and neurokinin A, and calcitonin gene-related peptide (CGRP). The tachykinins have been considered in a previous section for their effects after release from efferent autonomic nerves. They are synthesized in the dorsal route ganglia of sensory nerves together with somatostatin and galanin and transported to the nerve terminals in the gastrointestinal tract, skin, airways and cardiovascular system. Capsaicin causes their release and raises the plasma levels of these neuropeptides. Substance P and NKA cause the following responses:

1 They increase vascular permeability, resulting in plasma protein extravasation via NK_1 receptors.

2 Endothelium-dependent vasodilatation is produced via NK_1 receptors, probably involving release of NO and due to circulating tachykinins, although electrical stimulation of certain vessels causes release of CGRP and SP and vasorelaxation independent of the endothelium (porcine coronary arteries, rat mesentery). In other vessels, such as rabbit pulmonary artery, vasoconstriction occurs via NK_2 receptors.

3 Smooth muscle contraction is produced in the airways, gut and urinary tract via NK_1 and NK_2 receptors (see Table 12.2).

4   Mast cell degranulation is induced by high concentrations of SP, but it may be a mast cell primer at low concentrations without causing degranulation.

5   Recruitment or excitation of inflammatory cells occurs, for example stimulation of chemotaxis and of alveolar macrophages, suggesting an involvement in the immune response.

6   They stimulate secretions, including saliva, pancreatic secretions and secretions from seromucous glands of the airways. Tachykinins also exert a powerful mitogenic or stimulant action on various cell types including fibroblasts, synoviocytes, endothelial cells and macrophages which are involved in inflammation, the immune response and tissue repair. Their release during neurogenic inflammation is aimed at protecting the tissues from further injury and to promote tissue repair. The latter trophic role of the sensory neuropeptides is a physiological mechanism. They may, however, also become involved in pathophysiological processes. For example, in the airways a defect of the peptidases which break down peptides may result in enhanced levels. Also, in asthma, shedding of the airway epithelium may expose sensory nerve endings to excessive stimulation by airborne irritants (Maggi *et al.* 1993). There is an increase in SP-like immunoreactive nerves in asthmatic airways and consequently inhibition of neurogenic inflammation is a potential therapy for asthma (Barnes 1993b). Indeed, neurokinin A antagonists have been shown to reverse the bronchoconstriction induced in asthmatics by inhaled bradykinin.

These responses are mediated via the NK_1, NK_2 or NK_3 receptors described earlier and are inhibited by the tachykinin receptor antagonists shown in Table 12.3. Unlike tachykinins, which are involved in both efferent co-transmission and sensory neurogenic responses, CGRP is only involved in the latter processes. It has similar vascular effects to the tachykinins. In addition, CGRP enhances the vascular permeability changes induced by SP and increases the rate and force of cardiac contractions. It appears to be the major mediator of these responses to capsaicin in the heart and may be released during cardiac ischaemia. These effects are mediated via a separate class of CGRP receptors which may be readily desensitized by exposure to CGRP (Kupfermann 1991). Antagonists have been developed from the C-terminal fragments of CGRP, with $CGRP_{8-37}$ showing competitive blocking properties and selectivity for a $CGRP_1$ subtype (Watson & Girdlestone 1994).

The tachykinins released from sensory nerves also induce some of their inflammatory responses indirectly by mast cell degranulation and histamine release. SP also stimulates the release of proteases from mast cells which readily degrade CGRP. Conversely NKA causes vasodilatation only by a direct action and not through histamine release (Figure 12.4) (Kupfermann 1991, Dockray 1992).

Finally, the possibility of ATP being released from sensory nerves by antidromic stimulation was suggested as long ago as 1959 by Holton, who also showed its release into venous effluent from the rabbit ear and proposed that it mediated the vasodilator response. VIP is also present in the gut and airways in capsaicin-sensitive neurones suggesting that some of the VIP is contained in sensory nerves in these tissues (Maggi 1991a).

12.4 Dopaminergic Neurones

Noradrenaline is regarded as the primary neurotransmitter of sympathetic nerves, with dopamine serving as an intermediate in its synthesis. The conversion of dopamine to noradrenaline occurs in the storage vesicles by the action of the vesicular enzyme, dopamine β-hydroxylase (Figures 2.4 and 2.5, Chapter 2). Stimulation of sympathetic nerves therefore has the potential to release a small quantity of dopamine. Free dopamine in the plasma of humans under basal conditions accounts for ~20% of the free catecholamines. About 98% of circulating dopamine, however, is conjugated with sulphate (man) or glucuronides (rat) and as a consequence, the total dopamine in the blood represents 50% of the total circulating catecholamines. A substantial portion of conjugated dopamine arises from dietry sources, for example, after conversion from L-dopa. The remaining plasma dopamine is derived from sympathetic nerves, the adrenal medulla and the small intensely fluorescent (SIF) cells (see also Chapter 1). The latter cells are associated with sympathetic ganglia and are rich in dopamine which is released by M_2 receptor activation to modulate ganglionic transmission (Chapter 11). The baseline level of free dopamine in the plasma, although low, can be considered to represent the spontaneous activity of the sympathoadrenal system. Adrenalectomy, adrenal demedullation, denervation and chemical sympathectomy have been reported to decrease plasma dopamine levels. Whether the increases in sympathetic discharge under normal physiological conditions lead to elevated levels of dopamine is uncertain. Exercise and change in posture do not consistently induce increases in plasma dopamine (Van Loon 1983, De Feo 1990). More stressful stimuli such as haemorrhage or enforced restraint do cause more reliable increases in circulating dopamine (De Feo 1990).

Dopamine receptors are located in several peripheral tissues and exogenously applied dopamine exerts pharmacological effects in the kidneys, gastrointestinal tract and cardiovascular system. Some of these effects can be attributed to actions at α-adrenoceptors, such as the relaxation of the stomach (Guenaneche *et al.* 1991), or at β-adrenoceptors, such as the positive inotropic and chronotropic action on the heart (Lumley *et al.* 1977). Dopamine also has indirect sympathomimetic activity (Table 4.1, Chapter 4). Dopamine causes an increase in blood pressure which is also sensitive to α-adrenoceptor antagonism by phentolamine and therefore due to α-adrenoceptor-mediated vasoconstriction (Figure 5.2). In the presence of phentolamine, dopamine produces a fall in blood pressure which is not β-adrenoceptor-mediated but is attenuated by the dopamine receptor antagonist, haloperidol, indicating the involvement of dopamine receptors in this hypotensive response (Figure 5.2). Effects in the kidney and several vascular beds also appear to be due to stimulation of dopamine receptors since they are not antagonized by the adrenoceptor antagonists. The possibility therefore arises that dopamine may have an independent transmitter role in certain organs, which are innervated by neurones releasing primarily dopamine – dopaminergic neurones. The problem has been to separate such neurones from the noradrenergic sympathetic innervation.

Biochemical and histochemical studies have shown the presence of dopamine-containing neurones in the kidney and the canine paw pad and in the SIF cells of sympathetic ganglia. There is also strong evidence for nerves rich in dopamine in the gut and vas deferens. The kidney has been the most studied organ where dopaminergic sympathetic neurones have been shown to enter, together with the vasculature. A population of sympathetic nerves having a high dopa-decarboxylase immunoreactivity have been identified in a proportion of 5–20% of the total catecholamine-containing nerves (Harris *et al.* 1986). The neurones terminate in the renal cortex and electron microscopy has shown them to supply the juxtaglomerular blood vessels and tubular elements in the dog, rat and human kidney (Bell 1990). Further study of dopaminergic nerves has been made by analysis of the dopamine:noradrenaline ratio (DA:NA) in sympathetic ganglia which send projections to tissues such as the dog kidney, where dopaminergic neurones are thought to occur. The DA:NA ratio (DA being 8–10% of NA) in ganglia projecting to the kidney and hind limb vasculature was about twice that of neighbouring ganglia thought to project only noradrenergic neurones. Similarly, there is some evidence of ganglia in which neurones have a high immunoreactivity to dopa-decarboxylase but low dopamine-β-hydroxylase (DBH), which suggests high dopamine synthesis but low conversion to noradrenaline.

Also present in sympathetic ganglia and certain peripheral organs are detectable levels of the major metabolite of dopamine, 3,4-dihydroxyphenylacetic acid (DOPAC), which is not formed from noradrenaline or adrenaline (Figure 2.11, Chapter 2). Observations made in sympathetic ganglia are, however, limited by the possibility that the dopamine-containing characteristics originate in the SIF cells and not the cell bodies providing dopaminergic neurones. The evidence with neurones supplying the kidney is more convincing. Reserpine pretreatment depletes noradrenaline from sympathetic neurones, but interestingly it increases dopamine levels. This is because the vesicular conversion of dopamine to noradrenaline is impaired, and there is a compensatory induction of tyrosine hydroxylase activity (Chapter 6). Loading with dopa after reserpine pretreatment restores catecholamine fluorescence to some nerve fibres in the dog kidney, as would be expected of dopaminergic nerves (Hills & Jessen 1992).

Stimulation of the sympathetic renal nerves releases both noradrenaline and dopamine into the venous blood of anaesthetized dogs, the latter comprising ~8% of the total stimulus-evoked efflux (Petrovic *et al.* 1988). Administration of 6-OHDA in the presence of the selective inhibitor of dopamine uptake, GBR 12909, permits depletion of noradrenaline from noradrenergic neurones but protects the dopaminergic neurones. 6-OHDA alone abolished the renal efflux of dopamine and noradrenaline, in addition to the usual renal responses to sympathetic nerve stimulation. After GBR 12909 and 6-OHDA, the dopamine overflow was not attenuated (Bell & Sunn 1990). In the presence of DMI, which selectively prevents uptake into noradrenergic neurones, 6-OHDA causes depletion of dopamine (Chapter 6).

The predominant response to renal nerve stimulation is a vasoconstriction and reduced renal blood flow due to the α-vasoconstrictor effects of released noradrenaline. When this α-adrenoceptor-mediated response is antagonized by prazosin or phentolamine, a small inconsistent vasodilatation has been observed (Gomer & Zimmerman 1972, Bell & Lang 1973). Nerve stimulation after 6-OHDA and GBR 12909 treatment also permitted the appearance of a vasodilator response and diuretic action, although there was a small antinatriuresis (Bell & Sunn 1990). The vasodilator response is not always produced and it remains questionable whether dopaminergic neurones do induce renal vasodilatation. It is doubtful whether urinary dopamine originates from neuronal sources, most of it probably arising from synthesis in the tubular epithelium from L-dopa filtered from the blood.

Notwithstanding the lack of convincing vasodilator responses following stimulation of sympathetic nerves to the kidney, there is abundant evidence that the administration of exogenous dopamine exerts several pharmacological actions on the kidneys. Dopamine produces renal vasodilatation together with enhanced urine flow and natriuresis, which were initially thought to be secondary to the renal vasodilatation. It is now considered that dopamine has direct natriuretic effects on tubular function of the kidneys. An increase in sodium excretion can occur without an elevation of renal blood flow (Bell 1990). Furthermore, in isolated renal tubules, dopamine has been shown to decrease sodium and water transport across the epithelia of proximal renal tubules (Felder *et al.* 1984). Another renal effect of dopamine is to increase the circulating levels of renin and, as a result, also of angiotensin by the action of angiotensin converting enzyme (ACE). Whether this effect is a true stimulation of dopamine receptors located on the juxtaglomerular cells has yet to be convincingly demonstrated. The effect is certainly prevented by dopamine receptor antagonists, but may be a consequence of the renal vascular effects (Bell 1990).

12.4.1 *Dopamine Receptors*

Peripheral dopamine receptors were initially subclassified into DA_1 and DA_2 types. These are approximately equivalent in their pharmacological characteristics to the D_1 and D_2 receptors identified in the CNS. Receptor nomenclature does not now distinguish the central and peripheral dopamine receptor (Instructions to Authors 1993, Watson & Girdlestone 1994). Peripheral D_1 receptor-mediated responses are characterized by their antagonism by SCH 23390 and R-sulpiride and the selectivity of the agonists, fenoldopam and dopexamine, the latter also having β_2-adenoceptor-agonist properties (see Chapter 4, Table 4.5). D_2 receptors are preferentially blocked by domperidone and S-sulpiride; quinpirole and bromocriptine (Table 5.1) are selective agonists (Hilditch & Drew 1987, Brodde 1990) (Figure 12.5). D_1 receptors are classically coupled to stimulation of adenylyl cyclase and, in homogenates of rat renal cortex and medulla, dopamine increases levels of cAMP (Ricci *et al.* 1991). A D_1 receptor capable of activating phospholipase C leading to generation of inositol phosphates and diacylglycerol has also been suggested in the rat renal cortex. D_2 receptors, in contrast, are negatively linked to adenylyl cyclase and in the rat renal cortex membranes, quinpirole and bromocriptine elicit inhibitory effects upon cAMP generation (Ricci *et al.* 1991).

Subsequently, additional subtypes of dopamine receptor have been identified in the brain, and molecular biological approaches have resulted in the cloning of several D_1 receptor subtypes (D_{1A}, D_{1B} and D_5) which are known as D_1-like and all coupled to the stimulation of adenylyl cyclase. The D_2 receptor has two isoforms synthesized by alternative splicing. A shorter form composed of 415 amino acids is termed the $D_{2\ short}$ receptor. The long form ($D_{2\ long}$ receptor) is composed of 444 amino acids with a 29-amino acid insert in the third intracellular loop of the seven membrane spanning helical arrangement of the receptor (Chapter 13). Both receptors are coupled to the inhibition of adenylyl cyclase and display identical pharmacological profiles, but are expressed differently among brain areas. The extra 29 amino acids are not in ligand binding regions of the receptor. The D_5 receptor may be distinguished from the D_1 receptor by the 10-fold higher binding affinity of dopamine itself for this cloned receptor. D_3 and D_4 receptors have also been identified which are closely related to the D_2 receptor and the three receptors are referred to as D_2-like. D_3 and D_4 receptors, however, can be distinguished since they do not appear to be consistently linked to adenylyl cyclase. Furthermore, clozapine (Clozaril[R]) (Figure 12.5), the recently introduced antipsychotic drug for schizo-phrenia patients resistant to other conventional drugs, is a selective antagonist for D_4 receptors, although it also has α-adrenoceptor, 5-HT and histamine receptor blocking properties. Quinpirole is a selective D_3/D_4 agonist. Of the cloned receptors, the mRNA of the D_3 receptor has been reported to occur in the kidney. D_1 and D_2

Antagonists

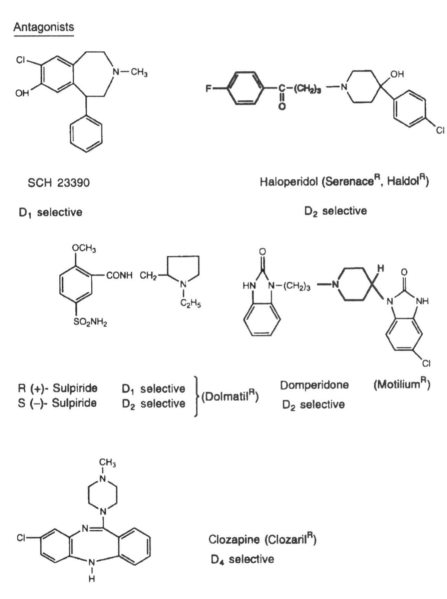

SCH 23390

D_1 selective

Haloperidol (Serenace[R], Haldol[R])

D_2 selective

R (+)- Sulpiride D_1 selective
S (−)- Sulpiride D_2 selective   }(Dolmatil[R])

Domperidone   (Motilium[R])

D_2 selective

Clozapine (Clozaril[R])

D_4 selective

Figure 12.5 Structures of some dopamine receptor antagonists, agonists and agonist prodrugs.

471

Agonists

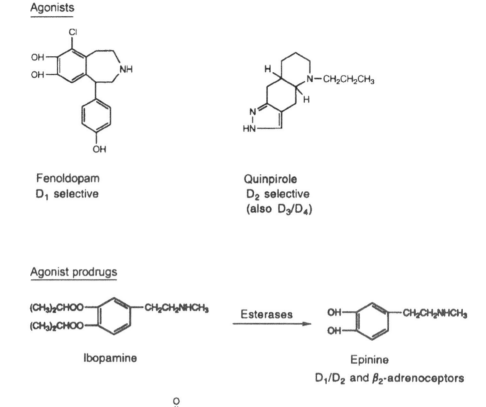

Fenoldopam
D$_1$ selective

Quinpirole
D$_2$ selective
(also D$_3$/D$_4$)

Agonist prodrugs

Ibopamine

Esterases

Epinine
D$_1$/D$_2$ and β$_2$-adrenoceptors

γ-Glutamyl-transpeptidase

Dopa

Dopa decarboxylase

N-L-γ-Glutamyl-dopa (gludopa)

Dopamine

Figure 12.5 *Continued*

receptors have been identified in the kidney by radioligand binding to tissue homogenates and by autoradiography. The D$_1$ receptors are located in the proximal tubules of the nephron and in the renal vasculature (Jose *et al.* 1992, Sokoloff & Schwartz 1995).

The renal vasodilatation, diuresis and increased electrolyte excretion (natriuresis) induced by dopamine are mediated via D$_1$ receptors. These responses are mimicked by the selective D$_1$ receptor agonists, fenoldopam and dopexamine, and are blocked by SCH 23390. Functional D$_2$ receptors also appear to be involved in the vasodilator response, being located on the prejunctional site of sympathetic neurones in the vascular smooth muscle. Stimulation of these prejunctional D$_2$ receptors by quinpirole exerts an inhibitory effect upon noradrenaline release (Chapter 6). Thus, renal vasoconstriction induced by sympathetic nerve activity is inhibited, resulting in

472

passive vasodilatation. D_2 receptors mediate prejunctional inhibitory effects upon noradrenaline release in several other tissues, including the cat nictitating membrane and the heart, and they mediate inhibition of transmitter release in the cholinergically innervated rabbit rectococcygeus muscle. They are also involved in inhibition of transmission in autonomic ganglia and inhibition of aldosterone release from the adrenal cortex. D_1 receptors mediate direct vasodilatation outside the kidney in the mesenteric, coronary and cerebral vascular beds. There is some indication that the D_1 receptors mediating vasodilatation may differ between vascular beds (Brodde 1990). It is worth remembering that the effect of dopamine itself on the vasculature is primarily vasoconstrictor via α-adrenoceptors, the D_1 receptor-mediated vasodilatation and fall in blood pressure only being observed after α-blockade (Figure 5.2). The selective D_1 receptor agonists, such as fenoldopam, lower blood pressure without α-adrenoceptor blockade.

12.4.2 Pathological Changes Associated with the Renal Dopaminergic System

A pathological role of defective dopaminergic mechanisms in the kidney has been proposed to explain certain forms of hypertension that are aggravated by sodium loading. In some hypertensives there is a decreased ability to increase renal dopamine production in response to a sodium load. A defective transduction of the dopamine receptor signal, possibly by reduced coupling to adenylyl cyclase in the proximal convoluted tubule, has also been suggested in animal models of hypertension, such as the SHR and in humans. This would result in a reduced ability of dopamine to inhibit Na^+/H^+ exchange in the renal brush-border membrane via D_1 receptors. This would impair its ability to promote sodium excretion; water retention and hypertension may then ensue (Jose *et al.* 1992).

12.4.3 Therapeutic Applications of the Peripheral Effects of Dopaminergic Agents

A potential therapeutic application of dopamine D_1 receptor agonists, such as fenoldopam and dopexamine, is in hypertension. Fenoldopam also has α-adrenoceptor antagonistic properties which may contribute to its blood pressure lowering activity. The disadvantages of fenoldopam are reflex tachycardia, vasoconstriction and renin release arising from the fall in blood pressure. The D_2 receptor agonists, bromocriptine (Parlodel[R]) and the ergot alkaloid mixture, co-dergocrine (Hydergine[R]) (see Chapter 5, ergot alkaloids), also lower blood pressure by inhibiting noradrenaline release. Such drugs, however, have the potential for causing centrally mediated nausea and vomiting, but they would not induce the reflex changes in heart rate or renin release of D_1 receptor agonism. The main therapeutic applications of the D_2 receptor agonists is to stimulate D_2 receptors on prolactin-secreting cells to inhibit prolactin secretion, and in the treatment of Parkinson's disease. Additionally, dopexamine (Dopacard[R]) and dopamine (Intropin[R]) are used in the acute management of low cardiac output states following cardiac surgery, where impaired renal perfusion and renal failure are complicating factors. Dopexamine has D_1 agonist activity combined with β_2-adrenoceptor agonist activity, which together increase cardiac performance by direct positive inotropy and reduced after-load, as well as improving renal blood flow. Dopexamine also induces a baroreceptor reflex cardiac stimulation arising from the hypotensive effect (Einstein *et al.* 1994).

Dopaminergic prodrugs have been developed to improve the delivery of the agent to its site of action and to achieve greater duration of action for chronic use in congestive heart failure (Rajfer & Davis 1990). This approach has utilized the catecholamine starting point of dopamine, which itself is poorly absorbed due to first-pass metabolism by COMT in the liver (Chapter 2). Retaining the dopamine moiety has advantages for dopaminergic activity, since it has limited access to the brain which avoids unwanted central side-effects. An example is ibopamine (Figure 12.5), which has been shown to be of benefit in clinical trials to many thousands of patients with congestive heart failure. Ibopamine is almost completely absorbed and is then rapidly hydrolysed by plasma esterases to yield the active drug, epinine, the free levels of which peak at ~1 hr after administration. The duration of action is 5–6 hr. Epinine shares the actions of dopamine (Table 4.1, Chapter 4) but is generally more potent at D_2 receptors and β_2-adrenoceptors and has less indirect sympathomimetic activity. Whether it is more potent at D_1 receptors than dopamine is less certain (van Woerkens *et al.* 1992). The beneficial effect in congestive heart failure may therefore be attributed to cardiac β_2-adrenoceptor stimulation and the reduced after-load from vasodilatation, and an improved renal function (see also Chapter 4). Additionally, D_2 receptor activation lowers plasma noradrenaline levels by reducing its release from sympathetic nerve endings. The sustained reduction of sympathetic nerve activity may have a favourable effect on the prognosis of heart failure patients in who excessive sympathetic activity may be a risk factor for their death (Rajfer & Davis 1990).

Protection of the amino group of dopamine has also been utilized in an attempt to overcome inactivation by MAO and to achieve specific site delivery. N-Acylation with amino acids results in compounds that are hydrolysed after absorption by peptide-cleaving enzymes, to a simple N-acyl amide that is resistant to further hydrolysis. γ-Glutamyldopa (gludopa) (Figure 12.5) was developed with the object of obtaining selective removal of the γ-glutamyl group to produce dopa by means of the enzyme, γ-glutamyl-transpeptidase, which is found predominantly in the kidney. This is further converted to dopamine by the activity of L-amino acid decarboxylase (dopa-decarboxylase) (Figure 2,4, Chapter 2), which also has high activity in the kidney. The local production of dopamine in the kidney, acting on D_1 receptors, can explain the improvements in renal perfusion and urine flow and the potential use of this drug in renal failure and essential hypertension (Casagrande & Santangelo 1990).

13

Receptors and Signal Transduction Pathways Involved in Autonomic Responses

13.1 Introduction

The previous chapters have described the receptors for the neurotransmitters, co-transmitters and neuromodulators of the autonomic nervous system and their

Table 13.1 G protein-coupled autonomic receptors and their second messengers

Receptor	Peptide length (human)	G protein	Cholera/ pertussis toxin sensitivity	Second messenger	Ion channels (direct coupling)	Response(s)
Adrenoceptors						
β_1	477	G_s	Cholera	cAMP↑		Cardiac positive inotropy
		G	No	—	Ca^{2+} open	
β_2	413	G_s	Cholera	cAMP↑		SM relaxation lipolysis
β_3	402	?	?	cAMP↑		
α_{1A}	—	G_q	No	IP$_3$/DAG↑ stim PLC, PLA$_2$, PLD		SM contraction
α_{1B}/α_{1C}	515/466	G_q	No			SM contraction (rat)
α_{1D}	560 (rat)	G_q	No			SM contraction
α_1		G	No		Ca^{2+} open	Prejunctional receptor EDRF release
$\alpha_{2A}/\alpha_{2B}/\alpha_{2C}$	450/450/461	G_i	Pertussis[a]	cAMP↓	—	
α_2		G	?	—	K^+ open	
		G	?	—	Ca^{2+} close	
Muscarinic						
M_1	460	G_q	No	IP$_3$/DAG↑	K^+ ($i_{K(M)}$) close	M current inactivation
		G	No			
M_2	466	G_i	Pertussis	cAMP↓	K^+ ($i_{K(ACh)}$)	Negative inotropy/ chronotropy
		G_K or G_0	Pertussis			
M_3	590	G_q	No	IP$_3$/DAG↑	Ca^{+2} open	SM contraction
		G	Pertussis			
M_4	479	G_i	Pertussis	cAMP↓		SM contraction
M_5	532	$G_{q/11}$	—	IP$_3$/DAG↑		SM contraction

	Amino acids	G protein	Toxin	Second messenger	Channel	Effect
Dopamine						
D_1	446	G_s	Cholera	cAMP↑	—	Vasodilatation
D_2	443 (large i_3)	G_i	Pertussis	cAMP↓	—	Prejunctional receptor
		G_0	Pertussis	—	K^+ open, Ca^{2+} close	
Purinoceptors						
P_1 A_1	326	G_i	Pertussis	cAMP↑	K^+ open	Negative inotropy/chronotropy (atria)
		G	Pertussis	—		
A_{2A}	409	G_s	Cholera	cAMP↑		SM relaxation
A_{2B}	328	G_s	Cholera	cAMP↑		SM relaxation
A_3	318	$G_{i/0}$	—	IP_3/DAG↑		Mast cell secretion
P_{2x}	(not identified as 7TM receptor)			—	Cation channel	SM contraction
P_{2y}	362 (chick)	G_q		IP_3/DAG↑		SM relaxation
P_{2u}	375	$G_{q/11}$		—		Vasodilatation
Neuropeptides						
NPY (Y_1)	384	G_i	Pertussis	cAMP↓		
NK_1, NK_2, NK_3	407/398/468	G_q	?	IP_3/DAG↑		SM contraction
VIP(VIP_1)	457	G_s	?	cAMP↑		

Notes: [a] α_2 Prejunctional receptor-mediated inhibition of noradrenaline release not pertussis toxin (PTx) sensitive.
Abbreviation: cAMP, Cyclic adenosine $3'5'$-monophosphate; DAG, 1,2-diacylglycerol; IP_3, inositol 1,4,5-triphosphate; NK_1, neurokinin$_1$; NPY, neuropeptide Y; SM, smooth muscle; VIP, vasoactive intestinal peptide.

classifications. With few exceptions, these receptors belong to a G protein-coupled superfamily which display common features. Many of the properties of these receptors have already been mentioned, including their coupling to G proteins, the second messengers involved, and features of their structures important for agonist and antagonist binding. This chapter is intended to assemble this information on the major receptor subtypes already considered. It is beyond the scope of this book to deal with each receptor type individually and all aspects of their structures, coupling and transduction mechanisms to the final tissue response. Instead, an attempt has been made to summarize the main generalizations that appear to apply to the majority of receptors already considered. Most of these receptors have now been cloned and their amino acid sequences determined and, as a consequence, the following common features have emerged.

1   They are located on the extracellular surface of the cell membrane with seven transmembrane spanning hydrophobic domains (G_m7).
2   They are coupled to a guanine nucleotide binding protein (G protein).
3   The pharmacological effects of agonist interactions are mediated through two second messenger systems: cyclic adenosine 3',5'-monophosphate (cAMP), levels of which may be elevated or supressed, or stimulation of phosphoinositide turnover to generate inositol triphosphate (IP_3) and diacylglycerol (DAG).
4   Some responses may also be produced by direct coupling through the G protein to ion channels without the intermediate second messenger.

The G proteins, second messengers and ion channels associated with each receptor and the known functional receptor subtypes are summarized in Table 13.1. The general features of the receptor structure, G protein coupling and the cAMP and phosphoinositide systems will therefore be considered. β-Adrenoceptors and muscarinic receptors will be used for illustrative purposes, since this is where most work has been done.

The two exceptions to the above generalizations are the nicotinic receptor and P_2 purinoceptors. Although the P_{2y} receptor has now been cloned and identified as having the classic seven membrane spanning domains, this does not currently apply to the P_{2x} receptor. P_{2x} receptors appear to be linked directly to a cation channel while the P_{2y} receptor is linked to second messengers, probably the stimulation of phosphoinositide turnover (Barnard *et al.* 1994). The nicotinic receptor of autonomic ganglia, the adrenal medulla and sensory neurones (N_N) is not linked to a G protein or second messenger, but forms an integral part of the sodium channel. Five subunits of the receptor are arranged around the ion channel, each one consisting of four, not seven, transmembrane domains formed into α-helices (Watson & Girdlestone 1994). The structure and function of the nicotinic receptor of the autonomic nervous system has been described thoroughly in Chapter 11 and will not be considered further.

13.2   Structure of G Protein-linked Receptors

The amino acid sequences of the G protein-linked receptors have been determined and shown to consist of single polypeptide chains. None of the receptors have been crystallized, so that the three-dimensional geometry of the receptors is not yet known. Bacteriorhodopsin is the purple-membrane protein of *Halobacterium halobium*, a halophilic (salt-loving) bacterium, which contains a retinal chromophore similar to rhodopsin, the visual pigment of mammalian eyes. This 25 kDa protein has a photosynthetic function in the absence of oxygen, to pump protons (H^+) from inside the bacterial cell to the outside, when it is exposed to light. As a result, ADP is converted to ATP. Bacteriorhodopsin is a membrane-spanning protein not coupled to a G protein, but probably making up the ion channel of the proton pump. It has been crystallized, making X-ray crystallography possible for elucidating the three-dimensional structure. High-resolution electron microscopy has shown this protein to consist of seven closely packed hydrophobic membrane-spanning α-helices extending nearly perpendicular to the plane of the membrane, some of the α-helices probably being tilted. Subsequently, the sequence of the visual pigment, bovine rhodopsin, was determined and its predicted structure was similar to that identified for bacteriorhodopsin. This model has served as the basis for the structural arrangement of all G protein-coupled receptors.

Analysis of the primary amino acid sequences indicates seven stretches of 20–30 hydrophobic amino acids which are believed to form the seven membrane-spanning sections that are arranged into α-helices. These are the transmembrane domains, TM1 to TM7 (Figure 13.1). There is a considerable degree of homology in the amino acid composition of the transmembrane domains between different receptors. Common

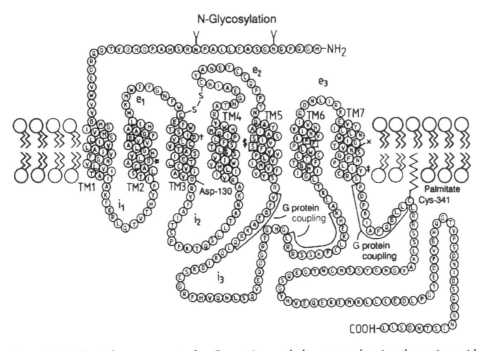

Figure 13.1 Typical arrangement of a G protein-coupled receptor showing the amino acid sequence of the human β_1-adrenoceptor. The transmembrane domains (TM1 to TM7) are arranged as α-helixes, with three extracellular loops (e_1 to e_3) and three intracellular loops (i_1 to i_3). The carboxyl terminal is anchored to the inside of the cell membrane lipid bilayer via a palmitate at Cys-341 (C). Two glycosylation sites occur at asparagine residues (N) in the amino terminal. The boxed residues are conserved in all catecholamine receptors. Residues that are important for agonist binding include aspartates (D) 79 (*) 113 (†) and 130, asparagine (N) 318 (X), tyrosine (Y) 326 and serines (S) 204 and 207 ($). Regions involved in G protein coupling are indicated by the bars alongside the chain in i_3 and the carboxyl terminal.

amino acids are found in certain positions, and proline residues, for example, may be involved in forming kinks in the α-helices to form them into a binding pocket for the agonist. The amino terminus is located extracellularly, while the carboxyl terminus is exposed on the cytoplasmic face of the cell membrane. This arrangement results in three extracellular hydrophilic loops and three intracellular loops. The third cytoplasmic loop between TM5 and TM6 is longer (Figure 13.1 and 13.2).

Based on the electron diffraction projection map of bacteriorhodopsin, the α-helices are arranged around a central pocket. The arrangement of bovine rhodopsin, which unlike bacteriorhodopsin is coupled to a G protein termed transducin, is also around a central pocket. The orientation of the α-helices, however, differs slightly, possibly as a result of the differences in the extraction and crystallization prior to electron crystallography of the bovine rhodopsin. The topography of the other G protein-coupled receptors remains to be determined experimentally but is currently modelled on that of bovine rhodopsin (Hoflack *et al.* 1994). Whether the α-helices are arranged clockwise or anticlockwise is unclear. In the model shown in Figure 13.2 they are presented in anticlockwise order after Kaumann (1991). A model predicted in association with this author from the calculation of bond forces between amino acid groups projecting into the receptor pocket also favours an anticlockwise orientation (Timms *et al.* 1992). Strosberg (1992), however, has presented a clockwise arrangement.

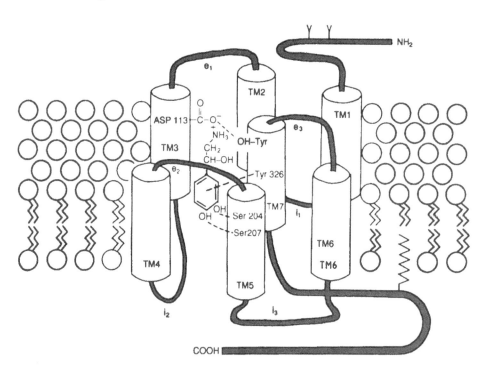

Figure 13.2 Three-dimensional model of the β_2-adrenoceptor. The seven transmembrane helices (TM1 to TM7) are shown as cylinders arranged anticlockwise in the cell membrane to form a pocket for the binding of noradrenaline. The positively charged amino group of noradrenaline interacts with the negatively charged Asp-113 on TM3 assisted by a tyrosine on TM7. The catechol hydroxyl groups of noradrenaline form hydrogen bonds with two serine amino acids (Ser-204 and Ser-207) of TM5.

Single cysteine residues that are thought to form disulphide bonds to stabilize the receptor structure are conserved among most of the receptors in the first two extracellular loops (Figure 13.1 and 13.2). The extracellular amino terminals vary considerably in length, from as few as seven amino acids in the adenosine A_2 receptor to 68 in the human cloned m_3 receptor. The amino terminal chain usually contains asparagine groups for N-glycosylation in a sequence, Asn-X-Ser/Thr. a_2-, β_1- and β_2-Adrenoceptors have several glycosylation sites, but they are absent in a_{2B} and adenosine receptors. Glycosylation involves the addition of a carbohydrate chain, which is always on the outside of the cell membrane and linked to a serine, threonine or asparagine amino acid, usually through N-acetylglucosamine or N-acetylgalactosamine.

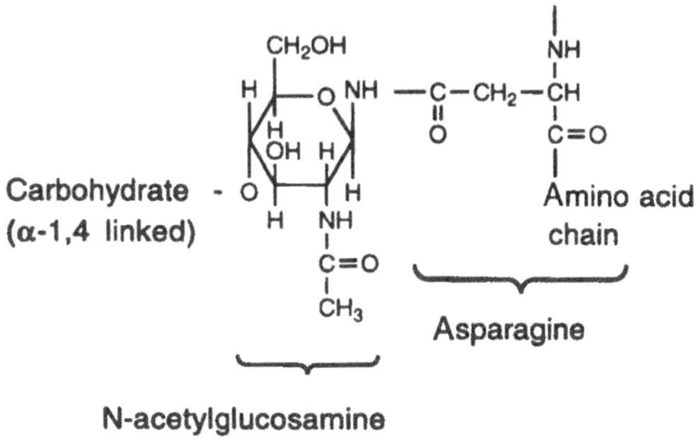

Carbohydrate - (α-1,4 linked)

N-acetylglucosamine

Asparagine

Amino acid chain

These carbohydrate residues are highly hydrophilic, allow for branching, and appear to assist in the orientation of the receptor glycoprotein in the cell membrane. They also appear to allow proper expression of the receptor; their deletion diminishes the level of receptor expression in the cell membrane but does not affect ligand binding (Harrison *et al.* 1991, Probst *et al.* 1992).

Another consistent feature is the presence of a cysteine residue (Cys-341) in the cytosolic carboxyl terminal, which is palmitoylated. In the case of the β-adrenoceptor, its substitution by glycine reduces agonist stimulation of adenylyl cyclase. The fatty acid chain of this palmitate group is expected to insert into the cell membrane, thereby anchoring the intracellular terminal, possibly for improved G protein coupling (Figure 13.1) (Probst *et al.* 1992).

The a-helices of the transmembrane domains contain 3.6 amino acid residues per helical turn. The arrangements of the residues in each a-helix have been predicted by computer-assisted modelling. These models predict that one side of the transmembrane domain has predominantly hydrophobic residues with hydrophilic residues on the other. The hydrophilic side of each domain faces inwards to form the pocket for polar ligand binding (Figure 13.2).

13.2.1 *Agonist Binding Sites*

The hydrophobic extracellular and intracellular domains are not directly involved in ligand binding, which occurs in the pocket formed by the transmembrane domains. In the a- and β-adrenoceptors, and dopamine and muscarinic receptors, a conserved aspartate in TM3 is crucial for agonist and antagonist binding. In the β-adrenoceptor site-directed mutagenesis studies suggest the importance of Asp-113 (Strader *et al.* 1988), while in the muscarinic receptor the distal aspartate of TM3 (Asp-105) has been implicated (Hulme *et al.* 1990, Caulfield 1993), and in dopamine D_1 and D_2 receptors, it is Asp-103 and Asp-114 (Civelli *et al.* 1991). More recently, Asp-147 on helix 3 has been proposed as the primary site of interaction of the quaternary head of Ach (Wess 1993). The carboxyl group of this aspartate residue is thought to form an ester linkage with the cationic head of the binding ligands. The aspartate residue of TM3 may form a cross-linkage with a tyrosine in TM7, by hydrogen bonding. Disruption of this hydrogen bond by the charged headgroup of Ach or noradrenaline could trigger a conformational change of the receptor. Thus, hydrogen bonding between Tyr and Asp may stabilize the carboxylate anion in the binding site. It may also polarize the tyrosine hydroxyl group, allowing a negative charge to develop on the oxygen, which could also be a site of interaction with the ligand cationic head. A conserved aromatic residue has also been identified in TM7, such as Tyr-326 in the β-adrenoceptor, which may interact with the aryl ring of adrenoceptor ligands.

Chimaeric receptor combinations of segments of the human β_2- and human platelet α_2-adrenoceptor have been produced by recombinant DNA technology. The fragments of DNA encoding the desired portion of each receptor were produced by means of specific restriction endonucleases, followed by splicing to produce chimaeric DNAs. The receptors may then be expressed in *Xenopus laevis* toad oocytes from the mRNA transcribed from injection of the recombinant DNA. The characteristics of receptors (natural wild-type or chimaeric) expressed in oocytes can be determined from radioligand binding with [^{125}I]cyanopindolol (β) and [^{3}H]yohimbine (α_2) and, in the case of β_2-adrenoceptors, also by stimulation of adenylyl cyclase. Exchange of TM7 in the β_2-adrenoceptor with the TM7 from the α_2-adrenoceptor caused a loss of β-adrenoceptor binding, but the appearance of weak α_2-adrenoceptor binding. Isoprenaline also became less potent than adrenaline at stimulating adenylyl cyclase (Kobilka *et al.* 1988). TM3 and TM7 therefore appear to be closely aligned around the receptor pocket and involved in the binding of the cationic head of agonist and antagonist ligands.

TM5 contains serine residues (Ser-204 and Ser-207) which may be involved in forming hydrogen bonds with the *meta* and *para* hydroxyl groups of adrenoceptor agonists at the β-adrenoceptor. Ser-204 of the α_2-adrenoceptor (equivalent to 207 in β) binds to the *para* hydroxyl group of adrenoceptor agonists and a corresponding serine residue appears in TM5 of all dopamine receptors (D_1, Ser-199 and D_2, Ser 202) (Civelli *et al.* 1991). In muscarinic receptors, a conserved asparagine in TM6, not found in other receptor subclasses, may be involved in interaction with the ester grouping of Ach (Probst *et al.* 1992).

13.2.2 Intracellular Coupling

13.2.2.1 G proteins

The cell surface receptors are coupled to various intracellular enzymes, ion channels or transporters by means of G proteins (Dolphin 1987). These are heterotrimeric proteins consisting of three individual protein units, each having separate genetic origins. They are composed of α-, β- and γ-subunits; the β- and γ-subunits generally do not vary between the different G proteins, but the α-subunits are variable (eg a_s, a_i and a_o) between G proteins (G_s, G_i and G_o). The G proteins are associated with their receptor at the intracellular face of the plasma membrane. The binding of an agonist to the receptor catalyses the replacement of GDP bound to the α-subunit of the G protein by GTP (Figure 13.3 and 13.4). This results in activation of the G protein and cleavage of the α-subunit which stimulates (or inhibits) the activity of the specific enzyme or channel that links receptor to response. The return of the α-subunit to the ground state and reassociation with the βγ complex is brought about by the intrinsic GTPase activity of the α-subunit. This converts the GTP-bound α-subunit back to the GDP-bound form, with liberation of P_i (see Figure 13.3 for adenylyl cyclase-linked receptors).

G proteins linking receptors to adenylyl cyclase are either stimulatory (G_s) or inhibitors (G_i). The α-subunit of G_s (G_{sa}) is the target for ADP-ribosylation by the ribosyl-transferase activity of cholera toxin, the diarrhoea-causing toxin from the bacterium,

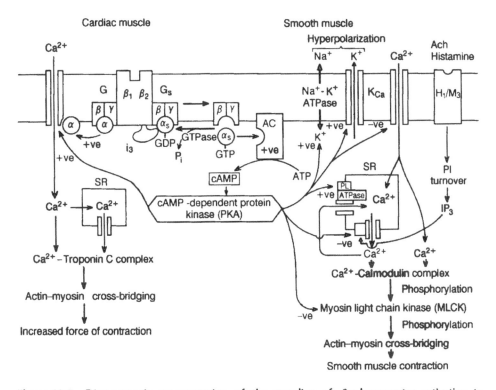

Figure 13.3 Diagrammatic representation of the coupling of β-adrenoceptor activation to increases in force of cardiac contraction (left side) and smooth muscle relaxation (right side). β-Adrenoceptor stimulation by agonist releases the GTP-bound a_s subunit of the guanine nucleotide binding protein (G_s) which activates adenylyl cyclase (AC). ATP is converted to cAMP which activates cAMP-dependent protein kinase (PKA). The activated catalytic subunits of PKA then phosphorylate a range of effector proteins to bring about cardiac stimulation (β_1) or smooth muscle relaxation (β_2). The primary target for cardiac contractility is the L-type Ca^{2+} channel allowing Ca^{2+} influx. Multiple targets are shown for inhibition of smooth muscle contractile tone induced by Ca^{2+} influx and by histamine or Ach via H_1 and M_3 receptors through increased phosphatidylinositol (PI) turnover stimulating release of Ca^{2+} from the sarcoplasmic reticulum (SR). These sites are described in the text. PL, Phospholamban.

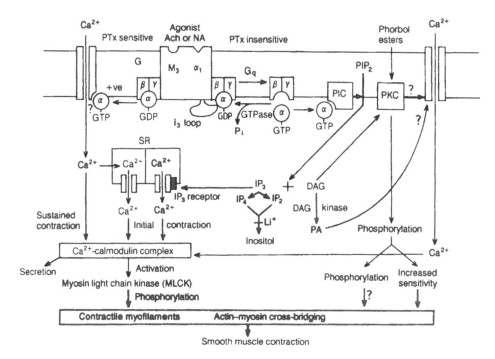

Figure 13.4 Diagrammatic representation of the coupling of M_3 muscarinic receptor or α_1-adrenoceptor stimulation by Ach or noradrenaline (NA) to smooth muscle contraction and glandular secretion. Agonist occupation of the receptor releases GTP-bound α-subunit of the guanine nucleotide binding protein (G_q). This activates phosphoinosidase C (PIC) which hydrolyses phosphatidylinositol 4,5-bisphosphate (PIP$_2$) to inositol 1,4,5-triphosphate (IP$_3$) and 1,2-diacylglycerol (DAG). IP$_3$ causes initial contraction of smooth muscle by release of Ca^{2+} from the sarcoplasmic reticulum (SR). DAG activates protein kinase C (PKC) which may contribute to the contraction by phosphorylation and opening Ca^{2+} channels. DAG is metabolized to phosphatidic acid (PA). Direct coupling of the receptor to Ca^{2+} channels via a pertussis toxin (PTx)-sensitive G protein may be responsible for the sustained contraction.

Vibrio cholerae. The toxin enters cells by interacting with a G_{M1} ganglioside on the cell surface. Once in the cell, the A_1 peptide unit of the toxin catalyses the transfer of an ADP-ribose unit from NAD^+ to an arginine residue on the G protein (G_{sa}).

This modification causes inhibition of the GTPase activity of G_s and hence maintains the G protein in a persistently activated GTP-bound state. Adenylyl cyclase remains active and cAMP levels increase. After long-term activation, however, adenylyl cyclase activity returns to normal, probably because of down-regulation of the $G_{s\alpha}$. The diarrhoea is caused by accumulation of cAMP in the intestinal epithelial cells causing excessive efflux of Na^+ and water into the gut.

The α-subunits of the G_i binding proteins are substrates for ADP-ribosylation catalysed by pertussis toxin (PTx). This toxin, also known as islet-activating protein, is produced by *Bordetella pertussis*, the causative organism of whooping cough. PTx blocks the receptor-mediated activation of the G_i protein and subsequent pharmacological responses mediated via inhibition of adenylyl cyclase and reduction of cAMP levels. For example, animals are pretreated with PTx (20 μg/kg iv) before removal of isolated tissues, or isolated tissues are incubated with PTx (1 μg/ml for 2 hr). This blocks the inhibitory effects of Ach mediated via M_2 receptors and of adenosine via A_1 receptors on left atrial tension or right atrial rate (Böhm *et al.* 1989b, Breslin & Docherty 1993). The effects of PTx on α_2-adrenoceptor-mediated responses linked to adenylyl cyclase are inconsistent. Generally, the α_2-mediated inhibitory effects on noradrenaline release from a range of peripheral tissues have not been prevented by PTx (Rump *et al.* 1992). The α_2-adrenoceptor-mediated endothelium-dependent relaxation of certain blood vessels are blocked, but the M_2 muscarinic receptor-mediated vasodilatation in rat aorta is not (Breslin & Docherty 1993).

The cardiac M_2 muscarinic receptor is linked to inhibition of adenylyl cyclase via the G_i regulatory protein and thereby exerts inhibition of cAMP-dependent positive inotropic and chronotropic responses, for example, the antiadrenergic effect (see Chapter 3). The M_2 muscarinic and adenosine A_1 receptors of atrial tissues also link directly to K^+ channels via a G protein (G_K or G_o) which is also PTx-sensitive (Hosey 1994). Activation of this G protein results in K^+ channel opening, efflux of K^+ and direct negative inotropic and chronotropic responses (Chapter 8, Urquhart *et al.* 1993). The muscarinic receptor-mediated *increase* in atrial contractility, in contrast, is not inhibited by PTx (Kenakin & Boselli 1990).

The G proteins linking receptors to phosphoinositide hydrolysis (G_q) are not PTx-sensitive. Thus, the M_3 receptor-mediated contractions of guinea-pig ileum, trachea, urinary bladder and oesophageal muscularis mucosae are not blocked by PTx pretreatment (Eglen *et al.* 1988a, 1994). The contractile responses of gastrointestinal smooth muscle are also mediated via muscarinic receptors coupling via G proteins to increases in cGMP and directly to inward Ca^{2+} channels, the latter by a PTx-sensitive G protein (Milligan & Green 1994). The $\beta\gamma$-subunit of the G protein probably has no physiological role in receptor coupling, although high concentrations have been shown to open K^+ channels and to activate phospholipase A_2 to produce arachidonic acid and lipoxygenase breakdown products (Hulme *et al.* 1990).

The question has been raised whether antagonists can have opposite inhibitory effects on G protein coupling without there being any agonist-induced coupling (Caulfield 1993). In the case of atropine at muscarinic receptors, there is some evidence to support this idea with the antagonist causing dissociation of the receptor–G protein complex in the absence of agonist. For this to be possible, G protein activation by muscarinic receptors would have to occur without agonist interaction. If the spontaneous formation of receptor–G protein complexes occurs, it is probably due to conformational changes in the receptor to an activated state. The presence of an agonist may then increase the population of this activated receptor (known as conformational selection). Alternatively, activation by the agonist may involve further conformational changes in the receptor which can only occur in the presence of an agonist; this is known as conformational induction. It is not clear whether an activated receptor of one type only couples to one G protein or whether it may couple to other G protein types. It is possible that when there is an overexpression of receptors then activation may spread to other G proteins (Kenakin 1995a). Another theoretical possibility is that several conformational forms of activated receptor may exist and that agonists may show structural selectivity in activating different forms. This may be a basis for the efficacy differences between agonists (Kenakin 1995b).

G proteins may be modified by several other agents. For example, N-ethylmaleimide (NEM) is an irreversible alkylating agent selective for sulphydryl groups which inactivates G_i. Incubation of tissues with this agent has been shown to impair α_2-adrenoceptor-mediated responses including the inhibition of noradrenaline release from field-stimulated sympathetic nerve endings (Rump *et al.* 1992). Earlier studies had shown that at low concentrations (0.1 mM) NEM could reduce β-adrenoceptor binding sites, but only when preincubated together with an agonist. This suggested that the agonist induced conformational changes in the receptor to expose sulphydryl groups for alkylation. Higher concentrations (1–10 mM) could prevent β-adrenoceptor coupling to adenylyl cyclase by alkylating the G protein (Andre *et al.* 1982). NEM was also shown to protect β-adrenoceptors against agonist-induced desensitization (Stadel & Lefkowitz 1979).

Protein kinase C (PKC) is activated by diacylglycerol, a product of phosphoinositide hydrolysis following muscarinic M_3 receptor stimulation and by phorbol esters (see later, Figure 13.4). PKC causes phosphorylation of several protein substrates and may *inactivate* G_i regulatory protein by phosphorylation, possibly in its GDP-bound state. Phorbol esters decrease the rate of activation of G_i by GTP (Houslay 1994). M_1 muscarinic receptor stimulation of PKC via the phosphoinositide pathway probably explains the enhancement of noradrenaline release from sympathetic nerves; it inhibits the G_i protein coupling to α_2-adrenoceptor-mediated inhibition of noradrenaline release (Costa *et al.* 1993). Inhibitors of PKC, such as polymyxin B, inhibit release of noradrenaline induced by high frequency sympathetic nerve stimulation (Musgrave & Majewski 1991, see also Chapter 6).

13.2.2.2 Receptor requirements for G protein coupling

Coupling of the receptor to the G protein occurs through identifiable regions of the receptor which have been determined by site-directed mutagenesis and deletion studies, whereby amino acid residues have been substituted or deleted, respectively, from the receptor. All of the intracellular domains are implicated in efficient G protein coupling, in particular the third cytoplasmic loop (i_3) and the carboxyl terminus. Examples include mutation of the highly conserved Asp-130, adjacent to TM3 in the second intracellular loop of the β-adrenoceptor, m_1 muscarinic receptors and α_{2A}-adrenoceptors, which produces receptors with high affinity ligand binding but with reduced or absent G protein coupling. Asp-79 in TM2 is highly conserved in nearly all G protein-coupled receptors and is necessary for agonist binding and for G protein activation.

Deletion experiments indicate that removal of a seven amino acid sequence in the third intracellular loop proximal to TM7 caused marked reduction in coupling of α_1-adrenoceptors to phospholipase C and phosphoinositide hydrolysis (Cotecchia *et al.* 1990). Deletion of carboxy terminal residues adjacent to TM7 also produced β-adrenoceptors with reduced ability to activate G proteins. Chimaeric receptors have been formed as hybrids of α_2- and β_2-adrenoceptors which are negatively and positively coupled to adenylyl cyclase through G_i and G_s proteins, respectively. Replacing the third intracellular loop (i_3) of the β_2-adrenoceptor with that of the α_2-adrenoceptor resulted in *inhibition* of adenylyl cyclase instead of stimulation (Kobilka *et al.* 1988). Similarly, substitution of the third intracellular loop of the D_2 dopamine receptor (negatively coupled to adenylyl cyclase) with that of the m_1 muscarinic receptor, which is linked to phospholipase C, resulted in chimaeras that hydrolysed phosphatidylinositol and mobilized Ca^{2+} (England *et al.* 1991). Adrenoceptor coupling is associated with a segment of the third intracellular loop adjacent to TM6, whereas sequences adjacent to TM5 control coupling selectivity between m_1 and m_2 muscarinic receptors. The first 16–21 amino acids of the third intracellular loop determine the coupling of m_2 and m_3 muscarinic receptors. Normally, m_2 gives modest phosphoinositide hydrolysis which is abolished by PTx (normally linked to adenylyl cyclase), whereas m_3 gives the expected robust phosphoinositide response that is insensitive to PTx. Exchange of the first 16–21 amino acids of i_3 reverses this profile. Additional substitution of the second intracellular loop (i_2) is necessary to abolish phosphoinositide turnover. The i_2 domain may not distinguish between the adenylyl cyclase and phosphoinositide responses of muscarinic receptors but for comparison between β-adrenoceptors and muscarinic receptors, it does come into play (Bonner 1992).

The G proteins linking receptors to their effectors are shown in Table 13.1. The effects of agonist activity at the autonomic receptors are mediated through two second messenger systems: the adenylyl cyclase system and the phosphoinositide system.

13.3   Receptors Linked to Phosphoinositide Hydrolysis

Stimulation of muscarinic M_1 and M_3 receptors, α_1-adrenoceptors and several other receptor types causes the G protein-mediated hydrolysis of the membrane phospholipid, phosphatidylinositol 4,5-bisphosphate (PIP_2) to inositol 1,4,5-triphosphate [$Ins(1,4,5)P_3$ or IP_3] and a 1,2-diacylglycerol (DAG) (Figure 13.4). This is catalysed by the enzyme, phosphoinosidase C (PIC) [also known as phospholipase C (PLC)], which is coupled to the receptor by G proteins of the G_q family (formerly termed G_p). These G proteins are not modified by pretreatment with either pertussis or cholera toxins. G_q and the related G_{11} probably activate a specific isoform of PIC, PIC-β. It is not yet clearly established that the ternary complex model of the G protein ($\alpha\beta\gamma$) applies to the PIC-linked receptors. PIC appears to hydrolyse all phosphoinositides, so that it can release IP_1 from phosphatidylinositide (PI), IP_2 from phosphatidylinositol 4-phosphate (PIP) and IP_3 from PIP_2 (Figure 13.4). IP_3 is a unique product of PIP_2 hydrolysis, whereas DAG can be derived from hydrolysis of other inositol or non-inositol glycerophosphates. Thus, monitoring total inositol phosphate accumulation as a measure of receptor activation of PIC may not be reliable for determining the nature of the primary transmembrane signalling event.

Muscarinic receptors activate phospholipases A_2 (PLA_2) and D (PLD) (Hosey 1994); the sites of their action on phospholipids are shown in Figure 13.5. They hydrolyse the release of fatty acids, PLA_2 in particular causing the release of arachidonic acid from phosphatidylcholine and phosphatidylethanolamine. Arachidonic acid is rapidly metabolized to oxygenated products by cyclo-oxygenase to prostaglandins (PGE_2, PGD_2 and $PGF_{2\alpha}$) and, via thromboxane synthase, to thromboxanes (TxB_2). Thus, muscarinic agonists may exert some of their pharmacological effects through increased production of prostaglandins. In the heart, ACh-induced increases in prostaglandin synthesis are mediated via M_2 receptors, being antagonized by methoctramine (Jaiswal & Malik 1991). Phospholipase A_2 is also activated by Ca^{2+} influx into the cell induced by cell injury and by calmodulin, which explains the release of arachidonic acid products during inflammatory processes. DAG, resulting from PLC activation, can also yield arachidonic acid by the action of diglyceride lipase and monoglyceride lipase.

IP_3 has been clearly identified as the second messenger for receptor-mediated release of intracellular Ca^{2+} from the sarcoplasmic reticulum (SR). Ca^{2+} is not released from the mitochondrial pool. The target is an IP_3 receptor which forms the Ca^{2+} channel spanning the membrane of the intracellular SR Ca^{2+} storage site. At least three separate IP_3 receptor genes have now been identified. Both IP_3-sensitive and IP_3-insensitive stores have been identified. The release of intracellular Ca^{2+} then initiates contractile responses of smooth muscle arising from α_1- or M_3 receptor stimulation and secretory responses such as enzyme secretion from the parotid salivary gland. The emptying of the intracellular Ca^{2+} pool by IP_3 may also trigger the influx of Ca^{2+} through sarcolemma membrane Ca^{2+} channels, as has been described recently for mast cells and other cell lines. The process of Ca^{2+} influx is said to be via a Ca^{2+}-release-activated channel (CRAC) and the current that flows through this channel is I_{CRAC}. Ca^{2+} influx via the CRAC does not necessarily require receptor occupancy, but can be triggered by other drugs or procedures that in common empty the intracellular Ca^{2+} stores. For example, the Ca^{2+} transporting ATPase inhibitor, thapsigargin, empties Ca^{2+} stores and causes elevated intracellular Ca^{2+} levels that are dependent upon extracellular Ca^{2+}. The possible link between the emptying of the SR by IP_3 receptor activation and Ca^{2+} influx via CRAC is a Ca^{2+} influx factor (CIF) released from the SR (Fasolato et al. 1994). The intracellular pool is subsequently replenished from the Ca^{2+} entering via Ca^{2+} influx. In addition to the IP_3 receptor mediating Ca^{2+} release from the SR, ryanodine receptors are also present. Ryanodine activates the release of Ca^{2+} from the intracellular stores and at higher concentrations ($>10\ \mu M$) inhibits the release. Caffeine also activates this receptor and induces smooth muscle contraction (Ehrlich et al. 1994).

Metabolism of IP_3 by ($InsP_3$)-5-phosphatase to produce $Ins(1,4)P_2$ (IP_2) is the major route of inactivation since IP_2 is not able to release Ca^{2+} from intracellular stores. A secondary route via a 3-kinase to IP_4 also occurs and, although this is less clearly understood, IP_4 may also be involved in regulation of Ca^{2+} influx or release (Figure 13.5). These products are in turn converted to inositol which is conserved by recycling back to phosphoinositides (Wojcikiewicz et al. 1994).

The other product of PIP_2 hydrolysis is DAG, which is known to activate PKC. This enzyme causes phosphorylation of a range of cellular proteins. PKC is also actively stimulated by the tumour-promoting phorbol esters, such as 12-O-tetradecanoylphorbol-13-acetate (TPA) or phorbol-12,13-dibutyrate, which are

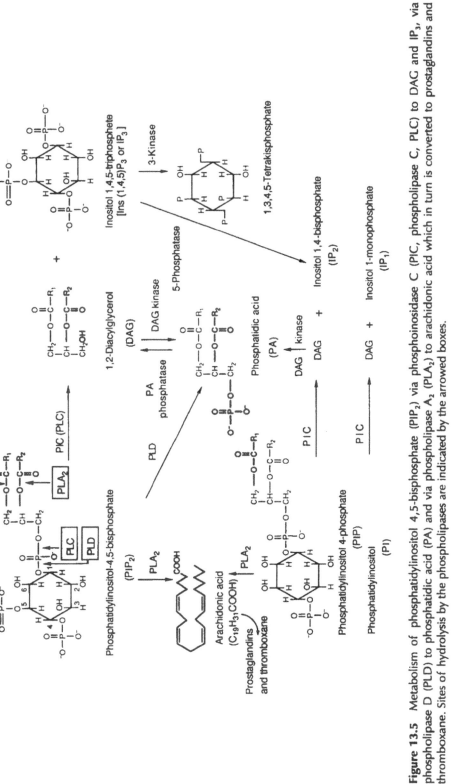

Figure 13.5 Metabolism of phosphatidylinositol 4,5-bisphosphate (PIP$_2$) via phosphoinosidase C (PIC, phospholipase C, PLC) to DAG and IP$_3$, via phospholipase D (PLD) to phosphatidic acid (PA) and via phospholipase A$_2$ (PLA$_2$) to arachidonic acid which in turn is converted to prostaglandins and thromboxane. Sites of hydrolysis by the phospholipases are indicated by the arrowed boxes.

capable of inducing slow contractions of smooth muscle (Urquhart & Broadley 1991). There are several isoforms of the PKC family of enzymes, of which PKC-β_I and PKC-β_{II} are Ca^{2+}-dependent. Ca^{2+} binds to the enzyme causing translocation of the enzyme to the plasma membrane, where it is activated following binding of DAG. The effect of DAG is only transient, since it is rapidly metabolized by both DAG kinase, which converts it to phosphatidic acid, and by DAG lipase. Phorbol esters bind directly to PKC with greater affinity than DAG and are more persistent, the contractile effects or elevation of intracellular Ca^{2+} peaking at >60 min with TPA (Wilkinson & Hallam 1994).

The role of DAG and activation of PKC in the contractile responses of smooth muscle to α_1-adrenoceptor and muscarinic M_3 receptor stimulation is less certain than for IP_3. To understand the possible effects of PKC activation, it is necessary to review the role of Ca^{2+} in smooth muscle contraction. An elevation of cytoplasmic free Ca^{2+} causes activation of the smooth muscle myofilaments involving actomyosin cross-bridge cycling (see Chapter 1). The cytoplasmic Ca^{2+} is believed to complex with the intracellular protein, calmodulin. This complex then activates myosin light chain kinase (MLCK), which in turn phosphorylates myosin light chain, the degree of which controls the actin–myosin cross-linkage and thus the smooth muscle contraction (Timmermans & Thoolen 1987). Intracellular Ca^{2+} levels are raised by influx of extracellular Ca^{2+} through voltage-operated (VOCs) or receptor-operated (ROCs) Ca^{2+} channels and by release of intracellular Ca^{2+} stored in the SR. The release of Ca^{2+} from the SR is induced by opening of IP_3-sensitive channels (see above) and through Ca^{2+}-activated channels that open when the cytoplasmic Ca^{2+} levels are raised, for example, by Ca^{2+} influx (Figure 13.4). The return of Ca^{2+} to the SR is driven by Ca^{2+},Mg^{2+}-ATPase which appears to be linked to the SR protein, phospholamban. Phosphorylation of phospholamban by elevated levels of cAMP or cGMP increases Ca^{2+},Mg^{2+}-ATPase activity and removal of Ca^{2+} from the cytoplasm. This, in part, explains the relaxation of smooth muscle by agents that raise levels of these two cyclic nucleotides (β_2 agonists and NO, respectively) (Giembycz & Raeburn 1991) (see later).

Agonist-induced Ca^{2+} influx via ROCs is more effective in raising intracellular cytoplasmic $[Ca^{2+}]_i$ than is Ca^{2+} influx via VOCs induced by depolarization through raising extracellular K^+. Agonists are also more effective contractile agents for an equivalent level of $[Ca^{2+}]_i$ because they enhance the sensitivity of the contractile myofilaments to Ca^{2+}. PKC activation by the agonist through DAG appears to have a role in this elevation of myofilament sensitivity. The contraction of rat aorta by agonists is associated with sensitization of myofilaments to Ca^{2+} through both PKC-dependent and PKC-independent pathways (Hori *et al.* 1993). The phorbol esters cause sustained smooth muscle contractions, but whether these contractions are dependent upon extracellular Ca^{2+} influx varies with the vessel and the species. The majority of evidence points to the contraction being at least partially dependent upon Ca^{2+} influx. The consequent increase in $[Ca^{2+}]_i$ appears to cause contraction by activation of MLCK through a Ca^{2+}–calmodulin complex, since the contractions of human pulmonary artery, at least, are inhibited by the calmodulin antagonist, trifluoperazine (see also Chapter 2) (Savineau *et al.* 1991). Phorbol esters also enhance the sensitivity of the contractile proteins by stimulating PKC.

The role of Ca^{2+} influx in smooth muscle contractile responses to α_1-adrenoceptor stimulation has been extensively examined by use of L-type Ca^{2+} channel antagonists. The results appear to depend upon the species, type of blood vessel and the agonist under study. In some cases the dihydropyridine L-type Ca^{2+} channel antagonist, such as nifedipine, reduces the contraction, whereas in other cases (eg rabbit aorta) the responses are quite resistant. The variability between agonists is not generally thought to be due to differences in receptor reserve but to utilization of the different Ca^{2+} processes between different agonists. Where inhibition of the response by a Ca^{2+} channel blocking agent occurs, it is thought to be due to blockade of ROCs. Ca^{2+} influx via VOCs may have a minor contribution to the contraction, as indicated by membrane depolarization associated with the contraction only under certain conditions, usually at higher agonist concentrations. The depolarization would then open VOCs. However, Ca^{2+} influx may occur without depolarization, indicating the involvement of ROCs (Timmermans & Thoolen 1987, Ruffolo *et al.* 1991).

The sequence of events in smooth muscle contraction induced by α_1-adrenoceptor stimulation therefore appears to be activation of the G protein (G_q), stimulation of PIC and release of IP_3 which causes mobilization of Ca^{2+} from the SR. The simultaneous release of DAG activates PKC which increases myofilament sensitivity to Ca^{2+} and may also be involved in regulation of Ca^{2+} influx through ROCs and a small elevation of myosin phosphorylation. PKC, however, generally phosphorylates substrates distinct from calcium-calmodulin-dependent kinases (Putney 1987) (Figure 13.4). It is of interest that G protein activation by the stable analogue, guanosine 5'-O-(3-thiotriphosphate) (GTP-γ-S), in skinned smooth muscle causes a tonic contraction in the presence of normal extracellular Ca^{2+}, which can be blocked by the PKC inhibitor, H7. GTP-γ-S also raises the sensitivity to noradrenaline-induced contractions (Nishimura *et al.* 1989). Thus, in contractions of vascular and tracheal smooth muscle via α_1 or M_3 muscarinic receptors, respectively, IP_3 generation appears to be involved in initiation of tone, whereas DAG accumulation and activation of PKC is involved in maintenance of smooth muscle tone (Giembycz & Raeburn 1991).

A further possibility is that the contraction of smooth muscle mediated via Ca^{2+} influx is due not to the DAG, but to the phosphatidic acid produced by DAG breakdown via DAG kinase (Figure 13.5). Phosphatidic acid has been recognized as a Ca^{2+} ionophore in many cell types. In support of this role, the DAG kinase inhibitor, R59022, impairs both the accumulation of phosphatidic acid and contractile responses of the taenia coli to muscarinic receptor stimulation (Chata *et al.* 1991). The possibility that phosphatidic acid is merely acting as a PLC activator should not, however, be discounted (Jackowski & Rock 1986).

The precise linking of the receptor activation through the G protein to Ca^{2+} influx via Ca^{2+} channels (ROCs) remains uncertain. In addition to the effects of DAG described above, it is possible that the receptor may be directly linked to the Ca^{2+} channel by a separate G protein. No second messengers would be involved in this type of ROC. In certain cells, Ca^{2+} influx is sensitive to pertussis and cholera toxins, whereas IP_3 generation and intracellular Ca^{2+} release are not. A further possibility is that the intracellular Ca^{2+} release by IP_3 triggers the influx of Ca^{2+}. Evidence against this, however, is that blockade of the IP_3 receptor by microinjection of heparin into smooth muscle cells leaves the influx of Ca^{2+} by muscarinic agonist intact. Finally, the second messenger itself, IP_3, could open the Ca^{2+} channel (Meldolesi *et al.* 1991, Fasolato *et al.* 1994).

The contraction of vascular smooth muscle by α_1-adrenoceptor stimulation is typically biphasic; a fast phasic component, usually completed in <30 sec, is followed by a slower sustained tonic response lasting 5–10 min. These two components can be related to the two processes of Ca^{2+} involvement in the contractile response outlined above. The early phasic contraction occurs in the absence of extracellular Ca^{2+}, but is a poorly maintained response. This is blocked by ryanodine, which releases Ca^{2+} from the SR. The response is therefore due to an immediate release of Ca^{2+} from the intracellular stores. The secondary tonic contraction is blocked in most vessels by the L-type Ca^{2+} channel blockers and is therefore due to the influx of extracellular Ca^{2+} (Timmermans & Thoolen 1987). The sustained contraction is eliminated by exclusion of Ca^{2+} from the bathing medium and the presence of 2 mM EGTA [ethylene glycol *bis*-(beta-aminoethylether) N,N,N,N-tetra-acetic acid] to chelate any residual Ca^{2+} on the extracellular cell surface (Rinaldi *et al.* 1991). The contractions of vascular smooth muscle mediated via a_2-adrenoceptors are also susceptible to Ca^{2+} channel blockade but, as will be seen later, they are negatively coupled to adenylyl cyclase and not to phosphoinositide turnover.

It has been suggested that the subtypes of a_1-adrenoceptor may be linked separately to the two Ca^{2+} mobilization processes. The a_{1A}-adrenoceptor-mediated contractions of the rat renal artery, rat portal vein and vas deferens are blocked by 5-methylurapidil and by the Ca^{2+} channel inhibitor, nifedipine. This suggests the involvement of extracellular Ca^{2+} influx through Ca^{2+} channels (Nichols & Ruffolo 1991; Blue *et al.* 1992). In contrast, the a_{1B}-adrenoceptor-mediated contractions of

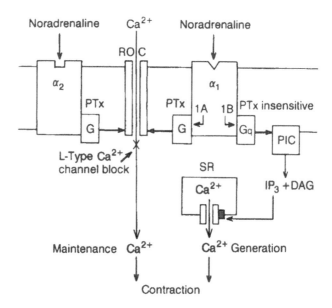

Figure 13.6 Summary of the coupling of a_1- and a_2-adrenoceptor coupling to Ca^{2+} mobilization in smooth muscle contraction. The a_{1B}-adrenoceptor is linked via a pertussis toxin (PTx)-insensitive guanine nucleotide regulatory protein (G_q) to activation of phosphoinosidase C (PIC) and increased production of inositol 1,4,5-triphosphate (IP_3) and 1,2-diacylglycerol (DAG). IP_3 causes Ca^{2+} release from the sarcoplasmic reticulum (SR) and generation of the muscle contraction. The a_{1A}- and a_2-adrenoceptors are linked via PTx-sensitive G proteins directly to the receptor operated Ca^{2+} channel (ROC), influx of Ca^{2+} causing maintenance of the contraction which is sensitive to L-type Ca^{2+} channel blockade.

492

the rat spleen and rat aorta are resistant to nifedipine blockade and are therefore assumed to be due to release of intracellular Ca^{2+} (Han *et al.* 1990, Minneman *et al.* 1991, Ruffolo *et al.* 1991) (see also Chapter 3 on α_1-adrenoceptor subtypes). The rat portal vein displays phasic myogenic activity and noradrenaline causes increased magnitude of these contractions and an increase in tone. Both responses are due to α_{1A}-adrenoceptor stimulation (Schwietert *et al.* 1991). The myogenic activity is due to activation of VOCs and spontaneous depolarizations and, like the increased tone, is prevented by Ca^{2+} antagonists (Timmermans & Thoolen 1987) (Figure 13.6). The activation of both α_{1A}- and α_{1B}-adrenoceptors appears to be linked to phosphoinositide turnover. In the heart, the increased cardiac contractility and action potential duration due to α_{1A}-adrenoceptor stimulation is also associated with increased levels of IP_3, which is not susceptible to PTx or the α_{1B}-adrenoceptor antagonist, chloroethylclonidine (CEC). There is also activation of PKC and the effect is probably due to an increase in sensitivity of the cardiac myofilaments to Ca^{2+} (Benfey 1993).

13.4 Receptors Linked to Adenylyl Cyclase

13.4.1 *Stimulation of Adenylyl Cyclase*

Most of our knowledge on receptors inducing stimulation of adenylyl cyclase has been obtained from studies on β-adrenoceptors. β-Adrenoceptors belong to a group of receptor types which are known to be coupled via a guanine nucleotide regulatory protein (G protein) to the adenylyl cyclase enzyme (EC 4.6.1.1) which generates the intracellular second messenger, cyclic adenosine 3',5'-monophosphate (cAMP). Receptors mediating their effects through adenylyl cyclase can be subdivided into two types: those which increase cAMP levels by stimulation of adenylyl cyclase by means of the stimulatory nucleotide regulatory protein (G_s) and those which decrease cAMP levels by inhibiting adenylyl cyclase through coupling to an inhibitory G protein (G_i). Thus, β_1-, β_2- and β_3-adrenoceptors, dopamine D_1 receptors, histamine H_2 receptors, A_2 purinoceptors and receptors for adrenocorticotrophic hormone (ACTH), thyroid stimulating hormone (TSH), follicle stimulating hormone (FSH) and vasoactive intestinal peptide (VIP) are all coupled to G_s and their occupation by specific agonists results in elevation of intracellular cAMP levels. In contrast, α_2-adrenoceptors, dopamine D_2 receptors, muscarinic M_2 and M_4 receptors, A_1 purinoceptors and receptors for somatostatin (SS_3 and SS_4), neuropeptide Y (Y_1) and μ and δ opioid receptors are coupled to G_i and inhibit cAMP production upon agonist occupation (Table 13.1) (Birnbaumer *et al.* 1990).

13.4.1.1 *cAMP as a second messenger*

The biochemical basis for the role of cAMP as the second messenger mediating the effects of β-adrenoceptor stimulation has its origins in the well known metabolic effect of adrenaline upon hepatic glycogenolysis: the conversion of glycogen to glucose-1-phosphate. This is a β_2-adrenoceptor-mediated biochemical process first described in detail by Sutherland (Sutherland & Rall 1960) (Figure 13.7). Adenylyl cyclase is coupled to the β-adrenoceptor by the G_s regulatory protein, activation of the enzyme being induced by agonist occupation. Adenylyl cyclase catalyses the

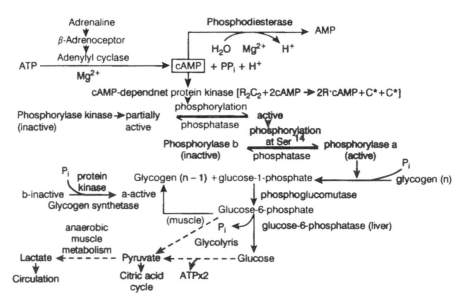

Figure 13.7 β-Adrenoceptor-mediated glycogenolysis cascade involving generation of second messenger cAMP. cAMP combines with the regulatory subunits (R$_2$) of cAMP-dependent protein kinase to release two active catalytic subunits (C*) of the enzyme.

convertion of adenosine triphosphate (ATP) to adenosine 3′,5′-monophosphate (cyclic AMP, cAMP) driven by the subsequent hydrolysis of the released pyrophosphate. Normally, cAMP is very stable but is destroyed by the presence of phosphodiesterases (PDE). PDEs convert cAMP to the inactive non-cyclic adenosine monophosphate and their pharmacological relevance will be considered in more detail later.

cAMP serves as a second messenger in many varied physiological events by activating cAMP-dependent protein kinase (PKA). This enzyme exists as a tetramer (R$_2$C$_2$) consisting of two regulatory subunits (R) and two catalytic subunits (C). Two molecules of cAMP bind to each regulatory subunit causing the dissociation of two activated catalytic subunits. These catalyse the phosphorylation of a wide range of proteins, and in glycogenolysis phosphorylase kinase is activated. This enzyme is also partially activated by a rise in intracellular Ca^{2+} which will accompany increases in cardiac muscle contractility induced by β-adrenoceptor stimulation and hence assist in energy provision. Activated phosphorylase kinase converts phosphorylase b (inactive) to phosphorylase a (active) by phosphorylation of a serine residue (Ser-14). This is the enzyme that controls the rate-limiting step in glycogenolysis, whereby glucose units are removed sequentially, as glucose-1-phosphate, from the glycogen polymer by phosphorolytic cleavage of the α-1,4 linkages. By the action of phosphoglucomutase, this in turn is converted to glucose-6-phosphate, which is utilized as a major energy source in muscles by incorporation into glycolysis pathways with the ultimate generation of ATP and pyruvate.

Cell membranes are impermeable to the phosphorylated glucoses, which are therefore not available as transportable energy sources. In the liver, however, glucose-6-phosphate is converted by glucose-6-phosphatase to glucose, which leaves the cell and increases blood sugar. This biochemical pathway therefore explains the hyperglycaemic response to adrenaline mediated via β_2-adrenoceptor stimulation (Figure 13.7). This cascade functions as an amplifying system, with each step having a reserve capacity. As a result, only a fraction of receptor occupancy can induce substantial generation of cAMP, activation of subsequent enzymes and the final response (see Figure 3.7, Chapter 3).

Sutherland first proposed that other β-adrenoceptor-mediated physiological responses could also be explained by this cascade and the stimulation of adenylyl cyclase to increase intracellular cAMP levels. He described the development of this now well-established concept in his Nobel Prize lecture (Sutherland 1972). At least six criteria were put forward to be satisfied for cAMP to be considered as the intracellular second messenger for the physiological effect of a hormone, drug or neurotransmitter (Robison *et al.* 1971).

1   The agonist under study should increase the cAMP levels in a time- and concentration-dependent manner consistent with the biological response. This has been clearly demonstrated to precede the pharmacological response in the heart and airways smooth muscle, both effects being prevented by β-adrenoceptor blockade.

2   The agonist should activate adenylyl cyclase in a cell-free system obtained from the tissue in question. In cardiac and lung membranes, catecholamines stimulate adenylyl cyclase in potency orders consistent with β_1- and β_2-adrenoceptors, respectively (Figure 3.2).

   Other agents are capable of stimulating adenylyl cyclase and in some cases also mimic the pharmacological response. Because these effects are not antagonized by β-adrenoceptor antagonists, however, the activation is independent of the β-adrenoceptor. An example is the diterpene, forskolin, obtained from the Indian herb, *Coleus forskohlii*. This compound is thought to stimulate the catalytic subunit of adenylyl cyclase directly and independently of either the regulatory G protein or the β-adrenoceptor (Seamon & Daly 1983). Forskolin can produce pharmacological effects identical to β-adrenoceptor agonists, including cardiac stimulation and relaxation of vascular, gastrointestinal and respiratory smooth muscle (Figure 13.8), which are associated with elevated levels of cAMP. These responses are not antagonized by β-adrenoceptor blocking doses of propranolol (Rodger & Shahid 1984). Fluoride ions, as NaF, also stimulate adenylyl cyclase, but this action only occurs in membrane fractions and not in intact cells. It does not therefore produce pharmacological responses comparable with β-adrenoceptor agonists. The action requires the presence of the regulatory G protein and is irreversible or only slowly reversible (Ross & Gilman 1980). Similarly, GTP or its stable analogue, Gpp(NH)p, stimulate adenylyl cyclase and increase cAMP levels in isolated membrane fractions by activating the G protein that couples the β-adrenoceptor to the enzyme. Membranes from the cyc⁻ mutant of S49 lymphoma cells, which lack the regulatory G protein but have β-adrenoceptors and adenylyl cyclase, do not generate cAMP in response to fluoride, Gpp(NH)p or β-adrenoceptor agonists (Johnson *et al.* 1980).

3   The biological response to the stimulation of the β-adrenoceptor should be mimicked by the putative second messenger, cAMP. This proved difficult to demonstrate initially because of the poor penetration of the cell membrane by cAMP and its rapid hydrolysis by PDE. Several non-hydrolysable cell permeable

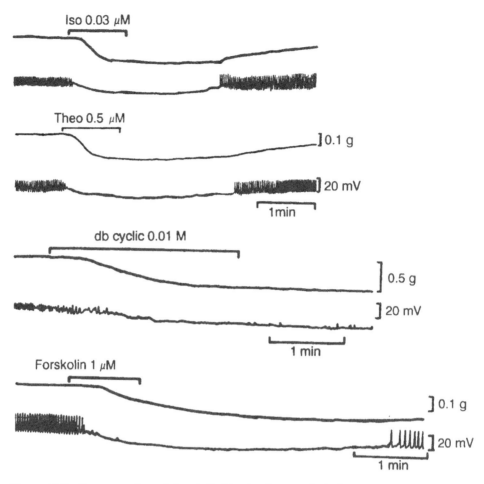

Figure 13.8 Demonstration of the similarities in pharmacological responses to agents acting through cAMP. Relaxation (top traces) and inhibition of spontaneous membrane potential changes recorded with intracellular microelectrodes (bottom traces) of guinea-pig tracheal smooth muscle are shown. The agents used are a β-adrenoceptor agonist, isoprenaline (Iso), a phosphodiesterase inhibitor, theophylline (Theo), the stable analogue of cAMP, dibutyryl cAMP (db cyclic) and the direct stimulant of adenylyl cyclase, forskolin. Adapted with permission from Honda et al. (1986).

analogues have been produced, including dibutyryl-cAMP and 8-bromo-cAMP, and these have been shown to induce tissue responses equivalent to β-adrenoceptor stimulation. These responses include increases in cardiac contractility, relaxation of airway smooth muscle (Giembycz & Raeburn 1991) and facilitation of noradrenaline release from sympathetic neurones (Johnston et al. 1987) (Figure 13.8).

4 The biological response should be mimicked and potentiated by inhibitors of the degradative enzyme, PDE. Several cAMP-dependent PDE inhibitors, including the non-selective 3-isobutyl-1-methylxanthine (IBMX) and selective Type III PDE inhibitors (eg siguazodan) have been widely shown to increase cardiac contractility and rate, relax airway and vascular smooth muscle, facilitate

noradrenaline release from sympathetic neurones and potentiate responses to isoprenaline (Nicholson *et al.* 1991) (Figure 13.8).

5   The intracellular receptor for cAMP, cAMP-dependent protein kinase (PKA), must be present in the effector cell and induce phosphorylation of substrates which alter their activity in a manner consistent with a functional biological response. This final criterion has been the most difficult to fulfil. In tracheal smooth muscle, stimulation of β-adrenoceptors results in activation of adenylyl cyclase and an increase in activity of PKA (Giembycz & Raeburn 1991).

The generation of cAMP and its activation of cAMP-dependent protein kinase is therefore a common result of β-adrenoceptor stimulation. Although the cascade of events leading from this step to the increased mobilization of glucose from the liver are well established, the steps leading to other pharmacological responses remain more speculative. The possible substrates that are phosphorylated by the increased activity of cAMP-dependent protein kinase will be considered later with regard to increases in cardiac contraction and in smooth muscle relaxation.

13.4.2   G Protein Coupling of β-adrenoceptors to Adenylyl Cyclase

Early radioligand binding experiments provided valuable information on the coupling of β-adrenoceptor occupancy with adenylyl cyclase. The competition binding curves for full agonists competing with antagonist radioligands, such as [^{3}H]DHA and [^{125}I]iodocyanopindolol, were shallow and biphasic (see Chapter 3). This indicated the existence of two affinity states for the β-adrenoceptor. When the binding assays were performed in the presence of guanine nucleotides, such as GTP or its non-hydrolysable analogue, Gpp(NH)p, monophasic displacement curves were obtained. This demonstrated that GTP converted the high-affinity state of the β-adrenoceptor to the low-affinity state. Conversely, the curves for antagonists in competition with radiolabelled antagonists were monophasic and not shifted to the right (to low affinity) by GTP. These findings led to the ternary complex model for stimulation of adenylyl cyclase by β-adrenoceptors (Kent *et al.* 1980) (Figure 13.9). The three components of this model are the receptor (R), the G protein (G$_s$, formerly termed N$_s$) and the catalytic moiety of adenylyl cyclase (C). Binding of the agonist or hormone (H) to the receptor forms a low affinity drug–receptor complex (H·R). This promotes the binding of the H·R complex with the regulatory G protein to form a transient high-affinity ternary complex (H·R·G$_s$). The formation of H·R·G$_s$ is associated with loss of tightly bound GDP from the G$_s$ binding protein and its replacement by GTP, which restores the H·R complex to its low-affinity state. The G$_s$–GTP complex then interacts with the catalytic unit of adenylyl cyclase stimulating the generation of cAMP. The association of the GTP-occupied G$_s$ protein with the catalytic subunit (G$_s$·GTP·C) is terminated by GTPase activity associated with the G$_s$ protein. This converts G$_s$·GTP·C back to the ground state (G$_s$·GDP), liberating P$_i$ and the catalytic unit (Lefkowitz *et al.* 1982, Harden 1983).

Since the original ternary complex model was proposed, the G protein coupled to β-adrenoceptors has been shown to exist as a heterotrimer comprising a, β and γ subunits (see earlier). In the current model for β-adrenoceptor-mediated activation of adenylyl cyclase, binding of the agonist to the receptor catalyses the replacement

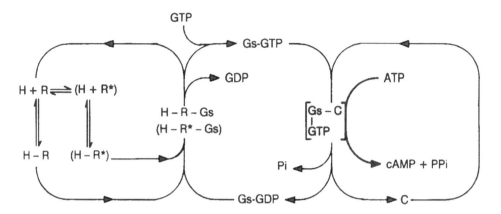

Figure 13.9 Ternary complex model for the activation of adenylyl cyclase by agonist or hormonal (H) stimulation of the β-adrenoceptor (R). The proposed interactions of the hormone–receptor complex (H–R), the guanine nucleotide binding protein (G_s) and the catalytic subunit of the adenylyl cyclase (C) are illustrated. The updated 'allosteric ternary complex model' is indicated within the left-hand loop by the additional step of conversion of the inactive conformation of the receptor (R) to an active conformer (R^*) by the presence of agonist or hormone. A detailed description is provided in the text.

of GDP by GTP on the 42 kDa α-subunit, which dissociates from the $\beta\gamma$-subunit of the G protein. The activated GTP-liganded α-subunit then stimulates adenylyl cyclase. The intrinsic GTPase activity of the α-subunit converts GTP to GDP; catecholamines facilitate this by stimulating GTPase (Ross & Gilman 1980). The α-subunit is thus returned to its ground state, allowing its reassociation with the $\beta\gamma$ complex (Strosberg 1992) (Figure 13.3).

Further modification of the ternary complex model has been required by the possibility that the receptor-G protein complex may be activated in the absence of an agonist. This proposal has arisen from the cloning of mutant receptors (constitutively active mutant receptors) by the interchange of key amino acids in the third intracellular loop of the β-adrenoceptors, which is involved in G protein coupling (see earlier). It is proposed that the natural wild-type receptor is normally conformationally constrained in the inactive form (R) but the presence of agonist releases this into the active conformer (R^*). This agonist-receptor complex (H·R^*) then catalyses the dissociation of the $\beta\gamma$-subunit from the G protein in a similar way to the classical ternary complex model (Figure 13.9). The modified allosteric ternary complex model allows for the spontaneous formation of the activated conformer of the receptor in mutant receptors. Whether spontaneous activation occurs with the wild-type receptor is uncertain, although there may be conditions under which some basal level of activity occurs. The consequence of such activity is that antagonists could exert negative basal activity (negative antagonist, inverse agonists) (Lefkowitz *et al.* 1993). The conventional view is that agonists with different efficacies will increase the proportion of receptors in the R^* conformer to varying degrees. However, an alternative possibility is that several activated conformers of the receptor may exist and that these have differing abilities to interact with the G protein (Kenakin 1995c).

13.4.3 Coupling of cAMP-dependent Protein Kinase to Tissue Responses

13.4.3.1 Cardiac muscle

The β-adrenoceptor-mediated increases in rate and force of cardiac contractions can be attributed to the activation of cAMP-dependent protein kinase by cAMP. The composition of the cAMP-dependent protein kinase is the same as that already described for glycogenolysis, the two cAMP-binding sites occurring on each regulatory subunit, both synergistically activating the protein kinase. The activated catalytic subunits then phosphorylate effector proteins of the cardiac myocyte. The primary site of phosphorylation appears to be the L-type Ca^{2+} channel, which allows Ca^{2+} influx. The increase in cardiac action potential height and duration due to β-adrenoceptor stimulation can be mimicked by injection of activated catalytic subunits. It is also possible that the activated α_s-subunit of the regulatory G protein can directly open the Ca^{2+} channel. This can occur 10 times faster than via activation of cAMP-dependent protein kinase and is also energy efficient, since it requires no ATP expenditure (Kaumann 1991). The influx of Ca^{2+} in turn releases intracellular Ca^{2+} from the SR. This increases the quantity of Ca^{2+} available for cardiac contractions through increased Ca^{2+}-calmodulin-regulated cross-bridging of the myofilaments (see under IP_3-mediated smooth muscle contractions) (Callewaert 1992).

Phosphorylation by activated cAMP-dependent protein kinase is also involved in the relaxation phase of the cardiac contraction. Phosphorylation of phospholamban increases its activity in regulating Ca^{2+} uptake into the SR (Figure 13.3). This accelerates removal of Ca^{2+} from the cytoplasm and favours relaxation of the muscle. It also enhances the provision of Ca^{2+} for release at subsequent contractions. Phosphorylation of troponin I also enhances relaxation. In cardiac muscle, as in skeletal muscle, troponin C serves as the Ca^{2+} binding protein, troponin I binds to actin and troponin T binds to tropomyosin. The released Ca^{2+} binding to troponin C allows conformational changes that are transmitted to tropomysin and thus to actin, which permits actin–myosin cross-bridging. The phosphorylation of troponin I by cAMP-dependent protein kinase reduces the sensitivity of troponin C for Ca^{2+} and facilitates cardiac muscle relaxation (Kaumann 1991) (Figure 13.3).

13.4.3.2 Smooth muscle relaxation

The relaxation of smooth muscle by cAMP-generating agonists is attributed to phosphorylation of several targets by the activation of cAMP-dependent protein kinase. A wide range of phosphorylation sites have been proposed but their relative importance has yet to be established (Figure 13.3). These targets probably vary between muscles, between the modes of tone induction (eg agonist-induced tone via phosphatidylinositol hydrolysis or depolarization-induced tone), and whether it is generation of muscle tone or its maintenance that is being inhibited by the muscle relaxant. The following are the more likely sites of phosphorylation by protein kinase-A. (PKA).

1. Interference with IP_3/DAG pathways. β_2-Adrenoceptor agonists reduce histamine-induced formation of IP_3 in airway smooth muscle but not M_3 receptor-mediated IP_3 accumulation. Thus, although this is a possible site of action, the situation is complex and may be influenced by other properties of the contractile agent. Phosphorylation of the IP_3 receptor on the SR at Ser-1589 by PKA has been demonstrated and this may *reduce* its affinity for IP_3 and therefore the ability to release Ca^{2+} into the cytosol.

2. Inhibition of Ca^{2+} influx. Ca^{2+} activation of the contractile myofilaments of smooth muscle occurs through formation of the Ca^{2+}–calmodulin complex which activates MLCK. This in turn phosphorylates the 20 kDa light chain of myosin (LC_{20}), which facilitates actin–myosin cross-bridging and contraction. There is a question whether LC_{20} phosphorylation is sustained and is involved in the mainten-ance of contractile tone since $[Ca^{2+}]_i$ is only just above basal level during this phase of contraction. Ca^{2+} appears to be rapidly extruded from the cell, being facilitated by calmodulin. The Ca^{2+} available for the initial contraction is through influx via the opening of L-type Ca^{2+} channels. This occurs by depolarization with high K^+ or by application of cathodal currents. Phosphorylation of these Ca^{2+} channels by activation of PKA with cAMP-elevating agents may reduce their Ca^{2+} transporting properties. An indirect mechanism on Ca^{2+} influx may be activation of sarcolemmal Na^+,K^+-ATPase. This drives Na^+ out of the cell in exchange for K^+ and maintains the resting membrane potential. cAMP-dependent activation of Na^+,K^+-ATPase by β-adrenoceptor agonists would lead to smooth muscle relaxation by hyperpolarizing the cell membrane. This would reduce the probability of voltage-operated Ca^{2+} channels opening, thereby reducing Ca^{2+} influx. Also, the enhanced Na^+/K^+ exchange would encourage Ca^{2+} loss from the cytosol through Na^+/Ca^{2+} exchange.

3. Sarcoplasmic reticulum (SR) Ca^{2+} uptake. The sequestering of Ca^{2+} from the cytosol into the SR is driven by the calmodulin-independent Ca^{2+},Mg^{2+}-ATPase of the SR membrane. The activity of this ATPase is enhanced by the phosphorylation of phospholamban by PKA. This probably represents a universal mechanism for Ca^{2+} sequestration and consequent relaxation of a wide range of smooth muscles and cardiac muscle.

4. K^+ channels. The limitation of depolarization is brought about by K^+ efflux through the opening of K^+ channels (outward rectification) and hyperpolarization. Thus, Ca^{2+}-dependent K^+ channel opening associated with hyperpolarization accompanies smooth muscle relaxation by β-adrenoceptor agonists. This may be mimicked by intracellular application of cAMP-dependent protein kinase. The relaxation of tracheal smooth muscle by β-agonists and by dibutyryl-cAMP has been shown to be blocked by the $K_{Ca^{2+}}$ channel inhibitor, charybdotoxin. This indicates an important role of cAMP-elevating agents and PKA on the K^+ channel in mediating smooth muscle relaxation. A direct coupling of β-adrenoceptors to ATP-sensitive K^+ channels, independent of the cAMP generation, may also explain part of vasorelaxation. Vascular responses to β-adrenoceptor agonists are inhibited by the K_{ATP}^+-channel antagonist, glibenclamide (Randall & McCulloch 1995).

5. Myosin light chain kinase (MLCK). Phosphorylation of MLCK by PKA may represent a site of inactivation of actin–myosin cross-bridging and contraction that is independent of Ca^{2+} levels. The normal contraction process of smooth muscle involves phosphorylation of MLCK. However, the cAMP-elevating agents, through activation of PKA, would double phosphorylate MLCK and this would reduce the ability to phosphorylate myosin light chain and thus reduce contractile tension.

According to this theory, the level of myosin light chain phosphorylation by MLCK is reduced. The catalytic subunit of PKA can reduce tension generation in skinned smooth muscle, an effect not due to altered Ca^{2+} mobilization, since these muscles cannot regulate Ca^{2+} influx.

In summary, the relaxation of smooth muscle by cAMP-generating agonists, such as β-adrenoceptor stimulants, is therefore due to the combined consequences of phosphorylation by PKA of various targets. These mechanisms lead to restriction of cytoplasmic Ca^{2+} levels either by removal to the SR or prevention of its influx, to membrane hyperpolarization or to a direct effect on the myofibrils (Giembycz & Raeburn 1991) (Figure 13.3).

13.4.4 *Receptor-mediated Inhibition of Adenylyl Cyclase*

The role of reduced levels of second messenger cAMP in the prejunctional α_2-adrenoceptor-mediated inhibition of transmitter release has already been described in Chapter 6. The contribution of falling levels of cAMP in the contractile responses of smooth muscle to agents known to be negatively coupled to adenylyl cyclase, such as α_2-adrenoceptor-mediated vasoconstriction (Chapter 3, Table 3.4) and M_2 muscarinic receptors in the ileum (Chapter 7), remains unclear.

α_2-Adrenoceptor-mediated vasoconstriction is blocked by L-type Ca^{2+} channel antagonists *in vivo* and in rat isolated tail artery *in vitro*. It is also virtually abolished in Ca^{2+}-free medium containing EGTA and is therefore almost entirely due to influx of extracellular Ca^{2+} via Ca^{2+} channels. Since there is little change in membrane potential during contractions in certain blood vessels, it is probably due to influx through ROCs rather than VOCs. While α_2-adrenoceptor activation is associated with a fall in cAMP in many tissues including platelets, postjunctional vascular smooth muscle, pancreatic islet cells and the renal cortex, the role of this in mediating Ca^{2+} influx and contraction of smooth muscle is unclear. The prejunctional α_2-adrenoceptor-mediated inhibition of transmitter release is associated with a *fall* in intracellular Ca^{2+}.

In many tissues, the α_2-adrenoceptor-mediated response is blocked by PTx, including lipolysis, insulin release from the pancreatic islets and vasoconstriction of pithed rats. This suggests the involvement of G_i protein coupled negatively to adenylyl cyclase in these responses. The mechanism of inhibition of adenylyl cyclase by the G_i regulatory protein is still unclear. Once possible mechanism may be that the liberated $\beta\gamma$-subunit is similar to that released from G_s and is therefore able to combine with the α_s-subunit released from G_s, that normally activates adenylyl cyclase. This would prevent G_s activation. An alternative proposal is that the α_i-subunit competes with α_s for binding to the catalytic subunit of adenylyl cyclase. These are probably not the sole mechanisms, however, since inhibition of adenylyl cyclase can occur in cells that do not contain G_s. A further possible mechanism is a direct inhibition of the catalytic subunit (Birnbaumer 1990).

Whether the α_2-adrenoceptor-mediated vasoconstriction can be attributed to reduced levels of cAMP, however, is questionable. It is difficult to imagine that there is sufficient endogenous cAMP generation under resting conditions for inhibition of its formation to induce a substantial contraction. For example, there would have to be basal activation of adenylyl cyclase by, for example, circulating adrenaline via β_2-adrenoceptors. The PTx sensitivity and dependence upon Ca^{2+}

influx suggests that the α_2-adrenoceptor, in common with the α_1-adrenoceptor, is directly linked to the sarcolemmal Ca^{2+} channel by a PTx-sensitive G protein (Ruffolo *et al.* 1991).

Reduced levels of cAMP do not therefore appear to play a role in the contractile response mediated via α_2-adrenoceptors. This is a similar situation to the contractions of smooth muscle, such as in the ileum, due to muscarinic receptor stimulation by Ach. M_2 receptors are also negatively coupled to adenylyl cyclase and have been demonstrated in the ileum by radioligand binding. However, the reduced cAMP levels may only serve a function in opposing the relaxation responses to β-adrenoceptor agonists. The contraction of the ileum is attributed to M_3 receptors linked to phosphatidylinositol hydrolysis (see Chapter 7) (Fernandes *et al.* 1992, Eglen *et al.* 1994).

13.5 Phosphodiesterases

The effects of cAMP as a second messenger are terminated by its hydrolysis to AMP (5'-adenosine monophosphate), which does not activate cAMP-dependent protein kinase (PKA), by the actions of phosphodiesterase (PDE, EC 3.1.4.17) (Figures 13.3 and 13.7). Inhibition of PDE will therefore allow intracellular cAMP levels to rise which is the basis of the pharmacological responses to PDE inhibitors. This enzyme is not a single entity but exists as several different isoenzymes. Confusion has occurred over different systems used for the naming of these isoenzymes. Initially, it was according to their order of elution from anion-exchange chromatography. This was not entirely satisfactory since the number of peaks of PDE activity can depend upon both the tissue and the chromatographic conditions, and a peak may include more than one isoenzyme. The classification now most widely adopted relates not only to the elution pattern, but also to the primary amino acid sequences of the enzyme and that deduced from cDNA. It also takes into account the kinetic properties of isoenzyme affinities for hydrolysis of cAMP or cGMP, susceptibility to inhibition by these cyclic nucleotides and by pharmacological agents.

At least five families of isoenzymes are recognized by these criteria, each of which is encoded by a separate gene or gene family. Most of these families contain two or more closely related subfamilies and many of the genes appear to generate two or more alternatively spliced mRNAs. In general, the amino acid sequence of members of one family share 20–25% sequence identity with members of other families. Much of this occurs at one domain, thought to be the catalytic centre of the enzyme. Most subfamily members are probably encoded by different, but highly homologous, genes (70–90% sequence identity). The term 'isoform' is often used to distinguish between members of a single family (Beavo & Reifsnyder 1990). The characteristics of the PDE families I to V are shown in Table 13.2.

The distribution of these isoenzymes differs between tissues, and the introduction of selective inhibitors of the isoenzymes provides the potential for elevating the cyclic nucleotide levels in specific target cells or organs. The airways contain predominantly PDEs III and IV, whereas in the heart, vascular smooth muscle and platelets PDE III predominates. Inflammatory cells contain a high proportion of PDE IV activity. The subcellular distribution of the PDE isoenzymes also differs between tissues. Cardiac, liver and adipose cells contain high levels (50–70%) of membrane-bound PDE in the particulate fraction of tissue homogenates. In contrast, smooth

Table 13.2 Phosphodiesterase isoenzymes (PDE) – their distribution and selective inhibitors

PDE	Characteristics	K_m (μM) cAMP	K_m (μM) cGMP	Distribution	Selective inhibitors
I	Ca^{2+}-calmodulin-dependent	2–70	2–20	Widespread soluble PDE; brain	Vinpocetin, 8-methoxymethyl-3-isobutyl-1-methylxanthine
II	cGMP-stimulated	30–45	10–15	Widespread soluble PDE; endothelium, mast cells, adrenal gland	None, but trequinsin and dipyridamole are active
III	cGMP-inhibited	0.1–0.5	0.1–0.5	Platelets, heart, airways and vascular smooth muscle; absent in brain	Amrinone, milrinone, enoximone, siguazodan, SK&F 94120
IV	cAMP-specific	0.5–2	>100	Inflammatory cells, airways smooth muscle, brain	Rolipram, Ro-20-1724, denbufylline
V	cGMP-specific	>100	1–5	Platelets, vascular smooth muscle	Zaprinast, dipyridamole

muscle, kidney and platelets have most of their PDE activity in the soluble cytosolic fraction (Nicholson *et al.* 1991).

13.5.1 Non-selective PDE Inhibitors

The methylxanthines, theophylline and caffeine, were the first inhibitors of PDE to be recognized. Other alkylxanthines, such as 3-isobutyl-1-methylxanthine (IBMX), oxypentifylline, enprofylline, the purine isosteres ICI 58,301 and etazolate, and the opium alkaloid, papaverine, are also non-selective inhibitors of the PDE isoenzymes (Table 13.3).

Caffeine is found in tea (*Thea sinensis*) and coffee (*Coffea arabica*), while cocoa contains theobromine and some caffeine. Another source of caffeine in the diet is from beverages based on cola flavouring, which contain extracts of cola nuts (*Cola acuminata*). The popularity of these caffeine-containing beverages is based upon the elevation of mood, decreased fatigue and increased capacity for work induced by the caffeine. Additionally, these non-selective PDE inhibitors have a wide range of peripheral pharmacological properties. They relax smooth muscle through elevation of cAMP and cGMP; these effects may be observed as bronchodilatation and vasodilatation of isolated blood vessels. Cardiac stimulation, diuresis and augmented secretion from exocrine and endocrine glands, particularly of gastric acid and pepsin, are further pharmacological actions of the methylxanthines. Blood pressure responses are variable but usually there is increased peripheral resistance and a rise in blood pressure, most likely due to release of catecholamines. The rise in blood pressure associated with caffeine consumption usually declines with chronic coffee

Table 13.3 Non-selective phosphodiesterase inhibitors

<u>Alkylxanthines</u>

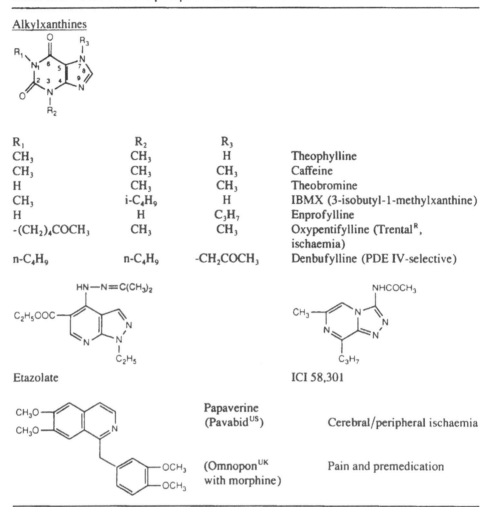

R_1	R_2	R_3	
CH_3	CH_3	H	Theophylline
CH_3	CH_3	CH_3	Caffeine
H	CH_3	CH_3	Theobromine
CH_3	i-C_4H_9	H	IBMX (3-isobutyl-1-methylxanthine)
H	H	C_3H_7	Enprofylline
-$(CH_2)_4COCH_3$	CH_3	CH_3	Oxypentifylline (Trental[R], ischaemia)
n-C_4H_9	n-C_4H_9	-CH_2COCH_3	Denbufylline (PDE IV-selective)

Etazolate

ICI 58,301

Papaverine
(Pavabid[US])   Cerebral/peripheral ischaemia

(Omnopon[UK]
with morphine)   Pain and premedication

drinkers. The pharmacological effects of the methylxanthines, as exemplified by theophylline, are not necessarily due to PDE inhibition and therefore a complete analysis of its mechanism of action is not warranted here. Additionally, theophylline is an adenosine receptor antagonist and is an inhibitor of intracellular Ca^{2+} mobilization.

Currently, theophylline remains the only agent with PDE inhibitory properties that has widespread use in the treatment of bronchial asthma. It is particularly useful for nocturnal asthma in sustained-release formulations, and is indicated as an adjunct to β_2-adrenoceptor agonist inhalation (when required) (see Chapter 4) and anti-inflammatory therapy in chronic asthma for children and adults. Administration of theophylline is by the oral route but can be associated with a high incidence of side-effects if the serum levels exceed 20 $\mu g/ml$. Monitoring of serum levels is therefore advised, and patients should not transfer to other formulations or brand names without a full clinical assessment and retitration of serum levels. The solubility of methylxanthines is low but is enhanced by complexing, for example, with ethylenediamine to form aminophylline (eg Phyllocontin Continus^R). Other salts in use include choline theophyllate (Choledyl^R); both preparations yield the parent methylxanthine when dissolved in aqueous solution.

The beneficial effects of theophylline in asthma are due to its bronchodilator action and to anti-inflammatory and anti-anaphylactic activity. It inhibits the release of mediators from mast cells and reduces both the early (mast cell-derived histamine-mediated) and late phases (inflammatory mediators released from the influx of eosinophils) of the bronchoconstrictor response to antigen challenge. There is little evidence in asthmatic subjects or animal models, however, that it has substantial effects on the bronchial hyperreactivity (Raeburn *et al.* 1993). Considerable debate has centred on the mechanism of action of theophylline in asthma, the two prime candidates being PDE inhibition and adenosine receptor antagonism. PDE inhibition has generally been discounted on the grounds of its low potency as a PDE inhibitor, which suggests that at therapeutic plasma levels it is unlikely to have PDE inhibitory activity. However, at therapeutically relevant doses, theophylline can attenuate the cardiovascular effects of adenosine. This would suggest that adenosine receptor antagonism may be important. Although adenosine normally relaxes airway smooth muscle, its inhalation by asthmatic patients causes bronchoconstriction and it has been argued that theophylline may antagonize this response. This would suggest a role for adenosine in the asthmatic response but this has still to be fully evaluated (Broadley 1995). The fact that the alkylxanthine, enprophylline, has bronchodilator activity and has been shown to have benefit in asthmatic patients, yet is relatively ineffective as an adenosine antagonist in most tissues, suggests that PDE inhibition rather than adenosine antagonism is responsible for the beneficial effects of theophylline. Recent experience with the selective PDE inhibitors, to be dealt with in the next section, also supports this view. Other alkylxanthines, such as IBMX, are potent non-selective PDE inhibitors with less adenosine receptor antagonistic properties. The adenosine antagonism is generally thought to be best avoided since it may contribute to the central side-effects of theophylline (Morley 1991).

13.5.2   PDE I – Ca^{2+}–Calmodulin Stimulated PDE

Stimulated by the Ca^{2+}-calmodulin complex, this isoenzyme has at least seven isoforms, a number of which are found in the brain. The two major isoforms, I$_\alpha$ and I$_\beta$, have different affinities for cAMP and cGMP. PDE I$_\alpha$ acts preferentially on cGMP, whereas PDE I$_\beta$ has equal potency on both substrates. In the presence of Ca^{2+}–calmodulin, the enzymes are activated, causing 2–4-fold reduction of the K_m. Selective inhibitors include the phenothiazines, such as trifluoroperazine, W-7 and the ginsenosides from ginseng (*Panax quinquefolium*) root. As the inhibition can be overcome by calmodulin, it appears that these agents are serving as calmodulin antagonists and probably bind to the calmodulin binding site. Vinpocetin and 8-methoxy-methyl-3-isobutyl-1-methylxanthine, however, inhibit the hydrolytic site on PDE I selectively, rather than the calmodulin binding site (Thompson 1991). These inhibitors relax airway and vascular smooth muscle and are also possible inhibitors of platelet aggregation (Nicholson *et al.* 1991).

13.5.3 PDE II – cGMP-stimulated PDE

PDE II hydrolyses both cAMP and cGMP, but has a slight preference for cGMP. The activity against cAMP is stimulated many-fold by physiological levels of cGMP (Torphy *et al.* 1993) and is therefore regarded as a cGMP-stimulated isoenzyme family. There is a widespread distribution of PDE II, in particular in adrenal glands, endothelial cells and rat mast cells. At the latter site, inhibition may be of value in preventing microvascular leakage and histamine release from mast cells associated with the late bronchoconstrictor phase of asthma (Giembycz 1992). No clearly defined selective inhibitor of PDE II, however, has been identified as yet and, as a result, functional roles have been poorly defined. Trequinsin (HL725) and dipyridamole (the adenosine transport inhibitor) have been classified as inhibitors of this enzyme (Beavo & Reifsnyder 1990, Thompson 1991).

13.5.4 PDE III – cGMP-inhibited PDE

This isoenzyme has the lowest K_m value for both cAMP and cGMP. It is inhibited by low concentrations of cGMP and activated by cAMP-dependent phosphorylation. The primary distribution of PDE III is in the heart, platelets and airway and vascular smooth muscle, although it is not found in the brain. Inhibition of the enzyme causes cAMP accumulation and positive inotropy, reduced aggregation of platelets, vasodilatation and bronchodilatation. The relaxation of vascular smooth muscle is partly due to accumulation of cGMP (Murray & England 1992).

Selective PDE III inhibitors have been developed for the treatment of congestive heart failure, their cardiotonic activity providing inotropic support to the heart and vasodilatation reducing pre- and after-load. Amrinone was the lead compound of this class and was the first to be approved in the USA for short-term intravenous administration in patients with severe heart failure refractory to other measures. Other agents of this type include milrinone (Primacor[R]), enoximone (Perfan[R]), imazodan, siguazodan, SK&F 94120 and potent derivatives, cilostamide and lixazinone (Table 13.4). In clinical trials these compounds have shown short-term haemodynamic improvement in patients with heart failure; however, long-term therapy with PDE III inhibitors has generally resulted in increased mortality and evidence of modification of the underlying disease process is lacking. As a result these agents are restricted to acute use in severe cardiac failure. They may also have the disadvantage of having an arrhythmic potential. Agents that combine PDE inhibitory activity with the ability to sensitize the contractile myofibrils to Ca^{2+} and *anti*-arrhythmic activity may prove more useful than those with PDE III inhibitory activity alone (eg Org 30029). Combination with β-adrenoceptor blocking activity, in the compound GI104313, may also have the advantage of preventing the undesirable cardiac arrythmias. The positive inotropy of PDE III inhibition would also offset the negative inotropy of the β-blockade (Shaffer *et al.* 1993). Although only 2% of the total cardiac cAMP hydrolytic activity resides in the SR fraction, it appears that PDE III inhibitors act at the level of the SR-PDE III (Nicholson *et al.* 1991, Thompson 1991).

The roles of PDE III and PDE IV in the myocardium may vary between species. In the rat heart, for example, the PDE III inhibitors, amrinone and milrinone, or the PDE IV inhibitors are ineffective as positive inotropes when administered alone.

However, when combined, they exert the expected increase in force of cardiac contraction. It is also reported that they are less effective in the failing human heart because of low basal cAMP levels. When combined with isoprenaline to raise cAMP levels, PDE III inhibitors produce full inotropic activity (Murray & England 1992).

Table 13.4 Selective inhibitors of phosphodiesterase (PDE)

PDE III inhibitors

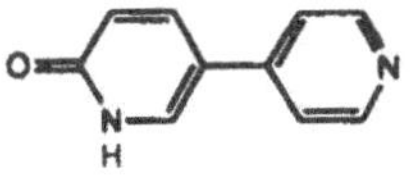

Amrinone

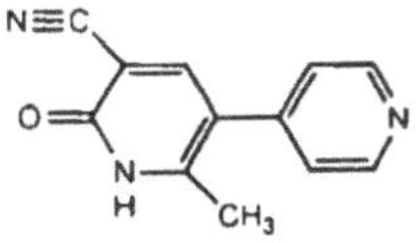

Milrinone (Primacor[R])

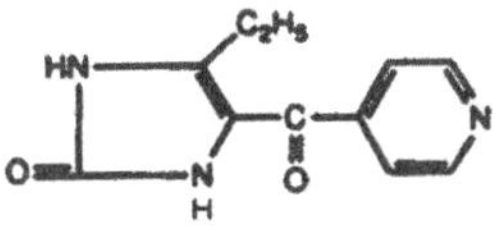

Piroximone (MDL 19,205)

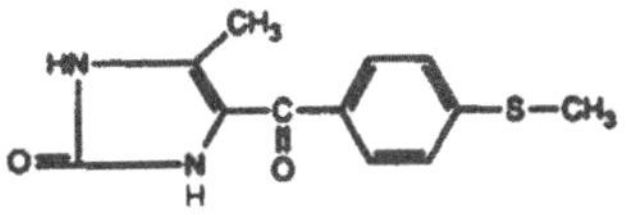

Enoximone (MDL 17,043) (Perfan[R])

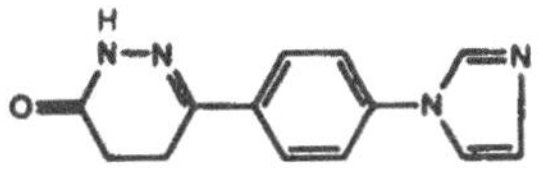

Imazodan

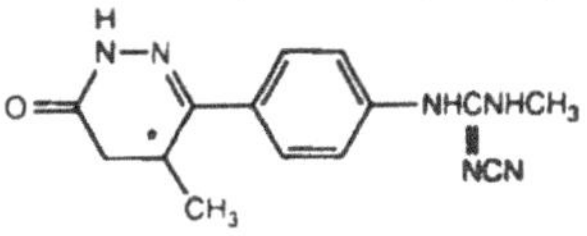

Siguazodan (SK&F 94836)

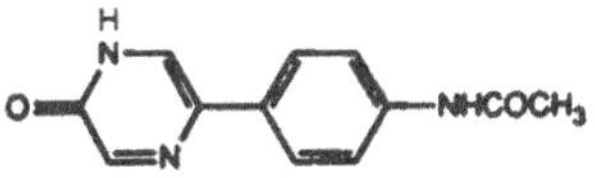

SK&F 94120

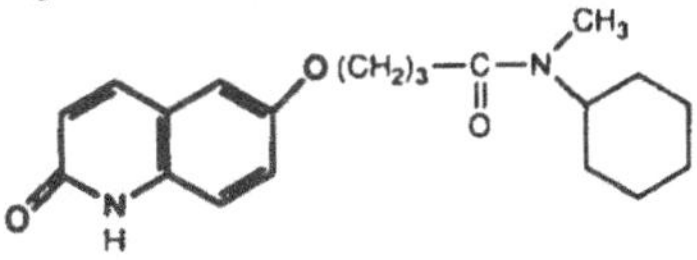

Cilostamide (OPC 3689)

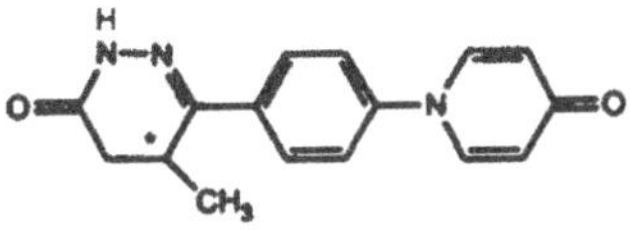

SK&F 95654

PDE IV inhibitors

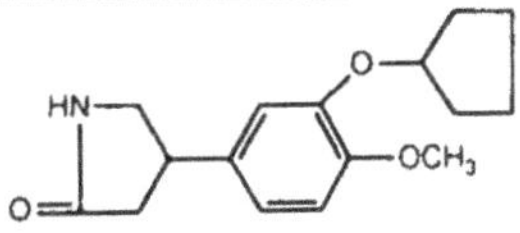

Rolipram

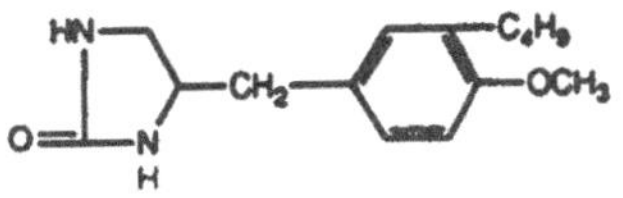

RO 20–1724

PDE V inhibitors

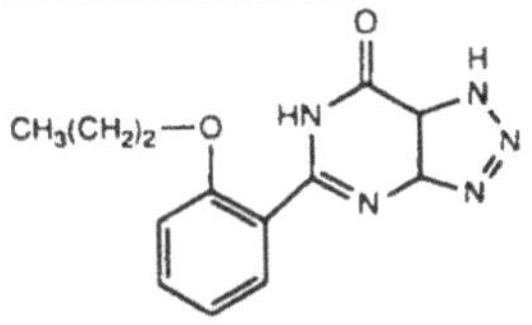

Zaprinast (M&B 22948)

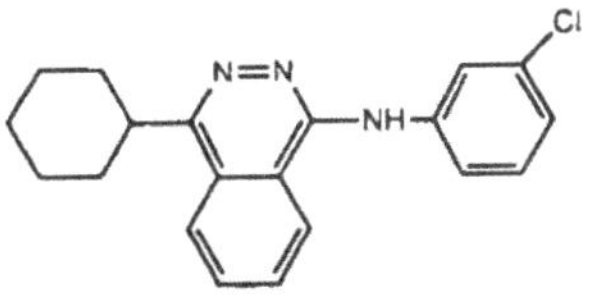

MY-5445

(*continued*)

Table 13.4 Selective inhibitors of phosphodiesterase (PDE) *(continued)*

PDE III/IV inhibitors

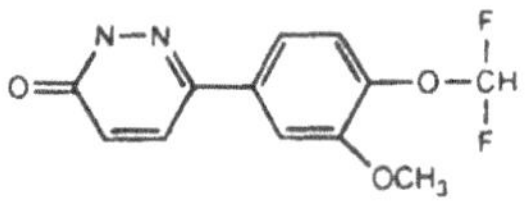

Zardaverine (MY-5445)

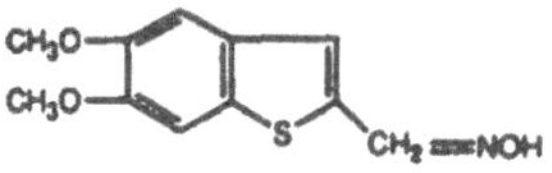

Org 30029

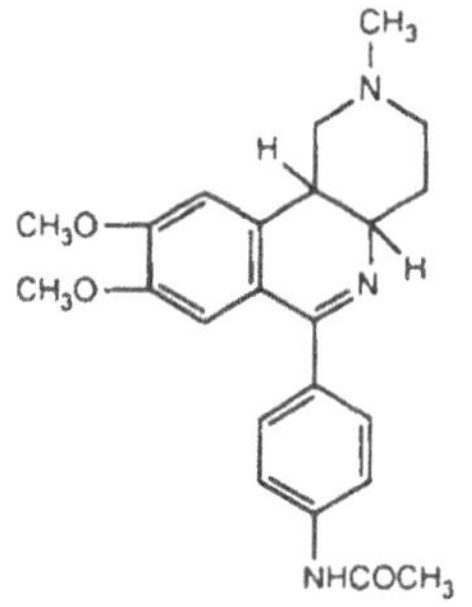

Benzafentrine (AH 21–132)

Platelets contain predominantly PDE III and therefore inhibition of platelet aggregation and antithrombotic therapy are potential targets for PDE III inhibitors, cilostazol (Pletaal[R]) having been marketed in Japan. The presence of PDE III in airway smooth muscle explains the bronchorelaxant activity of these compounds. In general, PDE III inhibitors such as siguadozan are less effective than the PDE IV inhibitors such as rolipram, indicating the relatively more important functional role of inhibiting PDE IV than PDE III in the airways. There appear to be species differences, however, PDE III being more important in the pig bronchus. Neither rolipram nor siguazodan are particularly effective in potentiating the bronchorelaxant effect of isoprenaline in human bronchus. This is probably due to a cAMP-independent component of the response, such as direct linkage of the β_2-adrenoceptor to the K^+ channel (Torphy *et al.* 1993). An interesting relationship has been found in airway smooth muscle between the β-adrenoceptor subtype and the relaxation induced by PDE III and IV inhibitors. Rolipram (IV) is a more effective relaxant of guinea-pig and bovine trachea (where β_2-adrenoceptors predominate) than in mouse and porcine trachea (where β_1-adrenoceptors predominate). Siguazodan (III) is more potent than rolipram in porcine trachea. A possible functional relationship is therefore suggested between the isoenzyme and the β-adrenoceptor subtype. Both PDE III and PDE IV, however, appear to be involved to some extent in regulating airway smooth muscle tone since both siguazodan and rolipram relax the airways, irrespective of the β-adrenoceptor subtype (Raeburn *et al.* 1993, Tomkinson *et al.* 1993).

13.5.5 *PDE IV – cAMP-specific PDE*

PDE IV is the only truly cAMP-specific PDE since it exhibits a higher affinity for cAMP than for cGMP (Table 13.2). There is a wide tissue distribution and in the brain it is the main cAMP-hydrolysing isoenzyme. Smooth muscle of the airways is an important location, but of current interest is the presence in inflammatory cells which has relevance to the use of selective inhibitors as anti-inflammatory agents in the treatment of asthma.

Selective inhibitors include rolipram, Ro 20-1724 (Table 13.4) and denbufylline (Table 13.3). Rolipram has undergone clinical evaluation of its antidepressant activity and has been found to be effective in the treatment of major depression. Mood elevation and neuroexcitability associated with enhanced transmitter release in the brain provide a basis for possible application in conditions associated with cerebral ischaemia. Denbufylline and other selective PDE IV inhibitors may have possible applications in senile dementia by elevating mood and improving cognitive function (Nicholson *et al.* 1991, Thompson 1991). Gastrointestinal side-effects occur, including an increase in gastric acid secretion. Also of major concern with selective PDE IV inhibitors is emesis arising from a central site of action (Murray & England 1992).

Airways smooth muscle is relaxed by the PDE IV inhibitors, rolipram and denbufylline, and PDE IV is the major isoenzyme present in human bronchus (Cortijo *et al.* 1993). This bronchodilator activity is therefore of potential value in the treatment of asthma, but also has important potential for inhibition of the anti-inflammatory component of bronchial asthma. The late phase of airway broncho-constriction to antigen provocation is associated with inflammatory cell influx into the airways and the consequent release of inflammatory mediators and free radicals, which promote all the sequelae of an acute asthma attack, including tissue damage, oedema and plasma protein extravasation.

Basophils, mast cells, eosinophils, neutrophils, lymphocytes and macrophages all contain mainly PDE IV. Inhibitors of PDE IV have been shown to inhibit the functions of these inflammatory cells. Rolipram and other PDE IV inhibitors reduce antigen-induced release of mediators including histamine and leukotriene C_4 (LTC_4) from mast cells and basophils (Torphy & Undem 1991). Macrophages release tissue necrosis factor (TNF) which facilitates the migration of eosinophils across the capillary wall into the sites of inflammation; this release is inhibited by rolipram (Schade & Schudt 1993). Eosinophils and neutrophils play major roles in the inflammatory response. Their activation in asthma is measured from the release of superoxide and H_2O_2 by FMLP, a synthetic analogue of the bacterial chemotactic peptide (see sensory neurones, Chapter 12) or by opsonized zymosan (known as a respiratory burst). This is inhibited by rolipram, although not opsonized zymosan-induced release in human neutrophils, suggesting that this bypasses the cAMP-dependent regulatory step for the respiratory burst. PDE IV inhibitors also impair other indicators of neutrophil activity such as FMLP-induced biosynthesis of PAF, LTB_4 and chemotaxis (Wright *et al.* 1990).

A further process that is suggested to be important in the inflammatory response of the airways is the contraction of endothelial cells of the pulmonary vasculature by inflammatory mediators such as PAF and leukotrienes. This results in the formation of physical pores between the cells, through which the proinflammatory cells and plasma proteins can pass, elevated levels of which appear in bron-choalveolar lavage (BAL) fluid. Exposure of endothelial cells in culture to H_2O_2 promotes a complex metabolic response including a fall in ATP and NADP, activation of the glutathione antioxidant cycle, increased hydrolysis of inositol phosphates and increased intracellular free Ca^{2+} levels which probably induces Ca^{2+}-dependent actin–myosin-based endothelial cell contraction. This contractile process is inhibited by PDE IV and PDE III inhibitors, probably by increasing cAMP activation of PKA and a reduction of Ca^{2+}-dependent MLCK activity (see Figure 13.3). The consequent relaxation of endothelial cells may therefore reduce microvascular leakage and airway oedema (Giembycz 1992). These anti-inflammatory effects of PDE IV inhibitors have been demonstrated *in vivo* as a reduction of leakage of inflammatory cells into lavage fluid and of bronchoconstriction after antigen challenge of sensitized animals (Underwood *et al*. 1993). There is also evidence of a reduction by PDE IV inhibitors of the airway hyperreactivity to spasmogens that occurs after antigen challenge in experimental animals (Howell *et al*. 1993, Raeburn *et al*. 1993).

From the foregoing account of the actions of PDE IV inhibitors in experimental situations, there is clear potential for their use in bronchial asthma. This activity may well be the basis of the therapeutic efficacy of theophylline which has widespread use in asthma treatment in spite of its potential for toxicity. The limitations of theophylline are the high incidence of adverse effects such as gastrointestinal upset and CNS activity including nausea, headache and insomnia, and its poor safety margin. These CNS side-effects, the emesis and gastrointestinal disturbances are also a potential problem with the selective PDE IV inhibitors.

13.5.6   PDE III and IV Inhibitors

Compounds have now been developed which inhibit PDE III and PDE IV but not the other isoenzymes. These include zardaverine, which has been developed as an anti-asthmatic drug, benzafentrine (AH 21-132), which has been discontinued, and Org 30029 (Table 13.4). Combined PDE III/IV inhibition could be an attractive proposition for the treatment of asthma, since it would combine the bronchodilator effects of PDE III inhibition with the anti-inflammatory properties of PDE IV inhibition. This combination may also act synergistically as airway smooth muscle relaxants. This may also apply to T lymphocytes, which have both PDE III and IV, so that dual inhibition will result in immunosupression. Such activity may be of benefit in severe asthma, but it would leave the patient immunocompromised and susceptible to infection. Several of these compounds have been shown to exert bronchodilator activity in human volunteers. There is debate however, whether a PDE IV or mixed PDE III/IV inhibitor would be preferable for asthma treatment. The disadvantage of PDE III inhibitory activity is the cardiovascular side-effects, including tachycardia, which is enhanced by the PDE IV activity as described above. These cardiovascular effects may be minimized by direct application to the airways via inhalation (Murray & England 1992, Raeburn *et al*. 1993).

13.5.7   PDE V – cGMP-specific PDE

PDE V is the only cGMP-selective isoenzyme and inhibitors of this PDE elevate the intracellular levels of cGMP rather than cAMP. PDE V is found predominantly in smooth muscle, platelets and the lungs. Selective inhibitors include zaprinast, dipyridamole, SK&F 96231 and MY5445. They relax smooth muscle by elevating intracellular cGMP and therefore potentiate the relaxant activity of sodium nitroprusside rather than isoprenaline. Zaprinast is a poor bronchodilator, but it inhibits histamine release from mast cells and may reduce endothelial cell permeability and plasma leakage. It inhibits exercise-induced asthma but not histamine-induced bronchospasm, suggesting an effect on neuronal pathways involved in bronchoconstriction (Murray & England 1992, Raeburn *et al.* 1993).

A potential application of PDE V inhibitors is in antithrombotic therapy to prevent platelet aggregation and arterial thrombosis. This property may be used in the management of intermittent claudication to improve circulation to ischaemic skeletal muscle. The inhibition of platelet aggregation would improve blood flow to the ischaemic areas without significant vasodilatation through PDE IV inhibition. The latter effect would otherwise divert blood flow away from the already underperfused ischaemic tissues (Nicholson *et al.* 1991). Non-selective PDE inhibitors of the alkylxanthine class, including oxypentifylline (Trental[R]), are currently in use for this purpose. Oxypentifylline is used for intermittent claudication due to chronic obstructive arterial disease and increases the distance that patients can walk without experiencing pain. It may also be of value in cerebrovascular disease. The beneficial action is probably not related to vasodilatation but to antithrombotic activity, to improved flow properties of blood and to a favourable tissue oxygen utilization profile (Angershach & Ochlich 1984).

PDE inhibitors selective for the different isoenzymes are therefore currently under active investigation for several therapeutic targets including asthma, cardiac failure, CNS disorders of mood and cognition and peripheral ischaemic conditions. Several agents are now available within each class and have received encouraging reports in clinical trials. Their true value, however, awaits the development of additional compounds with improved side-effect profiles.

Supersensitivity and Desensitization

14.1   Introduction

Throughout the preceding chapters of this book many examples of changes in sensitivity of tissues and organs both in *in vivo* and in isolated tissues have occurred. Indeed, the pharmacological evaluation of agonist activity of autonomic neurotransmitters is based upon the measurement of tissue responsiveness after drug treatments, surgical procedures or changes in environment. The basis of tissue sensitivity measurements, from the standpoint of receptor theory, was discussed in Chapter 3. The importance of concentration–response curves and determination of the EC_{50} values was stressed when examining changes in sensitivity induced by drugs and by alterations in the chemical, physical and physiological environment (Kenakin 1984, 1993). Factors controlling tissue sensitivity that have been dealt with in this book include the age of the subject (α-adrenoceptor sensitivity), sex and hormone status (eg thyroid hormone levels on adrenoceptor function) and disease state (eg β-adrenoceptors in asthma and heart failure) (Chapter 3).

A distinction can be made between acute and chronic changes in sensitivity. Furthermore, sensitivity of tissues and organs may be increased by a drug or procedure, when it is termed supersensitivity or sensitization, compared with a subsensitivity or desensitization. Acute and chronic changes in sensitivity may occur with the same drug treatment but in opposite directions. For example, the acute effect of a receptor antagonist is to block the agonist-induced responses of the target organ, thus dose–response curves are displaced to the right in its presence (eg β-blockade of cardiac responses, Figure 3.10). However, the chronic use of β-adrenoceptor antagonists can lead to an underlying up-regulation of β-adrenoceptors, so that when the antagonist is removed by abrupt termination of treatment there is an *increase* in sensitivity to β-agonists. This may precipitate angina or fatal arrhythmias in susceptible individuals due to the raised sensitivity to endogenous catecholamines (see Chapter 5).

14.2 Acute Sensitivity Changes

Acute changes in sensitivity to an agonist can be subdivided into five separate categories as follows.

1 *Presence of an antagonist.* Drugs which block the receptor competitively or irreversibly or at a postreceptor coupling site, such as the G protein or L-type Ca^{2+} channels (Chapter 13), can reduce agonist sensitivity.

2 *Changes in physical conditions of* in vitro *experiments.* The bathing environment of isolated tissues has an important impact on the sensitivity to agonists and must therefore remain constant both during an experiment and for comparisons between other agonists. Important factors are the temperature, pH and ionic concentration. Temperature has been shown to affect the thermodynamics of several drug–receptor interactions. More favourable interactions and therefore supersensitivity of β_1-adrenoceptor-mediated responses occurs as temperature is lowered (Broadley & Williams 1983). The pH of the bathing medium can affect the ionization of drugs or the charge on specific groups on the receptor and thereby change the ability to interact. The responses to noradrenaline of rat atria are increased by alkalosis and reduced by acidosis. pH can also affect the chemical stability of a drug and thus its concentration in the tissue bath. The best known example is the oxidation of catecholamines at alkaline pH. The presence of ascorbic acid prevents the degradation of noradrenaline (Kenakin 1984). An example of the effect of ionic concentration is the dependence of smooth muscle contractile responses upon adequate extracellular Ca^{2+} levels (Chapter 13).

3 *Concentration of agonist in the vicinity of the receptor.* The importance of equilibrium conditions for the measurement of tissue responses was emphasized in Chapter 3 (Table 3.1). When a tisssue possesses active removal mechanisms for the drug, then the response to that drug is governed by its steady-state concentration in the receptor compartment. This is controlled by the rate diffusion of the drug into the tissue receptor compartment and the rate of removal by tissue uptake or local metabolism. If the latter process exceeds the rate of diffusion, there could be a constant deficit of drug at the receptor compared with that in the tissue bath or in the circulation *in vivo*. When the uptake or metabolism pathways are inhibited by drugs, the concentration of

agonist in the receptor compartment rises and supersensitivity ensues. This is the situation with catecholamines that are substrates for neuronal uptake, inhibition of which by cocaine causes supersensitivity. This process was described in Chapter 3 (Figures 2.25 and 2.26). Similarly, inhibitors of synaptic metabolism of Ach or noradrenaline can potentiate the actions of agonists that are substrates. The responses to cholinesters are potentiated by the presence of an anticholinesterase, such as physostigmine (Figure 9.1, Chapter 10).

4   *Blockade of opposing influences.* This example of changes in sensitivity to an agonist occurs when the agonist has more than one site of action. Removal of an underlying mechanism that normally opposes the recorded response will then produce an apparent increase in sensitivity. There are numerous examples of this type of change in responsiveness throughout this book. Where an agonist interacts with two receptor subtypes that mediate opposing responses, blockade of one

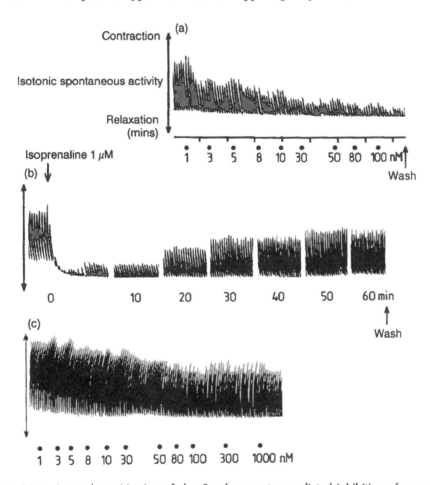

Figure 14.1 Acute desensitization of the β_1-adrenoceptor-mediated inhibition of myogenic activity of an isolated segment of rabbit jejunum recorded isotonically. A cumulative concentration–response curve for isoprenaline is constructed in (a) and (c). A single dose of isoprenaline is added in (b) and left in contact with the tissue for 60 min, during which time there is gradual recovery of myogenic activity (ie desensitization). After washout, the second curve for isoprenaline is substantially attenuated.

subtype will enhance the responses mediated via the unblocked receptor. For example, the α-adrenoceptor-mediated pressor response to adrenaline is enhanced by β-adrenoceptor blockade. Blockade of reflexes may also enhance the response to an agonist *in vivo*. Removal of vagal reflexes with atropine potentiates the tachycardia induced by noradrenaline. The α_1-adrenoceptor-mediated rise in blood pressure usually induces baroreceptor-mediated reflex vagal slowing of the heart which opposes the direct β-adrenoceptor-mediated tachycardia (Figure 5.2). Prejunctional and postjunctional locations of receptors also exert opposing actions on the responses to neurotransmitter released by nerve stimulation. For example, phentolamine blocks prejunctional α_2-adrenoceptors that normally exert negative feedback control of noradrenaline release and therefore potentiates the β-adrenoceptor-mediated responses to sympathetic nerve stimulation of isolated atria (Figure 6.2, Chapter 6). Indirectly acting sympathomimetic amines represent an example where sensitivity changes are the opposite to those of a directly acting agent (Chapter 4). An indirectly acting amine, such as tyramine, is inhibited by reserpine depletion whereas the sensitivity of directly acting amines such as noradrenaline are unaffected or even potentiated (Figure 6.8).

5 *Acute tachyphylaxis or desensitization.* The progressive loss of sensitivity to an agonist either with repeated exposure or in its continuous presence is desensitization or tachyphylaxis. The progressive reduction of tissue sensitivity to an indirectly acting sympathomimetic amine with repeated exposure is a special case of tachyphylaxis attributable to the loss of available stores of noradrenaline in superficial storage vesicles (Figure 4.1, Chapter 4). The term desensitization is usually applied to the phenomenon of a diminished functional responsiveness of the tissue to an agonist, despite the continued presence of that agonist (Figure 14.1). This process can occur rapidly within minutes or more slowly over a matter of hours or days. There has been considerable work on the process of desensitization of G protein-coupled receptors, much of which has centred on the β-adrenoceptor. The next sections of this chapter will therefore concentrate on the mechanisms of desensitization, dealing with both the acute and chronic phases of the process.

14.3 Desensitization at G Protein-coupled Receptors

The process of desensitization has been observed in a diverse range of biological systems, including chemotaxis responses of bacteria, pheromone responses of yeast and the adaptation of the eyes to light. One of the main model systems used to examine the mechanisms of receptor desensitization for functional pharmacological responses has been the β-adrenoceptor coupled to adenylyl cyclase.

The role of desensitization of adrenoceptor-mediated functional responses appears to be to limit the tissue responsiveness during continued neuronal firing or maintained circulating catecholamine levels. This process will thereby protect the tissues from excessive stimulation. The extremely short process of desensitization which involves uncoupling of the receptor from G protein activation occurs within seconds to minutes and may be a further means of turning off the agonist signal (Hausdorff *et al.* 1990). Termination of the agonist-induced response is brought about by the intrinsic ATPase activity of the GTP binding protein, which removes GTP from the α-subunit and allows the G protein trimer to reform in readiness for a further coupling of the activated receptor complex to the G protein (Figure 13.3).

The cyclic interaction between receptor, G protein and adenylyl cyclase may also be broken by the uncoupling of receptor from G protein, which would thus terminate the stimulation of adenylyl cyclase. These processes are therefore involved in the homeostasis of receptor sensitivity and their coupling to the functional response.

In homologous desensitization, only the responsiveness to the desensitizing agent is diminished and is usually specific for the receptor type in question. Heterologous desensitization is associated with a loss of responsiveness to a range of agents acting through the same second messenger pathway. The distinction is, however, not always clear since both types of desensitization may occur simultaneously (Harden 1983).

14.4 Mechanisms of Desensitization

14.4.1 β-Adrenoceptors

Considerably more work has been done to elucidate the mechanisms of desensitization by use of the β-adrenoceptor–adenylyl cyclase system because the structure of this receptor and its second messenger coupling has been well defined by molecular biological techniques. Knowledge has been assisted by the cloning of structurally modified β-adrenoceptors by site-directed mutagenesis and by the availability of β-adrenoceptor-containing cell lines lacking essential components of the receptor–response coupling. For example, the cyc⁻ mutant of the S49 lymphoma cell lines lacks the α_s-subunit of the G protein and the kin⁻ mutant lacks cAMP-dependent protein kinase (PKA). Both cell types can still undergo desensitization and phosphorylation (Strasser *et al.* 1986). This observation indicates that desensitization does not involve the G protein but the site is at the receptor itself.

Several mechanisms of β-adrenoceptor desensitization have been identified which occur under different experimental conditions depending upon the concentration of agonist and the duration of exposure. These have provided a useful model for other autonomic receptors but, as will be seen later, the processes are not identical for all receptors. Most studies have used isolated cell preparations and have measured the adenylyl cyclase activity from the accumulation of cAMP. In the continued presence of agonist, the intracellular cAMP levels generally plateau or even return to basal levels within a few minutes. The potential sites of this desensitization are therefore the receptor itself, the regulatory G protein (G_s for β-adrenoceptors) or adenylyl cyclase. The ability to stimulate adenylyl cyclase is generally unchanged since the responses to NaF, in cell fragments, or to forskolin, which stimulates the enzyme directly, are not reduced in short-term desensitization procedures. As noted above, the presence of G protein is not essential for demonstrating desensitization, which points to the β-adrenoceptor as the prime location for change during desensitization. Three processes have been described: uncoupling, sequestration and down-regulation (Hausdorff *et al.* 1990, Lefkowitz *et al.* 1990, Liggett & Lefkowitz 1994).

14.4.1.1 Uncoupling of the β-adrenoceptor from the G protein

The increase in cAMP levels caused by β-adrenoceptor activation in turn stimulates the activity of protein kinase A (PKA) which is responsible for the phosphorylation of a wide range of substrates involved in the pharmacological responses of smooth and cardiac muscle (Figure 13.3, Chapter 13). PKA also phosphorylates the receptor itself and is responsible for its uncoupling from G protein activation. Coupling of the receptor to the G protein is via the third intracellular loop of the receptor and sites of phosphorylation by PKA have been identified at residues 259–262 of the β_2-adrenoceptor intracellular loop. A mutated form of human β_2-adrenoceptor, expressed in a mammalian fibroblast cell line, in which the serine amino acids of this portion were replaced by alanines, showed a greatly impaired capacity for phosphorylation and desensitization. A second site of phosphorylation by PKA is at residues 343–346 of the proximal portion of the cytoplasmic tail (Figure 14.2). This process occurs within 15 min at nanomolar concentrations (10 nM) of isoprenaline and requires only the presence of cAMP. It may therefore be mimicked by other agents causing elevation of cAMP such as PDE inhibitors and dibutyryl cAMP. Furthermore, dibutyryl cAMP also phosphorylates other receptors linked via G_s to cAMP production, causing heterologous desensitization to, for example, PGE_1. Membranes from cell lines expressing the β_2-adrenoceptor but lacking the PKA catalytic subunit (S49 lymphoma kin⁻ variant) show only a partial desensitization indicating the role of PKA in this process. The residual desensitization of these cells is specific for the β-adrenoceptor agonist and indicates the presence of a second pathway of desensitization. This has been identified as being due to β-adrenoceptor kinase (β-ARK), which phosphorylates 11 serine and threonine residues in the

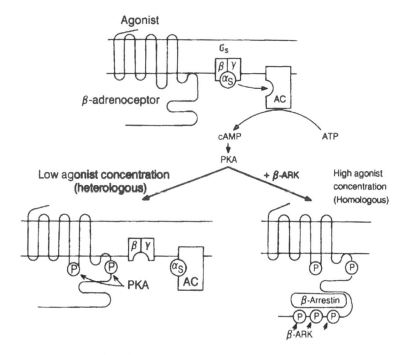

Figure 14.2 Agonist-induced desensitization of the β-adrenoceptor associated with uncoupling of the receptor from the guanine nucleotide regulatory protein (G_s). The receptor is phosphorylated (P) by protein kinase A (PKA) activated by the cAMP generated by the action of adenylyl cyclase (AC). At high agonist concentrations, β-adrenoceptor kinase (β-ARK) also phosphorylates the receptor at the carboxy terminus of the receptor. Uncoupling of the β-ARK-phosphorylated receptor also requires β-arrestin.

cytoplasmic tail of the β-adrenoceptor (Figure 14.2). Mutant β_2-adrenoceptors with these residues substituted or with a truncated carboxyl terminus showed impaired phosphorylation and desensitization, as measured by cAMP generation, although the G_s was normal. These two kinases (PKA and β-ARK) involved in phosphorylation of the β-adrenoceptor may be selectively inhibited. β-ARK is inhibited by polyanions such as heparin, which can be introduced into whole cells permeabilized with low concentrations of digitonin, whereas PKA is inhibited by protein kinase I, a 25 amino acid synthetic analogue of the naturally occurring inhibitor of the kinase.

Also present in crude preparations of β-ARK was a factor termed β-arrestin. This is analogous to arrestin, the 48 kDa protein that inactivates rhodopsin kinase in the rhodopsin photoreceptor system. β-ARK is comparable with rhodopsin kinase, the retinal enzyme that phosphorylates and inactivates only the light-bleached form of rhodopsin. Thus, arrestin inhibits transducin, the retinal G protein, from interacting with the phosphorylated rhodopsin. β-Arrestin interacts with the β-ARK phosphorylated receptor leading to enhanced uncoupling from the G_s regulatory protein. Therefore both β-ARK and β-arrestin are required for this agonist-specific phosphorylation and uncoupling process, leading to homologous desensitization. Human β-ARK has now been cloned and sequenced. It also interacts with the $\beta\gamma$-subunits of the G protein by the C-terminal portion of β-ARK. The β-ARK-$\beta\gamma$ complex is thereby anchored in the cell membrane and allows for more efficient receptor phosphorylation and desensitization. This would occur when the receptors are activated to release the $\beta\tau$-subunit and would thus facilitate termination of receptor activation through the phosphorylation process.

Phosphorylation and uncoupling by PKA and β-ARK occur under different concentration conditions. PKA activation requires only elevation of cAMP and occurs at low agonist concentrations (exposure to 10 nM isoprenaline for 15 min). Agonists at other receptors activating adenylyl cyclase and cAMP production, such as PGE_1, can also induce phosphorylation via PKA and heterologous desensitization. β-ARK, however, requires specific agonist occupation for these occupied receptors to be phosphorylated. Therefore desensitization with low concentrations of agonist which only occupy a fraction of the receptors (spare receptors) is mainly through PKA. At higher concentrations (eg 2 μM isoprenaline for 15 min), β-ARK has a significant role in desensitization together with PKA. Cloned β_1- and β_2-adrenoceptors show differences in their potential phosphorylation sites, with the β_1-adrenoceptor having only residues for PKA activity in the third cytoplasmic loop and 10 serine or threonine sites for potential β-ARK phosphorylation compared with 11 for the β_2-adrenoceptor. Otherwise, they both undergo phosphorylation and uncoupling with short-term exposure to agonist. The β_3-adrenoceptor (see Chapter 4) has no serine or threonine PKA phosphorylation sequences or β-ARK phosphorylation sites in the carboxyl tail. The β_3-adrenoceptor expressed in isolated cell lines does not undergo desensitization with exposure to concentrations of isoprenaline as high as 100 μM for 30 min or subsequent down-regulation.

14.4.1.2 Sequestration

This is the process of internalization of cell surface receptors to the intracellular compartment. It occurs along a similar time course (seconds to minutes) to that of uncoupling. Sequestration is associated with a loss of β-adrenoceptor binding sites measured with hydrophilic radioligands, such as CGP-12177 (see Chapter 5,

Table 5.3), which determines only cell surface receptors in intact cells. There is no change in total β-adrenoceptor binding assessed with hydrophobic ligands, such as dihydroalprenolol, that can penetrate the cell membrane. Thus, there is translocation of receptors into the cell during this phase of desensitization (Figure 14.3). Uncoupling precedes sequestration but phosphorylation is not a prerequisite for sequestration. Approximately 70% of cell surface β-adrenoceptors may be sequestered within 30 min of agonist exposure, but the question has been asked why they need to be sequestered when desensitization has already been effected by uncoupling. It is possible that sequestration must occur before resensitization can take place by dephosphorylation. Conconavalin A inhibits sequestration and has been shown to reduce functional resensitization.

14.4.1.3 Down-regulation

This term is used to describe a reduction in the total number of receptors irrespective of their location. This process has a much longer time course and occurs after several hours of agonist exposure. The receptor protein appears to break down, presumably by lysosomal pathways. Recovery from down-regulation requires the synthesis of new receptor protein and can take a matter of days. *In vitro* studies have shown that the recovery of β-adrenoceptors after long-term exposure of isolated cells to agonist (8 hr) is delayed by the protein synthesis inhibitor, cycloheximide. The desensitization after short-term exposure, however, shows rapid recovery which is not affected by cycloheximide. In this case, sequestered receptors merely reappear at the cell surface. Synthesis of new receptor protein after down-regulation appears to take place in the Golgi apparatus and may involve transport via microtubules to the cell surface. Receptors alkylated by the irreversible antagonists such as BAAM (see Chapter 5) are also replaced by similar resynthesis pathways and provide an index of the normal turnover rate of adrenoceptors. *In vivo* estimates provide $t_{\frac{1}{2}}$ values of 18 hr for β_2- and 45 hr for β_1-adrenoceptors (Mahan *et al.* 1987).

It is not certain whether sequestration is a prelude to receptor down-regulation, whereby the internalized receptors become degraded. However, while sequestration requires agonist exposure, down-regulation can be achieved by exposure to cAMP analogues alone. Furthermore, conconavalin A, which inhibits sequestration, does not significantly alter down-regulation. Studies with mutant β_2-adrenoceptors show that certain structural features of the receptor are required for down-regulation. There is some, but not an absolute requirement for G_s coupling and the presence of PKA phosphorylation sites.

With long-term exposure to the agonist over 12–24 hr, the down-regulation of β_2-adrenoceptors is associated with a decline in the levels of mRNA encoding the β_2-adrenoceptor. The half-life of mRNA is normally ~12 hr, but after down-regulation it declines to ~7 hr. It is concluded that there is no reduction in the rate of transcription from DNA, but a destabilization of the mRNA resulting in its degradative removal by a mRNA binding protein. This effect is mimicked by forskolin, but to a lesser extent, and is therefore mediated in part via the raised cAMP levels which probably activate the RNA binding protein (Bahouth & Malbon 1994). The possibility that the G protein is also a site of heterologous desensitization has been considered, since stimulation of G_s by PGE_1 for 3 hr in NG108–15 neuroblastoma cells results in a loss of the 45 kDa $G_{s\alpha}$ protein. Production of cAMP is not required since it is not mimicked by forskolin or dibutyryl cAMP. In S49

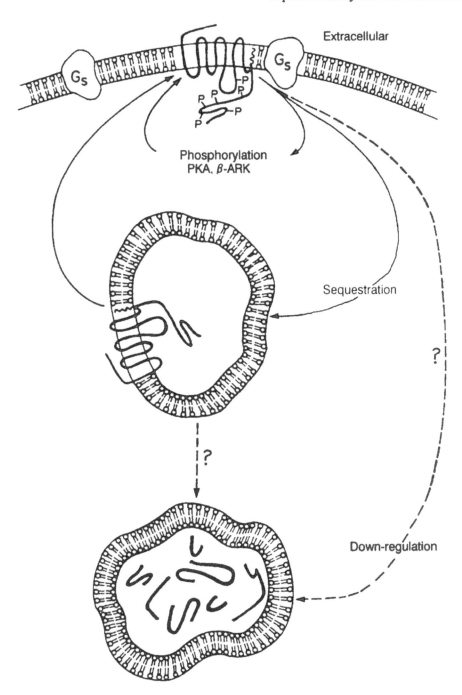

Figure 14.3 Mechanisms of agonist-induced β-adrenoceptor desensitization. *Uncoupling* of the β-adrenoceptor from the guanine nucleotide regulatory protein (G_s) follows phosphorylation by protein kinase A (PKA) and β-adrenoceptor kinase (β-ARK). *Sequestration* is a rapid intracellular translocation of the receptor. *Down-regulation* is a decrease in receptor number occurring after prolonged agonist exposure and associated with degradation of the receptor. Whether sequestration must precede down-regulation is unclear. Reproduced with permission from Hausdorff *et al.* (1990).

521

lymphoma cells, β-adrenoceptor activation causes translocation of $G_{s\alpha}$ from the membrane to the cytoplasmic fraction. Phosphorylation of the G protein also occurs, but is unlikely to contribute to the heterologous desensitization. The loss of $G_{s\alpha}$ may therefore be a common feature of longer term heterologous desensitization (Milligan & Green 1994).

14.4.2 α_2-Adrenoceptors

Although less work has been performed on the mechanisms of desensitization of receptors other than β-adrenoceptors, more information is now becoming available.

α_2-Adrenoceptors are negatively coupled to adenylyl cyclase (Chapters 3 and 13) and earlier studies were of limited value since inhibition of adenylyl cyclase is not easily measured and it was not always clear which subtypes of α_2-adrenoceptors were being examined. The mechanisms of desensitization appear to be similar to those of the β-adrenoceptor. Short-term agonist exposure causes phosphorylation at multiple serine and threonine residues of the third intracellular loop. Cell lines with mutant α_2-adrenoceptors lacking this vital segment do not undergo short-term desensitization as measured by the inhibition of forskolin-stimulated adenylyl cyclase by adrenaline. Sequestration also occurs within minutes, but is independent of phosphorylation. Conconavalin A blocks sequestration but not the short-term desensitization. It does, however, prevent long-term desensitization indicating that, unlike with β-adrenoceptors, sequestration is a prerequisite for down-regulation. Long-term desensitization of α_2-adrenoceptors does not require phosphorylation, but it is unclear how much down-regulation is involved (Liggett & Lefkowitz 1994). There is evidence that subtypes of α_2-adrenoceptors are down-regulated in different ways (Shreve *et al.* 1991).

Prolonged exposure to agonists that exert inhibitory effects on adenylyl cyclase can result in a compensatory increased activity of the enzyme. This is manifest as a supersensitivity or rebound increase in activity on withdrawal of the inhibitory agonist. An example is adipocytes removed from animals pretreated with adenosine receptor agonists, which inhibit adenylyl cyclase via A_1 receptors. Intact adipocytes show increases in basal and isoprenaline- or forskolin-stimulated cAMP accumulation. The precise mechanism is still not fully understood, but the chronic inhibition of cAMP may not be essential. A signal transduction mechanism mediated through the G protein (G_i or G_o) appears to induce an adaptive up-regulation of the catalytic subunit of adenylyl cyclase. This sensitization of adenylyl cyclase may explain the loss of activity (desensitization) that occurs with inhibitory agonists such as adenosine (Thomas & Hoffman 1994). A decrease in the α-subunit of G_i is also proposed. After 2 days of *in vivo* exposure there is no change in mRNA encoding for G_i, but a possible increase in G protein degradation (Milligan & Green 1994).

14.4.3 α_1-Adrenoceptors and Muscarinic M_3 Receptors

In contrast to the receptor types mentioned earlier, α_1-adrenoceptors and muscarinic M_3 receptors are coupled via the G_q regulatory protein to phosphoinositide turnover (Chapter 13). Desensitization occurs by internalization, with a $t_{1/2}$ $\leqslant$10 min. Upon removal of the agonist, receptors return to the cell surface with equal rapidity. The desensitization of muscarinic-receptor- and α_1-adrenoceptor-mediated IP$_3$ accumulation correlates well with the reduction of cell surface receptors. Internalization does not require phosphoinositide hydrolysis, since it still occurs when external Ca^{2+} is removed. Internalized muscarinic receptors, like those present in the sarcolemma, still appear to be able to bind agonists in a GTP-sensitive manner. This is in contrast to internalized β-adrenoceptors. They are not therefore uncoupled, although phosphorylation at the third intracellular loop may occur. After internalization the receptors are down-regulated.

Reduced numbers of muscarinic binding sites have been observed after incubation of various isolated cell lines with muscarinic agonists or after chronic *in vivo* treatment. The correlation with levels of mRNA encoding these receptors, however, is poor. For example, treatment of rats with an organophosphorus anticholinesterase resulted in reduced cardiac muscarinic receptor binding sites but no change in m$_2$ receptor mRNA (Caulfield 1993).

Agonist-induced IP$_3$ formation can be maintained for hour-long periods despite partial receptor desensitization. Therefore, desensitization can also occur downstream of phosphoinositidase-C (PIC) due to persistently elevated levels of IP$_3$ or DAG. The PKC that is activated by DAG can be down-regulated, which occurs with persistent activation by phorbol esters. The IP$_3$ receptor of the SR also appears to be down-regulated (Wojcikiewicz *et al.* 1994). Chronic activation of muscarinic m$_1$ or m$_3$ receptors with carbachol caused down-regulation of α-subunits of G$_q$. This requires agonist occupancy and is probably due to enhanced degradation (Milligan & Green 1994). Muscarinic receptors negatively coupled to adenylyl cyclase undergo heterologous desensitization requiring only activation of protein kinase C. These m$_2$ receptors are phosphorylated within 1–5 min of agonist exposure, which uncouples the receptor from the G protein. Phosphorylation occurs by the action of two protein kinases, PKC and a G protein-coupled receptor kinase (GRK) (Hosey 1994).

14.5 Desensitization of Functional Tissue Responses

So far we have considered the mechanisms of desensitization with regard to structural changes in the receptor and its coupling to the G protein. Such information has been generated essentially from radioligand binding and second messenger production (cAMP or IP$_3$ formation) in isolated cell lines. The functional relevance of this data, however, can only be assessed by extension to the *in vivo* situation or to isolated intact tissue responses.

Prolonged *in vitro* exposure of isolated tissues to an agonist usually shows a slowly developing loss of sensitivity. This desensitization may take an hour or more to develop, for example, the inhibition of spontaneous myogenic activity of the rabbit ileum is slowly reversed in the continued presence of the β-adrenoceptor agonist, isoprenaline. The response to isoprenaline after washout is then absent (Figure 14.1). Experiments with isolated cardiac tissues have shown that exposure to isoprenaline (10^{-6} M) for 4 hr is required to demonstrate a persistent desensitization. In these experiments the initial β-adrenoceptor sensitivity is assessed by constructing a dose–response curve to the agonist. The tissue is then incubated with isoprenaline and then washed thoroughly before testing the sensitivity by construction of a further dose–response curve. The desensitization is evident from the rightward shift of the dose–response curve and possible reduction of the maximum response, indicating loss of functional receptors (Herepath & Broadley 1990b). There is no cross-desensitization with histamine, suggesting that it is homologous and associated with down-regulation of receptors. It is not therefore of the rapid reversible uncoupling type, the function of which is probably to turn off the immediate agonist signal. The long-term slowly reversible desensitization is probably for protection of the effector from excessive stimulation with chronic agonist exposure.

Whether β_1- and β_2- adrenoceptors are equally susceptible to this long-term down-regulation is a matter of dispute. There are studies that show responses mediated via β_2-adrenoceptors to be down-regulated with chronic *in vitro* agonist exposure. Some of these can be criticized because of the experimental conditions, such as adequate washout of agonist before test of tissue sensitivity. If these factors are controlled, then under identical conditions β_1-adrenoceptors mediating atrial rate and force of contraction responses are desensitized whereas β_2-adrenoceptor-mediated relaxation of airway or vascular smooth muscle is not (Herepath & Broadley 1992). Incubation of isolated tissues such as guinea-pig ileum with muscarinic agonists also induces short-term desensitization characterized by rightward shift of the dose–response curve and depression of the maximum response (Siegel *et al.* 1984).

In vitro, α_1-adrenoceptor functional desensitization has been demonstrated in rabbit isolated aortic rings incubated with adrenaline for 7 hr. The dose–response curve for contractile responses to noradrenaline was displaced to the right without affecting histamine or 5-HT, characteristic of homologous desensitization. There was no change in the α_1-adrenoceptor binding sites or affinity, but phosphatidylinositol turnover was blunted, indicating an uncoupling mechanism (Lurie *et al.* 1985).

Chronic pretreatment of animals with agonists has also revealed an *ex vivo* loss of functional responses and radioligand binding in subsequently removed tissues. This has been found with cardiac β_1-adrenoceptors after isoprenaline infusions (Harden 1984, Brodde 1991), vascular α_1-adrenoceptors after adrenaline (Colucci *et al.* 1981b) and dopamine D_2 receptors of the renal vasculature after dopexamine infusions (Martin & Broadley 1994). The agonist-induced reduction in sensitivity and β-adrenoceptor binding characteristics of the airways and of the heart after chronic agonist administration to animals and humans have been discussed in Chapter 3. These changes occur over prolonged periods and the mechanisms are probably of the down-regulatory type.

β_2-Adrenoceptors have received considerable attention because of the relevance to tolerance to the bronchodilator effects of β-agonists with their chronic use in asthma therapy. Whether β_2-adrenoceptor functional responsiveness is reduced in vascular and airway smooth muscle after chronic agonist exposure is still unclear. There are reports of reduced β_2-adrenoceptor sensitivity in the heart (Brodde 1991), the vas deferens (May *et al.* 1986) and the vasculature, albeit substantially less than the β_1-adrenoceptor sensitivity (Cohen & Schenck 1987) after long-term β-agonist treatment. However, there are also reports of unaltered sensitivity of vascular β_2-adrenoceptors after chronic infusions of isoprenaline or adrenaline (Tsjujimoto & Hoffman 1985, Martin & Broadley 1994). Similarly, the β-adrenoceptor binding characteristics of the lungs were unchanged although β_2-adrenoceptor binding of lymphocytes from patients receiving chronic β_2-agonist treatment are down-regulated (Hauck *et al.* 1990). Indeed, the lymphocyte β_2-adrenoceptor has been widely used as an index of desensitization after chronic β-agonist administration to humans. Whether it is truly representative of desensitization of β-adrenoceptors, however, has been questioned (Michel *et al.* 1986, Hauck *et al.* 1990).

Down-regulation of receptors and associated desensitization may also occur when levels of endogenous agonists are chronically elevated. Phaeochromocytoma is an example. The adrenomedullary tumour causes elevated circulating levels of catecholamines. Tissues removed from animals with experimental phaeochromocytoma or patients show blunted adrenoceptor responsiveness (Hoffman 1987). Down-regulation and subsensitivity have also been demonstrated when the metabolism of endogenous amines is inhibited or when their removal by uptake processes is impaired by drugs. For example, muscarinic receptor desensitization has been demonstrated after chronic exposure of animals to the anticholinesterases, which raise the levels of Ach in the synapse. Chronic subacute daily exposure of animals to the organophosphorus compounds, DFP and sarin (Chapter 10), causes initial signs of Ach accumulation including salivation, urination and diarrhoea. After 5–6 days of treatment, however, these effects of anticholinesterase toxicity are attenuated or disappear. These changes are associated with a reduction in [^{3}H]QNB (Chapter 7) muscarinic binding sites in the trachea but only a small reduction in contractile sensitivity to a muscarinic agonist, bethanechol, that is not a substrate for AchE. The sensitivity of the trachea to Ach is still increased because of the AchE inhibition, although not as much as at 2 hr after commencing treatment (Mohan *et al.* 1988). The pupillary constriction in response to muscarinic agonists is impaired by chronic administration of anticholinesterases and by continuous light exposure, which causes maintained activation of parasympathetic pathways and release of ACh. These changes are related to a fall in muscarinic binding sites ([^{3}H]QNB). The pupillary constrictor response (miosis) to muscarinic agonists and to light is impaired in humans by chronic exposure to muscarinic agonists and anticholinesterases (Smith & Smith 1980). The pupil constriction is also reduced by continuous light exposure, which causes maintained activation of the parasympathetic pathways and release of Ach. These changes are related to a fall in muscarinic binding sites ([^{3}H]QNB). This loss of sensitivity with long-term agonist exposure does not appear to influence the intraocular pressure lowering efficacy of the anticholinesterases in glaucoma to a great extent. The development of tolerance is probably minimized by the multiple sites of action in lowering intraocular pressure (Massoulie & Toutant 1988) (Chapters 8 and 10). Chronic administration of the catecholamine neuronal uptake inhibitor, desmethylimipramine (Chapter 2), enhances synaptic noradrenaline levels and has been shown to reduce the density of β-adrenoceptors in the cerebral cortex of rats (Minneman *et al.* 1981).

14.6   Chronic Supersensitivity

A raised sensitivity to agonists occurs following chronic removal of receptor activation. Several examples of this phenomenon have been described throughout this book. Abrupt withdrawal of β-adrenoceptor antagonist therapy has been associated with an increased risk of angina and sudden death. This is attributed to an underlying up-regulation of β-adrenoceptors caused by their chronic blockade. When blockade is removed, the up-regulated β-adrenoceptors are exposed to circulating catecholamines and display enhanced sensitivity to the cardiac stimulant effects (Brodde 1991). Treatment of animals with propranolol results in enhanced sensitivity of isolated cardiac tissues removed from these animals and an increase in density of β-adrenoceptor binding sites (Chess-Williams & Broadley 1984). These processes and differential effects on β_1- and β_2-adrenoceptors are discussed in Chapter 3 (cardiac β-adrenoceptors in disease and after drug treatment) and Chapter 5 (properties of β-adrenoceptor antagonists).

Removal of endogenous neurotransmitter release from autonomic nerve endings by surgical or chemical disruption of nerve activity also causes a time-dependent increase in effector organ sensitivity to agonists. This form of supersensitivity has been demonstrated extensively after surgical sympathectomy of the rat vas deferens (Abel *et al.* 1985) and cat nictitating membrane, which show increases in sensitivity to the α_1-adrenoceptor-mediated contractions by noradrenaline after several days. If the preganglionic nerve to the nictitating membrane is severed (decentralization), then uptake mechanisms associated with the intact postganglionic neurone are unaffected and the enhanced sensitivity cannot be due to presynaptic changes. The postsynaptic increase in sensitivity is attributed to an up-regulation of α_1-adrenoceptors induced by the lack of tonic stimulation by endogenous noradrenaline (Figure 14.4) (Broadley 1990). Similarly, surgical interruption of the sympathetic innervation of the guinea-pig heart causes a slowly developing supersensitivity to β_1-adrenoceptor stimulation in subsequently removed isolated cardiac tissues (Goto *et al.* 1985).

Chemical sympathectomy has been achieved with noradrenaline depleting agents, such as reserpine and guanethidine, or with the neurotoxin, 6-hydroxydopamine (Chapter 6). Chronic treatment of animals with these agents has been shown to induce a supersensitivity of isolated tissues removed from these animals after several days. Increases in sensitivity have been observed for responses mediated via α_1-adrenoceptors of the vasculature, vas deferens, nictitating membrane and ileum, and β_1-adrenoceptors of cardiac tissues and the ileum (Grassby & Broadley 1986). The changes in sensitivity after 6-OHDA or reserpine are revealed as leftward shifts of the dose–response curves for directly acting sympathomimetic amines; indirectly acting amines are inhibited by the catecholamine depletion (Chapter 4). They are not due to inhibition of neuronal uptake, which may explain any immediate supersensitivity, since they occur with amines that are not substrates for uptake$_1$. The maximum responses of partial agonists are also raised, consistent with an increase in receptor number (Figure 14.5). In the heart there is no change in sensitivity to agonists not acting through β_1-adrenoceptors, such as histamine, although small increases in sensitivity to forskolin have been found, suggesting some change at the level of G$_s$ binding protein coupling to adenylyl cyclase (Broadley *et al.* 1984).

No change in affinity of agonists for the β-adrenoceptor has been determined in functional measurements of tissue responses by means of irreversible or functional antagonism (see Chapter 3) (Chess-Williams *et al.* 1985). Radioligand binding data has shown that there is a small but significant increase in the density of β-adrenoceptor binding sites in the heart after depletion by reserpine or 6-OHDA but no change in K_D, indicating that the mechanism of the supersensitivity is primarily due to an up-regulation of β-adrenoceptors (Chess-Williams *et al.* 1986, 1987a,b). Associated with the increase in β_1-adrenoceptor expression is an increase in mRNA in the heart and salivary glands after sympathectomy with reserpine or surgery. The stability of the mRNA is increased, the half-life extending from 3.5 to 8 hr. Therefore, the increased expression of β-adrenoceptors is not due to increased transcription of mRNA from DNA, but at a posttranscriptional level due to increased stability of the mRNA (Bahouth & Malbon 1994).

Interestingly, the cardiac α_1-adrenoceptors (Chapter 3) do not display supersensitivity or up-regulation of binding sites after 6-OHDA or reserpine pretreatment,

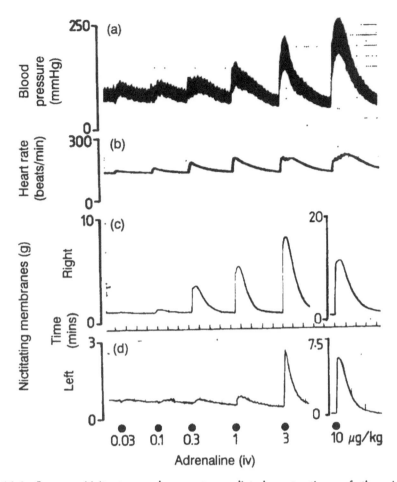

Figure 14.4 Supersensitivity to α-adrenoceptor-mediated contractions of the nictitating membrane induced by surgical denervation. The right superior cervical sympathetic nerve of the cat (2.91 kg) was cut preganglionically under halothane : nitrous oxide anaesthesia. Seven days later, a dose–response curve for intravenous adrenaline was obtained under chloralose (80 mg/kg iv) anaesthesia after induction with halothane : nitrous oxide on (a) blood pressure, (b) heart rate, (c) right nictitating membrane and (d) left nictitating membrane. Supersensitivity of the denervated membrane is indicated by its greater sensitivity to adrenaline. Note that the sensitivity range for the right membrane is over three-fold less than for the left membrane.

indicating that they are probably not under the tonic influence of the sympathetic innervation (Chess-Williams *et al.* 1987a,b). As indicated above, α_1-adrenoceptors are capable of showing supersensitivity; increases occur in other tissues. Also of interest is the observation that β_2-adrenoceptor-mediated relaxation responses of the lungs, uterus and vas deferens do not display supersensitivity after chronic reserpine or 6-OHDA treatment. This may indicate either that these β_2-adrenoceptors are not under the influences of the nervous innervation, or that β_2-adrenoceptors are not normally desensitized by endogenous catecholamines (Broadley *et al.* 1986, Grassby & Broadley 1986).

Cholinergic responses also display supersensitivity. The sensitivity of the iris to muscarinic stimulation is increased by ciliary ganglionectomy, by chronic HC-3

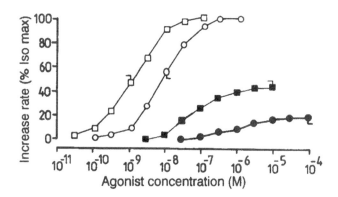

Figure 14.5 Reserpine-induced supersensitivity of the β-adrenoceptor-mediated increases in right atrial rate to isoprenaline (Iso, open symbols) and the partial agonist prenalterol (closed symbols). Right atria were removed from guinea-pigs that were untreated (○ ●, $n = 17$) or pretreated with reserpine (0.5 mg/kg daily for 7 days, ip) (□ ■, $n = 4$). Responses were measured as the increase in spontaneous rate of contraction and plotted as the mean percentage of the maximum response to isoprenaline. Supersensitivity is revealed as a shift of the concentration–response curve for isoprenaline to the left and an increase in maximum for the partial agonist.

treatment to prevent synthesis and release of Ach (Chapter 7), and by continuous exposure to dark, by covering the eye, which impairs nerve traffic. These functional changes are associated with increases in muscarinic receptor binding density (Massoulie & Toutant 1988). Chronic nicotine administration for 10 days causes an increase in the number of nicotinic binding sites ([^{3}H]ACh) but not muscarinic binding sites ([^{3}H]QNB) in rat cerebral cortex (Schwartz & Kellar 1983). This is probably due to chronic antagonism of these receptors; the link with smoking has been discussed in Chapter 11.

The supersensitivity to sympathomimetic amines induced by chronic treatment with reserpine or 6-OHDA is relatively selective and generally agreed to be the result of chronic depletion of catecholamines from the noradrenergic nerve terminals. This results in a loss of endogenous neurotransmitter release onto the adrenoceptors, which respond by up-regulation (Fleming *et al.* 1973, Fleming 1984). Rather than the supersensitivity being a positive up-regulation, it can be viewed as a passive loss of the normal desensitization by endogenous catecholamines and therefore a return to the true basal level of sensitivity. Thus, supersensitivity of this type probably only occurs if there is an underlying desensitization at rest.

Sympathetic denervation causes not only supersensitivity but also hypertrophy and hyperplasia of vascular smooth muscle, fibroblasts and cardiac muscle. Thus noradrenergic innervation appears to exert a repressive effect on the nuclear activity of effector cells. This does not appear to be due to the noradrenaline release, since the hypertrophy is not mimicked by depletion of noradrenaline with reserpine. Furthermore, it is not prevented by continuous infusions of noradrenaline. Instead, an adenosine antagonist has been shown to induce comparable hypertrophy in blood vessels and the heart. The repressive effect of the sympathetic innervation is therefore proposed to be exerted by adenosine, which is formed by the breakdown of the co-transmitter, ATP (see Chapter 12) (Osswald & Azevedo 1991).

In addition to the control of receptor function and tissue sensitivity by the tonic effects of neurotransmitter release, other endogenous factors also play an important role, for example the hormonal status. The effects of thyroid hormone on cardiac adrenoceptors and of oestrogens on uterine adrenoceptor balance have been described in Chapter 3. Also controlling receptor sensitivity are the corticosteroids (glucocorticoids). Oral administration of a glucocorticoid such as cortisone acetate, increases β_2-adrenoceptor density in neutrophils. The β_2-adrenoceptor density is also increased in rat lung after hydrocortisone (Mano *et al.* 1979) and in various cell types including smooth muscle in culture after prolonged exposure to dexamethasone. The β_2-adrenoceptor-mediated stimulation of adenylyl cyclase is also increased. The changes in receptor expression are preceded by an increase in mRNA. This is prevented by actinomycin D, an inhibitor of RNA polymerase II and mRNA transcription. The mechanism involves increased transcription of mRNA from DNA (Bahouth & Malbon 1994). Surprisingly, in these isolated cell studies, dexamethasone causes a down-regulation of β_1-adrenoceptors and their mRNA. *In vivo*, however, the effect of chronic stress has been shown to increase the sensitivity of isolated cardiac tissues to β_1-adrenoceptor stimulation, but only when the levels of glucocorticoids are markedly elevated. Thus, the potentiation of the β-adrenoceptor-mediated cardiac effects are attributed to the raised corticosteroids (Bassett *et al.* 1978).

The influence of corticosteroids on β_2-adrenoceptor function has particular relevance to the treatment of bronchial asthma. Steroids are of course used in their own right to relieve the inflammatory responses in asthma. Their use will also influence the β_2-adrenoceptor characteristics of the airways and of leucocytes. In asthma there is a reduced sensitivity of the airways to β-adrenoceptor stimulation and a down-regulation of lymphocyte β_2-adrenoceptors. The reduced airway responsiveness is probably due to uncoupling between receptor and adenylyl cyclase (Goldie *et al.* 1990) (Chapter 3). These effects are attributed to the elevation of catecholamines induced by the stress of an asthma attack. Additionally, the regular use of β_2-agonists as bronchodilators may further down-regulate β_2-adrenoceptors, as measured in lymphocytes from asthma patients treated with β-agonists such as terbutaline (Hauck *et al.* 1990). Administration of corticosteroids such as prednisone or beclomethasone (Becotide[R]), or of ketotifen, restores the lymphocyte β_2-adrenoceptor density to normal (Brodde *et al.* 1985, Svedmyr 1990). Thus, corticosteroids have a functional role in maintaining β-adrenoceptor sensitivity in the face of down-regulation by exogenous and endogenous agonists and have a permissive influence on β-adrenoceptor function (Davies & Lefkowitz 1984, Svedmyr 1990).

14.7 Reprise

The autonomic nervous system is involved in homeostasis of the body's internal environment and the constant maintenance of a steady state in the face of changes in external influences. Blood pressure, respiration, digestion, body temperature and less obvious factors such as metabolism, the immune response and haematological changes are all influenced by the autonomic nervous system. The mechanisms described in this chapter show that the receptor mechanisms of the autonomic nervous system are also self-regulating. This enables the body to compensate for too much or too little autonomic adaptation. The compensatory changes to over- or under-activity of the autonomic nervous system allows for fine tuning and maintenance of the status quo. With such efficient regulatory mechanisms occurring, it is therefore surprising that pharmacological changes can be induced in the intact individual by means of drugs. The number of drugs exerting their effects by interacting with the autonomic nervous system, either to increase or depress its function at selected sites, has increased steadily this century. The tables of chemical structures presented in this book are a testament to this fact and there appears to be no slowing of progress. Further step-by-step improvements in selectivity of these agents and consequent patient acceptability and the introduction of new classes of drugs can be envisaged. Pharmacology has its origins in the autonomic nervous system with the pioneering work of scientists such as Dale, and autonomic pharmacology continues to lead the subject forward to the next century.

'Finis Opus Coronat'

References

ABBS, E.T., BROADLEY, K.J. & ROBERTS, D.J. (1967) Inhibition of catechol-O-methyl transferase by some acid degradation products of adrenaline and noradrenaline. *Biochem. Pharmacol.*, **16**, 279–282.

ABEL, P.W., JOHNSON, R.D., MARTIN, T.J. & MINNEMAN, K.P. (1985) Sympathetic denervation does not alter the density or properties of *alpha-1* adrenergic receptors in rat vas deferens. *J. Pharmacol. Exp. Ther.*, **233**, 570–577.

ACHESON, G.H. & PEREIRA, S.A. (1946) The blocking effect of tetraethylammonium ion on the superior cervical ganglion of the cat. *J. Pharmacol. Exp. Ther.*, **87**, 273–280.

ADLER-GRASCHINSKY, E. & LANGER, S.Z. (1975) Possible role of a β-adrenoceptor in the regulation of noradrenaline release by nerve stimulation through a positive feed-back mechanism. *Br. J. Pharmacol.*, **53**, 43–50.

AGOSTON, D.V. (1988) Cholinergic co-transmission. In: *The Cholinergic synapse. Handbook of Experimental Pharmacology*, pp. 479–533, V.P. Whittacker (Ed.), Springer-Verlag, Berlin.

AHLQUIST, R.P. (1948) A study of adrenotropic receptors. *Am. J. Physiol.*, **153**, 586–600.

AHLQUIST, R.P. & LEVY, B. (1959) Adrenergic receptive mechanism of canine ileum. *J. Pharmacol. Exp. Ther.*, **127**, 146–149.

ALBERTS, P., BARTFAI, T. & STJÄRNE, L. (1982) The effects of atropine on [^{3}H] acetylcholine secretion from myenteric plexus evoked electrically or by high potassium. *J. Physiol.*, **329**, 93–112.

ALLEN, T.G.J. & BURNSTOCK, G. (1990) M_1 and M_2 muscarinic receptors mediate excitation and inhibition of guinea-pig intracardiac neurones in culture. *J. Physiol.*, **422**, 463–480.

ALSTER, P. & WENNMALM, A. (1983) Effect of nicotine on prostacyclin formation in rat aorta. *Eur. J. Pharmacol.*, **86**, 441–446.

AMBACHE, N. (1951) A further survey of the action of *Clostridium botulinum* toxin upon different types of autonomic nerve fibre. *J. Physiol.*, **113**, 1–17.

AMOBI, N.I.B. & SMITH, I.C.H. (1993) The relative importance of extracellular and intracellular calcium in the responses of the human vas deferens to noradrenaline and potassium: a study using Ca^{2+}-deprivation and Ca^{2+}-antagonists. *J. Auton. Pharmacol.*, **13**, 177–192.

ANDEN, N.-E. & GRABOWSKA-ANDEN, M. (1985) Synthesis and utilization of catecholamines in the rat superior cervical ganglion following changes in the nerve impulse flow. *J. Neural Transm.*, **64**, 81–92.

ANDERSSON, K.-E. (1988) Current concepts in the treatment of disorders of micturition. *Drugs*, 35, 477–494.

ANDRE, C., VAUQUELIN, G., SEVERNE, Y., DE BACKER, J-P. & STROSBERG, A.D. (1982) Dual effects of N-ethylmaleimide on agonist-mediated conformational changes of β-adrenergic receptors. *Biochem. Pharmacol.*, 31, 3657–3662.

ANDREWS, H.E., BRUCKDORFER, K.R., DUNN, R.C. & JACOBS, M. (1987) Low-density lipoproteins inhibit endothelium-dependent relaxation in rabbit aorta. *Nature*, 327, 237–239.

ANDREWS, J.S., JANSEN, J.H.M., LINDERS, S. & PRINCEN, A. (1993) Reversal of performance deficits induced by scopolamine or hemicholinium-3 (HC3). *Life Sci.*, 52, 593.

ANGELI, P., BRASILI, L., CINGOLANI, M.L., MARUCCI, G., PIGINI, P. & TONNINI, M.C. (1991) Pharmacological characterization of muscarinic receptor subtypes in rabbit isolated tissue preparations. *J. Auton. Pharmacol.*, 11, 315–321.

ANGERSBACH, D. & OCHLICH, P. (1984) The effect of 7-(2′-oxopropyl)-1,3-di-*n*-butyl-xanthine (BRL 30892) on ischaemic skeletal muscle PO_2, pH and contractility in cats and rats. *Arzneim.-Forsch. Drug Res.*, 34, 1274–1278.

ANTAL, L.C. & GOOD, C.S. (1980) Effects of oxprenolol on pistol shooting under stress. *The Practitioner*, 224, 755–760.

ANYUKHOVSKY, E.P. & ROSEN, M.R. (1991) Abnormal automatic rhythms in ischemic Purkinje fibers are modulated by a specific α_1-adrenoceptor subtype. *Circulation*, 83, 2076–2082.

ARCH, J.R.S., AINSWORTH, A.T. & CAWTHORNE, M.A. (1982) Thermogenic and anorectic effects of ephedrine and congeners in mice and rats. *Life Sci.*, 30, 1817–1826.

ARCH, J.R.S., AINSWORTH, A.T., CAWTHORNE, M.A., PIERCY, V., SENNITT, M.V., THODY, V.E., WILSON, C. & WILSON, S. (1984) Atypical β-adrenoceptor on brown adipocytes as targets for anti-obesity drugs. *Nature*, 309, 163–165.

ARCH, J.R.S., CAWTHORNE, M.A., CONEY, K.A., GUSTERSON, B.A., PIERCY, V., SENNITT, M.V., SMITH, S.A., WALLACE, J. & WILSON, S. (1991) β-Adrenoceptor-mediated control of thermogenesis, body composition and glucose homeostasis. In: *Obesity and Cachexia*, pp. 241–268, N.J. Rothwell & M.J. Stock (Eds), John Wiley, Chichester.

ARGYLL-ROBERTSON, D. (1863) The Calabar bean as a new agent in ophthalmic practice. *Edinburgh Med. J.*, 8, 815–820.

ARIËNS, E.J. (1954) Affinity and intrinsic activity in the theory of competitive inhibition. *Arch. Int. Pharmacodyn. Ther.*, 99, 32–49.

ARIËNS, E.J. (1960a) Various types of receptors for sympathomimetic drugs. In: *Ciba Foundation Symposium on Adrenergic Mechanisms*, pp. 264–274, J.R. Vane, G.E.W. Wolstenholme & M. O'Connor (Eds), Churchill, London.

ARIËNS, E.J. (1960b) Sympathomimetic drugs and their receptors. In: *Ciba Foundation Symposium on Adrenergic Mechanisms*, pp. 253–263, J.R. Vane, G.E.W. Wolstenholme & M. O'Connor (Eds), Churchill, London.

ARIËNS, E.J. & SIMONIS, A.M. (1983) Physiological and pharmacological aspects of adrenergic receptor classification. *Biochem. Pharmacol.*, 32, 1539–1545.

ARUNLAKSHANA, O. & SCHILD, H.O. (1959) Some quantitative uses of drug antagonists. *Br. J. Pharmacol. Chemother.*, 14, 48–58.

ASANO, M., MASUZAWA, K., MATSUDA, T. & ASANO, T. (1988) Reduced function of the stimulatory GTP-binding protein in *beta* adrenoceptor-adenylate cyclase system of femoral arteries isolated from spontaneously hypertensive rats. *J. Pharmacol. Exp. Ther.*, 246, 709–718.

ÅSTRAND, P., BROCK, J.A. & CUNNANE, T.C. (1988) Time course of transmitter action at the sympathetic neuroeffector junction in rodent vascular and non-vascular smooth muscle. *J. Physiol.*, 401, 657–670.

ATACK, J.R., QIAN, Y., SONCRANT, T.T., BROSSI, A. & RAPOPORT, S.I. (1989) Comparative inhibitory effects of various physostigmine analogs against acetyl- and butyrylcholinesterases. *J. Pharmacol. Exp. Ther.*, **249**, 194–202.

ATLAS, D. & BURSTEIN, Y. (1984) Isolation and partial purification of a clonidine-displacing brain substance. *Eur. J. Biochem.*, **144**, 287–293.

AUGUSTINE, G.J., CHARLTON, M.P. & SMITH, S.J. (1987) Calcium action in synaptic transmitter release. *Annu. Rev. Neurosci.*, **10**, 633–693.

AXELROD, J. (1971) Noradrenaline: fate and control of its biosynthesis. *Science*, **173**, 598–606.

BACH, A.W.J., LAN, N.C., JOHNSON, D.L., ABELL, C.W., BEMBENEK, M.E., KWAN, S.-W., SEEBURG, P.H. & SHIH, J.C. (1988) cDNA cloning of human liver monoamine oxidase A and B: molecular basis of differences in enzymatic properties. *Proc. Natl Acad. Sci. USA.*, **85**, 4934–4938.

BACQ, Z.M. (1975) *Chemical Transmission of Nerve Impulses*, Pergamon Press, Oxford.

BADEWITZ-DODD, L. (1991) Drugs and sport. *Media Medica, Chichester* (in collaboration with the Sports Council, London and IOC Medical Commission).

BAHOUTH, C. & MALBON, C.C. (1994) Genetic (transcriptional and post-transcriptional) regulation of G-protein-linked receptor expression. In: *Regulation of Cellular Signal Transduction Pathways by Desensitization and Amplification*, pp. 99–112, D.R. Sibley & M.D. Houslay (Eds), John Wiley, Chichester.

BALDWIN, H.A., DE SOUZA, R.J. SARNA, G.S., MURRAY, T.K., GREEN, A.R. & CROSS, A.J. (1991) Measurements of tacrine and monoamines in brain by *in vivo* microdialysis argue against release of monoamines by tacrine at therapeutic doses. *Br. J. Pharmacol.*, **103**, 1946–1950.

BALL, D.I., BRITTAIN, R.T., COLEMAN, R.A., DENYER, L.H., JACK, D., JOHNSON, M., LUNTS, L.H.C., NIALS, A.T., SHELDRICK, K.E. & SKIDMORE, I.F. (1991) Salmeterol, a novel long-acting β_2-adrenoceptor agonist: characterization of pharmacological activity *in vitro* and *in vivo*. *Br. J. Pharmacol.*, **104**, 665–671.

BARGER, G. & DALE, H.H. (1910) Chemical structure and sympathomimetic action of amines. *J. Physiol.*, **41**, 19–59.

BARLOW, R.B. & ING, H.R. (1948) Curare-like action of polymethylene *bis*-quaternary ammonium salts. *Br. J. Pharmacol. Chemother.*, **3**, 298–304.

BARLOW, R.B., BERRY, K.J., GLENTON, P.A.M., NIKOLAOU, N.M. & SOH, K.S. (1976) A comparison of affinity constants for muscarinic sensitive acetylcholine receptors in the guinea-pig atrial pacemaker cells at 29 °C and in ileum at 29 °C and 37 °C. *Br. J. Pharmacol.*, **58**, 613–620.

BARNARD, E.A., BURNSTOCK, G. & WEBB, T.E. (1994) G Protein-coupled receptors for ATP and other nucleotides: a new receptor family. *Trends Pharmacol. Sci.*, **15**, 67–70.

BARNES, P.J. (1993a) Muscarinic receptor subtypes in airways. *Life Sci.*, **52**, 521–527.

BARNES, P.J. (1993b) Autonomic pharmacology of the airways. In: *Pharmacology of the Respiratory Tract. Lung Biology in Health and Disease*, Vol. 67, pp. 415–455, K.F. Chung & P.J. Barnes (Eds), Marcel Dekker, New York.

BARNES, P.J. & CHUNG, K.F. (1992) Questions about inhaled β_2-adrenoceptor agonists in asthma. *Trends Pharmacol. Sci.*, **13**, 20–23.

BARNES, P.J., CHUNG, K.F. & PAGE, C.P. (1988) Inflammatory mediators and asthma. *Pharmacol. Rev.*, **40**, 49–84.

BARRÚS, M.T., MARIN, J. & BALFAGÓN, G. (1993) Presynaptic 5-hydroxytryptamine receptors modulating noradrenaline release in bovine cerebral arteries. *J. Auton. Pharmacol.*, **13**, 413–423.

BASSENGE, E. (1988) Cardiovascular actions of nicotine. In: *The Pharmacology of Nicotine*, pp. 117–137, M.J. Rand & K. Thurau (Eds), IRL Press, Oxford.

BASSETT, J.R., STRAND, F.L. & CAIRNCROSS, K.D. (1978) Glucocorticoids, adrenocorticotrophic hormone and related polypeptides on myocardial sensitivity to noradrenaline. *Eur. J. Pharmacol.*, **49**, 243–249.

BAUMGOLD, J., KARTON, Y., MALKA, N. & JACOBSON, K.A. (1992) High affinity acylating antagonists for muscarinic receptors. *Life Sci.*, **51**, 345–351.

BAXTER, G.F. & YELLON, D.M. (1993) Reversal of left ventricular hypertrophy: extending the aims of antihypertensive therapy. *Pharm. J.*, **250**, 60–63.

BAYLISS, W.M. (1901) On the origin from the spinal cord of the vasodilator fibres of the hind-limb, and on the nature of these fibres. *J. Physiol.*, **26**, 173–209.

BEAVO, J.A. & REIFSNYDER, D.H. (1990) Primary sequences suggest selective inhibition of individual cyclic nucleotide phosphodiesterase is possible. *Trends Pharmacol. Sci.*, **11**, 150–155.

BECKMANN, M.L., GERBER, J.G., BYYNY, R.L., LoVERDE, M. & NIES, A.S. (1988) Propranolol increases prostacyclin synthesis in patients with essential hypertension. *Hypertension*, **12**, 582–588.

BELFRAGE, P. (1985) Hormonal control of lipid degradation. In: *New Perspectives in Adipose Tissue: Structure, Function and Development*, pp. 121–144, A. Cryer & R.L.R. Van (Eds), Butterworths, London.

BELL, C. (1990) Dopamine and the kidney. In: *Peripheral Dopamine Pathophysiology*, pp. 87–98, F. Amenta (Ed.), CRC Press, Boca Raton, FL.

BELL, C. & LANG, W.J. (1973) Neural dopaminergic vasodilator control in the kidney. *Nature, New Biol.*, **246**, 27–29.

BELL, C. & SUNN, N. (1990) A functional role for renal dopaminergic nerves in the dog. *J. Auton. Pharmacol.*, **10**(Suppl. 1), s41–s45.

BELLEAU, B. (1960) Relationships between agonists, antagonists and receptor sites. In: *Ciba Foundation Symposium on Adrenergic Mechanisms*, pp. 223–245, J.R. Vane, G.E.W. Wolstenholme & M. O'Connor (Eds), Churchill, London.

BEMBILLA-PERROT, B. & TERRIER DE LA CHAISE, A. (1992) Provocation of supraventricular tachycardias by an intravenous class I antiarrhythmic drug. *Int. J. Cardiol.*, **34**, 189–198.

BENFEY, B.G. (1979) Cardiac inotropic beta-adrenoceptors are fully active at low temperatures. *J. Pharmacol. Exp. Ther.*, **210**, 429–432.

BENFEY, B.G. (1987) Function of myocardial α-adrenoceptors. *J. Appl. Cardiol.*, **2**, 49–70.

BENFEY, B.G. (1990) Function of myocardial α-adrenoceptors. *Life Sci.*, **46**, 743–757.

BENFEY, B.G. (1993) Functions of myocardial α-adrenoceptors. *J. Auton. Pharmacol.*, **13**, 351–372.

BENFIELD, P. & SORKIN, E.M. (1987) Esmolol. A preliminary review of its pharmacodynamic and pharmacokinetic properties, and therapeutic efficacy. *Drugs*, **33**, 392–412.

BENOWITZ, N.L. (1988a) Pharmacologic aspects of cigarette smoking and nicotine addiction. *N. Eng. J. Med.*, **319**, 1318–1330.

BENOWITZ, N.L. (1988b) Pharmacokinetics and pharmacodynamics of nicotine. In: *The Pharmacology of Nicotine*, pp. 3–18, M.J. Rand & K. Thurau (Eds), IRL Press, Oxford.

BERGOUGNAN, L., ROSENWEIG, P., DUCHIER, J., COURNOT, A., BERLIN, I. & MORSELLI, P.L. (1990) SL 84,0418: clinical and cardiovascular tolerance in healthy young volunteers of a new alpha-2 antagonist with anti-hyperglycaemic properties. *Eur. J. Pharmacol.*, **183**, 1015–1016.

BERNE, R.M. (1980) The role of adenosine in the regulation of coronary blood flow. *Circ. Res.*, **47**, 807–813.

BERRARD, S., BRUCE, A. & MALLET, J. (1989) Molecular genetic approach to study the mammalian choline acetyltransferase. *Brain Res. Bull.*, **22**, 147–153.

BERRY, J.L., SMALL, R.C., HUGHES, S.J., SMITH, R.D., MILLER, A.J., HOLLINGSWORTH, M., EDWARDS, G. & WESTON, A.H. (1992) Inhibition by adrenergic neurone blocking agents of the relaxation induced by BRL 38227 in vascular, intestinal and uterine smooth muscle. *Br. J. Pharmacol.*, **107**, 288–295.

BERTHELSEN, S. & PETTINGER, W.A. (1977) A functional basis for classification of α-adrenergic receptors. *Life Sci.*, **21**, 595–606.

BERTI, F. & SHORE, P.A. (1967) A kinetic analysis of drugs that inhibit the adrenergic neuronal membrane amine pump. *Biochem. Pharmacol.*, **16**, 2091–2094.

BESSE, J.C. & FURCHGOTT, R.F. (1976) Dissociation constants and relative efficacies of agonists acting on α-adrenergic receptors in rabbit aorta. *J. Pharmacol. Exp. Ther.*, **197**, 66–78.

BEVAN, J.A. & SU, C. (1973) Sympathetic mechanisms in blood vessels: nerve and muscle relationships. *Annu. Rev. Pharmacol.*, **13**, 269–285.

BEVAN, J.A., BEVAN, R.D., KITE, K. & ORIOWO, M.A. (1988) Species differences in sensitivity of aortae to norepinephrine are related to α-adrenoceptor affinity. *Trends Pharmacol. Sci.*, **9**, 87–89.

BHALLA, R.C., SHARMA, R.V. & ASHLY, T. (1978) Adenylate cyclase activity in myocardium of spontaneously hypertensive rat: effect of endogenous factors and solubilization. *Biochem. Biophys. Res. Commun.*, **82**, 273–280.

BHAT, M.B., MISHRA, S.K. & RAVIPRAKASH, V. (1989) Differential susceptibility of cholinergic and noncholinergic neurogenic responses to calcium channel blockers and low calcium medium in rat urinary bladder. *Br. J. Pharmacol.*, **96**, 837–842.

BIRDSALL, N.J.M., SPALDING, T.A., CORRIE, J.E.T., CURTIS, C.A.M. & HULME, E.C. (1993) Studies on muscarinic receptors using nitrogen mustards. *Life Sci.*, **52**, 561.

BIRMINGHAM, A.T. & WILSON, A.B. (1963) Preganglionic and postganglionic stimulation of the guinea-pig isolated vas deferens preparation. *Br. J. Pharmacol. Chemother.*, **21**, 569–580.

BIRNBAUMER, L. (1990) G Proteins and signal transduction. *Annu. Rev. Pharmacol. Toxicol.*, **30**, 675–705.

BIRNBAUMER, L., ABRAMOWITZ, J. & BROWN, A.M. (1990) Receptor–effector coupling by G proteins. *Biochim. Biophys. Acta*, **1031**, 163–224.

BLACK, J.W. & LEFF, P. (1983) Operational models of pharmacological agonism. *Proc. R. Soc. London, Ser. B.*, **220**, 141–162.

BLACK, J.W. & SHANKLEY, N.P. (1985) Pharmacological analysis of muscarinic receptors coupled to oxyntic cell secretion in the mouse stomach. *Br. J. Pharmacol.*, **86**, 601–607.

BLACK, J.W. & STEPHENSON, J.S. (1962) Pharmacology of a new adrenergic beta-receptor blocking compound (Nethalide). *Lancet*, **ii**, 311–314.

BLACK, J.W., GERSKOWITCH, V.P., RANDALL, P.J. & TRIST, D.G. (1981) Critical examination of the histamine–cimetidine interaction in guinea-pig heart and brain. *Br. J. Pharmacol.*, **74**, 978P–979P.

BLAKELEY, A.G.H. & CUNNANE, T.C. (1979) The packeted release of transmitter from the sympathetic nerves of the guinea-pig vas deferens: an electrophysiological study. *J. Physiol.*, **296**, 85–96.

BLASCHKO, H. (1939) The specific action of L-dopa-decarboxylase. *J. Physiol.*, **96**, 50P–51P.

BLUE, D.R., VIMONT, R.L. & CLARKE, D.E. (1992) Evidence for a noradrenergic innervation to α_{1A}-adrenoceptors in rat kidney. *Br. J. Pharmacol.*, **107**, 414–417.

BÖHM, M., DIET, F., FEILER, G., KEMKES, B. & ERDMANN, E. (1988) α-Adrenoceptors and α-adrenoceptor-mediated positive inotropic effects in failing human myocardium. *J. Cardiovasc. Pharmacol.*, **12**, 357–364.

BÖHM, M., PIESKE, B., SCHNABEL, P., SCHWINGER, R., KEMKES, B., KLOVE-KORN, W.-P. & ERDMANN, E. (1989a) Reduced effects of dopexamine on force of contraction in the failing human heart despite preserved β_2-adrenoceptor subpopulation. *J. Cardiovasc. Pharmacol.*, **14**, 549–559.

BÖHM, M., SCHMITZ, W., SCHOLZ, H. & WILKEN, A. (1989b) Pertussis toxin prevents adenosine receptor and m-cholinoceptor-mediated sinus rate slowing and AV conduction block in the guinea-pig heart. *Naunyn-Schmiedeberg's Arch. Pharmacol.*, **312**, 239–243.

BÖHM, M., GIERSCHIK, P., JAKOBS, K.-H., PIESKE, B., SCHNABEL, P., UNGERER, M. & ERDMANN, E. (1990) Increase in $G_{i\alpha}$ in human hearts with dilated but not ischemic cardiomyopathy. *Circulation*, **82**, 1249–1265.

BOJANIC, D. & NAHORSKI, S.R. (1983) Identification and subclassification of rat adipocyte beta-adrenoceptors using ($\pm$)-^{125}I-cyanopindolol. *Eur. J. Pharmacol.*, **93**, 235–243.

BOJANIC, D., JANSEN, J.D., NAHORSKI, S.R. & ZAAGSMA, J. (1985) Atypical characteristics of the β-adrenoceptor mediating cyclic AMP generation and lipolysis in the rat adipocyte. *Br. J. Pharmacol.*, **84**, 131–137.

BOLLI, R. (1992) Myocardial 'stunning' in man. *Circulation*, **86**, 1671–1691.

BOLLI, P., ERNE, P., KIOWSKI, W., JI, B.H., AMANN, F.W. & BUHLER, F.R. (1983) Important contribution of post-junctional alpha$_2$-adrenoceptor mediated vasoconstriction to arteriolar tone in man. *J. Hypertens.*, **1**(Suppl. 2), 257–259.

BOND, R.A. & CLARKE, D.E. (1988) Agonist and antagonist characterization of a putative adrenoceptor with distinct pharmacological properties from the α- and β-subtypes. *Br. J. Pharmacol.*, **95**, 723–734.

BÖNISCH, H. & MICHAEL-HEPP, J. (1990) ^{3}H-Desipramine binding to the uptake$_1$-carrier of bovine adrenal medulla. *J. Auton. Pharmacol.*, **10**, 10.

BÖNISCH, H. & TRENDELENBURG, U. (1988) The mechanism of action of indirectly acting sympathomimetic amines. In: *Catecholamines 1. Handbook of Experimental Pharmacology*, Vol. 90, pp. 247–277, U. Trendelenburg & N. Weiner (Eds), Springer-Verlag, Berlin.

BONNER, T.I. (1992) Domains of muscarinic acetylcholine receptors that confer specificity of G protein coupling. *Trends Pharmacol. Sci.*, **13**, 48–50.

BORKOWSKI, K.R. (1988) Pre- and postjunctional β-adrenoceptors and hypertension. *J. Auton. Pharmacol.*, **8**, 153–171.

BORTON, M., CONNAUGHTON, S. & DOCHERTY, J.R. (1991) Actions of 8-hydroxy-2-(N-dipropylamino)tetralin (8-OH-DPAT) at α_2-adrenoceptors. *J. Auton. Pharmacol.*, **11**, 247–253.

BOULANGER, C.M., MORRISON, K.J. & VANHOUTTE, P.M. (1994) Mediation by M$_3$-muscarinic receptors of both endothelium-dependent contraction and relaxation to acetylcholine in the aorta of the spontaneously hypertensive rat. *Br. J. Pharmacol.*, **112**, 519–524.

BOURA, A.L.A. & GREEN, A.F. (1965) Adrenergic neurone blocking agents. *Annu. Rev. Pharmacol.*, **5**, 183–212.

BOUSQUET, P., FELDMAN, J. & SCHWARTZ, J. (1984) Central cardiovascular effects of α-adrenergic drugs: differences between catecholamines and imidazolines. *J. Pharmacol. Exp. Ther.*, **230**, 232–236.

BOWMAN, W.C. (1981) Effects of adrenergic activators and inhibitors on the skeletal muscles. In: *Adrenergic Activators and Inhibitors. Handbook of Experimental Pharmacology*, Vol. 54, Pt. II, pp. 47–128, L. Szekeres (Ed.), Springer-Verlag, Berlin.

BOWMAN, W.C. & RAND, M.J. (1980) *Textbook of Pharmacology*, 2nd Edn, Blackwell Scientific, Oxford.

BOZLER, E. (1948) Conduction, automaticity and tonus of visceral muscles. *Experientia*, **4**, 213–218.

BRASCH, H. (1991) No influence of prejunctional α_2-adrenoceptors on the effects of nicotine and tyramine in guinea-pig atria. *J. Auton. Pharmacol.*, **11**, 37–44.

BRAVE, S.R., HOBBS, A.J., GIBSON, A. & TUCKER, J.F. (1991) The influence of L-N^G-nitro-arginine on field stimulation induced contractions and acetylcholine release in guinea-pig isolated tracheal smooth muscle. *Biochem. Biophys. Res. Commun.*, **179**, 1017–1022.

BRAVE, S.R., BHAT, S., HOBBS, A.J., TUCKER, J.F. & GIBSON, A. (1993) The influence of L-N^G-nitro-arginine on sympathetic nerve induced contraction and noradrenaline release in the rat isolated anococcygeus muscle. *J. Auton. Pharmacol.*, **13**, 219–225.

BRESLIN, E. & DOCHERTY, J.R. (1993) Endothelium-dependent relaxations of rat aorta to acetylcholine are not inhibited by pertussis toxin. *J. Auton. Pharmacol.*, **13**, 323–328.

536

BRISTOW, M.R., MINOBE, W., RASMUSSEN, R., HERSHBERGER, R.E. & HOFFMAN, B.B. (1988) Alpha-1 adrenergic receptors in the nonfailing and failing human heart. *J. Pharmacol. Exp. Ther.*, **247**, 1039–1045.

BRITTAIN, R.T., JACK, D. & RITCHIE, A.C. (1970) Recent β-adrenoceptor stimulants. *Adv. Drug Res.*, **5**, 197–253.

BROADLEY, K.J. (1980) The effect of temperature upon the responses of isolated cardiac muscle to sympathomimetic amines. *Meth. Find. Exp. Clin. Pharmacol.*, **2**, 181–193.

BROADLEY, K.J. (1982) Review. Cardiac adrenoreceptors. *J. Auton. Pharmacol.*, **2**, 119–145.

BROADLEY, K.J. (1990) Evidence from lack of decentralization-induced supersensitivity that β_2-adrenoceptors of the cat nictitating membrane are non-innervated. *Acta Physiol. Scand.*, **140**, 481–489.

BROADLEY, K.J. (1995) Purinoceptors. In: *Airways Smooth Muscle: A Reference Source*, Vol. III, *Ion Channels, Receptors and Signal Transduction*, D. Raeburn & M. Giembycz (Eds), Birkhäuser Verlag, Basel.

BROADLEY, K.J. & AL-ATTAR, T. (1990) Temperature-dependence of α-receptor subtypes. *J. Auton. Pharmacol.*, **10**, 12P.

BROADLEY, K.J. & GRASSBY, P.F. (1985) Alpha- and beta-adrenoceptor-mediated responses of the guinea-pig ileum and the effects of neuronal uptake inhibition. *Naunyn-Schmiedeberg's Arch. Pharmacol.*, **331**, 316–323.

BROADLEY, K.J. & HAWTHORN, M.H. (1983) Examination of cardiac β-adrenoceptor subtypes by pharmacological and radioligand binding techniques. *Br. J. Pharmacol.*, **78**, 136P.

BROADLEY, K.J. & NICHOLSON, C.D. (1981) Dissociation constants of isoprenaline and orciprenaline and their relative efficacies on guinea-pig isolated atria determined by use of an irreversible β-adrenoceptor antagonist. *Br. J. Pharmacol.*, **72**, 635–643.

BROADLEY, K.J. & PATON, D.M. (1990) Comparison of the uptake and O-methylation of isoprenaline by cardiac and respiratory tissues of guinea-pigs. *Pharmacol. Res.*, **22**, 573–585.

BROADLEY, K.J. & WILLIAMS, R.G. (1983) Temperature-induced changes in dissociation constants (K_A) of agonists at cardiac β-adrenoceptors determined by use of the irreversible antagonist Ro 03-7894. *Br. J. Pharmacol.*, **79**, 517–524.

BROADLEY, K.J., WILLIAMS, R.G., HAWTHORNE, M.H. & GRASSBY, P.F. (1984) Mechanisms of cardiac supersensitivity to sympathomimetic amines. *Meth. Find. Exp. Clin. Pharmacol.*, **6**, 179–186.

BROADLEY, K.J., BROOME, S. & PATON, D.M. (1985) Hypothermia-induced supersensitivity to adenosine for responses mediated via A_1 receptors but not A_2 receptors. *Br. J. Pharmacol.*, **84**, 407–415.

BROADLEY, K.J., CHESS-WILLIAMS, R.G. & GRASSBY, P.F. (1986) A physiological basis for subclassifying beta-adrenoceptors examined by chemical sympathectomy of guinea-pigs. *J. Physiol.*, **373**, 367–378.

BROCK, J.A. & CUNNANE, T.C. (1987) Relationship between the nerve action potential and transmitter release from sympathetic postganglionic nerve terminals. *Nature*, **326**, 605–607.

BROCK, J.A. & CUNNANE, T.C. (1988) Electrical activity at the sympathetic neuroeffector junction in the guinea-pig vas deferens. *J. Physiol.*, **399**, 607–632.

BROCK, J.A. & CUNNANE, T.C. (1992) Electrophysiology of neuroeffector transmission in smooth muscle. In: *Autonomic Neuroeffector Mechanisms*, pp. 121–213, G. Burnstock & C.H.V. Hoyle (Eds), Harwood Academic, Chur.

BROCK, J.A., CUNNANE, T.C., EVANS, R.J. & ZIOGAS, J. (1989) Inhibition of transmitter release from sympathetic nerve endings by w-conotoxin. *Clin. Exp. Pharmacol. Physiol.*, **16**, 333–339.

BRODDE, O.-E. (1987) Cardiac beta-adrenergic receptors. *ISI Atlas Sci.: Pharmacol.*, **1**, 107–112.

BRODDE, O.-E. (1990) Subclassification of peripheral dopamine receptors. *J. Auton. Pharmacol.*, **10**(Suppl. 1), s5–s10.

BRODDE, O.-E. (1991) β_1- and β_2-Adrenoceptors in the human heart: properties, function and alterations in chronic heart failure. *Pharmacol. Rev.*, **43**, 203–242.

BRODDE, O.-E. & MICHEL, M.C. (1992) Adrenergic receptors and their transduction mechanisms in hypertension. *J. Hypertens*, **10**(Suppl. 7), S133–S145.

BRODDE, O.-E., BRINKMAN, M., SCHEMUTH, R., O'HARA, N. & DAUL, A. (1985) Terbutaline-induced desensitization of human lymphocyte β_2-adrenoceptors. Accelerated restoration of β-adrenoceptor responsiveness by prednisone and ketotifen. *J. Clin. Invest.*, **76**, 1096–1101.

BRODDE, O.-E., SCHULER, S., KRETSCH, R., BRINKMAN, M., BORST, H.G., HETZER, R., REIDEMEISTER, J. CHR., WARNECKE, H. & ZERKOWSKI, H.-R. (1986) Regional distribution of β-adrenoceptors in the human heart: coexistence of functional β_1- and β_2-adrenoceptors in both atria and ventricles in severe congestive cardiomyopathy. *J. Cardiovasc. Pharmacol.*, **8**, 1235–1242.

BRODDE, O.-E., VAN TITS, L.J.H. & MICHEL, M.C. (1991) Acute immunomodulatory effects of beta-adrenoceptor agonists. In: *Adrenoceptors: Structure, Mechanisms, Function. Advances in Pharmacological Sciences*, pp. 237–244, E. Szabadi & C.M. Bradshaw (Eds), Birkhäuser-Verlag, Basel.

BROGDEN, R.N., HEEL, R.C., SPEIGHT, T.M. & AVERY, G.S. (1981) α-Methyl-*p*-tyrosine: a review of its pharmacology and clinical use. *Drugs*, **21**, 81–89.

BROWN, C.M., MACKINNON, A.C., REDFERN, W.S., HICKS, P.E., KILPATRICK, A.T., SMALL, C., RAMCHARAN, M., CLAGUE, R.U., CLARKE, R.D., MACFARLANE, C.B. & SPEDDING, M. (1993) The pharmacology of RS-15385-197, a potent and selective α_2-adrenoceptor antagonist. *Br. J. Pharmacol.*, **108**, 516–525.

BROWN, D.A. (1980) Locus and mechanism of action of ganglion blocking agents. In: *Pharmacology of Ganglionic Transmission. Handbook of Experimental Pharmacology*, Vol. 53, pp. 185–235, D.A. Kharkevich (Ed.), Springer-Verlag, Berlin.

BROWN, D.A. (1988) M Currents. In: *Ion Channels*, Vol. 1, pp. 55–94, T. Narahasi (Ed.), Plenum, New York.

BROWN, D.A. & CAULFIELD, M.P. (1979) Hyperpolarizing 'α_2'-adrenoceptors in rat sympathetic ganglia. *Br. J. Pharmacol.*, **65**, 435–445.

BROWN, D.A., GARTHWAITE, J., HAYASHI, E. & YAMADA, S. (1976) Action of surugatoxin on nicotinic receptors in the superior cervical ganglion of the rat. *Br. J. Pharmacol.*, **58**, 157–159.

BROWN, D.A., FATHERAZI, S., GARTHWAITE, J. & WHITE, R.D. (1980) Muscarinic receptors in rat sympathetic ganglia. *Br. J. Pharmacol.*, **70**, 577–592.

BROWN, D.A., DOCHERTY, J.R., FRENCH, A.M., MACDONALD, A., MCGRATH, J.C. & SCOTT, N.C. (1983) Separation of adrenergic and non-adrenergic contractions to field stimulation in the rat vas deferens. *Br. J. Pharmacol.*, **79**, 379–393.

BRYAN-LLUKA, L.J. & VUOCOLO, H.E. (1992) Evidence from guinea-pig trachealis that uptake$_2$ of isoprenaline is enhanced by hyperpolarization of the smooth muscle. *Naunyn-Schmiedeberg's Arch. Pharmacol.*, **346**, 399–404.

BUCKLEY, M.M.T., GOA, K.L. & CLISSOLD, S.P. (1990) Ocular betaxolol: a review of its pharmacological properties, and therapeutic efficacy in glaucoma and ocular hypertension. *Drugs*, **40**, 75–90.

BUCKLEY, N.J. & CAULFIELD, M. (1992) Transmission: acetylcholine. In: *Autonomic Neuroeffector Mechanisms*, pp. 257–322, G. Burnstock & C.H.V. Hoyle (Eds), Harwood Academic, Chur.

BUCKNER, C.K. & SAINI, R.K. (1975) On the use of functional antagonism to estimate dissociation constants for beta adrenergic receptor agonists in isolated guinea-pig trachea. *J. Pharmacol. Exp. Ther.*, **194**, 565–574.

BUHLER, F.R., LARAGH, J.H., BAER, L., VAUGHAN, D.E. & BRUNNER, H.R. (1972) Propranolol inhibition of renin secretion. *N. Engl. J. Med.*, **287**, 1209–1214.

BUKOWIECKI, L., FOLLÉA, N., PARADIS, A. & COLLET, A. (1980) Stereospecific stimulation of brown adipocyte respiration by catecholamines via β_1-adrenoceptors. *Am. J. Physiol.*, **238**, E552–E563.

BÜLTMAN, R. & STARKE, K. (1993) Choroethylclonidine: an irreversible agonist at prejunctional α_2-adrenoceptors in rat vas deferens. *Br. J. Pharmacol.*, **108**, 336–341.

BUNKER, C.B., TERENGHI, G., SPRINGALL, D.R., POLAK, J.M. & DOWD, P.M. (1990) Deficiency of calcitonin gene-related peptide in Raynaud's phenomenon. *Lancet*, **336/8730**, 1530–1533.

BURN, J.H. (1975) *The Autonomic Nervous System*, 5th Edn, Blackwell Scientific, Oxford.

BURN, J.H. & DALE, H.H. (1915) The action of certain quaternary ammonium bases. *J. Pharmacol. Exp. Ther.*, **6**, 417–438.

BURN, J.H. & RAND, M.J. (1958) The action of sympathomimetic amines in animals treated with reserpine. *J. Physiol.*, **144**, 314–336.

BURN, J.H. & RAND, M.J. (1959) Sympathetic postganglionic mechanism. *Nature*, **184**, 163–165.

BURN, J.H. & RAND, M.J. (1965) Acetylcholine in adrenergic transmission. *Annu. Rev. Pharmacol.*, **5**, 163–182.

BURN, J.H. & TAINTER, M.L. (1931) An analysis of the effect of cocaine on the action of adrenaline and tyramine. *J. Physiol.*, **71**, 169–193.

BURNSTOCK, G. (1972) Purinergic nerves. *Pharmacol. Rev.*, **24**, 509–581.

BURNSTOCK, G. & HOLMAN, M.E. (1962) Spontaneous potentials of sympathetic nerve endings in smooth muscle. *J. Physiol.*, **160**, 446–460.

BUTCHERS, P.R., VARDEY, C.J. & JOHNSON, M. (1991) Salmeterol: a potent and long-acting inhibitor of inflammatory mediator release from human lung. *Br. J. Pharmacol.*, **104**, 672–676.

BUXTON, B.F., JONES, C.R., MOLENAAR, P. & SUMMERS, R.J. (1987) Characterization and autoradiographic localization of β-adrenoceptor subtypes in human cardiac tissues. *Br. J. Pharmacol.*, **92**, 299–310.

BYLUND, D.B. (1987) Biochemistry and pharmacology of the alpha-1 adrenergic receptor. In: *The Alpha-1 Adrenergic Receptors*, pp. 19–69, R.R. Ruffolo, Jr (Ed.), Humana Press, Clifton, NJ.

BYLUND, D.B. (1992) Subtypes of α_1- and α_2- adrenergic receptors. *FASEB J.*, **6**, 832–839.

BYLUND, D.B., BLAXALL, H.S., MURPHY, T.J. & SIMMONEAUX, V. (1991) Pharmacological evidence for alpha-2C and alpha-2D adrenergic receptor subtypes. In: *Adrenoceptors: Structure, Mechanisms, Function*, pp. 27–36, E. Szabadi & C.M. Bradshaw (Eds), Birkhäuser-Verlag, Basel.

BYLUND, D.B., EIKENBERG, D.C., HIEBLE, J.P., LANGER, S.Z., LEFKOWITZ, R.J., MINNEMAN, K.P., MOLINOFF, P.B., RUFFOLO, R.R. & TRENDELENBURG, U. (1994) International Union of Pharmacology nomenclature of adrenoceptors. *Pharmacol. Rev.*, **46**, 121–136.

CAIRNS, S.P. & DULHUNTY, A.F. (1993) The effects of β-adrenoceptor activation on contraction in isolated fast- and slow twitch skeletal muscle fibres of the rat. *Br. J. Pharmacol.*, **110**, 1133–1141.

CALLEWAERT, G. (1992) Excitation–contraction coupling in mammalian cardiac cells. *Cardiovasc. Res.*, **26**, 923–932.

CALLINGHAM, B.A., HOLT, A. & ELLIOTT, J. (1991) Properties and functions of the semicarbazide-sensitive amine oxidases. *Biochem. Soc. Trans.*, **19**, 228–233.

CAMBRIDGE, D., WHITING, M.V., BUTTERFIELD, L.J. & ALLAN, G. (1992) The effects of combined angiotensin converting enzyme inhibition and β-adrenoceptor blockade on plasma renin activity in anaesthetized dogs. *Br. J. Pharmacol.*, **106**, 342–347.

CAMEJO, G., HURT, E., THUBRIKAR, M. & BONDJERS, G. (1991) Modification of low density lipoprotein association with arterial intima: a possible environment for the antiatherogenic action of β-blockers. *Circulation*, **84**(Suppl. 4), 4117–4122.

CAMERON, A.R., JOHNSTON, C.F., KIRKPATRICK, C.T. & KIRKPATRICK, M.C.A. (1983) The quest for the inhibitory neurotransmitter in bovine tracheal smooth muscle. *Quart. J. Exp. Physiol.*, **68**, 413–426.

CAMPBELL, T.J. (1987) Cellular electrophysiological effects of D- and DL-sotalol in guinea-pig sinoatrial node, atrium and ventricle and human atrium: differential tissue sensitivity. *Br. J. Pharmacol.*, **90**, 593–599.

CANDELL, L.M., YUN, S.H., TRAN, L.L.P. & EHLERT, F.J. (1990) Differential coupling of the muscarinic receptor to adenylate cyclase and phosphoinositide hydrolysis in the longitudinal muscle of the rat ileum. *Mol. Pharmacol.*, **38**, 689–697.

CANFIELD, P. & PARASKEVA, P. (1992) β-Adrenoceptor agonist stimulation of acid secretion by rat stomach *in vitro* is mediated by 'atypical' β-adrenoceptors. *Br. J. Pharmacol.*, **106**, 583–586.

CANNON, B. & NEDERGAARD, J. (1985) Brown adipose tissue: molecular mechanisms controlling activity and thermogenesis. In: *New Perspectives in Adipose Tissue*: *Structure, Function and Development*, pp. 223–270, A. Cryer & R.L.R. Van (Eds), Butterworths, London.

CANNON, W.B. (1929) Organisation for physiological homeostasis. *Physiol. Rev.*, **9**, 399–431.

CANNON, W.B. & ROSENBLUETH, A. (1933) Studies on conditions of activity in endocrine organs XXIX. Sympathin E and sympathin I. *Am. J. Physiol.*, **104**, 557–574.

CANNON, W.B. & URIDIL, J.E. (1921) Studies on the conditions of activity in endocrine glands. VIII. Some effects on the denervated heart of stimulating the nerves of the liver. *Am. J. Physiol.*, **58**, 353–354.

CARATI, C.J., CREED, K.E. & KEOGH, E.J. (1988) Vascular changes during penile erection in the dog. *J. Physiol.*, **400**, 75–88.

CARLSSON, E., ÅBLAD, B., BRANDSTROM, A. & CARLSSON, B. (1972) Differentiated blockade of the chronotropic effects of various adrenergic stimuli in the cat heart. *Life Sci.*, **11**, 953–958.

CARLSSON, E., DAHLÖF, C.-G., HEDBERG, A., PERSSON, H. & TANGSTRAND, B. (1977) Differentiation of cardiac chronotropic and inotropic effects of β-adrenoceptor agonists. *Naunyn-Schmiedeberg's Arch. Pharmacol.*, **300**, 101–105.

CARMICHAEL, S.W. (1986) *The Adrenal Medulla*, Vol. 4, Cambridge University Press, Cambridge.

CASAGRANDE, C. & SANTANGELO, F. (1990) Dopaminergic prodrugs. In: *Peripheral Dopamine Pathophysiology*, pp. 307–343, F. Amenta (Ed.), CRC Press, Boca Raton, FL.

CAULFIELD, M.P. (1993) Muscarinic receptors – characterization, coupling and function. *Pharmacol. Ther.*, **58**, 319–379.

CAULFIELD, M.P., JONES, S., VALLIS, Y., BUCKLEY, N.J., KIM, G.-D., MILLIGAN, G. & BROWN, D.A. (1994) Muscarinic M-current inhibition via $G_{aq}/11$ and α-adrenoceptor inhibition of Ca^{2+} current via G_{ao} in rat sympathetic neurones. *J. Physiol.*, **477**, 415–422.

CAUSSADE, F. & CLOAREC, A. (1993) Effect of tienoxolol, a new diuretic β-blocking agent, on urinary prostaglandin excretion in the rat. *Br. J. Pharmacol.*, **109**, 278–284.

CHAN, S.L.F., BROWN, C.A., SCARPELLO, K.E. & MORGAN, N.G. (1994) The imidazoline site involved in control of insulin secretion: characteristics that distinguish it from I_1- and I_2-sites. *Br. J. Pharmacol.*, **112**, 1065–1070.

CHANGEUX, J.-P., DEVILLERS-THIERY, A. & CHEMMUIVILLI, P. (1984) Acetylcholine receptor: an allosteric protein. *Science*, **25**, 1335–1345.

CHATA, H., NOBE, K. & MOMOSE, K. (1991) Role of phosphatidic acid in carbachol-induced contraction in guinea-pig *Taenia coli*. *Res. Commun. Chem. Pathol. Pharmacol.*, **71**, 59–72.

CHEN, K.K., WU, C.K. & HENRIKSEN, E. (1929) Relationship between the pharmacological action and chemical constitution and configuration of the optical isomers of ephedrine and related compounds. *J. Pharmacol. Exp. Ther.*, **36**, 363–400.

CHENG, Y.C. & PRUSOFF, W.H. (1973) Relationship between the inhibition constant (K_i) and the concentration of inhibitor which causes 50 per cent inhibition (I_{50}) of an enzymatic reaction. *Biochem. Pharmacol.*, **22**, 3099–3108.

CHESS-WILLIAMS, R.G. & BROADLEY, K.J. (1984) *Ex vivo* examination of β-adrenoceptor characteristics after propranolol withdrawal. *J. Cardiovasc. Pharmacol.*, **6**, 701–706.

CHESS-WILLIAMS, R.G., GRASSBY, P.F., CULLING, W., PENNY, W., BROADLEY, K.J. & SHERIDAN, D.J. (1985) Cardiac postjunctional supersensitivity to β-agonists after chronic chemical sympathectomy with 6-hydroxydopamine. *Naunyn-Schmiedeberg's Arch. Pharmacol.*, **329**, 162–166.

CHESS-WILLIAMS, R.G., BROADLEY, K.J. & SHERIDAN, D.J. (1986) Calculated and actual changes in beta-adrenoceptor number associated with increases in cardiac sensitivity. *J. Pharm. Pharmacol.*, **38**, 902–906.

CHESS-WILLIAMS, R.G., BROADLEY, K.J. & SHERIDAN, D.J. (1987a) Cardiac post-junctional alpha$_1$- and beta-adrenoceptors: effects of chronic chemical sympathectomy with 6-hydroxydopamine. *J. Receptor Res.*, **7**, 713–728.

CHESS-WILLIAMS, R.G., GRASSBY, P.F., BROADLEY, K.J. & SHERIDAN, D.J. (1987b) Cardiac alpha- and beta-adrenoceptor sensitivity and binding characteristics after chronic reserpine pretreatment. *Naunyn-Schmiedeberg's Arch. Pharmacol.*, **336**, 646–651.

CHESS-WILLIAMS, R.G., WILLIAMSON, K.L. & BROADLEY, K.J. (1990) Whether phenylephrine exerts inotropic effects through α- or β-adrenoceptors depends upon the relative populations. *Fund. Clin. Pharmacol.*, **4**, 25–37.

CHESS-WILLIAMS, R.G., DOUBLEDAY, B. & REYNOLDS, G.P. (1994) Differential regulation of cardiac α- and β-adrenoceptors by the sympathetic nervous system. *J. Auton. Pharmacol.*, **14**, 29–36.

CHOO, J.J., HORAN, M.A., LITTLE, R.A. & ROTHWELL, N.J. (1992) Anabolic effects of clenbuterol on skeletal muscle are mediated by β_2-adrenoceptor activation. *Am. J. Physiol.*, **263**, E50–E56.

CHRISTIE, J. & SATCHELL, D.G. (1980) Purine receptors in the trachea: is there a receptor for ATP? *Br. J. Pharmacol.*, **70**, 512–514.

CHRISTIOSON, R. (1855) On the properties of the ordeal bean of Old Calabar. *Mon. J. Med. (London)*, **20**, 193–204.

CIVELLI, O., BUNZOW, J.R., GRANDY, D.K., ZHON, Q.-Y. & VAN TOL, H.H.M. (1991) Molecular biology of the dopamine receptors. *Eur. J. Pharmacol. Mol. Pharmacol.*, **207**, 277–286.

CLAPHAM, D.E. & NEHER, E. (1984) Trifluoperazine reduces inward ionic currents and secretion by separate mechanisms in bovine chromaffin cells. *J. Physiol.*, **353**, 541–564.

CLARK, J.T., SMITH, E.R. & DAVIDSON, J.M. (1984) Enhancement of sexual motivation in male rats by yohimbine. *Science*, **225**, 847–849.

CLARKE, P.B.S. (1992) The fall and rise of neuronal α-bungarotoxin binding proteins. *Trends Pharmacol. Sci.*, **13**, 407–413.

CLARKE, P.B.S., REUBEN, M. & EL-BIZRI, H. (1994) Blockade of nicotinic responses by physostigmine, tacrine and other cholinesterase inhibitors in rat striatum. *Br. J. Pharmacol.*, **111**, 695–702.

CLAYTON, D.E., BUSSE, W.W. & BUCKNER, C.K. (1981) Contribution of vascular smooth muscle to contractile responses of guinea-pig isolated lung parenchymal strips. *Eur. J. Pharmacol.*, **70**, 311–320.

CLOW, A., HUSSAIN, T., GLOVER, V., SANDLER, M., DEXTER, D.T. & WALKER, M. (1991) (−)-Deprenyl can induce soluble superoxide dismutase in rat striata. *J. Neural Transm.*, **86**, 77–80.

COCKS, T.M. & ANGUS, J.A. (1983) Endothelium-dependent relaxation of coronary arteries by noradrenaline and serotonin. *Nature*, **305**, 627–630.

COHEN, M.L. & SCHENCK, K.W. (1987) Selective down regulation of vascular β_1-adrenergic receptors after prolonged isoproterenol infusion. *J. Cardiovasc. Pharmacol.*, 10, 365–368.

COLE, J.O. (1964) Therapeutic efficacy of antidepressant drugs. *J. Am. Med. Assoc.*, 190, 448–455.

COLEMAN, R.A. & LEVY, G.P. (1974) A non-adrenergic inhibitory nervous pathway in guinea-pig trachea. *Br. J. Pharmacol.*, 52, 167–174.

COLEMAN, R.A., DENYER, L.H. & SHELDRICK, K.E. (1987) β-Adrenoceptors in guinea-pig gastric fundus – are they the same as the 'atypical' β-adrenoceptors in rat adipocytes? *Br. J. Pharmacol.*, 90, 40P.

COLLIS, M.G. & HOURANI, S.M.O. (1993) Adenosine receptor subtypes. *Trends Pharmacol. Sci.*, 14, 360–366.

COLUCCI, W.S., ALEXANDER, R.W., WILLIAMS, G.H., RUDE, R.E., HOLMAN, B., KONSTAM, M., WYNNE, J., MUDGE, G.H. & BRAUNWALD, E. (1981a) Decreased lymphocyte beta-adrenergic receptor density in patients with heart failure and tolerance to the beta-adrenergic agonist pirbuterol. *N. Engl. J. Med.*, 305, 185–190.

COLUCCI, W.S., GIMBRONE, M.A. & ALEXANDER, R.W. (1981b) Regulation of the postsynaptic α-adrenergic receptor in rat mesenteric artery. Effects of chemical sympathectomy and epinephrine treatment. *Circ. Res.*, 48, 104–111.

CONNAUGHTON, S. & DOCHERTY, J.R. (1990) No evidence for differences between pre- and postjunctional α_2-adrenoceptors in the periphery. *Br. J. Pharmacol.*, 99, 97–102.

CONOLLY, M.E., TASHKIN, D.P., HUI, K.K.P., LITTNER, M.R. & WOLFE, R.N. (1982) Selective subsensitization of beta-adrenergic receptors in central airways of asthmatics and normal subjects during long-term therapy with inhaled salbutamol. *J. Allergy Clin. Immunol.*, 70, 423–431.

COOK, N.S. & UBBEN, D. (1990) Fibrinogen as a major risk factor in cardiovascular disease. *Trends Pharmacol. Sci.*, 11, 444–451.

COPPES, R.P., BROUWER, F., FREIE, I., SMIT, J. & ZAAGSMA, J. (1994) Sustained prejunctional facilitation of noradrenergic neurotransmission by adrenaline as a co-transmitter in the portal vein of freely moving rats. *Br. J. Pharmacol.*, 113, 342–344.

CORRIGAN, C.J. & KAY, A.B. (1992) Role of T-lymphocytes and lymphokines. *Asthma Br. Med. Bull.*, 48, 72–84.

CORTIJO, J., BOU, J., BELETA, J., CARDELÚS, I., LLENAS, J., MORCILLO, E. & GRISTWOOD, R.W. (1993) Investigation into the role of phosphodiesterase IV in bronchorelaxation, including studies with human bronchus. *Br. J. Pharmacol.*, 108, 562–568.

COSTA, M., FURNESS, J.B. & HUMPHREYS, C.M.S. (1986) Apamin distinguishes two types of relaxation mediated by enteric nerves in the guinea-pig gastrointestinal tract. *Naunyn-Schmiedeberg's Arch. Pharmacol.*, 332, 79–88.

COSTA, M., BARRINGTON, M. & MAJEWSKI, H. (1993) Evidence that M_1 muscarinic receptors enhance noradrenaline release in mouse atria by activating protein kinase C. *Br. J. Pharmacol.*, 110, 910–916.

COTE, M.G., PALAIC, D. & PANISSET, J.C. (1970) Changes in the number of vesicles and the sizes of sympathetic nerve terminals following nerve stimulation. *Rev. Can. Biol. Exp.*, 29, 111–114.

COTECCHIA, S., EXUM, S., CARON, M.G. & LEFKOWITZ, R.J. (1990) Regions of the α_1-adrenergic receptor involved in coupling to phosphatidylinositol hydrolysis and enhanced sensitivity of biological function. *Proc. Natl Acad. Sci. USA*, 87, 2896–2900.

COULDWELL, C., JACKSON, A., O'BRIEN, H. & CHESS-WILLIAMS, R.G. (1993) Characterization of the α_1-adrenoceptors of the rat prostate gland. *J. Pharm. Pharmacol.*, 45, 922–924.

COUPAR, I.M. & DE LUCA, A. (1994) Opiate and opiate antidiarrhoeal drug action on rat isolated intestine. *J. Auton. Pharmacol.*, 14, 69–78.

542

COUTURE, R., CUELLO, A.C. & HENRY, J.L. (1985) Trigeminal antidromic vasodilatation and plasma extravasation in the rat: effects of acetylcholine antagonists and cholinesterase inhibitors. *Br. J. Pharmacol.*, 84, 637–643.

COWAN, W.D. & DANIEL, E.E. (1983) Human female bladder and its non-cholinergic contractile function. *Can. J. Physiol. Pharmacol.*, 61, 1236–1246.

CRUICKSHANK. J.M. (1990) β-Blockers, plasma lipids, and coronary heart disease. *Circulation*, 82(Suppl. 2), II60–II65.

CULLING, W., PENNY, W.J., CUNLIFFE, G., FLORES, N.A. & SHERIDAN, D.J. (1987) Arrhythmogenic and electrophysiological effects of alpha adrenoceptor stimulation during myocardial ischaemia and reperfusion. *J. Mol. Cell. Cardiol.*, 19, 251–258.

CUNNANE, T.C., MUIR, T.C. & WARDLE, K.A. (1987) Is co-transmission involved in the excitatory responses of the rat anococcygeus muscle? *Br. J. Pharmacol.*, 92, 39–46.

DAHLOF, B., ANDREN, L., SVENSSON, A. & HANSSON, L. (1983) Antihypertensive mechanism of beta-adrenoceptor antagonism. The role of beta$_2$-blockade. *J. Hypertension*, 1(Suppl. 2), 112–115.

DAHLSTRÖM, A. (1965) Observations on the accumulation of noradrenaline in the proximal and distal parts of peripheral adrenergic nerves after compression. *J. Anat.*, 99, 677–689.

DALE, H.H. (1906) On some physiological actions of ergot. *J. Physiol.*, 34, 163–206.

DALE, H.H. (1914) The action of certain esters and ethers of choline, and their relation to muscarine. *J. Pharmacol. Exp. Ther.*, 6, 147–190.

DANIEL, E.E. (1985) Nonadrenergic noncholinergic (NANC) neuronal inhibitory interactions with smooth muscle. In: *Calcium and Contractility: Smooth Muscle*, pp. 385–426, A.K. Grover & E.E. Daniel (Eds), Humana Press, Clifton, NJ.

DANIELSSON, A., HENRIKSSON, R., LINDSTRÖM, P. & SEHLIN, J. (1982) The importance of an intact sympathetic innervation for the differentiation of the β-adrenoceptor subtypes in the rat parotid gland. *Acta Physiol. Scand.*, 115, 377–379.

DAVIES, A.O. & LEFKOWITZ, R.J. (1984) Regulation of β-adrenergic receptors by steroid hormones. *Annu. Rev. Physiol.*, 46, 119–130.

DAVIES, I.B. (1991) Adrenoceptors in autonomic failure. In: *Adrenoceptors: Structure, Mechanisms, Function. Advances in Pharmacological Sciences*, pp. 275–284, E. Szabadi & C.M. Bradshaw (Eds), Birkhäuser-Verlag, Basel.

DAWSON, G.R., BENTLEY, G., DRAPER, F., RYCROFT, W., IVERSEN, S.D. & PAGELLA, P.G. (1991) The behavioural effects of heptylphysostigmine a new cholinesterase inhibitor, in tests of long-term and working memory in rodents. *Pharmacol. Biochem. Behav.*, 39, 865–871.

DAY, M.D. (1967) The lack of crossed tachyphylaxis between tyramine and some other indirectly acting sympathomimetic amines. *Br. J. Pharmacol. Chemother.*, 30, 631–643.

DAY, M.D. (1979) *Autonomic Pharmacology. Experimental and Clinical Aspects*, Churchill Livingstone, Edinburgh.

DAY, M.D. & RAND, M.J. (1961) Effect of guanethidine in revealing cholinergic sympathetic fibres. *Br. J. Pharmacol. Chemother.*, 17, 245–260.

DAY, M.D. & ROACH, A.G. (1974) Central adrenoreceptors and the control of arterial blood pressure. *Clin. Exp. Pharmacol. Physiol.*, 1, 347–360.

DAY, M.D., OWEN, D.A.A. & WARREN, P.R. (1968) An adrenergic neurone blocking action of propranolol in isolated tissues. *J. Pharm. Pharmacol.*, 20, 130–134s.

DeBERNARDIS, J.F., WINN, M., KERKMAN, D.J., KYNCL, J.J., BUCKNER, S. & HORROM, B. (1987) A new nasal decongestant A-57219: a comparison with oxymetazoline. *J. Pharm. Pharmacol.*, 39, 760–763.

DE BLASI, A., LIPARTITI, M., PIRONE, F., ROCHAT, J.-F. & GARATTINI, S. (1986) Reduction of beta-adrenergic receptors by tertatolol: an additional mechanism for beta-adrenergic blockade. *Clin. Pharmacol. Ther.*, 39, 245–253.

DE CAMILLI, P. & JAHN, R. (1990) Pathways to regulated exocytosis in neurons. *Annu. Rev. Physiol.*, 52, 625–645.

DECLERMONT, P. (1854) Chimie organique – note sur la preparation de quelques esthers. *C. R. Acad. Sci.*, **39**, 338–341.

DE FEO, M.L. (1990) Plasma dopamine levels and their significance. In: *Peripheral Dopamine Pathophysiology*, pp. 15–25, F. Amenta (Ed.), CRC Press, Boca Raton, FL.

DE LUCA, A., LI, C.G., RAND, M.J., REID, J.J., THAINA, P. & WONG-DUSTING, H.K. (1990) Effects of *w*-conotoxin GVIA on autonomic neuroeffector transmission in various tissues. *Br. J. Pharmacol.*, **101**, 437–447.

DE MAN, J.G., BOECKXSTAENS, G.E., HERMAN, A.G. & PELCKMANS, P.A. (1994) Effect of potassium channel blockade and α_2-adrenoceptor activation on the release of nitric oxide from non-adrenergic non-cholinergic nerves. *Br. J. Pharmacol.*, **112**, 341–345.

DEMARINIS, R.M., WISE, M., HIEBLE, J.P. & RUFFOLO, R.R., JR (1987) Structure–activity relationships for alpha-1 adrenergic receptor agonists and antagonists. In: *The Alpha-1 Adrenergic Receptors*, pp. 211–265, R.R. Ruffolo, Jr (Ed.), Humana Press, Clifton, NJ.

DENERIS, E.S., CONNOLLY, J., ROGERS, S.W. & DUVOISIN, R. (1991) Pharmacological and functional diversity of neuronal nicotinic acetylcholine receptors. *Trends Pharmacol. Sci.*, **12**, 34–40.

DERRY, D.M., SCHÖNBAUM, E. & STEINER, G. (1969) Two sympathetic nerve supplies to brown adipose tissue of the rat. *Can. J. Physiol. Pharmacol.*, **47**, 57–63.

DESAI, K.M., WARNER, T.D., BISHOP, A.E., POLAK, J.M. & VANE, J.R. (1994) Nitric oxide, and not vasoactive intestinal peptide, as the main neurotransmitter of vagally induced relaxation of the guinea pig stomach. *Br. J. Pharmacol.*, **113**, 1197–1202.

DHEIN, S., TITZER, S., WALLSTEIN, M., MÜLLER, A., GERWIN, R., PANZNER, B. & KLAUS, W., (1992) Celiprolol exerts microvascular dilatation by activation of β_2-adrenoceptors. *Naunyn-Schmiedeberg's Arch. Pharmacol.*, **346**, 27–31.

DIFRANCESCO, D. (1993) Pacemaker mechanisms in cardiac tissue. *Annu. Rev. Physiol.*, **55**, 455–472.

DIXON, W.E. (1907) On the mode of action of drugs. *Med. Magazine (London)*, **16**, 454–457.

DOCHERTY, J.R. (1986) Aging and the cardiovascular system. *J. Auton. Pharmacol.*, **6**, 77–84.

DOCHERTY, J.R. (1989) The pharmacology of α_1- and α_2-adrenoceptor: evidence for and against a further subdivision. *Pharmacol. Ther.*, **44**, 241–284.

DOCKRAY, G.J. (1992) Transmission: peptides. In: *Autonomic Neuroeffector Mechanisms*, pp. 409–464, G. Burnstock & C.H.V. Hoyle (Eds), Harwood Academic, Chur.

DOGGRELL, S.A. (1990) Effects of bromoacetylalprenololmenthane (BAAM), an irreversible β-adrenoceptor antagonist, on the rat isolated left atria and portal vein. *J. Auton. Pharmacol.*, **10**, 333–344.

DOGGRELL, S.A. (1992) Further analysis of the inhibitory effects of dihydroergotamine, cycloheptadine and ketanserin on the responses of the rat aorta to 5-hydroxytryptamine. *J. Auton. Pharmacol.*, **12**, 223–236.

DOLPHIN, A.C. (1987) Nucleotide binding proteins in signal transduction and disease. *Trends Pharmacol. Sci.*, **10**, 53–57.

DOODS, H.N., QUIRION, R., MIHM, G., ENGEL, W., RUDOLF, K., ENTZEROTH, M., SCHIAVI, G.B., LADINSKY, H., BECHTEL, W.D., ENSINGER, H.A., MENDLA, K.D. & EBERLEIN, W. (1993) Therapeutic potential of CNS-active M_2 antagonists: novel structures and pharmacology. *Life Sci.*, **52**, 497–503.

DÖRJE, F., FRIEBE, T., TACKE, R., MUTSCHER, E. & LAMBRECHT, G. (1990) Novel pharmacological profile of muscarinic receptors mediating contraction of the guinea-pig uterus. *Naunyn-Schmiedeberg's Arch. Pharmacol.*, **342**, 284–289.

DOUGLAS, S.A., VICKERY-CLARK, L.M., LOUDEN, C., RUFFOLO, R.R., JR, FEUERSTEIN, G.Z. & OHLSTEIN, E.H. (1994) Acute pretreatment with carvedilol is sufficient for inhibition of neointima formation following rat carotid artery balloon angioplasty. *Pharmacol. Commun.*, **5**, 65–72.

DOWLING, G.P., MCDONOUGH, E.T. & BOST, R.O. (1987) 'Eve' and 'ecstacy'. A report of five deaths associated with the use of MDEA and MDMA. *J. Am. Med. Assoc.*, **257**, 1615–1617.

DOXEY, J.C. & ROACH, A.G. (1980) Presynaptic α-adrenoreceptors; *in vitro* methods and preparations utilised in evaluation of agonists and antagonists. *J. Auton. Pharmacol.*, **1**, 73–99.

DOXEY, J.C., LANE, A.C., ROACH, A.G. & VIRDEE, N.K. (1983) Comparison of the α-adrenoceptor antagonist profiles of idazoxan (Rx-781094), yohimbine, rauwolscine and corynanthine. *Naunyn-Schmiedeberg's Arch. Pharmacol.*, **325**, 136–144.

DRAPER, A.J., MEGHJI, S. & REDFERN, P.H. (1989) Enhanced presynaptic facilitation of vascular adrenergic neurotransmission in spontaneously hypertensive rats. *J. Auton. Pharmacol.*, **9**, 103–111.

DRAYER, D.E. (1987) Lipophilicity, hydrophilicity and the central nervous system side-effects of beta blockers. *Pharmacotherapy*, **7**, 87–91.

DREW, G.M. & WHITING, S.B. (1979) Evidence for two distinct types of postsynaptic receptor. *Br. J. Pharmacol.*, **67**, 207–215.

DREW, G.M., HILDITCH, A. & LEVY, G.P. (1978) Effect of labetalol on the uptake of [^{3}H]-noradrenaline into the isolated vas deferens of the rat. *Br. J. Pharmacol.*, **63**, 471–474.

DRUKARCH, B., KITS, K.S., LEYSEN, J.E., SCHEPENS, E. & STOOF, J.C. (1989) Restricted usefulness of tetraethylammonium and 4-aminopyridine for the characterization of receptor-operated K^+-channels. *Br. J. Pharmacol.*, **98**, 113–118.

DUCHAINE, D. (1992) Clenbuterol. In: *The Underground Steroid Handbook II (Update)*, pp. 31–44, HLR Technical Books, Venice, CA.

DUCIS, I. (1988) The high-affinity choline uptake system. In: *The Cholinergic Synapse. Handbook of Experimental Pharmacology*, Vol. 86, pp. 409–445, V.P. Whittaker (Ed.), Springer-Verlag, Berlin.

DUDLEY, M.W., HOWARD, B.D. & CHO, A.K. (1990) The interactions of the beta-haloethylbenzylamines, xylamine and DSP-4 with catecholaminergic neurons. *Annu. Rev. Pharmacol. Toxicol.*, **30**, 387–403.

DU PLOOY, W.J., HAY, L., KAHLER, C.P., SCHUTTE, P.J. & BRANDT, H.D. (1994) The dose-related hyper- and hypokalaemic effects of salbutamol and its arrhythmogenic potential. *Br. J. Pharmacol.*, **111**, 73–76.

DUSTING, G.J. & LI, D.M.F. (1986) Catecholamine release and potentiation of thromboxane A_2 production by nicotine in the greyhound. *Br. J. Pharmacol.*, **87**, 29–36.

DUVAL, N., ANGEL, I., OBLIN, A., EON, M.T. & LANGER, S.Z. (1991) Effect of a_2-adrenoceptor antagonists on norepinephrine release and inhibition of insulin secretion during pancreatic nerve stimulation. Interactions at pre- and postsynaptic sites. In: *Presynaptic Receptors and Neuronal Transporters. Advances in the Biosciences*, Vol. 82, pp. 25–26, S.Z. Langer, A.M. Galzin & J. Costentin (Eds), Pergamon Press, Oxford.

EASSON, L.H. & STEDMAN, E. (1933) CLXX. Studies on the relationship between chemical constitution and physiological action. V. Molecular dissymmetry and physiological activity. *Biochem. J.*, **27**, 1257–1266.

ECCLES, J.C. (1964) *The Physiology of Synapses*. Springer-Verlag, Berlin.

EGAN, T.M., NOBLE, D., NOBLE, S.J., POWELL, T. & TWIST, V.W. (1987) An isoprenaline activated sodium-dependent inward current in ventricular myocytes. *Nature*, **328**, 635–637.

EGLEN, R.M. & HARRIS, G.C. (1993) Selective inactivation of muscarinic M_2 and M_3 receptors in guinea-pig ileum and atria *in vitro*. *Br. J. Pharmacol.*, **109**, 946–952.

EGLEN, R.M. & WHITING, R.L. (1990) Heterogeneity of vascular muscarinic receptors. *J. Auton. Pharmacol.*, **10**, 233–245.

EGLEN, R.M., KENNY, B.A., MICHEL, A.D. & WHITING, R.L. (1987) Muscarinic activity of McN-A-343 and its value in muscarinic receptor classification. *Br. J. Pharmacol.*, 90, 693–700.

EGLEN, R.M., HUFF, M.M., MONTGOMERY, W.W. & WHITING, R.L. (1988a) Differential effects of pertussis toxin on muscarinic responses in isolated atria and smooth muscle. *J. Auton. Pharmacol.*, 8, 29–37.

EGLEN, R.M., MONTGOMERY, W.W., DAINTY, I.A., DUBUQUE, L.K. & WHITING, R.L. (1988b) The interaction of methoctramine and himbacine at atrial, smooth muscle and endothelial muscarinic receptors *in vitro*. *Br. J. Pharmacol.*, 95, 1031–1038.

EGLEN, R.M., REDDY, H., WATSON, N. & CHALLISS, R.A.J. (1994) Muscarinic acetylcholine receptor subtypes in smooth muscle. *Trends Pharmacol. Sci.*, 15, 114–119.

EHRLICH, B.E., KAFTAN, E., BEZPROZVANNAYA, S. & BEZPROZVANNY, I. (1994) The pharmacology of intracellular Ca^{2+}-release channels. *Trends Pharmacol. Sci.*, 15, 145–149.

EINSTEIN, R., ABDUL-HUSSEIN, N., WONG, T.-W., CHANG, D.H.-T., MATTHEWS, R. & RICHARDSON, D.P. (1994) Cardiovascular actions of dopexamine in anaesthetized and conscious dogs. *Br. J. Pharmacol.*, 111, 199–204.

EL AMRANI, A.I.K., LE CARPENTIER, Y., RIOU, B. & POURNY, J.C. (1989) Lusitropic effect and modification of contraction–relaxation coupling induced by α-adrenergic stimulation in rat left ventricular papillary muscle. *J. Mol. Cell. Cardiol.*, 21, 669–680.

ELBEIN, A.D. (1990) Hydrolases. In: *Comprehensive Medicinal Chemistry*, Vol. 2, *Enzymes and Other Molecular Targets*, pp. 365–389, P.G. Sammes (Ed.), Pergamon Press, Oxford.

ELLIOTT, T.R. (1905) The action of adrenalin. J. Physiol. 32, 401–467.

ELTZE, M., GONNE, S., RIEDEL, R., SCHLOTKE, B., SCHUDT, C. & SIMON, W.A. (1985) Pharmacological evidence for selective inhibition of gastric acid secretion by telenzepine, a new antimuscarinic drug. *Eur. J. Pharmacol.*, 112, 211–224.

EMORINE, L., BLIN, N. & STROSBERG, A.D. (1994) The human β_3-adrenoceptor: the search for a physiological function. *Trends Pharmacol. Sci.*, 15, 3–7.

EMORINE, L.J., FÈVE, B., PAIRAULT, J., BRIEND-SUTREN, M.-M., NAHMIAS, C., MARULLO, S., DELAVIER-KLUTCHKO, C. & STROSBERG, A.D. (1992) The human β_3-adrenergic receptor: relationship with atypical receptors. *Am. J. Clin. Nutr.*, 55, 215S–218S.

ENDOH, M., TAKANASHI, M. & NOROTA, I. (1992) Role of alpha$_{1A}$-adrenoceptor subtype in production of the positive inotropic effect mediated via myocardial alpha$_1$-adrenoceptors in the rabbit papillary muscle: influence of selective alpha$_{1A}$ subtype antagonists WB4101 and 5-methylurapidil. *Naunyn-Schmiedeberg's Arch. Pharmacol.*, 345, 578–585.

ENGEL, G. (1981) Subclasses of beta-adrenoceptors – a quantitative estimation of beta$_1$- and beta$_2$-adrenoceptors in guinea pig and human lung. *Postgrad. Med. J.*, 57(Suppl. 1), 77–83.

ENGEL, G., HOYER, D., BERTHOLD, R. & WAGNER, H. (1981) (±)[^{125}Iodo]cyano-pindolol, a new ligand for β-adrenoceptors: identification and quantitation of subclasses of β-adrenoceptors in guinea-pig. *Naunyn-Schmiedeberg's Arch. Pharmacol.*, 317, 277–285.

ENGLAND, B.P., ACKERMAN, M.S. & BARRETT, R.W. (1991) A chimeric D$_2$ dopamine/m$_1$ muscarinic receptor with D$_2$ binding specificity mobilizes intracellular calcium in response to dopamine. *FEBS Lett.*, 279, 87–90.

ERICKSON, J.D., EIDEN, L.E. & HOFFMAN, B.J. (1992) Expression cloning of a reserpine-sensitive vesicular monoamine transporter. *Proc. Natl Acad. Sci. USA*, 89, 10993–10997.

ERNSBERGER, P., MEELEY, M.P., MANN, J.J. & REIS, D.J. (1987) Clonidine binds to imidazole binding sites as well as α_2-adrenoceptors in the ventrolateral medulla. *Eur. J. Pharmacol.*, 134, 1–13.

EVANS, P.D. (1981) Multiple receptor types for octopamine in the locust. *J. Physiol.*, **318**, 99–122.

EVANS, R.G. & ANDERSON, W.P. (1995) Renal effects of infusion of rilmenidine and guanabenz in conscious dogs: contribution of peripheral and central nervous system α_2-adrenoceptors. *Br. J. Pharmacol.*, **116**, 1557–1570.

FALLON, J.R. (1991) Rounding up acetylcholine receptors. *Curr. Biol.*, **1**, 265–267.

FASOLATO, C., INNOCENTI, B. & POZZAN, T. (1994) Receptor-activated Ca^{2+} influx: how many mechanisms for how many channels? *Trends Pharmacol. Sci.*, **15**, 77–83.

FAULKNER, M. & SHARPEY-SCHAFER, E.P. (1959) Circulatory effects of trumpet playing. *Br. Med. J.*, **I**, 685–686.

FAVRE-MAURICE, R., DE HAUT, M., DALMAZ, Y. & PEYRIN, L. (1992) Differential effect of guanethidine on dopamine and norepinephrine in rat peripheral tissues. *J. Neural Transm.*, **88**, 115–126.

FDA REPORT (1991) Tacrine as a treatment for Alzheimer's disease: an interim report from the FDA. *N. Engl. J. Med.*, **324**, 349–352.

FEIGL, E.O. (1983) Coronary physiology. *Physiol. Rev.*, **63**, 1–205.

FEINSTEIN, A.R. (1988) Scientific standards in epidemiologic studies of the menace of daily life. *Science*, **242**, 1257–1263.

FELDER, R.A., BLECHER, M., CALCAGNO, P.L. & JOSE, P.A. (1984) Dopamine receptors in the proximal tubule of the rabbit. *Am. J. Physiol.*, **247**, F499–F505.

FELDMAN, A.M. & BRISTOW, M.R. (1990) The adrenergic pathway in the failing human heart: implications for inotropic therapy. *Cardiology*, **77**(Suppl. 1), 1–32.

FELDMAN, R.D. (1991) Physiological regulation of human beta adrenergic receptor responsiveness: effect of dietary sodium intake in normotensive and hypertensive subjects. In: *Adrenoceptors: Structure, Mechanism, Function. Advances in Pharmacological Sciences*, pp. 245–253, E. Szabadi & C.M. Bradshaw (Eds), Birkhäuser-Verlag, Basel.

FELDMAN, R.D. (1992) A low sodium diet corrects the defect in β-adrenergic response in older subjects. *Circulation*, **85**, 612–618.

FERNANDES, L.B., FRYER, A.D. & HIRSHMAN, C.A. (1992) M_2 Muscarinic receptors inhibit isoproterenol-induced relaxation of canine airway smooth muscle. *J. Pharmacol. Exp. Ther.*, **262**, 119–126.

FERRY, C.B. (1966) Cholinergic link hypothesis in adrenergic neuroeffector transmission. *Physiol. Rev.*, **46**, 420–456.

FIELD, J.L. & NEWBERRY, N.R. (1989) Methoctramine and hexahydrodifenidol antagonise two muscarinic responses on the rat superior cervical ganglion with opposite selectivity. *Neurosci. Lett.*, **100**, 254–258.

FILLENZ, M. (1990) *Noradrenergic Neurons*, Cambridge University Press, Cambridge.

FILLENZ, M. (1992) Transmission: noradrenaline. In: *Autonomic Neuroeffector Mechanisms*, Vol. 1, pp. 323–365, G. Burnstock & C.H.V. Hoyle (Eds), Harwood Academic, Chur.

FINKLEMAN, B. (1930) On the nature of inhibition in the intestine. *J. Physiol.*, **70**, 145–157.

FINNERTY, F.A. & BROGDEN, R.N. (1985) Guanadrel. A review of its pharmacodynamic and pharmacokinetic properties and therapeutic use in hypertension. *Drugs*, **30**, 22–31.

FLAVAHAN, N.A. & VANHOUTTE, P.M. (1987) Heterogeneity of alpha-adrenergic responsiveness in vascular smooth muscle: role of receptor subtypes and receptor reserve. In: *The Alpha-1 Adrenergic Receptors*, pp. 351–403, R.R. Ruffolo, Jr (Ed.), Humana Press, Clifton, NJ.

FLECKENSTEIN, A. & BURN, J.H. (1953) The effect of denervation on the action of sympathomimetic amines on the nictitating membrane. *Br. J. Pharmacol. Chemother.*, **8**, 69–78.

FLEMING, W.W. (1984) A review of postjunctional supersensitivity in cardiac muscle. In: *Neuronal and Extraneuronal Events in Autonomic Events*, pp. 205–219, W.W. Fleming, K.-H. Graefe, S.Z. Langer & N. Weiner (Eds), Raven Press, New York.

FLEMING, W.W., WESTFALL, D.P., DE LA LANDE, I.S. & JELLET, L.B. (1972) Lognormal distribution of equieffective doses of norepinephrine and acetylcholine in several tissues. *J. Pharmacol. Exp. Ther.*, **181**, 339–345.

FLEMING, W.W., McPHILLIPS, J.J. & WESTFALL, D.P. (1973) Post-junctional supersensitivity and subsensitivity of excitable tissues to drugs. *Ergeb. Physiol. Biol. Chem. Exp. Pharmakol.*, **68**, 55–119.

FLORES, N.A. & SHERIDAN, D.J. (1989) Electrophysiological and antiarrhythmic effects of UK 52,046-27 during ischaemia and reperfusion in the guinea-pig heart. *Br. J. Pharmacol.*, **96**, 670–674.

FOLKOW, B. (1982) Physiological aspects of primary hypertension. *Physiol. Rev.*, **62**, 347–504.

FOLKOW, B. & SVANBORG, A. (1993) Physiology of cardiovascular aging. *Physiol. Rev.*, **73**, 725–764.

FORD, A.P.D.W., TANEJA, D.T. & CLARKE, D.E. (1992) Atypical β-adrenoceptor-mediated relaxation and cyclic AMP formation in tunica muscularis mucosae (TMM) of the rat oesophagus. *Br. J. Pharmacol.*, **105**, 235P.

FORD, A.P.D.W., WILLIAMS, T.J., BLUE, D.R. & CLARKE, D.E. (1994) α_1-Adrenoceptor classification; sharpening Occam's razor. *Trends Pharmacol. Sci.*, **15**, 167–170.

FOZARD, J.R. & MIR, A.K. (1987) Are 5-HT receptors involved in the antihypertensive effects of urapidil? *Br. J. Pharmacol.*, **90**, 24P.

FRANCO-CERECEDA, A., LUNDBERG, J.M. & DAHLÖF, C. (1985) Neuropeptide Y and sympathetic control of heart contractility and coronary vascular tone. *Acta Physiol. Scand.*, **124**, 361–369.

FRASER, T.R. (1863) On the characters, actions and therapeutical uses of the ordeal bean of Calabar (*Physostigma venenosum*, Balfour). *Edinburgh Med. J.*, **9**, 36–56.

FREDHOLM, B.B. (1985) Nervous control of circulation and metabolism in white adipose tissue. In: *New Perspectives in Adipose Tissue: Structure, Function and Development*, pp. 45–64, A. Cryer & R.L.R. Van (Eds), Butterworths, London.

FREDHOLM, B.B., HU, P.-S., VAN DER PLOEG, J., DUNÉR-ENGSTRÄM, M. & DUNWIDDIE, T.V. (1991) Mechanisms of inhibition of transmitter release by adenosine analogs. In: *Presynaptic Receptors and Neuronal Transporters. Advances in the Biosciences*, Vol. 82, pp. 249–252, S.Z. Langer, A.M. Galzin & J. Costentin (Eds), Pergamon Press, Oxford.

FREEDMAN, S.B., HARLEY, E.A. & IVERSEN, L.L. (1988) Biochemical measurement of muscarinic receptor efficacy and its role in receptor regulation. *Trends Pharmacol. Sci.*, Suppl. 54–60.

FREEDMAN, S.B., DAWSON, G.R., IVERSEN, L.L., BAKER, R. & HARGREAVES, R.J. (1993) The design of novel muscarinic partial agonists that have functional selectivity in pharmacological preparations *in vitro* and reduced side-effect profile *in vivo*. *Life Sci.*, **52**, 489–495.

FRIED, G., TERENIUS, L., HÖKFELT, T. & GOLDSTEIN, M. (1985) Evidence for differential localization of noradrenaline and neuropeptide Y in neuronal storage vesicles isolated from rat vas deferens. *J. Neurosci.*, **5**, 450–458.

FRISHMAN, W.H. (1983) Pindolol: a new β-adrenoceptor antagonist with partial agonist activity. *N. Engl. J. Med.*, **308**, 940–944.

FRYER, A.D. & JACOBY, D.B. (1993) Effect of inflammatory cell mediators on M_2 muscarinic receptors in the lungs. *Life Sci.*, **52**, 529–536.

FUDER, H., MUSCHOLL, E. & SPEMANN, R. (1983) The determination of presynaptic pA_2 values of yohimbine and phentolamine on the perfused rat heart under conditions of negligible autoinhibition. *Br. J. Pharmacol.*, **79**, 109–119.

FULLER, R.W. (1982) Pharmacology of brain epinephrine neurons. *Annu. Rev. Pharmacol.*, **22**, 31–55.

FURCHGOTT, R.F. (1959) The receptors for epinephrine and norepinephrine (adrenergic receptors). *Pharmacol. Rev.*, **11**, 429–441.

FURCHGOTT, R.F. (1960) Receptors for sympathomimetic amines. In: *Ciba Foundation Symposium on Adrenergic Mechanisms*, pp. 246–252, J.R. Vane, G.E.W. Wolstenholme & C.M. O'Connor (Eds), Little, Brown, Boston, MA.

FURCHGOTT, R.F. (1966) The use of β-haloalkylamines in the differentiation of receptors and in the determination of dissociation constants of receptor–agonist complexes. In: *Advances in Drug Research*, Vol. 3, pp. 21–55, N.J. Harper & A.B. Simmonds (Eds), Academic Press, London.

FURCHGOTT, R.F. (1972) The classification of adrenoceptors (adrenergic receptors). An evaluation from the standpoint of receptor theory. In: *Handbook of Experimental Pharmacology*, Vol. 33, pp. 283–335, H. Blashko & E. Muscholl (Eds), Springer-Verlag, Berlin.

FURCHGOTT, R.F. (1976) Postsynaptic adrenergic receptor mechanisms in vascular smooth muscle. In: *Vascular Neuroeffector Mechanisms. 2nd International Symposium*, pp. 131–142, J.A. Bevan (Ed.), Karger, Basel.

FURCHGOTT, R.F. & BURSZTYN, P. (1967) Comparison of dissociation constants and of relative efficacies of selected agonists acting on parasympathetic receptors. *Ann. NY Acad. Sci.*, **144**, 882–899.

FURCHGOTT, R.F. & ZAWADZKI, J.V. (1980) The obligatory role of endothelial cells in the relaxation of arterial smooth muscle by acetylcholine. *Nature*, **288**, 373–376.

FURLONG, R. & BROGDEN, R.N. (1988) Xamoterol. A preliminary review of its pharmacodynamic and pharmacokinetic properties, and therapeutic use. *Drugs*, **36**, 455–474.

GABELLA, G. (1976) *Structure of the Autonomic Nervous System*, Chapman & Hall, London.

GABELLA, G. (1981) Structure of smooth muscles. In: *Smooth Muscle. An Assessment of Current Knowledge*, pp. 1–46, E. BÜLBRING, A.F. BRADING, A.W. JONES & T. TOMITA (Eds), Edward Arnold, London.

GABELLA, G. (1987) Innervation of airway smooth muscle: fine structure. *Annu. Rev. Physiol.*, **49**, 583–594.

GABELLA, G. (1992) Fine structure of post-ganglionic nerve fibres and autonomic neuro-effector junctions. In: *Autonomic Neuroeffector Mechanisms*, pp. 1–31, G. BURNSTOCK & C.H.V. HOYLE (Eds), Harwood Academic, Chur.

GADDUM, J.H. (1937) The quantitative effects of antagonistic drugs. *J. Physiol.*, **89**, 7P-9P.

GASCÓN, S., DIERSSEN, M., MARMOL, F., VIVAS, N.M. & BADIA, A. (1993) Effects of age on α_1-adrenoceptor subtypes in the heart ventriclar muscle of the rat. *J. Pharm. Pharmacol.*, **45**, 907–909.

GENTILINI, G., DI BELLO, M.G., RASPANTI, S., BINDI, D., MUGNAI, S. & ZILLETTI, L. (1994) Salmeterol inhibits anaphylactic histamine release from guinea-pig isolated mast cells. *J. Pharm. Pharmacol.*, **46**, 76–77.

GERSHON, M.D. (1981) The enteric nervous system. *Annu. Rev. Neurosci.*, **4**, 227–272.

GHADIMI, H., KUMAR, S. & ABACI, F. (1971) Studies on monosodium glutamate ingestion. 1. Biochemical explanation of Chinese restaurant syndrome. *Biochem. Med.*, **5**, 447–456.

GHELARDINI, C., GUALTIERI, F., BALDINI, M., DEI, S., ROMANELLI, M.N., GIOTTI, A. & BARTOLINI, A. (1993) R-(+)-hyoscyamine: the first selective antagonist for guinea-pig uterus muscarinic receptor subtype. *Life Sci.*, **52**, 569.

GIACOBINI, E. (1992) Cholinomimetic replacement of cholinergic function in Alzheimer disease. In: *Treatment of Dementias*, pp. 19–34, E.M. Meyer (Ed.) Plenum Press, New York.

GIEMBYCZ, M.A. (1992) Commentary. Could isoenzyme-selective phosphodiesterase inhibitors render bronchodilator therapy redundant in the treatment of bronchial asthma? *Biochem. Pharmacol.*, **43**, 2041–2051.

GIEMBYCZ, M.A. & RAEBURN, D. (1991) Putative substrates for cyclic nucleotide-dependent protein kinases and the control of airway smooth muscle tone. *J. Auton. Pharmacol.*, **11**, 365–398.

GLAVIN, G.B. & SMYTH, D.D. (1995) Effects of the selective I_1 imidazoline receptor agonist, moxonidine, on gastric secretion and gastric mucosal injury in rats. *Br. J. Pharmacol.*, **114**, 751–754.

GODFRAIND, T., MILLER, R. & WIBO, M. (1986) Calcium antagonism and calcium entry blockade. *Pharmacol. Rev.*, **38**, 321–416.

GOLDIE, R.G., PATERSON, J.W. & LULICH, K.M. (1990) Adrenoceptors in airway smooth muscle. *Pharmacol. Ther.*, **48**, 295–322.

GOLDSTEIN, D.S. (1980) Plasma norepinephrine in essential hypertension – a study of the studies. *Hypertension*, **3**, 48–52.

GOLENHOFEN, K. (1981) Differentiation of calcium activation processes in smooth muscle using selective antagonists. In: *Smooth muscle. An Assessment of Current Knowledge*, pp. 157–170, E. Bulbring, A.F. Brading, A.W. Jones & T. Tomita (Eds), Edward Arnold, London.

GOLENHOFEN, K., HERMSTEIN, N. & LAMMEL, E. (1973) Membrane potential and contraction of vascular smooth muscle (portal vein) during application of noradrenaline and high potassium, and selective inhibitory effects of iproveratril (verapamil) *Microvasc. Res.*, **5**, 73–80.

GOMER, S.K. & ZIMMERMAN, B.C. (1972) Determination of sympathetic vasodilator responses during renal nerve stimulation. *J. Pharmacol. Exp. Ther.*, **181**, 75–82.

GOODALL, McC. & KIRSHNER, N. (1958) Biosynthesis of epinephrine and norepinephrine by sympathetic nerves and ganglia. *Circulation*, **17**, 366–371.

GÖTHERT, M., NAWROTH, P. & NEUMEYER, H. (1979) Inhibitory effects of verapamil, prenylamine and D600 on Ca^{2+}-dependent noradrenaline release from the sympathetic nerves of isolated rabbit hearts. *Naunyn-Schmiedeberg's Arch. Pharmacol.*, **310**, 11–19.

GOTO, K., LONGHURST, P.A., CASSIS, L.A., HEAD, R.J., TAYLOR, D.A., PRICE, P.J. & FLEMING, W.W. (1985) Surgical sympathectomy of the heart in rodents and its effect on sensitivity to agonists. *J. Pharmacol. Exp. Ther.*, **234**, 280–287.

GOYAL, R.K. (1989) Muscarinic receptor subtypes. Physiology and clinical implications. *N. Engl. J. Med.*, **321**, 1022–1029.

GRAMMAS, P., DERESKI, M.O., DIGLIO, C., GIACOMELLI, F. & WIENER, J. (1989) Autonomic receptor interactions in isolated myocytes from hypertensive rats. *J. Mol. Cell. Cardiol.*, **21**, 807–815.

GRANT, J.A. & SCRUTTON, M.C. (1980) Interaction of selective α-adrenoceptor agonists and antagonists with human and rabbit blood platelets. *Br. J. Pharmacol.*, **71**, 121–134.

GRANT, W.M. (1969) Action of drugs on movement of occular fluids. *Annu. Rev. Pharmacol.*, **9**, 85–94.

GRASSBY, P.F. & BROADLEY, K.J. (1984) Characterization of β-adrenoceptors mediating relaxation of the guinea-pig ileum. *J. Pharm. Pharmacol.*, **36**, 602–607.

GRASSBY, P.F. & BROADLEY, K.J. (1986) Responses mediated via beta-1 adrenoceptors but not beta-2 adrenoceptors exhibit supersensitivity after chronic reserpine pretreatment. *J. Pharmacol. Exp. Ther.*, **237**, 950–958.

GRASSBY, P.F. & BROADLEY, K.J. (1987) Partial agonists at guinea-pig atrial β-adrenoceptors display relaxation responses in the guinea-pig ileum independent of β-adrenoceptor stimulation. *Gen. Pharmacol.*, **18**, 25–31.

GRASSBY, P.F. & MCNEILL, J.H. (1988) Sensitivity changes to inotropic agents in rabbit atria after chronic experimental diabetes. *Can. J. Physiol. Pharmacol.*, **66**, 1475–1480.

GRASSBY, P.F., ARCH, J.R.S., WILSON, C. & BROADLEY, K.J. (1987) Beta-adrenoceptor sensitivity of brown and white adipocytes after chronic pretreatment of rats with reserpine. *Biochem. Pharmacol.*, **36**, 155–162.

GRIFFIN, M.T., THOMAS, E.A. & EHLERT, F.J. (1993) Kinetics of solvolysis and *in vivo* muscarinic receptor binding of N-(α-bromoethyl)-4-piperidinyl diphenylacetate: an analogue of 4-DAMP mustard. *Life Sci.*, **52**, 587.

GRIMM, U., FUDER, H., MOSER, U., BÄUMERT, H.G., MUTSCHLER, E. & LAMBRECHT, G. (1994) Characterization of the prejunctional muscarinic receptors mediating inhibition of evoked release of endogenous noradrenaline in rabbit isolated vas deferens. *Naunyn-Schmiedeberg's Arch. Pharmacol.*, **349**, 1–10.

GROSS, N.J. (1992) Anticholinergic bronchodilators. In: *Asthma: Basic Mechanisms and Clinical Management*, 2nd Edn, pp. 555–566, P.J. BARNES, I.W. RODGER & N.C. THOMSON (Eds), Academic Press, London.

GROWCOTT, J.W., HOLLOWAY, B., GREEN, M. & WILSON, C. (1993) Zeneca ZD1114 acts as an antagonist at β_2-adrenoceptors in rat isolated ileum. *Br. J. Pharmacol.*, **110**, 1375–1380.

GRUNBERG, N.E. (1988) Nicotine and body weight: behavioral and biological mechanisms. In: *The Pharmacology of Nicotine*, pp. 97–110, M.J. Rand & K. Thurau (Eds), IRL Press, Oxford.

GUENANECHE, F., SCHUURKES, J.A.J. & LEFEBVRE, R.A. (1991) Influence of fenoldopam and quinpirole in the guinea-pig stomach. *J. Auton. Pharmacol.*, **11**, 221–235.

GULDBERG, H.C. & MARSDEN, C.A. (1975) Catechol-O-methyltransferase: Pharmacological aspects and physiological role. *Pharmacol. Rev.*, **27**, 135–206.

GUNDLACH, A.L., RUTHERFURD, S.D. & LOUIS, W.J. (1990) Increase in galanin and neuropeptide Y mRNA in locus coeruleus following acute reserpine treatment. *Eur. J. Pharmacol.*, **184**, 163–167.

GURIN, S. & DELLUVA, A.M. (1947) The biological synthesis of radioactive adrenalin from phenylalanine. *J. Biol. Chem.*, **170**, 545–550.

GURNEY, A.M. & RANG, H.P. (1984) The channel-blocking action of methonium compounds on rat submandibular ganglion cells. *Br. J. Pharmacol.*, **82**, 623–642.

GYERMEK, L. (1980) Methods for the examination of ganglion-blocking activity. In: *Pharmacology of Ganglionic Transmission. Handbook of Experimental Pharmacology.*, Vol. 53, pp. 63–121, D.A. KHARKEVICH, (Ed.), Springer-Verlag, Berlin.

HAAS, M., CHENG, B., RICHARDT, G., LANG, R.E. & SCHÖMIG, A. (1989) Characterization and presynaptic modulation of stimulation-evoked exocytotic co-release of noradrenaline and neuropeptide Y in guinea-pig heart. *Naunyn-Schmiedeberg's Arch. Pharmacol.*, **339**, 71–78.

HAEFELY, W.E. (1980) Non-nicotinic chemical stimulation of autonomic ganglia. In: *Pharmacology of Ganglionic Transmission. Handbook of Experimental Pharmacology*, Vol. 53, pp. 313–357, D.A. KHARKEVICH (Ed.), Springer-Verlag, Berlin.

HAINSWORTH, R. (1991) Reflexes from the heart. *Physiol. Rev.*, **71**, 617–658.

HAMBERGER, B., MALMFORS, T., NORBERG, K.-A. & SACHS, C. (1964) Uptake and accumulation of catecholamines in peripheral adrenergic neurons of reserpinized animals, studied with a histochemical method. *Biochem. Pharmacol.*, **13**, 841–844.

HAMILTON, C.A., DALRYMPLE, H. & REID, J. (1982) Recovery *in vivo* and *in vitro* of α-adrenoceptor responses and radioligand binding after phenoxybenzamine. *J. Cardiovasc. Pharmacol.*, **4** (Suppl.), S125–S128.

HAMILTON, C.A., REID, J.R. & YAKUBA, M.A. (1988) [^{3}H]-Yohimbine and [^{3}H]-idazoxan bind to different sites on rabbit forebrain and kidney membrane. *Eur. J. Pharmacol.*, **146**, 345–348.

HAMMER, R. & GIACHETTI, A. (1982) Muscarinic receptor subtypes: M_1 and M_2 biochemical and functional characterization. *Life Sci.*, **31**, 2991–2998.

HAMMER, R., BERRIE, C.P., BIRDSALL, N.J.M., BURGEN, A.S.V. & HULME, E.C. (1980) Pirenzepine distinguishes between different subclasses of muscarinic receptors. *Nature*, **283**, 90–92.

HAMMER, R., GIRALDO, E., SCHIAVI, G.B. MONTERINI, E. & LADINSKY, H. (1986) Binding profile of a novel cardioselective muscarinic receptor antagonist, AF-DX 116, to membranes of peripheral tissues and brain in the rat. *Life Sci.*, **38**, 1653–1662.

HAN, C., ABEL, P.W. & MINNEMAN, K.P. (1987) a_1-Adrenoceptor subtypes linked to different mechanisms for increasing intracellular Ca^{2+} in smooth muscle. *Nature*, **329**, 333–335.

HAN, C., LI, J. & MINNEMAN, K.P. (1990) Subtypes of a_1-adrenoceptors in rat blood vessels. *Eur. J. Pharmacol.*, **190**, 97–104.

HANFT, G. & GROSS, G. (1989) Subclassification of a_1-adrenoceptor recognition sites by urapidil derivatives and other selective antagonists. *Br. J. Pharmacol.*, **97**, 691–700.

HARDEN, T.K. (1983) Agonist-induced desensitization of the β-adrenergic receptor-linked adenylate cyclase. *Pharmacol. Rev.*, **35**, 5–32.

HARE, M.L.C. (1928) Tyramine oxidase. 1. A new enzyme system in liver. *Biochem. J.*, **22**, 968–979.

HARGREAVES, R.J., McKNIGHT, A.T., SCHOLEY, K., NEWBERRY, N.R., STREET, L.J., HUTSON, P.H., SEMARK, J.E., HARLEY, E.A., PATEL, S. & FREEDMAN, S.B. (1992) L-689,660, a novel cholinomimetic with functional selectivity for M_1 and M_3 muscarinic receptors. *Br. J. Pharmacol.*, **107**, 494–501.

HARRIS, T., MULLER, B.D., COTTON, R.G.H., BORRI VOLTATTORNI, C. & BELL, C. (1986) Dopaminergic and noradrenergic sympathetic nerves of the dog have different DOPA decarboxylase activities. *Neurosci. Lett.*, **65**, 155–160.

HARRISON, J.K., PEARSON, W.R. & LYNCH, K.R. (1991) Molecular characterization of a_1- and a_2-adrenoceptors. *Trends Pharmacol. Sci.*, **12**, 62–67.

HASHIMOTO, H., HAYASHI, M., NAKAHARA, Y., NIWAGUCHI, T. & ISHII, H. (1977) Actions of D-lysergic acid diethylamide (LSD) and its derivatives on 5-hydroxytryptamine receptors in the isolated uterine smooth muscle of the rat. *Eur. J. Pharmacol.*, **45**, 341–348.

HAUCK, R.W., BÖHM, M., GENGENBACH, S., SUNDER-PLASSMANN, L., FRUHMANN, G. & ERDMANN, E. (1990) β_2-Adrenoceptors in human lung and peripheral mononuclear leukocytes of untreated and terbutaline-treated patients. *Chest*, **98**, 376–381.

HAUSDORFF, W.P., CARON, M.C. & LEFKOWITZ, R.J. (1990) Turning off the signal: desensitization of β-adrenergic receptor function. *FASEB J.*, **4**, 2881–2889.

HAUSSINGER, D., BRODDE, O-E. & STARKE, K. (1987) Alpha-adrenoceptor antagonistic action of amiloride. *Biochem. Pharmacol.*, **36**, 3509–3515.

HAVARD, C.W. & FONSECA, V. (1990) New treatment approaches to myasthenia gravis. *Drugs*, **39**, 66–73.

HAWTHORN, M.H. & BROADLEY, K.J. (1982) Evidence from use of neuronal uptake inhibition that β_1-adrenoceptors, but not β_2-adrenoceptors, are innervated. *J. Pharm. Pharmacol.*, **34**, 664–666.

HAWTHORN, M.H. & BROADLEY, K.J. (1984) Reserpine-induced supersensitivity occurs for β-adrenoceptor-mediated responses of heart and trachea but not of the uterus and lung. *Eur. J. Pharmacol.*, **105**, 245–255.

HAWTHORN, M.H., BROADLEY, K.J. & GIBBON, C.J. (1985) Examination of the bronchoconstrictor response of guinea-pig isolated lung to β-phenylethylamine. *Gen. Pharmacol.*, **16**, 371–378.

HEAD, R.J., IRVINE, R.J., BARONE, S., STITZEL, R.E. & DE LA LANDE, I.S. (1985) Nonintracellular, cell-associated O-methylation of isoproterenol in the isolated rabbit aorta. *J. Pharmacol. Exp. Ther.*, **234**, 184–189.

HEARSE, D.J. & DENNIS, S.C. (1982) Myocardial ischaemia and metabolic changes associated with the genesis of ventricular arrhythmias. In: *Early Arrhythmias Resulting from Myocardial Ischaemia*, pp. 15–35, J.R. PARRATT (Ed.), Macmillan Press, London.

HEDBERG, A., MINNEMAN, K.P. & MOLINOFF, P.B. (1980) Differential distribution of beta-1 and beta-2 adrenergic receptors in cat and guinea-pig heart. *J. Pharmacol. Exp. Ther.*, **213**, 503–508.

HEDBERG, A., GERBER, J.G., NIES, A.S., WOLFE, B.B. & MOLINOFF, P.B. (1986) Effects of pindolol and propranolol on beta adrenergic receptors on human lymphocytes. *J. Pharmacol. Exp. Ther.*, **239**, 117–123.

HEDQVIST, P. (1977) Basic mechanisms of prostaglandin action on autonomic neurotransmission. *Annu. Rev. Pharmacol. Toxicol.*, **17**, 259–279.

HEIJNIS, J.B. & VAN ZWIETEN, P.A. (1992) Enhanced inotropic responsiveness to α_1-adrenoceptor stimulation in isolated working hearts from diabetic rats. *J. Cardiovasc. Pharmacol.*, **20**, 559–562.

HENNING, M. (1969) Interaction of DOPA decarboxylase inhibitors with the effect of α-methyldopa on blood pressure and tissue monoamines in rats. *Acta Physiol. Scand.*, **27**, 135–148.

HENNING, M. & VAN ZWIETEN, P.A. (1968) Central hypotensive effect of α-methyldopa. *J. Pharm. Pharmacol.*, **20**, 409–417.

HEREPATH, M.L. & BROADLEY, K.J. (1990a) Affinities of full agonists for cardiac β-adrenoceptors calculated by use of *in vitro* desensitization. *Naunyn-Schmiedeberg's Arch. Pharmacol.*, **341**, 525–533.

HEREPATH, M.L. & BROADLEY, K.J. (1990b) Desensitization of β-adrenoceptor-mediated functional responses of guinea-pig atria by *in vitro* incubation with isoprenaline. *J. Cardiovasc. Pharmacol.*, **15**, 259–268.

HEREPATH, M.L. & BROADLEY, K.J. (1992) Resistance of β_2-adrenoceptor-mediated responses of lung strips to desensitization by long-term agonist exposure – comparison with atrial β_1-adrenoceptor-mediated responses. *Eur. J. Pharmacol.*, **215**, 209–219.

HERTTING, G. & AXELROD, J. (1961) Fate of tritiated noradrenaline at the sympathetic nerve endings. *Nature*, **192**, 172–173.

HIEBLE, J.P. & KOLPAK, D.C. (1993) Mediation of the hypotensive action of systemic clonidine in the rat by α_2-adrenoceptors. *Br. J. Pharmacol.*, **110**, 1635-1639.

HIEBLE, J.P. & RUFFOLO, R.R. Jr. (1987) Therapeutic application of agents interacting with alpha$_1$-adrenergic receptors. In: *The Alpha-1 Adrenergic Receptors*, pp. 477-500, R.R. RUFFOLO, Jr (Ed.), Humana Press, Clifton, NJ.

HIGGINS, B., VATNER, S.F. & BRAUNWALD, E. (1973) Parasympathetic control of the heart. *Pharmacol. Rev.*, **25**, 119–155.

HIGGINS, T.L. & CHERNOW, B. (1987) Pharmacotherapy of circulatory shock. *Dis. Mon.*, **33**, 309–361.

HILDITCH, A. & DREW, G.M. (1987) Subclassification of peripheral dopamine receptors. *Clin. Exp. Hypertens.*, A9(Suppl. 5 & 6), 853–872.

HILL, A.V. (1913) The combinations of haemoglobin with oxygen and with carbon monoxide I. *Biochem. J.*, **7**, 471–480.

HILLS, J.M. & JESSEN, K.R. (1992) Transmission: τ-aminobutyric acid (GABA), 5-hydroxytryptamine (5-HT) and dopamine. In: *Autonomic Neuroeffector Mechanisms*, pp. 465-507, G. BURNSTOCK & C.H.V. HOYLE, (Eds), Harwood Academic, Chur.

HIMMS-HAGEN, J. (1985) The role of brown adipose tissue thermogenesis in energy balance. In: *New Perspectives in Adipose Tissue: Structure, Function and Development*, pp. 199–222, A. CRYER & R.L.R. VAN, (Eds), Butterworths, London.

HINSTRIDGE, V. & SPEIGHT, T.M. (1991) An overview of therapeutic interventions in myocardial infarction. Emphasis on secondary prevention. *Drugs*, **42**, (Suppl. 2), 8–20.

HIRSCH, L.S. (1939) Controlling appetite in obesity. *J. Med. Cinncinnati*, **20**, 84–85.

HIRSCHOWITZ, B.I. & MOLINA, E. (1984) Classification of muscarinic effects on gastric secretion and heart rate in intact dogs. *Trends Pharmacol. Sci.*, Suppl. 69–73.

HJEMDAHL, P. & LINDE, B. (1983) Influence of circulating norepinephrine and epinephrine on adipose tissue vascular resistance and lipolysis in humans. *Am. J. Physiol.*, **245**, H447–H452.

HOFFMAN, B.B. (1987) Regulation of alpha-1 adrenergic receptors. In: *The Alpha-1 Adrenergic Receptors*, pp. 457–474, R.R. Ruffolo, Jr (Ed.), Humana Press, Clifton, NJ.

HOFLACK, J., TRUMPP-KALLMEYER, S. & HIBERT, M. (1994) Re-evaluation of bacteriorhodopsin as a model for G protein-coupled receptors. *Trends Pharmacol. Sci.*, **15**, 7–9.

HOJNACKI, J., MULLIGAN, J., CLUETTE-BROWN, J., IGOE, F. & OSMOLSKI, T. (1986) Oral nicotine impairs clearance of plasma low density lipoproteins. *Proc. Soc. Exp. Biol. Med.*, **182**, 414–418.

HOLMAN, M.E. (1970) Junction potentials in smooth muscle. In: *Smooth Muscle*, pp. 244–288, E. BÜLBRING, A.F. BRADING, A.W. JONES & T. TOMITA (Eds), Edward Arnold, London.

HOLMES, B., BROGDEN, R.N., HEEL, R.C., SPEIGHT, T.M. & AVERY, G.S. (1983) Guanabenz: a review of its pharmacodynamic properties and therapeutic efficacy in hypertension. *Drugs*, **26**, 212–229.

HOLMSTEDT, B. (1959) Pharmacology of organophosphorus cholinesterase inhibitors. *Pharmacol. Rev.*, **11**, 567–688.

HOLMSTEDT, B. (1988) Toxicity of nicotine and related compounds. In: *The Pharmacology of Nicotine*, pp. 61–88, M.J. RAND & K. THURAU (Eds), IRL Press, Oxford.

HOLTON, P. (1959) The liberation of adenosine triphosphate on antidromic stimulation of sensory nerves. *J. Physiol.* **145**, 494–504.

HONDA, K., SATAKE, T., TAKAGI, K. & TOMITA, T. (1986) Effects of relaxants on electrical and mechanical activities in the guinea-pig tracheal muscle. *Br. J. Pharmacol.*, **87**, 665–671.

HORI, M., SATO, K., MIYAMOTO, S., OZAKI, H. & KARAKI, H. (1993) Different pathways of calcium sensitization activated by receptor agonists and phorbol esters in vascular smooth muscle. *Br. J. Pharmacol.*, **110**, 1527–1531.

HOSEY, M.M. (1994) Desensitization of muscarinic cholinergic receptors and the role of protein phosphorylation. In: *Regulation of Cellular Signal Transduction Pathways by Desensitization and Amplification*, pp. 113–128. D.R. SIBLEY & M.D. HOUSLAY (Eds), John Wiley, Chichester.

HOUGEN, H.P., THYGESEN, P., CHRISTENSEN, H.B., RYGAARD, J., SVENDSEN, O. & JUUL, P. (1992) Effect of immunosuppressive agents on the guanethidine-induced sympathectomy in athymic and euthymic rats. *Int. J. Immunopharmacol.*, **14**, 1113–1123.

HOUSLAY, M.D. (1994) Protein kinase C and the modulation of G-protein-controlled adenylyl cyclase and other signal transmission systems. In: *Regulation of Cellular Signal Transduction Pathways by Desensitization and Amplification*, pp. 129–168, D.R. SIBLEY & M.D. HOUSLAY (Eds), John Wiley, Chichester.

HOWELL, R.E., SICKELS, B.D. & WOEPPEL, S.L. (1993) Pulmonary antiallergic and bronchodilator effects of isozyme-selective phosphodiesterase inhibitors in guinea pigs. *J. Pharmacol. Exp. Ther.*, **264**, 609–615.

HOYLE, C.H.V. (1992) Transmission: purines. In: *Autonomic Neuroeffector Mechanisms*, pp. 367–407, G. BURNSTOCK & C.H.V. HOYLE, (Eds), Harwood Academic, Chur.

HOYLE, C.H.V. & BURNSTOCK, G. (1985) Atropine-resistant excitatory junction potentials in rabbit bladder are blocked by α,β-methylene ATP. *Eur. J. Pharmacol.*, **114**, 239–240.

HOYLE, C.H.V., KNIGHT, G.E. & BURNSTOCK, G. (1990) Suramin antagonizes responses to P_2-purinoceptor agonists and purinergic nerve stimulation in the guinea-pig urinary bladder and *Taenia coli*. *Br. J. Pharmacol.*, **99**, 617–621.

HUCHO, F., JÄRV, J. & WEISE, C. (1991) Substrate-binding sites in acetylcholinesterase. *Trends Pharmacol. Sci.*, **12**, 422–426.

HULME, E.C., CURTIS, C.A.M., WHEATLEY, M., AITKEN, A. & HARRIS, A.C. (1989) Localization and structure of the muscarinic receptor ligand binding site. *Trends Pharmacol. Sci.*, Suppl. 22–25.

HULME, E.C., BIRDSALL, N.J.M. & BUCKLEY, N.J. (1990) Muscarinic receptor subtypes. *Annu. Rev. Pharmacol. Toxicol.*, **30**, 633–673.

HUNTER, A.J., MURRAY, T.K., JONES, J.A., CROSS, A.J. & GREEN, A.R. (1989) The cholinergic pharmacology of tetrahydroaminoacridine *in vivo* and *in vitro*. *Br. J. Pharmacol.*, **98**, 79–86.

HURVITZ, L.M., KAUFMAN, P.L., ROBIN, A.L., WEINREB, R.N., CRAWFORD, K. & SHAW, B. (1991) New developments in the treatment of glaucoma. *Drugs*, **41**, 514–532.

HUTTNER, W.B., BENEDUM, V.M. & ROSA, P. (1988) Biosynthesis, structure and function of the secretogranins/chromogranins. In: *Molecular Mechanisms in Secretion*, pp. 380–389. N.A. THORN, M. TREIMAN & O.H. PETERSEN (Eds), Munksgaard, Copenhagen.

ICHINOSE, M. & BARNES, P.J. (1989) Inhibitory histamine H_3-receptors on cholinergic nerves in human airways. *Eur. J. Pharmacol.*, **163**, 383–386.

IGNARRO, L.J. (1990) Nitric Oxide: a novel signal transduction mechanism for transcellular communication. *Hypertension*, **16**, 477–483.

INSTRUCTIONS TO AUTHORS. Receptor Nomenclature (1993) *Br. J. Pharmacol.*, **108**, 282–284.

INTERNATIONAL CONSENSUS REPORT ON DIAGNOSIS AND MANAGEMENT OF ASTHMA (1992) *Clin. Exp. Allergy*, **22**, (Suppl. 1).

ISHAC, E.J.N., PENNEFATHER, J.N. & HANDBERG, G.M. (1983) Effects of changes in thyroid state on atrial α- and β-adrenoceptors, adenylate cyclase activity, and catecholamine levels in the rat. *J. Cardiovasc. Pharmacol.*, **5**, 396–405.

ISHIKAWA, S. & SPERELAKIS, N. (1987) A novel class (H_3) of histamine receptors on perivascular nerve terminals. *Nature*, **327**, 158–160.

IVERSEN, L.L. (1963) The uptake of noradrenaline by the isolated perfused rat heart. *Br. J. Pharmacol.*, **21**, 523–537.

IVERSEN, L.L. (1965) The uptake of catechol amines at high perfusion concentrations in the isolated rat heart: A novel catechol amine uptake process. *Br. J. Pharmacol.*, **25** 18–33.

IVERSEN, L.L. (1967) *The Uptake and Storage of Noradrenaline in Sympathetic Nerves*, Cambridge University Press, Cambridge.

IVERSEN, L.L. (1971) Role of transmitter uptake mechanisms in synaptic neurotransmission. *Br. J. Pharmacol.*, **41**, 571–591.

JACKOWSKI, S. & ROCK, C.O. (1989) Stimulation of phosphatidylinositol 4,5-bisphosphate-phospholipase C activity by phosphatidic acid. *Arch. Biochem. Biophys.*, **268**, 516–524.

JAISWAL, N. & MALIK, K.U. (1991) Methoctramine, a cardioselective antagonist: muscarinic receptor mediating prostaglandin synthesis in isolated rabbit heart. *Eur. J. Pharmacol.*, **192**, 63–70.

JAMES, I.M., GRIFFITH, D.N.W., PEARSON, R.M. & NEWBURY, P. (1977) Effect of oxprenolol on stage-fright in musicians. *Lancet*, ii, 952–954.

JARROTT, B. & VAJDA, F.J.E. (1987) The current status of monoamine oxidase and its inhibitors. *Med. J. Aust.*, **146**, 634–638.

JEFFERY, P.K. (1992) Pathology of asthma. In: *Asthma.* pp. 23–39, P.J. BARNES (Ed.) (British Medical Bulletin 48), Churchill Livingstone, London.

JERUSALINSKY, D. & HARVEY, A.L. (1994) Toxins from mamba venoms: small proteins with selectivities for different subtypes of muscarinic acetylcholine receptors. *Trends Pharmacol. Sci.*, **15**, 424–430.

JOHANSSON, L.-H. & PERSSON, H. (1983) β_2-Adrenoceptors in guinea-pig atria. *J. Pharm. Pharmacol.*, **35**, 804–807.

JOHANSSON, S.R.M., ANDERSSON, R.G.G. & WIKBERG, J.E.S. (1980) Comparison of β_1- and β_2-receptor stimulation in oestrogen or progesterone dominated rat uterus. *Acta Pharmacol. Toxicol.*, **47**, 252–258.

JOHNSON, G.L., KASLOW, H.R., FARFEL, Z. & BOURNE, H.R. (1980) Genetic analysis of hormone-sensitive adenylate cyclase. *Adv. Cyclic Nucleotide Res.*, **13**, 1–37.

JOHNSTON, H., MAJEWSKI, H. & MUSGRAVE, I.F. (1987) Involvement of cyclic nucleotides in prejunctional modulation of noradrenaline release in mouse atria. *Br. J. Pharmacol.*, **91**, 773–781.

JOHNSTON, J.P. (1968) Some observations upon a new inhibitor of monoamine oxidase in brain tissue. *Biochem. Pharmacol.*, **17**, 1285–1297.

JOSE, P.A., RAYMOND, J.R., BATES, M.D., APERIA, A., FELDER, R.A. & CAREY, R.M. (1992) The renal dopamine receptors. *J. Am. Soc. Nephrol.*, **2**, 1265–1278.

KAISER, C., COLELLA, D.F., SCHWARTZ, M.S., GARVEY, E. & WARDELL, J.R. (1974) Adrenergic Agents. 1. Synthesis and potential β-adrenergic agonist activity of some catecholamine analogs bearing a substituted amino functionality in the meta position. *J. Med. Chem.*, **17**, 49–57.

KALSNER, S. (1974) A new approach to the measurement and classification of forms of supersensitivity of autonomic effector responses. *Br. J. Pharmacol.*, **51**, 427–434.

KALSNER, S. (1980) The effects of (+)- and (−)-propranolol on ^{3}H-transmitter efflux in guinea-pig atria and the presynaptic β-adrenoceptor hypothesis. *Br. J. Pharmacol.*, **70**, 491–498.

KAMEYAMA, M., HOFMANN, F. & TRAUTWEIN, W. (1985) On the mechanism of β-adrenergic regulation of the Ca channel in the guinea-pig heart. *Pflügers Arch.*, **405**, 285–293.

KANTHAKUMAR, K., CUNDELL, D.R., JOHNSON, M., WILLS, P.J., TAYLOR, G.W., COLE, P.J. & WILSON, R. (1994) Effect of salmeterol on human nasal epithelial cell ciliary beating: inhibition of the ciliotoxin, pyocyanin. *Br. J. Pharmacol.*, **112**, 493–498.

KARAKI, H. (1987) Use of tension measurements to delineate the mode of action of vasodilators. *J. Pharmacol. Meth.*, **18**, 1–21.

KARCZMAR, A.G. (1970) History of the research with anticholinesterase agents. In: *Anticholinesterase Agents*, Vol. 1, *International Encyclopaedia of Pharmacology and Therapeutics*, pp. 1–44, A.G. KARCZMAR (Ed.), Pergamon Press, Oxford.

KASAKOV, L., BELAI, A., VLASKOVSKA, M. & BURNSTOCK, G. (1994) Noradrenergic–nitrergic interactions in the rat anococcygeus muscle: evidence for postjunctional modulation by nitric oxide. *Br. J. Pharmacol.*, **112**, 403–410.

KAUFMAN, P.L., WEIDMAN, T. & ROBINSON, J.R. (1984) Cholinergics. In: *Pharmacology of the Eye. Handbook of Experimental Pharmacology*, Vol. 69, pp. 149–191, M.L. SEARS, (Ed.), Springer-Verlag, Berlin.

KAUMANN, A.J. (1972) Potentiation of the effect of isoprenaline and noradrenaline by hydrocortisone in cat heart muscle. *Naunyn-Schmiedeberg's Arch. Pharmacol.*, **273**, 134–153.

KAUMANN, A.J. (1977) Relaxation of heart muscle by catecholamines and by dibutyryl cyclic adenosine 3'5'-monophosphate. *Naunyn-Schmiedeberg's Arch. Pharmacol.*, **296**, 205–215.

KAUMANN, A.J. (1981) In kitten ventricular myocardium, the inotropic potency of an agonist is determined by both its intrinsic activity for adenylyl cyclase and its affinity for the β-adrenoceptors. *Naunyn-Schmiedeberg's Arch. Pharmacol.*, **317**, 13–18.

KAUMANN, A.J. (1987) Are receptor classifications derived from cellular, membrane, and molecular assays compatible? In: *Perspectives on Receptor Classification*, pp. 207–219, J.W. BLACK, D.H. JENKINSON & V.P. GERSKOWITCH (Eds), Alan R. Liss, New York.

KAUMANN, A.J. (1989) Is there a third heart β-adrenoceptor? *Trends Pharmacol. Sci.*, **10**, 316–320.

KAUMANN, A.J. (1991) Some aspects of heart beta adrenoceptor function. *Cardiovasc. Drugs Therapy*, **5**, 549–560.

KAWASAKI, H., CLINE, W.H. JR. & SU, C. (1982) Enhanced presynaptic *beta* adrenoceptor-mediated modulation of vascular adrenergic neurotransmission in spontaneously hypertensive rats. *J. Pharmacol. Exp. Ther.*, **223**, 721–735.

KELLAR, K.J., GIBLIN, B.A. & MARTINO-BARROWS, A. (1988) Nicotinic cholinergic receptor agonist recognition sites in brain. In: *The Pharmacology of Nicotine*, pp. 193–206, M.J. RAND & K. THURAU (Eds), IRL Press, Oxford.

KENAKIN, T.P. (1980) Errors in the measurement of agonist potency-ratios produced by uptake processes. A general model applied to β-adrenoceptor agonists. *Br. J. Pharmacol.*, **71**, 407–417.

KENAKIN, T.P. (1981) An *in vitro* quantitative analysis of the α-adrenoceptor partial agonist activity of dobutamine and its relevance to inotropic selectivity. *J. Pharmacol. Exp. Ther.*, **216**, 210–219.

KENAKIN, T.P. (1984) The classification of drugs and drug receptors in isolated tissues. *Pharmacol. Rev.*, **36**, 165–222.

KENAKIN, T.P. (1987) *Pharmacologic Analysis of Drug–Receptor Interaction*, Raven Press, New York.

KENAKIN, T.P. (1993) *Pharmacologic Analysis of Drug–Receptor Interaction*, 2nd Edn, Raven Press, New York.

KENAKIN, T. (1995a) Agonist–receptor efficacy I: mechanisms of efficacy and receptor promiscuity. *Trends Pharmacol. Sci.*, **16**, 188–192.

KENAKIN, T. (1995b) Agonist-receptor efficacy II: agonist trafficking of receptor signals. *Trends Pharmacol. Sci.*, **16**, 232–238.

KENAKIN, T. (1995c) Pharmacological proteus. *Trends Pharmacol. Sci.*, **16**, 256–258.

KENAKIN, T.P. & BEEK, D. (1982a) Is prenalterol (H133/80) really a selective beta-1 adrenoceptor agonist? Tissue selectivity resulting from differences in stimulus-response relationships. *J. Pharmacol. Exp. Ther.*, **213**, 406–412.

KENAKIN, T.P. & BEEK, D. (1982b) *In vitro* studies on the cardiac activity of prenalterol with reference to use in congestive heart failure. *J. Pharmacol. Exp. Ther.*, **220**, 77–85.

KENAKIN, T.P. & BLACK, J.W. (1977) Can chloropractolol alkylate β-adrenoceptors? *Nature*, **265**, 365–366.

KENAKIN, T.P. & BOSELLI, C. (1990) Promiscuous or heterogeneous muscarinic receptors in rat atria? I. Schild analysis with simple competitive antagonists. *Eur. J. Pharmacol.*, **191**, 39–48.

KENNEDY, C., SAVILLE, V.L. & BURNSTOCK, G. (1986) The contributions of noradrenaline and ATP to the responses of the rabbit central ear artery to sympathetic nerve stimulation depend on the parameters of stimulation. *Eur. J. Pharmacol.*, **122**, 291–300.

KENNY, B.A., CHALMERS, D.H., PHILPOTT, P.C. & NAYLOR, A.M. (1995) Characterization of an α_{1D}-adrenoceptor mediating contractile responses of rat aorta to noradrenaline. *Br. J. Pharmacol.*, **115**, 981–986.

KENT, R.S., DE LEAN, A. & LEFKOWITZ, R.J. (1980) A quantitative analysis of beta-adrenergic receptor interactions: resolution of high and low affinity states of the receptor by computer modeling of ligand binding data. *Mol. Pharmacol.*, **17**, 14–23.

KERSHBAUM, A. & BELLET, S. (1964) Cigarette smoking and blood lipids. *J. Am. Med. Assoc.*, **187**, 32–36.

KILBINGER, H. (1984) Facilitation and inhibition by muscarinic agonists of acetylcholine release from guinea pig myenteric neurones: mediation through different types of neuronal muscarinic receptors. *Trends Pharmacol. Sci.*, Suppl. 49–52.

KILBINGER, H., WOLF, D. & D'AGOSTINO, G. (1991) Pre- and postjunctional muscarinic receptors in the guinea-pig trachea. In: *Presynaptic Receptors and Neuronal Transporters. Advances in Biosciences*, Vol. 82, pp. 7–10, S.Z. Langer, A.M. Galzin & J. Costentin (Eds), Pergamon Press, Oxford.

KINOSHITA, S., OHLSTEIN, E.H. & FELDER, R.A. (1990) Dopamine-1 receptors in rat proximal convoluted tubule: regulation by intrarenal dopamine. *Am. J. Physiol.*, **258**, F1068–F1074.

KIRKPATRICK, K. & BURNSTOCK, G. (1987) Sympathetic nerve-mediated release of ATP from the guinea-pig vas deferens is unaffected by reserpine. *Eur. J. Pharmacol.*, **138**, 207–214.

KIRSHNER, N. (1959) Biosynthesis of adrenaline and noradrenaline. *Pharmacol. Rev.*, **11**, 350–357.

KJELDSEN, S.E., EIDE, I., AAKESSON, I. & LEREN, P. (1983) Increased arterial catecholamine concentration in 50-year-old men with essential hypertension. *Scand. J. Clin. Lab. Invest.*, **43**, 343–349.

KLEIN, R.L. (1982) Chemical composition of the large noradrenergic vesicles. In: *Neurotransmitter Vesicles*, pp. 133–174, R.L. KLEIN, H. LAGERCRANTZ & H. ZIMMERMAN (Eds), Academic Press, New York.

KO, F-N., GUH, J-H., YU, S-M., HOU, Y-S., WU, Y-C & TENG, C-M. (1994)(−)-Discretamine, a selective α_{1D}-adrenoceptor antagonist, isolated from *Fissistigma glaucescens. Br. J. Pharmacol.*, **112**, 1174–1180.

KOBILKA, B.K., KOBILKA, T.S., DANIEL, K., REGAN, J.W., CARON, M.G. & LEFKOWITZ, R.J. (1988) Chimeric α_2-, β_2-adrenergic receptors: delineation of domains involved in effector coupling and ligand binding specificity. *Science*, **240**, 1310–1316.

KOBINGER, W. & WALLAND, A. (1967) Investigation into the mechanism of the hypotensive effect of 2-(2,6-dichlorphenylamino)-2-imidazoline-HCL. *Eur. J. Pharmacol.*, **2**, 155–162.

KOHI, M., NOROTA, I., TAKANASHI, M. & ENDOH, M. (1993) On the mechanism of action of the *beta*-1 partial agonist denopamine in regulation of myocardial contractility: effects on myocardial *alpha* adrenoceptors and intracellular Ca^{++} transients. *J. Pharmacol. Exp. Ther.*, **265**, 1292–1300.

KOHN, A.N., MOSS, A.P., HARGETT, N.A., RITCH, R., SMITH, H. & PODOS, S.M. (1979) Clinical comparison of dipivalyl epinephrine and epinephrine on the treatment of glaucoma. *Am. J. Ophthalmol.*, **87**, 196–201.

KOLTZENBURG, M. & MCMAHON, S.B. (1991) The enigmatic role of the sympathetic nervous system in chronic pain. *Trends Pharmacol. Sci.*, **12**, 399–402.

KOMORI, S. & OHASHI, H. (1984) Presynaptic, muscarinic inhibition of non-adrenergic, non-cholinergic neuromuscular transmission in the chicken rectum. *Br. J. Pharmacol.*, **82**, 73–84.

KOPIN, I.J. (1968) False adrenergic transmitters. *Annu. Rev. Pharmacol. Toxicol.*, **8**, 377–394.

KOSTRZEWA, R.M. & JACOBOWITZ, D.M. (1974) Pharmacological actions of 6-hydroxydopamine. *Pharmacol. Rev.*, **26**, 199–288.

KOYANAGAWA, H., MUSHA, T., KANDA, A., KIMURA, T. & SATOH, S. (1989) Inhibition of vagal transmission by cardiac sympathetic nerve stimulation in the dog: possible involvement of opioid receptor. *J. Pharmacol. Exp. Ther.*, **250**, 1092–1096.

KRISHNAN, G. (1976) Oxprenolol in the treatment of examination stress. *Curr. Med. Res. Opin.*, **4**, 241–243.

KROMER, W. (1988) Endogenous and exogenous opioids in the control of gastrointestinal motility and secretion. *Pharmacol. Rev.*, **40**, 121–162.

KROMER, W. & ELTZE, M. (1991) Is field (vagal) stimulation of gastric acid secretion mediated by M_1 or non-M_1 muscarinic receptors? A methodological problem exemplified in the mouse stomach *in vitro. J. Auton. Pharmacol.*, **11**, 337–342.

KRSTEW, E., MCPHERSON, G.A., MALTA, E. & RAPER, C. (1984) Is Ro 03–7894 an irreversible antagonist at β-adrenoceptor sites? *Br. J. Pharmacol.*, **82**, 501–508.

KUNOS, G. & NICKERSON, M. (1977) Effects of sympathetic innervation and temperature on the properties of rat heart adrenoceptors. *Br. J. Pharmacol.*, **59**, 603–614.

KUNOS, G., VERMES-KUNOS, I. & NICKERSON, M. (1974) Effects of thyroid state on adrenoceptor properties. *Nature*, **250**, 779–781.

KUNOS, G., ROBERTSON, B., KAN, W.H., PREIKSAITIS, H. & MUCCI, L. (1978) Adrenergic reactivity of the myocardium in hypertension. *Life Sci.*, **22**, 847–854.

KUPFER, L.E., BILEZIKIAN, J.P. & ROBINSON, R.B. (1986) Regulation of alpha and beta adrenergic receptors by triiodothyronine in cultured rat myocardial cells. *Naunyn-Schmiedeberg's Arch. Phrmacol.*, **334**, 275–281.

KUPFERMANN, I. (1991) Functional studies of cotransmission. *Physiol. Rev.*, **71**, 683–732.

KURIYAMA, H. (1970) Effects of ions and drugs on the electrical activity of smooth muscle. In: *Smooth Muscle*, pp. 366–395, E. BÜLBRING, A.F. BRADING, A.W. JONES & T. TOMITA (Eds), Edward Arnold, London.

KYBURZ, E. (1990) New developments in the field of MAO inhibitors. *Drugs News Perspect.*, **3**, 592–599.

LAMBRECHT, G., FEIFEL, R., MOSER, U., WAGNER-RÒDER, M., CHOO, L.K., CAMUS, J., TASTENOY, M., WAELBROECK, M., STROHMANN, C., TACKE, R., RODRIGUES DE MIRANDA, J.F., CHRISTOPHE, J. & MUTSCHLER, E. (1989) Pharmacology of hexahydro-difenidol, hexahydro-sila-difenidol and related selective muscarinic antagonists. *Trends Pharmacol. Sci.*, Suppl. 60–64.

LANDS, A.M., ARNOLD, A., McAULIFF, J.P., LUDUENA, F.P. & BROWN, T.G. JR. (1967) Differentiation of receptor systems activated by sympathomimetic amines. *Nature*, **214**, 597–598.

LANG, M., HÜMMER, B. & HAHN, H.-L. (1988) Effects of multiple applications of nicotine on mucus secretion and on circulatory and ventilatory variables. In: *The Pharmacology of Nicotine*, pp. 178–179, M.J. RAND & K. THURAU (Eds), IRL Press, Oxford.

LANGE, W. & VON KRUEGER, G. (1932) Über Ester der Monofluorphosphorsäure. *Ber. Dtsch. Chem. Ges.*, **65**, 1598–1601.

LANGER, S.Z. (1974) Selective metabolic pathways for noradrenaline in the peripheral and in the central nervous system. *Med. Biol.*, **52**, 372–383.

LANGER, S.Z. (1977) Presynaptic receptors and their role in the regulation of transmitter release. *Br. J. Pharmacol.*, **60**, 481–497.

LANGER, S.Z. (1981) Presynaptic regulation of the release of catecholamines. *Pharmacol. Rev.*, **32**, 337–362.

LANGER, S.Z. & HICKS, P.E. (1984) Alpha-adrenoreceptor subtypes in blood vessels: Physiology and pharmacology. *J. Cardiovasc. Pharmacol.*, **6**, S547–S558.

LANGER, S.Z., ALDER-GRASHINSKY, E. & GIORGI, O. (1977) Physiological significance of α-adrenoceptor-mediated negative feedback mechanism regulating noradrenaline release during nerve stimulation. *Nature*, **265**, 648–650.

LANGLEY, J.N. (1892) On the origin from the spinal cord of the cervical and upper thoracic sympathetic fibres with some observations on white and gray rami communicantes. *Philos. Trans. R. Soc. London, Ser. B*, **183**, 85–124.

LANGLEY, J.N. (1901) Observations on the physiological action of extracts of the supra-renal bodies. *J. Physiol.*, **27**, 237–256.

LANGLEY, J.N. (1906) Croonian lecture, on nerve endings and a special excitable substances in cells. *Proc. R. Soc. London Ser. B*, **78**, 170–194.

LANGLEY, J.N. (1921) *The Autonomic Nervous System*, Pt. 1, Heffer, Cambridge.

LANGLEY, J.N. & DICKINSON, W.L. (1889) On the local paralysis of peripheral ganglia, and on the connexion of different classes of nerve fibres with them. *Proc. R. Soc. London Ser. B*, **46**, 423–431.

LANGTRY, H.D. & McTAVISH, D. (1990) Terodiline. A review of its pharmacological properties and therapeutic use in the treatment of urinary incontinence. *Drugs*, **40**, 748–761.

LARSEN, A.A. & LISH, P.M. (1964) A new bio-isostere: alkylsulphonamido-phenethanolamines. *Nature*, **203**, 1283–1284.

LASAGNA, L. (1988) *Phenylpropanolamine – A review.* John Wiley, New York.

LAUTT, W.W. (1980) Hepatic nerves: a review of their functions and effects. *Can. J. Physiol. Pharmacol.,* **58**, 105–123.

LAVIE, C.J., VENTURA, H.O. & MESSERLI, F.H. (1991) Regression of increased left ventricular mass by antihypertensives. *Drugs,* **42**, 945–961.

LAVIN, T.M., HEALD, S.L., JEFFS, P.W., SHORR, R.G., LEFKOWITZ, R.J. & CARON, M.G. (1981) Photoaffinity labeling of the β-adrenergic receptor. *J. Biol. Chem.,* **256**, 11944–11950.

LAZARENO, S., BUCKLEY, N.J. & ROBERTS, F.F. (1990) Characterization of muscarinic M_4 binding sites in rabbit lung, chicken heart, and NG 108–15 cells. *Mol. Pharmacol.,* **38**, 805–815.

LEBLOND, F.A., MORISSET, J. & LEBEL, D. (1989) Alterations of pancreatic growth and of GP-2 content in the reserpinized rat model of cystic fibrosis. *Pediatr. Res.,* **25**, 478–481.

LEFEBVRE, R.A. (1986) Study on the possible neurotransmitter of the non-adrenergic non-cholinergic innervation of the rat gastric fundus. *Arch. Int. Pharmacodyn. Ther.,* **280**(Suppl.), 110–136.

LEFEBVRE, R.A. & DEVREESE, K. (1991) Effect of co-dergocrine in the autoperfused superior mesenteric vascular bed of the rat. *J. Auton. Pharmacol.,* **11**, 155–165.

LEFEBVRE, R.A., DE BEURME, F.A. & SAS, S. (1991) Effect of apamin on the responses to VIP, ATP and NANC neurone stimulation in the rat and cat gastric fundus. *J. Auton. Pharmacol.,* **11**, 73–83.

LEFF, P., MARTIN, D. & MORSE, J.M. (1985) Application of the operational model of agonism to establish conditions when functional antagonism may be used to estimate agonist dissociation constants. *Br. J. Pharmacol.,* **85**, 655–663.

LEFF, P., PRENTICE, D.J., GILES, H., MARTIN, G.R. & WOOD, J. (1990) Estimation of agonist affinity and efficacy by direct, operational model-fitting. *J. Pharmacol. Meth.,* **23**, 225–237.

LEFKOWITZ, R.J., CARON, M.G., MICHEL, T. & STADEL, J.M. (1982) Mechanisms of hormone receptor-effector coupling: the β-adrenergic receptor and adenylate cyclase. *Fed. Proc.,* **41**, 2664–2670.

LEFKOWITZ, R.J., HAUSDORFF, W.P. & CARON, M.G. (1990) Role of phosphorylation in desensitization of the β-adrenoceptor. *Trends Pharmacol. Sci.,* **11**, 190–194.

LEFKOWITZ, R.J., COTECCHIA, S., SAMAMA, P. & COSTA, T. (1993) Constitutive activity of receptors coupled to guanine nucleotide regulatory proteins. *Trends Pharmacol. Sci.,* **14**, 303–307.

LEMOINE, H., TENG, K.J., SLEE, S.J. & KAUMANN, A.J. (1989) On minimum cyclic AMP formation rates associated with positive inotropic effects mediated through β_1-adrenoceptors in kitten myocardium. β_1-Specific and non-adrenergic stimulant effects of denopamine. *Naunyn-Schmiedeberg's Arch. Pharmacol.,* **339**, 113–128.

LESAR, T.S. (1987) Comparison of ophthalmic β-blocking agents. *Clin. Pharm.,* **6**, 451–463.

LEVEY, A.I. (1993) Immunological localization of m1–m5 muscarinic acetylcholine receptors in peripheral tissues and brain. *Life Sci.,* **52**, 441–448.

LEVI-MONTALCINI, R. (1972) The morphological effects of immunosympathectomy. In: *Immunosympathectomy,* pp. 55–78, G. Steiner & E. Schönbaum (Eds), Elsevier, Amsterdam.

LIEDEBERG, F., KANNISTO, P., OWMAN, CH. & SCHMIDT, G. (1993) Effects of vasoactive intestinal polypeptide (VIP) on the neuromuscular complex in the bovine ovarian follicle wall. *J. Auton. Pharmacol.,* **13**, 201–209.

LIGGETT, S.B. & LEFKOWITZ, R.J. (1994) Adrenergic receptor-coupled adenylyl cyclase systems: regulation of receptor function by phosphorylation, sequestration and downregulation. In: *Regulation of Cellular Signal Transduction Pathways by Desensitization and Amplification,* pp. 71–97, D.R. SIBLEY & M.D. HOUSLAY (Eds), John Wiley, Chichester.

LIMAS, C.J., LIMAS, C. & GOLDENBERG, I.F. (1989) Intracellular distribution of adrenoceptors in the failing human myocardium. *Am. Heart J.*, **117**, 1310–1316.

LIMBERGER, N., TRENDELENBURG, A-U. & STARKE, K. (1992) Pharmacological characterization of presynaptic α_2-autoreceptors in rat submaxillary gland and heart atrium. *Br. J. Pharmacol.*, **107**, 246–255.

LJUNG, B., BEVAN, J.A. & SU, C. (1973) Evidence for uneven alpha-receptor distribution in the rat portal vein. *Circ. Res.*, **32**, 556–563.

LOEWI, O. (1921) Uebertragbarkeit der herznervenwirkung. *Pflügers Arch. Gesamte Physiol. Menshen Tiere*, **189**, 238–242.

LOEWI, O. & NAVRATIL, E. (1926) Uber humorale übertrag barkeit der herznervenwirkung. X Uber das schicksal des vagusstoffes. *Pflügers Arch. Gesamte Physiol. Menschen Tiere*, **214**, 678–688.

LORENZO, P.S. & ADLER-GRASCHINSKY, E. (1992) Gangliosides prevent the dimethyl sulphoxide-induced increases in [^{3}H]-noradrenaline release from rat isolated atria. *J. Auton. Pharmacol.*, **12**, 349–357.

LORING, R.H. & ZIGMOND, R.E. (1988) Characterization of neuronal nicotinic receptors by snake venom neurotoxins. *Trends Neurosci.*, **11**, 73–78.

LOTTI, M. & MORETTO, A. (1995) Cholinergic symptoms and Gulf War Syndrome. *Nature Medicine*, **1**, 1225–1226.

LOTTI, V.J., LEDOUAREC, J.C. & STONE, C.A. (1984) Autonomic nervous system: adrenergic antagonists. In: *Pharmacology of the Eye. Handbook of Experimental Pharmacology*, pp. 247–277, M.L. SEARS (Ed.), Springer-Verlag, Berlin.

LOU, Y.P., FRANCO-CERECEDA, A. & LUNDBERG, J.M. (1991) Omega conotoxin inhibits CGRP release and bronchoconstriction evoked by a low concentration of capsaicin. *Acta Physiol. Scand.*, **141**, 135–136.

LUCCHELLI, A., GRANA, E. & SANTAGOSTINO-BARBONE, M.G. (1992) Influence of lithium pretreatment and of cooling on the responsiveness of the rat isolated jejunum and urinary bladder to muscarinic agonists. *J. Auton. Pharmacol.*, **12**, 61–72.

LUFF, S.E. & MCLACHLAN, E.M. (1989) Frequency of neuromuscular junctions on arteries of different dimensions in the rabbit, guinea-pig and rat. *Blood Vessels*, **26**, 95–106.

LULICH, K.M., GOLDIE, R.G. & PATERSON, J.W. (1988) Beta-adrenoceptor function in asthmatic bronchial smooth muscle. *Gen. Pharmacol.*, **19**, 307–311.

LUMLEY, P. & BROADLEY, K.J. (1975) Differential blockade of guinea-pig atrial rate and force responses to (−)-noradrenaline by practolol – an uptake phenomenon. *Eur. J. Pharmacol.*, **34**, 207–217.

LUMLEY, P. & BROADLEY, K.J. (1977) Evidence from agonist and antagonist studies to suggest that the β_1-adrenoceptors subserving the positive inotropic and chronotropic responses of the heart do not belong to two separate subgroups. *J. Pharm. Pharmacol.*, **29**, 598–604.

LUMLEY, P., BROADLEY, K.J. & LEVY, G.P. (1977) Analysis of the inotropic: chronotropic selectivity of dobutamine and dopamine in anaesthetized dogs and guinea-pig isolated atria. *Cardiovasc. Res.*, **11**, 17–25.

LUNDBERG, J.M. & HOKFELT, T. (1983) Coexistence of peptides and classical neurotransmitters. *Trends Neurosci.*, **6**, 325–333.

LUNDBERG, J.M., ÅNGGÅRD, A., FAHRENKRUG, J., LUNGREN, G. & HOLMSTEDT, B. (1982) Corelease of VIP and acetylcholine in relation to blood flow and salivary secretion in cat submandibular salivary gland. *Acta Physiol. Scand.*, **115**, 525–528.

LURIE, K., TSUJIMOTO, G. & HOFFMAN, B. (1985) Desensitization of α_1-adrenergic receptor-mediated vascular smooth muscle contraction. *J. Pharmacol. Exp. Ther.*, **234**, 147–152.

MACKAY, D. (1978) How should values of pA_2 and affinity constants for pharmacological competitive antagonists be estimated? *J. Pharm. Pharmacol.*, **30**, 312–313.

MACKAY, D. (1982) Dose–response curves and mechanisms of drug action. *Trends Pharmacol. Sci.*, **3**, 496–499.

MACENZIE, I., BURNSTOCK, G. & DOLLY, J.O. (1982) The effects of purified botulinum neurotoxin type A on cholinergic, adrenergic and non-cholinergic, non-adrenergic, atropine-resistant autonomic neuromuscular transmission. *Neuroscience*, 7, 997–1006.

MACKINNON, A.C., SPEDDING, M. & BROWN, C.M. (1994) α_2-Adrenoceptors: more subtypes but fewer functional differences. *Trends Pharmacol. Sci.*, 15, 119–123.

MAELICKE, A. (1988) Structure and function of the nicotinic acetylcholine receptor. In: *The Cholinergic Synapse. Handbook of Experimental Pharmacology*, pp. 267–313, V.P. WHITTAKER (Ed.), Springer-Verlag, Berlin.

MAGGI, C.A. (1991a) The pharmacology of the efferent function of sensory nerves. *J. Auton. Pharmacol.*, 11, 173–208.

MAGGI, C.A. (1991b) Omega conotoxin and prejunctional modulation of the biphasic response of the rat isolated urinary bladder to single pulse electrical field stimulation. *J. Auton. Pharmacol.*, 11, 295–304.

MAGGI, C.A., MELI, A. & SANTICIOLI, P. (1987) Neuroeffector mechanisms in the voiding cycle of the guinea-pig urinary bladder. *J. Auton. Pharmacol.*, 7, 295–308.

MAGGI, C.A., COY, D.H. & GIULIANI, S. (1992) Effect of [D-Phe[6]] bombesin (6–13) methylester, a bombesin receptor antagonist, towards bombesin-induced contractions in the guinea-pig and rat isolated urinary bladder. *J. Auton. Pharmacol.*, 12, 215–222.

MAGGI, C.A., PATACCHINI, R., ROVERO, P. & GIACHETTI, A. (1993) Tachykinin receptors and tachykinin receptor antagonists. *J. Auton. Pharmacol.*, 13, 23–93.

MAGGI, C.A., HOLZER, P. & GIULIANI, S. (1994a) Effect of w-conotoxin on cholinergic and tachykininergic excitatory neurotransmission to the circular muscle of the guinea-pig colon. *Naunyn-Schmiedeberg's Arch. Pharmacol.*, 350, 529–536.

MAGGI, C.A., ASTOLFI, M., GIULIANI, S., GOSO, C., MANZINI, S., MEINI, S., PALACCHINI, R., PAVONE, V., PEDONE, C., QUARTARA, L., RENZETTI, A.R. & GIACHETTI, A. (1994b) MEN 10,627, a novel polycyclic peptide antagonist of tachykinin NK$_2$ receptors. *J. Pharmacol. Exp. Ther.*, 271, 1489–1500.

MAGGI, C.A., PATACCHINI, R., MEINI, S. & GIULIANI, S. (1994c) Effects of longitudinal muscle-myenteric plexus removal and indomethacin on the responses to tachykinin NK-2 and NK-3 receptor agonists in the circular muscle of the guinea-pig ileum. *J. Auton. Pharmacol.*, 14, 49–60.

MAHAN, L.C., MCKERNAN, R.M. & Insel, P.A. (1987) Metabolism of alpha- and beta-adrenergic receptors *in vitro* and *in vivo*. *Annu. Rev. Pharmacol. Toxicol.*, 27, 215–235.

MAIN, B.G. (1990) β-Adrenergic receptors: In: *Comprehensive Medicinal Chemistry*, Vol. 3, *Membranes and Receptors*, pp. 187–228, P.G. SAMMES & J.B. TAYLOR (Eds), Pergamon Press, Oxford.

MAJEWSKI, H. (1983) Modulation of noradrenaline release through activation of presynaptic β-adrenoceptors. *J. Auton. Pharmacol.*, 3, 47–60.

MAKHLOUF, G.M. (1984) Regulation of gastrin and somatostatin secretion by gastric intramural neurones. *Trends Pharmacol. Sci.*, Suppl. 63–65.

MALIK, K.U. (1970) Potentiation by anticholinesterases of the response of rat mesenteric arteries to sympathetic postganglionic nerve stimulation. *Circ. Res.*, 27, 647–655.

MALLARD, N.J., MARSHALL, R.W. & SPRIGGS, T.L.B. (1992) Neurotransmitter feedback is not important in modulating the noradrenergic component of responses of rat vas deferens to twin pulse electrical field stimulation. *J. Auton. Pharmacol.*, 12, 165–174.

MAN IN'T VELD, A.J., VAN DEN MEIRACKER, A.H. & SCHALEKAMP, M.A. (1988) Do beta-blockers really increase peripheral vascular resistance? Review of the literature on new observations under basal conditions. *Am. J. Hypertens.*, 1, 91–96.

MANO, K., AKBARZADEH, A. & TOWNLEY, R.G. (1979) Effect of hydrocortisone on beta-adrenergic receptors in lung membranes. *Life Sci.*, 25, 1925–1930.

MANZINI, S., MAGGI, C.A. & MELI, A. (1985) Further evidence for involvement of adenosine-5'-triphosphate in non-adrenergic non-cholinergic relaxation of the isolated rat duodenum. *Eur. J. Pharmacol.*, 113, 399–408.

MARÍN, J. (1993) Mechanisms involved in the increased vascular resistance in hypertension. *J. Auton. Pharmacol.*, **13**, 127–176.

MARÍN, J. & RODRÍGUEZ-MARTÍNEZ, M.A. (1995) Nitic oxide, oxygen-derived free radicals and vascular endothelium. *J. Auton. Pharmacol.* **15**, 279–307.

MARKS, M.J., BURCH, J.B. & COLLINS, A.C. (1983) Effects of chronic nicotine infusion on tolerance development and nicotinic receptors. *J. Pharmacol. Exp. Ther.*, **226**, 817–825.

MARSHALL, C.R. (1913) Studies on the pharmaceutical action of tetra-alkyl-ammonium compounds. *Trans. R. Soc. Edinburgh.*, **1**, 17–40.

MARSHALL, I., BURT, R.P. & CHAPPLE, C.R. (1995) Noradrenaline contractions of human prostate mediated by α_{1A}-(α_{1c}) adrenoceptor subtype. *Br. J. Pharmacol.*, **115**, 781–786.

MARSHALL, J.M. (1973) Effects of catecholamines on the smooth muscle of the female reproductive tract. *Annu. Rev. Pharmacol.*, **13**, 19–32.

MARTIN, S.W. & BROADLEY, K.J. (1994) Effects of chronic intravenous infusions of dopexamine and isoprenaline to rats on D_1-, β_1- and β_2-receptor-mediated responses. *Br. J. Pharmacol.*, **112**, 595–603.

MARTINDALE (1989) *The Extra Pharmacopoea* 29th Edn, J.E.F. REYNOLDS (Ed.), The Pharmaceutical Press, London.

MARTINEAU, L., HORAN, M.A., ROTHWELL, N.J. & LITTLE, R.A. (1992) Salbutamol, a β_2-adrenoceptor agonist, increases skeletal muscle strength in young men. *Clin. Sci.*, **83**, 615–621.

MARTINEZ, T.T. & MCNEILL, J.H. (1977) The effect of temperature on cardiac adrenoceptors. *J. Pharmacol. Exp. Ther.*, **203**, 457–466.

MASINI, I., PORCIATTI, F., BOREA, P.A., BARBIERI, M., CERBAI, E. & MUGELLI, A. (1991) Cardiac beta-adrenoceptors in the normal and failing heart: electrophysiological aspects. *Pharmacol. Res.*, **24**(Suppl. 1), 21–27.

MASSOUDI, M. & MILLER, D.S. (1977) Ephedrine, a thermogenic and potential slimming drug. *Proc. Nutr. Soc.*, **36**, 135A.

MASSOULIÉ, J. & TOUTANT, J.-P. (1988) Vertebrate cholinesterases: structure and types of interaction. In: *The Cholinergic Synapse. Handbook of Experimental Pharmacology*, pp. 167–224, V.P. WHITTAKER (Ed.), Springer-Verlag, Berlin.

MASSOULIÉ, J., PEZZEMENTI, L., BON, S., KREJCI, E. & VALLETTE, F.-M. (1993) Molecular and cellular biology of cholinesterases. *Prog. Neurobiol.*, **41**, 31–91.

MATUCCI, R., BIANCHI, B., MAGGI, M., DI MICHELE, N. & GIOTTI, A. (1993) Estrogen modulation of cholinergic receptors in guinea-pig myometrium. *Life Sci.*, **52**, 581.

MAXWELL, R.A., MULL, R.P. & PLUMMER, A.J. (1959) [2-(Octahydro-1-azocinyl)-ethyl]-guanidine sulphate (CIBA 5864-Su), a new synthetic antihypertensive agent. *Experientia*, **15**, 267.

MAXWELL, R.A., FERRIS, R.M. & BURCSU, J.E. (1976) Structural requirements for inhibition of noradrenaline uptake by phenylethylamine derivatives, desipramine, cocaine, and other compounds. In: *The Mechanism of Neuronal and Extraneuronal Transport of Catecholamines*, pp. 95–153, D.M. PATON (Ed.), Raven Press, New York.

MAY, J.M., ABEL, P.W. & MINNEMAN, K.P. (1986) Regulation of β-adrenoceptor density and function in rat vas deferens. *Eur. J. Pharmacol.*, **122**, 221–229.

MCCLESKEY, E.W., FOX, A.P., FELDMAN, D.H., CRUZ, L.J., OLIVERA, R.M., TSIEN, R.W. & YOSHIKAMI, D. (1987) w-Conotoxin: direct and persistent blockade of specific types of calcium channels in neurons but not muscle. *Proc. Natl Acad. Sci. USA*, **84**, 4327–4331.

MCGRATH, J.C. (1982) Evidence for more than one type of post-junctional alpha-adrenoceptor. *Biochem. Pharmacol.* **31**, 467–484.

MCGRATH, J.C. (1984) Alpha-adrenoceptor antagonism by apoyohimbine and some observations on the pharmacology of alpha-adrenoceptors in the rat anococcygeus and vas deferens. *Br. J. Pharmacol.*, **82**, 769–781.

MCGUIRE, M.C., NOGUEIRA, C.F., BARTES, C.F., LIGHTSTONE, H., MAJRA, A., VAN DER SPEK, A.F.L., LOCKRIDGE, O. & LADU, B.N. (1989) Identification of the structural mutation responsible for the dibucaine resistant variant form of human serum cholinesterase. *Proc. Natl. Acad. Sci. USA*, **86**, 953–957.

MCNEILL, J.H. (1985) Endocrine dysfunction and cardiac performance. *Can. J. Physiol. Pharmacol.*, **63**, 1–8.

MCPHERSON, G.A., MOLENAAR, P. & MALTA, E. (1985) The affinity and efficacy of naturally occurring catecholamines at β-adrenoceptor subtypes. *J. Pharm. Pharmacol.*, **37**, 499–501.

MCTAVISH, D., CAMPOLI-RICHARDS, D. & SORKIN, E.M. (1993) Carvedilol; a review of its pharmacodynamic and pharmacokinetic properties, and therapeutic efficacy. *Drugs*, **45**, 232–258.

MEDGETT, I.C. & LANGER, S.Z. (1984) Heterogeneity of smooth muscle alpha-adrenoceptors in rat tail artery *in vitro*. *J. Pharmacol. Exp. Ther.*, **229**, 823–830.

MEKORI, Y.A., BLICKSTEIN, D., BARAM, D., ALTER, A., RADNAY, J., ROZENSZAJN, L.A. & RAVID, M. (1989) Characterization of the interference of T cell activation by reserpine. *Cell Immunol.*, **124**, 308–319.

MELANDER, A., SUNDLER, F. & WESTGREN, V. (1973) Intrathyroidal amines and the synthesis of thyroid hormone. *Endocrinology*, **93**, 193–200.

MELANDER, A., ERICSON, L.E., SUNDLER, F. & INGBAR, S.H. (1974) Sympathetic innervation of the mouse thyroid and its significance in thyroid hormone secretion. *Endocrinology*, **94**, 959–966.

MELCHIORRE, C., MINARINI, A., ANGELI, P., GIARDINA, D., GULINI, U. & QUAGLIA, W. (1989) Polymethylene tetraamines as muscarinic receptor probes. *Trends Pharmacol. Sci.*, Suppl. 55–59.

MELDOLESI, J., CLEMENTI, E., FASOLATO, C., ZACHETTI, D. & POZZAN, T. (1991) Ca^{2+} Influx following receptor activation. *Trends Pharmacol. Sci.*, **12**, 289–292.

MENDEL, B. & RUDNEY, H. (1943) Studies on cholinesterase. I. Cholinesterase and pseudo-cholinesterase. *Biochem. J.*, **37**, 59–63.

MENKVELD, G.J. & TIMMERMAN, H. (1991) A rapid *in vitro* assay of the histamine H_3-receptor: inhibition of electrically evoked contractions of guinea-pig ileum preparations. In: *Presynaptic Receptors and Neuronal Transporters. Advances in the Biosciences*, Vol. 82. pp. 37–38. S.Z. LANGER, A.M. GALZIN & J. COSTENTIN (Eds), Pergamon Press, Oxford.

MEYER, U.A. (1994) The molecular basis of genetic polymorphism of drug metabolism. *J. Pharm. Pharmacol.*, **46**(Suppl. 1), 409–415.

MICHEL, M.C., BECKERINGH, J.J., IKEZONO, K., KRETSCH, R. & BRODDE, O-E. (1986) Lymphocyte beta$_2$-adrenoceptors mirror precisely beta$_2$-adrenoceptor; but poorly beta$_1$-adrenoceptor changes in the human heart. *J. Hypertens.*, **4**, (Suppl. 6), S215–S218.

MICHEL, M.C., PINGSMAN, A., BECKERINGH, J.J., ZERKOWSKI, H.R., DOETSCH, N. & BRODDE, O-E. (1988) Selective regulation of β_1- and β_2-adrenoceptors in the human heart by chronic β-adrenoceptor antagonist treatment. *Br. J. Pharmacol.*, **94**, 685–692.

MICHEL, M.C., HANFT, G. & GROSS, G. (1994) Functional studies on α_1-adrenoceptor subtypes mediating inotropic effects in rat right ventricle. *Br. J. Pharmacol.*, **111**, 539–546.

MICHEL, A.D., LOURY, D.N. & WHITING, R.L. (1989) Differences between the α_2-adrenoceptor in rat submaxillary gland and the α_{2A}- and α_{2B}-adrenoceptor subtypes. *Br. J. Pharmacol.*, **98**, 890–897.

MIHARA, S. & NISHI, S. (1989) Muscarinic excitation and inhibition of neurons in the submucous plexus of the guinea-pig caecum. *Neuroscience*, **31**, 247–257.

MILLER, V.M. & VANHOUTTE, P.M. (1989) Is nitric oxide the only endothelium-derived relaxing factor in canine femoral veins? *Am. J. Physiol.*, **257**, H1910–H1916.

MILLIGAN, G. & GREEN, A. (1994) Receptor-mediated downregulation of cellular G-protein levels as a mechanism for the development of sustained heterologous desensitization. In: *Regulation of Cellular Signal Transduction Pathways by*

Desensitization and Amplification, pp. 233–248, D.R. SIBLEY & M.D. HOUSLAY (Eds), John Wiley, Chichester.

MIMRAN, A. & DUCAILAR, G. (1988) Systemic and regional haemodynamic profile of diuretics and alpha- and beta-blockers. A review comparing acute and chronic effects. *Drugs*, **35**(Suppl. 6), 60–69.

MINNEMAN, K.P. (1988) α_1-Adrenergic receptor subtypes, inositol phosphates, and sources of cell Ca^{2+}. *Pharmacol. Rev.*, **40**, 87–119.

MINNEMAN, K.P. & ABEL, P.W. (1984) Relationship between α_1-adrenoceptor density and functional response of rat vas deferens. Studies with phenoxybenzamine. *Naunyn-Schmiedeberg's Arch. Pharmacol.*, **327**, 238–246.

MINNEMAN, K.P. & MOWRY, C.B. (1986) Interactions of putatively irreversible antagonists with β_1- and β_2-adrenergic receptors. *Biochem. Pharmacol.*, **35**, 857–864.

MINNEMAN, K.P., HEGSTRAND, L.R. & MOLINOFF, P.B. (1979) The pharmacological specificity of beta-1 and beta-2 adrenergic receptors in rat heart and lung *in vitro*. *Mol. Pharmacol.*, **16**, 21–33.

MINNEMAN, K.P., PITTMAN, R.N. & MOLINOFF, P.B. (1981) β-Adrenergic receptor subtypes: properties, distribution, and regulation. *Annu. Rev. Neurosci.*, **4**, 419–461.

MINNEMAN, K.P., WILSON, K.M. & HAN. C. (1991) Alpha-1-adrenergic receptor subtypes: pharmacology and signal transduction. In: *Adrenoceptors: Structure, Mechanisms, Function. Advances in Pharmacological Sciences*, pp. 15–25, E. SZABADI & C.M. BRADSHAW (Eds), Birkhäuser-Verlag, Basel.

MISU, Y. & KUBO, T. (1986) Presynaptic β-adrenoceptors. *Med. Res. Rev.*, **6**, 197–225.

MIYAMOTO, M.D. (1978) The actions of cholinergic drugs on motor nerve terminals. *Pharmacol. Rev.*, **29**, 221–247.

MOCHIDA, S., MIZOBE, F., FISHER, A., KAWANISHI, G. & KOBAYASHI, H. (1988) Dual synaptic effects of activating M_1-muscarinic receptors, in superior cervical ganglia of rabbits. *Brain Res.*, **455**, 9–17.

MOHAN, P.M., SAUNDERS, H.M.H., YANG, C-M., DWYER, T.M. & FARLEY, J.M. (1988) Contractile responses of tracheal smooth muscle in organophosphate-treated swine: 1. Agonist changes. *J. Auton. Pharmacol.*, **8**, 93–106.

MOLDERINGS, G.J., COLLING, E., LIKUNGU, J., JAKSCHIK, J. & GÖTHERT, M. (1994) Modulation of noradrenaline release from the sympathetic nerves of the human saphenous vein and pulmonary artery by presynaptic EP_3- and DP-receptors. *Br. J. Pharmacol.*, **111**, 733–738.

MOLENAAR, P. & MALTA, E. (1986) Analysis of agonist dissociation constants as assessed by functional antagonism in guinea pig left atria. *J. Pharmacol. Meth.*, **15**, 105–117.

MOLENAAR, P. & SUMMERS, R.J. (1987) Characterization of beta-1 and beta-2 adrenoceptors in guinea pig atrium: functional and receptor binding studies. *J. Pharmacol. Exp. Ther.*, **241**, 1041–1047.

MONCADA, S., PALMER, R.M.J. & HIGGS, E.A. (1991) Nitric oxide: physiology, pathophysiology and pharmacology. *Pharmacol. Rev.*, **43**, 109–142.

MORAN, N.C. & PERKINS, M.E. (1958) Adrenergic blockade of the mammalian heart by a dichloro analogue of isoproterenol. *J. Pharmacol. Exp. Ther.*, **124**, 223–237.

MORLEY, J. (1991) New drug developments for asthma. In: *Preventive Therapy in Asthma*, pp. 253–273, J. MORLEY (Ed.), Academic Press, London.

MORRIS, J.L. & GIBBINS, I.L. (1992) Co-transmission and neuromodulation. In: *Autonomic Neuroeffector Mechanisms*, pp. 33–119, G. BURNSTOCK & C.H.V. HOYLE (Eds), Harwood Academic, Chur.

MORRIS, T.H. & KAUMANN, A.J. (1984) Different steric characteristics of β_1- and β_2-adrenoceptors. *Naunyn-Schmiedeberg's Arch. Pharmacol.*, **327**, 176–179.

MORRISON, K.J. & VANHOUTTE, P.M. (1992) Characterization of muscarinic receptors that mediate contraction of guinea-pig isolated trachea to choline esters: effect of removing epithelium. *Br. J. Pharmacol.*, **106**, 672–676.

MORROW, A.L. & CREESE, I. (1986) Characterization of α_1-adrenergic subtypes in rat brain: a reevaluation of [^{3}H]-WB4101 and [^{3}H]-prazosin binding. *Mol. Pharmacol.*, **29**, 321–329.

MOSER, U., LAMBRECHT, G., WAGNER, M., WESS, J. & MUTSCHLER, E. (1989) Structure–activity relationships of new analogues of arecaidine propargyl ester at muscarinic M_1 and M_2 receptor subtypes. *Br. J. Pharmacol.*, **96**, 319–324.

MOTOMURA, S., ZERKOWSKI, H-R., DAUL, A. & BRODDE, O-E. (1990) On the physiologic role of beta-2 adrenoceptors in the human heart: *in vitro* and *in vivo* studies. *Am. Heart J.*, **119**, 608–619.

MOTTRAM, D.R. (1988) *Drugs in Sport*, E. & F.N. Spon, London.

MUELLER, R.A., THOENEN, H. & AXELROD, J. (1969a) Increase in tyrosine hydroxylase activity after reserpine administration. *J. Pharmacol. Exp. Ther.*, **169**, 74–79.

MUELLER, R.A., THOENEN, H. & AXELROD, J. (1969b) Inhibition of trans-synaptically increased tyrosine hydroxylase activity by cycloheximide and actinomycin D. *Mol. Pharmacol.*, **5**, 463–469.

MUELLER, R.A., THOENEN, H. & AXELROD, J. (1970) Inhibition of neuronally induced tyrosine hydroxylase by nicotinic receptor blockade. *Eur. J. Pharmacol.*, **10**, 51–56.

MUKHERJEE, S., LEVER, J.D., NORMAN, D., SYMONS, D., SPRIGGS, T.L.B. & JUNG, R.T. (1989) A comparison of the effects of 6-hydroxydopamine and reserpine on noradrenergic and peptidergic nerves in rat brown adipose tissue. *J. Anat.*, **167**, 189–193.

MURAMATSU, I., OHMURA, T. & OSHITA, M. (1989) Comparison between sympathetic adrenergic and purinergic transmission in the dog mesenteric artery. *J. Physiol.*, **411**, 227–243.

MURAMATSU, I., KIGOSHI, S. & OSHITA, M. (1990) Two distinct α_1-adrenoceptor subtypes involved in noradrenaline contraction of the rabbit thoracic aorta. *Br. J. Pharmacol.*, **101**, 662–666.

MURRAY, K.J. & ENGLAND, P.J. (1992) Inhibitors of cyclic nucleotide phosphodiesterases as therapeutic agents. *Biochem. Soc. Trans.*, **20**, 460–464.

MUSCHOLL, E., HABERMEIER-MUTH, A., ALTES, U. & FORSYTH, K. (1991) Cholinergic–adrenergic presynaptic interactions in the heart and characterization of the receptors involved. In: *Presynaptic Receptors and Neuronal Transporters. Advances in Biosciences*, Vol. 82, pp. 15–18, S.Z. LANGER, A.M. GALZIN & J. COSTENTIN (Eds), Pergamon Press, Oxford.

MUSGRAVE, I., MARLEY, P. & MAJEWSKI, H. (1987) Pertussis toxin does not attenuate α_2-adrenoceptor mediated inhibition of noradrenaline release in mouse atria. *Naunyn-Schmiedeberg's Arch. Pharmacol.*, **336**, 280–286.

MUSGRAVE, I.F. & MAJEWSKI, H. (1991) Protein kinase C and modulation of neurotransmission: studies with protein kinase inhibitors. In: *Presynaptic Receptors and Neuronal Transporters. Advances in Biosciences*, Vol. 82. pp. 279–280, S.Z. LANGER, A.M. GALZIN & J. COSTENTIN (Eds), Pergamon Press, Oxford.

NADELMANN, J. & FRISHMAN, W.H. (1990) Clinical use of β-adrenoceptor blockade in systemic hypertension. *Drugs*, **39**, 862–876.

NAGAMANI, M., KELVER, M.E. & SMITH, E.R. (1987) Treatment of menopausal hot flushes with transdermal administration of clonidine. *Am. J. Obstet. Gynecol.*, **156**, 561–565.

NAKAZATO, Y., OHGA, A., OLESHANSKY, M., TOMITA, U. & YAMADA, Y. (1988) Voltage-independent catecholamine release mediated by the activation of muscarinic receptors in guinea-pig adrenal glands. *Br. J. Pharmacol.*, **93**, 101–109.

NETTER, F.H. (1974) *The Ciba Collection of Medical Illustrations*, Vol. 1., *Nervous System*, Ciba Medical Education Division, New Jersey.

NEWBERRY, N.R. & PRIESTLEY, T. (1987) Pharmacological differences between two muscarinic responses of the rat superior cervical ganglion *in vitro*. *Br. J. Pharmacol.*, **92**, 817–826.

NEWHOUSE, P.A., POTTER, A., CORWIN, J. & LENOX, R. (1992) Acute nicotinic blockade produces cognitive impairment in normal humans. *Psychopharmacology*, 108, 480–484.

NEWTON, M.W. & JENDEN, D.J. (1986) False transmitters as presynaptic probes for cholinergic mechanisms and function. *Trends Pharmacol. Sci.*, 7, 316–320.

NG, N.L. & MALTA, E. (1989) Organ bath studies using the irreversible β-adrenoreceptor antagonist bromoacetylalprenololmenthane (BAAM) *J. Auton. Pharmacol.*, 9, 189–200.

NICHOLS, A.J. & RUFFOLO, R.R., JR. (1991) Alpha-adrenoceptors and Ca^{2+} translocation in vascular smooth muscle. In: *Adrenoceptors: Structure, Mechanisms, Function. Advances in Pharmacological Sciences*, pp. 139–148, E. SZABADI & C.M. BRADSHAW (Eds), Birkhäuser-Verlag, Basel.

NICHOLSON, C.D. & BROADLEY, K.J. (1978) Irreversible β-adrenoceptor blockade of atrial rate and tension responses. *Eur. J. Pharmacol.*, 52, 259–269.

NICHOLSON, C.D., CHALLIS, R.A.J. & SHAHID, M. (1991) Differential modulation of tissue function and therapeutic potential of selective inhibitors of cyclic nucleotide phosphodiesterase isoenzymes. *Trends Pharmacol. Sci.*, 12, 19–27.

NICKERSON, M. (1949) The pharmacology of adrenergic blockade. *Pharmacol. Rev.*, 1, 27–101.

NICKERSON, M. & GOODMAN, L.S. (1947) Pharmacological properties of a new adrenergic blocking agent; N,N-dibenzyl-β-chloroethylamine (dibenamine) *J. Pharmacol. Exp. Ther.*, 89, 167–185.

NICKERSON, M. & GUMP, W.S. (1949) The chemical basis for adrenergic blocking in compounds related to dibenamine. *J. Pharmacol. Exp. Ther.*, 97, 25–47.

NISHIMURA, J., KHALIL, R.A. & VAN BREEMEN, C. (1989) Agonist-induced vascular tone. *Hypertension*, 13, 835–844.

NOBLE, D. (1979) *The Initiation of the Heartbeat.*, 2nd Edn, Clarendon Press, Oxford.

NOBLE, D. (1984) The surprising heart: a review of recent progress in cardiac electrophysiology. *J. Physiol.*, 353, 1–50.

NOREL, X., ANGRISANI, M., LABAT, C., GORENNE, I., DULMET, E., ROSSI, F. & BRINK, C. (1993) Degradation of acetylcholine in human airways: role of butyrylcholinesterase. *Br. J. Pharmacol.*, 108, 914–919.

NORTH, R.A. (1986) Mechanisms of autonomic integration. In: *Handbook of Physiology. Section 1. The Nervous System*, Vol. IV, *Intrinsic Regulatory Systems of the Brain*, pp. 115–153, American Physiological Society, Bethesda, MD.

NORTH, R.A., SLACK, B.E. & SURPRENANT, A-M. (1985) Muscarinic M_1 and M_2 receptors mediate depolarization and presynaptic inhibition in guinea-pig enteric nervous system. *J. Physiol.*, 368, 435–452.

OATES, J.A., GILLESPIE, L., UDENFRIEND, S. & SJOERDSMA, A. (1960) Decarboxylase inhibition and blood pressure reduction by alpha methyl-3,4-dihydroxy-DL-phenylamine. *Science*, 131, 1890–1891.

O'DONNELL, S.R. (1991) Pharmacology of β-adrenoceptors in lung disease. In: *Adrenoceptors: Structure, Mechanisms, Function. Advances in Pharmacological Sciences*, pp. 265–274, E. SZABADI & C.M. BRADSHAW (Eds), Birkhäuser-Verlag, Basel.

O'DONNELL, S.R. & WANSTALL, J.C. (1974) Potency and selectivity *in vitro* of compounds related to isoprenaline and orciprenaline on β-adrenoceptors in the guinea-pig. *Br. J. Pharmacol.*, 52, 407–417.

O'DONNELL, S.R. & WANSTALL, J.C. (1979a) pA$_2$ Values of selective β-adrenoceptor antagonists on isolated atria demonstrate a species difference in the β-adrenoceptor populations mediating chronotropic responses in cat and guinea-pig. *J. Pharm. Pharmacol.*, 31, 686–690.

O'DONNELL, S.R. & WANSTALL, J.C. (1979b) The importance of choice of agonist in studies designed to predict $\beta_2:\beta_1$ adrenoceptor selectivity of antagonists from pA$_2$ values in guinea-pig trachea and atria. *Naunyn-Schmiedeberg's Arch. Pharmacol.*, 308, 183–190.

OHMURA, T., OSHITA, M., KIGOSHI, S. & MURAMATSU, I. (1992) Identification of α_1-adrenoceptor subtypes in the rat vas deferens: binding and functional studies. *Br. J. Pharmacol.*, **107**, 697–704.

OSHITA, M., KIGOSHI, S. & MURAMATSU, I. (1989) Selective potentiation of extracellular Ca^{2+}-dependent contraction by neuropeptide Y in rabbit mesenteric arteries. *Gen. Pharmacol.*, **20**, 363–367.

OSSWALD, W. & AZEVEDO, I. (1991) Role of adenosine in the trophic effects of sympathetic innervation. *Trends Pharmacol. Sci.*, **12**, 442–443.

OTSUKA, M. & YOSHIOKA, K. (1993) Neurotransmitter functions of mammalian tachykinins. *Physiol. Rev.*, **73**, 229–308.

PACHOLCZYK, T., BLAKELY, R.D. & AMARA, S.G. (1991) Expression cloning of a cocaine- and antidepressant-sensitive human noradrenaline transporter. *Nature*, **350**, 350–354.

PAGE, C.P. (1991) One explanation of the asthma paradox: inhibition of natural anti-inflammatory mechanism by β_2-agonists. *Lancet*, **337**, 717–720.

PALMER, J.B.D., CUSS, F.M.C. & BARNES, P.J. (1986) VIP and PHM and their role in nonadrenergic inhibitory responses in isolated human airways. *J. Appl. Physiol.*, **61**, 1322–1328.

PAPPANO, A.J. (1971) Propranolol-insensitive effects of epinephrine on action potential repolarization in electrically driven atria of the guinea-pig. *J. Pharmacol. Exp. Ther.*, **177**, 85–95.

PAREKH, A.B. & BRADING, A.F. (1992) The M_3 muscarinic receptor links to three different transduction mechanisms with different efficacies in circular muscle of guinea-pig stomach. *Br. J. Pharmacol.*, **106**, 639–643.

PARK, K.H., LONG, J.P. & CANNON, J.G. (1991) Evaluation of the central and peripheral components for induction of postural hypotension by guanethidine, clonidine, dopamine 2 receptor agonists and 5-hydroxytryptamine 1A receptor agonists. *J. Pharmacol. Exp. Ther.*, **259**, 1221–1230.

PARRAMÓN, M., GONZÁLEZ, M.P. & OSET-GASQUE, M.J. (1995) A reassessment of the modulatory role of cyclic AMP in catecholamine secretion by chromaffin cells. *Br. J. Pharmacol.*, **114**, 517–523.

PARRATT, J.R. (1982) Inhibitors of the slow calcium current and early ventricular arrhythmias. In: *Early Arrhythmias Resulting from Myocardial Ischaemia*, pp. 329–346, J.R. PARRATT (Ed.), Macmillan Press, London.

PATIL, P.N. & JACOBOWITZ, D. (1968) Steric aspects of adrenergic drugs. IX. Pharmacologic and histochemical studies on isomers of cobefrin (α-methylnorepinephrine) *J. Pharmacol. Exp. Ther.*, **161**, 279–295.

PATIL, P.N., PATEL, D.G. & KRELL, R.D. (1971) Steric aspects of adrenergic drugs. XV. Use of isomeric activity as a criterion to differentiate adrenergic receptors. *J. Pharmacol. Exp. Ther.*, **176**, 622–633.

PATIL, P.N., FUDGE, K. & JACOBOWITZ, D. (1972) Steric aspects of adrenergic drugs. XVIII. Alpha-adrenergic receptors of mammalian aorta. *Eur. J. Pharmacol.*, **19**, 79–87.

PATON, D.M. (1976a) Characteristics of uptake of noradrenaline by adrenergic neurons. In: *The Mechanism of Neuronal and Extraneuronal Transport of Catecholamines*, pp. 49–66, D.M. PATON (Ed.), Raven Press, New York.

PATON, D.M. (1976b) Characteristics of efflux of noradrenaline from adrenergic neurons. In: *The Mechanism of Neuronal and Extraneuronal Transport of Catecholamines*, pp. 155–174, D.M. PATON, (Ed.), Raven Press, New York.

PATON, W.D.M. (1957) The action of morphine and related substances on contraction and on acetylcholine output of coaxially stimulated guinea-pig ileum. *Br. J. Pharmacol. Chemother.*, **11**, 119–127.

PATON, W.D.M. & ZAIMIS, E.J. (1949) The pharmacological actions of polymethylene bistrimethylammonium salts. *Br. J. Pharmacol. Chemother.*, **4**, 381–400.

PATON, W.D.M. & ZAR, M.A. (1968) The origin of acetylcholine released from guinea-pig intestine and longitudinal muscle strips. *J. Physiol.*, **194**, 13–33.

PATRA, P.B., WADSWORTH, R.M., HAY, D.W.P. & ZEITLIN, I.J. (1990) The effect of inhibitors of prostaglandin formation on contraction of the rat, rabbit and human vas deferens. *J. Auton. Pharmacol.*, **10**, 55–64.

PENNER, S.B. & SMYTH, D.D. (1994) Sodium excretion following central administration of an I_1 imidazoline receptor agonist, moxonidine. *Br. J. Pharmacol.*, **112**, 1089–1094.

PETERSEN, O.H. & FINDLAY, I. (1987) Electrophysiology of the pancreas. *Physiol. Rev.*, **67**, 1054–1116.

PETERSON, J.S., PATTON, A.J. & NORONHA-BLOB, L. (1990) Mini-pig urinary bladder function: comparisons of *in vitro* anticholinergic responses and *in vivo* cystometry with drugs indicated for urinary incontinence. *J. Auton. Pharmacol.*, **10**, 65–73.

PETROFSKY, J.S. & LIND, A.R. (1975) Isometric strength, endurance and the blood pressure and heart rate responses during isometric exercise in healthy men and women, with special reference to age and body fat content. *Pflügers Arch.*, **360**, 49–61.

PETROVIC, T., HARRIS, P.J. & BELL, C. (1988) Comparison of resting and stimulus evoked catecholamine release from the femoral and renal vasculature. *J. Auton. Nerv. Syst.*, **25**, 195–203.

PIERCY, V. (1987) The β-adrenoreceptors mediating uterine relaxation throughout the oestrus cycle of the rat are predominantly of the β_2-subtype. *J. Auton. Pharmacol.*, **8**, 11–18.

PILEBLAD, E., MAGNUSSON, T. & FORNSTEDT, B. (1989) Reduction of brain glutathione by L-buthionine sulfoximine potentiates the dopamine-depleting action of 6-hydroxydopamine in rat striatum. *J. Neurochem.*, **52**, 1978–1980.

PIRES, J.G.P. & RAMAGE, A.G. (1990) Evidence suggesting that the 5-HT$_2$ antagonist ICI 169,369 activates vagal afferents and in addition has a central hypotensive action in anaesthetized rats. *J. Auton. Pharmacol.*, **10**, 345–351.

PLOSKER, G.L. & CLISSOLD, S.P. (1992) Controlled release metoprolol formulations. A review of their pharmacodynamic and pharmacokinetic properties and therapeutic use in hypertension and ischaemic heart disease. *Drugs*, **43**, 382–414.

POLLARD, R.M., FILLENZ, M. & KELLY, P. (1982) Parallel changes in ultrastructure and noradrenaline content of nerve terminals in rat vas deferens following transmitter release. *Neuroscience*, **7**, 1623–1629.

PORTE, D. & ROBERTSON, R.P. (1973) Control of insulin secretion by catecholamines, stress, and the sympathetic nervous system. *Fed. Proc.*, **32**, 1792–1796.

POWELL, C.E. & SLATER, I.H. (1958) Blocking of inhibitory adrenergic receptors by a dichloro analogue of isoproterenol. *J. Pharmacol. Exp. Ther.*, **122**, 480–488.

POWIS, G. (1973) Binding of catecholamines to connective tissue and the effect upon the responses of blood vessels to noradrenaline and to nerve stimulation. *J. Physiol.*, **234**, 145–162.

PRETOLANI, M. & VARGAFTIG, B.B. (1993) From lung hypersensitivity to bronchial hyperreactivity. What can we learn from studies on animal models? *Biochem. Pharmacol.*, **45**, 791–800.

PRICE, N.M., SCHMITT, L.G., McGUIRE, J., SHAW, J.E. & TROBOUGH, G. (1981) Transdermal scopolamine in the prevention of motion sickness at sea. *Clin. Pharmacol. Ther.*, **29**, 414–419.

PRIOR, C., MARSHALL, I.G. & PARSONS, S.M. (1992) The pharmacology of vesamicol: an inhibitor of the vesicular acetylcholine transporter. *Gen. Pharmacol.*, **23**, 1017–1022.

PRITCHARD, B.N.C. & GILLAM, P.M.S. (1964) Use of propranolol (Inderal) in treatment of hypertension. *Br. Med. J.*, **2**, 725–727.

PROBST, W.C., SNYDER, L.A., SCHUSTER, D.I., BROSIUS, J. & SEALFON, S.C. (1992) Sequence alignment of the G-protein coupled superfamily. *DNA and Cell Biol.*, **11**, 1–20.

PRUNEAU, D. & ANGUS, J.A. (1990) *w*-Conotoxin GVIA is a potent inhibitor of sympathetic neurogenic responses in rat small mesenteric arteries. *Br. J. Pharmacol.*, **100**, 180–184.

PUECH, A.J. LECRUBIER, Y. & SIMON, P. (1979) Are alpha and beta presynaptic receptors involved in mood regulation? Pharmacological and clinical data. In: *Presynaptic Receptors*, pp. 359–362, S.Z. LANGER, K. STARKE & M.L. DUBOCOVICH (Eds), Pergamon Press, Oxford.

PUTNEY, J.W., JR. (1987) Phosphoinositides and alpha-1 adrenergic receptors. In: *The Alpha-1 Adrenergic Receptors*, pp. 189–208, R.R. Ruffolo (Ed.), Humana Press, Clifton, NJ.

QUIRION, R., AUBERT, I., LAPCHAK, P.A., SCHAUM, R.P., TEOLIS, S., GAUTHIER, S. & ARAUJO, D.M. (1989) Muscarinic receptor subtypes in human neurodegenerative disorders: focus on Alzheimer's disease. *Trends Pharmacol. Sci.*, Suppl. 80–84.

RAEBURN, D., SOUNESS, J.E., TOMKINSON, A. & KARLSSON, J-A. (1993) Isozyme-selective cyclic nucleotide phosphodiesterase inhibitors: Biochemistry, pharmacology and therapeutic potential in asthma. *Prog. Drug Res.*, **40**, 9–32.

RAJFER, J., ARONSON, W.J., BUSH, P.A., DOREY, F.J. & IGNARRO, L.J. (1992) Nitric oxide as a mediator of relaxation of the corpus cavernosum in response to nonadrenergic, noncholinergic neurotransmission. *N. Engl. J. Med.*, **326**, 90–94.

RAJFER, S.I. & DAVIS, F.R. (1990) Role of dopamine receptors and the utility of dopamine agonists in heart failure. *Circulation*, **82**, (Suppl.I), 97–102.

RAMAGE, A.G. (1986) A comparison of the effects of doxazosin and alfuzosin with those of urapidil on preganglionic sympathetic nerve activity in anaesthetized cats. *Eur. J. Pharmacol.*, **129**, 307–314.

RAMNARINE, S.I., HIRYAMA, Y., BARNES, P.J. & ROGERS, D.F. (1994) 'Sensory-efferent' neural control of mucus secretion: characterization using tachykinin receptor antagonists in ferret trachea *in vitro*. *Br. J. Pharmacol.*, **113**, 1183–1190.

RAND, M.J. & LI, C.G. (1993) Modulation of acetylcholine-induced contractions of the rat anococcygeus muscle by activation of nitrergic nerves. *Br. J. Pharmacol.*, **110**, 1479–1482.

RAND, M.J., MCCULLOCH, M.W. & STORY, D.F. (1980) Catecholamine receptors on nerve terminals. In: *Handbook of Experimental Pharmacology. Adrenergic Activators and Inhibitors*, Vol. 54/1, pp. 223–266, L. SZEKERES (Ed.), Springer-Verlag, Berlin.

RANDALL, M.D. & MCCULLOCH, A.I. (1995) The involvement of ATP-sensitive potassium channels in β-adrenoceptor-mediated vasorelaxation in the rat isolated mesenteric arterial bed. *Br. J. Pharmacol.*, **115**, 607–612.

REES, D.D., PALMER, R.M., HODSON, H.F. & MONCADA, S. (1989) A specific inhibitor of nitric oxide formation from L-arginine attenuates endothelium-dependent relaxation. *Br. J. Pharmacol.*, **96**, 418–424.

REGOLI, D., BOUDON, A. & FAUCHÉRE, J-L. (1994) Receptors and antagonists for substance P and related peptides. *Pharmacol. Rev.*, **46**, 551–599.

REID, D., MORALES, A., HARRIS, C., SURRIDGE, D.H.C., CONDRA, M. & OWEN, J. (1987) Double-blind trial of yohimbine in treatment of psychogenic impotence. *Lancet*, **2**, 421–423.

REID, J.J., WONG-DUSTING, H.K. & RAND, M.J. (1989) The effect of endothelin on noradrenergic transmission in rat and guinea-pig atria. *Eur. J. Pharmacol.*, **168**, 93–96.

REUTER, H. (1983) Calcium channel modulation by neurotransmitters, enzymes and drugs. *Nature*, **301**, 569–574.

REVERTE, M., GARCIA-BARRADO, M.J. & MORATINOS, J. (1991) Role of α-adrenoceptors in control of plasma potassium in conscious rabbits. *J. Auton. Pharmacol.*, **11**, 305–313.

RIBEIRO, J.A. (1991) Purinergic regulation of transmitter release. In: *Adenosine and Adenine Nucleotides as Regulators of Cellular Function*, pp. 155–167, J.W. PHILLIS (Ed.), CRC Press, Boca Raton, FL.

RICCI, A., COLLIER, W.L., ROSSDODIVITA, I. & AMENTA, F. (1991) Dopamine receptors mediating inhibition of the cyclic adenosine monophosphate generating system in the rat renal cortex. *J. Auton. Pharmacol.*, **11**, 121–127.

RICHARDS, J.G. & DA PRADA, M. (1977) Uranaffin reaction: a new cytochemical technique for the localization of adenine nucleotides in organelles storing biogenic amines. *J. Histochem. Cytochem*, **25**, 1322–1336.

RICHARDSON, J.B. (1979) State of the art: nerve supply to the lungs. *Am. Rev. Resp. Dis.*, **119**, 785–802.

RINALDI, G.J., GRAND, C. & CINGOLANI, H.E. (1991) Characteristics of the fast and slow components of the response to noradrenaline in rat aorta. *Can. J. Cardiol.*, **7**, 316–322.

RIVA, E., MENNINI, T. & LATINI, R. (1991) The α- and β-adrenoceptor blocking activities of labetolol and its RR–SR (50:50) stereoisomers. *Br. J. Pharmacol.*, **104**, 823–828.

ROATH, S. (1989) *Raunaud's: A Guide for Health Professionals*, Chapman & Hall, London.

ROBBERECHT, P., WINAND, J., CHATELAIN, P., POLECZEK, P., CAMUS, J., DE NEFF, P. & CHRISTOPHE, J. (1981) Comparison of β-adrenergic receptors and the adenylate cyclase system with muscarinic receptors and guanylate cyclase activities in the hearts of spontaneously hypertensive rats. *Biochem. Pharmacol.*, **30**, 385–387.

ROBERTSON, R.P. & PORTE, D. (1973) Adrenergic modulation of basal insulin secretion in man. *Diabetes* **22**, 1–8.

ROBISON, G.A., BUTCHER, R.W. & SUTHERLAND, E.W. (1971) Cyclic AMP and hormone action. In: *Cyclic AMP*, pp. 17–47, G.A. ROBISON, (Ed.), Academic Press, New York.

RODGER, I.W. & SHAHID, M. (1984) Forskolin, cyclic nucleotides and positive inotropism in isolated papillary muscles of the rabbit. *Br. J. Pharmacol.*, **81**, 151–159.

ROESLER, J.M., McCAFFERTY, J.P. DeMARINIS, R.M., MATTHEWS, W.D. & HEIBLE, J.P. (1986) Characterization of the antihypertensive activity of SK&F 86466, a selective α_2-antagonist in the rat. *J. Pharmacol. Exp. Ther.*, **236**, 1–7.

ROFFEL, A.F., MEURS, H., ELZINGA, C.R.S. & ZAAGSMA, J. (1995) No evidence for a role of muscarinic M_2 receptors in functional antagonism in bovine trachea. *Br. J. Pharmacol.*, **115**, 665–671.

ROSS, E.M. & GILMAN, A.G. (1980) Biochemical properties of hormone-sensitive adenylate cyclase. *Annu. Rev. Biochem.*, **49**, 533–564.

ROSS, P.J., LEWIS, M.J., SHERIDAN, D.J. & HENDERSON, A.H. (1981) Adrenergic hypersensitivity after beta-blocker withdrawal. *Br. Heart J.*, **45**, 637–642.

ROSS, S.B. (1976) Structural requirements for uptake into catecholamine neurons. In: *The Mechanism of Neuronal and Extraneuronal Transport of Catecholamines*, pp. 67–93, D.M. PATON (Ed.), Raven Press, New York.

ROSZKOWSKI, A.P. (1961) An unusual type of sympathetic ganglionic stimulant. *J. Pharmacol. Exp. Ther.*, **132**, 156–170.

RUBINSTEIN, R. & COHEN, S. (1992) Lack of agonist activity of McN-A-343 may be circumstantial. *J. Auton. Pharmacol.*, **12**, 1–4.

RUBIO, M.C. & LANGER, S.Z. (1973) Effects of noradrenaline metabolites on tyrosine hydroxylase activity in guinea-pig atria. *Naunyn-Schmiedeberg's Arch. Pharmacol.*, **280**, 315–330.

RUFFOLO, R.R., JR. (1982) Review. Important concepts of receptor theory. *J. Auton. Pharmacol.*, **2**, 277–295.

RUFFOLO, R.R. JR., NICHOLS, A.J., STADEL, J.M. & HIEBLE, J.P. (1991) Structure and function of α-adrenoceptors. *Pharmacol. Rev.*, **43**, 475–505.

RUFFOLO, R.R. JR., NICHOLS, A.J., STADEL, J.M. & HIEBLE, J.P. (1993) Pharmacologic and therapeutic applications of α_2-adrenoceptor subtypes. *Annu. Rev. Pharmacol. Toxicol.*, **32**, 243–279.

RUFFOLO, R.R., RICE, P.J., PATIL, P.N., HAMADA, A. & MILLER, D.D. (1983) Differences in the applicability of the Easson–Stedman Hypothesis to the alpha$_1$- and alpha₂-adrenergic effects of phenylethylamines and imidazolines. *Eur. J. Pharmacol.*, **86**, 471–475.

RUFFOLO, R.R. JR., ROSING, E.L. & WADDELL, J.E. (1979) Receptor interactions of imidazolines. I. Affinity and efficacy for alpha adrenergic receptors in rat aorta. *J. Pharmacol. Exp. Ther.*, **209**, 429–436.

RUFFOLO, R.R., TUROWSKI, B.S. & PATIL, P.N. (1977) Lack of cross-desensitization between structurally dissimilar alpha-adrenoceptor agonists. *J. Pharm. Pharmacol.*, **29**, 378–380.

RUMP, L.C., WOLK, V., RUFF, G. & SCHOLLMEYER, P. (1992) Activation of α_1- and α_2-adrenoceptors inhibit noradrenaline release in rabbit renal arteries: effect of pertussis toxin and N-ethylmaleimide. *J. Auton. Pharmacol.*, **12**, 97–108.

RYALL, R.W. (1961) Effects of cocaine and antidepressant drugs on the nictitating membrane of the cat. *Br. J. Pharmacol. Chemother.*, **17**, 339–357.

SADOSHIMA, J., TOKUTOMI, N. & AKAIKE, N. (1988) Effects of neostigmine and physostigmine on the acetylcholine receptor–ionophore complex in frog isolated sympathetic neurones. *Br. J. Pharmacol.*, **94**, 620–624.

SAGRADA, A., DURANTI, P., GIUDICI, L. & SCHIAVONE, A. (1993) Himbacine discriminates between two M_1 receptor-mediated responses. *Life Sci.*, **52**, 574.

SANTICIOLI, P. & MAGGI, C.A. (1994) Inhibitory transmitter action of calcitonin gene-related peptide in guinea-pig ureter via activation of glibenclamide-sensitive K channels. *Br. J. Pharmacol.*, **113**, 588–592

SANTTI, E., HUUPPONEN, R., ROURU, J., HÄNNINEN, V., PESONEN, U., JHANWAR-UNIYAL, M. & KOULU, M. (1994) Potentiation of the anti-obesity effect of the selective β_3-adrenoceptor agonist BRL 35135 in obese Zucker rats by exercise. *Br. J. Pharmacol.*, **113**, 1231–1236.

SARANTOS-LASKA, C., McCULLOCH, M.W., RAND, M.J. & STORY, D.F. (1981) Nicotine release of noradrenaline in the presence of tetrodotoxin from sympathetic nerve terminals in rabbit isolated atria. *J. Pharm. Pharmacol.*, **33**, 315–316.

SATOH, T., MORIYAMA, T., KURIKI, H. & KARAKI, H. (1992) Ca^{2+} Channel blocker-like activity of reserpine in smooth muscle. *Jap. J. Pharmacol.*, **60**, 291–293.

SAVINEAU, J-P., MARTHAN, R. & CREVEL, H. (1991) Contraction of vascular smooth muscle induced by phorbol 12,13 dibutyrate in human and rat pulmonary arteries. *Br. J. Pharmacol.*, **104**, 639–644.

SAXENA, P.R. & FERRARI, M.D. (1989) 5-HT-like receptor agonists and the pathophysiology of migraine. *Trends Pharmacol. Sci.*, **10**, 200–204.

SAYET, I., NEUILLY, G., RAKOTOARISOA, L., MIRONNEAU, C. & MIRONNEAU, J. (1993) Relation between α_1-adrenoceptor subtypes and noradrenaline-induced contraction in rat portal vein smooth muscle. *Br. J. Pharmacol.*, **110**, 207–212.

SCATCHARD, G. (1949) The attraction of proteins for small molecules and ions. *Ann. NY Acad. Sci.*, **51**, 660–672.

SCHADE, F.U. & SCHUDT, C. (1993) The specific type-III and IV phosphodiesterase inhibitor zardaverine suppresses formation of Tumor Necrosis Factor by macrophages. *Eur. J. Pharmacol.*, **230**, 9–14.

SCHILD, H.O. (1947) pA, A new scale for the measurement of drug antagonism. *Br. J. Pharmacol. Chemother.*, **2**, 189–206.

SCHMIEDEBERG, O. & KOPPE, R. (1869) *Das Muscarin, das giftige Alkaloid des Fliegenpilzes*, F.C.W. VOGEL, LEIPZIG.

SCHOLZ, H. (1980) Effects of beta- and alpha-adrenoceptor activators and adrenergic transmitter releasing agents on the mechanical activity of the heart. In: *Handbook of Experimental Pharmacology*, Vol. 54, Pt. 1, pp. 651–733, L. SZEKERES L. (Ed.), Springer-Verlag, Berlin.

SCHÖMIG, A. (1988) Adrenergic mechanisms in myocardial infarction: cardiac and systemic catecholamine release. *J. Cardiovasc. Pharmacol.*, **12**(Suppl. 1), S1–S7.

SCHRADER, G. (1952) *Die Entwicklung neuer Insektizide auf Grundlage von Organischen Fluorund Phosphorverbindungen*, Monographie No. 62, 2 Aufl., Verlag Chemie, Weinhein.

SCHÜMANN, H.J. (1983) What role do α- and β-adrenoceptors play in the regulation of the heart? *Eur. Heart J.*, **4**(Suppl. A), 55–60.

SCHWARTZ, R.D. & KELLAR, K.J. (1983) Nicotinic cholinergic receptor binding sites in the brain. Regulation *in vivo. Science*, **220**, 214–216.

SCHWARTZ, R.D. & KELLAR, K.J. (1985) In vivo regulation of (^{3}H) acetylcholine recognition sites in brain by nicotinic cholinergic drugs. *J. Neurochem.*, **45**, 427–433.

SCHWIETERT, H.R., GOUW, M.A.M., WILHELM, D., WILFFERT, B. & VAN ZWIETEN, P.A. (1991) The role of α_1-adrenoceptor subtypes in the phasic and tonic responses to phenylephrine in the longitudinal smooth muscle of the rat portal vein. *Naunyn-Schmiedeberg's Arch. Pharmacol.*, **343**, 463–471.

SCHWIETZER, E.S. & KELLY, R.B. (1982) ATP Release from cholinergic synapses. *Soc. Neurosci. Abstr.*, **8**, 493.

SCRIABINE, A. (1979) β-Adrenoceptor blocking drugs in hypertension. *Annu. Rev. Pharmacol.*, **19**, 269–284.

SEAMON, K.B. & DALY, J.W. (1983) Forskolin, cyclic AMP and cellular physiology. *Trends Pharmacol. Sci.*, **4**, 120–123.

SEKIZAWA, K., FUKUSHIMA, T., IKARASHI, Y., MARUYAMA, Y. & SASAKI, H. (1993) The role of nitric oxide in cholinergic neurotransmission in rat trachea. *Br. J. Pharmacol.*, **110**, 816–820.

SHAFFER, J.E., GRIZZLE, M.K., ANDERSON, D.K. & WHEELER, T.N. (1993) The inotropic and beta blocking effects of a chimeric molecule that putatively inhibits both type III phosphodiesterase and beta adrenoceptors in anesthetized dogs. *J. Pharmacol. Exp. Ther.*, **265**, 1105–1112.

SHAND, D.G. & WOOD, A.J.J. (1978) Propranolol withdrawal syndrome – why? *Circulation*, **58**, 202–203.

SHARMA, R.K. & JEFFERY, P.K. (1990) Airway β-adrenoceptor number in cystic fibrosis and asthma. *Clin. Sci.*, **78**, 409–417.

SHEPPARD, D. (1987) Physiology of the parasympathetic nervous system of the lung. *Postgrad. Med. J.*, **63**, (Suppl. 1), 21–27.

SHERIDAN, D.J., PENKOSKE, P.A., SOBEL, B.E. & CORR, P.B. (1980) Alpha adrenergic contributions to dysrhythmia during myocardial ischemia and reperfusion in cats. *J. Clin. Invest.*, **65**, 161–171.

SHERMAN, G.P., GREGA, G.J., WOODS, R.J. & BUCKLEY, J.P. (1968) Evidence for a central hypotensive mechanism of 2-(2,6-dichlorophenylamino)-2-imidazoline (Catapresan, St-155) *Eur. J. Pharmacol.*, **2**, 326–328.

SHREVE, P.E., TOEWS, M.L. & BYLUND, D.B. (1991) α_{2A}- and α_{2C}-Adrenoceptors are differentially down-regulated by norepinephrine. *Eur. J. Pharmacol. Mol. Pharmacol.*, **207**, 275–276.

SIDELL, F.R. (1992) Clinical considerations in nerve agent intoxication. In: *Chemical Warfare Agents*, pp. 155–194, S.M. SOMANI (Ed.), Academic Press, London.

SIEGEL, H., JIM, K., BOLGER, G.T. & TRIGGLE, D.J. (1984) Specific and non-specific desensitization of guinea-pig ileal smooth muscle. *J. Auton. Pharmacol.*, **4**, 109–126.

SINGH, B.N., DEEDWANIA, P., KOONLAWEE, N., WARD, A. & SORKIN, E.M. (1987) Sotalol. A review of its pharmacodynamic and pharmacokinetic properties and therapeutic use. *Drugs*, **34**, 311–349.

SKOK, V.I. (1980) Ganglionic transmission: morphology and physiology. In: *Pharmacology of Ganglionic Transmission. Handbook of Experimental Pharmacology*, Vol. 53, pp. 9–39, D.A. KHARKEVICH (Ed.), Springer-Verlag, Berlin.

SKOMEDAL, T., AASS, H., OSNES, J.B., FJELD, N.B., KLINGEN, G., LANGSLET, A. & SEMB, G. (1985) Demonstration of an *alpha* adrenoceptor-mediated inotropic effect of norepinephrine in human atria. *J. Pharmacol. Exp. Ther.*, **233**, 441–446.

SMITH, G.W. & O'CONNOR, S.E. (1988) An introduction to the pharmacological properties of Dopacard (dopexamine hydrochloride) *Am. J. Cardiol.*, **62**, 9C–17C.

SMITH, K., CONNOUGHTON, S. & DOCHERTY, J.R. (1992) Investigations of prejunctional α_2-adrenoceptors in rat atrium, vas deferens and submandibular gland. *Eur. J. Pharmacol.*, **211**, 251–256.

SMITH, S.A. & SMITH, S.E. (1980) Subsensitivity to cholinoceptor stimulation of the human iris sphincter *in situ* following acute and chronic administration of cholinomimetic miotic drugs. *Br. J. Pharmacol.*, **69**, 513–518.

SMYTH, D.D., UMEMURA, S. & PETTINGER, W.A. (1985) Renal nerve stimulation causes α_1-adrenoceptor-mediated sodium retention but not α_2-adrenoceptor antagonism of vasopressin. *Circ. Res.*, **57**, 304–311.

SNEDDON, P. & GRAHAM, A. (1992) Role of nitric oxide in the autonomic innervation of smooth muscle. *J. Auton. Pharmacol.*, **12**, 445–456.

SNEDDON, P. & MACHALY, M. (1992) Regional variation in purinergic and adrenergic responses in isolated vas deferens of rat, rabbit and guinea-pig. *J. Auton. Pharmacol.*, **12**, 421–428.

SOKOLOFF, P. & SCHWARTZ, J-C. (1995) Novel dopamine receptors half a decade later. *Trends Pharmacol. Sci.*, **16**, 270–275.

SOLL, A.H. (1984) Fundic mucosal muscarinic receptors modulating acid secretion. In: *Subtypes of Muscarinic Receptors, Trends Pharmacol. Sci.*, Suppl. pp. 60–62, B.I. HIRSCHOWITZ, R. HAMMER, A. GIACHETTI, J.J. KEIRNS & R.R. LEVINE (Eds), Elsevier, Cambridge.

SOMANI, S.M., SOLANA, R.P. & DUBE, S.N. (1992) Toxicodynamics of nerve agents. In: *Chemical Warfare Agents*, pp. 67–123, S.M. SOMANI (Ed.) Academic Press, London.

SORKIN, E.M. & HEEL, R.C. (1986) Guanfacine: a review of its pharmacodynamic and pharmacokinetic properties, and therapeutic efficacy in the treatment of hypertension. *Drugs*, **31**, 301–336.

SOURKES, T.L. (1954) Inhibition of dihydroxyphenylalanine decarboxylase by derivatives of phenylalanine. *Arch. Biochem.*, **51**, 444–456.

SPECTOR, S., SJOERDSMA, A. & UDENFRIEND, S. (1965) Blockade of endogenous norepinephrine synthesis by α-methyl-tyrosine, an inhibitor of tyrosine hydroxylase. *J. Pharmacol. Exp. Ther.*, **147**, 86–95.

SPRIGGS, T.L.B., MALLARD, N.J., MARSHALL, R.W. & SITHERS, A.J. (1991) Functional discrimination of α_{1A} and α_{1B} adrenoceptors in rat vas deferens. *Br. J. Pharmacol.*, **102**, 17P.

STADEL, J.M. & LEFKOWITZ, R.J. (1979) Multiple reactive sulfhydryl groups modulate the function of adenylate cyclase coupled *beta*-adrenergic receptors. *Mol. Pharmacol.*, **16**, 709–718.

STADEL, J.M. & LEFKOWITZ, R.J. (1991) Beta-adrenergic receptors. Identification and characterization by radioligand binding studies. In: *The Beta-Adrenergic Receptors*, pp. 1–40, J.P. PERKINS (Ed.), Humana Press, Clifton, NJ.

STAEHELIN, M., SIMONS, P., JAEGG, K. & WIGGER, N. (1983) CGP-12177. A hydrophilic β-adrenergic receptor radioligand reveals high affinity binding of agonists to intact cells. *J. Biol. Chem.*, **258**, 3496–3502.

STANASZEK, W.F., KELLERMAN, D., BROGDEN, R.N. & ROMANKIEWICZ, J.A. (1983) Prazosin update. A review of its pharmacological properties and therapeutic use in hypertension and congestive heart failure. *Drugs*, **25**, 339–384.

STANFIELD, P.R. (1983) Tetraethylammonium ions and the potassium permeability of excitable cells. *Rev. Physiol. Biochem. Pharmacol.*, **97**, 1–67.

STANTON, B., PUGLISI, E. & GELLAI, M. (1987) Localization of α_2-adrenoceptor mediated increase in renal Na^+, K^+ and water excretion. *Am. J. Physiol.*, **252**, F1016–F1021.

STARKE, K., MONTEL, H. & SCHÜMANN, H.J. (1971) Influence of cocaine and phenoxybenzamine on noradrenaline uptake and release. *Naunyn-Schmiedeberg's Arch. Pharmacol.*, **270**, 210–214.

STARKE, K., GÖTHERT, M. & KILBINGER, H. (1989) Modulation of neurotransmitter release by presynaptic autoreceptors. *Physiol. Rev.*, **69**, 864–989.

STEDMAN, E., STEDMAN, E. & EASSON, L.H. (1932) Choline-esterase. An enzyme present in the blood-serum of the horse. *Biochem. J.*, **26**, 2056–2066.

STEINFATH, M., DANIELSEN, W., VON DER LEYEN, H., MENDE, U., MEYER, W., NEUMANN, J., NOSE, M., REICH, T., SCHMITZ, W., SCHOLZ, H., STARBATTY, J., STEIN, B., DORING, V., KALMAR, P. & HAVERICH, A. (1992) Reduced α_1 and β_2-adrenoceptor-mediated positive inotropic effects in human end-stage heart failure. *Br. J. Pharmacol.*, **105**, 463–469.

STENE-LARSEN, G. & HELLE, K.B. (1978) Cardiac β_2-adrenoceptor in the frog. *Comp. Biochem. Physiol.*, **80**, 165–173.

STEPHENSON, R.P. (1956) A modification of receptor theory. *Br. J. Pharmacol. Chemother.*, **11**, 379–393.

STILES, G.L., CARON, M.G. & LEFKOWITZ, R.J. (1984) β-Adrenergic receptors: biochemical mechanisms of physiological regulation. *Physiol. Rev.*, **64**, 661–743.

STOLERMAN, I.P. & SHOAIB, M. (1991) The neurobiology of tobacco addiction. *Trends Pharmacol. Sci.*, **12**, 467–473.

STONE, T.W. (1991) Receptors for adenosine and adenine nucleotides. *Gen. Pharmacol.*, **22**, 25–31.

STRADER, C.D., SIGAL, I.S., CANDELORE, M.R., RANDS, E., HILL, W.S. & DIXON, R.A.F. (1988) Conserved aspartate residues 79 and 113 of the β-adrenergic receptor have different roles in receptor function. *J. Biol. Chem.*, **263**, 10267–10271.

STRASSER, R.H., SIBLEY, D.R. & LEFKOWITZ, R.J. (1986) A novel catecholamine-activated adenosine cyclic 3',5'-phosphate independent pathway for β-adrenergic receptor phosphorylation in wild-type and mutant S_{49} lymphoma cells: mechanism of homologous desensitization of adenylate cyclase. *Biochemistry*, **25**, 1371–1377.

STROESCU, V., FULGA, I.G., TURTUROIU, D. & TOMA, A. (1991) Experimental research about the influence of pentazocine and reserpine on morphine addiction. *Rev. Roum. Physiol.*, **28**, 3–11.

STROLIN BENEDETTI, M. & DOSTERT, PH. (1989) Monoamine oxidase, brain ageing and degenerative diseases. *Biochem. Pharmacol.*, **38**, 555–561.

STROSBERG, A.D. (1992) Biotechnology of β-adrenergic receptors. *Mol. Neurobiol.*, **4**, 211–250.

STRUTHERS, A.D., BROWN, D.C., BROWN, M.J., SCHUMER, B. & BLOOM, S.R. (1985) Selective α_2-receptor blockade facilitates the insulin response to adrenaline but not to glucose in man. *Clin. Endocrinol.*, **23**, 539–546.

SUMMERS, R.J., RUSSELL, F.D. & MOLENAAR, P. (1991) Autoradiographic characterization of β-adrenoceptor subtypes. In: *Adrenoceptors: Structure, Mechanism, Function. Advances in Pharmacological Sciences*, pp. 37–46. E. SZABADI & C.M. BRADSHAW (Eds), Berkhäuser-Verlag, Basel.

SUNN, N., HARRIS, P.J. & BELL, C. (1990) Effects on renal sympathetic axons in dog of acute 6-hydroxydopamine treatment in combination with selective neuronal uptake inhibitors. *Br. J. Pharmacol.*, **99**, 655–660.

SUTHERLAND, E.W. (1972) Studies on the mechanism of hormone action. *Science*, **177**, 401–408.

SUTHERLAND, E.W. & RALL, T.W. (1960) The relation of adenosine-3',5'-phosphate and phosphorylase to the actions of catecholamines and other hormones. *Pharmacol. Rev.*, **12**, 265–299.

SVEDMYR, N. (1990) Action of corticosteroids on beta-adrenergic receptors. *Am. Rev. Resp. Dis.*, **141**, S31–S38.

SYNETOS, D., MANOLOPOULOS, V.G., ATLAS, D. & PIPILI-SYNETOS, E. (1991) Human plasma-derived material with clonidine displacing substance (CDS)-like properties contracts the isolated rat aorta. *J. Auton. Pharmacol.*, **11**, 343–351.

SZEKERES, L. & PAPP, J. GY. (1980) Effect of adrenergic activators and inhibitors on the electrical activity of the heart. In: *Handbook of Experimental Pharmacology*, Vol. 54, Pt 1, pp. 597–650, L. SZEKERES (Ed.), Springer-Verlag, Berlin.

SZELENYI, I. (1982) Does pirenzepine distinguish between 'subtypes' of muscarinic receptors? *Br. J. Pharmacol.*, **77**, 567–569.

SZENTIVANYI, A. (1968) The beta-adrenergic theory of the atopic abnormality in bronchial asthma. *J. Allergy*, **42**, 203–232.

TAINTER, M.L. & CHANG, D.K. (1927) The antagonism of the pressor action of tyramine by cocaine. *J. Pharmacol. Exp. Ther.*, **30**, 193–207.

TANAKA, H., MANITA, S., MATSUDA, T., ADACHI, M. & SHIGENOBU, K. (1995) Sustained negative inotropism mediated by α-adrenoceptors in adult mouse myocardia: developmental conversion from positive response in the neonate. *Br. J. Pharmacol.*, **114**, 673–677.

TARKOVÁCS, G., BLANDIZZI, C. & VIZI, E.S. (1994) Functional evidence that α_{2A}-adrenoceptors are responsible for antilipolysis in human abdominal fat cells. *Naunyn-Schmiedeberg's Arch. Pharmacol.*, **349**, 34–41.

TATTERSALL, J.E.H. (1993) Ion channel blockade by oximes and recovery of diaphragm muscle from soman poisoning *in vitro*. *Br. J. Pharmacol.*, **108**, 1006–1015.

TATTERSFIELD, A.E. (1992) Bronchodilators: new developments. In: *Asthma*, pp. 190–204, P.J. BARNES (Ed.), (British Medical Bulletin 48), Churchill Livingstone, London.

TAYLOR, S.H. & MEERAN, M.K. (1972) The cardiovascular response to some environmental stresses and their modification by oxprenolol. In: *New Perspectives in Beta-Blockade. An International Symposium*, Aarhus, Denmark. pp. 293–306, D.M. BURLEY, J.H. FRIER, R.K. RONDEL & S.H. TAYLOR (Eds), Ciba Laboratories, Horsham, UK.

TECLE, H., LAUFFER, D.J., MIRZADEGAN, T., MOOS, W.H., MORELAND, D.W., PAVIA, M.R., SCHWARZ, R.D. & DAVIS, R.E. (1993) Synthesis and SAR of bulky 1-azabicyclo[2.2.1]-3-one oximes as muscarinic receptor subtype selective agonists. *Life Sci.*, **52**, 505–511.

TESCH, P.A. (1985) Exercise performance and beta-blockade. *Sports Med.*, **2**, 389–412.

TFELT-HANSEN, P. (1986) Efficacy of β-blockers in migraine. A critical review. *Cephalalgia*, **6**(Suppl. 5) 15–24.

THAINA, P., NOTT, M.W. & RAND, M.J. (1993) Inhibition by α_2-adrenoceptor agonists of contractions of rabbit isolated colon elicited by pelvic nerve stimulation. *J. Auton. Pharmacol.*, **13**, 115–126.

THIEMERMANN, C. & VANE, J.R. (1990) Inhibition of nitric oxide synthesis reduces the hypotension induced by bacterial lipopolysaccharide in the anaesthetized rat. *Eur. J. Pharmacol.*, **182**, 591–595.

THOENEN, H. (1970) Induction of tyrosine hydroxylase in peripheral and central adrenergic neurones by cold-exposure of rat. *Nature*, **228**, 861–862.

THOMAS, J.M. & HOFFMAN, B.B. (1994) Sensitivity of adenylyl cyclase by prior activation of inhibitory receptors. In: *Regulation of Cellular Transduction Pathways by Desensitization and Amplification*, pp. 193–215, D.R. SIBLEY & M.D. HOUSLAY (Eds), John Wiley, Chichester.

THOMPSON, W.J. (1991) Cyclic nucleotide phosphodiesterases: pharmacology, biochemistry and function. *Pharmacol. Ther.*, **51**, 13–33.

THURESON-KLEIN, Å. (1982) Ultrastructural preservation of vesicles in sympathetic nervous tissue. In: *Neurotransmitter Vesicles*, pp. 65–87, R.L. KLEIN, H. LAGERCRANTZ & H. ZIMMERMAN (Eds), Academic Press, New York.

TIMMERMANS, P.B.M.W.M. & THOOLEN, M.J.M.C. (1987) Ca^{2+} Utilization in signal transformation of alpha-1 adrenergic receptors. In: *The Alpha-1 Adrenergic Receptors*, pp. 113–187, R.R. RUFFOLO (Ed.), Humana Press, Clifton, NJ.

TIMMERMANS, P.M.W.M. & VAN ZWIETEN, P.A. (1980) Postsynaptic a_1- and a_2-adrenoceptors in the circulatory system of the pithed rat: selective stimulation of the a_2-type by B-HT933. *Eur. J. Pharmacol.*, **63**, 199–202.

TIMMERMANS, P.B.M.W.M., CHIU, A.T. & THOOLEN, M.J.M.C. (1990) α-Adrenergic receptors. In: *Comprehensive Medicinal Chemistry*, Vol. 3, *Membranes and Receptors*, pp. 133–185, P.G. SAMMES & J.B. TAYLOR (Eds), Pergamon Press, Oxford.

TIMMS, D., WILKINSON, A.J., KELLY, D.R., BROADLEY, K.J. & DAVIES, R.H. (1992) Interactions of tyr^{377} in a ligand-activated model of signal transmission through β_1-adrenoceptor α-helices. *Int. J. Quant. Chem.*, **19**, 197–215.

TIPTON, K.F. (1990) The design and behaviour of selective monoamine oxidase inhibitors. In: *Antidepressants: 30 Years On*, pp. 195–203, B. LEONARD & P. SPENCER (Eds), CNS (Clinical Neuroscience) Publishers, London.

TITELER, M. (1989) Receptor binding theory and methodology. In: *Receptor Pharmacology and Function*, pp. 17–45, M. WILLIAMS, R.A. GLENNON & P.B.M.W.M. TIMMERMANS (Eds), Marcel Dekker, New York.

TITMARSH, S. & MONK, J.P. (1987) Terazosin. A review of its pharmacodynamic and pharmacokinetic properties; and therapeutic efficacy in essential hypertension. *Drugs*, **33**, 461–477.

TOMITA, T. (1981) Electrical activity (spikes and slow waves) in gastrointestinal smooth muscles. In: *Smooth Muscle. An Assessment of Current Knowledge*, pp. 127–156, E. BÜLBRING, A.F. BRADING, A.W. JONES & T. TOMITA (Eds), Edward Arnold, London.

TOMITA, T. (1989) Electrical properties of airway smooth muscle. In: *Airway Smooth Muscle in Health and Disease*, pp. 151–167, R.F. COBURN (Ed.), Plenum, New York.

TOMITA, T. & WATANABE, H. (1973) Factors controlling myogenic activity in smooth muscle. *Phil. Trans. R. Soc.*, **265**, 73–85.

TOMKINSON, A., KARLSSON, J-A & RAEBURN, D. (1993) Comparison of phosphodiesterase type III and IV in airway smooth muscle with differing β-adrenoceptor subtypes. *Br. J. Pharmacol.*, **108**, 57–61.

TOMLINSON, P.R., WILSON, J.W. & STEWART, A.G. (1994) Inhibition by salbutamol of the proliferation of human airway smooth muscle cells grown in culture. *Br. J. Pharmacol.*, **111**, 641–647.

TORPHY, T.J. & UNDEM, B.J. (1991) Phosphodiesterase inhibitors: new opportunities for the treatment of asthma. *Thorax*, **46**, 512–523.

TORPHY, T.J., UNDEM, B.J., CIESLINSKI, L.B., LUTTMAN, M.A., REEVES, M.L. & HAY, D.W.P. (1993) Identification, characterisation and functional role of phosphodiesterase isoenzymes in human airway smooth muscle. *J. Pharmacol. Exp. Ther.*, **265**, 1213–1223.

TOUTANT, J. -P. & MASSOULIE, J. (1988) Cholinesterases: tissue and cellular distribution of molecular forms and their physiological regulation. In: *The Cholinergic Synapse. Handbook of Experimental Pharmacology*, pp. 225–265, V.P. WHITTAKER (Ed.), Springer-Verlag, Berlin.

TRENDELENBURG, P. (1917) Physiologische und pharmakologische versuche über die dunndarm peristaltik. *Arch. Exp. Pathol. Pharmakol.*, **81**, 55–129.

TRENDELENBURG, U. (1972) Classification of sympathomimetic amines. In: *Catecholamines, Handbook of Experimental Pharmacology*, Vol. 33, pp. 336–362, H. BLASCHKO & E. MUSCHOLL (Eds), Springer-Verlag, Berlin.

TRENDELENBURG, U. (1976) The extraneuronal uptake and metabolism of catecholamines in the heart. In: *The Mechanism of Neuronal and Extraneuronal Transport of Catecholamines*, pp. 259–280, D.M. PATON (Ed.), Raven Press, New York.

TRENDELENBURG, U. (1991) Functional aspects of the neuronal uptake of noradrenaline. *Trends Pharmacol. Sci.*, **12**, 334–337.

TRIGGLE, D.J. (1965) 2-Halogenoethylamines and receptor analysis. *Adv. Drug Res.*, **2** 173–189.

TRIGGLE, D.J. (1976) Structure–activity relationships. In: *Chemical Pharmacology of the Synapse*, pp. 233–430, D.J. TRIGGLE & C.R. TRIGGLE (Eds), Academic Press, London.

TRIGGLE, D.J. & TRIGGLE, C.R. (1976) *Chemical Pharmacology of the Synapse*, Academic Press, London.

TRIMBLE, W.S., LINIAL, M. & SCHELLER, R.H. (1991) Cellular and molecular biology of the presynaptic nerve terminal. *Annu. Rev. Neurosci.*, **14**, 93–122.

TSIEN, R.W. (1977) Cyclic AMP and contractile activity in heart. In: *Advances in Cyclic Nucleotide Research*, Vol. 8, pp. 363–420, P. GREENGARD & G. ROBISON (Eds), Raven Press, New York.

TSIEN, R.W., ELLINOR, P.T. & HORNE, W.A. (1991) Molecular diversity of voltage-dependent Ca^{2+} channels. *Trends Pharmacol. Sci.*, **12**, 349–354.

TSJUJIMOTO, G. & HOFFMAN, B.B. (1985) Desensitization of β-adrenergic-mediated vascular smooth muscle relaxation. *Mol. Pharmacol.*, **27**, 210–217.

TUČEK, S. (1988) Choline acetyltransferase and the synthesis of acetylcholine. In: *The Cholinergic Synapse. Handbook of Experimental Pharmacology*, Vol. 86, pp. 125–165, V.P. WHITTAKER (Ed.), Springer-Verlag, Berlin.

TUČEK, S. & PROŠKA, J. (1995) Allosteric modulation of muscarinic acetylcholine receptors. *Trends Pharmacol. Sci.*, **16**, 205–212.

TUCKER, G.T. (1994) Clinical implications of genetic polymorphism in drug metabolism. *J. Pharm. Pharmacol.*, **46**(Suppl. 1), 417–424.

TUNG, C-S, GOLDBERG, M.R., HOLLISTER, A.S., SWEETMAN, B.J. & ROBERTSON, D. (1988) Depletion of brainstem epinephrine stores by α-methyldopa: possible relation to attenuated sympathetic outflow. *Life Sci.*, **42**, 2365–2371.

TYTELL, M. & STADLER, H. (1988) Axonal transport in cholinergic neurons. In: *The Cholinergic Synapse. Handbook of Experimental Pharmacology*, Vol. 86, pp. 399–407, V.P. WHITTAKER (Ed.), Springer-Verlag, Berlin.

UDENFRIEND, S. (1966) Tyrosine hydroxylase. *Pharmacol. Rev.*, **18**, 43–51.

UEDA, H., GOSHIMA, Y. & MISU, Y. (1983) Presynaptic mediation by α_2, β_1- and β_2-adrenoceptors of endogenous noradrenaline and dopamine release from slices of rat hypothalamus. *Life Sci.*, **33**, 371–376.

UNDERWOOD, D.C., OSBORN, R.R., NOVAK, L.B. MATHEWS, J.K., NEWSHOLME, S.J., UNDEM, B.J., HAND, J.M. & TORPHY, T.J. (1993) Inhibition of antigen-induced bronchoconstriction and eosinophil infiltration in the guinea pig by the cyclic AMP-specific phosphodiesterase inhibitor, rolipram. *J. Pharmacol. Exp. Ther.*, **266**, 306–313.

UNWIN, N., TOYOSHIMA, C. & KUBALEK, E. (1988) Arrangement of the acetylcholine receptor subunits in the resting and desensitized states, determined by cryoelectron microscopy of crystallized *Torpedo* postsynaptic membranes. *J. Cell. Biol.*, **107**, 1123–1138.

U'PRITCHARD, D.C., GREENBERG, D.A., SHEEHAN, P.O. & SNYDER, S.H. (1978) Tricyclic antidepressants: therapeutic properties and affinities for α-noradrenergic receptor binding sites in the brain. *Science*, **199**, 197–198.

URQUHART, R.A. & BROADLEY, K.J. (1991) Comparison of adenosine receptor agonists with other vasodilators on noradrenaline-, potassium- and phorbol ester-contracted rabbit aorta. *Eur. J. Pharmacol.*, **200**, 35–45.

URQUHART, R.A. & BROADLEY, K.J. (1992) The effect of P_1- and muscarinic-receptor agonists upon cAMP-dependent and independent inotropic responses of guinea-pig cardiac preparations. *Gen. Pharmacol.*, **23**, 619–226.

URQUHART, R.A., FORD, W.R. & BROADLEY, K.J. (1993) Potassium channel blockade of atrial negative inotropic responses to P_1-purinoceptor and muscarinic receptor agonists and to cromakalim. *J. Cardiovasc. Pharmacol.*, **21**, 279-288.

VALVO, E., GAMMARO, L., OLDRIZZI, L., TESSITORE, N. & MASCHIO, G. (1982) Long-term treatment with atenolol in essential and renal hypertension: effect on blood pressure, systemic hemodynamics and renal function. *Curr. Ther. Res.*, **31**, 564–572.

VAN BAAK, M.A. (1988) β-Adrenoceptor blockade and exercise. An update. *Sports Med.*, **5**, 209–225.

VAN BREEMEN, C. & SAIDA, K. (1989) Cellular mechanisms regulating $[Ca^{2+}]_i$ smooth muscle. *Annu. Rev. Physiol.*, **51**, 315–329.

VAN DER VLIET, A., RADEMAKER, B. & BAST, A. (1990) A beta adrenoceptor with atypical characteristics is involved in the relaxation of the rat small intestine. *J. Pharmacol. Exp. Ther.*, **255**, 218–226.

VANHOUTTE, P.M., RUBANYI, G.M., MILLER, V.M. & HOUSTON, D.S. (1986) Modulation of vascular smooth muscle contraction by the endothelium. *Annu. Rev. Physiol.*, **48**, 307–320.

VAN ROSSUM, J.M. (1962) Classification and molecular pharmacology of ganglion blocking agents. II. Mode of action of competitive and noncompetitive ganglionic blocking agents. *Int. J. Neuropharmacol.*, **1**, 403–421.

VAN WOERKENS, L.J., BOOMSMA, F., MAN IN'T VELD, A.J., BEVERS, M.M. & VERDOUW, P.D. (1992) Differential cardiovascular and neuroendocrine effects of epinine and dopamine in conscious pigs before and after adrenoceptor blockade. *Br. J. Pharmacol.*, **107**, 303–310.

VAN ZWIETEN, P.A. (1988) Antihypertensive drugs interacting with α- and β-adrenoceptors. A review of basic pharmacology. *Drugs*, **35**(Suppl. 6), 6–19.

VARAGIĆ, V. & KRSTIĆ, M. (1966) Adrenergic activation by anticholinesterases. *Pharmacol. Rev.*, **18**, 799–800.

VARAGIĆ, V.M. & STOJANOVIĆ, V. (1987) The effect of enkephalins and of β-endorphin on the hypertensive response to physostigmine in the rat. *Br. J. Pharmacol.*, **92**, 197–202.

VATNER, D.E., LEE, D.L., SCHWARZ, K.R., LONGABAUGH, J.P., FUJII, A.M., VATNER, S.F. & HOMCY, C.J. (1988) Impaired cardiac muscarinic receptor function in dogs with heart failure. *J. Clin. Invest.*, **81**, 1836–1842.

VELLOM, D.C., RADIĆ, Z., LI, Y., PICKERING, N.A., CAMP, S. & TAYLOR, P. (1993) Amino acid residues controlling acetylcholinesterase and butyrylcholinesterase specificity. *Biochemistry*, **32**, 12–17.

VERNON, R.G. & CLEGG, R.A. (1985) The metabolism of white adipose tissue *in vivo* and *in vitro*. In: *New Perspectives in Adipose Tissue: Structure, Function and Development*, pp. 65–86, A. CRYER & R.L.R. VAN (Eds), Butterworths, London.

VILA, E., BADIA, A. & JANÉ, F. (1982) A preferential blocking effect of oxprenolol on α_1-adrenoceptors in the rat. *J. Pharm. Pharmacol.*, **34**, 671–673.

VINCENT, H.H., MANIN'T VELD, A.J., BOOMSMA, F., DERKX, F. & SCHALEKAMP, M.A.D.H. (1985) Compound ICI 118,551, a β_2-adrenoceptor antagonist, lowers blood pressure. *J. Hypertens*, **3**,(Suppl. 3), S247–S249.

VLIETSTRA, R.E. & BLINKS, J.R. (1976) Heterogeneity of cardiac beta-adrenoceptors. *Fed. Proc.*, **35**, 210.

VOLLE, R.L. (1966) Modification by drugs of synaptic mechanisms in autonomic ganglia. *Pharmacol. Rev.*, **18**, 839–869.

VOLLE, R.L. (1980) Ganglionic actions of anticholinesterase agents, catecholamines, neuro-muscular blocking agents, and local anaesthetics. In: *Pharmacology of Ganglionic Transmission. Handbook of Experimental Pharmacology*, Vol. 53, pp. 385–410 D.A., KHARKEVICH (Ed.), Springer-Verlag, Berlin.

VON EULER, U.S. (1946) A specific sympathomimetic ergone in adrenergic nerve fibres (sympathin) and its relations to adrenaline and noradrenaline. *Acta Physiol. Scand.*, **12**, 73–97.

VON KÜGELGEN, I. (1994) Purinoceptors modulating the release of noradrenaline. *J. Auton. Pharmacol.*, **14**, 11–12.

VU, H.Q., BUDAI, D & DUCKLES, S.P. (1989) Neuropeptide Y preferentially potentiates responses to adrenergic nerve stimulation by increasing rate of contraction. *J. Pharmacol. Exp. Ther.*, **251**, 852–857.

WADWORTH, A.N., MURDOCH, D. & BROGDEN, R.N. (1991) Atenolol. A reappraisal of its pharmacological properties and therapeutic use in cardiovascular disorders. *Drugs*, **42**, 468–510.

WAELBROECK, M., TASTENOY, M., CAMUS, J., FEIFEL, R., MUTSCHLER, E., STROHMANN, C., TACKE, R., LAMBRECHT, G. & CHRISTOPHE, J. (1989) Stereoselectivity of the interaction of muscarinic antagonists with their receptors. *Trends Pharmacol. Sci.* (Suppl.), 65–69.

WAELBROECK, M., CAMUS, J., TASTENOY, M. & CHRISTOPHE, J. (1992) Binding properties of nine 4-diphenyl-acetoxy-N-methyl-piperidine (4-DAMP) analogues to M_1, M_2, M_3 and putative M_4 muscarinic receptor subtypes. *Br. J. Pharmacol.*, **105**, 97–102.

WAKADE, A.R. & WAKADE, T.D. (1982) Biochemical evidence for re-use of noradrenergic storage vesicles in the guinea-pig heart. *J. Physiol.*, **327**, 337–362.

WAKADE, A.R., MALHOTRA, R.K. & WAKADE, T.D. (1990) Co-transmission in the rat adrenal medulla. *J. Auton. Pharmacol.*, **10**, 8.

WALDECK, B. (1976) An *in vitro* method for the study of β-receptor mediated effects on slow contracting skeletal muscle. *J. Pharm. Pharmacol.*, **28**, 434–436.

WALLE, T., WEBB, J.G., BAGWELL, E.E., WALLE, U.K., DANIELL, G.B. & GAFFNEY, T.E. (1988) Stereoselective delivery and actions of beta receptor antagonists. *Biochem. Pharmacol.*, **37**, 115–124.

WALLIN, J.D. & FRISHMAN, W.H. (1989) Dilevalol: a selective beta-2 adrenergic agonist vasodilator with beta adrenergic blocking activity. *J. Clin. Pharmacol.*, **29**, 1057–1068.

WALLIS, D. (1989) Interaction of 5-hydroxytryptamine with autonomic and sensory neurones. In: *The Peripheral Actions of 5-Hydroxytryptamine*, pp. 220–246, J.R. FOZARD (Ed.), Oxford Medical Publications, New York.

WALSH, L.K.B., BEGENISICH, T. & KASS, R.S. (1989) β-Adrenergic modulation of cardiac ion channels. Differential temperature sensitivity of potassium and calcium currents. *J. Gen. Physiol.*, **93**, 841–854.

WALTER, M., LEMOINE, H. & KAUMANN, A.J. (1984) Stimulant and blocking effects of optical isomers of pindolol on the sinoatrial node and trachea of guinea pig. Role of β-adrenoceptor subtypes in the dissociation between blockade and stimulation. *Naunyn-Schmiedeberg's Arch. Pharmacol.*, **327**, 159–175.

WATANABE, A.M., JONES, L.R., MANALAN, A.S. & BESCH, H.R. JR. (1982) Cardiac autonomic receptors. Recent concepts from radiolabeled ligand-binding studies. *Circ. Res.*, **50**, 161–174.

WATSON, N., BARNES, P.J. & MACLAGAN, J. (1992) Actions of methoctramine, a muscarinic M_2 receptor antagonist, on muscarinic and nicotinic cholinoceptors in guinea-pig airways *in vivo* and *in vitro*. *Br. J. Pharmacol.*, **105**, 107–112.

WATSON, S. & GIRDLESTONE, D. (1994) Receptor nomenclature supplement (5th Edn) *Trends Pharmacol. Sci.*

WEINSHILBOUM, R.M. (1979) Serum dopamine β-hydroxylase. *Pharmacol. Rev.*, **30** 133–166.

WELLSTEIN, A. & PITSCHNER, H.F. (1988) Complex dose–response curves of atropine in man explained by different functions of M_1- and M_2-cholinoceptors. *Naunyn-Schmiedeberg's Arch. Pharmacol.*, **338**, 19–27.

WELSH, N.J., SHANKLEY, N.P. & BLACK, J.W. (1994) Comparative analysis of the vagal stimulation of gastric acid secretion in rodent isolated stomach preparations. *Br. J. Pharmacol.*, **112**, 93–96.

WESS, J. (1993) Molecular basis of muscarinic acetylcholine receptor function. *Trends Pharmacol. Sci.*, **14**, 308–313.

WHEELDON, N.M., MCDEVITT, D.G. & LIPWORTH, B.J. (1992) Selectivity of antagonist and partial agonist activity of celiprolol in normal subjects. *Br. J. Clin. Pharmacol.*, **34**, 337–343.

WIDDICOMBE, J.G. (1990) The NANC system and airway vasculature. *Arch. Int. Pharmacodyn.*, **303**, 83–99.

WILFFERT, B., TIMMERMANS, P.B.M.W.M. & VAN ZWIETEN, P.A. (1982) Extrasynaptic location of alpha-2 and noninnervated beta-2 adrenoceptors in the vascular system of the pithed normotensive rat. *J. Pharmacol. Exp. Ther.*, **221**, 762–768.

WILKINSON, S.E. & HALLAM, T.J. (1994) Protein kinase C: is its pivotal role in cellular activation over-stated? *Trends Pharmacol. Sci.*, **15**, 53–57.

WILLEMS, J.L., BUYLAERT, W.A., LEFEBVRE, R.A. & BOGAERT, M.G. (1985) Neuronal dopamine receptors on autonomic ganglia and sympathetic nerves and dopamine receptors in the gastrointestinal system. *Pharmacol. Rev.*, **37**, 165–216.

WILLIAMS, L.T. & LEFKOWITZ, R.J. (1978) *Receptor Binding Studies in Adrenergic Pharmacology*, Raven Press, New York.

WILLIAMS, L.T., LEFKOWITZ, R.J., WATANABE, A.M., HATHAWAY, D.R. & BESCH, H.R. JR. (1977) Thyroid hormone regulation of β- adrenergic receptor number. *J. Biol. Chem.*, **252**, 2787–2789.

WILLIAMS, R.G. & BROADLEY, K.J. (1982) Responses mediated via β_1- but not β_2-adrenoceptors exhibit hypothermia-induced supersensitivity. *Life Sci.*, **31**, 2977–2983.

WILLIAMS, R.G. & BROADLEY, K.J. (1983) Determination of agonist affinity for cardiac β-adrenoceptors during reserpine-induced supersensitivity. *Eur. J. Pharmacol.*, **87**, 95–105.

WILLIAMS, R.J., URQUHART, C.J., WILSON, K.A., DOWNING, O.A., DETTMAR, P.W. & ROACH, A.G. (1990) α_2-Adrenoceptor antisecretory responses in the rat jejunum. *J. Auton. Pharmacol.*, **10**, 109–118.

WILLIAMSON, K.L. & BROADLEY, K.J. (1987) Characterization of the α-adrenoceptors mediating positive inotropy of rat left atria by use of selective agonists and antagonists. *Arch. Int. Pharmacodyn. Ther.*, **285**, 181–198.

WILLIAMSON, K.L. & BROADLEY, K.J. (1989) Do both adrenaline and noradrenaline stimulate cardiac α-adrenoceptors to induce positive inotropy of rat atria? *Br. J. Pharmacol.*, **98**, 597–611.

WILLS, M. & DOUGLAS, J.S. (1988) Aging and cholinergic responses in bovine trachealis muscle. *Br. J. Pharmacol.*, **93**, 918–924.

WILLS-KARP, M. & GAVETT, S. (1993) Modulation of muscarinic agonist binding by cations and guanine nucleotides during aging. *Life Sci.*, **52**, 566.

WILSON, C. & LINCOLN, C. (1984) β-Adrenoceptor subtypes in human, rat, guinea pig, and rabbit atria. *J. Cardiovasc. Pharmacol.*, **6**, 1216–1221.

WILSON, I.B. & HARRISON, M.A. (1961) Turnover number of acetylcholinesterase. *J. Biol. Chem.*, **236**, 2292–2295.

WILSON, V.G., BROWN, C.M. & MCGRATH, J.C. (1991) Are there more than two types of α-adrenoceptors involved in physiological responses? *Exp. Physiol.*, **76**, 317–346.

WINNIFORD, M.D., WHEELAN, K.R., KREMERS, M.S., UGOLINI, V., VAN DEN BERG, E., NIGGEMAN, E., JANSEN, D.E. & HILLIS, L.D. (1986) Smoking-induced coronary vasoconstriction in patients with atherosclerotic coronary artery disease: evidence for adrenergically mediated alterations in coronary artery tone. *Circulation*, **73**, 662–667.

WINSLOW, J.B. (1732) *Exposition Anatomique de la Structure du Corps Humain*, G. Desprez, Paris.

WOJCIKIEWICZ, R.J.H., TOBIN, A.B., SAFRANY, S.T. & NAHORSKI, S.R. (1994) Regulation of phosphoinositidase C-linked receptors and phosphoinositide-mediated cell signalling. In: *Regulation of Cellular Signal Transduction Pathways by Desensitization and Amplification*, pp. 169–190, D.R. SIBLEY & M.D. HOUSLAY (Eds), John Wiley, Chichester.

WONG-DUSTING, H.K. & RAND, M.J. (1985) Effect of [D-Ala^2, Met^5]-enkephalinamide and [D-Ala^2, D-Leu^5]-enkephalin on cholinergic and noradrenergic neurotransmission in isolated atria. *Eur. J. Pharmacol.*, **111**, 65–72.

WONNACOTT, S. & DAJAS, F. (1994) Neurotoxins: Nature's untapped bounty. *Trends Pharmacol. Sci.*, **15**, 1–3.

WOOD, P.B. & ROBINSON, M.L. (1981) An investigation of the comparative liposolubilities of β-adrenoceptor blocking agents. *J. Pharm. Pharmacol.*, **33**, 172–173.

WOODS, S.W., KOSTER, K., KRYSTAL, J.K., SMITH, E.O., ZUBAL, I.G., HOFFER, P.B. & CHARNEY, D.S. (1988) Yohimbine alters regional cerebral blood flow in panic disorder. *Lancet*, **2**, 678.

WOOLVERTON, W.L. & JOHNSON, K.M. (1992) Neurobiology of cocaine abuse. *Trends Pharmacol. Sci.*, **13**, 193–200.

WRIGHT, C.D., KUIPERS, P.J., KOBYLARZ-SINGER, D., DEVALL, L.J., KLINKEFUS, B.A. & WEISHAAR, R.E. (1990) Differential inhibition of human neutrophil function. Role of cyclic AMP-specific, cyclic GMP-insensitive phosphodiesterase. *Biochem. Pharmacol.*, **40**, 699–707.

YAMADA, S., ISHIMA, T., TOMITA, T., HAYASHI, M., OKADA, T. & HAYASHI, E. (1984) Alterations in cardiac alpha and beta adrenoceptors during the development of spontaneous hypertension. *J. Pharmacol. Exp. Ther.*, **228**, 454–460.

YAMAGUCHI, I. & KOPIN, I.J. (1980) Blood pressure, plasma catecholamines and sympathetic outflow in pithed SHR and WKY rats. *Am. J. Physiol.* **238**, H365–H377.

YANG, C.M. (1991) Characterization of muscarinic receptors in dog tracheal smooth muscle cells. *J. Auton. Pharmacol.*, **11**, 51–61.

YAZAWA, K. & KAMEYAMA, M. (1990) Mechanism of receptor-mediated modulation of the delayed outward potassium current in guinea-pig ventricular myocytes. *J. Physiol.*, **421**, 135–150.

YEOMANS, N.D. (1988) Gastrointestinal actions of nicotine. In: *The Pharmacology of Nicotine*, pp. 91–96, M.J. RAND & K. THURAU (Eds), IRL Press, Oxford.

YOUDIM, M.B.H. & FINBERG, J.P.M. (1990) Monoamine oxidase A and B inhibitors and substrates as antidepressants. In: *Antidepressants: 30 Years On*, pp. 204–211, B. LEONARD & P. SPENCER (Eds), CNS (Clinical Neuroscience) Publishers, London.

YOUDIM, M.B.H. & FINBERG, J.P.M. (1991) New directions in monoamine oxidase A and B selective inhibitors and substrates. *Biochem. Pharmacol.*, **41**, 155–162.

YOUNG, R.A. & BROGDEN, R.N. (1988) Doxazosin. A review of its pharmacodynamic and pharmacokinetic properties, and therapeutic efficacy in mild or moderate hypertension. *Drugs*, **35**, 525–541.

YU-HONG, L., BARNES, P.J. & ROGERS, D.F. (1993) Regulation of NANC neural bronchoconstriction *in vivo* in the guinea-pig: involvement of nitric oxide, vasoactive intestinal peptide and soluble guanylyl cyclase. *Br. J. Pharmacol.*, **108**, 228–235.

ZAAGSMA, J., VAN DER HEIJDEN, P.J.C.M., VAN DER SCHAAR, M.W.G. & BANK, C.M.C. (1983) Comparison of functional β-adrenoceptor heterogeneity in central and peripheral airway smooth muscle of guinea pig and man. *J. Receptor Res.*, **3**, 89–106.

ZABLUDOWSKI, J.R., BALL, S.G. & ROBERTSON, J.I.S. (1985) Ketanserin and α_1-adrenergic antagonism in man. *J. Cardiovasc. Pharmacol.*, **7**,(Suppl. 7), S123–S125.

ZELLER, E.A. & BARSKY, A.J. (1952) *In vivo* inhibition of liver and brain monoamine oxidase by 1-isonicotinyl-2-isopropylhydrazine. *Proc. Soc. Exp. Biol. Med.*, **81**, 459–461.

ZHANG, L., HOROWITZ, B. & BUXTON, I.L.O. (1991) Muscarinic receptors in canine colonic circular smooth muscle. I. Coexistence of M_2 and M_3 subtypes. *Mol. Pharmacol.*, **40**, 943–951.

ZHANG, X., XIAO, W-B. & NORDBERG, A. (1993) 1,2,3,4-Tetrahydro-9-aminoacridine (THA) may interact with cholinergic presynaptic receptors to regulate *in vivo* acetylcholine release in the striatum of anaesthetized rats. *Life Sci.*, **52**, 587.

ZIEGLER, M.G., LAKE, C.R. & KOPIN, I.J. (1976) Plasma noradrenaline increases with age. *Nature*, **261**, 333–335.

ZIMMERMANN, H. (1988) Cholinergic synaptic vesicles. In: *The Cholinergic Synapse. Handbook of Experimental Pharmacology*, Vol. 86, pp. 349–382, V.P. WHITTAKER (Ed.), Springer-Verlag, Berlin.

582

ZIOGAS, J., STORY, D.F. & RAND, M.J. (1984) Effects of locally generated angiotensin II on noradrenergic transmission in guinea-pig isolated atria. *Eur. J. Pharmacol.*, **106**, 11–18.

ZÜRCHER, G., KELLER, H.H., KETTLER, R., BORGULYA, J., BONETTI, E.P., EIGENMANN, R. & DA PRADA, M. (1990) Ro 40–7592, a novel, very potent and orally active inhibitor of catechol-O-methyltransferase: a pharmacological study in rats. In: *Advances in Neurology*, Vol. 53, *Parkinson's Disease: Anatomy Pathology and Therapy*, pp. 497-503, M.B. STREIFLER, A.D. KORCZYN, E. MELAMED & M.B.H. YOUDIM (Eds), Academic Press, New York.

Index

glaucoma 188, 232, 242, 281, 338, 344, 345, 368, 373, 407
Glaucostat^{FR} (aceclidine) 342, 344−5
glossopharyngeal nerve 13, 30, 32
glucagon 26
glucose metabolism 198
N-L-γ-Glutamyl-dopa (gludopa) 472
glutathione antioxidant system 289
glycogenolysis (liver) 24, 26, 123, 142, 182, 199, 235, 243, 245−6, 493−5
glycopyrrolate (Robinul^R) 354, 377
glycopyrronium 406
Golgi apparatus 9, 453, 520
green mamba toxin 312
guanabenz (Wytensin^{US}) 150, 179, 264, 266, 269
guanadrel (Hylorel^R) 276, 280
guanethidine (Ismelin^R) 37, 54, 82, 84−7, 168, 238, 276, 449, 462−3
 in glaucoma 281
 indirect sympathomimetic activity 168, 278, 279
 mechanisms of neurone blockade 278−9
 pharmacology 276−8
 supersensitivity after chronic treatment 526
 uses 280
guanfacin (Tenex^{US}) 179, 264, 266, 269
guanine nucleotide (G) binding proteins 308, 323, 476−8, 482−7
 receptor requirements 479, 486−7
guanine nucleotide (G_i) binding proteins 131, 262, 324, 476−7, 485
guanine nucleotide (G_q) binding proteins 476−7, 484
guanine nucleotide (G_s) binding proteins 131, 482, 497−8, 517−19
guanoclor 276
guanosine 5'-triphosphate (GTP) 99, 321
guanoxan 150, 276, 280−1
guanyl-5'yl-imidotriphosphate (Gpp(NH)p) 99, 495, 497
guanylyl-cyclase 331, 462
guinea-pig
 atrium
  negative inotropic response, M_2 receptors 313, 315, 357, 362
  noradrenaline release from 167, 258
  pA_2 values for muscarinic antagonists 315, 357
 electrically stimulated superfused tracheal strip 194
 gall bladder 315, 447
 heart ventricular membranes 97
 ileum 121, 139, 307, 313, 315, 404, 456
  effects of ganglion stimulants 430−1, 442
  muscarinic M_3 receptors 362, 365, 371

pA_2 values for muscarinic antagonists 352, 356, 358, 365
 isolated perfused hearts (Langendorff heart) 159, 329, 365
 isolated soleus muscle (skeletal muscle) 184
 isolated spontaneously beating atria 270
 left atria, β-agonists on 78, 105, 111, 115, 118−9
 lung parenchymal strips 76, 120
 perfused lung 175
 right atria, β-agonists on 105, 113, 115, 118−9
  negative chronotropic response, M_2 315
 stomach 141
 taenia coli 447, 449, 452
 trachea 89, 120, 133, 305, 318, 462, 463
 transmurally stimulated ileum (Paton preparation) 38, 139−40, 263, 306−7, 366, 405
 urinary bladder 447
 uterus 312−3
 vas deferens 277
gynaecomastia (breast enlargement) 267, 287

H35/25 227, 230
haemorrhoids (piles) 187
haemostatis 186
Hagedorn compound, HI−6 411
hair tonics 340
Haldol^R (Serenace^R, haloperidol) 231, 468, 471
β-haloalkylamines 206−12, 248, 290, 362, 392
 pharmacology and uses 211−12
 reversal of α-blockade 211
Halobacterium halobium 478
haloperidol (Serenace^R, Haldol^R) 231, 468, 471
heart 22, 122, 145, 157, 315
 effects of parasympathomimetic amines 328−31
 effects of sympathomimetic amines 180−182
 rate 181, 234, 243, 364, 415, 424, 527
  of the pithed rat 142
  response to exercise 131
α-helixes
 of β-adrenoceptors 230, 479, 481−2
 of muscarinic receptor 311, 358, 363
 of nicotinic (N_N) receptor pentamers 292
 of nicotinic receptor subunits 417
 of receptor structure 478−82
hemicholinium−3 (HC−3) 54, 296, 299, 370, 416
 effect of chronic treatment 527
hemlock (*Conium maculatum*) 429

*For Product Safety Concerns and Information please contact
our EU representative GPSR@taylorandfrancis.com Taylor & Francis
Verlag GmbH, Kaufingerstraße 24, 80331 München, Germany*

T - #0070 - 160425 - C0 - 254/178/33 [35] - CB - 9780748405565 - Gloss Lamination